Theory and Design of Railway Truss Bridges

Railroad bridges are a critical component of the infrastructure and economy of many countries, and many of these bridges are nearing the end of their useful service life. *Theory and Design of Railway Truss Bridges* provides comprehensive coupled information regarding the structural analysis and design of steel railway truss spans. Most books cover either analysis or design of structures, but none cover both the analysis and design of railway trusses. Further, the book presents technical information on the analysis of railway trusses currently unavailable in other modern books. It also provides readers with up-to-date information concerning the modern methods of design recommended by the American Railway Engineering and Maintenance-of-Way Association (AREMA):

- Includes detailed information on the analysis of trusses for moving loads.
- Presents information on topics specific to railway trusses such as loading effects, secondary stresses and stress reversal.
- Includes information on the history of railway truss analysis, design and construction.
- Covers methods for the analysis of statically indeterminate spans.
- Describes methods to determine the displacement of truss spans.
- Provides up-to-date theory and design methods based on current AREMA recommendations.

Theory and Design of Railway Truss Bridges

John F. Unsworth

CRC Press
Taylor & Francis Group
Boca Raton London New York

CRC Press is an imprint of the
Taylor & Francis Group, an **informa** business

Designed cover image: Shutterstock

First edition published 2025
by CRC Press
2385 NW Executive Center Drive, Suite 320, Boca Raton FL 33431

and by CRC Press
4 Park Square, Milton Park, Abingdon, Oxon, OX14 4RN

CRC Press is an imprint of Taylor & Francis Group, LLC

ISBN: 978-1-032-73740-9 (hbk)
ISBN: 978-1-032-74582-4 (pbk)
ISBN: 978-1-003-46997-1 (ebk)

DOI: 10.1201/9781003469971

Typeset in Times
by Apex CoVantage, LLC

To Johanna, whose intellectual proficiency and efforts in STEM are inspirational to those who love her and are portents of a meaningful and fulfilling life.

Contents

PART B: Structural Design of Railway Truss Bridges

Preface

Railway truss spans are critical components of the North American freight railroad system infrastructure. Most existing railway truss spans have served the North American railroad industry safely and reliably for over a century but have experienced substantial increases in the magnitude and frequency of freight rail traffic throughout their useful life. The assessment, maintenance and rehabilitation of railway truss spans are fundamental aspects of a sustainable, safe and reliable transportation system. However, in some cases, due to functional and/or structural obsolescence, the replacement of railway truss spans is required.

Consequently, this book is written as assistance to practicing engineers and students concerning the rehabilitation and replacement design of railway truss spans. It is intended to complement the recommended practices of Chapter 15—Steel Structures of the American Railway Engineering and Maintenance-of-way Association (AREMA) Manual for Railway Engineering (MRE). The AREMA MRE is updated annually by committees of active railway bridge owners, engineers, consultants, fabricators, constructors, academics and researchers. The reader is recommended to consult the most recent edition of the AREMA MRE when engaged in the assessment, maintenance, rehabilitation and replacement of railway truss bridges. Nevertheless, much of the information contained herein will remain valid through subsequent editions of the AREMA MRE.

This book is divided into twelve chapters of which, following the first chapter concerning the history of railway truss bridges, seven focus on theory and structural analysis, and four have an emphasis on structural design.

The early history of structural engineering is the history of steel railway bridges. Accordingly, Chapter 1 provides a brief history of the design and construction of steel railway truss bridges throughout the world. It outlines the development of wooden, iron and steel railway trusses and the associated development of structural mechanics and specifications.

Chapter 2 considers the fundamental issue of statical determinacy. The internal and external statical determinacy of truss spans are studied as the initial stage of structural analysis.

Structural analysis requires the input of the loads and forces on the truss span. Chapter 3 outlines the appropriate dead, live and environmental loads to consider in the structural analysis of railway truss spans.

Most railway truss spans are statically determinate. Therefore, Chapter 4 focuses on the theory and structural analysis of statically determinate truss spans. Freight train loads are moving loads, and it is imperative that the locations of live loads providing the maximum and minimum forces in truss members be established. Therefore, determination of uniformly distributed and concentrated live load positions for maximum and minimum member forces, and the magnitude of the member forces, in horizontal parallel chord, inclined chord and subdivided panel truss spans is outlined in considerable detail. The chapter includes over fifty examples to illustrate the nuances of the theory and analytical methods applied on a variety of truss forms.

The theory and analysis of statically determinate truss spans is followed, in Chapter 5, by concepts and analytical techniques for statically indeterminate spans. The behavior and analysis of statically indeterminate truss spans is best understood through the classical energy methods of analysis employed in this chapter.

In Chapter 6, the classical energy methods of analysis are further employed to determine the displacement of truss spans. Truss displacements are important for camber design and to ensure AREMA serviceability requirements are maintained during freight rail operations.

Secondary bending in ordinary truss members is typically of little significance. However, for cases where secondary bending is considerable and requires consideration, Chapter 7 describes methods of determining the secondary flexural stresses in railway truss members due to joint rigidity, eccentricity, floorbeam rotation, transverse forces, rotationally fixed bearings and member elongation.

Chapter 8 is the last chapter of Part A concerning the theory and structural analysis of railway truss spans; it concerns the lateral and longitudinal loads and forces on railway truss spans. Lateral and longitudinal forces result from live loads and environmental forces. This chapter presents information regarding the lateral force-resisting systems used in railway truss spans and the distribution of lateral and longitudinal forces within those systems. Lateral and longitudinal loads and forces on loaded and unloaded through- and deck-truss spans are included.

Chapter 9 is the first chapter of Part B concerning the structural design of railway truss members and connections. This chapter outlines design criteria related to the strength and fatigue design of tension members and the strength and stability design of compression members. Combined axial forces and bending moments in truss members are considered along with relevant topics about secondary bending, web member overload and stress reversal. The chapter concludes with a review and examples regarding load combinations for member and connection design.

Chapter 10 encompasses the detailed strength and fatigue design of railway truss tension members and concludes with practical advice and examples concerning railway truss tension member design.

Chapter 11 involves the detailed strength and stability design of railway truss compression members and finishes with practical advice concerning built-up railway truss compression member design.

Chapter 12 is the concluding chapter of the book and contains the important subject of connection design. In particular, the design of gusset plates has elicited considerable interest in recent years. Accordingly, member end connection, splice and gusset plate design are examined in the chapter. Gusset plate design for continuous and discontinuous chord members are deliberated and illustrated with examples of each.

This book is an endeavor to provide fundamental information regarding the design of ordinary modern steel railway truss spans and does not purport to be a definitive text on the subject. Other books, manuals, handbooks, codes, guides, specifications, reports and papers are essential for the safe and reliable design of railway truss bridges. Such resources, as available to the author, were used in the preparation of this book. The author gratefully appreciates any corrections or recommendations that are drawn to his attention.

John F. Unsworth
Cochrane, Alberta, Canada

About the Author

John F. Unsworth is Professional Engineer (P.Eng.) who holds a bachelor of engineering (B.Eng.) degree with distinction in civil engineering from Concordia University (Montreal) and a master of engineering (M.Eng.) degree in structural engineering from the University of British Columbia (Vancouver). He retired as Chief Bridge Engineer at the Canadian Pacific Railway in 2018 and was subsequently employed as a consultant with AECOM until his retirement in 2022. He is a Fellow of the Canadian Society for Civil Engineering.

He is a former President of the American Railway Engineering and Maintenance-of-way Association (AREMA) and has served as the Chair of AREMA Committee 15—Steel Structures and the Association of American Railroads (AAR) Bridge Research Advisory Committees. As recognition of his service to those organizations, he is an Emeritus Member of AREMA Committee 15 and Chair Emeritus of the AAR Bridge Research Advisory Committee. He is also a former member of the National Academy of Sciences, Engineering and Medicine Transportation Research Board where he served on the Steel Bridges Committee for over a decade. He continues to provide service to the industry through the American Institute of Steel Construction (AISC) National Steel Bridge Alliance (NSBA), AREMA, AAR and the International Heavy Haul Association (IHHA).

He has contributed to the fifth edition of the *Structural Steel Designer's Handbook* and the IHHA *Best Practices* books. He is the author of the second edition of the CRC Press book *Design and Construction of Modern Steel Railway Bridges.*

Figures

Tables

1 History of Railway Truss Bridges

1.1 INTRODUCTION

The truss form is not a phenomenon of nature[1] but an ingenious invention of mankind at the end of the Renaissance period of human history. Wooden trusses have been used since the 16th century when truss forms were employed by the Italian architect and builder Andrea Palladio[2] for roofs and pedestrian bridges. Nevertheless, wooden trusses were not widely used for almost another 300 years until German[3] and early American bridge builders such as Timothy Palmer,[4] Theodore Burr,[5] Enoch Hale and Lewis Wernwag[6] constructed timber truss bridges to convey pedestrian, equestrian and horse-drawn wagon traffic. The advent of the railroad in the early 19th century required that wooden trusses carry steam locomotives.[7]

1.2 WOODEN BRIDGE TRUSSES

Wooden trusses were patented by Ithiel Town in 1820 and Stephen Long in 1830. The Town truss consisted of top and bottom chord members of parallel timbers connected by a web of diagonal boards and was an early prototype of the iron lattice trusses used by railroads later in the 19th century. The Long truss consisted of top and bottom chord members of three timbers connected by a web of double post and diagonal members. Many of the early timber bridges, particularly those of longer spans, were arch-stiffened trusses or trussed arches.

The first wooden bridge carrying railroad traffic was constructed by Lewis Wernwag for the Baltimore and Ohio Railroad (B&O) in 1830. Subsequently, several wooden truss bridges carrying railroads in New England across the Susquehanna River were built in the early 1830s. One of the wooden truss bridges crossing the Susquehanna River was a 1304 m (4277 ft) long bridge consisting of twenty-three 55 m (180 ft) spans, each with four lines of double lattice wood trusses.[8] In 1834, Theodore Burr constructed a wooden truss bridge over the Schuylkill River at Philadelphia which served the Columbia Railroad, and its successor the Philadelphia and Reading Railroad, for 50 years until it was replaced with an iron truss bridge. Benjamin H. Latrobe of the B&O Railroad designed and built wooden trusses in the 1830s that were like future Howe trusses[9] but with rigid counterbraces. Town truss railroad bridges were built at New York and Philadelphia for the Harlem River and B&O Railroad, respectively. Seven 55 m (180 ft) Long trusses were used for the 384 m (1260 ft) crossing of the Connecticut River in 1844.[10] Figure 1.1 shows wooden Town, Long and Burr trusses typically used to carry railroad traffic in the early nineteenth century.

Herman Haupt[11] and William Howe[12] were granted patents for bridges proposed to carry railroad traffic in 1839 and 1840, respectively. In 1840, seven 55 m (180 ft) long Howe truss spans were used to convey railroad traffic across the Connecticut River.[13] Howe wood trusses were also used in the 1067 m (3500 ft) long bridge – consisting of thirteen spans up to 76 m (250 ft) long – crossing the Susquehanna River for the Philadelphia, Baltimore and Washington Railroad. Figure 1.2 shows a timber Howe truss railroad span designed by C.C. Schneider and built in 1885 for the Canadian Pacific Railway.[14]

In 1844, Caleb Pratt patented a truss that, for balanced loads, had vertical web members in compression and diagonal web members in tension. Very few wood Pratt trusses were built, but the form became very prevalent when metal[15] trusses began to replace wooden trusses in the latter part of the 19th century.

A considerable number of timber railroad bridges with truss spans of up to 67 m (220 ft) continued to be built up to the end of the 19th century. However, many railroads ceased timber truss construction in the middle of the 19th century due to concerns over fire and maintenance. On some railroads, however,

DOI: 10.1201/9781003469971-1

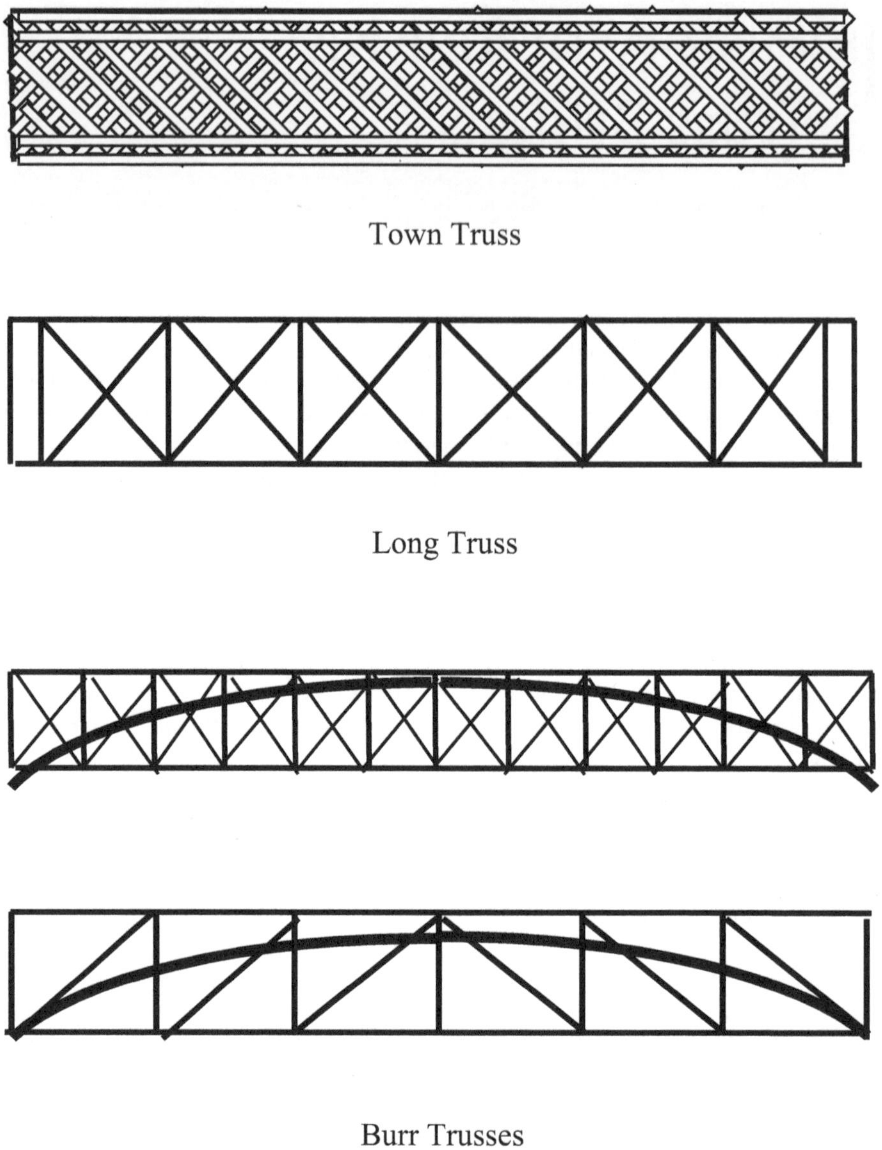

FIGURE 1.1 Wooden Town, Long and Burr truss spans.

timber truss span construction, particularly using Howe truss spans, continued into the late 19th century.[16] Howe trusses with wood compression members[17] and wrought iron tension members were used extensively in early American railroad bridge construction for almost 50 years before being supplanted by iron truss bridges. Since early steam locomotives were relatively light, the use of wood truss spans continued until the rapid railroad expansion in North America and Europe occurred in the latter half of the 19th century. Railroad growth required heavier locomotives and stronger bridges. In response, railroad bridge engineers embraced new materials (cast and wrought iron) and new truss forms.

1.3 IRON BRIDGE TRUSSES

As early as 1786 Thomas Paine promoted the use of cast iron for arch spans carrying roadway traffic.[18] The first iron railway bridge on a chartered railroad was built in 1823 by George Stephenson

FIGURE 1.2 Wooden bridge with timber Howe truss spans.

Source: **Courtesy of Canadian Pacific Kansas City Railway.**

FIGURE 1.3 Gaunless River bridge of the Stockton and Darlington Railway built in 1825 at West Auckland, UK. From Chris Lloyd, The Northern Echo, Darlington.

on the Stockton to Darlington Railway in England (Figure 1.3). The bridge consisted of 3.8 m (12.5 ft) long lenticular truss spans[19] in a trestle arrangement. Wendall Bollman used a form of iron truss with inclined tension members at each joint in the bottom chord such that the wooden bottom chord carried no stress. This form of the Bollman truss was used on the B&O Railroad during the 1840s.

The iron Bollman truss, patented in 1852 and used by the B&O and other railroads until 1873, was an example of the innovative[20] use of iron in American railway bridge construction.

The first railway bridge in North America constructed entirely of metal was an iron Howe truss, built by the Philadelphia and Reading Railroad in 1845, with cast iron compression members and wrought iron tension members. Another B&O Railroad engineer, Albert Fink, improved[21] Bollman's truss design in 1850, and the Fink truss was used extensively for the next 25 years on the B&O Railroad. Like the Howe truss, the Bollman and Fink trusses typically had cast iron compression members and wrought iron tension members. It is noteworthy that following the collapse of an iron truss bridge in 1850 on the Erie Railroad, some American railroads dismantled their iron trusses and replaced them with wood trusses. Nevertheless, in 1859 the New York Central Railroad[22] built the first bridge constructed entirely of wrought iron[23] in North America. Likewise, the practice of constructing iron truss bridges did not cease on North America's first commercial railroad, the B&O Railroad.[24]

Bollman and Fink continued to use their own patented cast and wrought iron trusses extensively between 1840 and 1875.[25] Iron Howe trusses were also constructed by the Boston and Albany Railroad in 1847 near Pittsfield, Massachusetts, and on the Harlem and Erie Railroad in 1850. Subsequently, iron truss bridges, particularly of the Whipple[26] and Pratt truss forms, became increasingly common as American railroads continued their rapid expansion.

Early examples of Pratt truss use were the Pennsylvania Railroad's cast and wrought iron arch-stiffened Pratt truss bridges of the 1850s. An iron bowstring truss,[27] also utilizing cast iron compression members and wrought iron tension members, was designed by Squire Whipple[28] for the Rensselaer & Saratoga Railway in 1852. Notably, iron trusses were also built by the North Pennsylvania Railroad in 1856 (a Whipple truss) and the Catasauqua and Fogelsville Railroad in 1857. In 1859, the Lehigh Valley Railroad built the first completely pin-connected iron truss. The Royal Albert Railway Bridge across the Tamar River, also completed in 1859, is a significant example of an Isambard Kingdom Brunel wrought iron railway bridge using large lenticular trusses (Figure 1.4).

In 1861, the Pennsylvania Railroad established the use of forged eyebars in a pin-connected truss over the Schuylkill River.[29] After this, many North American railway bridges were constructed with pin-connected eyebars while European practice still favored the use of riveted construction.[30] Riveted construction was considered superior, but pin-connected construction enabled the economical and rapid erection of railway truss span bridges in remote areas of North America. The principal exception was the New York Central Railroad, which used riveted construction exclusively for its iron railway truss spans.

In 1863, the Pennsylvania Railroad successfully crossed the Ohio River using a 98 m (320 ft) long iron truss span.[31] The Pennsylvania Railroad typically used the relatively rigid Whipple truss for such long spans. This bridge construction encouraged greater use of longer-span iron trusses to carry heavy freight railroad traffic in the United States. Another notable wrought iron railway truss was the 119 m (390 ft) span built by the B&O Railroad at Louisville, KY, in 1869.

The Erie Railroad pioneered the use of iron Post truss bridges in 1865, and they remained a standard of construction on the B&O Railroad for the next fifteen years. Post trusses were also used by many other railroads between 1865 and 1880. The web post members are inclined at an angle to the vertical of arctan (H/2P) (horizontal run of 50% of, P, the panel length), and the tension strut members are inclined at a vertical angle of arctan (3H/2P) (horizontal run of 150% of, P, the panel length), where H is the height of the truss. The inclined web members made fabrication difficult and more costly than other truss forms, such as Whipple and Pratt trusses. Consequently, the Post truss was not used after 1880.

The Pratt truss (patented in 1844) became predominant for short- and medium-span iron railway bridges in North America after 1870. Also, by 1870 the use of cast iron[32] was abandoned by most American railroads. The Pratt truss form is well suited for use in wrought iron bridges with the diagonal web members in tension under balanced load conditions.[33] Pratt trusses with an even number of panels and web members that can resist both tension and compression[34] are statically determinate.[35] Pratt trusses with web members that can resist only tension[36] are also statically determinate[37] using

FIGURE 1.4 The Royal Albert Bridge built in 1859 over the Tamar River at Saltash, UK.

Source: **From Owen Dunn, http://en.wikipedia, June 2005.**

counter members.[38] Pratt trusses with an odd number of panels and web members that can resist both tension and compression are statically indeterminate.[39] Wrought iron Whipple, Warren and Post trusses were also used by American railroads in the decade commencing in 1870.

Baltimore, Petit or Pennsylvania truss spans were often used for longer wrought iron railway bridge spans.[40] The first use of a Baltimore truss (a Pratt truss with subdivided panels) was on the Pennsylvania Railroad in 1871. Truss forms typically used by North American railroads in the 19th and 20th centuries are shown in Figures 1.5a and 1.5b.

An iron Howe truss collapsed under a train at Tariffville, Connecticut, in 1867, and a similar event occurred a decade later at Chattsworth, Illinois. However, the most significant railway bridge failure, due to the considerable loss of life associated with the incident, was the collapse of the cast iron Howe deck-truss span on the Lake Shore & Michigan Southern Railroad at Ashtabula, Ohio, in 1876 (Figures 1.6a and 1.6b). The Ashtabula bridge failure provided further evidence that cast iron was not appropriate for heavy railway moving loads and caused all American railroad companies to abandon the use of cast iron elements for bridges.

Railroad construction was also advancing in Germany in the latter part of the 19th century, and the failures of American railroad trusses interested Karl Culmann during the construction of some major bridges for the Royal Bavarian Railroad. He considered that American engineers used allowable stresses that were too high to preclude fatigue failures of iron truss railway bridges, and he recognized the issue of truss compression chord instability. A new emphasis on truss analysis and elastic stability was developing in response to the iron railway bridge failures and wrought iron trusses continued to be built in conjunction with the rapid railroad expansion of the late 19th

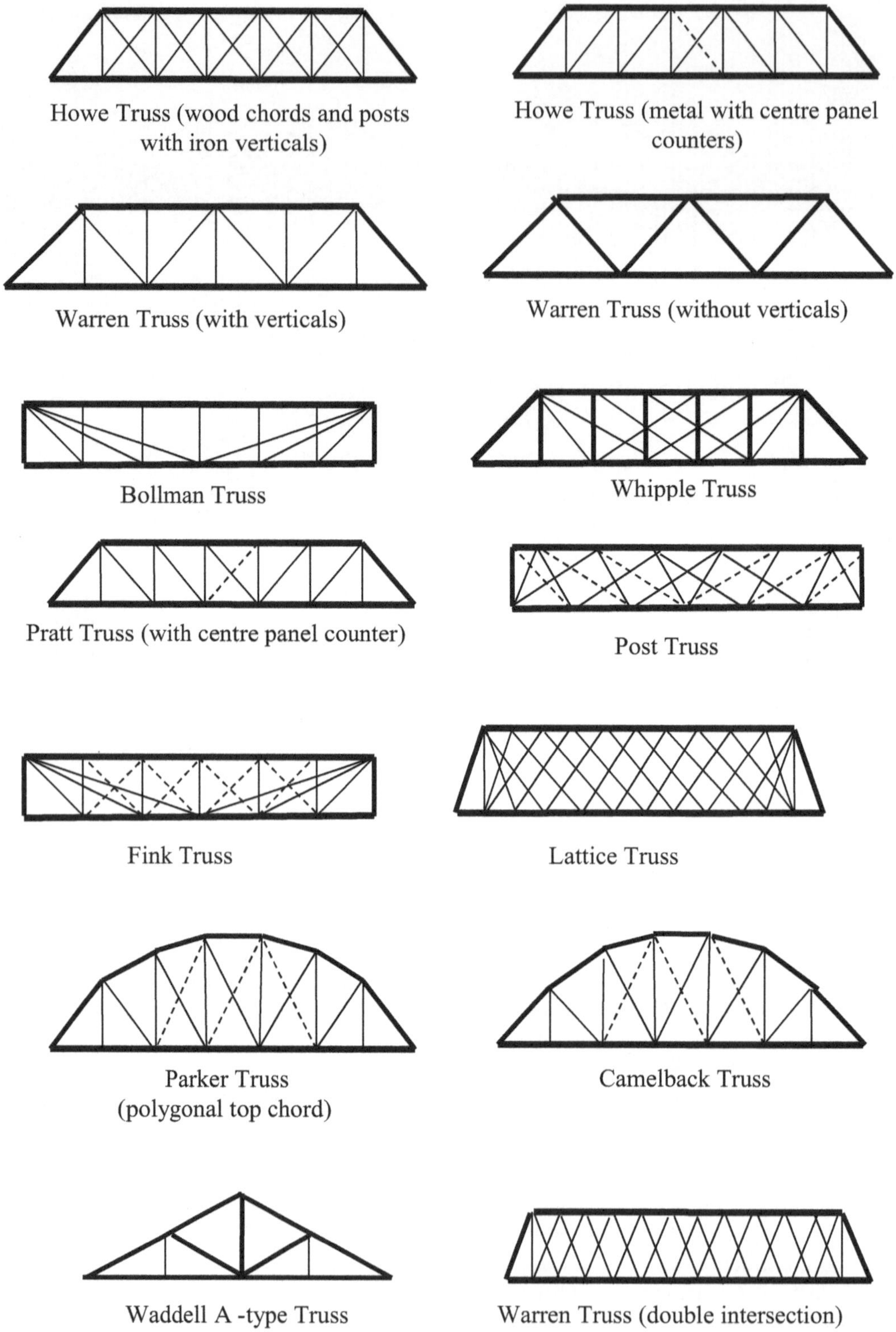

FIGURE 1.5a Short- and medium-span truss forms used by North American railroads.

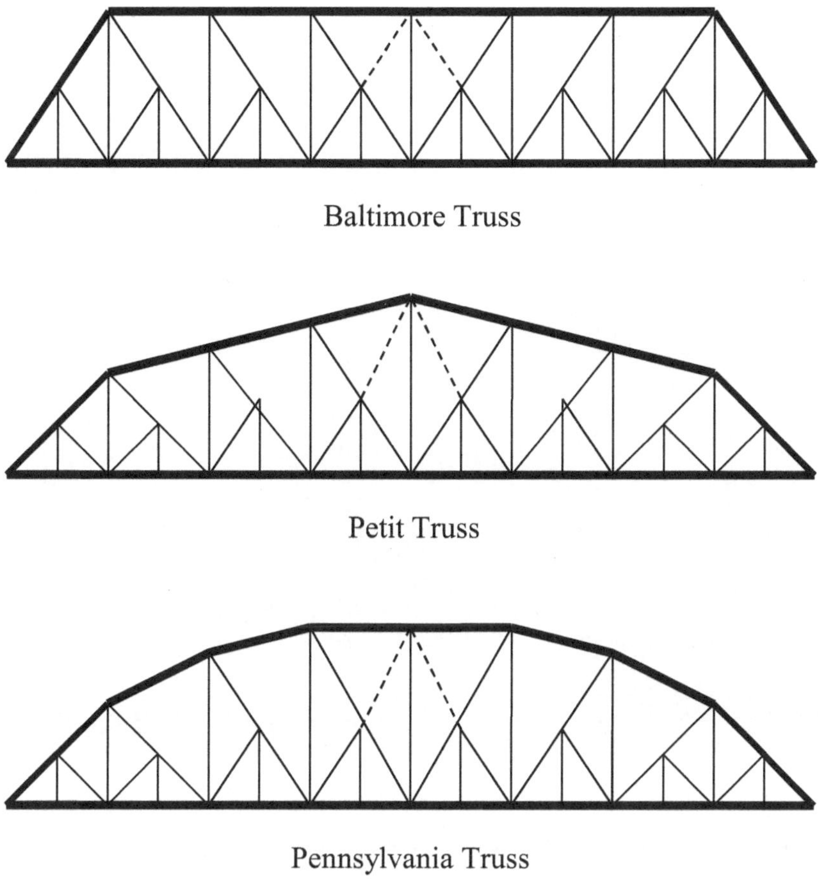

FIGURE 1.5b Long-span subdivided truss forms used by North American railroads.

century. Table 1.1 indicates some notable cast and wrought iron truss span bridges constructed between 1823 and 1877. Nevertheless, railroad bridge engineers embraced the new material of steel, and steel began to replace iron for the design and construction of railway bridge trusses through the second half of the 19th century.[41]

1.4 STEEL BRIDGE TRUSSES

The development of steel, with higher strength and ductility than iron, intensified steel railway truss design and construction. This elicited the construction of the first entirely steel railway bridge (comprising Whipple trusses) by the Chicago and Alton Railway in 1879 in Glasgow, Missouri. Steel trusses using the same forms as those used in iron spans, particularly the Whipple and Pratt types, became a standard of construction on many North American railroads in the early years of the 1880s. Table 1.2 indicates some notable steel simple-span truss bridges constructed between 1879 and 1907. Steel also permitted bridge engineers to design longer-truss spans required for major river crossings with difficult substructure conditions.[42] Consequently, commencing in the last quarter of the 19th century, many steel cantilever and continuous span trusses were designed and built.

FIGURE 1.6a The Ashtabula Bridge, OH, before the 1876 collapse.

Source: **Courtesy of Ashtabula Railway Historical Foundation.**

FIGURE 1.6b The Ashtabula Bridge, OH, after the 1876 collapse.

Source: **Courtesy of Ashtabula Railway Historical Foundation.**

TABLE 1.1
Iron Truss Simple-Span Railway Bridges Constructed between 1823 and 1877

Location	Railroad	Engineer	Year	Truss Type	Length m (ft)
West Auckland, UK (Figure 1.3)	Stockton to Darlington	G. Stephenson	1823	Lenticular	3.8 (12.5)
Ireland	Dublin & Drogheda	G. Smart	1824	Lattice	25.5 (84)
Manayunk, PA	Philadelphia and Reading	R. Osborne	1845	Howe	10.5 (34)
Pittsfield, MA	Boston & Albany	-	1847	Howe	9 (30)
Windsor, UK	Great Western	I. K. Brunel	1849	Bowstring	57 (187)
Newcastle, UK	Northwestern	I. K. Brunel	1849	Bowstring	38 (125)
New York	Harlem & Erie	-	1850	Howe	-
Various	Pennsylvania	H. Haupt	1850s	Pratt—arch stiffened	-
Harper's Ferry, W. VA	Baltimore & Ohio	W. Bollman	1852	Bollman	38 (124)
Fairmont, W. VA	Baltimore & Ohio	A. Fink	1852	Fink	62.5 (205)
-	Rensselaer & Saratoga	S. Whipple	1852	Whipple	-
Newark Dyke, UK	Great Northern	C. Wild	1853	Warren	79 (259)
-	North Pennsylvania	-	1856	Whipple	-
Guth, PA, Jordan Creek	Catasauqua & Fogelsville	F. C. Lowthorp	1857	-	33.5 (110)
Phillipsburg, NJ	Lehigh Valley	J. W. Murphy	1859	Whipple (pin-connected)	50 (165)
Plymouth, UK (Figure 1.4)	Cornish (Great Western)	I. K. Brunel	1859	Lenticular	139 (455)
Frankfort, Germany	-	-	1859	Lenticular	105 (345)
Various	New York Central	H. Carroll	1859	Lattice	27.5 (90)
Kehl River, Germany	Baden State	Keller	1860	Lattice	60 (197)
Schuylkill River	Pennsylvania	J. H. Linville	1861	Whipple	58.5 (192)
Steubenville, OH	Pennsylvania	J. H. Linville	1863	Murphy-Whipple	97.5 (320)
Mauch Chunk, PA.	Lehigh Valley	J. W. Murphy	1863	-	-
Liverpool, UK	London & Northwestern	W. Baker	1863	-	93 (305)
Blackfriars Bridge, UK	-	Kennard	1864	Lattice	-
Orival, France	Western	-	~1865	Lattice	51 (167)
Various	Baltimore & Ohio	S. S. Post	1865	Post	-
Lockport, IL	Chicago & Alton	S. S. Post	~1865	Post	-
Schuylkill River	Connecting Railway of Philadelphia	J. H. Linville	1865	Linville	-
Burlington, IA	Chicago, Burlington & Quincy	M. Hjorstberg	1868	-	-
Dubuque, IA	Illinois Central	J. H. Linville	1868	Linville	76 (250)
Quincy, IL	Chicago, Burlington & Quincy	T. C. Clarke	1868	-	76 (250)
Kansas City, MO.	Chicago, Burlington & Quincy	J. H. Linville & O. Chanute	1869	-	71 (234)
Louisville, KY	Baltimore & Ohio	A. Fink	1869	Subdivided Warren & Fink	119 (390)

(Continued)

TABLE 1.1 (*Continued*)
Iron Truss Simple-Span Railway Bridges Constructed between 1823 and 1877

Location	Railroad	Engineer	Year	Truss Type	Length m (ft)
Parkersburg & Benwood, WV	Baltimore & Ohio	J. H. Linville	1870	Bollman	106 (348)
Hannibal, MO	-	-	1871	-	-
Cincinnati, OH	Cincinnati Southern	J. H. Linville & G. L. F. Bouscaren	1876	Linville	157 (515)
Tay River, Scotland	-	Sir T. Bouch	1877	Lattice	-

TABLE 1.2
Steel Simple-Span Truss Railway Bridges Constructed between 1879 and 1907

Location	Railroad	Engineer	Year	Type	Length m (ft)
Glasgow, MO	Chicago & Alton	-	1879	Whipple	732 (2402)
Bismark, ND	-	G. S. Morison & C. C. Schneider	1882	Whipple	-
Hannibal, MO	-	-	1886	-	-
Tay River, Scotland	-	-	1887	-	-
Sioux City, IA	-	-	1888	-	122 (400)
Cincinnati, OH	-	W. H. Burr	1888	-	168 (550)
Benares, India	-	-	1888	Lattice	109 (356)
Hawkesbury, Australia	-	-	1889	-	127 (416)
Henderson Bridge	Louisville & Nashville	-	~1889	Subdivided Warren	160 (525)
Cairo, IL	Illinois Central		1889	-	-
Ceredo RR Bridge (W. VA)	-	Doane & Thomson	~1890	-	159 (521)
Merchant's Bridge, St. Louis, MO	-	G. S. Morison	1890	Petit	158 (517)
Kansas City, MO	-	-	1891	-	-
Burlington (2), IA	Chicago, Burlington & Quincy	G.S. Morison	1892	-	-
Louisville, KY	-	-	1893	Petit	168 (550)
Nebraska City, NE	-	G. S. Morison	1895	Whipple	122 (400)
Sioux City, IA	-	-	1896	-	149 (490)
Montreal, QC	Grand Trunk	-	1897	-	106 (348)
Kansas City, MO	Kansas City Southern	J. A. L. Waddell	1900	Pratt	-
Rumford, ON	Canadian Pacific	C. N. Monsarrat	1907	Subdivided Warren	126 (412)

1.5 STEEL CANTILEVER BRIDGE TRUSSES

North American and European engineers favoured steel cantilever trusses for long-span railway bridges, which due to their rigidity were considered to best resist the effects of vibration and heavy concentrated moving railway loads. The structural and construction efficacy of cantilever-type bridges led to the erection of many long-span steel railway bridges of cantilever truss design after 1876.

The Cincinnati Southern Railway constructed the first cantilever, or Gerber[43] type, steel-truss railway bridge in the United States over the Kentucky River in 1877[44] (Figure 1.7). In 1883, the

FIGURE 1.7 Steel cantilever truss railway bridge over the Kentucky River.

Source: **From https://alchetron.com/High-Bridge-of-Kentucky.**

Michigan Central and Canada South Railway completed the construction of a counterbalanced cantilever deck-truss bridge[45] across the Niagara Gorge parallel to Roebling's 1854 railway suspension bridge, which was later replaced by a steel spandrel braced hinged arch in 1897[46] (Figure 1.8). Shortly afterward, in 1884, the Canadian Pacific Railway crossed the Fraser River in British Columbia with the first balanced cantilever steel deck truss (Figure 1.9). The Canadian Pacific Railway also constructed a cantilever through-truss bridge in 1885 over the St John River in New Brunswick. It was an unusual configuration consisting of cantilever arms, each side of the suspended through-truss span, of different length, making the bridge asymmetrical (Figure 1.10). Cantilever bridges became customary for long-span railway bridge construction as they provided the rigidity required to resist large dynamic train loads, could be made statically determinate and required no main-span falsework to erect.

Benjamin Baker reviewed precedent[47] cantilever bridges constructed in North America (particularly, those on the Canadian Pacific Railway) and proposed a steel cantilever through-truss bridge for the Firth of Forth railway bridge crossing in Scotland. It was a monumental undertaking completed in 1890 and is an example of steel cantilever-truss railway bridge construction on an imposing scale with cantilever arms of 207 m (680 ft) supporting a 107 m (350 ft) suspended span (Figure 1.11). The bridge is very stiff, and the 90 mm (3½ in) deflection, measured by designer Baker under the heaviest locomotives available on the North British Railway, compared well with his estimate of 100 mm (4 in). The bridge was further tested under extreme wind conditions with two long heavy coal trains and the cantilever tip deflection was less than 180 mm (7 in).

FIGURE 1.8 Steel counterbalanced cantilever deck-truss bridge across the Niagara Gorge.

Source: From https://civildigital.com.

However, the Forth Railway Bridge used a large quantity of steel (due to the large tubular members) and was costly to construct. This prompted engineers such as Theodore Cooper, who had worked with James Eads on the 1869–1874 construction of St. Louis Bridge across the Mississippi River, and Ralph Modjeski to consider cantilever truss construction with different truss span types using relatively smaller members. Two such statically determinate cantilever-type railway bridges were the 205 m (671 ft) main span of the bridge crossing the Mississippi River at Thebes, Illinois, and the 549 m (1800 ft) main span of the Quebec Bridge.

The Thebes bridge, constructed in 1905, consists of five pin-connected through truss spans; of which two spans are 159 m (521 ft) fixed double anchor spans (anchoring four 46.5 m (152.5 ft) cantilever arms) and three contain 111.5 m (366 ft) suspended spans (Figure 1.12). The Quebec Bridge, an example of economical long-span steel cantilevered truss construction for railroad loads, was completed in 1917 after two construction failures (Figures 1.13a and 1.13b). The initial 1907 failure was likely due to calculation error in determining dead load compressive stresses in the bottom chord members during construction as the cantilever arms were increased in length. The bridge was redesigned,[48] and a new material, nickel steel,[49] was used in the reconstruction. In 1916, the suspended-span truss fell while being hoisted into place. It was quickly rebuilt, and the Quebec Bridge was opened to railway traffic in 1917 (Figure 1.13c). Another major cantilever-type bridge was not to be constructed until after 1930. The Quebec Bridge remains the longest-span cantilever bridge in the world. Table 1.3 indicates some notable steel cantilever truss spans built from 1876 to 1917.

FIGURE 1.9 Steel-balanced cantilever deck-truss bridge across the Fraser River.

Source: **Courtesy of Canadian Pacific Kansas City Railway.**

FIGURE 1.10 Steel cantilever through-truss bridge over the St John River in New Brunswick.

Source: **Courtesy of Canadian Pacific Kansas City Railway.**

FIGURE 1.11 The Forth Rail Bridge over the Firth of Forth, Scotland.

Source: **From GFDL Andrew Bell, http://en.wikipedia, January 2005.**

FIGURE 1.12 The Thebes Bridge over the Mississippi River at Thebes, IL.

Source: **US Geological Survey.**

FIGURE 1.13a The 1907 Quebec Bridge collapse.

Source: **Courtesy of Carlton University Civil Engineering Exhibits.**

FIGURE 1.13b The 1916 Quebec Bridge collapse.

Source: **Courtesy of A. A. Chesterfield, Library and Archives Canada.**

FIGURE 1.13c The Quebec Bridge over the St Lawrence River, completed in 1917.

Source: **Courtesy of Carlton University Civil Engineering Exhibits.**

1.6 STEEL CONTINUOUS-SPAN TRUSSES

Continuous truss spans were frequently used for long-span steel railway bridge construction in Europe but seldom in North America due to the practice of avoiding statically indeterminate structures. The first long-span continuous steel truss railway bridge in North America was built by the Canadian Pacific Railway over the St. Lawrence River at Montreal in 1886 (Figure 1.14).[50] The 124.5 m (408 ft) main spans were erected by the cantilever method with careful consideration of the deflections and stresses in the bottom chords of the truss. These were controlled during the cantilever erection with cables and adjustment screws attached to the partially completed truss supported on the center pier of the continuous span (Werry, 1997).

In 1917, the Chesapeake & Ohio Railroad completed construction of two 236.5 m (775 ft) span continuous steel trusses across the Ohio River at Sciotoville, Ohio (Figure 1.15). This bridge remains the longest continuous-span bridge in the world. Riveting was used extensively in Europe but only became a standard of American long-span steel railway bridge fabrication after construction of the Sciotoville bridge. Table 1.4 indicates some notable steel continuous truss spans built from 1886 to 1929.

Many iron and steel railway bridges were replaced in the first decades of the 20th century due to the development of substantially more powerful and heavier locomotives.[51]

TABLE 1.3
Steel Cantilever Truss Spans Constructed between 1876 and 1917

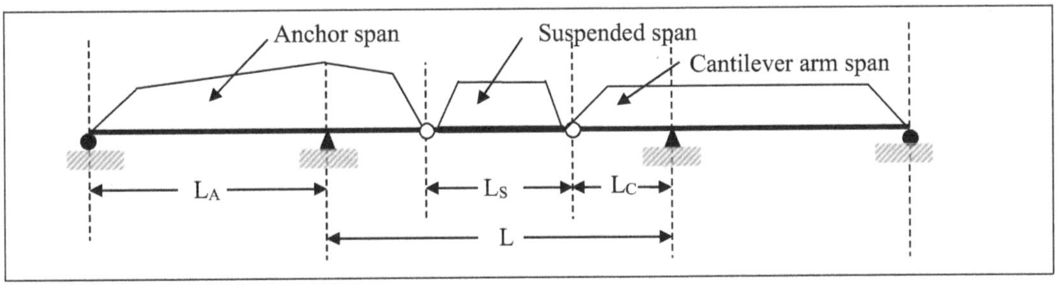

The part of the line from left solid circle to first open circle represents anchor span, the part between two open circles represents suspended span, and part from the right open circle to solid circle on the right represents cantilever arm span. The anchor span has a support at a distance of L subscript A from the left. The cantilever arm span has a support at distance of L subscript C from open circle on the right. The distance between two supports is L and the length of the suspended span is L subscript S.

				Length of Cantilever Truss Spans m (ft)			
Location	**Railroad**	**Engineer**	**Year**	L_A	L_C	L_S	L
Posen, Poland	-	-	1876	23 (74)	-	-	45 (148)
Dixville, KY (Figure 1.7)	Cincinnati Southern	C. Shaler Smith & G. Bouscaren	1877	~73 (~240)	49.5 (162.5)	0	99 (325)
St Paul, MN	Chicago, Milwaukee & St Paul	C. Shaler Smith	1880	74 (243)	49.5 (162)	0	99 (324)
Niagara Gorge, NY (Figure 1.8)	Michigan Central	C. C. Schneider	1883	63 (207.5)	57 (187.5)	36.5 (120)	151 (495)
Fraser River, BC (Figure 1.9)	Canadian Pacific	C. C. Schneider	1884	32 (105)	32 (105)	32 (105)	96 (315)
St John, NB (Figure 1.10)	Canadian Pacific	G. H. Duggan	1885	43.5 (143) & 58 (190)	43.5 (143) & 58 (190)	43.5 (143)	145 (476)
Louisville, KY	-	-	1886	55 (180)	49 (160)	49 (160)	146 (480)
Point Pleasant, WV	-	-	1888	73 (240)	43.5 (142.5)	61 (200)	148 (485)
Tyrone, KY	Louisville and Southern	J.W. MacLeod	1889	~68.5 (~225)	-	-	551
Poughkeepsie, NY	Central New England	-	1889	80 (262.5)	~49 (~160)	~69.5 (~228)	167 (548)
Hoogly, India	East India	Sir B. Leslie	1890	-	-	-	-
Firth of Forth, Scotland (Figure 1.11)	North British	Sir B. Baker & Sir J. Fowler	1890	207 (680)	207 (680)	107 (350)	521 (1710)
Pecos River	Southern Pacific	A. Bonzano	1891	-	-	-	~61 (~200)
Red Rock, CO	-	J.A.L. Waddell	1892	50 (165)	50 (165)	100 (330)	200 (660)
Callao, Peru	Lima and Oroya	L.L. Buck	~1892	-	-	-	265
Cernavoda, Romania	-	-	~1892	71 (233.5)	50 (164)	90 (295)	190 (623)

(Continued)

TABLE 1.3 (*Continued*)
Steel Cantilever Truss Spans Constructed between 1876 and 1917

Location	Railroad	Engineer	Year	Length of Cantilever Truss Spans m (ft)			
				L_A	L_C	L_S	L
Memphis, TN	-	G. S. Morison	1892	69 (226) & 94.5 (310)	52 (170)	137 (450)	241 (790.5)
Ottawa, ON	Canadian Pacific	G. H. Duggan	1900	75 (247)	37.5 (123.5)	94 (308)	169 (555)
Loch Etive, Scotland	-	Sir J. W. Barry	1903	42.5 (139.5)	44.5 (146)	71 (232)	160 (524)
Pittsburgh, PA	Wabash	-	1904	106 (346)	69 (226)	110 (360)	248 (812)
Mingo Junction, OH	Wabash	-	1904	91 (298)	-	-	213 (700)
Thebes, IL (Figure 1.12)	-	A. Noble & R. Modjeski	1905	79.5 (260.5) (½ of span)	46.5 (152.5)	111.5 (366)	204.5 (671)
Blackwell's Island (Queensboro), NY	City of New York (light rail)	G. Lindenthal	1907	143 (469.5) & 192 (630)	180 (591)	0	360 (1182)
Khushalgarth, India	-	Rendel & Robertson	1908	-	-	-	-
Westerburg, Prussia	Prussian State	-	1908	-	-	33.5 (110)	-
Daumer Bridge, China	Yunnan	-	1909	37.5 (123)	27.5 (90)	51 (168)	106 (348)
Beaver, PA	Pittsburgh & Lake Erie	-	1910	97.5 (320)	74 (242)	87 (285)	234.5 (769)
Quebec, QC (Figure 1.13)	Government of Canada	Duggan, Vautelet, Monsarrat, Modjeski, Schneider	1917	157 (515)	177 (580)	195 (640)	549 (1800)

1.7 RAILWAY TRUSS THEORY AND DESIGN

Timber and iron railway trusses were failing at a distressing rate, in both the United States and England, in the latter half of the 19th century. For example, it was estimated that between 1875 and 1888, 25% of American railroad bridges were failing annually. A rational and rigorous scientific means of truss analysis was imperative following these railway truss failures. Also, heavier locomotives, the need for long-span bridges and an increased focus on public safety made a scientific approach to the analysis and design of railway bridges essential.

Early developments in elasticity and strength of materials theories supported the objective of a scientific approach to truss design. The early work of Robert Hooke (1678) concerning the elastic force and deformation relationship, Leonard Euler (1759) and C. A. Coulomb (1773) concerning the elastic stability of compression members and Louis Navier (1826) on the theory of elasticity laid the foundation for the rational analysis of trusses. France led in the development of elasticity theory and mechanics of materials in the 18th and early 19th centuries. French schools of bridge engineering produced engineers who, in many cases, became the leaders of American railway truss engineering practice.[52]

FIGURE 1.14 The continuous-span steel truss bridge built over the St Lawrence River at Montreal in 1886.

Source: **Courtesy Canadian Pacific Kansas City Railway.**

FIGURE 1.15 The continuous-span steel truss bridge built over the Ohio River at Sciotoville, Ohio, in 1917.

Source: **Sciotoville Railroad Bridge—HistoricBridges.org.**

TABLE 1.4
Continuous Truss Span Railway Bridges Constructed between 1850 and 1929

Location	Railroad	Engineer	Year	Type	Longest Span m (ft)
Torksey, UK	-	J. Fowler	1850	3-span continuous tubular girder	40 (130)
Britannia Bridge, Menai Straits, UK	London-Chester-Holyhead	R. Stephenson	1850	4-span continuous tubular	140 (460)
Montreal, QC	Grand Trunk	R. Stephenson	1860	25-span continuous tubular	100 (330)
Montreal, QC (Figure 1.14)	Canadian Pacific	C. Shaler Smith	1886	4-span continuous trusses	124 (408)
Sciotoville, OH (Figure 1.15)	Chesapeake & Ohio	G. Lindenthal & D. B. Steinman	1917	2-span continuous truss	236 (775)
Allegheny River, PA	Bessemer & Lake Erie	-	1918	3-span continuous truss	158.5 (520)
Nelson River, PA	Bessemer & Lake Erie	-	1918	3-span continuous truss	122 (400)
Cincinnati, OH	C.N.O. & T.P.	-	1922	3-span continuous truss	157 (516)
Cincinnati, OH	Cincinnati & Ohio	-	1929	3-span continuous truss	206 (675)

1.7.1 Wood Railway Truss Theory and Design

American railway wood truss engineering practice was principally experiential and based on the use of proven truss forms. Early wooden Town, Long and Howe railway trusses were constructed using empirical rules without an understanding of the distribution and magnitude of forces in the truss members. Wood truss designs evolved by experience[53] through iterative revisions of member sizes. Therefore, the timber railway bridges were built and operated over without assurances of adequate safety. Timber railway truss spans were typically replaced due to member failures or fire.

1.7.2 Metal Railway Truss Theory and Design

Like wood trusses, early iron railway trusses were constructed without knowledge of the distribution and magnitude of forces in the truss members. Early iron truss designs were typically based on wood truss forms and evolved by experience through iterative revisions of truss forms, member sizes and materials. The absence of scientific approaches to truss design was exposed by the many failures of American iron railway bridge trusses between 1850 and 1870.

American engineers developed a great interest in truss analysis because of the extensive use of iron trusses on American railroads. Squire Whipple published the first rational treatment of statically determinate trusses (the method of joints) in 1847. At this juncture, European engineers were also interested in the problems of truss analysis and compression member elastic stability.

Concurrently, the rapid advancement of elasticity theory and engineering mechanics in Europe in the mid-nineteenth century encouraged French and German engineers to design iron railway trusses using scientific methods. Following Whipple's treatise on truss analysis, two European railway bridge engineers, Dmitrii Zhuravskii (aka D.J. Jourawski) and Karl Culmann, provided significant contributions to the theory of truss analysis for iron railway bridges. Culmann, an engineer of the Royal Bavarian Railroad, was a strong and early proponent of the mathematical analysis of trusses. He presented, in 1851, an analysis of the Howe truss commonly used in the United States and analyzed the statically indeterminate Long, Town and Burr trusses using approximate methods.

The Warren truss was developed in 1846[54] and by 1850 William Blood had developed a method of analysis for triangular trusses. William Fairbairn also considered the effects of moving loads on determinate trusses as early as 1857. Johann Schwedler, a German engineer, also made a substantial contribution to truss analysis by introducing the method of sections in 1862. Concurrently, August Ritter improved truss analysis by simplifying the method of sections through development of the equilibrium equation at the intersection of two truss members.

James Clerk Maxwell and Culmann both published graphical methods for truss analysis between 1864 and 1866. Truss graphical analysis methods were further developed and improved by Maxwell and Otto Mohr between 1864 and 1874. In the decade following 1860, railway trusses were typically designed for a uniformly distributed live load of between 30 kN/m and 60 kN/m (between 1 ton/ft and 2 tons/ft). Developed by Emil Winkler in 1867, bridge engineers were also given the powerful tool of influence lines for moving load analysis.[55] Between 1885 and 1889, Friedrich Engesser, a German railway bridge engineer, further developed practical compression member stability analysis.

Railway truss engineering profited from the scientific developments in the latter half of the 19th century by promoting standards for the design, fabrication and construction of iron and steel truss span railway bridges in the late 19th and early 20th centuries. Accordingly, early standard specifications for steel railway truss spans included provisions concerning specified materials, truss types and geometry, loads, member design (allowable stresses and proportioning of parts), details, fabrication (workmanship and painting) and inspection (mill and shop).

1.7.3 Railway Truss Design Specifications

Prior to 1870, railway truss spans were designed and fabricated by specialty bridge companies. Almost 40 bridges (50% of them iron) were collapsing annually in the United States during the decade following 1870. This was alarming as the failing bridges comprised 25% of the entire American railway bridge inventory at that time. In the decade between 1876 and 1886, almost 200 bridge spans collapsed in the United States.

In many cases, proprietary railway bridges were failing due to a lack of rigidity and lateral stability. Most of these spans were built by bridge companies without the benefit of independent engineering design, and while some bridge companies had good specifications for design and construction, others did not. Therefore, without independent engineering analysis and design, railroad company officials required a good knowledge of bridge engineering to ensure public safety. This was not always the case, as demonstrated by the Ashtabula bridge collapse, where it was learned in the subsequent inquiry that the proprietary bridge design had been approved by a railroad company executive without bridge design experience.[56] Consequently, to preclude further failures, American engineers were proposing the development and implementation of railroad company specifications that all bridge fabricators would build in accordance with. Developments in the fields of materials and structural mechanics had supplied the tools for scientific bridge design and provided the basis on which to establish specifications for iron and steel railway bridges.

The first specification for iron railway bridges was made by the Clarke, Reeves and Company (later the Phoenix Bridge Co.) in 1871. This was followed in 1873 by George S. Morison's[57] "Specifications for Iron Bridges" for the Erie Railroad (formerly the New York, Lake Erie & Western Railroad). Louis F. G. Bouscaren of the Cincinnati Southern Railroad published the first specifications with concentrated wheel loads in 1875.[58]

By 1876, the practice of bridge truss design by consulting engineers working on behalf of the railroad companies had become more prevalent in conjunction with the expanding railroad business. Cooper's publications concerning railway loads, design specifications and construction were particularly significant contributions to the development of a rational basis for the design of steel railway bridges. Cooper produced specifications for iron and steel railway bridges in 1884 that were intended for use by all railroad companies. These specifications for iron and steel railway bridges were updated until 1888.

In 1890, Cooper delivered his first specification for steel railway bridges, which was updated until 1905, when his specifications were superseded by the development, in 1906, of general specifications for steel railway bridges by the American Railway Engineering & Maintenance-of-way Association (AREMA). This latter specification has been continuously updated and is the current recommended practice on which most North American railroad company design requirements are based. Other significant milestones in the development of general specifications for iron and steel railway bridges were:

- 1867 St. Louis Bridge Co. specifications for Eads' steel arch[59]
- 1873 Chicago & Atchison Railroad Co.
- 1877 Chicago, Milwaukee & St. Paul Railway Co. (Charles Shaler Smith)
- 1877 Lake Shore & Michigan Southern Railway (Charles Hilton)
- 1877 Western Union Railroad Co.
- 1879 Erie Railroad (Theodore Cooper for Octave Chanute)
- 1880 Quebec Government Railways
- 1880 New York, Pennsylvania & Ohio Railroad
- 1887 Pencoyd Iron Works (Charles C. Schneider)
- 1895 Baltimore & Ohio Railroad

The large-magnitude dynamic loads imposed on bridges by railroad traffic created a need for scientific design to ensure safe, reliable and economical[60] construction. Accordingly, railroad engineers and consulting engineers engaged in iron and steel railway bridge design were the leaders in the development of structural engineering practice.[61]

In early specifications, the allowable stresses for materials were based on generally conservative estimates of the tensile and compressive strength of steel. Nevertheless, these early specifications permitted greater allowable stresses for truss members where secondary stresses[62] had been determined. Allowable stresses for steel and fasteners have been continuously modified in subsequent editions of the AREMA MRE, reflecting the latest materials research and bridge engineering research.

Design loads also evolved in response to existent railroad train loads. In the decade following 1860, iron railway trusses were typically designed for a uniformly distributed live load of between 30 kN/m and 60 kN/m (1 ton/ft and 2 tons/ft). By 1875, 100 kN (12 ton) axle loads were being specified, and in 1877 100 kN (24 kips) axle loads with a uniformly distributed live load of 45 kN/m (3 kips/ft) were stipulated. The 1879 specification of the Erie Railroad, prepared by Theodore Cooper, provided a design live load equivalent to Cooper's subsequent 1884 live load recommendation of E22. AREMA MRE design loads in 1906 were specified as Cooper's E40. The design loads increased in 1920, 1935 and 1968 to Cooper's E60, E72 and E80, respectively. In addition to the specified Cooper's design live loads were recommendations for alternate live load[63] and impact factor equations.[64] The 2024 AREMA MRE indicates a minimum Cooper's E80 (SI equivalent is Cooper's EM360) live load with an alternate load consisting of four 445 kN (100 kips) axles.

Various dynamic increment (impact) formulae were used for steel railway truss design prior to the 1906 AREMA MRE (Figure 1.16). Since 1906, the MRE has specified formulae for calculating steam and diesel locomotive impact forces in various editions of the AREMA and American Railway Engineering Association (AREA)[65] specifications and recommended practices. Most North American freight railroads discontinued steam locomotive use in the early 1960s. The diesel and diesel-electric locomotives that followed, in conjunction with improved track design, construction and maintenance practices, have allowed bridge designers to use smaller-impact loads for design.[66] Figure 1.17 outlines the recommended dynamic increment (impact) for truss spans in the AREMA and AREA specifications and recommended practices of 1906, 1920, 1935 and 1968.[67]

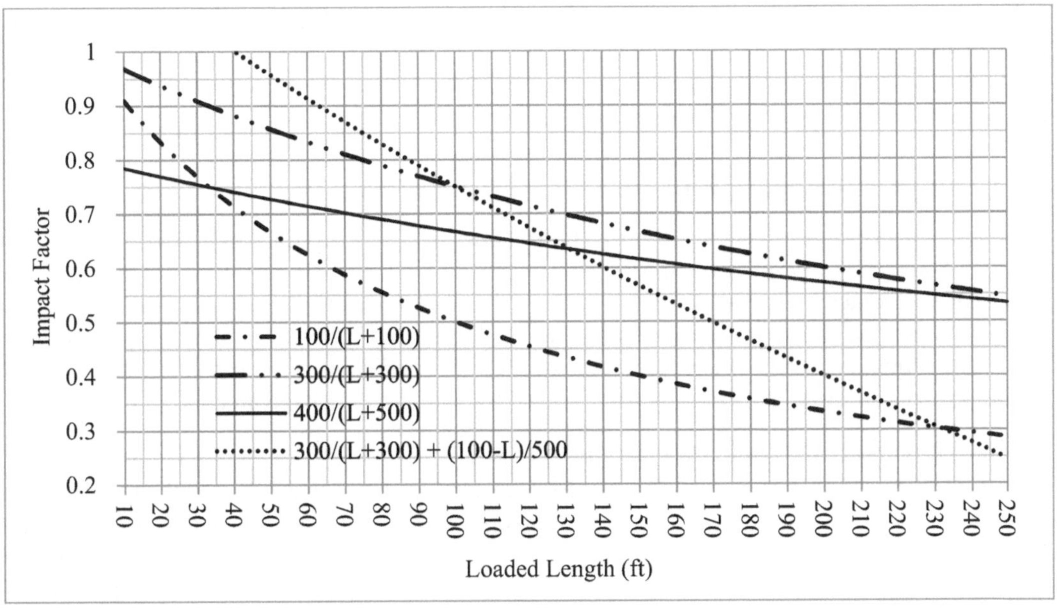

FIGURE 1.16 Dynamic increment (impact) used for steel railway truss design prior to the 1906 AREMA specification.

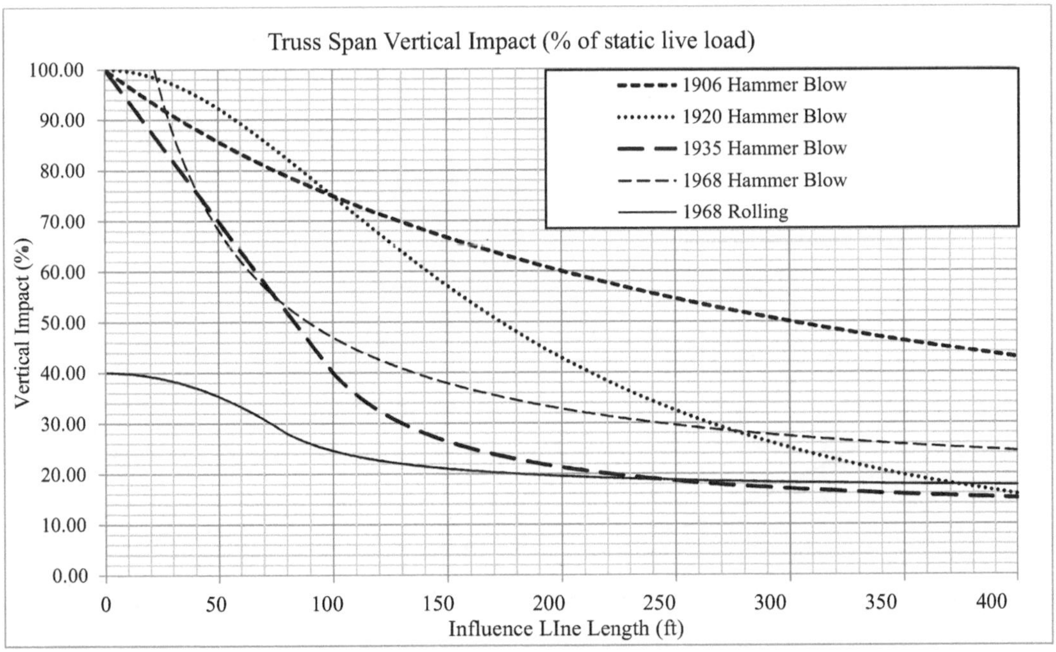

FIGURE 1.17 AREMA and AREA impact loads for truss spans since 1906.

Well-maintained steel railway bridges designed prior to 1960 with conservative allowable stresses[68] for heavy steam locomotives[69] and large dynamic increment (steam locomotive impact) continue to safely and reliably carry modern railway traffic.[70] Many of these bridges are over 100 years old, providing evidence of the exceptional design, fabrication and construction skills of early steel railway bridge engineers and the scientific methods and specifications that guided their work. Modern steel railway bridge design practice continues this record of safety and reliability using cost-effective materials, analysis and design methods based on updated design specifications and codes, such as AREMA MRE Chapter 15—Steel Structures.

1.7.4 North American Steel Bridge Design Specifications

Various aspects of steel railway truss design have evolved since the original 1906 AREMA *General Specifications for Steel Railroad Bridges*. The design live load, a major component of the AREMA and AREA specifications, was revised in 1920, 1935 and 1968. Other provisions have also been updated as the specification is reviewed and updated annually.[71] This evolution has occurred due to the dedication and hard work of engineers striving for continuous improvement in the scientific methods associated with steel railway truss bridge design, fabrication and erection.

1.7.4.1 1906 AREMA Steel Truss Specifications

Cooper's E40 live load (maximum axle load of 178 kN (40 kips)) with two 222 kN (50 kips) axle loads at 2.1 m (7 ft) spacing with impact percentage as 30,000/(L+300) with loaded length L in ft, as shown in Figure 1.17, was specified in the 1906 AREMA specifications. An equivalent uniformly distributed live load of $\frac{T}{2\sqrt{S}}$, where T is the driving axle force (lbs.) and S is the span length in ft was also considered and is plotted in Figure 1.18 for typical driving axle force T (lbs) for various span lengths S (ft).

Truss width and depth (height) restrictions in terms of span length were specified. A truss width greater than span/20 was specified to prevent overturning due to lateral forces. A truss depth (height) greater than span/10 was directed for deflection control. If a shallower truss was used, the truss members had to be designed so that the truss deflection at midspan would not be greater than it would be for a truss with depth (height) of span/10.

The hip vertical (hanger) and similar members (for example, subvertical members in subdivided trusses) and the two end panels of the bottom chord were specified to be rigid (rivetted construction)

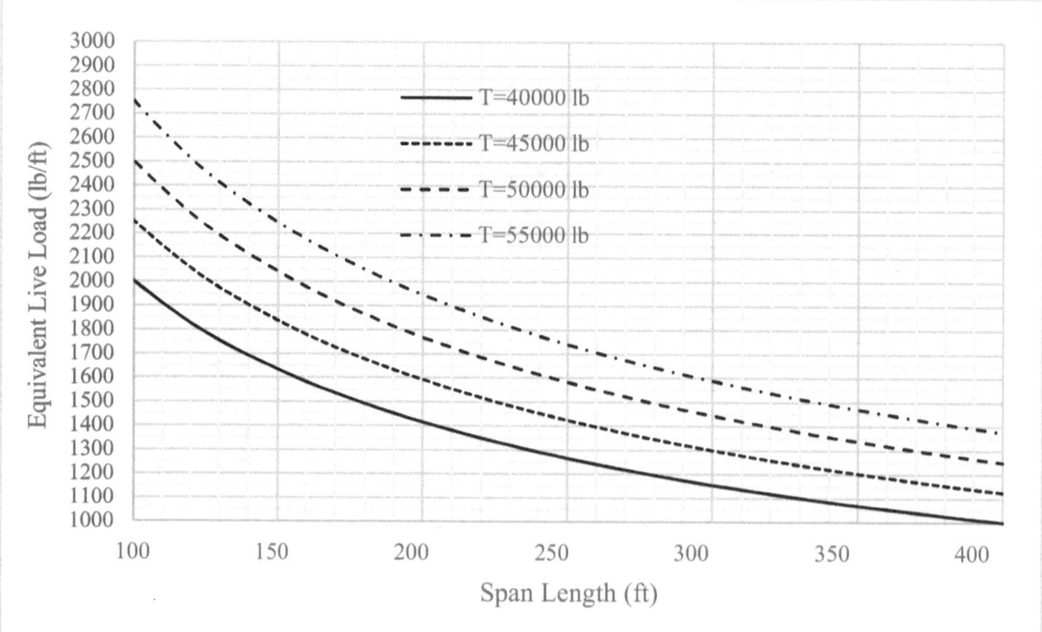

FIGURE 1.18 Equivalent uniformly distributed live load, T (lb) (1906).

in single-track pin-connected through-truss spans not longer than 90 m (300 ft). The hip vertical (hanger) members required rigidity to resist secondary out-of-plane and in-plane bending stresses. Rigidity of the bottom chord members was required to resist compression from lateral and live load rocking forces in bottom chord members lightly stressed by primary tension forces in end panels.

Members that experienced stress reversal had to be designed for both tension and compression with connections designed for the algebraic sum of the tension and compression forces. Rigid counter members were also preferred and rivetted to chord members where strain (stress) reversal occurred. If adjustable counter members were employed, they had to have turnbuckles for maintenance tightening because of loosening due to strain reversal.

The 1906 specification indicates that Pony trusses, without top chord lateral bracing, had to be riveted with double-webbed chord members to provide lateral stiffness to resist top chord instability. Truss camber was specified by increasing the top chord panel length by 3 mm (1/8 in) per 3.05 m (10 ft) of span length. In relation to this, the geometrical effects of reducing the contact surfaces at compression member joints of inclined chord trusses were deliberated upon. In addition, intermediate transverse (sway) frame and portal bracing requirements were outlined.

Secondary stresses were a concern, and provisions limited the inclination of eyebars to maintain secondary stresses at low levels in tension members. Also identified were secondary stresses in truss post members from eccentricities due to floorbeam deflection (particularly for shallow floorbeams). The specification also acknowledged that secondary bending stresses in truss chord members, due to bracing and floor system arrangements, could occur when one track is loaded on double track bridges. Nevertheless, the secondary strains were not accurately determined, and the recommendation was to disregard secondary bending stresses in typical or ordinary truss spans. There was no mention of secondary stresses in truss members due to truss deformation.[72]

Secondary bending stresses in truss members were considered only by cautioning designers that the secondary bending stresses are more important in compression members, where the secondary strains cause further deflection of the member, than for tension members where primary axial tension strains resist further deflection of the member. Truss members are subjected to axial forces and bending moments due to axial force eccentricities (often unintentional and related to connection eccentricities, member out-of-straightness and/or secondary deflection effects) and when axial members are laterally loaded (typically by self -weight and/or wind). These normal axial and flexural stresses must be combined. Axial tension combined with bending is generally of lesser concern than axial compression combined with bending due to the potential for instability of slender compression members. Therefore, compression member design had to consider slenderness in relation to elastic buckling and the inelastic effects related to load eccentricities and initial out-of-straightness. The compression member slenderness ratio for truss members was specified as less than 100 to avoid elastic buckling. Inelastic effects were represented by the Johnson formula (Equation 1.1) proposed in 1905, but a straight-line approximation (Equation 1.2) was used in the 1906 specification:

$$F_{cr} = F_y - B\left(\frac{L}{r}\right)^2 = 16000 - 0.70\left(\frac{L}{r}\right)^2 \tag{1.1}$$

$$F_{cr} = F_y - B\left(\frac{L}{r}\right) = 16000 - 70\left(\frac{L}{r}\right) \tag{1.2}$$

where F_{cr} is the critical buckling stress (psi) and F_y is the yield stress (psi).

L/r is the slenderness ratio where L is the length of the member and r is the radius of gyration in consistent units.

1.7.4.2 1920 AREA Steel Truss Specifications

A rapid increase in locomotive weight occurred in the three decades from 1890 to 1920, which precipitated a 50% increase in the design live load in the 1920 AREA specifications.[73] There were also deliberations between 1914 and 1919 regarding the effects that the rapidly increasing loads would have on existing truss spans built for lower design live loads. The 1920 specifications were appropriate for riveted trusses from 30.5 to 91.5 m (100 to 300 ft) long and pin-connected trusses from 45.7 to 91.5 m (150 to 300 ft) long.

Cooper's E60 live load (maximum axle load of 267 kN (60 kips)) with two 334 kN (75 kips) axle loads at 2.1 m (7 ft) spacing with impact percentage as $30{,}000/((L^2/100) + 300)$ with loaded length, L, in ft as shown in Figure 1.17 was specified in the 1920 AREA specifications. The 1920 specification also contained general rules for trusses, many of which remain appropriate for the design of modern steel railway truss spans. The specification indicated the use of single-intersection web systems, inclined end post members and top chord and end post member cross-sections with side components, cover plates and lacing bars. Symmetrical cross-sections for the bottom chord and web members of rivetted trusses was recommended in conjunction with the specification that gusset plates were to be proportioned for the stress transferred[74] but be no less than 13 mm (1/2″ inch) thick. Riveted truss shop assembly was also specified with the requirement that all rivet holes be drilled or reamed, and field connections were to be assembled in the shop or with metal templates.

Truss width and depth (height) restrictions in terms of span were specified. A truss width greater than span/15 was specified to prevent overturning due to lateral forces. A truss depth (height) greater than span/10 was indicated for deflection control with the same criteria for the design of shallower trusses as the 1906 specification. A further stipulation that the panel length must not exceed 150% of the truss width was added to the 1920 specification. The specification also contained provisions relating to deck truss sway, end post and end bent (portal)[75] bracing. The hip vertical (hanger) and similar members (for example, subvertical members in subdivided trusses) and the two end panels of the bottom chord were specified to be rigid (rivetted construction) in single-track pin-connected through-truss spans.

Members that experience reversal must be designed to resist both tension and compression. The 1920 specification indicated that increased resultant tensile and compressive strains be determined as the sum of the resultant tensile or compressive strain plus the smaller of either 50% of the resultant tensile or compressive strain. The connections of truss members that undergo stress reversal were specified to be designed for the sum of the tension and compression forces in the member. These criteria relating to stress reversal are shown in Example 1.1. Members that experienced stress reversal and counter members were preferred to be rivetted. If adjustable counters were employed, they had to have turnbuckles for maintenance tightening because of loosening due to strain reversal.

Example 1.1a (SI Units)

Determine the design forces specified by the 1920 AREA specification for a truss member subjected to stress reversal and the following forces:

DL = −100 kN
LL + I = −1160 kN
Reversal LL + I = 735 kN

Determine the design forces for this member subject to stress reversal:

Total force in member = −100 − 1160 = −1260 kN
Total reversal force in member = −100 + 735 = 635 kN

50% of the lesser total force = 0.50(635) = 317.5 kN
Compression design force = −1260 − 317.5 = −1577.5 kN
Tension design force = 635 + 317.5 = 952.5 kN
Connection design force = −(−1260) + 635 = 1895 kN

Example 1.1b (US Customary and Imperial Units)

Determine the design forces specified by the 1920 AREA specification for a truss member subjected to stress reversal and the following forces:

DL = −22.3 kips
LL + I = −260.0 kips
Reversal LL + I = 165.1 kips

Determine the design forces for this member subject to stress reversal:

Total force in member = −22.3 − 260.0 = −282.3 kips
Total reversal force in member = −22.3 + 165.1 = 142.8 kips
50% of the lesser total force = 0.50(142.8) = 71.40 kips
Compression design force = −282.3 − 71.4 = −353.7 kips
Tension design force = 142.8 + 71.4 = 214.2 kips
Connection design force = −(−282.3) + 142.8 = 425.1 kips

Truss camber was specified by changing truss member lengths to accommodate the deflection due to dead load and live load (without impact). This required that truss deflections and corresponding member changes in length under dead and live load be calculated.[76]

Intermediate transverse (sway) frames were specified at each panel point with a minimum depth of 1.8 m (6 ft) deep if clearances allowed. Otherwise, sway bracing with struts as deep as chord members and knee braces was specified. End portal bracing with knee bracing requirements were outlined. The 1920 specification also indicated that lateral bracing must be arranged to avoid bending stresses in truss members. Furthermore, the specification indicated that both systems of double-bracing arrangements were to be considered active if the members met the requirements for both tension and compression members.

Secondary stresses in ordinary trusses were deemed to not be of concern provided that the width of members in the plane of the truss did not exceed 10% of the member length. Secondary stresses due to truss deflections in trusses with subpanels or trusses with members with width greater than 10% of member length had to be determined. AREA Committee 15 recognized that calculating secondary stresses was not common practice and, in 1914, presented a method of determining secondary stresses in truss members due to truss deformation; in vertical post members due to floorbeam deflection from frame action and lateral bending stresses of floorbeams due to longitudinal deformation of the truss chords. Eyebar cross-section[77] and packing rules were provided to maintain secondary stresses at low levels in tension members.

Axial tension combined with bending is generally of lesser concern than axial compression combined with bending due to the potential for instability of slender compression members. Superimposed tension stresses from combined axial and bending forces in truss members[78] were specified to be less than the allowable tension stress.[79] It was also specified that if the truss chord member subjected to combined axial and bending forces was continuous over a panel point, only 75% of the bending stress had to be added to the axial stress. The compression member slenderness ratio for truss members was specified as less than 100 to avoid elastic buckling. Inelastic effects were represented by a straight-line approximation (Equation 1.3):

$$F_{cr} = 15000 - 50\left(\frac{L}{r}\right) \tag{1.3}$$

Of great concern to railway bridge engineers was the large increase in locomotive weight over a few decades that precipitated a 50% increase in the design live load. The effects that the rapidly increasing loads would have on existing truss spans[80] built for lower design live loads was investigated. It was determined that as the live load increases, some members reach capacity before other members. Analyses of typical railway truss spans built in the first two decades of the 20th century revealed that the largest percentage increase in stress for live loads heavier than the design live load occurs in web members near the center of the truss. Example 1.2 demonstrates the issue for a 36.6 m (120 ft) six-panel Pratt truss built without counters to Cooper's EM178 (Cooper's E40) design live load.

Example 1.2a (SI units)

A 7.62 m high and 36.59 m long six-panel Pratt through truss span was designed in 1908 for a dead load of 17.5 kN/m and live load of Cooper's EM178. The impact factors were 71% for truss members except for hip vertical members (hangers) where impact was 94% (Figure 1.17).

By 1920, the specified design live load was Cooper's EM267, and the railroad wanted to understand how the bridge would perform under the updated design live load of Cooper's EM267. The impact factors were 68% for truss members except for hip vertical members (hangers) where impact was 99% (Figure 1.17). Dead load was assumed to increase by 33% due to heavier deck ties (sleepers). The forces in the truss members with Cooper's EM178 and Cooper's EM267 live load are shown in Table E1.2a

TABLE E1.2a
Forces in Truss Members with Cooper's EM178 and Cooper's EM267 Live Load

Member		DL + LL1 + I1 (EM178) (kN)	DL + LL3+ I3 (EM267) (kN)	(DL + LL3+ I3) / (DL + LL1 + I1)
End post chord	L0-U1	−1482.1	−2159.5	1.46
Top chord	U1-U2	−1448.7	−2110.1	1.46
Top chord	U2-U3	−1591.5	−2317.9	1.46
Bottom chord	L0-L1	925.2	1348.2	1.46
Bottom chord	L1-L2	925.2	1348.2	1.46
Bottom chord	L2-L3	1448.7	2110.1	1.46
Hanger (hip vertical) web	U1-L1	619.6	942.1	1.52
Vertical web	U2-L2 (LLC)	−431.0	−628.5	1.46
	U2-L2 (LLT)	165.0	246.9	1.50
Vertical web	U3-L3	−17.8	−23.6	1.33
Diagonal web	U1-L2 (LLC)	36.5	39.1	1.07
	U1-L2 (LLT)	967.0	1410.5	1.46
Diagonal web	U2-L3 (LLC)	−210.4	−314.9	1.50
	U2-L3 (LLT)	527.1	771.7	1.46

Example 1.2b (Imperial/US Customary units)

A 25 ft high and 120 ft long six-panel Pratt through-truss span was designed in 1908 for a dead load of 1200 lb/ft and live load of Cooper's E40. The impact factors were 71% for truss members except for hip vertical members (hangers), where impact was 94% (Figure 1.17).

By 1920, the specified design live load was Cooper's E60 and the railroad wanted to understand how the bridge would perform under the updated design live load of Cooper's E60. The impact factors were 68% for truss members except for hip vertical members (hangers), where impact was 99% (Figure 1.17). Dead load was assumed to increase by 33% due to heavier deck ties (sleepers). The forces in the truss members with Cooper's E40 and Cooper's E60 live load are shown in Table E1.2b

TABLE E1.2b
Forces in Truss Members with Cooper's E40 and Cooper's E60 Live Load

Member		DL + LL1 + I1 (E40) (kips)	DL + LL3+ I3 (E60) (kips)	(DL + LL3+ I3) / (DL + LL1 + I1)
End post chord	L0-U1	−333.2	−485.5	1.46
Top chord	U1-U2	−325.7	−474.4	1.46
Top chord	U2-U3	−357.8	−521.1	1.46
Bottom chord	L0-L1	208.0	303.1	1.46
Bottom chord	L1-L2	208.0	303.1	1.46
Bottom chord	L2-L3	325.7	474.4	1.46
Hanger (hip vertical) web	U1-L1	139.3	211.8	1.52
Vertical web	U2-L2 (LLC)	−96.9	−141.3	1.46
	U2-L2 (LLT)	37.1	55.5	1.50
Vertical web	U3-L3	−4.0	−5.3	1.33
Diagonal web	U1-L2 (LLC)	8.2	8.8	1.07
	U1-L2 (LLT)	217.4	317.1	1.46
Diagonal web	U2-L3 (LLC)	−47.3	−70.8	1.50
	U2-L3 (LLT)	118.5	173.5	1.46

Members with an increase in axial force greater than that for chord members were considered at risk of overload prior to the chord members under loads heavier than the design live load. These overloaded members were web members in the trusses analyzed. Consequently, many trusses were removed from service due to weaker web members and concerns about overloading. Essentially, many truss spans were removed from service due to the difficulty of strengthening[81] web members, which constitute only a small portion of the total weight of the span, or the complexity of adding counter members if panel shear reversal developed under heavier loads.[82] To avoid web member overloading that would be associated with future live load increases, railway truss design engineers needed a design method to ensure that no members of the truss reached their overload capacity before the chord members, and they considered three solutions: a high design live load with high allowable stresses (presuming higher allowable stresses in future); the usual design load but with higher loads anticipated by using lower live load allowable stresses; or a design load for web members that is associated with an increase in axial force in chord members under loads heavier than the design live load.

The first was a typical modern approach, but reluctance to estimate future live loading[83] based on the substantial live load increases since the beginning of the 20th century made this option problematic.[84] The second option of using lower allowable stresses for live loads was also not satisfactory due to the complexity it brought to design load combinations and inconsistency with material strength.[85] The third option was adopted of designing truss web members for a live load associated with an increase in axial live load force in chord members that increased the total stress in the chord by an overload amount.

Nevertheless, in 1919, the method was implemented by proportioning web members of trusses for two thirds of the dead load plus one and one sixth times the live load stress, including impact. This was applied when the resulting stress was greater than the sum of the dead load and live load stresses including impact. There was considerable deliberation regarding the inappropriateness of estimating overload capacity using low allowable stresses and some support for designing truss members to carry maximum load with maximum allowable stress. Yet the 1920 AREA specification retained the proposed two-thirds dead load plus one and one-sixth times live load rule pending further investigation.

By 1924, the specification committee was recommending that web members be proportioned such that an increase of live load which increases the total stresses in the chords by 50%, will produce total stresses in the web members not more than 50% greater than the allowable stresses. The 50% criterion, shown in Figure 1.19 appears to align with the Cooper's E60 to Cooper's E40 axle load ratio of 1.50.

The proportioning of truss web members for a live load associated with an increase in axial force in chord members that increases stress in the chord by an overload amount continued to be deliberated upon as it was believed that the 50% criterion was too severe. Accordingly, the 1935 specification provided for proportioning of web members such that an increase of live load that increased the total stresses in the chords by 33% would produce total stresses in the web members not more than 33% greater than the allowable stresses.

This recommendation for overload design of truss web members remains in the 2024 edition of the AREMA MRE, is shown in Figure 1.20 and is outlined in Chapter 9, where the overload design force, F_{ol}, is shown to be

$$F_{ol} = \frac{DL_w + K_c\left(LL + I + CF\right)_w}{1.33} \tag{1.4a}$$

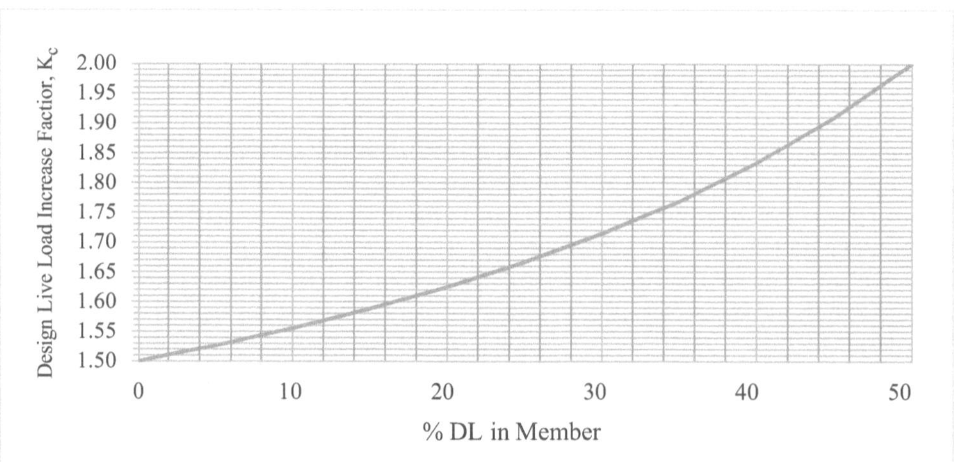

FIGURE 1.19 Design live load increase factor for 50% criterion of 1924.

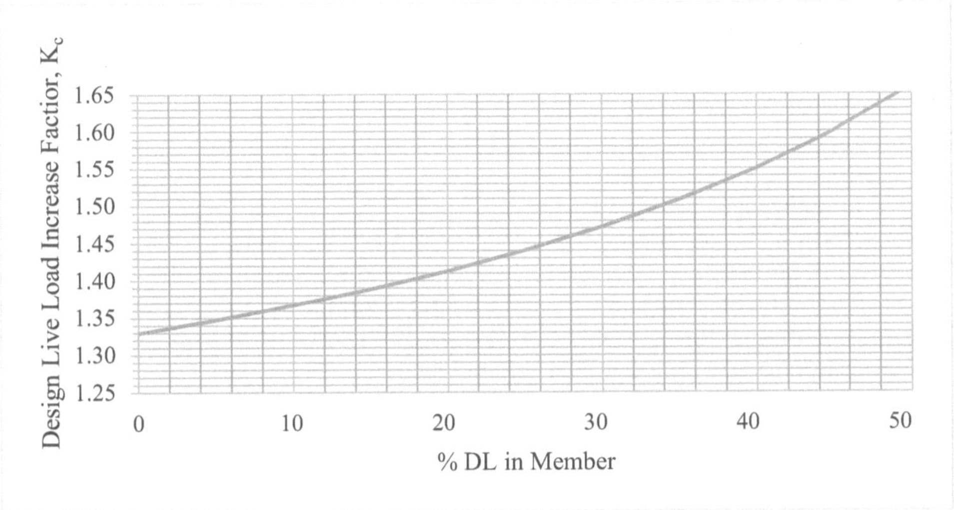

FIGURE 1.20 Design live load increase factor for 33% criterion of 1935.

where DL_W = dead load in web member under consideration.

$$K_c = \frac{133 - \%DL_c}{100 - \%DL_c} \tag{1.4b}$$

$\%DL_c$ = the percentage of total load in the truss chord attributed to dead load = $100(DL_c/TL_c)$
DL_c = dead load in most highly stresses chord member
TL_c = total load (dead load, live load plus impact, and centrifugal force) in most highly stressed chord member

1.7.4.3 1920 Canadian Engineering Standards for Steel Railway Truss Bridges

The Canadian specifications relating to steel railway truss span design differ from the AREA specifications in some respects.

The specification was appropriate for trusses up to 152 m (500 ft) long. Cooper's series live load with two axle loads at 2.1 m (7 ft) spacing with impact percentage as $3 \times 10^6/(L^2 + 30 \times 10^3)$ with loaded length, L, in ft was specified with axle load magnitudes as directed by the engineer. Truss spacing was specified as that necessary to resist overturning from lateral loads but not less than span/20.

Members that experienced stress reversal had to be designed for[86] dead load plus live load and impact of the same sign and two-thirds of the dead load plus live load and impact of the opposite sign. Web members subject to stress reversals at each application of live load also had to be designed for dead load plus live load and impact of each sign plus 50% of the smaller. The connections of truss members that undergo stress reversal were specified to be designed for the sum of the tension and compression forces in the member. This criterion for the design of members subject to stress reversal is shown in Example 1.3. Members that experienced stress reversal and counter members were preferred to be rivetted. If adjustable counters were employed, they had to have turnbuckles for maintenance tightening because of loosening due to strain reversal.

Example 1.3 (US Customary and Imperial Units)

Determine the design forces specified by the 1920 Canadian specification for a truss member subjected to stress reversal and the following forces:

DL = −22.3 kips
LL + I = −260.0 kips
Reversal LL + I = 165.1 kips

Determine the design forces for members subject to stress reversal:

Member compression design force = −22.3 − 260.0 = −282.3 kips
Member tension design force = 0.67(−22.3 + 165.1) = 95.7 kips
Web member compression design force = −353.7 kips (Example 1.1b)
Web member tension design force = 214.2 kips (Example 1.1b)
Compression design force = −282.3 − 71.4 = −353.7 kips
Tension design force = 142.8 + 71.4 = 214.2 kips

The member compression design force of −282.3 kips and member tension design force of 95.7 kips would apply to chord members that experience stress reversal:

Member connection design force = −(−282.3) + 95.7 = 378.0 kips
Web member connection design force = −(−22.3 − 260.0) + (−22.3 + 165.1) = 282.3 + 142.8 = 425.1 kips (Example 1.1b)

Truss camber was specified by either increasing the length of top chord members by 3 mm in 3.0 m (1/8 in. in 10 ft) or by modification of truss member lengths such that the bottom chord would be level when loaded by dead, live and impact loads. The latter required calculating truss distortion and corresponding member length changes under dead and live loads.

The specification also indicated that bracing arrangements were to be designed as axial tension or compression members considering bending due to self-weight. Secondary stresses due to truss deflections from dead, live and impact loads were required to be determined if directed by the engineer. The combined primary axial and secondary bending stresses were not to exceed the allowable axial stress[87] by more than 33%. Combined primary axial, secondary bending stresses and stresses due to wind, temperature and/or longitudinal forces due to braking or traction were not to exceed the allowable axial stress by more than 50%. A maximum eyebar inclination of 1 mm in 190 mm (1 in in 16 ft) was specified to maintain secondary stresses at low levels in tension members.

Members subject to combined axial tension or compression and bending stresses were to be designed such that allowable axial tension or compression stresses, respectively, were not exceeded. If bending in the member was due to self-weight or force eccentricities at joints, then the combined axial and bending stresses were not to exceed the allowable axial tension or compression stresses by 10%. It was also specified that if the truss member subjected to combined axial and bending forces was continuous over a panel point, only 75% of the bending stress must be added to the axial stress. The compression member slenderness ratio for truss members was specified as less than 175 to avoid elastic buckling. Inelastic effects were represented by a parabolic approximation (Equation 1.4):

$$F_{cr} = 12,500 - 0.25\left(\frac{L}{r}\right)^2 \qquad (1.4)$$

The specification of designing truss web members for a live load associated with an increase in axial live load force in chord members that increases the total stress in the chord by an overload amount was not considered in the Canadian specification of 1920. The Canadian specification of 1950 was similar to the AREA specification of 1948 with some revised clauses and no consideration for truss web member design for overload.

1.7.4.4 1935 AREA Steel Truss Specifications

The 1935 specifications were appropriate for riveted trusses greater than 30.5 m (100 ft) long and pin-connected trusses longer than 91.5 m (300 ft) long. Consequently, by 1935, most steel railway truss spans in North America were designed, fabricated and erected with rivetted joints.

Cooper's E72 live load (maximum axle load of 320 kN (72 kips)) with two 400 kN (90 kips) axle loads at 2.1 m (7 ft) spacing with impact percentage as 100—0.6L for loaded length, L, in ft less than 100 ft and (1800/(L – 40)) + 10 for loaded length, L, in ft greater than 100 ft as shown in Figure 1.17 was specified in the 1935 AREA MRE. The 1935 specification increased the impact percentage by adding a lurching (rocking) effect percentage of 100/S, where S is the spacing of longitudinal members (trusses, stringers) and length of transverse members (floorbeams).

The 1935 specification also contained general rules for trusses, many of which were the same as specified in the 1920 specification. Truss width and depth (height) restrictions in terms of span were specified. A through-truss width greater than span/20 and a deck truss width greater than span/15 were specified to prevent overturning due to lateral forces. A truss depth (height) greater than span/10 was indicated for deflection control. Even though the 1935 specification applied to pin-connected through-truss spans greater than 91.5 m (300 ft), hanger and similar members, and the two end panels of the bottom chord were specified to be rigid.

Members that experience reversal must be designed to resist both tension and compression. The 1935 specification indicated that increased resultant tensile and compressive strains be determined as the sum of the resultant tensile or compressive strain and the smaller of either 50% of the resultant tensile or compressive strain. The connections of truss members that undergo stress reversal were specified to be designed for the algebraic sum of the tension and compression forces in the member. Members that experience stress reversal and counter members were preferred to be rivetted. If adjustable counters were employed, they had to have turnbuckles for maintenance tightening because of loosening due to strain reversal.

Truss camber was specified to accommodate the deflection due to dead load plus 43.8 kN/m/track (3.0 kips/ft/track). This required that truss deflections and corresponding member changes in length be calculated. However, using an equivalent uniformly distributed live load reduced the calculation effort required.

Intermediate transverse (sway) frames were specified at each panel point with a minimum depth of 1.8 m (6 ft) deep if clearances allow. Otherwise, sway bracing with struts as deep as chord and knee braces was specified. End portal bracing with knee bracing requirements were outlined. The specification also indicated that both systems of double bracing arrangements were to be considered active if the members met the requirements for both tension and compression members.

Secondary stresses due to truss distortion and floorbeam deflection were to be kept to a minimum and were not of concern in ordinary trusses provided that the width of members parallel to the plane of the member deformation did not exceed 10% of the member length. The 1935 specification indicated that secondary stresses exceeding 27.6 MPa (4 ksi) in tension members and 20.7 MPa (3 ksi) in compression members be treated as primary axial stresses.

Superimposed stresses from combined axial and bending forces in truss members were specified to be less than the allowable axial stress. It was also specified that if the truss member subjected to combined axial and bending forces was continuous over a panel point, only 75% of the bending stress was to be added to the axial stress. The compression member slenderness ratio for truss members was specified as less than 100 to avoid elastic buckling. Inelastic effects were represented by a parabolic approximation for rivetted (Equation 1.5a) and pin-connected (Equation 1.5b) truss members:

$$F_{cr} = 15000 - 0.25\left(\frac{L}{r}\right)^2 \tag{1.5a}$$

$$F_{cr} = 15000 - 0.33\left(\frac{L}{r}\right)^2 \tag{1.5b}$$

The effects increasing loads on existing truss spans remained of concern.[88] Higher allowable stresses were permitted for secondary stresses and load combinations, but the concern was that as live loads increased, truss web members would reach capacity prior to the chord members. Therefore, AREA maintained a truss web member design criterion that used a live load associated with an increase in axial live load force in chord members that increases the total stress in the chord by an overload amount. The design criterion was to proportion web members such that increasing live load which increases the total stresses in the chords by 33% would produce total stresses in the web members not more than 33% greater than the allowable stresses. The 33% factor was related to the ratio of allowable stresses under usual loads and those proposed for overload conditions. The allowable tensile stress in the 1935 specification was 124 MPa (18 ksi), and it was proposed to review chord stresses under higher loads at 165 MPa (24 ksi).[89]

1.7.4.5 North American Railway Truss Design Specifications 1935 to 2024

North American, and particularly US, railway bridge engineers have engaged in considerable research, testing and assiduous standards development from 1935 to the present. Accordingly, Canadian railway bridge engineers favored the continuously updated AREA recommended practices over the 1950 Canadian Standard for Steel Railway Bridges (CSA-S1). Therefore, innately, the AREA specification became the model for North American railway company bridge design standards and continues in this role through the efforts of AREMA Committee 15—Steel Structures. This committee, consisting of engineers engaged in academia, regulation, research, design, fabrication and construction of steel railway bridges in North America, annually reviews and updates AREMA Chapter 15.

NOTES

1 Other structural forms such as beams, columns, arches and cables are found in nature.
2 Palladio built bridges and roofs of truss forms in the latter half of the 16th century. His truss forms consisted of top and bottom chord members with perpendicular vertical members and either single or double diagonal web members. He also wrote a treatise on truss architecture.
3 Wooden bridges were built over the Rhine River in the middle of the 18th century.
4 Palmer constructed a bridge over the Schuylkill River at Philadelphia in 1804. It was later replaced to carry railroad loads.
5 Burr built a four-span wood bridge over the Hudson River in 1804. His truss had parallel horizontal chord members, vertical post members and double diagonal web members. It was reinforced by a timber arch. This truss form was used extensively to carry American highway loads at many locations for another 50 years.
6 Wernwag built a bridge over the Schuylkill River at Philadelphia in 1812. Its clear span of almost 104 m (340 ft) was a hallmark of American bridge building in the early 19th century.
7 For example, locomotive axle loads were about 50 kN (11 kips) on the Baltimore and Ohio (B&O) Railroad in 1835 (about 15% of modern locomotive axle loads).
8 The Susquehanna River bridge burned in 1844.
9 The Howe truss was patented in the United States by William Howe in 1840.
10 Replicating American practice, many wooden Long trusses were used in the construction of the St. Petersburg and Moscow Railroad in the mid-19th century.
11 Haupt wrote a book titled *General Theory of Bridge Construction* in 1851.
12 The Howe truss is like the earlier wooden Burr truss but using iron rod vertical members and excluding the wood stiffening arch.
13 This bridge was replaced with wrought iron truss spans in 1874.
14 This bridge was replaced with a steel-trussed arch in 1893.
15 Progressively, cast iron, wrought iron and steel were used to construct railway bridges.

16 For example, many temporary timber Howe truss bridges were constructed in the late 1880s during construction of the Canadian Pacific Railway to expedite construction and meet contractual obligations to the Government of Canada.
17 The wood compression members typically framed into the wood chord members with cast iron blocks.
18 Highway traffic consisting of pedestrian, equestrian and horse-drawn cart traffic. Paine developed a model of the arch span, but the US Congress built no such bridges. The model was later sold to pay Paine's creditors.
19 Also referred to as Pauli spans.
20 Bollman trusses used wrought iron tension members and cast-iron compression members. The redundant nature of the truss form reduced the possibility of catastrophic failure.
21 The Fink truss used fewer inclined tensioned members and a wrought iron bottom chord that carried live load.
22 The New York Central Railroad also initiated the use of iron stringers (instead of the typically used wood stringers) in railway trusses in the 1860s.
23 The bridge was a lattice truss span.
24 The B&O Railroad was chartered in 1827 and built the first railroad bridge in North America, an arch-stiffened wooden truss, in 1830.
25 The first all-iron trusses on the B&O were designed by Fink in 1853.
26 The Whipple truss, a double-intersection truss form, enables the subdivision of long panels in a Pratt truss but is statically indeterminate (cannot determine member forces from only static equilibrium).
27 The double intersection tension ties in the web extend over two truss panels. The truss members are pin connected at the top and bottom chords.
28 In 1847, Whipple published *A Treatise on Bridge Building*, the first book on the scientific or mathematical analysis of trusses.
29 The truss was designed by Jacob. H. Linville.
30 Many of the pin-connected truss spans in Canada were replaced with riveted truss spans in the late 19th and early 20th centuries. However, pin-connected spans in the United States remained in service for many decades after the turn of the century.
31 John W. Murphy designed the Whipple truss using all wrought iron members (cast iron was retained for the joint blocks). The double-intersection Whipple truss using all wrought iron members is sometimes referred to as the Murphy truss or Murphy–Whipple truss.
32 With the exception of cast-iron bearing blocks at ends of truss compression members.
33 The earliest Pratt trusses had iron diagonal members in tension and vertical post members of wood in compression.
34 Web members that resist both tension and compression forces are riveted built-up members.
35 Statically determinate trusses are those in which the internal member forces may be calculated based on only force equilibrium at joints or sections of the truss.
36 Web members that resist tension forces only are eyebar members.
37 The truss form is statically determinate if only one member in each panel is active in tension. However, the truss form will change as the live load traverses the span.
38 Counter members provide for a truss form such that web members are not subject to stress reversal created by positive and negative panel shear forces, particularly in the central panels.
39 Statically indeterminate truss forms have redundant members such that the equations of force equilibrium (statics) are insufficient to determine internal member forces. In such cases, additional equations corresponding to the deformation of each redundant member are required. These additional equations determine the relationship between the deformation of the redundant members and the statically determinate truss with the redundant members removed. Consequently, to avoid the difficulties associated with the analysis of statically indeterminate truss forms, an approximate analysis was typically conducted assuming that the shear force in a panel was equally distributed to each member with one member in compression and the other in tension.
40 These truss forms had subdivided panels to accommodate long spans. The Petit truss was used extensively by American railroad companies.
41 In 1895, steel completely replaced wrought iron in the production of manufactured structural shapes.
42 Typically, where fast-flowing, wide or deep watercourses must be traversed, and substructure construction cost is reduced by use of longer spans. For example, in 1906, Canadian Pacific Railway constructed a 126 m (413 ft) rivetted subdivided Warren truss span to avoid a 27 m (90 ft) deep river crossing pier. The span was one of the longest simply supported steel railway truss spans in the world.

43 This type of bridge design and construction is attributed to the German engineer Heinrich Gerber, who patented and constructed the first cantilever type bridge in 1867.
44 At the location of an uncompleted suspension bridge by John Roebling.
45 This was the first use of cantilever construction using a suspended span.
46 The 1897 steel arch crossing of the Niagara Gorge was designed by Leffert L. Buck and detailed in his 1899 report to the joint Board of Directors of the Niagara Falls Suspension Bridge Company and the Niagara Falls International Bridge Company. The report is based on the paper by Richard S. Buck in the Transactions of the American Society of Civil Engineers, Vol. xi, p. 125.
47 American engineers Charles Shaler Smith and Charles C. Schneider had designed and constructed cantilever railway bridges in North America.
48 The original designers were Theodore Cooper and Peter Szlapka (Phoenixville Bridge Company). Following the collapse, a design was submitted by Henri. E. Vautelet, but the redesign of the bridge was tendered to various bridge companies and carried out by George H. Duggan (St. Lawrence Bridge Company) under the direction of Charles C. Schneider, Ralph Modjeski and Charles N. Monsarrat.
49 Alloy nickel steel was first used in 1909 on the Blackwell's Island (now Queensboro) Bridge in New York. Nickel steel was also used extensively by J. A. L. Waddell for long-span railway bridge designs. A. N. Talbot conducted tests of nickel steel connections for the Quebec bridge reconstruction.
50 These spans were replaced in 1912 due to concern over performance under heavier train loads. The lead ends of the simple-span replacement trusses were supported by falsework on a movable barge during installation on an adjacent alignment. A thorough historical review of the St. Lawrence River bridge is provided by S.D. Werry in the *Canadian Journal of Civil Engineering*, Vol. 24: p. 480–488 (1997).
51 Locomotive weights were typically about 40 tons in 1860, 70 tons in 1880, 100 tons in 1890, 125 tons in 1900 and 150 tons in 1910.
52 Charles Ellet (1830), Ralph Modjeski (1855), Louis F.G. Bouscaren, Chief Engineer of the Cincinnati Southern Railroad (1873) and Henri. E. Vautelet, Bridge Engineer of the Canadian Pacific Railway (ca. 1876) were graduates of early French engineering schools.
53 Often through failures of trusses.
54 The Warren truss, invented by James Warren and Willoughby Monzani, was first used in a railway bridge in 1853 on the Great Northern Railway in England.
55 Influence lines are used in algorithms for modern digital truss analysis.
56 There were also material quality issues with the cast iron compression blocks that were not discovered as the testing arranged by the Lake Shore & Michigan Southern Railroad company was inadequate.
57 Morison and Octave Chanute (an engineer educated in France) designed the 249 m (818 ft) long wrought iron viaduct for the Erie Railroad at Portage, New York, in 1875. This bridge remained in service with extensive strengthening until it was replaced in 2018 with a trussed-steel arch designed by consulting engineering firm Modjeski and Masters for the Norfolk Southern Railway. The design and construction of the 2018 bridge is outlined in the paper by Johns, Greco and Swanson presented at the AREMA Annual Conference in 2018.
58 However, it appears that the first use of concentrated wheel loads for bridge design was by the New York Central Railroad in 1862. Impact of 10% to 30%, depending on the truss member, was included in these Cincinnati Southern Railroad specifications.
59 This was not a general specification but was the first use of specification documents in the design and construction of railway bridges in the United States. The specification also included the first requirements for the inspection of material.
60 This can be a critical consideration as most railway bridge construction projects are privately funded by railroad companies.
61 The advanced state of steel design and construction knowledge possessed by railway bridge engineers made them greatly in demand by architects from about 1880 to 1900 during the rebuilding of Chicago after the Great Fire.
62 Secondary stresses in truss members are typically flexural stresses due to fabricated eccentricities at joints and flexural stresses created due to joint rigidity and truss deformation.
63 The alternate load consisted of two axles spaced at 2.1 m (7 ft), each axle with magnitudes of 223, 334 and 400 kN (50, 75 and 90 kips) in 1906, 1920 and 1935, respectively.
64 The changes to the specified impact factors for truss spans in 1906, 1920, 1935 and 1968 are shown in Figure 1.16.
65 From 1912 to 1998, AREMA was known as AREA.
66 Steam locomotive impacts were very large due to eccentric reciprocating wheel motion or "hammer blow".

67 Figure 1.17 is shown in US customary or imperial units only as the impact formulae of these older specifications and recommended practices were provided only in those units. The 2024 AREMA recommendations for impact loads in Chapter 15 are the same as the 1968 recommendations.
68 Particularly for bridges designed in the early part of the 20th century.
69 Steam locomotives used in the early part of the 20th century weighed about the same as modern diesel locomotives.
70 Modern rail car axle loads are typically not greater than modern diesel locomotive or older steam locomotive combined static and dynamic loads. However, older bridge design specifications did not consider fatigue as a design limit state (and did not need to due to the light rail cars pulled by few heavy locomotives). Nevertheless, older railway bridges generally perform well in the modern cyclical railway live load environment due to low allowable design stresses, internal redundancy of riveted connections and the use of modern methods of fatigue life evaluation. Modern steel bridges must be designed considering fatigue due to the large number of high-magnitude tensile stress ranges experienced by some steel superstructure members (typically members with short influence lines) and details.
71 For example, allowable stresses, impact formulae, column formulas and many other provisions of the specifications have been reviewed and revised since the first edition of the AREMA *General Specifications for Steel Railroad Bridges* in 1906.
72 This is likely because pin-connected joints were not considered as rigid.
73 In the late 19th and early 20th centuries, many bridge companies, railroads and consulting engineers were already using Cooper's E50 or other larger design live loads than the Cooper's E40 specified in the 1906 AREMA specifications.
74 Gusset plate design guidance was very limited. Gusset plate design was advanced significantly with Whitmore's experiments of the early 1950s.
75 To transfer top lateral bracing system reactions to the substructure.
76 Usually, by virtual work methods. Using an equivalent uniformly distributed live load reduced the calculation effort required.
77 Cross-section rules for the eyebar shaft and head were provided.
78 Including bending from floorbeam deflections.
79 110 MPa (16,000 psi) in the 1920 MRE.
80 Typically, less than 30 years old.
81 Members in tension, such as U1-L2 in Example 1.2, may be eyebars that are difficult to strengthen at congested chord joints.
82 Member U2-L3 in Example 1.2 would be in tension only with a counter present. For member U2-L3 and the counter in the 3rd panel, the increase in axial force is about the same as that for the chord members, thereby reducing the concerns with overloading.
83 Typically for about 50 years in the future.
84 This is less problematic for modern truss design as it is unlikely that axle loads will exceed even 25% of modern locomotive and freight car axle loads due to many other constraints related to railroad infrastructure and operations. However, the substantial increases of the early 20th century affected primarily bridges and not other railroad infrastructure.
85 This method effectively imposed different factors to the allowable stresses for dead and live loads, which is analogous to load factor design methods.
86 By 1950, the Canadian Standards Association specification were indicating that members that experience stress reversal must be proportioned for the maximum combined (typically, combined dead, live and impact loads) force of each sign, each being increased by 50% of the smaller.
87 Allowable axial tensile stress = 110 MPa (16 ksi). Allowable axial compression stress by Equation 1.4.
88 The design live load increased by 20% since the previous 1920 design load specification.
89 Proposed in 1934 by Shortridge Hardesty. Hardesty also indicated that when the live to dead load ratio was 5, the web member design live load would increase by about 40%.

BIBLIOGRAPHY

Akesson, B., 2008, *Understanding Bridge Collapses*, Taylor & Francis, London.

AREA Specifications for Iron and Steel Railway Bridges, 1906, American Railway Engineering Association, Chicago, IL.

AREMA Manual for Railway Engineering—Steel Structures, 1968, American Railway Engineering and Maintenance-of-way Association, Washington, DC.

AREMA Manual for Railway Engineering—Steel Structures, 2024, American Railway Engineering and Maintenance-of-way Association, Lanham, MD.

AREMA Specifications for Steel Railway Bridges, 1920, American Railway Engineering and Maintenance-of-way Association, Chicago, IL.

AREMA Specifications for Steel Railway Bridges, 1935, American Railway Engineering and Maintenance-of-way Association, Chicago, IL.

Baker, B., 1862, *Long-Span Railway Bridges*, Reprint from Original, BiblioBazaar, Charleston, SC.

Bennett, R., and Skinner, T., 1996, *Bridge Failures, Recent and Past Lessons for the Future*, American Railway Bridge and Building Association, Homewood, IL.

Buck, L.L., 1898, *The Niagara Railway Arch*, Vol. xi, Transactions, American Society of Civil Engineers, New York, p. 125.

CESA Standard Specifications for Steel Railway Bridges, 1920, Canadian Engineering Standards Association, Ottawa, ON.

Chatterjee, S., 1991, *The Design of Modern Steel Bridges*, BSP Professional Books, Oxford.

Clark, J.G., 1939, *Specifications for Iron and Steel Railroad Bridges Prior to 1905*, Published by Author, Urbana, IL.

Clary, J.N., undated, *History of Early Bridge Specifications*, Virginia Department of Transportation, Structure and Bridge Division, Charlottesville, VA.

Cooper, T., 1889, *American Railroad Bridges*, Engineering News, New York.

CSA Specification for Steel Railway Bridges, 1950, Canadian Standards Association, Ottawa, ON.

Gasparini, D.A., and Fields, M., 1993, "Collapse of Ashtabula Bridge on December 29, 1876", *Journal of Performance of Constructed Facilities, ASCE*, 7(2), 109–125.

Ghosh, U.K., 2006, *Design and Construction of Steel Bridges*, Taylor and Francis, London.

Griggs, F.E., 2002, "Kentucky High River Bridge", *Journal of Bridge Engineering, ASCE*, 7(2), 73–84.

Griggs, F.E., 2006, "Evolution of the Continuous Truss Bridge", *Journal of Bridge Engineering, ASCE*, 12(1), 105–119.

Griggs, F.E., 2015, *Historic Structures: The Quebec Bridge*, Structure Magazine, Algonguin, IL.

Haupt, H., 1851, *General Theory of Bridge Construction*, D. Appleton & Co., New York.

Jacobs, D.W. and Malla, R.B., 2013, *Review of Live Load Impact Factor for Existing Truss Railroad Bridges in the United States*, Proceedings of the Joint Rail Conference, Knoxville, TN.

Johns, K.W., Greco, N., and Swanson, H.C., 2018, *Design and Construction of the New Portageville Railroad Bridge*, Proceedings of the AREMA 2018 Annual Conference & Exposition, Chicago, IL.

Kurrer, K.-E., 2008, *The History of the Theory of Structures*, Ernst & Sohn Verlag, Berlin.

Kuzmanovic, B.O., 1977, "History of the Theory of Bridge Structures", *Journal of the Structural Division, ASCE*, ST5, 1095–1111.

Marianos, W.N., 2008, "G. S. Morison and the Development of Bridge Engineering", *Journal of Bridge Engineering, ASCE*, 13(3), 291–298.

Middleton, W.D., 2001, *The Bridge at Quebec*, Indiana University Press, Bloomington, IN.

Motley, P.B., 1914, *Double Tracking of the CPR's St. Lawrence River Bridge*, Canadian Railway and Marine World, Montreal, QC.

Petroski, H., 1996, *Engineers of Dreams*, Random House, New York.

Plowden, D., 2002, *Bridges: The Spans of North America*, W.W. Norton & Co., New York.

Ryall, M.J., Parke, G.A.R., and Harding, J.E., 2000, *Manual of Bridge Engineering*, Thomas Telford, London.

Timoshenko, S.P., 1983, *History of Strength of Materials*, Dover Publications, New York.

Todhunter, I., 1885, *A History of the Theory of Elasticity and of the Strength of Materials*, Vol. I, Dover Publications, Reprint 1960.

Todhunter, I., and Pearson, K., 1893, *A History of the Theory of Elasticity and of the Strength of Materials*, Vol. II, Dover Publications, Reprint 1960.

Tyrrell, H.G., 1911, *History of Bridge Engineering*, H. G. Tyrell, Chicago, IL.

Waddell, J.A.L., 1898, *De Pontibus*, John Wiley & Sons, New York.

Waddell, J.A.L., 1916, *Bridge Engineering*, Vol. 1, John Wiley & Sons, New York.

Waddell, J.A.L., 1916, *Bridge Engineering*, Vol. 2, John Wiley & Sons, New York.

Werry, S.D., 1997, *Rails Across the River: The Story of the St. Lawrence River Bridge (1881–1915)*, Canadian Journal of Civil Engineering, Volume 24, National Research Council of Canada, Ottawa, ON.

Whipple, S., 1873, *Treatise on Bridge Building*, Reprint from Original 2nd Edition, University of Michigan, Ann Arbor, MI.

Willis, R., 1849, *Application of Iron to Railway Structures*, Commissioner's Report, William Clowes, London.

Part A

Theory and Structural Analysis of Railway Truss Bridges

2 Stability and Determinacy of Railway Truss Spans

2.1 INTRODUCTION—CLASSIFICATION OF TRUSS SPANS

Trusses may be classified as simple, compound or complex spans. Simple trusses are built from three non-concurrent joints connected by three members to form a triangle with subsequent joints connected with non-concurrent members. A simple truss may be supported in the same manner as a beam. Compound trusses are constructed through the connection of certain joints in simple trusses by three non-parallel and non-concurrent links. Compound trusses may also be constructed with links and hinges at common joints. A compound truss may be supported in the same manner as a simple truss. Trusses that cannot be classified as simple or compound trusses are complex trusses. Complex truss member forces may not be readily ascertained through investigation of static equilibrium at joints or sections of the truss. Railway truss spans are typically constructed from simple and stable statically determinate[1] planar trusses.

2.2 PLANAR TRUSS SPANS FOR RAILWAY BRIDGES

Railway bridge truss spans are typically designed as vertical planar deck- and through-trusses connected with horizontal and vertical lateral bracing systems. The lateral bracing systems are often planar trusses.

Planar trusses are framed structures comprising straight members lying in the same plane with axes that intersect at joints, typically mathematically considered as frictionless pins,[2] resisting forces generally acting in the same plane as the members. Typical railway truss spans are modeled and designed as an assemblage of two-dimensional planar trusses.[3] The planar trusses may be components of through- or deck-[4] truss span configurations as shown in Figures 2.1a and 2.1b, respectively.[5]

2.3 SIMPLY SUPPORTED PLANAR TRUSS SPANS FOR RAILWAY BRIDGES

Modern moderate length railway through- and deck-truss spans are typically designed with simple vertical planar Pratt, Warren and Howe trusses (Figure 2.2a).[6] Longer-span truss designs for railway loads often use compound vertical trusses such as Baltimore (subdivided Pratt), Pennsylvania or Petit (Baltimore truss with inclined top chords) or subdivided Warren trusses (Figure 2.2b). These vertical trusses are subdivided by subtie or substrut members within the truss panels.[7] Horizontal bracing, resisting lateral forces in modern railway through- and deck- truss superstructures, is typically designed as Pratt or Warren trusses. Depending on the number and arrangement of members, vertical and horizontal planar trusses may be internally unstable, internally stable and statically determinate or internally stable and statically indeterminate. The planar truss forms in Figures 2.2a and 2.2b are internally stable and statically determinate.

The vertical trusses in railway superstructures are supported on bearings that transmit reactions from the superstructures to the substructures (abutments and piers). Depending on bearing[8] and substructure[9] configurations, the vertical planar truss spans may be externally unstable, externally stable and statically determinate or externally stable and statically indeterminate.

Therefore, stable planar truss spans may be statically determinate or statically indeterminate based on internal (member forces) and/or external (reactions) conditions. Statically unstable planar

DOI: 10.1201/9781003469971-3

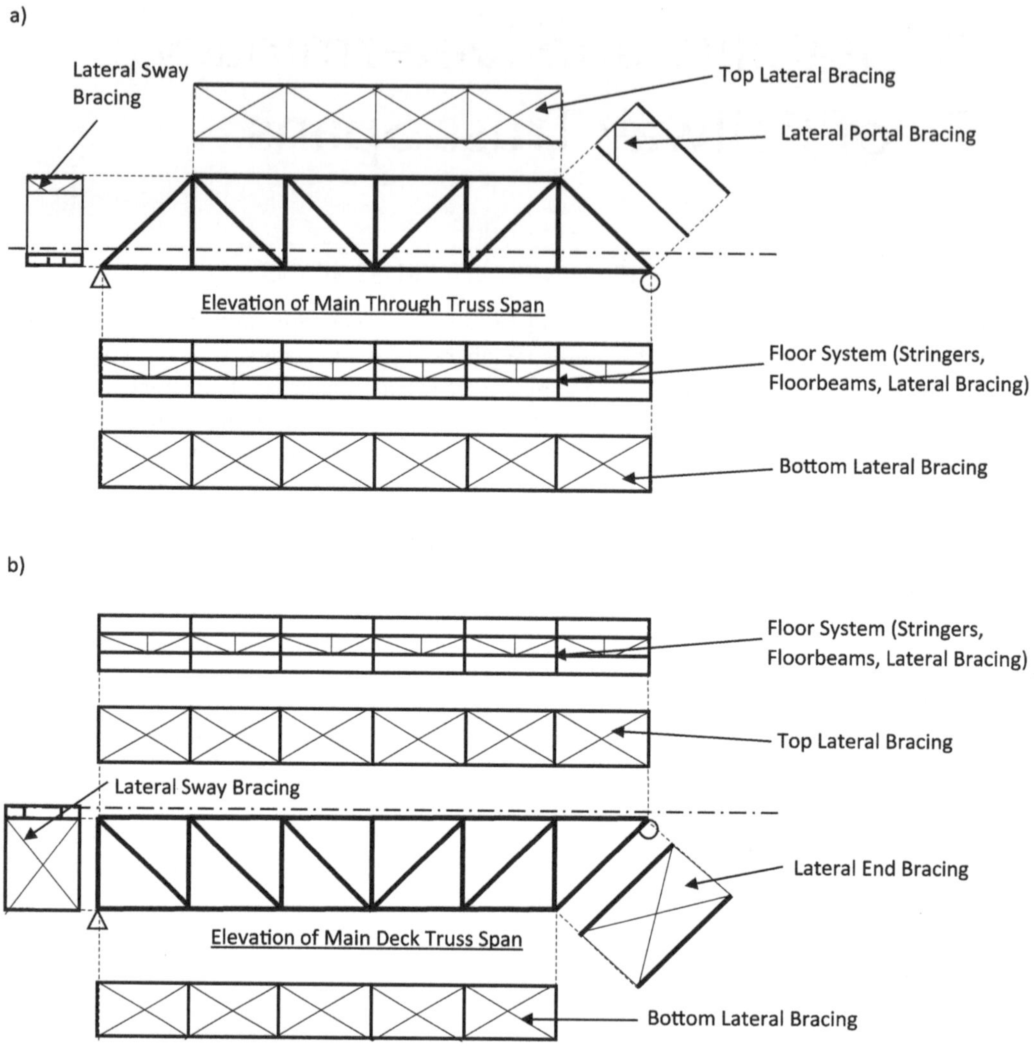

FIGURE 2.1 Principal components of through- and deck-truss spans.

trusses will collapse as a mechanism (internally unstable) or be subject to rigid body movement[10] (externally unstable). Truss spans without redundant members (internally statically determinate) and three support restraints[11] (externally statically determinate) are statically determinate. Trusses with redundant members[12] are internally statically indeterminate. Truss spans with more than three reactions are externally statically indeterminate. Statically indeterminate truss spans are generally objectionable because of analytical difficulty (force equilibrium with deformation compatibility requiring knowledge of truss member section properties)[13] and because the truss member forces are affected by temperature changes, member fit[14] and support movement (affects deformation compatibility of truss members).

Hence, most vertical trusses in railway superstructures are designed, fabricated and erected as statically determinate trusses. Nevertheless, statically indeterminate trusses[15] are typically more rigid under moving loads and, particularly for long spans, may be more economical than statically determinate trusses. Consequently, defining planar truss span internal and external determinacy is an essential initial step in the structural analysis of railway truss superstructures.

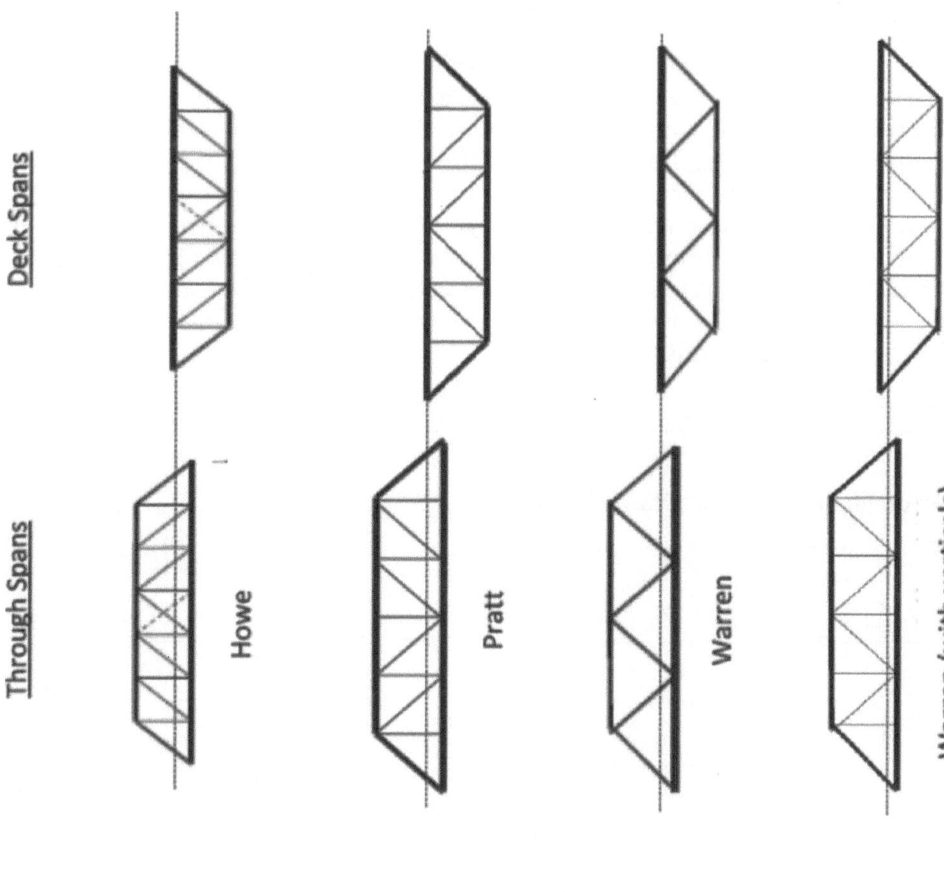

FIGURE 2.2a Typical moderate-length railway truss spans.

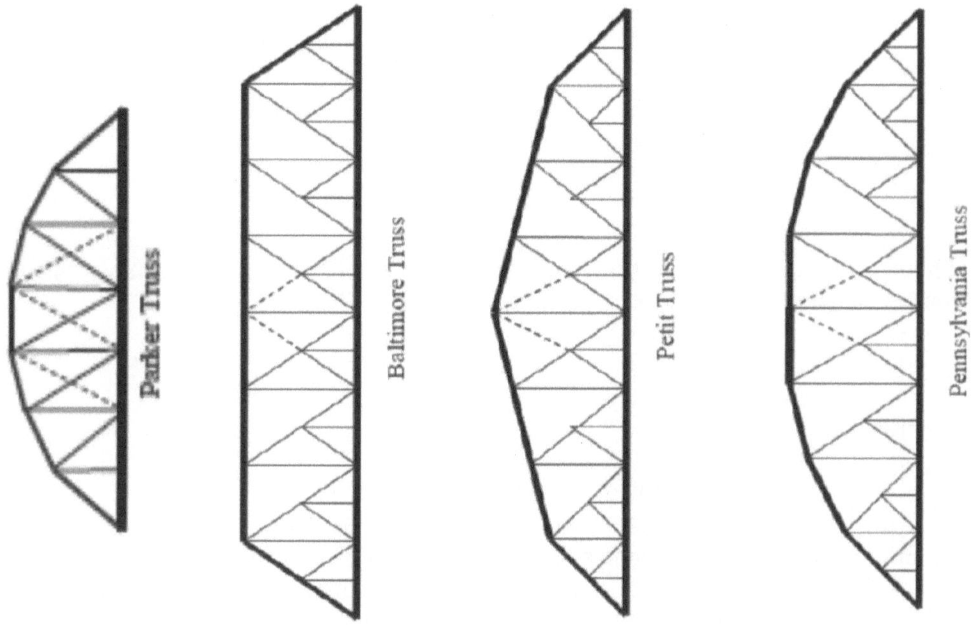

FIGURE 2.2b Typical long railway truss spans.

2.4 INTERNAL DETERMINACY OF PLANAR TRUSSES

The inherently stable triangle, with pinned connections between members, is the building block of geometrically stable[16] planar trusses. Building a geometrically stable planar truss from the triangle involves adding two members and one joint to maintain triangulation. Therefore, the total number of members in a planar truss that maintains triangulation as it is built, NM, is equal to the original three members plus two additional members with each additional joint, NJ_a. which is expressed as

$$NM = 3 + 2NJ_a \tag{2.1}$$

The total number of joints in a planar truss that maintains triangulation as it is built, NJ, is equal to the additional number of joints, NJ_a, plus the original three joints, or

$$NJ = NJ_a + 3 \tag{2.2}$$

Substituting Equation 2.2 into Equation 2.1 yields

$$NM = 3 + 2(NJ - 3) = 2NJ - 3 \tag{2.3}$$

NM is the minimum number of members required for a stable internally statically determinate planar truss with NJ joints. Building a planar truss with fewer members than NM will result in a geometrically unstable truss. Building a planar truss with more members than NM (redundant members) will result in an internally statically indeterminate truss. The degree of internal indeterminacy, ID_I, imposed by a planar truss form can be calculated from Equation 2.3 as

$$ID_I = NM + 3 - 2NJ \tag{2.4}$$

Some vertical planar truss forms typically used in railway spans and their degree of internal indeterminacy, ID_I, are shown in Table 2.1 and Table 2.2. Trusses a, b and g in Table 2.1 (Long, wooden Howe and Whipple trusses, respectively) illustrate that many of the early truss forms were highly internally statically indeterminate.[17] Trusses d, e, f and i in Table 2.1 (Howe, Warren and Pratt trusses) are internally statically determinate. Trusses c and h in Table 2.1 (Howe and Pratt trusses with an odd number of panels) are internally statically indeterminate to the first degree. Trusses a, b, c, h and j in Table 2.1 are also internally statically indeterminate to a degree equal to the number of doubly braced panels. Trusses a, b and c in Table 2.2 (Baltimore, Petit and Pennsylvania trusses) are all internally statically determinate.

Example 2.1 illustrates stable internally statically determinate, stable internally statically indeterminate and unstable planar truss forms.

TABLE 2.1
Internal Redundancy of Simple Planar Truss Spans

	Planar Simple Span Truss Forms—Internal Redundancy	NM	NJ	ID_I
a	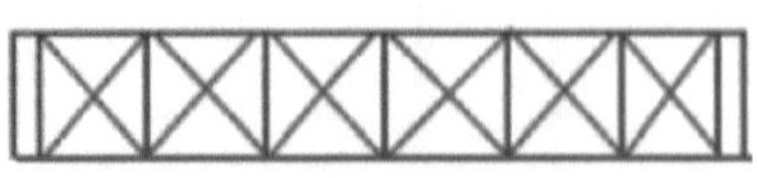 Long Truss	43	20	6
b	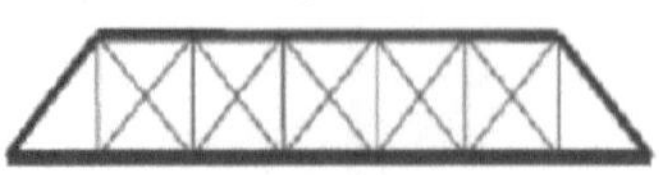 Howe Truss (wood with iron verticals)	40	19	5

TABLE 2.1 ***(Continued)***
Internal Redundancy of Simple Planar Truss Spans

	Planar Simple Span Truss Forms—Internal Redundancy	NM	NJ	ID_I
c	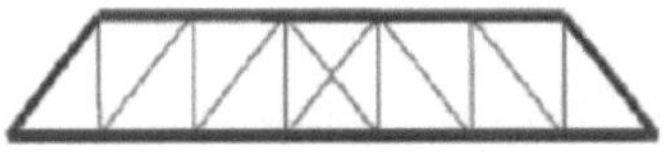 Howe Truss	28	15	1
d	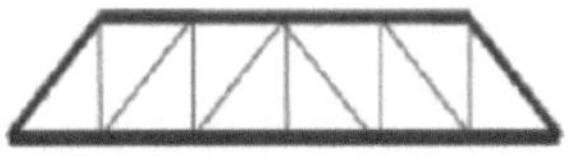 Howe Truss (iron)	21	12	0
e	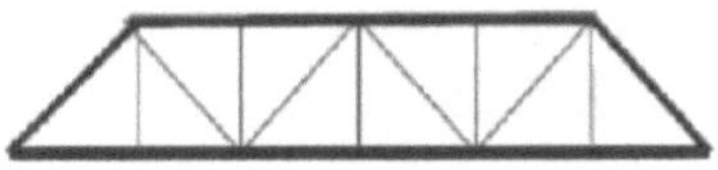 Warren Truss (with verticals)	21	12	0
f	 Warren Truss (without verticals)	11	7	0
g	 Whipple Truss	46	18	13
h	 Pratt Truss	28	15	1
i	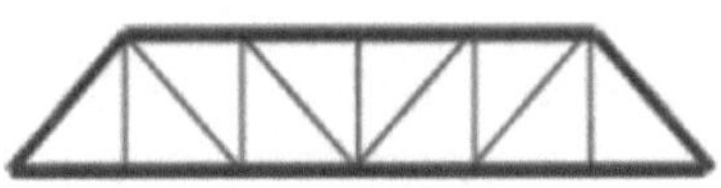 Pratt Truss	21	12	0
j	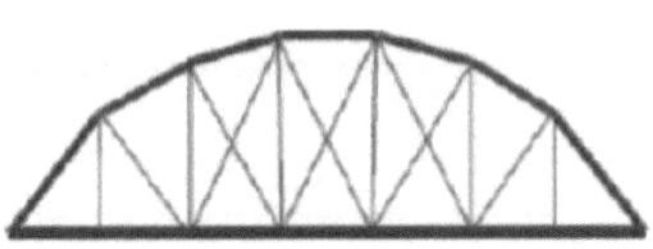 Parker Truss (polygonal top chord)	34	17	3

TABLE 2.2
Internal Redundancy of Compound Planar Truss Spans

	Planar Simply Supported Compound Truss Forms—Internal Redundancy	NM	NJ	ID_I
a	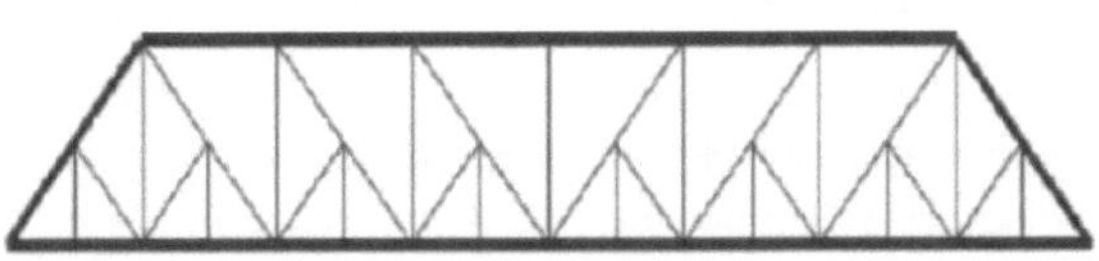 Baltimore Truss	61	32	0

(Continued)

TABLE 2.2 *(Continued)*
Internal Redundancy of Compound Planar Truss Spans

Planar Simply Supported Compound Truss Forms—Internal Redundancy	NM	NJ	ID_i
b 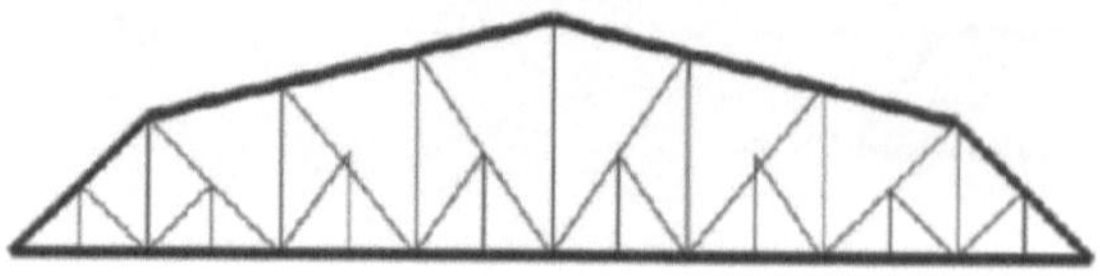 Petit Truss	61	32	0
c 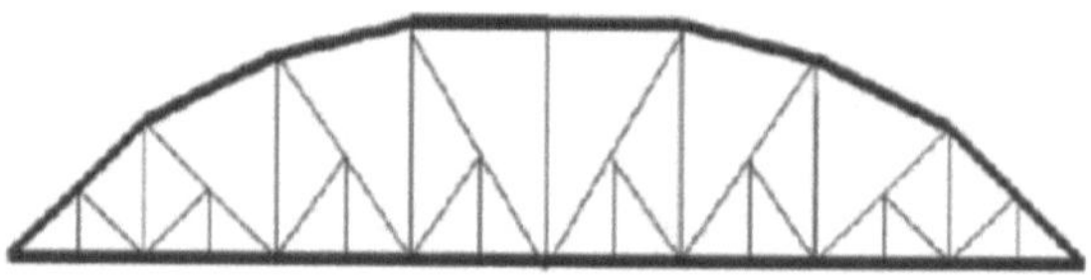 Pennsylvania Truss	61	32	0

Example 2.1

Establish the internal indeterminacy of the trusses shown in Figure E2.1.

The planar truss in Figure E2.1a has NM = 21 and NJ = 12; $ID_i = 21 + 3 - 2(12) = 0$. It is stable and internally statically determinate.

The planar truss in Figure E2.1b has NM = 27 and NJ = 14; $ID_i = 27 + 3 - 2(14) = 2$. It is stable and internally statically indeterminate to the second degree.

The planar truss in Figure E2.1c has NM = 23 and NJ = 13; $ID_i = 23 + 3 - 2(13) = 0$. It is statically determinate, but it cannot resist panel shear and is internally unstable. An attempt at structural analysis will reveal this.

The planar truss in Figure E2.1d has NM = 20 and NJ = 12; $ID_i = 20 + 3 - 2(12) = -1$. It is internally unstable.

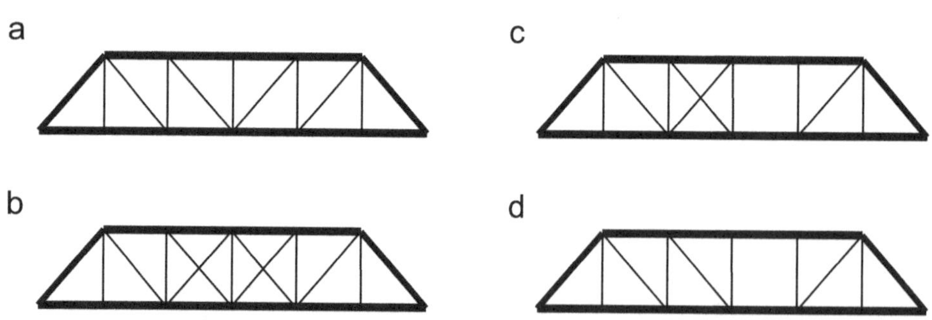

FIGURE E2.1 Determinacy of Pratt truss spans.

It is interesting to note that many trusses in the latter part of the 19th century were parallel-chord Warren trusses with multiple web member systems[18] to subdivide panels and obtain longer spans. The Whipple truss invented in 1852 by Squire Whipple (see Chapter 1) is a double-intersection truss that combines the internally statically determinate Pratt truss with additional web members to create an internally statically indeterminate truss span.

Whipple assumed that each web system acted independently, which enabled an approximate structural analysis. However, some loaded joints belong to both web member systems. It was assumed that such joint loads were resisted by either web member system to provide the maximum force in any member, shared by each system or that the shear in any panel is equally divided by the two diagonal members in the panel. However, the web member systems do not act independently, and the structural analyses were inherently approximate for the Whipple and other multiple intersection truss spans. A quadruple-intersection Warren truss is shown in Figure 2.3.

2.5 EXTERNAL DETERMINACY OF PLANAR TRUSS SPANS

The number and location of supports will affect the number of reactions, NR, and the external determinacy, ID_E, of a truss span bridge. An externally stable planar truss span requires a minimum of three reactions for external determinacy,[19] and therefore,

$$ID_E = NR - 3 = (2(NS_F) + (NS\text{-}NS_F)) - 3 = NS + NS_F - 3 \tag{2.5}$$

where NS is the number of supports and NS_F is the number of fixed supports.

A statically determinate simply supported truss span (NS = 2) consisting of a fixed ($NS_F = 1$) support[20] and an expansion[21] support, provides three independent reactions (two vertical and one horizontal) yielding $ID_E = 0$. However, for continuous bridges[22] with multiple supports, NS, Equation 2.5 indicates that ID_E may be large. Consequently, in addition to the potential for detrimental stresses in some members due to excessive temperature changes and/or support displacements (see Chapter 5), the analytical complexity[23] of externally statically indeterminate bridge spans increases. The internal and external indeterminacy of some simply supported truss spans is shown in Table 2.3.[24] The external indeterminacy of some continuous-truss span bridges is shown in Table 2.4.

ID_E decreases with the introduction of various construction conditions, NC, such as hinges, rockers (vertical links pinned at each end), rollers, sliding surfaces and, where appropriate, the

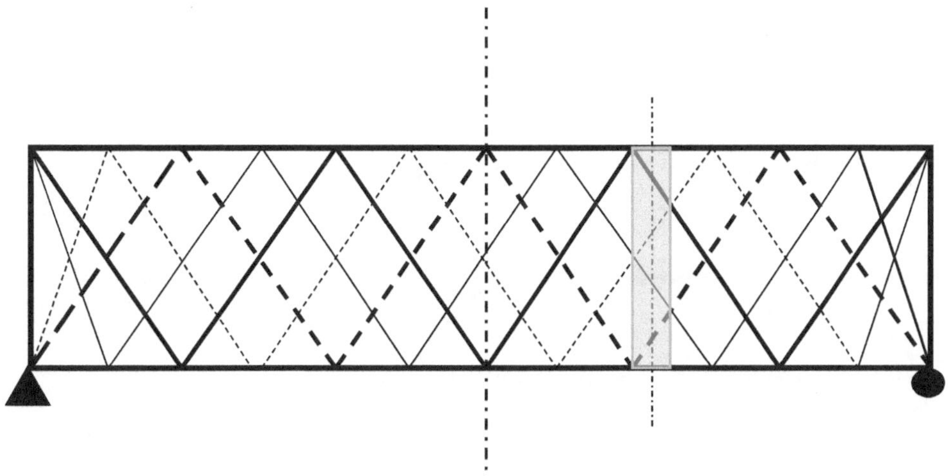

FIGURE 2.3 A quadruple-intersection Warren truss span.

TABLE 2.3
Determinacy of Simply Supported Truss Spans

	Simply Supported Truss Spans—Internal and External Redundancies	NM	NJ	ID_I	NR	ID_E
a	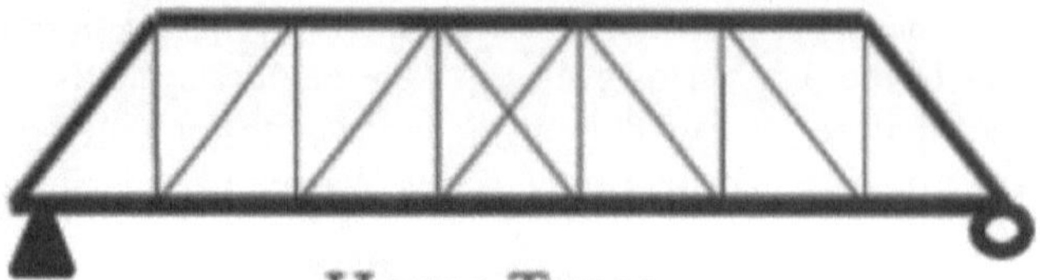 Howe Truss	28	15	1	3	0
b	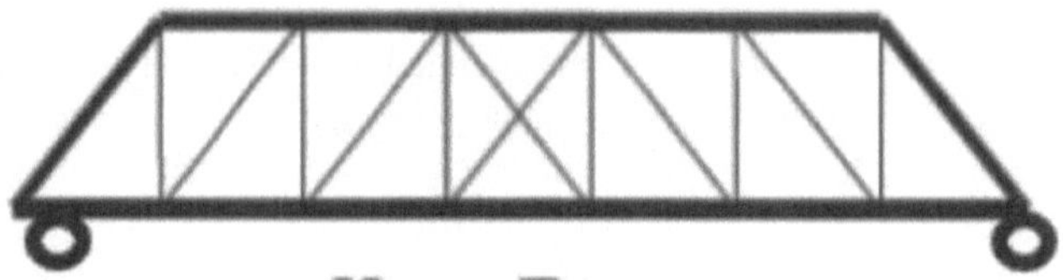 Howe Truss	28	15	1	2	−1
c	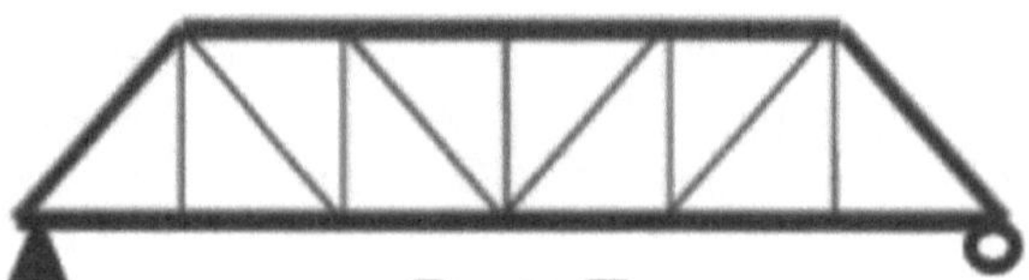 Pratt Truss	21	12	0	3	0
d	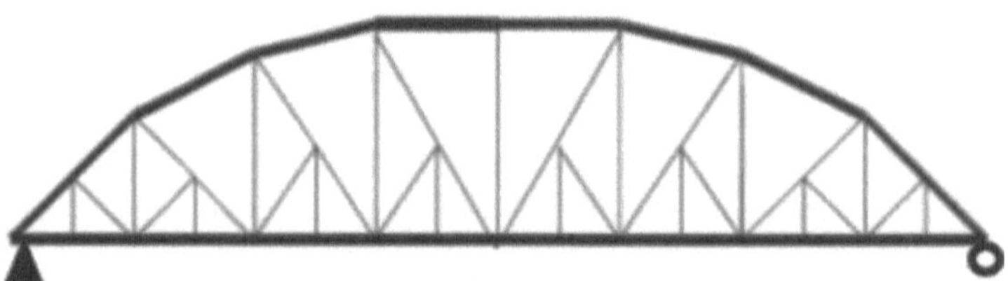 Pennsylvania Truss	61	32	0	3	0
e	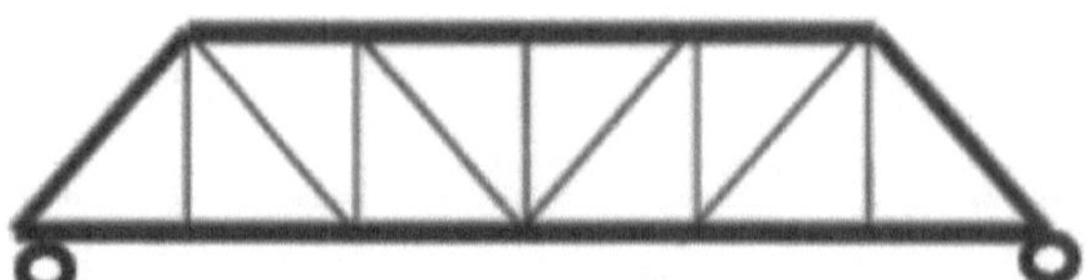 Pratt Truss	21	12	0	2	−1
f	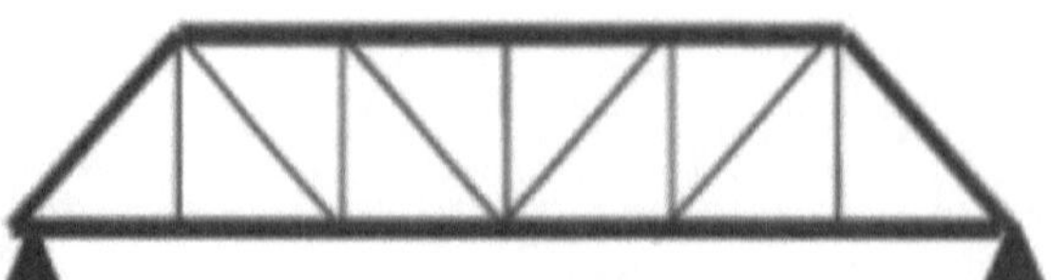 Pratt Truss	21	12	0	4	1

TABLE 2.4
External Determinacy of Continuous-Truss Spans

Continuous Truss Spans—External Redundancy		NS	NS_F	NR	ID_E
a	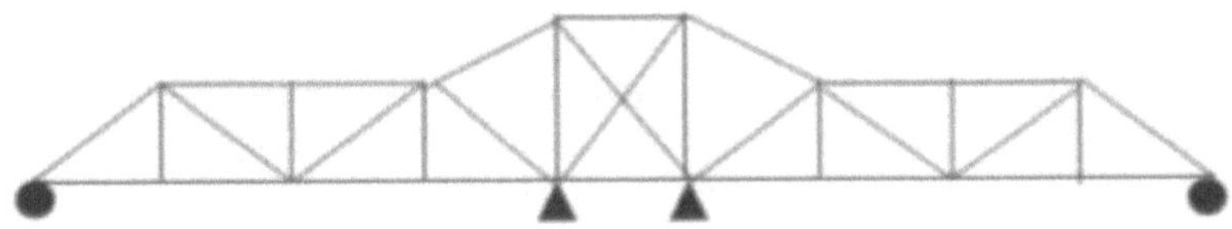	4	2	6	3
b	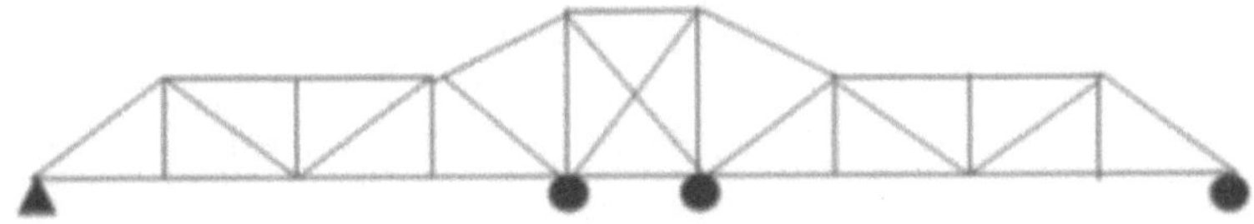	4	1	5	2
c	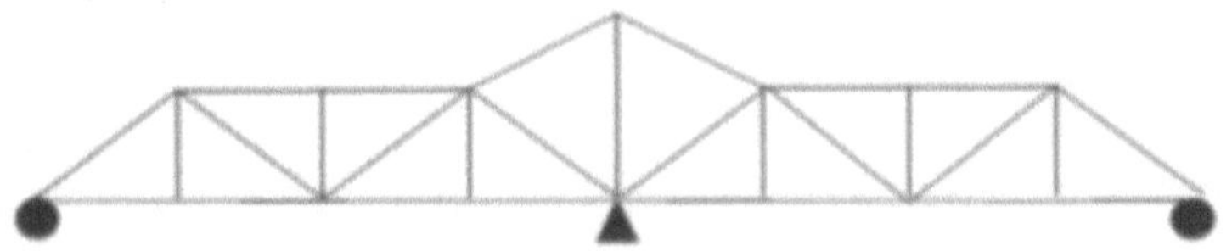	3	1	4	1
d	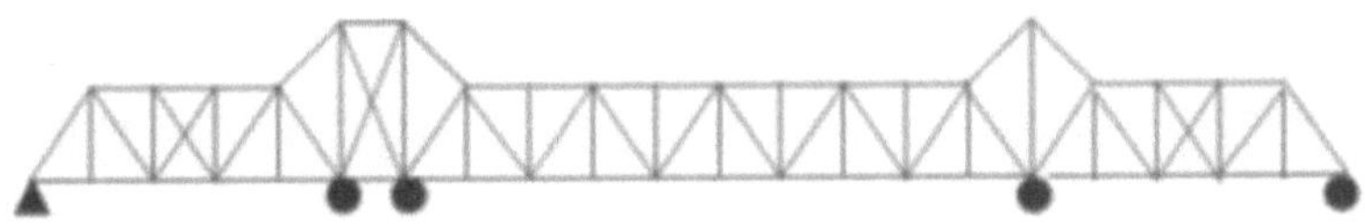	5	1	6	3

use of light bracing[25] between the vertical posts over supports. Hinges provide that $\Sigma M = 0$ on the free body diagram on either side of the hinge. Rollers and sliding surfaces ensure that $\Sigma F_x = 0$ on the free body diagram on either side of the device. Rockers provide that both $\Sigma M = 0$ and $\Sigma F_x = 0$ on the free body diagram on either side of the rocker. The use of light bracing between the vertical posts over supports provides that the panel over the supports cannot transfer shear ($\Sigma F_y = 0$) such that the condition of equal moments at the supports is imposed.[26] If NC are considered, Equation 2.5 becomes

$$ID_E = NS + NS_F - 3 - NC \qquad (2.6)$$

Continuous-span bridges that effectively utilize construction conditions to reduce or eliminate ID_E are cantilever-span bridges.[27] Cantilever-truss bridges are continuous bridges with both analytical and erection advantages. Cantilever bridges have truss spans that extend beyond intermediate supports of the bridge to carry truss spans at the ends of the cantilever arms.

Cantilever-truss bridges can be erected by the cantilever method of erection,[28] precluding the need for falsework over deep and/or navigable channels. The anchor span trusses may be erected conventionally on falsework, or by lateral,[29] longitudinal[30] or vertical[31] installation (Unsworth, 2018). The cantilever arm trusses may then be built out toward the center of the main span using small derricks and cranes.[32] Following completion of cantilever arm truss erection, the suspended span may be built out toward the center of the main span[33] from each end.

Temporary members between the ends of the suspended and cantilever arm spans are installed to allow the suspended span cantilever behavior. With the suspended-span erection completed, the temporary members are removed following the installation of pins at one end ($\Sigma M = 0$) and a rocker or sliding arrangement ($\Sigma M = \Sigma F_x = 0$) at the other end of the suspended span to effect a simply supported suspended span. The suspended span members must be designed with due consideration of the cantilever erection stresses unless the suspended span is lifted into place from barges.[34]

Table 2.5 shows some cantilever bridge forms and their ID_E under various NC incorporated into the bridge to reduce ID_E. The ID_E is also shown for three existing renowned cantilever railway bridges. The cantilever-truss bridge in Table 2.5a is externally statically determinate using four construction conditions provided by light bracing over a pier ($\Sigma F_y = 0$), a link between a cantilever arm and suspended span ($\Sigma Fx = 0$ and $\Sigma M = 0$) and a hinge between a cantilever arm and suspended span ($\Sigma M = 0$). The cantilever truss bridge in Table 2.5c is externally statically determinate using three construction provided by two hinges and a sliding or rolling device. The bridge over the Mississippi River at Thebes, IL, is externally statically determinate due to the use of three rockers and one hinge to provide seven construction conditions and their accompanying equilibrium equations.

TABLE 2.5
External Determinacy of Cantilever-Truss Spans

Cantilever Truss Spans—External Redundancy	NS	NS_F	NR	NC	ID_E
a 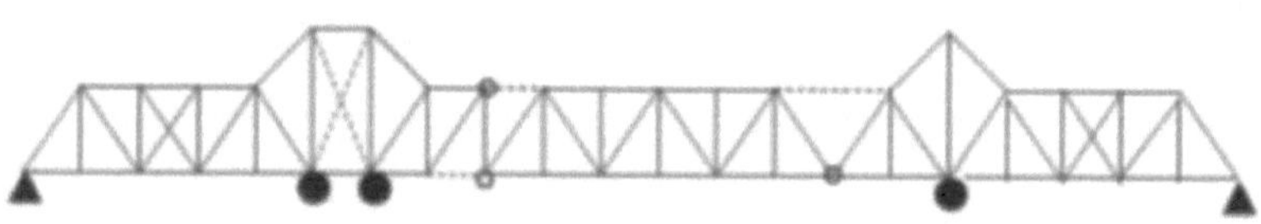	5	2	7	4	0
b 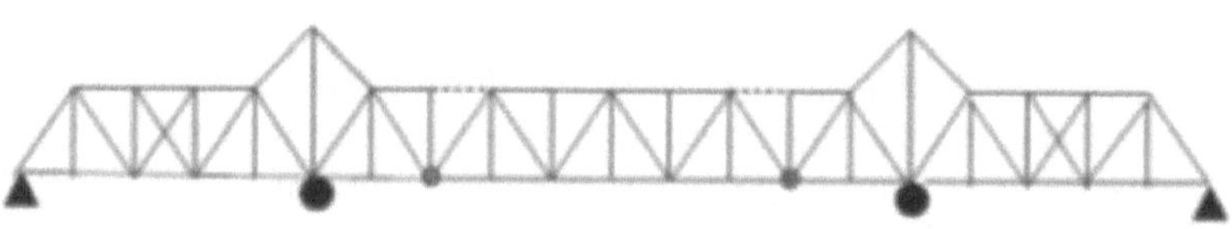	4	2	6	2	1
c 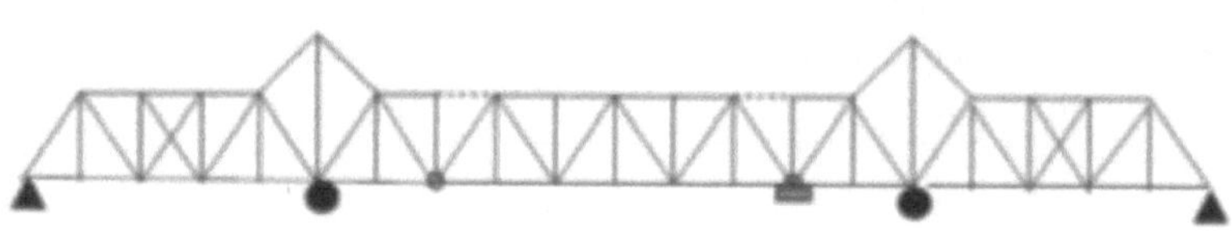	4	2	6	3	0
d 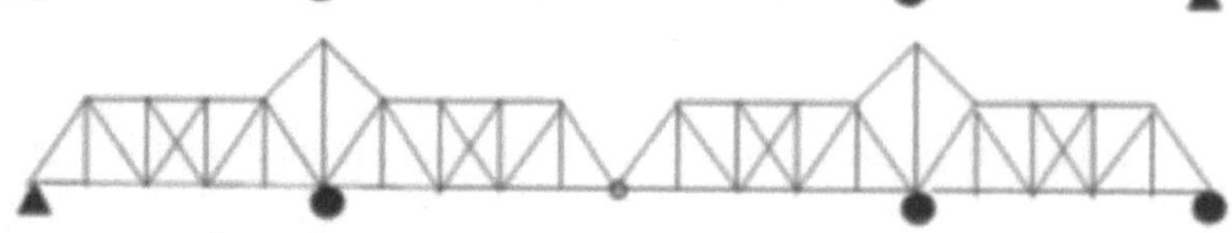	4	1	5	1	1
e Firth of Forth, Scotland (see Chapter 1)	8	3	11	6	2
f Thebes, IL, Mississippi River (see Chapter 1)	6	4	10	7	0
g Beaver, PA, Ohio River	4	2	6	3	0

2.6 STRUCTURAL ANALYSIS OF PLANAR TRUSS SPANS IN BRIDGES

The angles between truss members loaded at idealized pin connections change very little to accommodate changes in truss member lengths. However, more rigid connections between truss members cannot allow the angles between members to change unrestrained, and adjacent members will bend slightly. Therefore, since typical bolted gusset plate connections between truss members are more rigid than pins, some secondary normal stresses due to bending and associated shear stresses will occur in truss members adjacent to the gusset plate connections. These secondary normal and shear stresses may not appreciably affect the magnitude of the normal axial stress in the member and can be calculated, if necessary, by separate analyses (see Chapter 7). Therefore, statically determinate and indeterminate trusses can be typically analyzed with sufficient accuracy for member forces and joint deflections assuming pinned joints.

Statically determinate pin-jointed planar truss span member forces are readily obtained through the three equations of static equilibrium, $\Sigma F_x = 0$, $\Sigma F_y = 0$ and $\Sigma M = 0$, applied to free body segments of the truss span by isolating joints (method of joints) or sections (method of sections) of the truss span (see Chapter 4). Statically indeterminate pin-jointed planar truss span member forces can be determined by approximate methods or via energy methods that consider deformation compatibility (see Chapter 5).[35] Statically determinate and indeterminate pin-connected planar truss span joint deflections can also be determined by energy methods (see Chapter 6).[36] However, modern digital computer applications based on the stiffness (displacement)[37] and flexibility (force) methods have rendered the calculation of truss span deformations by classical and energy methods of structural analysis unnecessary. Nevertheless, an understanding of the classical and energy methods of truss span analysis is essential for a thorough understanding of truss span behavior and for the practical verification of the results of computer application analyses.

2.7 LOADS AND FORCES ON RAILWAY PLANAR TRUSS SPANS

Vertical loads and forces are resisted by the horizontal floor system (typically deck ties, stringers and/or floorbeams) and transferred to the main vertical planar truss span joints. The internal member forces in truss span floor systems and main vertical planar trusses[38] are due to

- dead load (imposed and self-weight);
- live load (from moving trains);
- impact load (dynamic increment of live load due to vibration);
- longitudinal forces (due to traction and braking of trains);
- centrifugal forces (from moving trains on curved track);
- wind forces (lateral and longitudinal on the loaded and unloaded span);
- lateral forces from equipment (truck or wheelset hunting or nosing of locomotives and cars);
- thermal forces (in superstructure and continuous welded rail on superstructure); and
- seismic forces (if applicable by bridge type, configuration, and geographical location).

Lateral loads and forces are resisted by the horizontal floor system and horizontal and vertical planar bracing trusses and frames. Lateral bracing of truss superstructures generally occurs in horizontal planes coincident with the top[39] and bottom chords of the vertical truss[40] and in vertical planes between the vertical truss panel points.[41] Additional lateral bracing in the horizontal and vertical planes is typically required between stringers[42] in truss span floor systems. Internal member forces in truss span bracing systems are generally due to

- wind forces (lateral on the loaded and unloaded span);
- longitudinal forces (may require traction bracing design);
- centrifugal forces;
- lateral forces from equipment; and
- derailment loads (an extreme event loading).

NOTES

1. Member forces can be determined by considering only the equations of static equilibrium.
2. This assumption is reasonable for relatively long and slender truss members connected with modern bolted gusset plate connections. However, in some cases, the secondary stresses in members due to joint rigidity require explicit consideration (see Chapter 7).
3. Truss spans typically have vertical planar trusses resisting gravity loads and forces and horizontal planar trusses resisting lateral loads and forces.
4. Some deck truss spans do not have floor systems supporting the track but have cross-ties (sleepers) supported directly on the top chord of the deck truss. Such cases require analyzing combined bending and axial compression in the top chord of the deck truss.
5. Large and complex railway truss spans may warrant a three-dimensional space frame analysis. Space frame analysis using the classical methods of structural analysis is onerous and usually impractical. However, the modern matrix methods of structural analysis combined with the computational efficiency of digital computers makes space frame analysis of large and complex railway truss spans feasible.
6. Pratt and Warren trusses are most prevalent due to the inefficiencies of Howe trusses (see Chapter 4).
7. Typically, substrut members are preferable from the perspective of secondary stresses in the chord members (see Chapter 7).
8. Related to the number of restraints to translational or rotational movements.
9. Related to the number and location of substructures supporting the truss superstructure.
10. Rigid body movement of a truss span will occur when there are fewer than three support restraints.
11. Three reactions are required to satisfy the three equilibrium conditions concerning translation and rotation at supports. Expansion and fixed bearings at the end of simply supported truss spans provide two vertical and one horizontal reaction.
12. Trusses are internally statically indeterminate when the number of members exceeds that required for internal static equilibrium. The members that exceed the number of members required for internal static equilibrium are redundant members.
13. Modern computer methods based on stiffness (finite element method) and flexibility make the analysis of statically indeterminate trusses no longer onerous. Nevertheless, in statically indeterminate truss analyses, member section properties must be estimated to investigate deformation compatibility.
14. Typically due to fabrication error.
15. For example the vertical trusses in some cantilever and continuous truss bridges.
16. Geometric instability occurs when a planar truss is triangulated such that it cannot carry panel shear. Geometrically unstable trusses create a mechanism that will collapse under a load condition. Geometrically stable trusses have no relative movement of the truss elements other than those created by elastic deformations.
17. These trusses were often used for railway trusses before more rational analyses were enabled in the mid-19th century (see Chapter 1).
18. The multiple web member systems provided double-, triple- or quadruple-intersection Warren truss spans. A quadruple-intersection truss span contains four web member systems with four diagonal members within each subpanel of the span.
19. A minimum of three reactions ensuring the equilibrium of horizontal forces ($\Sigma F_x = 0$), vertical forces ($\Sigma F_y = 0$) and moments ($\Sigma M_z = 0$) are required.
20. Translation restrained with free rotation.
21. Free translation and rotation.
22. Continuous construction is typically economical when spans are longer than about 750 ft. Nevertheless, relatively short continuous spans ensue for center and rim-bearing movable swing spans in the closed position.
23. Member sizes are required for statically indeterminate structural analysis, and preliminary member sizes must be established based on experience, previous similar designs or approximate analyses. Structures with many redundant members (large ID_I) and/or reactions (large ID_E) are onerous to analyze (see Chapter 5).
24. $ID_E = -1$ in Table 2.3b and Table 2.3e is due to the instability of rigid body motion.
25. Another technique for preventing shear transfer is to provide links between the vertical posts over supports and the top chord of the truss.
26. This construction condition is often used in rim-bearing swing span bridges to reduce the external indeterminacy by one.

27 Typically economical with main spans up to about 2000 ft. Also, typically economical for long spans with relatively large dead load (e.g. ballasted deck bridges) due to reduced moments and better distribution of dead load at supports.
28 Simple-span trusses may also be erected using cantilever erection with end launching.
29 Typically by erection on adjacent temporary support bents or frames and lateral sliding.
30 Typically by erection and launching with or without temporary intermediate supports (e.g., frames and/or barges).
31 Typically by erection and installation using cranes.
32 The truss members must be designed considering the erection sequence and associated loads. However, the erection stresses in the cantilever arm truss members are typically within the service stress levels.
33 Like the erection of the cantilever arm trusses.
34 The suspended spans of the Quebec Bridge (following an initial erection failure) and other cantilever-span bridges have been lifted into position from barges in the main channel.
35 Statically indeterminate truss member forces can be calculated with Castigliano's Second Theorem or by superposition of deflections and the Principle of Virtual Work.
36 Truss joint displacements can be calculated with Castigliano's Second Theorem or with the Principle of Virtual Work.
37 Essentially, the finite element method for one-dimensional members in framed structures.
38 Primarily axial forces in truss members and bending and shear forces in stringers and floorbeams.
39 Pony trusses are relatively low-profile through-trusses without top lateral bracing. Compression chords are typically laterally braced by knee braces on either the track or field side of the pony truss.
40 Lateral bracing in the horizontal plane of through- and deck-truss spans typically consists of either single- or double-diagonal truss-type bracing in the plane of the top and bottom chords. Lateral bracing in the horizontal plane of pony truss spans typically consists of either single- or double-diagonal truss-type bracing in the plane of the bottom chord.
41 Lateral bracing in the vertical plane of through-truss spans typically consists of end or portal frame bracing and intermediate or sway frame bracing between the main trusses at panel points. Lateral bracing in the vertical plane of deck truss spans supported at the bottom chord typically consists of end bracing trusses and intermediate or sway bracing trusses between the main trusses at panel points. End lateral bracing trusses are in the inclined plane of deck truss spans supported at the top chord. Lateral bracing in the vertical plane of pony truss spans typically consists of knee bracing trusses at the main trusses at panel points.
42 Typically required in the horizontal plane to resist lateral forces and in the vertical plane against lateral-torsional buckling of the top flange and provide load distribution in accordance with requirements in the AREMA MRE Chapter 15.

REFERENCE

Unsworth, J.F., 2018, *Design and Construction of Modern Steel Railway Bridges*, 2nd Edition, CRC Press, Boca Raton, FL.

BIBLIOGRAPHY

Armenakas, A.E., 1988, *Classical Structural Analysis: A Modern Approach*, McGraw-Hill, New York.

Ghali, A., Neville, A.M., and Brown, T.G., 2007, *Structural Analysis: A Unified Classical and Matrix Approach*, Spon Press, London.

Norris, C.H., Wilbur, J.B., and Utku, S., 1991, *Elementary Structural Analysis*, 4th edition, McGraw-Hill, New York.

3 Loads and Forces on Railway Truss Bridges

3.1 INTRODUCTION

The loads and forces on railway trusses are vertical (gravity), longitudinal or lateral in nature. Gravity loads that comprise dead, live and impact loads are the principal loads to consider for railway truss design. Live load impact (dynamic effect) is included due to the relatively rapid application of railway live loads. However, longitudinal and lateral forces (due to live load, wind forces, thermal forces and/or due to seismic activity) also warrant careful consideration in railway truss design. Railway trusses are subjected to specific forces related to railroad moving loads. These are live load impacts from vertical[1] and rocking[2] effects, longitudinal forces from acceleration or deceleration of railroad equipment, lateral forces caused by irregularities at the wheel-to-rail interface (commonly referred to as 'truck hunting' or 'nosing') and centrifugal forces due to track curvature.

3.2 DEAD LOADS

Superstructure dead load consists of the weight of the trusses, floor system, track, deck (open[3] or ballasted), utilities (conduits, pipes, and cables), walkways,[4] permanent formwork, snow, ice and anticipated future dead loads (for example, larger deck ties, increases in ballast depth, additional utilities). However, snow and ice loads are generally excluded from consideration due to their relatively low magnitude and transient nature. For open-deck railway trusses, dead load is often a comparatively small proportion of the total load in the members. However, for ballasted deck railway trusses, dead load is often a considerable proportion of the total load in the members.

At the commencement of design, dead load must be estimated from experience or review of similar railway truss span designs. This estimated design dead load must be reviewed against the actual dead load calculated after final design of the truss span. Small differences between the estimated and actual dead load are not important provided that the dead load is a reasonably small constituent of the total design load. Railway bridge engineers will often include an allowance of 10% to 15% of estimated steel truss weight to account for bolts, gusset plates, stiffeners and other appurtenant steel components. Dead loads typically used for ordinary steel railway truss design are shown in Table 3.1. If possible, temporary construction dead loads should also be considered during design.[5]

3.3 RAILWAY LIVE LOADS

Railroad locomotives and equipment (box and flat cars, commodity gondolas, hopper and tank cars) vary greatly with respect to weight, number of axles and axle spacing. Modern freight locomotives have two three-axle sets (trucks or bolsters) with spacing between axles of between 1.96 m (6.42 ft) and 2.08 m (6.83 ft) and spacing between axle sets of between 13.91 m (45.62 ft) and 16.66 m (54.63 ft). These modern-generation locomotives weigh up to 200,000 kg (435,000 lb). There are, however, many four- and six-axle locomotives that weigh between 115,000 kg (250,000 lb) and 180,000 kg (400,000 lb) with lengths between 15 m (50 ft) and 25 m (80 ft) operating on the railroad infrastructure.

Axle spacing is typically between 1.52 m (5 ft) and 1.78 m (5.83 ft) for North American four-axle freight car equipment. Truck spacing varies from 5 m (17 ft) to 20 m (66 ft) (Dick, 2002). Gross car weights up to 130,000 kg (286,000 lb) are common on North American railroads[6] and some railroad lines carry 145,000 kg (315,000 lb) cars. This variability in railroad equipment weight and geometry

DOI: 10.1201/9781003469971-4

TABLE 3.1
Dead Loads Used for Ordinary Steel Railway Truss Design

Item	Dead Load	
Track (rails and fastenings)	200 lb/ft	300 kg/m
Steel	490 lb/cu ft	7850 kg/cu m
Reinforced and prestressed concrete	150 lb/cu ft	2400 kg/cu m
Plain (unreinforced) concrete	145 lb/cu ft	2320 kg/cu m
Timber	35 to 60 lb/cu ft	560 to 960 kg/cu m
Sand and gravel, compacted (railroad ballast)	120 lb/cu ft	1920 kg/cu m
Sand and gravel, loose	100 lb/cu ft	1600 kg/cu m
Permanent formwork (incl. concrete in valleys)	15 lb/sq ft	75 kg/sq m
Waterproofing on decks	10 lb/sq ft	50 kg/sq m

requires a representative live load model for design that provides a safe and reliable estimate of railroad operating equipment characteristics within the design life of the bridge.

3.3.1 Railway Freight Train Live Load

The railway bridge design live load currently recommended in AREMA (2024) is Cooper's EM360[7] (Cooper's E80) load. This design load is based on two consolidation-type steam locomotives with trailing cars represented by a uniformly distributed load (Cooper, 1894).[8] The maximum locomotive axle load is 360 kN (80 kips), and freight equipment is represented by a uniformly distributed load of 120 kN per m of track (8 kips per ft of track). An alternate live load consisting of four 445 kN (100 kips) axles is also recommended in AREMA (2024, Chapter 15) to represent the stress range effects of adjacent heavy rail cars on short spans. These design live loads are shown in Figure 3.1.

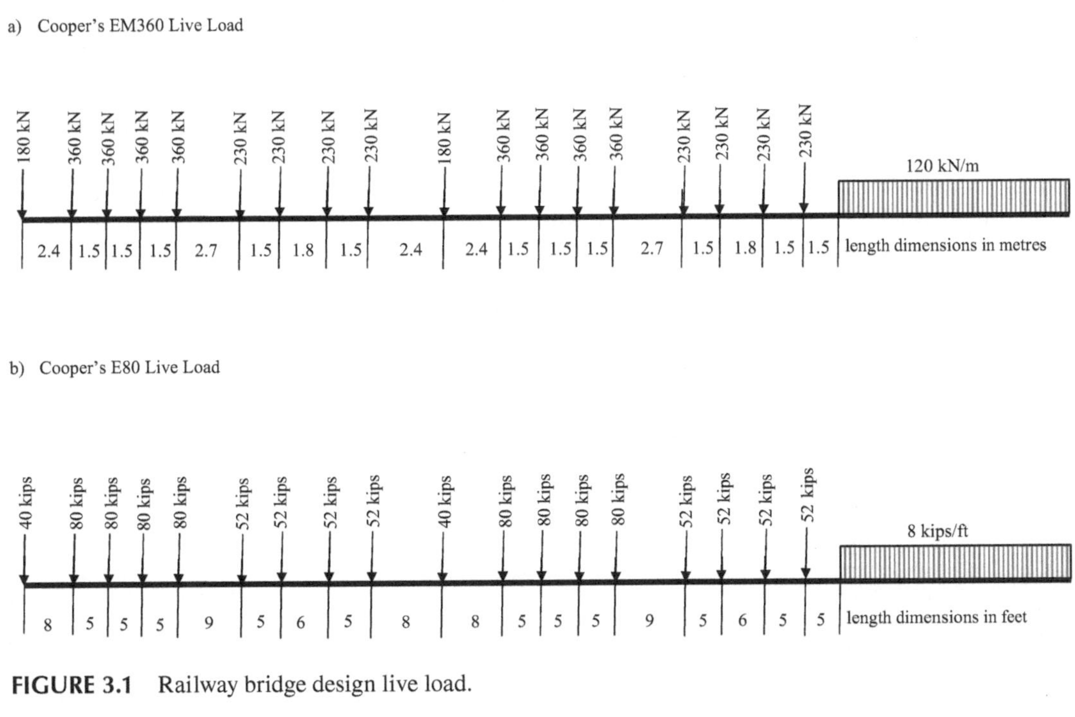

FIGURE 3.1 Railway bridge design live load.

a) Alternate Live Load – SI Units

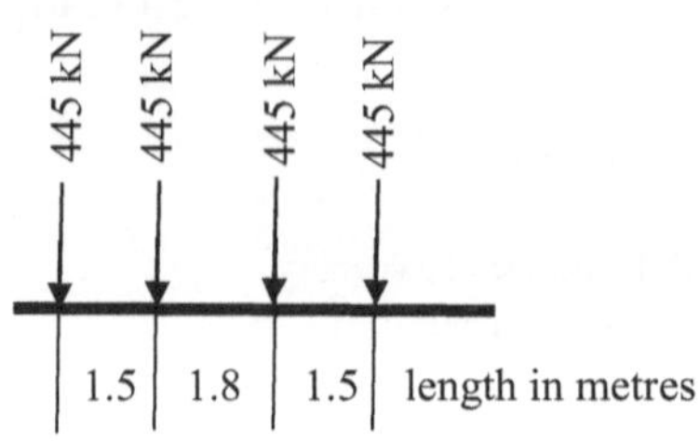

b) Alternate Live Load – US customary and imperial units

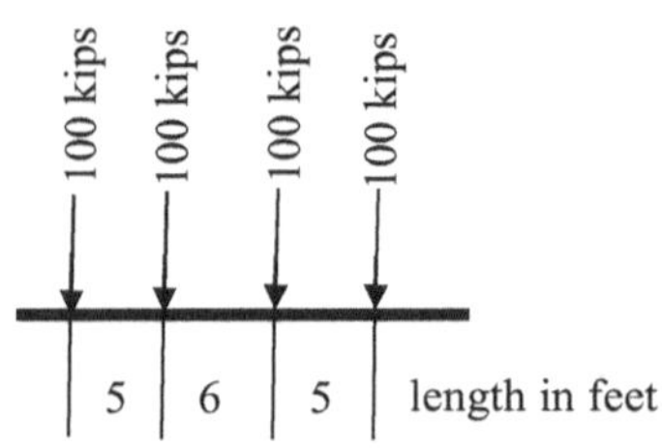

FIGURE 3.1 (Continued)

The positions of concentrated and uniformly distributed live loads on a truss span for maximum chord member forces, as well as maximum and minimum web member forces, must be calculated. However, as shown in Chapter 4, calculating the positions of live loads on the truss span that create maximum chord member forces,[9] and develop maximum and minimum web member forces,[10] is less onerous for uniformly distributed loads than for concentrated loads. Accordingly, equivalent uniformly distributed loads for Cooper's and other locomotive and train live loads were presented often in early railway bridge design literature (Waddell, 1916; Ketchum, 1924; Johnson, Bryan and Turneaure, 1946). The equivalent uniformly distributed live load for maximum forces in web members due to shear, V_i, in panel i is calculated as shown in Figure 3.2.

From Figure 3.2,

$$V_i = \frac{w_{ev}}{2L}\left((a_i + x_i)^2 - nx_i^2\right) \tag{3.1a}$$

The maximum V_i occurs where

$$\frac{dV_i}{dx} = 0 = \frac{w_{ev}}{L}(a_i + x_i - nx_i) \tag{3.1b}$$

such that

$$x_i = \frac{a_i}{(n-1)} \tag{3.1c}$$

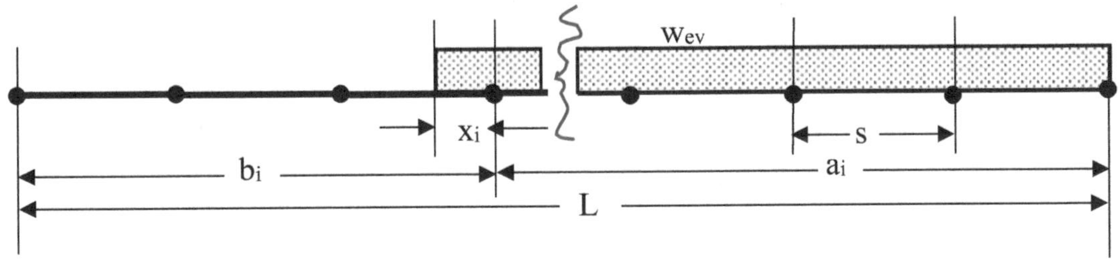

FIGURE 3.2 Uniformly distributed load on a truss span for maximum panel shear force.

Substitution of Equation 3.1c in Equation 3.1a yields

$$V_i^{max} = \frac{w_{ev}a_i^2}{2L}\left(\frac{n}{n-1}\right) = \frac{w_{ev}a_i^2}{2(L-s)} \tag{3.1d}$$

where V_i = the shear in panel i and n = the number of equal length panels of length, s, in the truss span = L/s.

From Figure 3.3,

$$V_i = \frac{P_T x_T}{L} - P_F - \frac{P_i X_i}{s} = \frac{\sum_{i=1}^{n} P_i\left(X_i + (n-i)s\right)}{L} - \sum_{j=1}^{i-1} P_j - \frac{P_i X_i}{s} \tag{3.2}$$

where

P_T = the resultant of all loads in panels on the span = $\sum_{i=1}^{n} P_i$

$P_T X_T$ = the moment about joint n (joints = 0 to n) from all loads on the span = $\sum_{i=1}^{n} P_i\left(X_i + (n-i)s\right)$

P_F = the resultant of loads from panel1 to panel i − 1 = $\sum_{j=1}^{i-1} P_j$

P_i = the resultant of loads in panel i
X_i = the distance from load P_i to joint i
X_F = the distance from load P_F to end joint i-1
X_T = the distance from load P_T to end joint n

Equating Equations 3.1d and 3.2 provides the following:

$$w_{ev}^{max} = \frac{2(L-s)\left(\frac{P_T X_T}{L} - P_F - \frac{P_i X_i}{s}\right)}{a_i^2} = \frac{2L\left(\frac{P_T X_T}{L} - P_F - \frac{P_i X_i}{s}\right)}{a_i^2}\left(\frac{n-1}{n}\right) \tag{3.3}$$

Equation 3.3 can be plotted for P_T and P_F, which depend on the load configuration on the span, and P_i, which is contingent on the load configuration within panel, i, and a_i, which reflects the panel under consideration, for truss spans of n panels. The equivalent uniformly distributed live load for maximum forces in chord members due to moment, M_j, at panel point j is calculated as in Figure 3.4.

From Figure 3.4,

$$M_j = \frac{w_{em} a_j}{2}\left(L - a_j\right) \tag{3.4}$$

From Figure 3.3,

$$M_j = M_T - M_F$$

where

$$M_T = \frac{P_T X_T \left(L - a_j\right)}{L} = \frac{\sum_{i=1}^{n} P_i \left(X_i + (n-i)s\right)\left(L - a_j\right)}{L}$$

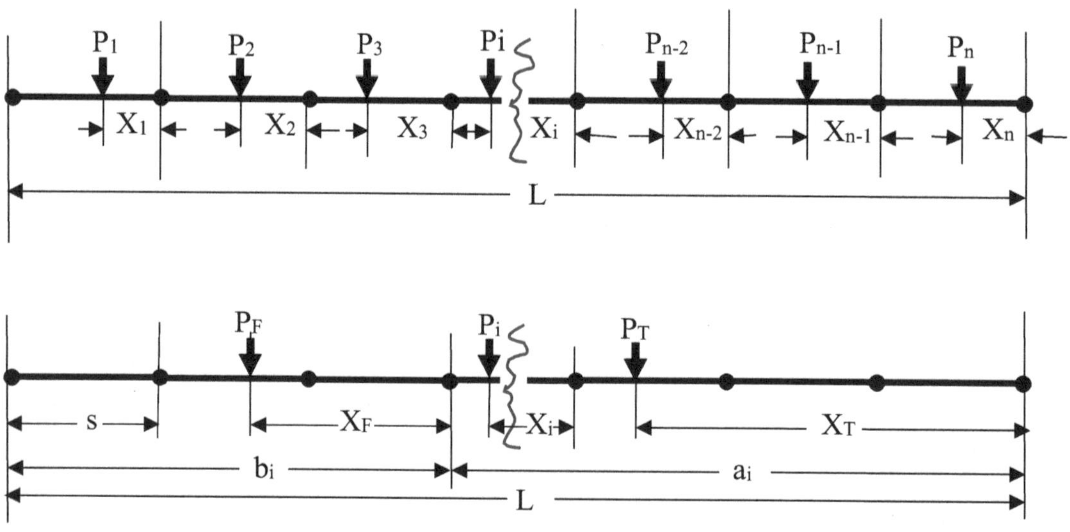

FIGURE 3.3 Concentrated loads on a truss span.

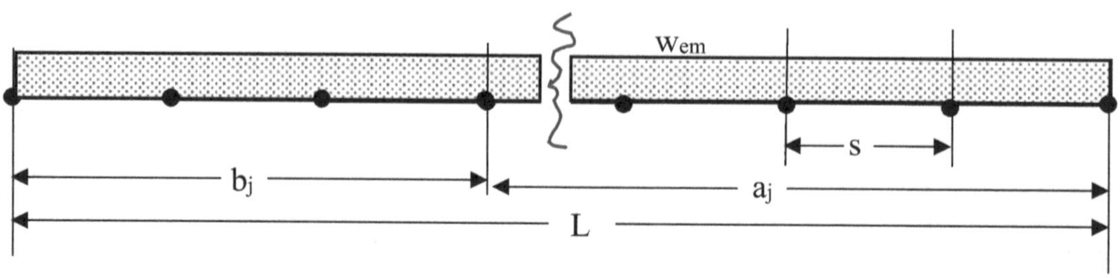

FIGURE 3.4 Uniformly distributed load on a truss span for maximum panel point moment.

$$M_F = P_F x_F = \sum_{j=1}^{i-1} P_j \left(X_j + (i-1-j)s \right)$$

and i = panel under consideration.

Equating Equations 3.3 and 3.4 provides Equation 3.5:

$$w_{em} = \left(\frac{2}{a_j \left(L - a_j \right)} \right) \left(\frac{P_T X_T \left(L - a_j \right)}{L} - P_F x_F \right) \quad (3.5)$$

Equation 3.5 can be plotted for P_T and P_F, which depend on the load configuration on the span, and a_j, which reflects the panel point under consideration, for truss spans of n panels.

However, Equation 3.3 and Equation 3.5 require that the shear forces and bending moments for the concentrated load system be calculated to determine the equivalent uniformly distributed loads. This would make calculating equivalent uniformly distributed loads redundant for a particular truss design project. However, equivalent uniformly distributed load charts may be produced for a concentrated load system such as the Cooper's configuration for various locations on many spans. Subsequently, the equivalent uniformly distributed load charts are a useful tool for calculating panel shear forces and panel point bending moments. Nevertheless, the use of equivalent uniformly distributed live loads for bridge design has not gained favor among North American railway bridge engineers, who prefer to use concentrated loads that more closely reflect actual locomotive and train wheel load configurations.[11] Consequently, modern truss design in North America is typically based on Cooper's E live load configuration shown in Figure 3.1.

The Cooper's E configuration design load appears antiquated, particularly with respect to the use of steam locomotive geometry. However, it is a reasonably good representation of the maximum load effects of modern freight traffic. Therefore, Cooper's EM360 (Cooper's E80) design live load appears to be generally adequate for current equipment weights and load control practices (Unsworth, 2018).

Nevertheless, actual loads can exceed design loads during the life of a bridge. Actual rail cars with 25% higher individual axle loads are not uncommon due to flat wheels, rail and track conditions, overloading and/or unbalanced loading (Tobias, Foutch and Choros, 1996). These large axle loads can significantly influence the maximum effects on members with short influence lines (e.g., stringers in floor systems). Also, in addition to reflecting current equipment weight and overloads, the design live load must consider anticipated future freight equipment weight.[12]

Increasing the AREMA design live load was recommended by Tobias, Foutch and Choros (1996) following a statistical study of the load spectra from over 500 North American freight trains (Unsworth, 2018). Accordingly, some railroad companies vary from the Cooper's EM360 (Cooper's E80) bridge design load based on their current and/or projected operating practice.[13] Nevertheless, the cyclical stress ranges created by Cooper's design load do not necessarily accurately reflect the cyclical stress ranges created by modern railway freight equipment on truss spans. Therefore, the allowable fatigue stress ranges, S_{Rfat}, recommended for design in AREMA (2024) are based on equivalent constant amplitude stress cycles from projected variable amplitude stress cycles due to typical railroad traffic. The service life of steel railway trusses designed for the appropriate loads and correctly fabricated with suitable materials is generally considered to be about 80 years (AREMA, 2024). However, steel trusses that are well maintained after erection can safely and reliably carry live loads for considerably longer service lives.[14]

3.3.2 Railway Equipment Live Load for Fatigue Design

As equipment (locomotives and rail cars) traverses the length of an influence line or loaded length of a span, cyclical variable amplitude stress ranges are induced in the truss members carrying the live load. If large enough, the cyclical[15] nature of the tensile stress ranges can induce progressive fatigue damage that results in cracking, particularly in members of the truss with details subject to stress concentrations.[16]

The AREMA (2024) fatigue design methodology uses the Cooper's EM360 (Cooper's E80) live load as the model for a fatigue design load.[17] The stress-life approach, recommended for the design of steel trusses by AREMA (2024), requires that the variable-amplitude cyclical railway live load be developed as an effective, or equivalent, constant amplitude cyclical design load. This is necessary because fatigue strength is established by constant amplitude cyclical stress testing of typical steel superstructure member and detail specimens. This equivalent constant-amplitude cyclical fatigue design load must accumulate the same damage as the variable-amplitude cyclical load over the total number of stress range cycles to failure. The stress ratio, $R = S_{remin}/S_{remax}$ and stress range, $\Delta S_{re} = S_{remax}-S_{remin}$, describes constant amplitude loading. The mean stress[18] is, $S_{remean} = (S_{remax} + S_{remin})/2$.

However, actual freight rail traffic stress ranges measured on in-service bridges are often complex. The variable-amplitude load and stress cycles from actual freight trains are typically irregular and require a method of reducing the complex spectra to simple cyclic loads and stresses. There are many techniques for counting the significant cycles of variable amplitude load or stress range spectra. For many structures, rainflow cycle counting provides very good results (Dowling, 1999), counting the number of full cycles and their load or stress range magnitudes from half cycles. A frequency distribution histogram for the numbers of stress range cycles, n_i, can be developed from rainflow cycle counting.

AREMA (2024) recommends counting all live load stress range cycles as a complete tensile cycle (even those with a compressive component due to stress reversal) instead of counting only the tensile portion of stress range cycles. This is appropriate because near flaws and details, a member may be subjected to a fully effective tensile stress range cycle due to the superposition of tensile residual stresses. This is analogous to raising the mean stress such that the entire stress range cycle is in tension.

The equivalent constant amplitude stress range, $\Delta S_{re,}$ that accumulates the same damage over the total number of stress cycles to failure as the frequency distribution histogram for the actual variable amplitude stress range cycles can be calculated by considering a damage accumulation rule with crack growth behavior models. A damage accumulation rule is required for the number, $n_{i,}$ of stress range cycles, ΔS_i, of the frequency distribution histogram. There are many damage accumulation rules, but it is usual to apply the Palmgren–Miner (Miner, 1945) linear damage accumulation rule because, although load cycle sequence and interaction effects are not accounted for, the linear damage rule provides good agreement with test results (Stephens, Fatemi, Stephens and Fuchs, 2001). The rule is also independent of the stress magnitude. Also, where residual stresses are high[19] (typical of modern steel railway truss fabrications), mean stress effects are negligible[20] and the stress range magnitude is of principal importance. The Palmgren–Miner linear damage accumulation rule[21] is

$$\Sigma \frac{n_i}{N_i} = 1.0 \tag{3.6}$$

where n_i = the number of cycles at stress range level, ΔS_i and N_i = the number of cycles to failure at stress range level, ΔS_i. A log-log straight line relationship exists between N_i and ΔS_i (Basquin, 1910). This relationship is also observed in constant amplitude fatigue testing of members and details (Kulak and Smith, 1995).

Crack growth behaviour, as defined by the Paris–Erdogan power law,[22] can be used to establish a relationship between stress range and number of cycles to failure. The crack growth rate is

$$\frac{da}{dN} = C\Delta K^{m} \tag{3.7}$$

where a = the crack length, N = the total number of constant amplitude stress range cycles, m = a material constant established from regression analysis of test data (m = 3 for structural steel), C = a material constant established from regression analysis of test data, $\Delta K = C_K \Delta S_{re} \sqrt{\pi a}$ = the change in stress intensity factor for the equivalent constant amplitude stress range, ΔS_{re}, and C_K = constant depending on shape and size of crack, edge conditions, stress concentration and residual stresses (Pilkey, 1997).

Integrating Equation 3.7 yields

$$N = \frac{1}{C}\int_{a_i}^{a_f} \frac{da}{\Delta K^{m}} \tag{3.8}$$

where a_i = the initial crack length and a_f = the final crack length. Substituting $\Delta K = C_K \Delta S_{re} \sqrt{\pi a}$ into Equation 3.8 provides

$$N = \frac{\left(\sqrt{\pi}\Delta S_{re}\right)^{-m}}{C}\int_{a_i}^{a_f} \frac{da}{\left(C_K\sqrt{a}\right)^{m}} \tag{3.9}$$

Since C, C_K and a = a_i ($a_f >> a_i$, therefore neglect terms with a_f because of –m power) are constants (Kulak and Smith, 1995),

$$N = A\left(\Delta S_{re}^{-m}\right) \tag{3.10}$$

where A = a constant depending on detail and established from regression analysis of material strength test data.

Equation 3.10 illustrates that the number of cycles to failure, N, for steel truss members or details is very sensitive to the equivalent constant amplitude stress range, ΔS_{re}. Equation 3.10 also provides the number of cycles at failure, N_i, at stress range level, ΔS_i, as

$$N_i = A\left(\Delta S_i^{-m}\right) \tag{3.11}$$

Substitution of Equation 3.11 into Equation 3.6 yields

$$\Sigma\frac{n_i}{N_i} = \Sigma\frac{n_i}{A\left(\Delta S_i^{-m}\right)} = \Sigma\frac{\lambda_i N}{A\left(\Delta S_i^{-m}\right)} = 1 \tag{3.12}$$

where $\lambda_i = n_i/\Sigma n_i = n_i/N$

and substituting Equation 3.11 into Equation 3.12 provides

$$\sum \frac{\lambda_i \Delta S_{re}^{-m}}{\Delta S_i^{-m}} = 1 \quad (3.13a)$$

or

$$\Delta S_{re} = \left(\sum \lambda_i \Delta S_i^m \right)^{1/m} \quad (3.13b)$$

Equation 3.13 with m = 3 is the root mean cube (RMC) probability density function describing ΔS_{re}, which causes the same amount of fatigue damage as the variable-amplitude stress range spectrum. No factor of safety is applied since the Palmgren–Miner linear damage accumulation rule is considered relatively accurate for service level railway live loads (Fisher, 1984). Equation 3.13 indicates that the railway fatigue design load can be expressed in terms of number of cycles and magnitude of load. The fatigue design load recommended by AREMA (2024) is based on analyses of continuous unit freight trains typical of grain, coal and other bulk commodity traffic on North American and other heavy haul railways.

The AREMA (2024) fatigue design load postulates locomotives and equipment with maximum axle loads of 360 kN (80,000 lbs. or 80 kips). In addition, for loaded lengths or spans greater than 30.5 m (100 ft), it is based on a maximum equivalent uniform load of 90 kN/m (6000 lb/ft). These live loads are characteristic of modern train traffic that typically create forces equivalent to those induced by live loads between Cooper's EM220 (Cooper's E50) and Cooper's EM360 (Cooper's E80).[23] Therefore, since cycles corresponding to the typical characteristic load geometry and loaded length are considered, the maximum Cooper's EM360 (Cooper's E80) load can be used to calculate the fatigue design stress range with the number of design cycles adjusted for the characteristic load magnitude.

The recommended N is based on an analysis of loaded lengths for various member types and lengths subjected to a 110-car unit train[24] with 350 kN (78.75 kip) axle loads (AREMA, 2024). The AREMA (2024) recommendations assume variable-amplitude stress range cycles estimated from 1.75×10^6 trains (60 trains per day over a design life of 80 years), to provide infinite life for loaded lengths or spans less than 30.5 m (100 ft) long (Table 3.2). It may be necessary to increase the number cycles shown in Table 3.2 for spans greater than 23 m (75 ft) long to account for specific load patterns used in accordance with railroad operating practices.[25] The analyses also considered the cyclical loading based on orientation and number of tracks for transverse members (typically, floorbeams) and the effects on longitudinal members by transverse loads applied directly at, or within, panel points (typically, web members). For spans or loaded lengths greater than 90 m (300 ft), a more detailed analysis by influence lines or using modern structural analysis computer software[26] may be required.

The adjusted equivalent N considering a Cooper's EM270 (Cooper's E60) characteristic load magnitude is

$$N = N_v \left(\frac{(\Delta S_{E60})}{\Delta S_{E80}} \right)^m = N_v \left(\frac{(270)}{360} \right)^3 = N_v \left(\frac{(60)}{80} \right)^3 = 0.42 N_v \quad (3.14)$$

where N_v = the total number of variable amplitude load cycles, ΔS_{EM270} (ΔS_{E60}) = stress range from Cooper's EM270 (Cooper's E60) load (characteristic of modern freight train loads), and ΔS_{EM360} (ΔS_{E80}) = stress range from Cooper's EM360 (Cooper's E80) design load.

Table 3.2 (based on 1.75×10^6 trains over the bridge design life) and Equation 3.14 provide the adjusted number of equivalent constant amplitude stress range cycles over the truss design life as shown in Table 3.3.

AREMA (2024) recommends reductions in the fatigue design live load as a means of considering the lower number of cycles on lightly travelled railway lines. Therefore, on railway lines with less than 2.8 MGMT (million gross metric tons) per kilometer (5 MGT (million gross tons) per mile) per year, AREMA (2024) recommends a fatigue design load based on Cooper's EM180 (Cooper's E40). On railway lines with 2.8 to 8.5 MGMT per kilometer (5 to 15 MGT per mile) per year, stress ranges from a Cooper's EM290 (Cooper's E65) live load are recommended. Therefore, for all traffic levels, the number of effective or equivalent constant amplitude live load stress ranges in Table 3.3 can be used to develop appropriate allowable fatigue stress ranges for steel railway truss members and details.

3.3.3 Dynamic Freight Train Live Load

Equipment (locomotives and rail cars) traversing the length of an influence line or span create actions in the longitudinal, lateral and vertical directions related to the movement of the train. Longitudinal forces and pitching rotations (rotations around an axis perpendicular to longitudinal axis of the bridge) are caused by applied train braking and traction forces. Lateral forces are caused by wheel and truck yawing ('hunting' or 'nosing'). Lateral centrifugal forces are also created on bridges with curved track. Rocking (rotations around an axis parallel to longitudinal axis of the bridge) and vertical dynamic forces are created by vehicle–bridge interaction (VBI).[27]

TABLE 3.2
Variable-Amplitude Stress Range Cycles per Train

Span Length, L (m)	Span Length, L (ft)	Variable Amplitude Stress Range Cycles per Train	Total Variable Amplitude Stress Range Cycles, N_v
L>30	L>100	3	5.3×10^6
30 ≥ L>23	100 ≥ L>75	6	10.5×10^6
23 ≥ L>15	75 ≥ L>50	55	96.3×10^6
15 ≥ L	50 ≥ L	110	192.5×10^6

TABLE 3.3
Constant-Amplitude Stress Range Cycles

Span Length, L (m)	Span Length, L (ft)	Equivalent Constant Amplitude Stress Range Cycles Over Member or Detail Life, N
L > 30	L > 100	2.2×10^6
30 ≥ L > 23	100 ≥ L > 75	4.4×10^6
23 ≥ L > 15	75 ≥ L > 50	40.7×10^6
15 ≥ L	50 ≥ L	81.3×10^6

3.3.3.1 Rocking and Vertical Dynamic Forces

Lateral rocking of moving trains will amplify the vertical wheel loads. This load augmentation will increase stresses in members supporting the track, and AREMA (2024) includes this rocking load effect, RE, as a component of the impact load, I_F.

VBI dynamics generate amplification of the vertical moving loads that typically increase deflections and stresses in members supporting the track. The vertical dynamic effect, I_v, at location, x, on a span at time, t, may be expressed as,

$$I_v = \frac{R_d - R_s}{R_s} \tag{3.15}$$

where R_d = dynamic response at location x at time t and R_s = static response at location, x, at t. In terms of deflection dynamic response, the vertical impact factor is $I_v = \frac{y(x,t) - y_s(x,t)}{y_s(x,t)}$ where y(x, t) and $y_s(x, t)$ are the dynamic and static vertical deflections, respectively. Vertical dynamic shear and/or moment responses can also be considered to calculate a vertical impact factor.[28]

AREMA (2024) includes the vertical dynamic effect, I_v, as a component of the impact load, I_F. Analytical calculation of I_v, for even simply supported superstructures, is complex. Therefore, AREMA (2024) provides an empirical I_v based only on member influence line length[29] to provide deterministic values for vertical impact design. The impact factor is

$$I_F = RE + I_v \tag{3.16}$$

The dynamic load effect is

$$LE_D = (1 + I_F)(LE_S) \tag{3.17}$$

where LE_D = design load effect (includes dynamic and static responses for the linear elastic system) and LE_S = static load effect (static response of the linear elastic system).

3.3.3.1.1 Rocking Effects

Railroad freight equipment will rock or sway in a lateral direction due to wind forces, rail profile variances and equipment spring stiffness differences. Rocking due to rail and equipment conditions will affect the magnitude of equipment axle loads and is considered as a dynamic increment of the static axle load by AREMA (2024). Rocking effects are independent of train speed (AREA, 1949; Ruble, 1955).

RE is calculated for each member supporting the track as a percentage of the vertical live load. The applied rocking effect, as recommended in AREMA (2024), is the force couple of an upward force on one rail and downward force on the other rail equal to 20% of the design wheel load, or 0.20W, where W = wheel load (1/2 of axle load). The applied force couple is resisted by the members engaged on each side of the track centerline.

3.3.3.1.2 Vertical Dynamic Effects on Simply Supported Truss Spans

Due to VBI, dynamic responses (increased deflections, shear forces and bending moments) are induced by moving vehicle (locomotives and cars) suspension systems as the wheels traverse a railway bridge with surface irregularities.[30]

In VBI analyses, locomotives and railcars are modeled as moving multiple-degree-of-freedom masses connected by suspension systems modeled with linear spring stiffness and viscous damping elements. The wheel–rail interface is the contact between the moving vehicle and the superstructure.

The superstructure can be modeled as a Euler–Bernoulli beam with distributed mass, m_s, and flexural stiffness, EI_s[31] supporting a deck[32] with irregular track.

Modeling the dynamic behavior of VBI interaction is complex but can be simplified for some train (weight, suspension system, speed) and superstructure (rail, deck, length) characteristics (Unsworth, 2018) by neglecting certain features, for instance the superstructure mass compared with vehicle mass, although it should be noted that superstructure mass considerably less than vehicle weight is not typical of modern steel railway truss bridges.[33] It is also possible to ignore the inertia effects of the stationary mass and the vehicle suspension dynamics for superstructures carrying freight rail traffic at relatively slow vehicle speed, as well as all vibration modes with exception of the first mode of vibration, which is sufficient for simply supported spans (Veletsos and Huang, 1970). Other features that can be ignored in modeling VBI interaction are damping for structures with light damping, which is generally the case for steel railway bridges supporting relatively slow speed heavy freight traffic, dynamic effects of the load (vehicle suspension system dynamics), and the effects of surface irregularities.

It is evident that many of the parameters affecting the dynamic behavior of a steel railway bridge are complex and stochastic in nature. Deterministic solutions are difficult, even with many simplifying assumptions. Modern dynamic finite element analysis (FEA) methods and numerical analysis software enable the incremental, modal superposition and frequency domain analysis of structures. FEA is of particular use in the dynamic analysis of long span, continuous and complex superstructures.[34] However, for routine bridge design, the dynamic effects of moving concentrated live loads are most effectively developed using empirical data.[35] The parameters which affect the dynamic behavior of steel railway bridges are (Byers, 1970; Taly, 1998; Heywood, Roberts and Boully, 2001):

- dynamic characteristics of the live load (mass, vehicle suspension stiffness, natural frequencies and damping);
- train speed (a significant parameter);
- train handling (causing pitching accelerations);
- dynamic characteristics of the bridge (mass, stiffness, natural frequencies, damping);
- span length and continuity (increased impact due to higher natural frequencies of short span bridges);
- deck and track geometry irregularities on bridge (surface roughness) (a significant parameter);
- track geometry irregularities approaching the bridge;
- rail joints and flat or out-of-round wheel conditions (a significant parameter of particular importance for short spans);
- bridge supports (alignment and elevation);
- bridge layout (member arrangement, skewed, curved); and
- probability of attaining maximum dynamic effect concurrently with maximum load.

Many of these parameters are non-deterministic and difficult to assess. Therefore, as with most highway bridge design procedures, ordinary steel railway bridges are designed for dynamic allowance based on empirical equations developed from service load testing. AREMA (2024) provides deterministic values for design impact that are considered adequate, with an estimated probability of exceedance of 1% or less for an 80-year service life based on in-service railway bridge testing (Ruble, 1955; AREMA, 2024). AREMA's (2024) recommended impacts due to vertical effects for simply supported open deck steel bridges are shown in Figure 3.5. The impact load for ballasted deck steel bridges may be reduced to 90% of the total impact load determined for open deck steel bridges (AREA, 1966).

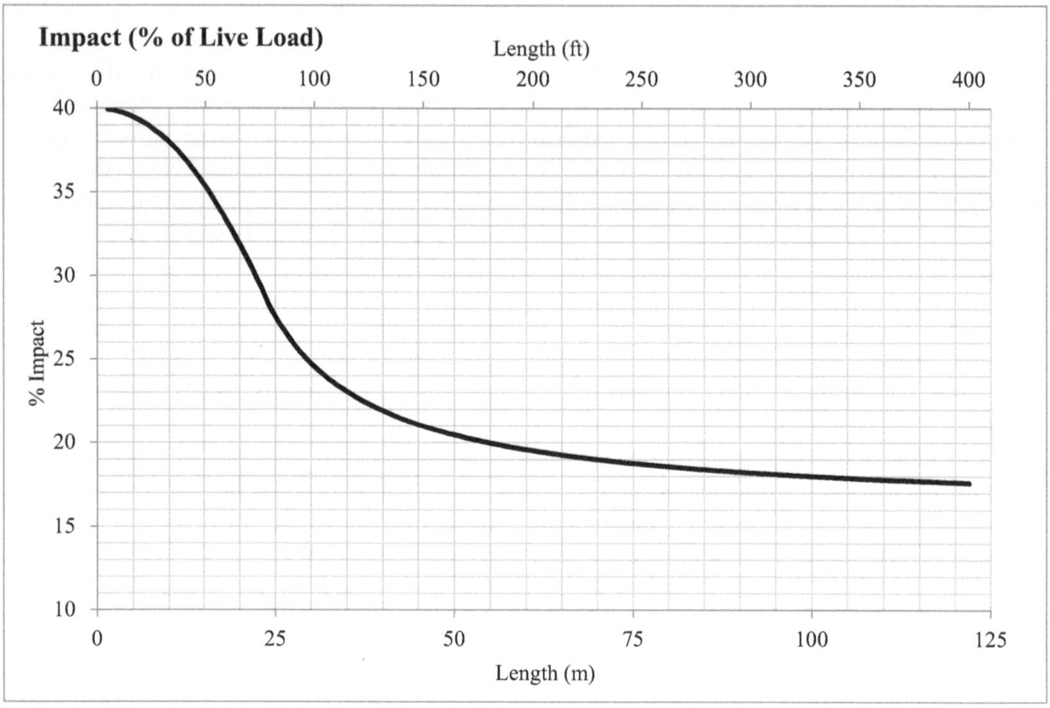

FIGURE 3.5 Impact due to vertical effects on simply supported open deck steel spans.

TABLE 3.4
Mean Impacts for Fatigue Design

Member	Percentage of Total Impact Load
Truss Members	65
Stringers and floorbeams with loaded length ≤ 3 m (10 ft)	65
Stringers and floorbeams with loaded length > 3 m (10 ft)	35

A statistical investigation of steel railway bridge impact (Byers, 1970) revealed that the test data (AREA, 1960) followed a normal frequency distribution with means and standard deviations increasing with increasing speed and decreasing with increasing span length. The study also indicated that track irregularity effects is a relatively large component of the total impact and that short-span impacts are more sensitive to speed effects than those of longer spans.

Also, based on this same statistical investigation of the test data, the mean dynamic amplification (impact) values are presented in AREMA (2024) for fatigue design based on Cooper's EM360 (Cooper's E80) stress ranges. Fatigue is sensitive to detail, and the criteria are given in Table 3.4 for various truss members as a percentage of the design impact load. However, these reductions may not be appropriate for fatigue design of members where consistently poor track or wheel conditions exist.

3.3.3.2 Design Impact Load

The total impact load is the sum of the impact load due to rocking and vertical effects as shown in Equation 3.16.

Modern bridge codes have attempted to formulate dynamic load allowance as a function of fundamental frequency. However, the great number of random parameters generally leads research and development in the direction of simplification for ordinary bridge design. Therefore, typical modern highway bridge design codes represent dynamic load allowance or impact as a simple function of length or specify a constant value within various span ranges. The AREMA (2024) recommendations provide simple equations for dynamic load allowance (impact) based on span type and length.[36] Impact for direct fixation of track to the bridge, or where track discontinuities exist (i.e. movable bridge joints), can be very large and can require refined dynamic analyses and special design considerations for support and damping.

3.3.4 Longitudinal Forces due to Locomotive Traction and Train Braking

Longitudinal forces, due to train braking (acting at the center of gravity of the live load) and locomotive tractive effort (acting at the freight equipment drawbars or couplers), are considerable for modern railway freight equipment. Longitudinal forces from railway live load exhibit the following characteristics (Otter, Sweeney and Dick, 2000):

- tractive effort and dynamic braking forces are greatest when accelerating and decelerating, respectively, at low train speeds;
- span length does not affect the relative magnitude of braking forces due to the distributed nature of emergency train braking systems;
- traction forces from locomotives may affect a smaller length of bridge;
- participation of the rails is relatively small (particularly when the bridge and approaches are loaded) due to the relatively stiff elastic rail fastenings used in modern bridge deck construction;
- the ability of the approach embankments to resist longitudinal forces at the bearings is reduced when the bridge and approaches are loaded; and
- grade-related traction is relatively insignificant for modern high-adhesion locomotives.

The locomotive and car wheels are modeled as accelerating or decelerating rolling[37] masses that do not slide (complete adhesion)[38] as they traverse the bridge superstructure. This provides a dynamic differential equation that allows for the numerical solution of longitudinal force (Unsworth, 2018). The longitudinal forces described by the dynamic differential equation (including the effects of axle bearing, wheel rim friction, air resistance, rolling friction and other effects) have been observed and recorded in field testing in both Europe and the United States. Nonetheless, numerical solutions are too arduous and costly for ordinary railway bridge design.

However, a static analysis may be performed because the longitudinal forces exhibit almost static behavior as maximum tractive and braking forces occur at low speeds when starting and at the end of braking. Accordingly, for a static longitudinal force analysis, the rails may be modeled as a series of longitudinal elastic bars (with independent longitudinal and flexural deformations) on horizontal elastic decks represented[39] by equivalent horizontal springs. The superstructures may also be modeled as elastic bars supported on expansion bearings at one end and fixed at the other end. These models provide a system of differential equations from consideration of static longitudinal force equilibrium of the approach rails, rails on the bridge and superstructure. However, even relatively simple bridge models will involve considerable calculation to determine the longitudinal displacements and forces in the superstructure (Unsworth, 2018).

Accordingly, extensive testing and analytical work has been performed (Foutch, Tobias and Otter, 1996; Foutch, Tobias, Otter, LoPresti and Uppal, 1997; LoPresti, Otter, Tobias and Foutch, 1998; LoPresti and Otter, 1998; Otter, LoPresti, Foutch and Tobias, 1996; Otter, LoPresti, Foutch and Tobias, 1997; Otter, Joy and LoPresti, 1999, Otter, Sweeney and Dick, 2000; Tobias, Otter and LoPresti, 1998; Tobias, Foutch, Lee, Otter and LoPresti, 1999; Uppal, Otter and Joy, 2001) to overcome the theoretical model complexities and numerical modeling efforts. This work has established relationships for braking and traction dependent on the length of the portion of the bridge under consideration. Testing in the United States has provided longitudinal forces for Cooper's EM360 (Cooper's E80) design live loads that are shown in Figure 3.6, Equation 3.18 and Equation 3.19. For loaded lengths less than 110 m (350 ft), longitudinal force due to traction governs. However, locomotive traction occurs over a relatively small length and braking forces in each span from a loaded length consisting of the entire bridge may exceed the tractive effort. The force due to traction governs for short and medium length bridges.

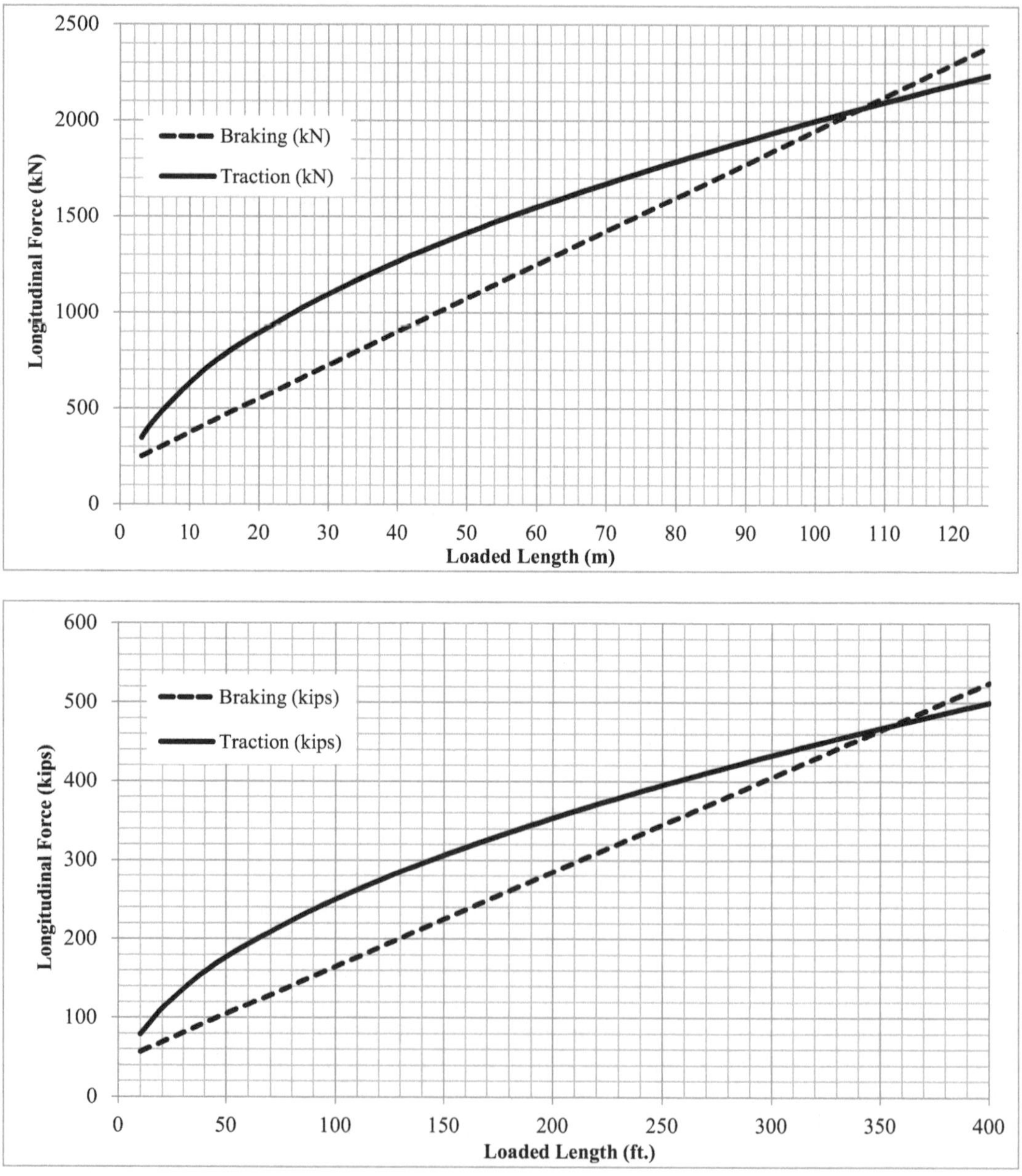

FIGURE 3.6 Longitudinal forces for Cooper's EM360 (Cooper's E80) design live load.

The longitudinal braking, LF_B, and traction, LF_T, forces for Cooper's EM360 (Cooper's E80) design live load are

$$LF_B = 200 + 17.5L \text{ (SI units)} \tag{3.18a}$$

$$LF_T = 200\sqrt{L} \text{ (SI units)} \tag{3.18b}$$

where LF_B = longitudinal force due to train braking (kN), LF_T = longitudinal force due to locomotive traction (kN), and L = length of the portion of bridge under consideration (m).

$$LF_B = 45 + 1.2L \text{ (US customary or imperial units)} \tag{3.19a}$$

$$LF_T = 25\sqrt{L} \text{ (US customary or imperial units)} \tag{3.19b}$$

where LF_B = longitudinal force due to train braking (kips), LF_T = longitudinal force due to locomotive traction (kips), and L = length of the portion of bridge under consideration (ft).

Nevertheless, while estimates of the magnitude of the applied longitudinal traction and braking forces appropriate for design are readily available from Equations 3.18 and 3.19, the distribution of longitudinal forces (the longitudinal force path from rails to bearings) for superstructure members may require careful consideration (see Chapter 8). Due to diaphragm behavior, the resistance of steel deck plates in relatively short-ballasted truss spans may preclude the need for bracing elements to carry longitudinal forces to the main trusses. However, in many moderate- and long-truss spans, specific consideration of the lateral bracing (traction bracing) requirements is necessary to adequately transfer longitudinal forces to the main trusses for transmission to the substructures through the bearings. To preclude the torsional and/or lateral bending of floorbeams that might result from longitudinal forces transmitted by floor systems without connection to the lateral bracing, traction bracing[40] is used (see Chapter 8).

Traction bracing is provided through connection of the stringers to the lateral bracing and addition of a new transverse member between the stringers at the bracing connections. Provided the main truss fixed bearings are adequate to transfer the longitudinal forces to the substructure, the traction bracing truss will avoid lateral loading of floorbeams since the stringers can carry no longitudinal force. Other traction bracing arrangements may be used in a similar manner for long single- and multiple-track truss spans to properly transmit longitudinal traction and braking forces to the bearings. For very long or complex steel truss bridges, a numerical analysis using FEA software that can model longitudinal loads due to locomotive traction and train braking should be used.

3.3.5 Centrifugal Forces

Centrifugal forces acting horizontally at the vehicle center of gravity, recommended as 2.4 m (8 ft) above top of rail in AREMA (2024), act on the moving live load as it traverses curved track on a bridge, as shown in Figure 3.7.

The centrifugal force corresponding to each axle load is

$$CF_A = \frac{m_A V^2}{R} \tag{3.20}$$

where $m_A = A/g$, A = axle load, g = acceleration due to gravity, V = speed of train, and R = radius of curve, which in SI units is expressed as

$$CF_A = \left(4.42 \times 10^{-6}\right) AV^2 D \text{ (SI units)} \tag{3.21a}$$

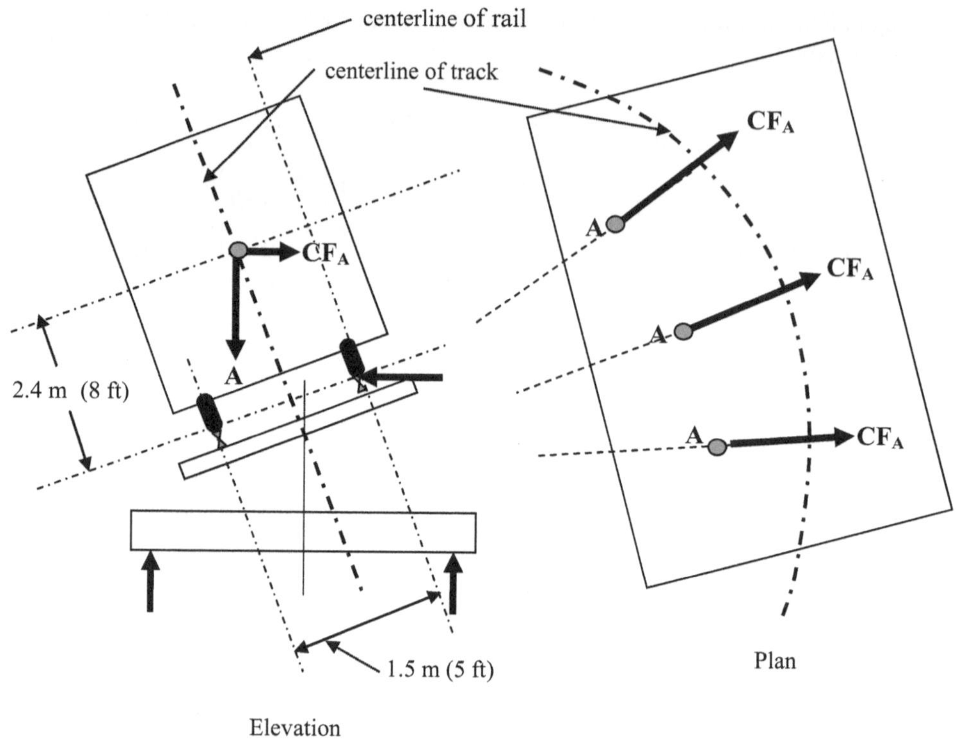

FIGURE 3.7 Centrifugal forces on the moving live load traversing a span supporting curved track.

where CF_A = centrifugal force at each axle, kN; A = axle load, kN; V = speed of train, km/h; R = 1747.5/D, m; and D = degree of curve (the central angle subtended by a 30.5 m chord).

In US customary or imperial units, this is expressed as

$$CF_A = \left(11.7 \times 10^{-6}\right) AV^2 D \text{ (US customary or imperial units)} \tag{3.21b}$$

where CF_A = centrifugal force at each axle, kips; A = axle load, kips; V = speed of train, mph; R = 5730/D, ft; and D = degree of curve (the central angle subtended by a 100 ft chord). Due to the rail–wheel interface contact as trains traverse through curved track, the entire centrifugal force will be transmitted at the outside or high rail, and therefore, centrifugal effects are not considered for longitudinal superstructure members inside the curved track.

3.3.6 Lateral Forces from Moving Freight Equipment

In addition to the centrifugal lateral forces due to track curvature, lateral forces caused by track irregularities at the wheel/rail interface must be considered in the design of steel railway bridges. The differential equation for lateral deflection is obtainable (Unsworth, 2018) but analytical solutions are often unsuitable for routine or ordinary design, and it is most suitable to determine design lateral forces from tests conducted on in-service bridges. Tests concerning the dynamic lateral forces on in-service bridges (Otter, Doe and Belport, 2005) have confirmed that the AREMA (2024) design recommendation of a single moving lateral force of 25% of the heaviest axle of Cooper's EM360 (Cooper's E80) load is a suitable representation of these effects.

The magnitude of lateral forces is of particular importance regarding the design of truss span lateral and cross bracing.[41] Therefore, in addition to the recommendation of a single moving lateral force of 25% of the heaviest axle of the Cooper's EM360 (Cooper's E80) live load, a notional vibration load of 2.9 kN/m (200 lb/ft) applied to the loaded chord and 2.2 kN/m (150 lb/ft) on the unloaded chord is recommended. This notional vibration load is recommended by AREMA (2024) to ensure a minimum lateral bracing stiffness to resist vibration from live load. It is not to be combined with other loads and forces (Waddell, 1916) but is to be applied to the lateral bracing of the loaded chord and unloaded chord as an alternative to the wind load on a loaded railway bridge.

3.3.7 Distribution of Live Load

Unlike highway loads, which may move laterally across the bridge deck, railway live loads are generally fixed in lateral position. However, they are a longitudinal series of large magnitude concentrated wheel loads, and longitudinal and lateral distribution to the deck and supporting members must be considered.

3.3.7.1 Distribution of Live Load for Open Deck Steel Bridges

For open-deck truss spans, no longitudinal distribution is made, and lateral distribution to supporting members is based on span cross-section geometry and type of lateral bracing system. Lateral bracing between longitudinal beams (stringers) should be made with cross-frames, or for spans with shallow beams, rolled beams and/or close beam spacing, solid diaphragms.[42] The cross frames and diaphragms should not have a spacing exceeding 5.5 m (18 ft). In some cases, AREMA (2024) recommends that diaphragms and cross bracing be fastened to the beam flanges. When the lateral bracing system meets these criteria and is properly designed for the lateral forces, all beams supporting the track are considered equally loaded.

3.3.7.2 Distribution of Live Load for Ballasted Deck Steel Bridges

For ballasted deck truss spans, the longitudinal and lateral distribution of design live load to the deck, and the longitudinal (stringers) and transverse (floorbeams) members supporting the deck, is based on tests performed by the Association of American Railroads (Sanders and Munse, 1969).

For deck design, axle loads are distributed as shown in Figure 3.8. The longitudinal deck distribution width, 915 mm + d_b + h (3 ft + d_b + h), should not exceed 1525 mm (5 ft) nor the minimum axle spacing of the design live load. The lateral deck distribution width, (length of tie + d_b +h), should not exceed 4.25 m (14 ft), or the distance between adjacent track centerlines or width of the deck. Timber, concrete and steel decks of thickness h supporting rock ballast with depth d_b should be designed in accordance with the recommended practices of AREMA (2024) Chapters 7, 8 and 15, respectively.

For longitudinal beam (stringer) design, axle loads are distributed as shown in Figure 3.9. Axle loads are distributed equally to all beams or girders within the lateral distribution width, Length of tie + 2(d_b + h). No longitudinal distribution of live load is made for longitudinal members spaced equally along the centerline of the track.

3.3.7.3 Distribution of Live Load for Direct Fixation Deck Steel Bridges

For longitudinal members supporting the deck, axle loads are distributed laterally based on a structural analysis of the deck and supporting beam system considering the location of wheel loads, the location and spacing of supporting members, and deck and longitudinal member materials and properties. No longitudinal distribution of live load is made for longitudinal beams supporting direct-fixation decks.

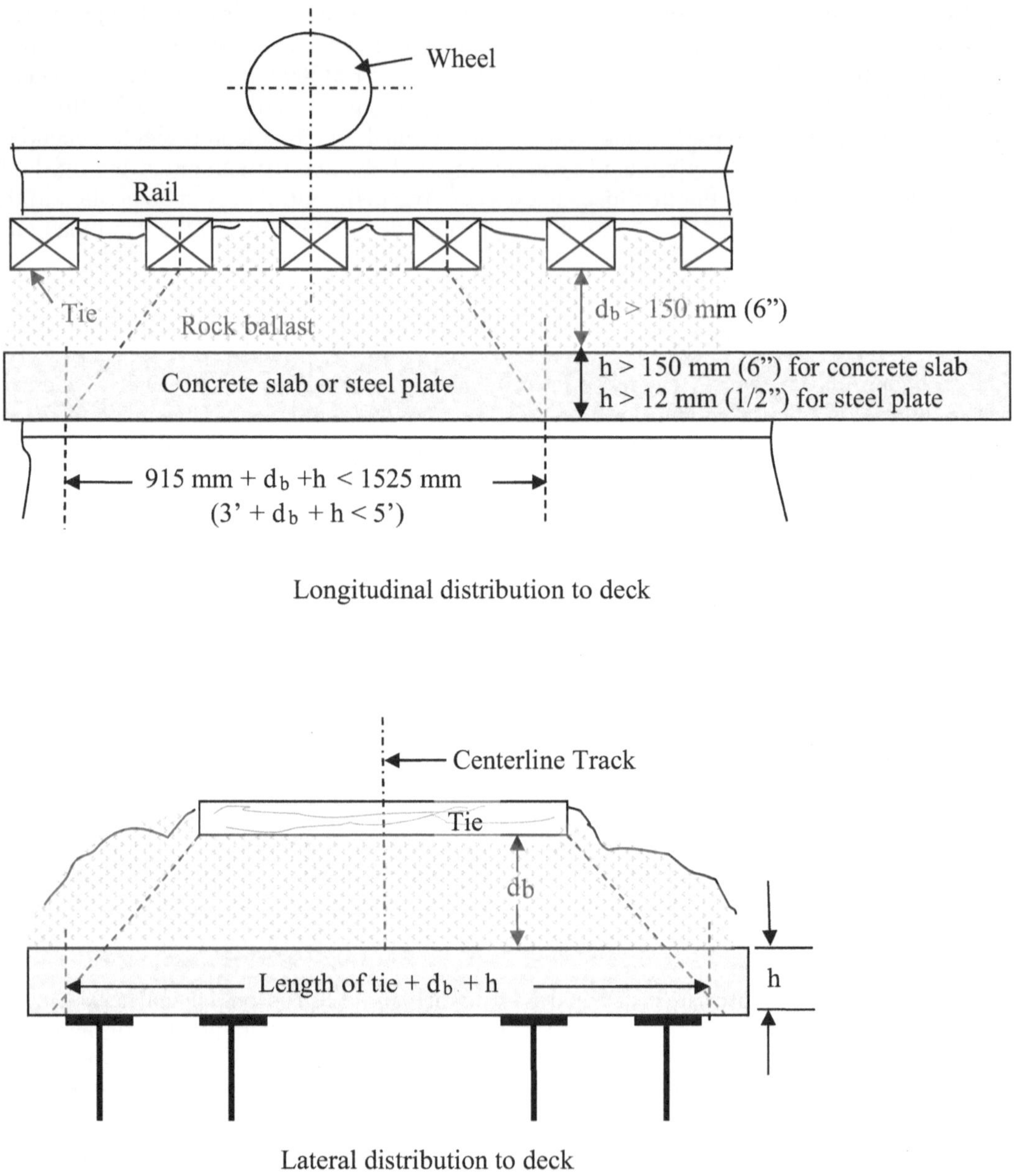

FIGURE 3.8 Lateral and longitudinal distribution of axle loads to deck.

3.4 ENVIRONMENTAL AND OTHER STEEL RAILWAY BRIDGE DESIGN FORCES

In addition to dead and live loads, environmental forces (wind, thermal and seismic) and other miscellaneous forces related to serviceability and overall stability criteria must be considered in the design of steel railway bridges.

3.4.1 Wind Forces on Steel Railway Bridges

In contrast to long-span flexible bridges (such as suspension or cable-stayed bridges) typical steel railway truss superstructures need not consider aerodynamic effects[43] of the wind.[44] However, the

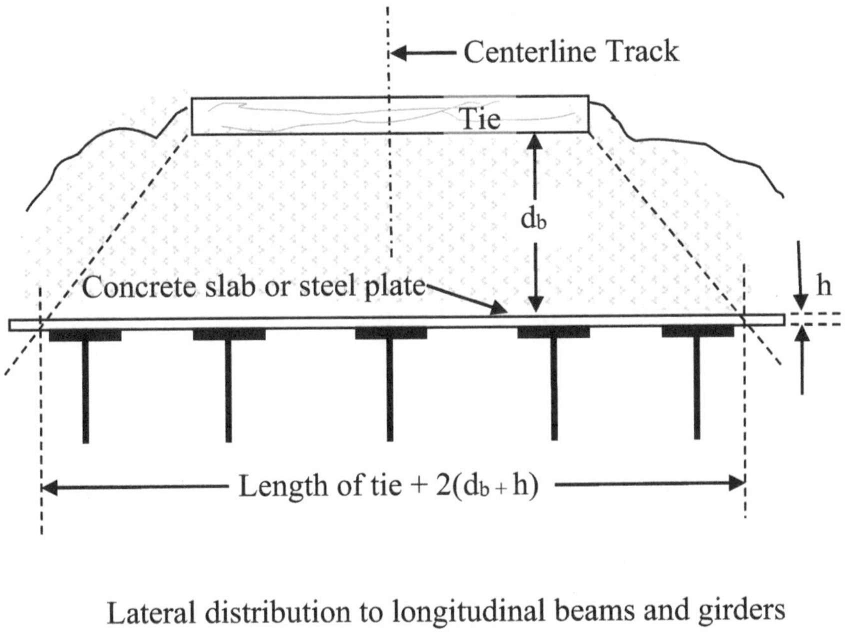

FIGURE 3.9 Lateral distribution of axle loads to stringers.

aerostatic effects of the wind on the superstructure and moving train must be considered, particularly for lateral bracing design.

A steady wind with uniform upstream velocity, V_u, flowing past a bluff body will create a maximum steady state local or dynamic pressure, p_m, in accordance with Bernoulli's fluid mechanics equation as follows:

$$p_m = p_{amb} + \frac{1}{2}\rho V_u^2 \tag{3.22}$$

where p_{amb} = ambient air pressure = 0 at atmospheric pressure, ρ = air density, and V_u = upstream air speed.

However, the average dynamic pressure on the bridge span will be less than the maximum dynamic pressure given by Equation 3.22. Therefore, the dynamic pressure, p, at any point on the bluff body can be expressed as,

$$p = C_p p_m = C_p\left(\frac{1}{2}\rho V_u^2\right) \tag{3.23}$$

where C_p = a dimensionless mean pressure coefficient which depends on the shape of the obstruction. For example, if we assume a 160 km/h (100 mph) wind speed (such as during gale and hurricane winds), Equation 3.22 yields a maximum dynamic pressure of 1.13 kPa (23.7 psf).

Design wind forces must be based on average dynamic wind pressures (i.e. reduced by use of an appropriate pressure coefficient) calculated over an appropriate cross-sectional area. The design wind forces must also consider the effects of wind gusts.[45] It is beneficial, from a design perspective, to calculate design wind forces based on the maximum dynamic pressure, a characteristic area and a dimensionless coefficient that includes the effects of bridge cross-sectional shape as well as the wind flow characteristics.[46] These coefficients are determined from tests and applied to the design process. If the dynamic pressure, p, is integrated over the surface of the bluff body, it will create a force, F, and moment, M. F is resolved into horizontal (drag), F_D, and vertical (lift), F_L, forces. The equations for the forces and moments can then be expressed in a form like Equation 3.23 as

$$F_D = C_D\left(\frac{1}{2}\rho V_u^2\right)A_{RD} \tag{3.24a}$$

$$F_L = C_L\left(\frac{1}{2}\rho V_u^2\right)A_{RL} \tag{3.24b}$$

$$M = C_M\left(\frac{1}{2}\rho V_u^2\right)A_{RM}^2 \tag{3.24c}$$

where C_D = dimensionless drag coefficient that depends on span geometry and Reynolds number, R_e. The Reynolds number indicates wind flow patterns related to inertial effects (R_e large and C_D small) and viscous effects (R_e small and C_D large). R_e is

$$R_e = \frac{\rho V_u L_D}{\mu} \tag{3.25}$$

where L_D = a characteristic length of the bridge or object for drag, C_L = a dimensionless lift coefficient, C_M = a dimensionless moment coefficient, A_{RD} = a characteristic area of the bridge or object for drag, A_{RL} = a characteristic area of the bridge or object for lift, A_{RM} = a characteristic area of the bridge or object for moment, and μ = dynamic wind viscosity.

The drag force or total wind thrust on a bluff body, such as a bridge cross section, created by the wind flow is of primary interest for the design of bridges. Therefore, drag coefficients are established by wind tunnel tests, which incorporate the effects of geometry and flow characteristics (as described by the Reynolds number), which may be used for design purposes. Drag coefficients for a truss are typically 1.70 (Simiu and Scanlon, 1986). The characteristic dimension of the effective area is generally taken as the area projected onto a plane normal to the wind flow. The solidity ratio, ϕ, is defined as the ratio of the effective area to the gross area. The effects of the usual pairing of trusses in steel railway bridges must also be considered. Figure 3.10 illustrates the typical relationship between the drag coefficient relating to the total wind force on two trusses, C_{DT}, to the drag coefficient for a single truss, C_D, in terms of the solidity ratio, ϕ, for spans with truss spacing, s, no greater than the truss height, h.

The AREMA (2024) design recommendations for lateral wind force on a loaded steel railway bridge superstructure assumes that the maximum wind velocity at which trains can safely operate[47] will produce a wind pressure of 1.44 kPa (30 psf). A notional lateral vibration load of 2.9 kN/m (200 lb/ft) applied to the loaded chord and 2.2 kN/m (150 lb/ft) applied to the unloaded chord is recommended in AREMA (2024) as an alternative to the wind pressure of 1.44 kPa (30 psf) on a loaded railway truss.[48] The AREMA (2024) design recommendations for lateral wind load on an unloaded steel railway bridge superstructure assumes a maximum wind velocity corresponding to a typical hurricane event with wind pressure of 2.39 kPa (50 psf). These lateral wind pressures produce wind forces that are considered moving forces.[49] To account for the effects of paired or multiple trusses, these lateral wind pressures are to be applied to the projected surface area of the windward truss plus the projected surface area of the leeward truss not shielded by the floor system.

The AREMA (2024) design recommendations also indicate the lateral wind force on the moving train is to be taken as 4.38 kN/m (300 lb/ft) at a distance 2.4 m (8 ft) above the top of rail. Designers

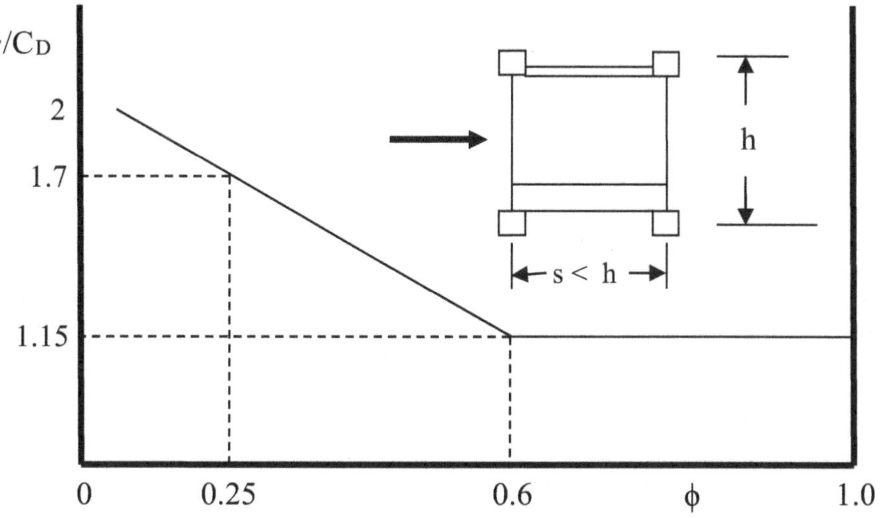

FIGURE 3.10 Typical relationship between the drag coefficient on two trusses and the drag coefficient for a single truss.

of railway bridges that carry high loads (e.g., double-stack rail cars) should review this recommendation. Longitudinal wind forces must be considered as 50% of the lateral wind forces.

3.4.2 Thermal Forces from Continuously Welded Rail on Steel Railway Bridges

Continuously welded rail (CWR) is used in modern track construction because it diminishes dynamic effects (no impact forces due to joints in the rail), provides a smoother ride and results in reduced rail maintenance and increased tie life. The rail can be fastened to the deck to provide lateral and longitudinal restraint.[50] The deck is also fastened to the superstructure to provide lateral and longitudinal restraint.[51]

The longitudinal forces[52] generated due to restraint of thermal expansion and contraction of the rail and superstructure may need to be considered in the design of some steel railway superstructures. Longitudinal forces in the rail are due to thermal expansion and contraction restrained by rail-to-deck interface mechanical fasteners and the stone ballast. However, on open deck bridges, there are additional longitudinal forces created by thermal expansion and contraction of the superstructure with magnitude dependent on the stiffness of rail-to-deck and deck-to-superstructure longitudinal restraint. The distribution of longitudinal forces through the superstructure may be of little concern for some superstructure types. However, for other types of superstructures and long spans, the longitudinal force path from rails to bearings may be of importance (for example, floorbeams with direct fixation of track and some span floor systems). In addition to thermal actions, the CWR may experience internal stresses due to bridge span movements from live load bending. The magnitude of the CWR-superstructure thermal interaction is governed by the following conditions:

- movement of the bridge spans, in particular the maximum span length which may freely expand in the bridge;
- the rail-laying temperature (neutral temperature) and ambient temperature extremes at the bridge (the temperature ranges experienced by the rail and superstructure depend on neutral temperature, and maximum and minimum ambient temperatures);
- the type of bridge (open deck, ballasted)[53] and bridge materials;

- the connection at the rail-to-deck and deck-to-superstructure interfaces;
- the cross-sectional area of the rail and coefficient of thermal expansion of the bridge; and
- the location of fixed and expansion bearings (spans with adjacent expansion bearings on the same pier create a long expansion length and generally provide the governing condition for design).

The longitudinal restraint at the rail-to-deck and deck-to-superstructure interfaces affects the stress in the CWR, relative displacements between rail-to-deck and deck-to-superstructure at the interfaces and the extent of rail travel (resulting in a gap in the rail) upon rail fracture at cold temperatures. The partial differential equation of horizontal motion from force equilibrium on a simply supported span bridge of constant mass and stiffness may be developed to determine the magnitude of longitudinal forces due to thermal effects on CWR (Unsworth, 2018). The partial differential equation is solved considering the following failure criteria:

- safe rail gap (separation) on a bridge after fracture of CWR;
- safe stress in the CWR to preclude buckling;
- acceptable relative displacement between rail-to-deck and deck-to-superstructure to preclude damage to the deck and/or fasteners; and
- bearing component damage.

For long open deck spans, there are conflicting design requirements that the rail-to-deck-to-superstructure connection be flexible enough to avoid excessive compressive stress in CWR (that could precipitate buckling) and rigid enough to reduce rail separation (and the associated high strain rate forces suddenly imparted to the superstructure) and relative displacements at rail-to-deck and deck-to-superstructure interfaces. The CWR can be anchored to only a portion of the deck length to accommodate these conflicting design requirements. The portion of the deck length to which the CWR is anchored should be adjacent to the fixed bearings to allow the necessary movement between rail and superstructure (anchored CWR over expansion bearing areas will resist the thermal movements of the span).

Based on similar considerations, AREMA (2024) and many railway companies established standard practice for longitudinal anchoring of CWR to long open deck steel spans. In general, the recommended practice is to use longitudinal rail anchors or elastic rail fasteners on approaches, and near fixed ends of spans, allowing some movement near expansion ends of spans.[54] For very long or complex steel bridges, a numerical analysis using FEA software that can model thermal loads in rails, deck and superstructure should be used. This software may also accommodate the analysis of longitudinal loads due to locomotive traction and train braking.

3.4.3 Seismic Forces on Steel Railway Bridges

The superstructure must transmit seismic forces to the bearings by lateral bracing, diaphragm action of deck plates and end cross frames. The seismic forces carried to the bearings are then transmitted to the substructures. The seismic forces, determined based on seismic risk and seismic hazard, must be considered from a performance perspective by conceptual and detail deliberations, and by dynamic structural analysis. Three levels of performance criteria or limit states are considered by AREMA (2024) as outlined in Table 3.5.

The average return period for the ground motion levels is based on the seismic risk and bridge importance considerations. The seismic risk for ground motion level (GML) 1 concerns the safety of continued train operations. The seismic risk for GML 2 and GML 3 relates to economic and passenger service considerations. AREMA (2024) indicates the following issues with respect to seismic risk or bridge importance used to determine the appropriate seismic return period for each limit state: immediate safety, immediate value, and replacement value of the bridge. The GML associated

TABLE 3.5
Performance Criteria for Seismic Analysis and Design

Performance Criterion or Limit State		Ground Motion Level of Seismic Event (GML)	Frequency of Seismic Event	Average Return Period of Seismic Event (years)	Design Considerations
Serviceability	• Elastic response of critical structural members • Moderate damage allowing traffic to continue at reduced speed • No permanent deformations	1	Occasional	50–100	Conceptual and Dynamic Structural Analysis
Ultimate	• Strength and stability of structural members not compromised • Inelastic response with traffic arrested until bridge inspection • Damage easily detected and repaired • No damage that affects structural integrity due to deformations	2	Rare	200–475	Conceptual and Details
Survivability	• Extensive structural damage but no structural collapse • Structure unusable by traffic	3	Very rare	1000–2475	Conceptual and Details

with each limit state reflects the seismic hazard level at the bridge site. It is defined by the base acceleration coefficients for peak ground acceleration (PGA), short-period spectral acceleration (S_S) and long-period spectral acceleration (S_L) at the bridge site.

Considering the importance of a bridge (its seismic risk), linear interpolation can be used to calculate an appropriate seismic return period that falls between the minimum and maximum average seismic return periods for each limit state.[55] Accordingly, AREMA (2024) provides equations with which to calculate the PGA, S_S and S_L base acceleration coefficients for return periods other than the maximum average return periods for GML 1, GML 2 and GML 3 of 100, 475 and 2475 years, respectively.

The level of seismic dynamic analysis required depends on the location and characteristics of the bridge. Railway bridges require an appropriate conceptual design for all limit states, structural analysis at GML 1 seismic events and detail considerations for GML 2 and GML 3 seismic events. However, for bridge sites with low ground motion levels, conceptual and detail considerations are unnecessary if the bridge remains elastic in a structural analysis at GML 3 loads.

The conceptual design required for all limit states includes consideration of the bridge configuration (alignment, substructure conditions); superstructure characteristics (support conditions, weight, span length); and substructure features (bridge seats, bent cap types, column configurations). The detail design of truss spans at GML 2 and GML 3 limit states must consider[56] continuity (transmission of lateral forces from superstructure to the ground); superstructure characteristics (load path to bearings by lateral bracing, deck diaphragm behavior, end cross frames); redundancy (bearing seat support width, bearing shear connectors, span ties, foundation rocking at GML 3, CWR load path); and member ductility. The joints in main members carrying lateral seismic forces must be designed for 30% greater than the member's yield capacity or GML 3 load levels.

Structural analysis required for GML 1 limit state is based on response spectrum analysis that determines the modal seismic response coefficient based on site factors, base acceleration coefficients, damping and the modal period of vibration. Site factors are based on the site class[57] and base acceleration coefficient. The base acceleration coefficients are determined from information provided by the US Geological Survey, Geological Survey of Canada or other applicable sources.

The response spectrum is the seismic response coefficient over a range of modal vibration periods; it exhibits a perfectly rigid region at vibration periods less than 0.03 seconds, a low period linear transition region, a constant acceleration region defined by the short period spectral response coefficient and a region that decreases proportionally to the inverse of the modal period of vibration and defined by the long period spectral response acceleration coefficient. However, due to the rigidity of typical railroad bridges, the AREMA (2024) equation for the modal seismic response coefficient does not include a low period seismic response. Nevertheless, should the bridge be considered flexible,[58] AREMA (2024) provides an equation for the seismic response coefficient for modal vibration periods less than 0.03 seconds based on PGA and an equation for the seismic response coefficient for modal vibration periods less than the modal vibration period at the commencement of constant acceleration[59] based on PGA and S_S.

Single-span bridges require no seismic analysis. Two-span or regular multi-span bridges can be analyzed by equivalent lateral force (ELF) analysis or modal analysis (MA); multi-span irregular[60] bridges must undergo MA. The ELF procedure considers the modal period of vibration equal to the natural period of vibration to determine the seismic response coefficient in the lateral and longitudinal directions. The distributed seismic force in the lateral and longitudinal directions is determined based on the seismic response coefficient and the distributed weight of the bridge. The seismic force in the lateral and longitudinal directions is distributed to members of the bridge based on stiffness and support conditions. The seismic design forces are combined using the 100%-30% rule for a static structural analysis.

The elastic response spectrum is determined based on the modal seismic response coefficient and the modal period of vibration. A dynamic modal structural analysis in the lateral and longitudinal directions to determine mode shapes, frequencies and member forces based on the bridge mass, stiffness and support conditions is typically conducted with finite element software.[61] The seismic loads in members in the lateral and longitudinal directions may be combined with the 100%-30% rule or by the square root of the sum of the squares in each direction.

3.4.4 Loads Relating to Overall Stability of the Superstructure

The overall stability of the superstructure against wind, nosing and centrifugal forces must also be ensured. The stability of truss spans should be calculated using a live load, without impact, of 18 kN/m (1200 lb/ft).[62] On multiple-track bridges, this live load should be placed on the most leeward track on the bridge. A 50% increase in allowable stress is permissible when determining stresses in members resisting overall instability of the span (such as anchor rods at bearings).

3.4.5 Pedestrian Loads

Typical walkways for steel railway bridges consist of a steel grating or other system with non-slip surfaces. The walkway components are designed for a load of 4.1 kPa (85 psf) and a maximum deflection of 1/160 of the walkway span length between supports. Guardrails (handrails and posts) for pedestrian walkways are typically designed for railing and post loads of 890 N (200 lb) applied laterally or vertically at the location of maximum effect.

3.4.6 Load and Force Combinations for Design of Steel Railway Truss Superstructures

Table 3.6 indicates the allowable stress factors for main truss members for some load combinations.

TABLE 3.6
Allowable Stress Factors for Truss Members

Load Combination		Truss Members	Allowable Stress Factor
LC1A	$DL + LL + I + CF$	Chord and web members	1.00
LC1B	$LC1A + W_L + LF + N + CWR + OF$	Chord, web and bracing members	1.25
LC2	$DL + (LL + I + CF)_{in}$	Web members	1.33
LC3	$LL + I_f + CF$	Chord and web members with tensile force range from live load or live load force reversal	S_{rfat}
LC4	W_U	Chord and bracing members	1.00
LC5	$B_R + N + W_L + CWR_L + CF$	Bracing members between compression chord members	1.25
LC6	$DL + EQ$	All members in active seismic zones	1.50
LC7	$SL + N + CF$	Members resisting overall instability	1.50

The internal member forces in Table 3.6 are due to

DL = dead load (vertically applied external load)
LL = live load (concentrated and/or uniformly distributed vertically applied external load)
I = maximum impact or dynamic increment (vertically applied external load)
CF = the centrifugal effect of live load on curved track (laterally applied external load)
W_L = wind on the loaded truss span (laterally and longitudinally applied external forces from wind on the structure and wind on the live load)
LF = the effect of live load traction or braking (longitudinally applied external loads)
N = nosing or truck hunting (laterally applied external load)
CWR = the forces due to restraint of continuously welded rail on tangent track (longitudinally applied external force)
OF = other forces such as those related to statical indeterminacy
$(LL + I + CF)_{in}$ = the live load + maximum impact load + centrifugal effect forces in web members for the strength overload condition.
$LL + I_f$ = live load + fatigue impact force range (typically 0.65I)
B_R = Notional force for compression chord bracing members (laterally applied notional force)
CWR_L = the forces due to restraint of continuously welded rail on curved track (laterally applied external force)
W_U = wind on the unloaded truss span (laterally and longitudinally applied external forces from wind on the structure)
EQ = seismic load on members from earthquake ground motions
SL = live load on leeward track of 18 kN/m (1200 lb/ft) without impact

In addition, members of the truss stressed during construction have allowable stress factors of 1.25 for erection loads (EL) and 1.33 for erection loads combined with wind on the unloaded structure ($EL + W_U$). Members of the truss stressed during maintenance jacking have allowable stress factors of 1.50 for dead loads (DL).

Load case LC1A applies to truss chord and web members subjected to live load; LC1B applies to all truss members subjected to live load, forces from CWR and lateral and longitudinal forces due to live load and wind. LC2 applies to truss web members and concerns web member strength overload. As outlined in Section 1.7.4.4, LC2 is based on the AREMA (2024) recommendation that truss web members and their connections be designed for a live load that increases the total stress by 33% over the design stress in the most highly stressed chord of a truss. This live load ensures that web members attain their safe capacity at about the same increased live load as other truss members because in steel railway trusses, the web members typically reach capacity prior to other members in the truss (Hardesty, 1935).[63]

LC3 concerns the fatigue design of chord and web members with tensile force ranges from cyclical live load or live load force reversal. LC4 provides the load combination to consider for the strength design of lateral bracing members (includes main truss chord members) to resist the wind forces on the unloaded bridge. LC5 relates to the strength of bracing members between compression chord members to resist lateral forces. LC6 concerns the strength of members to resist forces from seismic activity.[64] LC7 outlines the load combinations to consider causing the overall instability of the superstructure.

NOTES

1 Augmentation of static vertical load due to vibrations.
2 Lateral sway due to wind forces and differences in rail profile and equipment suspension stiffness.
3 To maintain a relatively low dead load, many truss spans are designed with open decks (ties (sleepers) supported on longitudinal beams (stringers) of the floor system).
4 Some engineers include walkway live load as a component of superstructure dead load. AREMA (2024) recommends a walkway live load of 4.1 kPa (85 psf).
5 The designer and erector will have to be able to communicate early in the design. Otherwise, truss member strengthening may be required to accommodate construction method and temporary construction loads.
6 The Association of American Railroads considers 130,000 kg (286,000 lb) rail cars free interchange cars.
7 Cooper's EM360 is the recommended live load in units of the International System (Système International or SI), the modern metric system. AREMA (2024) Chapter 15—Steel Structures does not explicitly indicate Cooper's EM360; however, AREMA (2024) Chapter 8—Concrete Structures and Foundations does outline the Cooper's EM360 live load as shown in Figure 3.1.
8 Alternate live loads have been proposed, such as the M live load system proposed by D. B. Steinman (1922), but Cooper's E live load has been maintained as the railway bridge design load by most railroads to date.
9 Maximum chord member forces are due to maximum panel point moments. Since any load added to a simply supported span will increase bending moments, it is intuitive that maximum chord member forces occur with the uniformly distributed load on the entire span. This is mathematically illustrated in Chapter 4.
10 Maximum and minimum web member forces occur due to maximum positive and negative shear forces in the truss panels. It is intuitive that maximum and minimum shear forces in panels occur with the uniformly distributed load partially on the span. This is mathematically illustrated in Chapter 4.
11 As shown in Chapter 4, charts for Cooper's E load configuration can be prepared to effectively calculate panel shear forces and panel point moments in truss spans.
12 In particular four-axle rail car weights, which have been increasing at a greater rate than locomotive weights.
13 Some North American railroads currently design new bridges for Cooper's EM400 (Cooper's E90) live load.
14 Many steel railway truss spans are over 100 years old, and some are over 125 years old.
15 Due to wheel configuration, typical railway traffic creates substantially more cycles on shorter spans than on longer spans.
16 At a microscopic level, cyclical stresses can precipitate movement of atomic dislocations and create slip lines resulting in surface stress raisers and crack initiation. However, in typical steel superstructures, stress concentrations occur due to poor design details and/or fabrication practices.
17 The Cooper's EM360 (Cooper's E80) live load is the current basis for the AREMA (2024) fatigue design load. However, a specific fatigue design load has been proposed (Dick, Otter and Connor, 2011; Dick, Connor, Korkmaz, Sarment and Maginot, 2022).
18 Since constant amplitude cyclical fatigue testing of members and details includes the effects of stress concentrations and residual stresses (from rolling, forming, fabricating and welding operations) the fatigue life is not influenced by mean stress effects, and the range of stress is significant for fatigue.
19 Typically at yield stress level.
20 Dead load is also unimportant since mean stress effects are negligible.

21 May be interpreted as the proportion of total fatigue life consumed at stress range level, ΔS_i.

22 This is a log-log linear relationship. At crack growth rates below log-log linear behavior, a threshold exists below which cracks will not propagate. At crack growth rates above log-log linear behavior, fracture occurs at critical stress intensity equal to the fracture toughness of steel (Barsom and Rolfe, 1987; Anderson, 2005).

23 For longer spans (greater than about 15 or 23 m (50 or 75 ft) depending on car lengths), modern unit freight train traffic typically creates forces equivalent to about Cooper's EM220 to Cooper's EM270 (Cooper's E50 to Cooper's E60). For shorter spans, modern unit freight train traffic can generate forces equivalent to about Cooper's EM270 to Cooper's EM360 (Cooper's E60 to Cooper's E80).

24 This design train consist was developed before some modern freight railroads started running trains with almost 200 cars.

25 For example, it is theoretically possible to generate 55 cycles on spans almost 30.5 m (100 ft) long with a repeating load pattern of two loaded and two unloaded rail cars (AREMA, 2024).

26 The software algorithms are often based on the influence line methodology.

27 For example VBI related to train suspension, deck and superstructure stiffness and damping.

28 Nevertheless, deflection is often used to establish dynamic increments for both shear and bending stresses.

29 Which appears to be reasonable based on the loaded simply supported fundamental frequency of free vibration $\omega_1 = \left(\frac{\pi^2}{L^2}\sqrt{\frac{EI}{m}}\right)$ where L = span length, EI = flexural rigidity of span, m = mass per unit length of span and ψ is a constant depending on weight of load and L (ψ = 1.00 for unloaded spans).

30 Surface irregularities such as flat wheels, rail joints, poor track geometry and even lesser aberrations such as rail undulation from bending between ties can excite vehicle and superstructure vibrations. Flat wheels and rail joints are of particular concern on short-span bridges with high natural frequency. Surface irregularities, particularly on short spans, are of considerable importance in relation to railway live load dynamic effects (Byers, 1970).

31 Particularly for simply supported steel truss spans with relatively constant cross sections.

32 Typically modeled as an elastic layer with uniform springs for open and ballasted decks.

33 Particularly modern truss superstructures with ballasted decks.

34 Up to three modes of vibration should be considered for continuous and cantilever bridges (Veletsos and Huang, 1970).

35 Although not often used in modern railway bridge design, impact equations for steam locomotives are provided in AREMA (2024) in addition to those recommended for modern diesel and diesel-electric powered locomotives.

36 Shown in Figure 3.5.

37 Rolling is the superposition of translation and rotation (Beer and Johnston, 1976). Axle bearing and wheel rim friction are very small in comparison with rolling friction and are typically neglected in analyses of longitudinal forces due to locomotive traction and train braking.

38 Non-uniform speed (acceleration for starting and deceleration for braking) and adhesion must exist between the wheel and rail interface to start and stop trains.

39 Other bridge deck and approach track models that incorporate different longitudinal restraint at the rail-to-deck and deck-to-superstructure can be used to provide greater accuracy.

40 It is typical to install traction bracing (thrust frames) in about every third panel to transfer longitudinal forces to the chord members without creating excessive lateral loads on floorbeams.

41 Lateral loads from freight rail equipment are considered applied directly to bracing members (see Chapter 8) without producing lateral bending of supporting member flanges or chords.

42 Channels, coped wide flange shapes and plate/angle fabrications, are often used for diaphragms between longitudinal beams. Plates alone are generally not used due to the absence of flanges and low bending strength.

43 The effects from dynamic behavior and buffeting.

44 An equivalent static wind pressure is appropriate since the natural or fundamental frequency of the superstructure is substantially greater than the frequency of localized gust effects.

45 Gust factors are generally between 2 and 3 for tall structures (Liu, 1991).

46 Wind flow characteristics are described by the Reynolds number on a characteristic geometry, which is dependent on wind velocity and viscosity.

47 To avoid the overturning of empty cars.

48 The notional vibration load can occur only when the bridge is loaded by live load.

49 The recommendation that the wind forces be considered as moving loads does not seem intuitive. However, the intent is to consider the wind forces as moving to ensure maximum member forces due to wind are considered (see Chapter 8).

50 Longitudinal restraint by elastic hold-down fasteners, friction and/or rail anchors applied at the base-of-rail against the ties.
51 Ballasted decks are generally rigidly connected to the superstructure. However, open deck spans have various degrees of longitudinal restraint depending on deck-to-superstructure connection. Open decks are often fastened to the superstructure with bolts or "hook bolts" installed at regular intervals (for example, every third tie).
52 These longitudinal forces caused by the thermal expansion and contraction of restrained rails and superstructure are in addition to the longitudinal forces in the rails and superstructures due to traction and braking.
53 For ordinary ballast deck bridges, the differential thermal movements are generally accommodated by the relatively flexible ballast section between the rail and superstructure.
54 Movement at the expansion end of spans is accommodated by unanchored rail or zero-longitudinal-force elastic fasteners. Rail expansion joints can be required for very long or complex bridges, or bridges with unusual bearing configurations (i.e. adjacent expansion bearings on long spans).
55 In some cases, designers conservatively consider the appropriate seismic return period as the maximum average seismic return period for each limit state.
56 Detail design provisions for reinforced concrete structures are also outlined in AREMA (2024).
57 The site class is determined based on soil stiffness in the upper 30 m (100 ft) of the soil profile measured by the average shear wave velocity.
58 Based on calculation of the effective moment of inertia for reinforced concrete structure, inclusion of foundation rocking to determine the vibration period and consideration of lateral flexibility of superstructures in the seismic response.
59 The modal vibration period at the commencement of constant acceleration is based on the short and long period spectral acceleration coefficients and site factors.
60 Bridges with a significantly irregular configuration or substructure stiffness.
61 An adequate number of modes must be included in the analysis to ensure lateral and longitudinal responses that have a minimum of 90% mass participation. For some bridges, this may require many modes in the analysis to capture the significant mass participation occurring at high frequencies.
62 This represents a uniform load of empty rail cars.
63 However, in parametric studies of some recent railway bridge designs, load combination LC2 did not govern web member design (Conway, 2003), and when tensile stress ranges or stress reversals are present, fatigue criteria often governs truss web member design, in any case.
64 LC6 assumes no live load on the bridge during a significant earthquake event. However, for unusually long viaducts, the presence of live load during a seismic event is a remote possibility.

REFERENCES

American Railway Engineering Association (AREA), 1949, "Test Results on Relation of Impact to Speed", *AREA Proceedings*, 50, 432–443.

American Railway Engineering Association (AREA), 1960, "Summary of Tests on Steel Girder Spans", *AREA Proceedings*, 61, 151–178.

American Railway Engineering Association (AREA), 1966, "Reduction of Impact Forces on Ballasted Deck Bridges", *AREA Proceedings*, 67, 699.

American Railway Engineering and Maintenance-of-Way Association (AREMA), 2024, *Manual for Railway Engineering*, Chapter 15, Steel Structures, Lanham, MD.

Anderson, T.L., 2005, *Fracture Mechanics*, 3rd edition, Taylor & Francis, Boca Raton, FL.

Barsom, J.M., and Rolfe, S.T., 1987, *Fatigue and Fracture Control in Steel Structures*, 2nd edition, Prentice-Hall, Englewood Cliffs, NJ.

Basquin, O.H., 1910, *Exponential Law of Endurance Tests*, Proc. American Society for Testing Materials, Vol. 10, Part 2, ASTM, West Conshohocken, PA.

Beer, F.P., and Johnston, E.R., 1976, *Statics and Dynamics*, 3rd edition, McGraw-Hill, New York.

Byers, W.G., 1970, "Impact from Railway Loading on Steel Girder Spans", *Journal of the Structural Division, ASCE*, 96(ST6), 1093–1103.

Conway, W.B., 2003, *AREMA, Chapter 15, Article 1.3.16*, Communication of February 13, 2003 with AREMA Committee 15, Minutes of Spring 2003 Committee 15 Meeting.

Cooper, T., 1894, "Train Loadings for Railroad Bridges", *Transactions of the American Society of Civil Engineers*, 31, 174–184.

Dick, S.M., 2002, *Bending Moment Approximation Analysis for use in Fatigue Life Evaluation of Steel Railway Girder Bridges*, PhD Thesis, University of Kansas.

Dick, S.M., Connor, R.J., Korkmaz, C., Sarment, M., and Maginot, C., 2022, *Advancements in Fatigue Evaluation and Design for Steel Railway Girder Bridges*, Proceedings of the 2022 AREMA Annual Conference, Lanham, MD.

Dick, S.M., Otter, D.E., and Connor, R.J., 2011, *Comparison of Railcar and Bridge Design Loadings for Development of a Railroad Bridge Fatigue Loading*, Proceedings of the 2011 AREMA Annual Conference, Lanham, MD.

Dowling, N.E., 1999, *Mechanical Behavior of Materials*, 2nd edition, Prentice Hall, Upper Saddle River, NJ.

Fisher, J.W., 1984, *Fatigue and Fracture in Steel Bridges*, John Wiley and Sons, New York.

Foutch, D.A., Tobias, D.H., and Otter, D.E., 1996, *Analytical Investigation of the Longitudinal Loads in an Open-Deck Through-Plate-Girder Bridge*, Report R-894, Association of American Railroads.

Foutch, D.A., Tobias, D.H., Otter, D.E., LoPresti, J.A., and Uppal, A.S., 1997, *Experimental and Analytical Investigation of the Longitudinal Loads in an Open-Deck Plate-Girder Railway Bridge*, Report R-905, Association of American Railroads.

Hardesty, S., 1935, "Live Loads and Unit Stresses", *AREA Proceedings*, 36.

Heywood, R., Roberts, W., and Boully, G., 2001, *Dynamic Loading of Bridges*, Journal of the Transportation Research Board, National Academy Press, Washington, DC.

Johnson, J.B., Bryan, C.W., and Turneaure, F.E., 1946, *The Theory and Practice of Modern Framed Structures, Part 1*, John Wiley & Sons, New York.

Ketchum, M.S., 1924, *Structural Engineer's Handbook*, McGraw-Hill, New York.

Kulak, G.L., and Smith, I.F.C., 1995, *Analysis and Design of Fabricated Steel Structures for Fatigue*, University of Alberta Structural Engineering Report No. 190, Edmonton, AB.

LoPresti, J.A., and Otter, D.A., 1998, *Longitudinal Forces in a Two-Span Open-Deck Steel Bridge at FAST*, Technology Digest 98–020, Association of American Railroads.

LoPresti, J.A., Otter, D.A., Tobias, D.H., and Foutch, D.A., 1998, *Longitudinal Forces in an Open-Deck Steel Bridge*, Technology Digest 98–007, Association of American Railroads.

Otter, D.E., Doe, B., and Belport, S., 2005, *Rail Car Lateral Forces for Bridge Design and Rating*, Technology Digest 05–002, Association of American Railroads.

Otter, D.E., Joy, R., and LoPresti, J.A., 1999, *Longitudinal Forces in a Single-Span, Ballasted-Deck, Plate-Girder Bridge*, Report R-935, Association of American Railroads.

Otter, D.E., LoPresti, J., Foutch, D.A., and Tobias, D.H., 1996, *Longitudinal Forces in an Open-Deck Steel Plate-Girder Bridge*, Technology Digest 96–024, Association of American Railroads.

Otter, D.E., LoPresti, J., Foutch, D.A., and Tobias, D.H., 1997, "Longitudinal Forces in an Open-Deck Steel Plate-Girder Bridge", *AREA Proceedings*, 98, 101–105.

Otter, D.E., Sweeney, R.A.P., and Dick, S.M., 2000, *Development of Design Guidelines for Longitudinal Forces in Bridges*, Technology Digest 00–018, Association of American Railroads.

Pilkey, W.D., 1997, *Stress Concentration Factors*, 2nd edition, John Wiley and Sons, New York.

Ruble, E.J., 1955, "Impact in Railroad Bridges", *Proceedings of the American Society of Civil Engineers*, 736, 1–36.

Sanders, W.W., and Munse, W.H., 1969, "Load Distribution in Steel Railway Bridges", *Journal of the Structural Division, ASCE*, 95(ST12), 2763–2781.

Simiu, E., and Scanlon, R.H., 1986, *Wind Effects on Structures*, John Wiley & Sons, New York.

Steinman, D.B., 1922, *Locomotive Loadings for Railway Bridges*, Transactions, Paper No. 1520, ASCE, New York.

Stephens, R.I., Fatemi, A., Stephens, R.R., and Fuchs, H.O., 2001, *Metal Fatigue in Engineering*, 2nd edition, John Wiley and Sons, New York.

Taly, N., 1998, *Design of Modern Highway Bridges*, McGraw-Hill, New York.

Tobias, D.H., Foutch, D.A., and Choros, J., 1996, "Loading Spectra for Railway Bridges under Current Operating Conditions", *Journal of Bridge Engineering, ASCE*, 1(4), 127–134, New York.

Tobias, D.H., Foutch, D.A., Lee, K., Otter, D.E., and LoPresti, J.A., 1999, *Experimental and Analytical Investigation of the Longitudinal Loads in a Multi-span Railway Bridge*, Report R-927, Association of American Railroads.

Tobias, D.H., Otter, D.E., and LoPresti, J.A., 1998, *Longitudinal Forces in Three Open-Deck Steel Bridges*, Proceedings of the 1998 AREMA Technical Conference, Lanham, MD.

Unsworth, J.F., 2018, *Design and Construction of Modern Steel Railway Bridges*, 2nd edition, CRC Press, Taylor and Francis, Boca Raton, FL.

Uppal, A.S., Otter, D.E., and Joy, R.B., 2001, *Longitudinal Forces in Bridges Due to Revenue Service*, Report R-950, Association of American Railroads.

Veletsos, A.S., and Huang, T., 1970, "Analysis of Dynamic Response of Highway Bridges", *Journal of Engineering Mechanics, ASCE*, 96(EM5).

Waddell, J.A.L., 1916, *Bridge Engineering*, Vols. 1 & 2, John Wiley and Sons, New York.

BIBLIOGRAPHY

Biggs, J.M., 1964, *Introduction to Structural Dynamics*, McGraw-Hill, New York.

Carnahan, B., Luther, H.H., and Wilkes, J.O., 1969, *Applied Numerical Methods*, John Wiley & Sons, New York.

Chatterjee, P.K., Datta, T.K., and Surana, C.S., 1994, "Vibration of Continuous Bridges under Moving Vehicles", *Journal of Sound and Vibration, Academic Press*, 169(5).

Chopra, A.K., 2004, *Dynamics of Structures*, 2nd edition, Prentice-Hall, NJ.

Clough, R.W., and Penzien, J., 1975, *Dynamics of Structures*, McGraw-Hill, New York.

Dick, S.M., 2006, *Estimation of Cycles for Railroad Girder Fatigue Life Assessment*, Bridge Structures, Taylor & Francis.

Fryba, L., 1972, *Vibration of Solids and Structures under Moving Loads*, Noordoff International, Groningen, Netherlands.

Fryba, L., 1996, *Dynamics of Railway Bridges*, Thomas Telford, London.

Huang, D., Wang, T.L., and Shahawy, M., 1992, "Impact Analysis of Continuous Multigirder Bridges Due to Moving Vehicles", *Journal of Structural Engineering, ASCE*, 118(12).

Inglis, C.E., 1928, *Impact in Railway-Bridges*, Minutes of Proceedings of the Institution of Civil Engineers, London.

Jin, Z., Pei, S., Li, X., and Qiang, S., 2016, "Vehicle-Induced Lateral Vibration of Railway Bridges: An Analytical-Solution Approach", *Journal of Bridge Engineering, ASCE*, 21(2).

Joy, R., Read, D., and Otter, D., 2007, *Continuous Welded Rail Restraint on an Open-Deck Girder Bridge*, Technology Digest 07–026, Association of American Railroads, Pueblo, CO.

Joy, R., Read, D., Otter, D., and Daniels, L., 2009, *Thermal Forces on Open Deck Steel Bridges*, Research Report R-996, Association of American Railroads, Pueblo, CO.

Kreyszig, E., 1972, *Advanced Engineering Mathematics*, John Wiley & Sons, New York.

Lin, J.H., 2006, "Response of a Bridge to a Moving Vehicle Load", *Canadian Journal of Civil Engineering*, 33(1).

Liu, H., 1991, *Wind Engineering—A Handbook for Structural Engineers*, Prentice-Hall, Englewood Cliffs, NJ.

McLean, W.G., and Nelson, E.W., 1962, *Engineering Mechanics*, McGraw-Hill, New York.

Miner, M.A., 1945, "Cumulative Damage in Fatigue", *Journal of Applied Mechanics*, 67, New York.

Priestley, M.J.N., Seible, F., and Calvi, G.M., 1996, *Seismic Design and Retrofit of Bridges*, John Wiley & Sons, New York.

Sorgenfrei, D.F., Marianos, W.N., and Sweeney, R.A.P., 2014, "Railroad Bridge Design Specifications", in *Bridge Engineering Handbook*, 2nd edition (Eds. Chen, W.F., and Duan, L.), CRC Press, Boca Raton, FL.

Sweeney, R.A.P., 2015, *Resistance and Loading for Steel Bridge Fatigue Evaluation*, Proceedings of the 2015 AREMA Conference, Lanham, MD.

Sweeney, R.A.P., Oommen, G., and Le, H., 1997, *Impact of Site Measurements on the Evaluation of Steel Railway Bridges*, International Association for Bridge and Structural Engineering Reports, Vol. 76, Zurich, Switzerland.

Uppal, A.S., Otter, D.E., and Doe, B.E., 2003, *Impact Loads in Railroad Short Steel Bridge Spans*, Report R-964, Association of American Railroads.

Yang, Y.B., Liao, S.S., and Lin, B.H., 1995, "Impact Formulas for Vehicles Moving Over Simple Span Continuous Beams", *Journal of Structural Engineering, ASCE*, 121(11).

Yang, Y.B., Yau, J.D., and Wu, Y.S., 2004, *Vehicle-Bridge Interaction Dynamics*, World Scientific, Singapore.

Yashinsky, M., and Karshenas, M.J., 2003, *Fundamentals of Seismic Protection for Bridges*, Earthquake Engineering Research Institute, Oakland, CA.

4 Statically Determinate Railway Truss Span Analysis

4.1 INTRODUCTION

Simple[1] through-truss spans are commonly used on railroads where spans longer than 50 m (175 ft) must span obstructions requiring adequate under clearance.[2] Simple deck-truss spans are generally of considerable depth but are often economical for high-level crossings requiring longer spans.[3] Whether through- or deck-truss spans are used to traverse the obstruction depends on site conditions[4] and the economics of fabrication and construction.

Single-span simply supported trusses are economically designed for railway live loads with span-to-depth ratios of 5 to 7. Span-to-depth ratios greater than 10 should be avoided. Furthermore, efficient truss design directs that truss diagonal members should be at a 45° orientation within each panel.[5] Depending on truss span length, economical panel lengths for railway live loads are typically not greater than 9 m (30 ft) to 15 m (50 ft). Economical preliminary truss designs are typically established with parametric studies of various simple planar truss types,[6] panel lengths[7] and truss depth. Considering the parametric relationships between span-to-depth ratio and truss panel length to total steel weight provides information on optimum span-to-depth ratios and panel lengths (Eisenbeis, 2010). Truss height must be based on minimum applicable vertical clearances[8] with diagonal web members at an economical orientation.

Short truss spans less than 35 m (115 ft) can be constructed using horizontal parallel-chord Warren trusses without vertical web members,[9] but moderate-length spans of between 45 m (150 ft) and 75 m (250 ft) used for railway bridge superstructures typically require simple horizontal parallel- or inclined-chord Warren trusses with vertical members,[10] Howe or Pratt trusses. Many railroad truss spans are of moderate length with horizontal parallel chords, inclined end members and single diagonal web members,[11] as shown in Figure 4.1a. Howe trusses are rarely constructed due to web member behavior[12] and relatively large magnitude forces in vertical web members.[13] Pratt and Warren truss forms are most prevalent for modern truss design and fabrication. Warren through-truss forms with vertical web members can generate higher secondary[14] flexural stresses in top and bottom chord members than experienced by Pratt truss forms.

Long simple-truss spans typically require inclined-chord or subdivided panel forms. Spans longer than 60 m (200 ft) are economically designed, fabricated and erected as inclined- or curved-chord truss spans. The Parker through-truss in Figure 4.1b is an inclined top chord Pratt through-truss commonly used for longer railway spans. For very long simple-truss spans greater than 90 m (300 ft), subdivided panel trusses such as the Baltimore,[15] Petit or Pennsylvania[16] trusses shown in Figure 4.1b are frequently economical. Subdivided panel truss forms can create high secondary stresses in chord members. Bridges incorporating very long spans are typically constructed using compound trusses,[17] which may be statically indeterminate, or with continuous trusses, which are statically indeterminate. Continuous-truss spans are relatively rare and usually used only where foundations are unyielding and temperature effects can be accommodated.

DOI: 10.1201/9781003469971-5

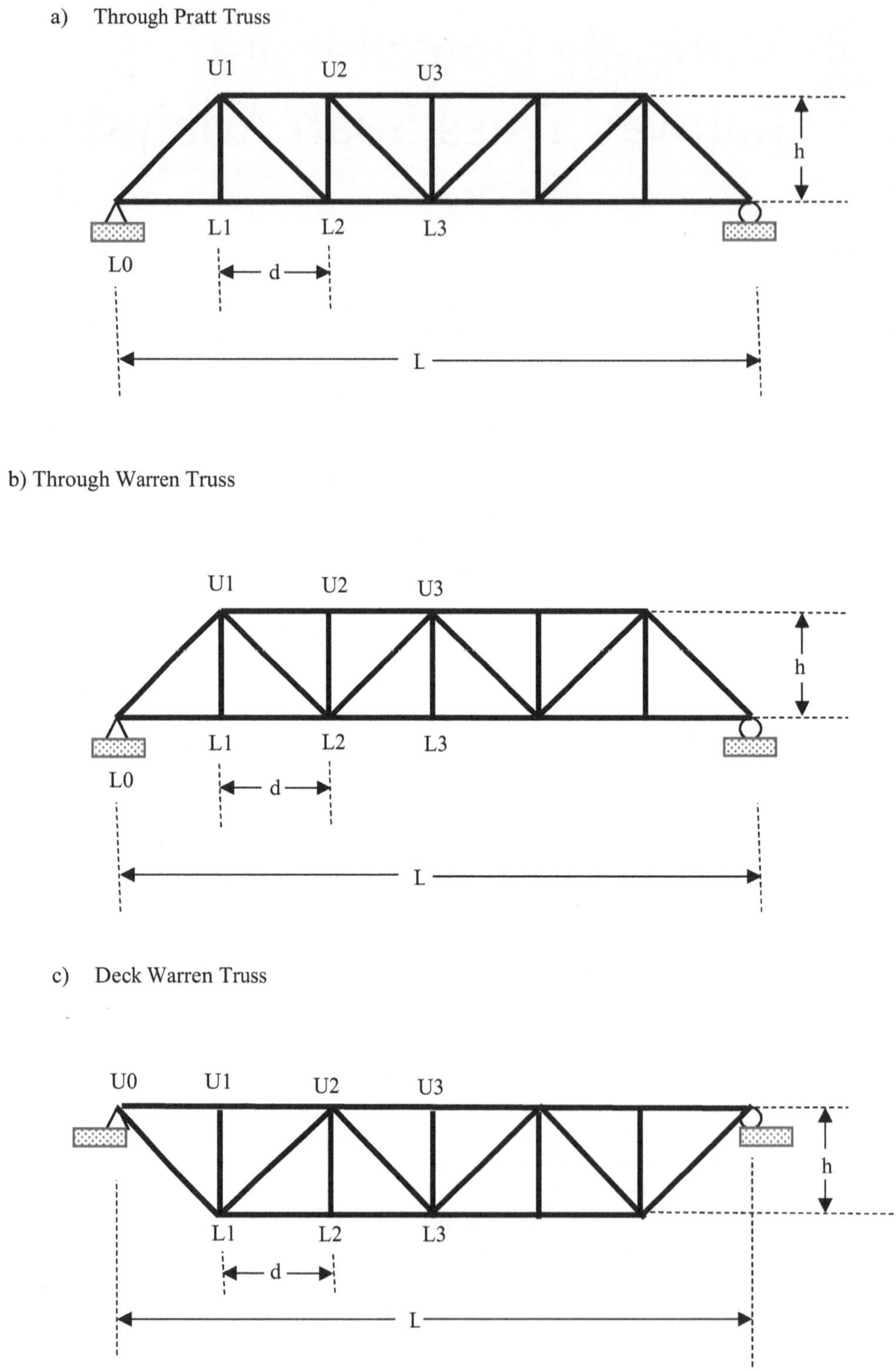

FIGURE 4.1a a) Pratt through-, b) Warren through-, c) Warren deck-, d) Howe through- and e) "K" through-truss spans.

d) Through Howe Truss

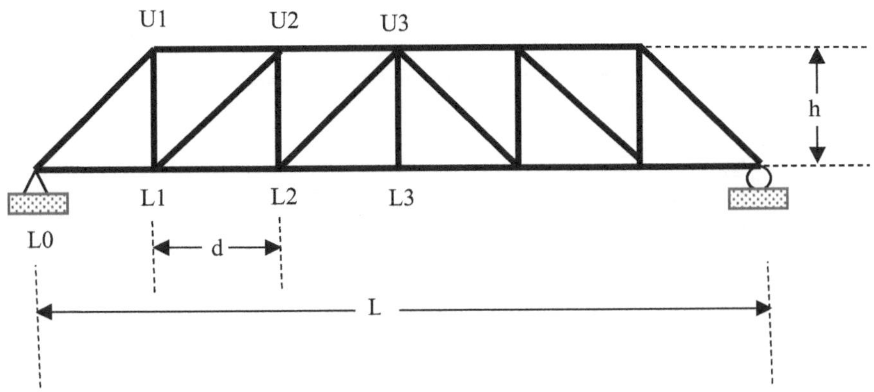

k) Through "K" Truss

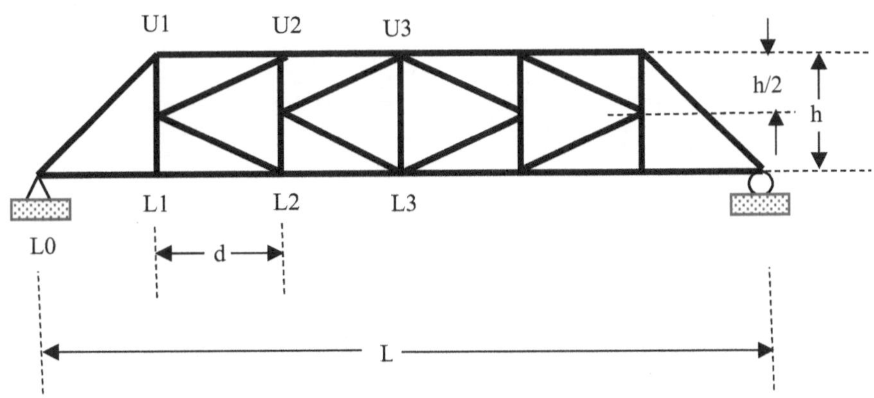

FIGURE 4.1a (Continued)

4.2 STRUCTURAL ANALYSIS FOR MEMBER FORCES

Following preliminary design to establish the appropriate truss type and configuration,[18] the detailed strength and serviceability design of planar trusses requires structural analyses for member forces and joint displacements due to primary loads,[19] secondary forces[20] and other external effects.[21] The complexity of the required structural analysis depends predominantly on whether the truss is statically determinate or statically indeterminate.

Railway truss spans can be structurally analyzed via classical or modern matrix methods. As outlined in Chapter 2, modern matrix methods of structural analysis are computationally laborious but well suited to analysis by digital computer. In particular, the direct stiffness method provides the basis for efficient truss analysis with finite element analysis (FEA) software. FEA modeling efforts are

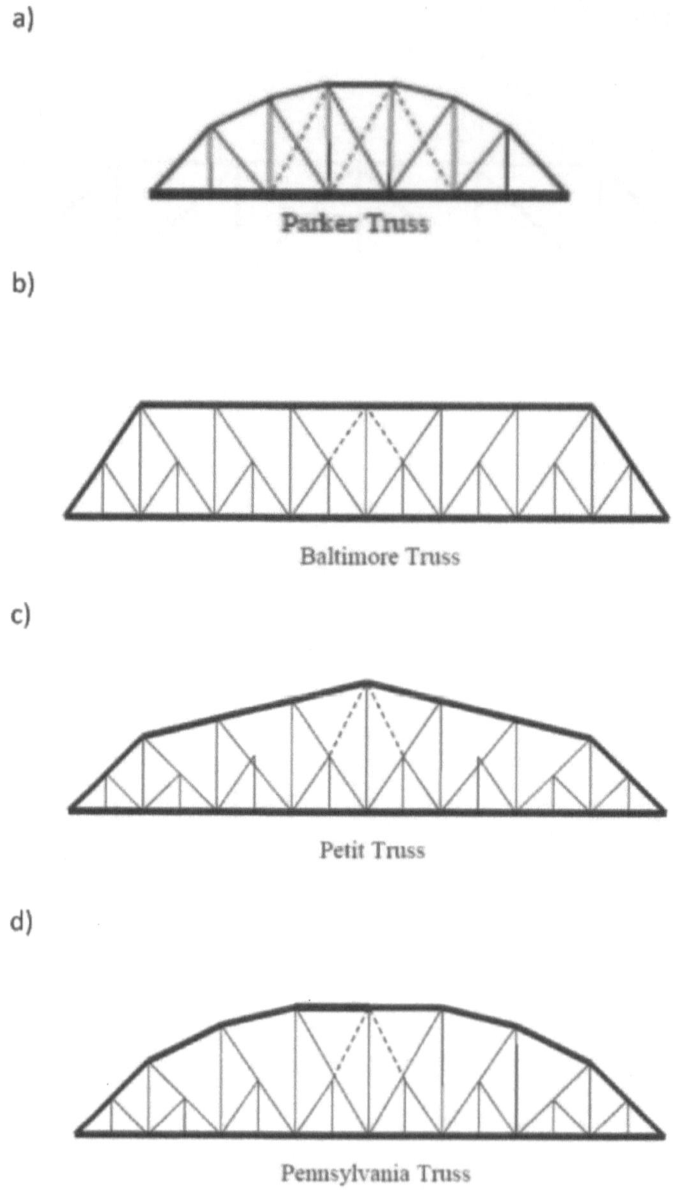

FIGURE 4.1b a) Parker (inclined top chord Pratt truss), b) Baltimore, c) Petit and d) Pennsylvania through-truss spans.

typically similar for simple statically determinate and indeterminate trusses. However, member properties are required for statically indeterminate truss analysis, thereby precipitating iterative analyses.[22]

Modeling errors in the systematic analysis of statically determinate trusses are readily detected by analytical checks of member forces using static equilibrium.[23] Correspondingly, the classical energy methods of structural analysis allow for promptly identifying modeling mistakes in the systematic analysis of member forces in statically indeterminate trusses and joint displacements in statically determinate and indeterminate trusses by analytical checks. Consequently, assessments using classical methods of structural analysis will improve the accuracy and timeliness of the truss analysis and design process. In addition, the classical methods of structural analysis improve the design engineer's comprehension of structural behavior and, when necessary, provide the analytical tools for the rapid preliminary evaluation of trusses.[24]

Calculating the member forces in statically indeterminate trusses requires additional equations available through consideration of displacement compatibility, as outlined in Chapter 5. The displacement of statically determinate and indeterminate truss joints due to external actions can be determined by energy methods[25] such as Castigliano's Second Theorem and the method of virtual work, as outlined in Chapter 6. However, the member forces in statically determinate trusses can be determined with merely the equations of static equilibrium.

A truss is statically determinate when the unknown member forces and unknown support reactions can be determined by considering only the equations of static equilibrium. If each joint of a truss is isolated, linear equations of translational equilibrium can be developed for $\Sigma F_x = 0$ and $\Sigma F_y = 0$, thereby providing 2NJ linear equations of equilibrium where NJ is the number of joints in the truss span. That is, a truss span is statically determinate when NM + NR = 2NJ where NM is the number of unknown members and NR is the number of unknown reactions. If NM + NR > 2NJ, there are too many unknowns, NR + NM, to determine from the simultaneous solution of 2NJ linear equations of equilibrium and the truss span is statically indeterminate. Therefore, calculating the member forces in statically indeterminate truss spans requires additional equations (displacement compatibility equations) to supplement the 2NJ linear equations of equilibrium.[26] Nevertheless, statically determinate and statically indeterminate trusses are not necessarily statically (geometrically) stable.[27]

Truss support reactions and member forces in statically determinate trusses can be calculated by solving the 2NJ simultaneous linear equations developed by investigating the horizontal and vertical translational equilibrium at each joint. However, to avoid solving many simultaneous linear equations, classical structural analysis of planar truss spans is effectively achieved by investigating the static equilibrium of free body diagrams[28] at successive truss joints (method of joints) and truss sections (method of sections). The method of joints and method of sections are uncomplicated for static load analyses but require calculating the locations of the live loads for maximum and minimum member forces, whether by mathematical criteria or by influence lines.[29] Influence lines also provide a useful visual indication of the positions of live loads for maximum and minimum member forces.

4.3 STATICALLY DETERMINATE PLANAR TRUSS MEMBER FORCES BY STATIC EQUILIBRIUM (METHOD OF JOINTS AND METHOD OF SECTIONS)

A truss in static equilibrium must also have its isolated parts (joints or sections) in static equilibrium. Therefore, static equilibrium of a statically determinate stable planar truss requires the following:

$$\Sigma F_x = 0 \tag{4.1a}$$

$$\Sigma F_y = 0 \tag{4.1b}$$

$$\Sigma M_z = 0 \tag{4.1c}$$

where F_x = the resultant force at a joint or section in the horizontal or the x-direction (translational static equilibrium), F_y = the resultant force at a joint or section in the vertical or the y-direction (translational static equilibrium) and M_z = the resultant moment of all forces at a section about an axis normal to the plane of the truss (rotational static equilibrium).

An isolated joint (method of joints) can be acted upon by only concurrent forces (member forces, reaction forces or applied loads), and therefore, only translational Equations 4.1a and 4.1b are required for static equilibrium. However, an isolated section (method of sections) of the truss with two or more joints will be acted upon by nonconcurrent forces, and Equations 4.1a, 4.1b and 4.1c must be satisfied for static equilibrium of the truss section. The simultaneous solution of Equations 4.1a and 4.1b (at joints) or 4.1a, 4.1b and 4.1c (at sections) requires that all forces (applied loads, reactions, construction conditions[30] and internal member forces) be included at the truss joint or section being considered. A free body diagram ensures that all forces are considered at the joint or section[31] being analyzed. If only a few member forces are required, the method of sections is typically an expedient analysis. The effective analysis of all member forces in trusses utilizes both the method of joints and the method of sections.

4.3.1 The Method of Joints

After the external reactions are calculated at the supports of simple trusses, the method of joints is used to determine internal member forces. The method of joints is particularly effective for the analysis of simple trusses if the joints are analyzed sequentially in the reverse order of the simple truss assembly. The translational forces on a typical joint in a simple truss are shown in the free body diagram in Figure 4.2a.

Static equilibrium of the joint in Figure 4.2a indicates the following:

$$\Sigma F_x = 0 = F_4 - F_1 - F_2 \cos\varnothing \tag{4.2a}$$

$$\Sigma F_y = 0 = F_2 \sin\varnothing + F_3 - W \tag{4.2b}$$

Successive application of the two equilibrium equations at all joints in the truss will yield the member forces. However, these calculations are cumbersome, and the method of sections is a more efficient approach to calculating member forces.

4.3.2 The Method of Sections

After the external reactions at the supports of simple trusses are calculated, the method of sections is used to calculate internal member forces. The method of sections is particularly effective for the analysis of simple trusses if the sections are carefully chosen such that no more than three members are intersected at the section.

The forces acting on a section through horizontal parallel-chord Pratt truss panel BC-ED are shown in the free body diagram in Figure 4.2b providing the following:

$$\Sigma F_x = 0 = F_{BE} + F_{CD} + F_{BD} \cos\varnothing \tag{4.2c}$$

$$\Sigma F_y = 0 = (R_0 - W_0 - W_1 - W_2) - F_{BD} \sin\varnothing = V_{BC-ED} - F_{BD} \sin\varnothing \tag{4.2d}$$

$$F_{BD} = \frac{V_{BC-ED}}{\sin\varnothing} \tag{4.2e}$$

$$\text{where } V_{BC-ED} = R_0 - W_0 - W_1 - W_2 \tag{4.2f}$$

Equation 4.2e reveals that the vertical component of the force, F_{BD}, in diagonal web member BD is equal to the positive[32] magnitude of the shear force, $V_{BC\text{-}ED}$, in panel BC-ED.

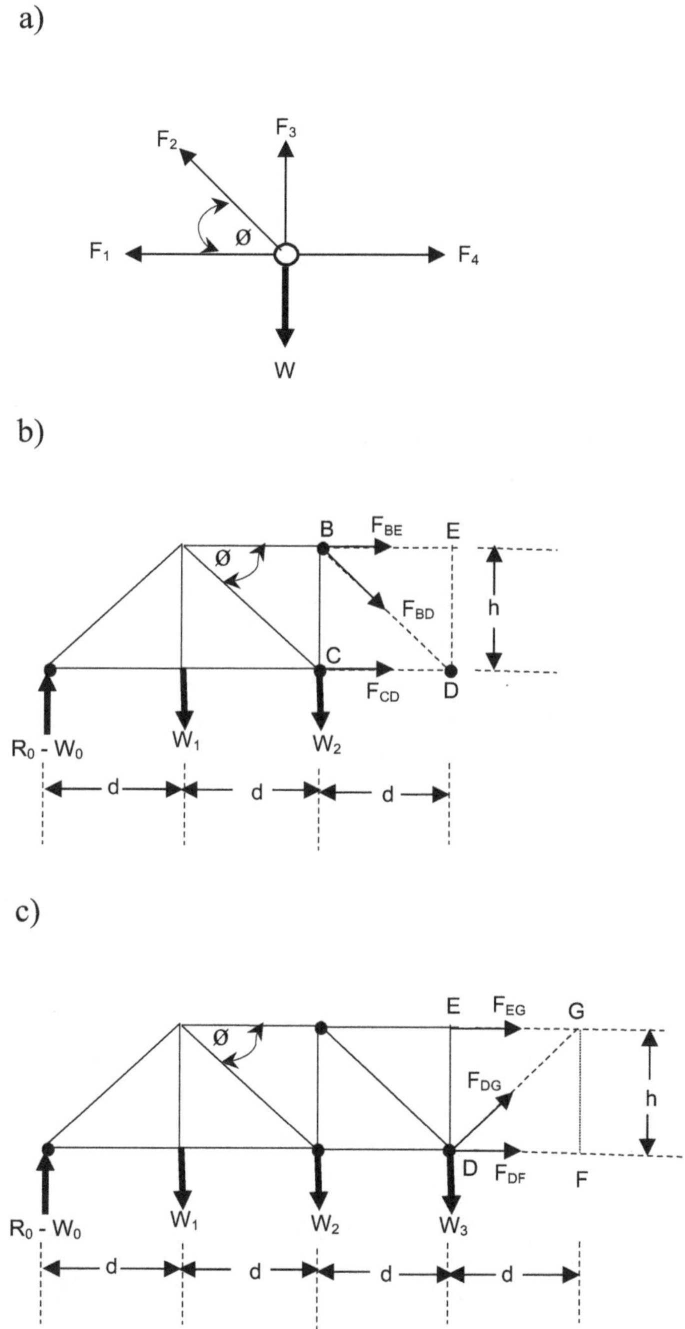

FIGURE 4.2 Free body diagrams of truss joint and sections.

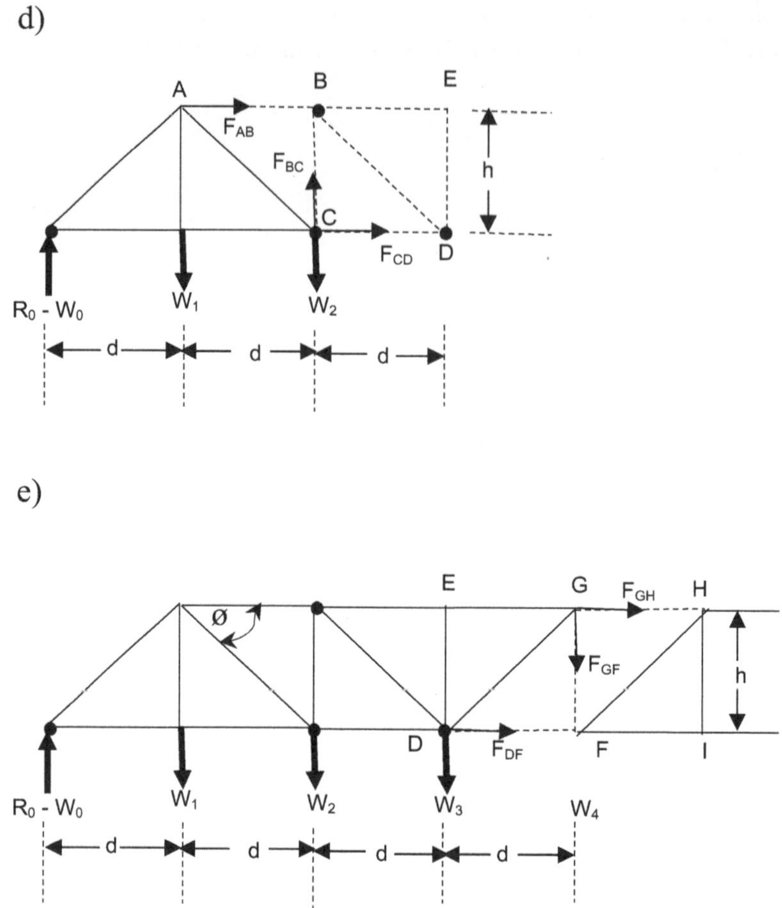

FIGURE 4.2 (Continued)

The forces acting on a section through horizontal parallel-chord Pratt truss panel ED-GF are shown in the free body diagram in Figure 4.2c providing the following:

$$\Sigma F_x = 0 = F_{EG} + F_{DF} + F_{DG}\cos\varnothing \tag{4.2g}$$

$$\Sigma F_y = 0 = (R_0 - W_0 - W_1 - W_2 - W_3) + F_{DG}\sin\varnothing = V_{ED-GF} + F_{DG}\sin\varnothing \tag{4.2h}$$

$$F_{DG} = \frac{-V_{ED-GF}}{\sin\varnothing} \tag{4.2i}$$

$$\text{where } V_{ED\text{-}GF} = R_0 - W_0 - W_1 - W_2 - W_3 = V_{BC\text{-}ED} - W_3 \tag{4.2j}$$

and since $R_0 + R_6 = W_0 + W_1 + W_2 + W_3 + W_4 + W_5 + W_6$,

$$V_{ED\text{-}GF} = -(R_6 - W_4 - W_5 - W_6) \tag{4.2k}$$

Equation 4.2i reveals that the vertical component of the force, F_{DG}, in diagonal web member DG is equal to the negative magnitude of the shear force, $V_{ED\text{-}GF}$, in panel ED-GF. Comparison of Equations 4.2f and 4.2k reveals that if joints to the right of joint D in Figure 4.2b are loaded, the shear in panel BC-ED is equal to $V_{ED\text{-}GF}$. This illustrates that depending on the direction of live loads, shear in a panel can be positive or negative.

The forces acting on a section of a horizontal parallel-chord Pratt through-truss panel BC-ED are shown in the free body diagram in Figure 4.2d providing the following:

$$\Sigma F_x = 0 = F_{AB} + F_{CD} \tag{4.2l}$$

$$\Sigma F_y = 0 = R_0 - W_0 - W_1 - W_2 + F_{BC} = V_{BC-ED} + F_{BC} \tag{4.2m}$$

$$F_{BC} = -V_{BC-ED} \tag{4.2n}$$

Equation 4.2l reveals that the force, F_{AB}, in top chord member AB is equal in magnitude but opposite in attitude[33] to the magnitude of the force, F_{CD}, in bottom chord member CD. Equation 4.2n reveals that the force, F_{BC}, in vertical web member BC is equal to the negative magnitude of the shear force, $V_{BC\text{-}ED}$, in panel BC-ED.[34]

The forces acting on a section of a horizontal parallel-chord Pratt through-truss panel GF-HI are shown in the free body diagram in Figure 4.2e providing the following:

$$\Sigma F_x = 0 = F_{GH} + F_{DF} \tag{4.2o}$$

$$\Sigma F_y = 0 = R_0 - W_0 - W_1 - W_2 - W_3 - F_{GF} = V_{ED-GF} - F_{GF} \tag{4.2p}$$

$$F_{GF} = V_{ED-GF} \tag{4.2q}$$

Equation 4.2o reveals that F_{GH} in top chord member GH is equal in magnitude but opposite in attitude to F_{DF} in bottom chord member DF. Equation 4.2q reveals that F_{GF} in vertical web member GF is equal to the positive magnitude of shear force $V_{ED\text{-}GF}$ in panel ED-GF. Comparison of Equation 4.2n and Equation 4.2q reveals that if joints to the right of joint C in Figure 4.2d are loaded, the force in vertical member BC is equal to the positive magnitude of $V_{BC\text{-}ED}$ in panel BC-ED. Once again, this illustrates that depending on the direction of live loads, shear in a panel can be positive or negative.

The forces in chord members of the horizontal parallel-chord Pratt truss can be calculated via moments at panel points[35] such that $\Sigma M_z = 0$, which taken about joints B and D in Figure 4.2b, yields the following:

$$\Sigma M_{z(B)} = 0 = \left(2(R_0 - W_0) - W_1\right)(d) - F_{CD}(h) \tag{4.2r}$$

$$\Sigma M_{z(D)} = 0 = \left(3(R_0 - W_0) - 2W_1 - W_2\right)(d) + F_{BE}(h) \tag{4.2s}$$

Equations 4.2r and 4.2s do not include diagonal member force F_{BD} but represent the moment of chord member force F_{CD} (at joint C) about panel point B and the moment of F_{BE} (at joint B) about panel point D, respectively.

Equations 4.2e and 4.2i illustrate that the maximum and minimum forces in the diagonal web members of horizontal parallel-chord Pratt truss spans may be determined from positive and negative panel shear forces. Equations 4.2n and 4.2q illustrate that the maximum and minimum forces in the vertical web members of horizontal parallel-chord Pratt truss spans can be determined from positive and negative panel shear forces. Equations 4.2l, 4.2o, 4.2r and 4.2s illustrate that maximum chord member forces in horizontal parallel-chord Pratt truss spans can be determined from panel point moments.

Overall, Equations 4.2 indicate that with regard to horizontal parallel-chord Pratt truss spans, maximum chord member forces are obtained via panel point moments and maximum and minimum diagonal and vertical web member forces are obtained via panel shear forces. Similar method of sections analysis of horizontal parallel-chord Warren truss spans reveals that maximum chord member forces are obtained via panel point moments, that maximum and minimum diagonal web member forces are obtained via panel shear forces and that maximum and minimum vertical web member forces are obtained directly as floorbeam reactions or are zero-force members. The method of sections in conjunction with judicious selection of equilibrium equations provides efficiency in calculating member forces at truss sections in panels with single diagonal members.

4.4 STATICALLY DETERMINATE HORIZONTAL PARALLEL CHORD TRUSS RAILWAY SPANS

Moderate-length truss spans[36] for railway bridge superstructures are economically designed, fabricated and erected with horizontal parallel-chord Pratt, Warren and Howe trusses with inclined end members. These truss spans with an odd number of panels require double-diagonal web members, each oriented in differing direction in central panels, rendering the span internally statically indeterminate. However, a horizontal parallel-chord Pratt, Warren or Howe truss span with an even number of panels is internally statically determinate with single-diagonal web members in each truss panel. For this reason, many modern moderate-length railway truss spans are designed, fabricated and erected with an even number of panels.[37]

Examples 4.1, 4.2 and 4.3 use the method of sections to determine member forces through the examination of static equilibrium of sections through panels of horizontal parallel-chord Pratt, Warren and Howe through-truss spans with inclined end post members.[38]

Example 4.1

Find the forces in members a, b, c, d, e and f of the horizontal parallel-chord Pratt through truss with inclined end post members sectioned through the third and second panels shown in Figure E4.1a and Figure E4.1b, respectively.

Taking a section through the third panel of the Pratt truss provides the free body diagram in Figure E4.1a, which indicates

$$\Sigma F_x = 0 = F_b\left(\frac{d}{\sqrt{d^2 + h^2}}\right) + F_a + F_c$$

$$\Sigma F_y = 0 = R_0 - W_0 - W_1 - W_2 - F_b\left(\frac{h}{\sqrt{d^2 + h^2}}\right)$$

$$\Sigma M_z \text{ about joint at top chord} = 0 = -F_c h - W_1 d + 2(R_0 - W_0)d$$

These three equations of equilibrium give

$$F_a = -(3(R_0 - W_0) - 2W_1 - W_2)\frac{d}{h}$$

$$F_b = (R_0 - W_0 - W_1 - W_2)\frac{\sqrt{h^2 + d^2}}{h}$$

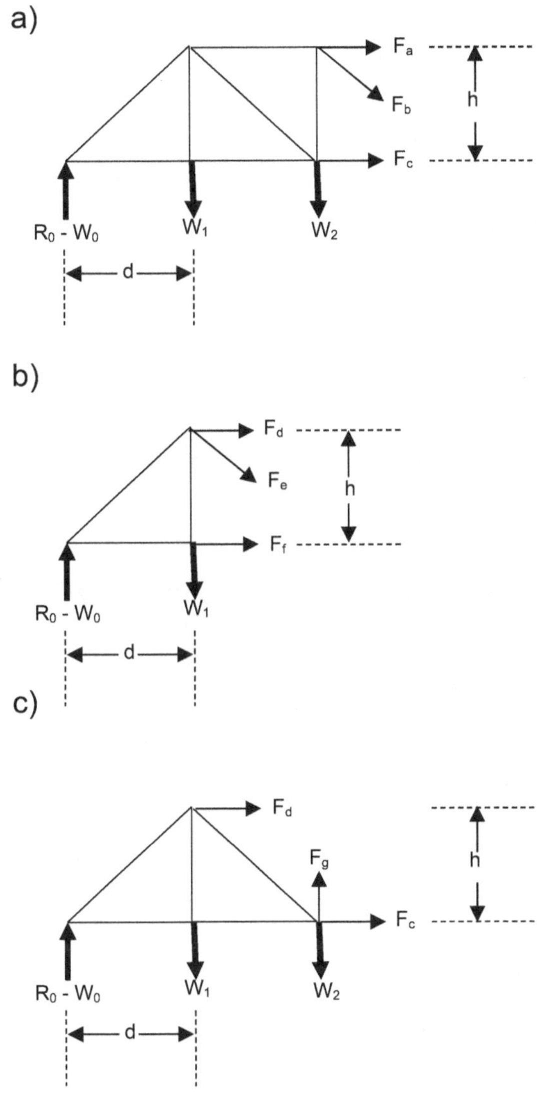

FIGURE E4.1 Free body diagrams of sections through a horizontal parallel-chord Pratt truss.

$$F_c = \left(2(R_0 - W_0) - W_1\right)\frac{d}{h}$$

Taking a section through the second panel of the Pratt truss provides the free body diagram in Figure E4.1b, which indicates

$$\Sigma F_x = 0 = F_e\left(\frac{d}{\sqrt{d^2 + h^2}}\right) + F_d + F_f$$

$$\Sigma F_y = 0 = R_0 - W_0 - W_1 - F_e\left(\frac{h}{\sqrt{d^2 + h^2}}\right)$$

ΣM_z about joint at top chord=0= $-F_f h + (R_0 - W_0)d$

and these three equations of equilibrium give

$$F_d = -(2(R_0 - W_0) - W_1)\frac{d}{h}$$

$$F_e = (R_0 - W_0 - W_1)\frac{\sqrt{h^2 + d^2}}{h}$$

$$F_f = (R_0 - W_0)\frac{d}{h}$$

Taking a section through the panels of the Pratt truss shown in Figure 4.1c provides the free body diagram that indicates

$$F_g = -(R_0 - W_0 - W_1 - W_2)$$

Example 4.2

Find the forces in members a, b, c, d, e and f of the horizontal parallel-chord Warren through-truss with inclined end post members in Figures E4.2a and E4.2b.

Taking a section through the third panel of the Warren truss provides the free body diagrams in Figure E4.2a, which indicate

$$\Sigma F_x = 0 = F_b\left(\frac{d}{\sqrt{d^2 + h^2}}\right) + F_a + F_c$$

$$\Sigma F_y = 0 = R_0 - W_0 - W_1 - W_2 + F_b\left(\frac{h}{\sqrt{d^2 + h^2}}\right)$$

$$\Sigma M_{z \text{ about joint at top chord}} = 0 = -F_c(h) - F_b\left(\frac{d}{\sqrt{d^2 + h^2}}\right)(h) - W_1(d) + (R_0 - W_0)(2d)$$

These three equations of equilibrium give

$$F_a = -(2(R_0 - W_0) - W_1)\frac{d}{h}$$

$$F_b = -(R_0 - W_0 - W_1 - W_2)\frac{\sqrt{h^2 + d^2}}{h}$$

$$F_c = (3(R_0 - W_0) - 2W_1 - W_2)\frac{d}{h}$$

The free body diagram in Figure E4.2b provides the same member forces F_d, F_e and F_f as the diagram in Figure E4.1b intersecting the second panel of a Pratt through-truss. Therefore,

$$F_d = -(2(R_0 - W_0) - W_1)\frac{d}{h}$$

$$F_e = (R_0 - W_0 - W_1)\frac{\sqrt{h^2 + d^2}}{h}$$

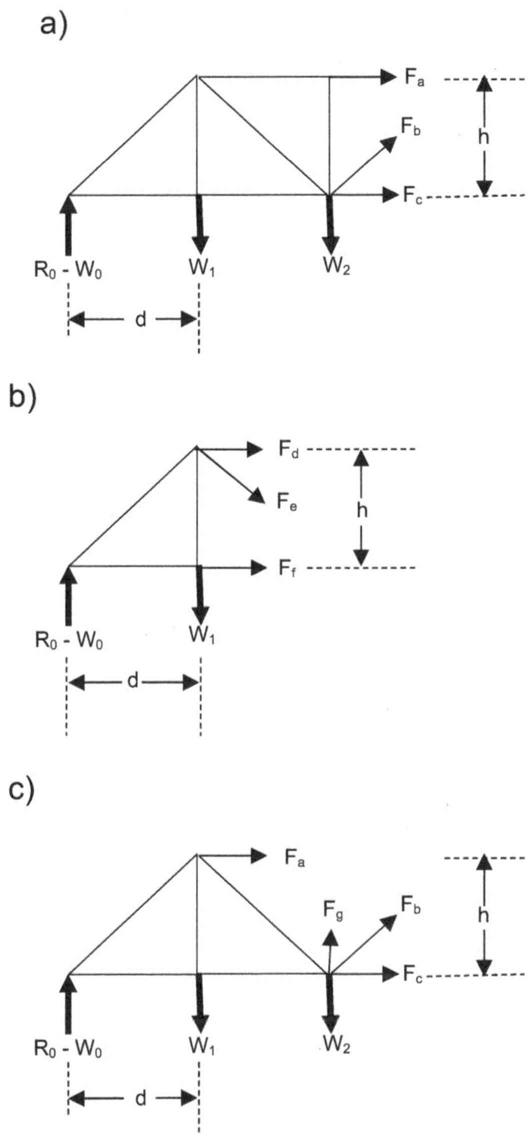

FIGURE E4.2 Free body diagrams of sections through a horizontal parallel-chord Warren truss.

$$F_f = (R_0 - W_0)\frac{d}{h}$$

Taking a section through the panels of the Warren truss shown in Figure 4.2c provides the free body diagram that indicates

$$F_g = -F_b\left(\frac{h}{\sqrt{h^2 + d^2}}\right) - (R_0 - W_0 - W_1 - W_2) = (R_0 - W_0 - W_1 - W_2)$$

$$\frac{\sqrt{h^2 + d^2}}{h}\left(\frac{h}{\sqrt{h^2 + d^2}}\right) - (R_0 - W_0 - W_1 - W_2) = 0$$

Example 4.3

Find the forces in members a, b, c, d, e and f of the horizontal parallel chord Howe through truss with inclined end post members in Figures E4.3a and E4.3b.

The free body diagram in Figure E4.3a provides the same member forces, F_a, F_b and F_c as the diagram in Figure E4.2a intersecting the third panel of a Warren through truss. Therefore,

$$F_a = -\left(2(R_0 - W_0) - W_1\right)\frac{d}{h}$$

$$F_b = -\left(R_0 - W_0 - W_1 - W_2\right)\frac{\sqrt{h^2 + d^2}}{h}$$

$$F_c = \left(3(R_0 - W_0) - 2W_1 - W_2\right)\frac{d}{h}$$

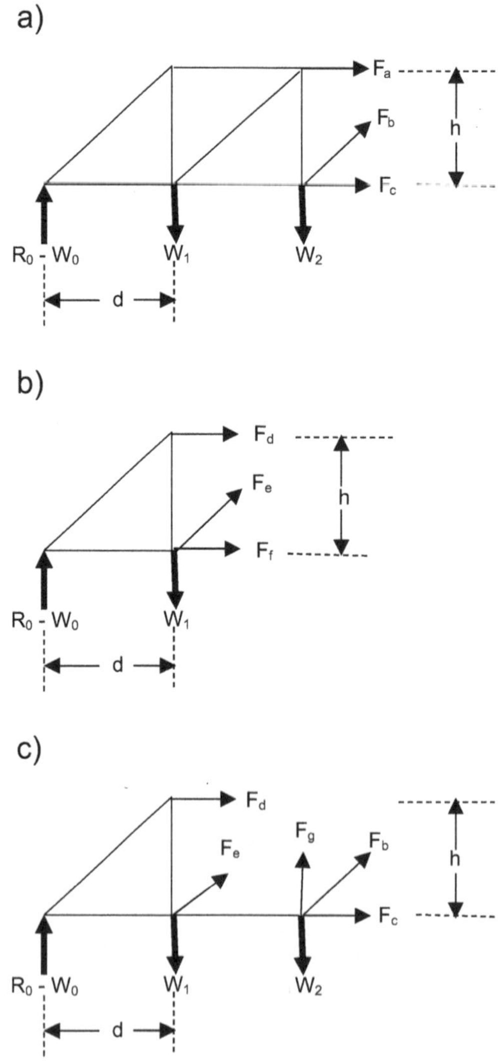

FIGURE E4.3 Free body diagrams of sections through a horizontal parallel-chord Howe truss.

Taking a section through the second panel of the Howe truss provides the free body diagram in Figure E4.3b, which shows that

$$\Sigma F_x = 0 = F_e\left(\frac{d}{\sqrt{d^2+h^2}}\right) + F_d + F_f$$

$$\Sigma F_y = 0 = R_0 - W_0 - W_1 + F_e\left(\frac{h}{\sqrt{d^2+h^2}}\right)$$

$$\Sigma M_{z\text{ (about joint at top chord)}} = 0 = -F_f(h) - F_e\left(\frac{d}{\sqrt{d^2+h^2}}\right)(h) + (R_0 - W_0)(d)$$

These three equations of equilibrium give

$$F_d = -(R_0 - W_0)\frac{d}{h}$$

$$F_e = -(R_0 - W_0 - W_1)\frac{\sqrt{h^2+d^2}}{h}$$

$$F_f = (2(R_0 - W_0) - W_1)\frac{d}{h}$$

Taking a section through the panels of the Howe truss shown in Figure E4.3c provides the free body diagram that reflects the following:

$$\begin{aligned} F_g &= -R_0 + W_0 + W_1 + W_2 - F_e\left(\frac{h}{\sqrt{h^2+d^2}}\right) - F_b\left(\frac{h}{\sqrt{h^2+d^2}}\right) \\ &= -R_0 + W_0 + W_1 + W_2 + R_0 - W_0 - W_1 + (R_0 - W_0 - W_1 - W_2) = R_0 - W_0 - W_1 \end{aligned}$$

Example 4.4 uses the equations of Examples 4.1, 4.2 and 4.3, developed from the method of sections in conjunction with the method of joints, to calculate the member forces due to uniformly distributed loads on the entire span of horizontal parallel-chord Pratt, Warren and Howe truss spans with inclined end post members.[39]

Example 4.4

Calculate the forces in the members of the 150 ft long (six 25 ft panels) and 35 ft high horizontal parallel-chord Pratt, Warren and Howe through-truss spans using the equations of Examples 4.1, 4.2 and 4.3. The truss spans are loaded with a uniformly distributed load of 5.5 kips/ft across the entire span:

$$W = 5.5(25) = 137.5 \text{ kips}$$

$$R_0 = R_6 = 6W/2 = 412.5 \text{ kips}$$

Use the equations of Example 4.1 to calculate the forces in chord and diagonal web members of the horizontal parallel-chord Pratt through-truss:

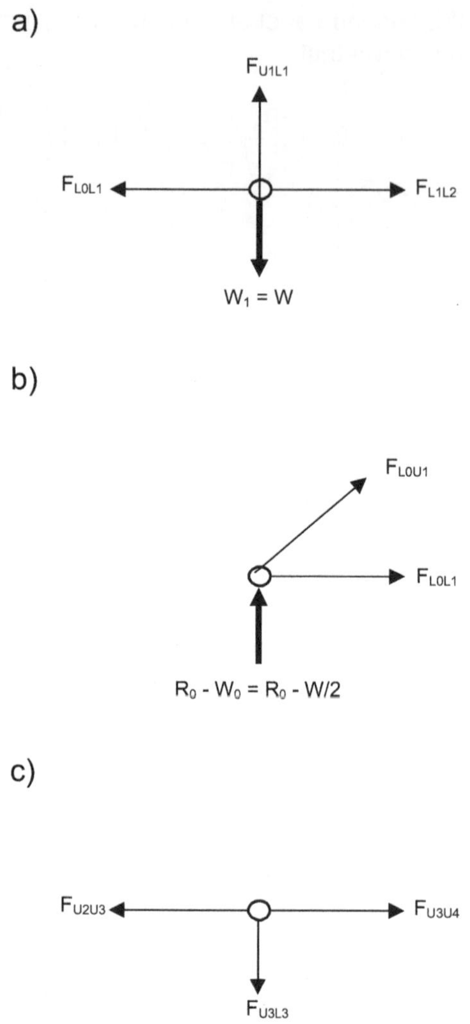

FIGURE E4.4 Free body diagrams and equilibrium at joints L1, L0 and U3.

$$F_{U2U3} = -3(R_0 - 1.5W)\frac{d}{h} = -3(412.5 - 206.25)(25/35) = -441.96 \text{ kips (compression)}$$

$$F_{U2L3} = (R_0 - 2.5W)\frac{\sqrt{h^2 + d^2}}{h} = (412.5 - 2.5(137.50))(1.229) = 84.49 \text{ kips(tension)}$$

$$F_{L2L3} = (2R_0 - 2W)\frac{d}{h} = (2(412.5) - 275.0)(0.714) = 392.86 \text{ kips(tension)}$$

$$F_{U1U2} = -(2R_0 - 2W)\frac{d}{h} = -(2(412.5) - 275.0)(0.714) = -392.86 \text{ kips(compression)}$$

$$F_{U1L2} = (R_0 - 1.5W)\frac{\sqrt{h^2 + d^2}}{h} = (412.5 - 206.25)(1.229) = 253.46 \text{ kips(tension)}$$

$$F_{L1L2} = (R_0 - W_0)\frac{d}{h} = 343.75(0.714) = 245.54 \text{ kips(tension)}$$

$$F_{U2L2} = -(R_0 - W_0 - W_1 - W_2) = -(343.75 - 137.50 - 137.50) = -68.75 \text{ kips (compression)}$$

Forces in members L0L1, L0U1, U1L1 and U3L3 of the Pratt through-truss are calculated by investigating equilibrium of forces at joints L0 (Figure E4.4b), L1 (Figure E4.4a) and U3 (Figure E4.4c).

$$\Sigma F_x = 0 : F_{L0L1} = F_{L1L2} = 245.54 \text{ kips (tension)}$$

$$\Sigma F_y = 0 : F_{U1L1} = W = 137.50 \text{ kips (tension)}$$

$$\Sigma F_x = 0 : F_{L0U1} = -F_{L0L1}\left(\frac{\sqrt{35^2 + 25^2}}{25}\right) = -245.54(1.720) = -422.33 \text{ kips (compression)}$$

Alternatively, from

$$\Sigma F_y = 0 : F_{L0U1} = -(R_0 - W_0)\left(\frac{\sqrt{35^2 + 25^2}}{35}\right) = -343.75(1.229) = -422.33 \text{ kips (compression)}$$

$$\Sigma F_x = 0 : F_{L2L3} = F_{L1L2} + F_{U1L2}(0.581) = 245.54 + 253.46(0.581) = 392.86 \text{ kips (tension)}$$

$$\Sigma F_x = 0 : F_{U3U4} = F_{U2U3} = -441.96 \text{ kips(compression)}$$

$$\Sigma F_y = 0 : F_{U3L3} = 0$$

The equations of Examples 4.2 and 4.3, based on the method of sections and the method of joints, can be used to calculate the forces in members of horizontal parallel-chord Warren and Howe through-trusses with inclined end post members, respectively. Table E4.4 is a summary of the member forces for the horizontal parallel-chord Pratt, Warren and Howe through-truss spans in Figures 4.1aa, 4.1ab and 4.1ad with a 5.5 kip/ft uniformly distributed load across the entire span.

TABLE E4.4
Member Forces for Horizontal Parallel-Chord Pratt, Warren and Howe Through-Truss Spans

Uniform Load of 5.5 kips/ft on Entire Span (W = 137.5 kips)				
Parallel-Chord Through-Truss Member		**Pratt Truss**	**Warren Truss**	**Howe Truss**
L0L1 & L5L6	Bottom chord	245.54	245.54	245.54
L1L2 & L4L5	Bottom chord	245.54	245.54	392.86
L2L3 & L3L4	Bottom chord	392.86	441.96	441.96
L0U1 & U5L6	End post	−422.33	−422.33	−422.33
U1U2 & U4U5	Top chord	−392.86	−392.86	−245.54
U2U3 & U3U4	Top chord	−441.96	−392.86	−392.86
U1L1 & U5L5	Vertical	137.50	137.50	245.54
U2L2 & U4L4	Vertical	−68.75	0	206.25

(Continued)

TABLE E4.4 (*Continued*)
Member Forces for Horizontal Parallel-Chord Pratt, Warren and Howe Through-Truss Spans

Uniform Load of 5.5 kips/ft on Entire Span (W = 137.5 kips)				
Parallel-Chord Through-Truss Member		**Pratt Truss**	**Warren Truss**	**Howe Truss**
U3L3	Vertical	0	137.50	137.50
Panels 1-2 & 4-5	Diagonal	253.46	253.46	−253.46
Panels 2-3 & 3-4	Diagonal	84.49	−84.49	−84.49

Examples 4.1, 4.2 and 4.3 reveal that chord member forces F_a, F_c, F_d and F_f, in horizontal parallel-chord trusses with single-diagonal web members can be calculated from opposite panel point moments and that web member forces F_b, F_e and F_g can be evaluated by considering shear forces in truss panels. These examples also indicate that the magnitude of the chord member forces in horizontal parallel-chord Pratt, Warren and Howe truss panels depends on the orientation of the diagonal member in the truss panel. Furthermore, these examples reveal equal-magnitude diagonal web member forces in corresponding panels of horizontal parallel-chord Pratt, Warren and Howe truss panels with attitude (axial compression or axial tension) dependent on orientation of the diagonal web member. Vertical web member forces, except vertical web members directly loaded by floorbeam reactions, depend on the truss form.

Examples 4.1, 4.2, 4.3 and 4.4 use the method of sections in each panel of horizontal parallel-chord Pratt, Warren and Howe trusses with inclined end post members to determine chord and web member forces. However, for long truss spans with many panels, a more general methodology is useful for calculating chord member, single-diagonal web member and vertical web member forces. Howe through-trusses were well suited for early railway span use.[40] Nevertheless as illustrated in Example 4.4, the large vertical web member forces[41] in Howe truss spans promoted the use of either horizontal parallel-chord Pratt or Warren trusses for moderate-length spans.

Equations 4.2 identify that the forces in chord members can be determined from panel point moments at the opposite panel point.[42] Equations 4.2 also indicate that there are equal but opposite-attitude shear forces on each side of the section through-truss panels. Accordingly, all chord and web member forces in panels $(j-1)-(j)$,[43] where $j \le n/2$, of statically determinate Pratt and Warren trusses with inclined end post members can be determined by the method of sections as shown in Figure 4.3. Determination of chord and web member forces in half the truss is sufficient for trusses with symmetrical geometry and loading[44] about the centerline of the truss span.

The end reactions in Figure 4.3 are calculated as

$$R_0 = \sum_{i=0}^{n}\left(\frac{n-i}{n}\right)W_i \tag{4.3a}$$

$$R_n = \sum_{i=0}^{n}W_i - R_0 = \sum_{i=0}^{n}W_i - \left(\sum_{i=0}^{n}\left(\frac{n-i}{n}\right)W_i\right) = \sum_{i=0}^{n}\left(\frac{i}{n}\right)W_i \tag{4.3b}$$

where n = the number of panels in the truss span and W_i = the load at panel point (i). Chord member forces in Figures 4.3a and 4.3c are calculated via panel point moments.

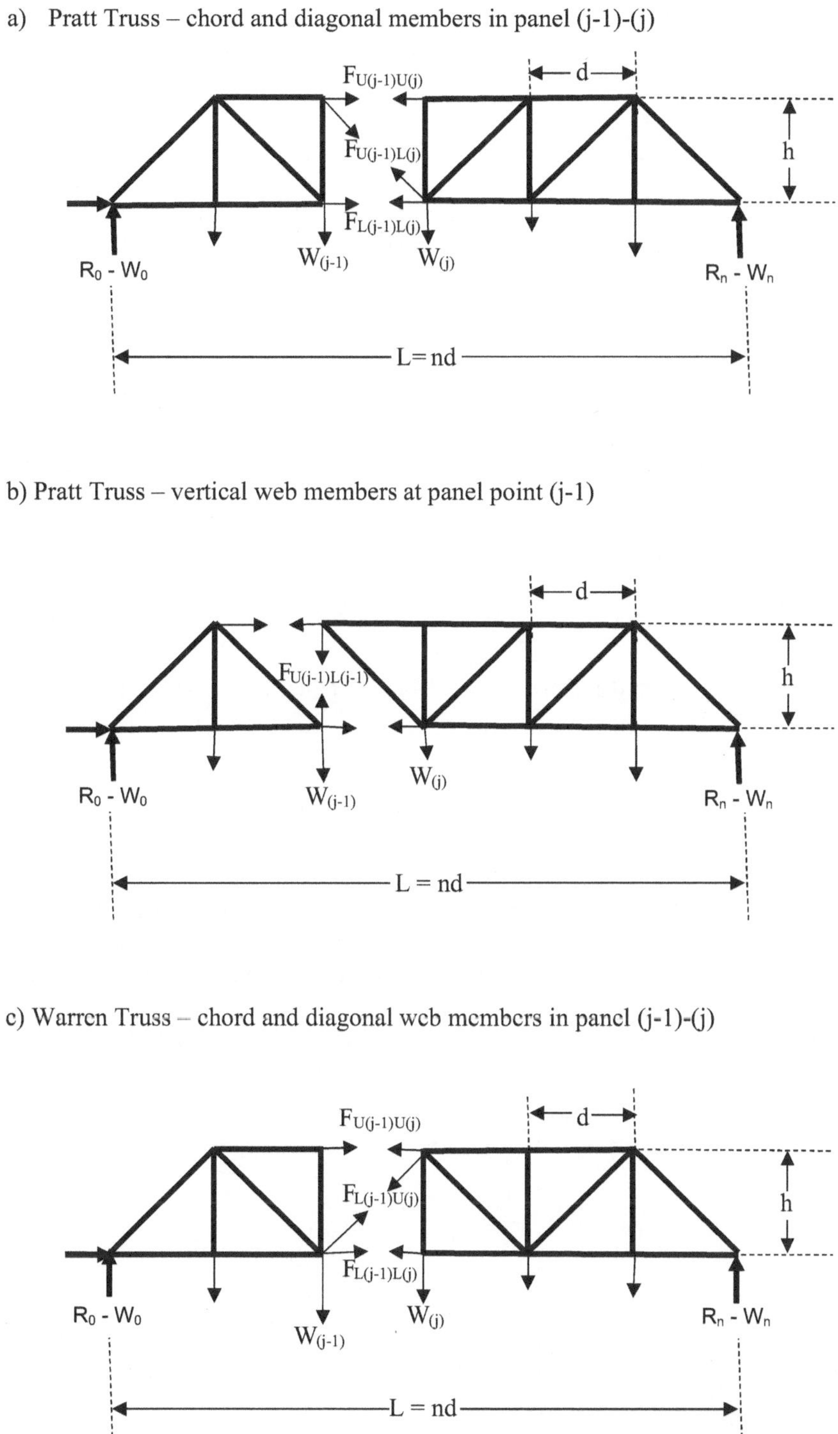

FIGURE 4.3 Chord and diagonal web member forces in panel (j – 1) – (j) of horizontal parallel chord Pratt and Warren trusses, and vertical web member forces at panel point (j – 1) or (j), where j < n/2, of horizontal parallel chord Pratt trusses.

The force in the top chord member U(j – 1)U(j) of panel (j – 1) – (j), in which the diagonal member is U(j – 1)L(j),[45] is

$$F^{P}_{U(j-1)U(j)} = -\frac{\Sigma M_{L(j)}}{h} = -\left(jR_0 - \sum_{i=0}^{j-1}(j-i)W_i\right)\frac{d}{h} \tag{4.3c}$$

The force in the bottom chord member L(j – 1)L(j) of panel (j – 1) – (j), in which the diagonal member is U(j – 1)L(j), is

$$F^{P}_{L(j-1)L(j)} = \frac{\Sigma M_{U(j-1)}}{h} = \left((j-1)R_0 - \sum_{i=0}^{j-1}(j-i-1)W_i\right)\frac{d}{h} \tag{4.3d}$$

The force in the top chord member U(j – 1)U(j) of panel (j – 1) – (j), in which the diagonal member is L(j – 1)U(j),[46] is

$$F^{W}_{U(j-1)U(j)} = -\frac{\Sigma M_{L(j-1)}}{h} = -\left((j-1)R_0 - \sum_{i=0}^{j-1}(j-i-1)W_i\right)\frac{d}{h} = -F^{P}_{L(j-1)L(j)} \tag{4.3e}$$

The force in the bottom chord member L(j – 1)L(j) of panel (j – 1) – (j), in which the diagonal member is L(j – 1)U (j), is

$$F^{W}_{L(j-1)L(j)} = \frac{\Sigma M_{U(j)}}{h} = \left(jR_0 - \sum_{i=0}^{j-1}(j-i)W_i\right)\frac{d}{h} = -F^{P}_{U(j-1)U(j)} \tag{4.3f}$$

A section in panel 0-1 of trusses with inclined end post members contains only end post member L0U1 and bottom chord member L0L1. The force in the bottom chord member L0L1, in which the diagonal member in panel 1-2 is U1L2, is determined using the method of joints by investigation of equilibrium at joint L0 or L1.

Diagonal web member forces in Figure 4.3a and 4.3c are determined by panel shear forces:

$$\begin{aligned} V^{+}_{(j-1)-(j)} &= R_0 - \sum_{i=0}^{j-1} W_i = \sum_{i=0}^{n}\left(\frac{n-i}{n}\right)W_i - \sum_{i=0}^{j-1} W_i = \sum_{i=j}^{n} W_i - \sum_{i=0}^{n}\left(\frac{i}{n}\right)W_i \\ &= -R_n + \sum_{i=j}^{n} W_i = -V^{-}_{(j-1)-(j)} \end{aligned} \tag{4.3g}$$

$$V^{+}_{(j-1)-(j)} = R_0 - \sum_{i=0}^{j-1} W_i \quad \text{= the positive shear}^{47} \text{ in panel } (j-1)-(j), \text{ where } j \le n/2 \tag{4.3h}$$

$$V^{-}_{(j-1)-(j)} = R_n - \sum_{i=j}^{n} W_i \quad \text{= the negative}^{48} \text{ shear in panel } (j-1)-(j), \text{ where } j \le n/2 \tag{4.3i}$$

The force in diagonal web members U(j – 1)L(j) due to positive panel shear in panel (j – 1) – (j), as shown in Figure 4.3a is

$$F^{+P}_{U(j-1)L(j)} = V^{+}_{(j-1)-(j)}\left(\frac{\sqrt{h^2+d^2}}{h}\right) = \left(R_0 - \sum_{i=0}^{j-1} W_i\right)\left(\frac{\sqrt{h^2+d^2}}{h}\right) \quad (4.3j)$$

The force in diagonal web members U(j-1)L(j) due to negative panel shear in panel (j – 1) – (j), as shown in Figure 4.3a, is

$$F^{-P}_{U(j-1)L(j)} = -V^{-}_{(j-1)-(j)}\left(\frac{\sqrt{h^2+d^2}}{h}\right) = -\left(R_n - \sum_{i=j}^{n} W_i\right)\left(\frac{\sqrt{h^2+d^2}}{h}\right) \quad (4.3k)$$

The force in diagonal web member L(j – 1)U(j) due to positive shear in panel (j – 1) – (j), as shown in Figure 4.3c, is

$$F^{+W}_{L(j-1)U(j)} = -V^{+}_{(j-1)-(j)}\left(\frac{\sqrt{h^2+d^2}}{h}\right) = -\left(R_0 - \sum_{i=0}^{j-1} W_i\right)\left(\frac{\sqrt{h^2+d^2}}{h}\right) = -F^{+P}_{U(j-1)L(j)} \quad (4.3l)$$

The force in diagonal web member L(j – 1)U(j) due to negative shear in panel (j – 1) – (j), as shown in Figure 4.3c, is

$$F^{-W}_{L(j-1)U(j)} = V^{-}_{(j-1)-(j)}\left(\frac{\sqrt{h^2+d^2}}{h}\right) = \left(R_n - \sum_{i=j}^{n} W_i\right)\left(\frac{\sqrt{h^2+d^2}}{h}\right) = -F^{-P}_{U(j-1)L(j)} \quad (4.3m)$$

The forces in vertical web members of horizontal parallel chord Pratt trusses that are not directly loaded by floorbeams or are not zero-force members can be calculated from the panel shear forces at the section through the vertical member and its adjacent chord members. This section through the vertical member and its adjacent chord members does not intersect any diagonal web members, so no vertical component of force from a diagonal web member exists at the section, as shown in Figure 4.3b.

The force in the vertical web member U(k)L(k), where 1 < k < n/2, of a Pratt truss due to positive shear in panel (k) – (k + 1), in which the diagonal member is U(k)L(k + 1), as shown in Figure 4.3b where k = j – 1, is[49]

$$F^{+P}_{U(k)L(k)} = F^{+P}_{U(j-1)L(j-1)} = -V^{+}_{(j-1)-(j)} = -\left(R_0 - \sum_{i=0}^{j-1} W_i\right) \quad (4.3n)$$

The force in the vertical web member U(k)L(k), where 1 < k < n/2, of a Pratt truss, due to negative shear in panel (k) – (k + 1), in which the diagonal member is U(k)L(k + 1), as shown in Figure 4.3b where k = j – 1, is

$$F^{-P}_{U(k)L(k)} = F^{-P}_{U(j-1)L(j-1)} = V^{-}_{(j-1)-(j)} = \left(R_n - \sum_{i=j}^{n} W_i\right) \quad (4.3o)$$

If k = j,[50] the vertical member in the Pratt truss is U(j)L(j):

$$F^{+P}_{U(k)L(k)} = F^{+P}_{U(j)L(j)} = -V^{+}_{(j)-(j+1)} = -\left(R_0 - \sum_{i=0}^{j} W_i\right) \quad (4.3p)$$

and

$$F^{-P}_{U(k)L(k)} = F^{-P}_{U(j)L(j)} = V^{-}_{(j)-(j+1)} = \left(R_n - \sum_{i=j+1}^{n} W_i \right) \tag{4.3q}$$

The force in vertical member U1L1 of a Pratt or Warren truss can be calculated with the method of joints by reviewing equilibrium at joint L1. Therefore, the force in vertical member U1L1 is equal to the load (floorbeam reaction) at panel point L1. The force in vertical member U(n/2)L(n/2) of a Pratt truss can be calculated with the method of joints by reviewing equilibrium at joint U(n/2), which for horizontal parallel chord trusses is zero.

The alternating orientation of the diagonal members in adjacent panels of horizontal parallel chord Warren trusses cannot be omitted from a section through a panel (Figures 4.1b and 4.1c), and panel shear cannot be used to determine the forces in vertical web members U(k)L(k) of Warren truss spans without determination of the forces in diagonal members. However, vertical members U(j – 1)L(j – 1) in a Warren truss panel (j – 1) – (j) with horizontal parallel-chord members and diagonal members L(j – 1)U(j) are zero-force members in accordance with equilibrium at panel point (joint) U(j – 1). Vertical members U(j)L(j) in a Warren truss panel (j – 1) – (j) with horizontal chord members and diagonal members L(j – 1)U(j) are directly loaded by the floorbeam reaction at L(j).

A section in panel 0-1 of Pratt and Warren trusses with inclined end post members contains only end post member L0U1 and bottom chord member L0L1. The force in end post member L0U1 is determined using the method of joints by investigation of equilibrium at joint L0.

Railway truss spans are subjected to static (dead) loads and moving (live) loads. The live loads are typically modeled as uniformly distributed and concentrated loads. Therefore, the analysis of horizontal parallel chord truss railway spans for maximum and minimum member forces requires that member dead and live loads be considered.

4.4.1 Dead Load Analysis

Dead load[51] is typically adequately modeled as a static uniformly distributed load[52] across the truss span allocated equally at the panel points.[53] The dead load may be apportioned between the top and bottom panel points of the truss based on estimates of deck, floor system, bracing and truss weight considering the location of their respective centers of gravity. Accordingly, dead load for through-truss spans may be apportioned such that 1/3 is assigned to the top chord panel points and 2/3 to the bottom chord panel points. Likewise, dead load for deck truss spans may be apportioned such that 2/3 is assigned to the top chord panel points and 1/3 to the bottom chord panel points. Nevertheless, the dead load forces in horizontal chord members and diagonal web members will remain unchanged by the proportioning of dead load between top and bottom chord panel points. However, even if vertical web members in compression will endure an increase in compression equal to the amount of dead load assigned to the top chord panel point and vertical web members in tension will experience a decrease in tension equal to the dead load distributed to the top chord panel point, the dead load analysis of typical moderate length railway spans is often appropriately conducted with the dead load apportioned entirely to the loaded chord[54] of the truss span.

4.4.2 Live Load Analysis

Live loads may be modeled as distributed and concentrated moving loads. The Cooper's E series design live load recommended by AREMA (2024) consists of both concentrated[55] and uniformly distributed[56] loads. The moving distributed and concentrated loads are typically transferred from the floor system to the loaded chord at the panel points in truss spans.[57] Therefore, as the distributed and concentrated live loads traverse the floor system of a simply supported truss span, chord and web member forces will vary. The end post and chord members in simply supported truss spans experience live load forces of the same attitude (axial compression or tension) as the dead load forces. Therefore, the minimum force in simply supported truss end post and chord members is due to dead load, and the maximum force in end post and truss chord members is due to combined dead and live load.

However, the web members in simply supported truss spans experience live load forces not necessarily of the same attitude as the dead load forces. Consequently, calculating the maximum and minimum forces in web members must include consideration of dead load as well as the maximum positive and negative shear forces within truss panels created by moving (live) loads. Furthermore, since the truss is in equilibrium, positive and negative shear forces at a section are equal in magnitude but of opposite direction, as shown by Equation 4.3g. Therefore, web members in truss panels may be subject to a range of live load forces that might precipitate force reversals in some web members.[58] The maximum magnitude of live load force reversal occurs when the minimum force magnitude is greatest and of opposite attitude to the maximum force in the web member.

Modern structural engineering software used for bridge analysis can perform moving load calculations that provide joint displacements and forces in the members as distributed and concentrated live loads traverse the truss span. Consequently, modern bridge analysis software has made investigating the effects of distributed and concentrated moving loads a routine component of the truss analysis and design process. Still, it is often necessary that only individual members of a truss span be investigated[59] or that relatively simple truss spans be designed. In such cases, and in general, a basic comprehension of classical moving load analysis techniques is of benefit to the design engineer.[60] The positions of concentrated and uniformly distributed live loads that create the maximum and minimum forces in truss members can be ascertained through mathematical criteria and, subsequently, the member forces determined by static equilibrium or by using influence lines.

4.4.2.1 Uniformly Distributed Live Load Analysis

Uniformly distributed live loads used in railway truss span analyses are typically the trailing portion of the Cooper's E design load or an equivalent uniformly distributed load corresponding to the bending moment and shear force effects of the Cooper's E design load for the member under consideration.[61]

The position of uniformly distributed live loads that create maximum and minimum member forces for design may be ascertained through mathematical criteria. A uniformly distributed load across the entire truss span may be used to determine dead load forces in chord and web members. However, as shown in Figures 4.4a and 4.4b, the uniformly distributed live load may be only partially distributed across the span length as it traverses the bridge and, thereby, only partially distributed within panel (j − 1) − (j). Therefore, the position of the uniformly distributed live load for maximum chord member forces[62], and maximum and minimum web member forces[63] is required.

The moment $M_{(k)}$ at truss panel point (k = j) in Figure 4.4a is

$$M_{(k)} = (kd)R_0 - d\sum_{i=0}^{k-1}(k-i)W_i = k\left(\frac{w\left((n-k)d + x^{+}_{(k-1)-(k)}\right)^2}{2n}\right) - \frac{w\left(x^{+}_{(k-1)-(k)}\right)^2}{2} \qquad (4.4a)$$

Since any load placed on a beam will increase the positive bending moment, the maximum moment at panel point (k) will occur when $x^{+}_{(k-1)-(k)} = kd$.

Substituting $x^{+}_{(k-1)-(k)} = kd$ into Equation 4.4a yields

$$M^{max}_{(k)} = \frac{wa_{(k)}}{2}\left(L - a_{(k)}\right) = \frac{wk(n-k)d^2}{2} \qquad (4.4b)$$

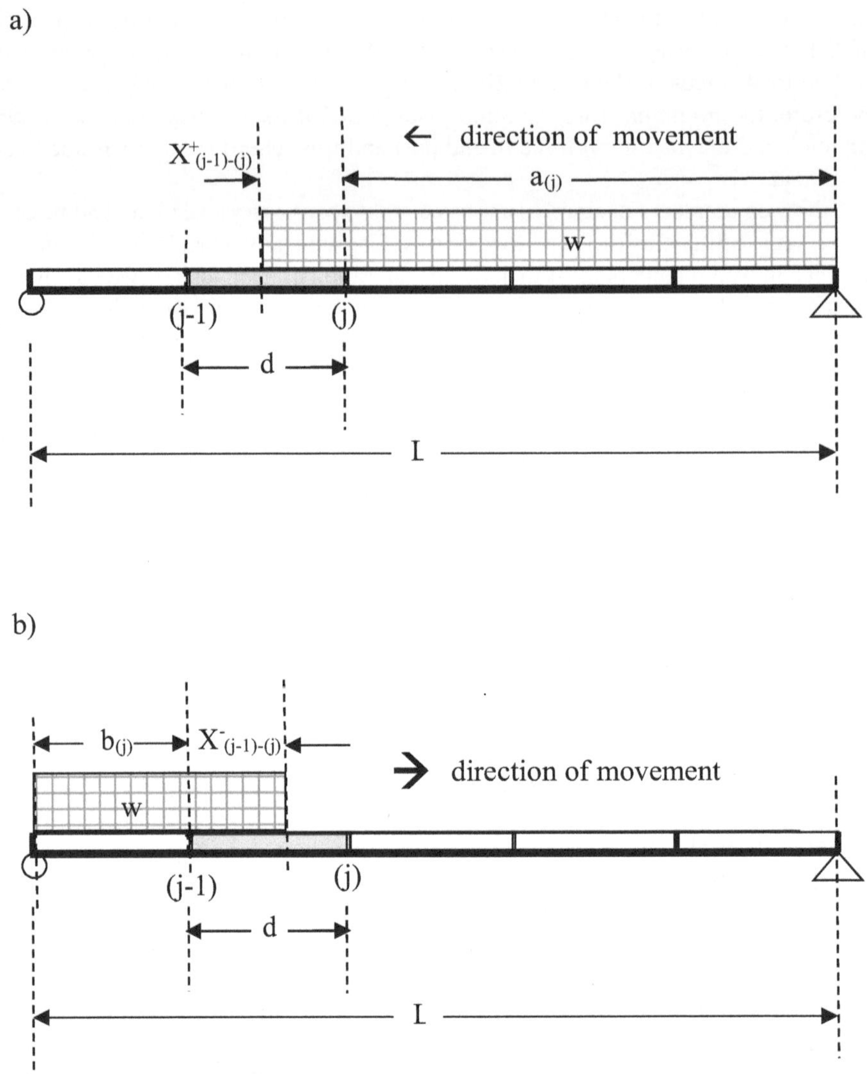

FIGURE 4.4 Uniformly distributed live load on a truss span causing positive and negative shear forces in panel (j – 1) – (j).

The force in chord members is equal to the sum of the moments of the external loads at the truss section about the opposite chord panel point.

For truss panels with single diagonal web members U(j – 1)L(j),

$$F^{max}_{U(j-1)U(j)} = \frac{-M^{max}_{(j)}}{h} = \frac{-wj(n-j)d^2}{2h} \tag{4.4c}$$

$$F^{max}_{L(j-1)L(j)} = \frac{M^{max}_{(j-1)}}{h} = \frac{w(j-1)(n-j+1)d^2}{2h} \tag{4.4d}$$

For truss panels with single diagonal web members L(j – 1)U(j),

$$F^{max}_{U(j-1)U(j)} = \frac{-M^{max}_{(j-1)}}{h} = \frac{-w(j-1)(n-j+1)d^2}{2h} \quad (4.4e)$$

$$F^{max}_{L(j-1)L(j)} = \frac{M^{max}_{(j)}}{h} = \frac{wj(n-j)d^2}{2h} \quad (4.4f)$$

When the uniformly distributed live load is entirely on the span,[64] the shear force, $V_{(j-1)-(j)}$, in truss panel (j-1)-(j), where j ≤ n/2,[65] is

$$V_{(j-1)-(j)} = \frac{w(L-d)}{2} - (j-1)wd = \frac{wd}{2}(n-2j+1) \quad (4.4g)$$

and the force in single diagonal web member U(j-1)L(j) is

$$F_{U(j-1)L(j)} = V_{(j-1)-(j)}\left(\frac{\sqrt{h^2+d^2}}{h}\right) = \left(\frac{wd}{2}(n-2j+1)\right)\left(\frac{\sqrt{h^2+d^2}}{h}\right) \quad (4.4h)$$

When the uniformly distributed live load is entirely on the span, the force in single diagonal web member L(j – 1)U(j) in panel (j – 1) – (j), where j ≤ n/2, is

$$F_{L(j-1)U(j)} = -V_{(j-1)-(j)}\left(\frac{\sqrt{h^2+d^2}}{h}\right) = -\left(\frac{wd}{2}(n-2j+1)\right)\left(\frac{\sqrt{h^2+d^2}}{h}\right) \quad (4.4i)$$

When the uniformly distributed live load is entirely on the span, the force in vertical web member U(j-1)L(j-1) of the Pratt truss, where the diagonal member in panel (j – 1) – (j) is U(j – 1)L(j), and where j ≤ n/2, is[66]

$$F_{U(j-1)L(j-1)} = -V_{(j-1)-(j)} = \frac{-wd}{2}(n-2j+1) \quad (4.4j)$$

The vertical web members U(j – 1)L(j – 1) of the Warren truss shown in Figure 4.3c are zero-force members. The vertical web members U(j)L(j) in Figure 4.3c are directly loaded by floorbeam reactions.

Example 4.5 uses Equation 4.4 to determine chord member forces in the Pratt truss in Figure 4.1aa loaded entirely with a uniformly distributed live load.

Example 4.5

The six-panel horizontal top chord Pratt truss in Figure 4.1aa is 150 ft long with h = 35 ft. It is subjected to a uniformly distributed dead load of 1500 lb/ft and a uniformly distributed live load of 5500 lb/ft. Live load forces in the Pratt truss members are required when the span is fully loaded by the uniformly distributed live load:

$$W_{DL} = (1.5)(25) = 37.50 \text{ kips}$$
$$W_{LL} = (5.5)(25) = 137.50 \text{ kips}$$

Maximum forces in the top chord members are determined in accordance with Equation 4.4c.

		Diagonal Member is U(j − 1)L(j) (Equation 4.4c) (kips)	
Top Chord Member	j	Dead Load Force	Live Load Force
U1U2	2	−107.14	−392.86
U2U3	3	−120.54	−441.96

Maximum forces in the bottom chord members are determined in accordance with Equation 4.4d.

		Diagonal Member is U(j − 1)L(j) (Equation 4.4d) (kips)	
Bottom Chord Member	j	Dead Load Force	Live Load Force
L1L2	2	66.96	245.54
L2L3	3	107.14	392.86

Equilibrium of forces at joint L1 indicates $F_{L0L1} = F_{L1L2}$

Equilibrium of forces at joint L0 indicates

$$F_{L0U1} = -F_{L0L1}\frac{\sqrt{h^2+d^2}}{d} = -1.720F_{L0L1} = -1.720F_{L1L2}$$
$$= -1.720(66.96+245.54) = -115.17-422.33 = -537.50 \text{ kips}$$

Member	Dead Load Force (kips)	Live Load Force (kips)
L0L1	66.96	245.54
L0U1	−115.17	−422.33

Forces in the diagonal web members are determined in accordance with Equation 4.4h.

		Diagonal Member is U(j-1)L(j) (Equation 4.4h) (kips)	
Diagonal Web Member	j	Dead Load Force	Live Load Force
U1L2	2	69.13	253.46
U2L3	3	23.04	84.49

Forces in the vertical web members are determined in accordance with Equation 4.4j.

		Diagonal Member Is U(j − 1)L(j) (Equation 4.4j) (kips)	
Vertical Web Member	j	Dead Load Force	Live Load Force
U2L2	3	−18.75	−68.75

Forces in members U1L1 and U3L3 are readily determined by considering the equilibrium of free body diagrams at truss joints L1 and U3, respectively, as

$$F_{U1L1(DL)} = 37.5 \text{ kips}$$

$$F_{U1L1(LL)} = 137.5 \text{ kips}$$

$$F_{U3L3(DL)} = F_{U3L3(LL)} = 0$$

Dead and live load member forces are summarized in Table E4.5.

TABLE E4.5
Member Forces in Horizontal Parallel-Chord Pratt Through Truss due to Uniformly Distributed Live Load across the Entire Span

Member	Dead Load Force (kips)	Live Load Force (kips)	Dead + Live Load Force (kips)
L0L1	66.96	245.54	312.50
L1L2	66.96	245.54	312.50
L2L3	107.14	392.86	500.00
L0U1	−115.17	−422.33	−537.50
U1U2	−107.14	−392.86	−500.00
U2U3	−120.54	−441.96	−562.50
U1L1	37.50	137.50	175.00
U2L2	−18.75	−68.75	−87.50
U3L3	0	0	0
U1L2	69.13	253.46	322.59
U2L3	23.04	84.49	107.53

However, the forces due to live load in diagonal and vertical web members of the Pratt truss in Example 4.5 are not maximum when the span is entirely loaded by the uniformly distributed live load because this load position does not maximize truss panel shear. As uniformly distributed live loads traverse the span, they produce positive and negative shear forces in the truss panels, which generate maximum positive and maximum negative web member forces. Therefore, as shown in Figures 4.4a and 4.4b, the uniformly distributed live load may be only partially distributed within panel (j-1)-(j) to generate maximum positive and maximum negative shear forces in the panel.[67] Consequently, at a unique position within panel (j − 1) − (j), the partially distributed live load will generate the maximum positive and maximum negative shear forces in panel (j − 1) − (j) and thus the maximum and minimum forces in the web members of trusses.

The positive shear force, $V^{+}_{(j-1)-(j)}$, in truss panel (j − 1) − (j), where j ≤ n/2, is

$$V^{+}_{(j-1)-(j)} = R_0 - \sum_{i=0}^{j-1} W_i = R_0 - W_{(j-1)} = \left(\frac{w\left(a_{(j)} + x^{+}_{(j-1)-(j)}\right)^2}{2L} \right) - \frac{w\left(x^{+}_{(j-1)-(j)}\right)^2}{2d} \tag{4.5a}$$

The maximum positive shear force $V^{+max}_{(j-1)-(j)}$ in panel (j − 1) − (j), is determined from

$$\frac{d\left(V^{+}_{(j-1)-(j)}\right)}{dx} = 0 = \frac{\left(a_{(j)} + x^{+}_{(j-1)-(j)}\right)}{L} - \frac{x^{+}_{(j-1)-(j)}}{d} \tag{4.5b}$$

which yields

$$x^{+}_{(j-1)-(j)} = \frac{a_{(j)}d}{(L-d)} = \frac{a_{(j)}}{(n-1)} = \frac{(n-j)d}{(n-1)} \tag{4.5c}$$

Substituting Equation 4.5c into Equation 4.5a yields the maximum positive shear force, $V^{+max}_{(j-1)-(j)}$, in panel (j − 1) − (j) as

$$V^{+max}_{(j-1)-(j)} = \frac{wa^2_{(j)}}{2(L-d)} = \frac{w\left((n-j)d\right)^2}{2d(n-1)} \tag{4.5d}$$

The negative shear force $V^{-}_{(j-1)-(j)}$ in truss panel (j – 1) – (j), where $j \le n/2$, is

$$V^{-}_{(j-1)-(j)} = R_n - \sum_{i=j}^{n} W_i = R_n - W_{(j)} = \left(\frac{w\left(b_{(j)} + x^{-}_{(j-1)-(j)}\right)^2}{2L}\right) - \frac{w\left(x^{-}_{(j-1)-(j)}\right)^2}{2d} \tag{4.5e}$$

The maximum negative shear force $V^{-max}_{(j-1)-(j)}$ in panel (j – 1) – (j) is determined from

$$\frac{d\left(V^{-}_{(j-1)-(j)}\right)}{dx} = 0 = \frac{\left(b_{(j)} + x^{-}_{(j-1)-(j)}\right)}{L} - \frac{x^{-}_{(j-1)-(j)}}{d} \tag{4.5f}$$

which yields

$$x^{-}_{(j-1)-(j)} = \frac{b_{(j)}d}{(L-d)} = \frac{b_{(j)}}{(n-1)} = \frac{(j-1)d}{(n-1)} = d - x^{+}_{(j-1)-(j)} \tag{4.5g}$$

Equation 4.5g is expected due to the form of the shear force diagram for a uniformly distributed load in panel (j) – (j – 1).[68] Substituting Equation 4.5g into Equation 4.5e yields the maximum negative shear force $V^{-max}_{(j-1)-(j)}$ in panel (j – 1) – (j) as follows:

$$V^{-max}_{(j-1)-(j)} = \frac{wb^2_{(j)}}{2(L-d)} = \frac{w\left((j-1)d\right)^2}{2d(n-1)} \tag{4.5h}$$

Therefore, the maximum force $F^{+max}_{U(j-1)L(j)}$ in diagonal web member U(j – 1)L(j) due to positive panel shear, in accordance with Equation 4.3j, is the following:

$$F^{+max}_{U(j-1)L(j)} = \left(\frac{w\left((n-j)d\right)^2}{2d(n-1)}\right)\left(\frac{\sqrt{h^2+d^2}}{h}\right) \tag{4.5k}$$

and the maximum force $F^{-max}_{U(j-1)L(j)}$ in diagonal web member U(j – 1)L(j) due to negative panel shear, in accordance with Equation 4.3k, is

$$F^{-max}_{U(j-1)L(j)} = -\left(\frac{w\left((j-1)d\right)^2}{2d(n-1)}\right)\left(\frac{\sqrt{h^2+d^2}}{h}\right) \tag{4.5l}$$

The maximum force $F^{+max}_{L(j-1)U(j)}$ in diagonal web member L(j – 1)U(j) due to positive panel shear, in accordance with Equation 4.3l, is

$$F^{+max}_{L(j-1)U(j)} = -\left(\frac{w\left((n-j)d\right)^2}{2d(n-1)}\right)\left(\frac{\sqrt{h^2+d^2}}{h}\right) = -F^{+max}_{U(j-1)L(j)} \tag{4.6a}$$

and the maximum force $F^{-max}_{L(j-1)U(j)}$ in diagonal web member L(j – 1)U(j) due to negative panel shear, in accordance with Equation 4.3m, is

$$F^{-max}_{L(j-1)U(j)} = \left(\frac{w((j-1)d)^2}{2d(n-1)}\right)\left(\frac{\sqrt{h^2+d^2}}{h}\right) = -F^{-max}_{U(j-1)L(j)} \tag{4.6b}$$

The maximum force $F^{+max}_{U(j-1)L(j-1)}$ in vertical web member U(j – 1)L(j – 1) of a Pratt truss, where the diagonal web member is U(j – 1)L(j), that is not directly loaded by floorbeam reaction,[69] due to positive panel shear, in accordance with Equation 4.3n, is[70]

$$F^{+max}_{U(j-1)L(j-1)} = -\left(\frac{w((n-j)d)^2}{2d(n-1)}\right) \tag{4.7a}$$

Additionally, the maximum force, $F^{-max}_{U(j-1)L(j-1)}$, in vertical web member U(j – 1)L(j – 1) of a Pratt truss, where the diagonal web member is U(j – 1)L(j), that is not directly loaded by floorbeam reaction, due to negative panel shear, in accordance with Equation 4.3o, is

$$F^{-max}_{U(j-1)L(j-1)} = \left(\frac{w((j-1)d)^2}{2d(n-1)}\right) \tag{4.7b}$$

If the vertical member in a Pratt truss that is not directly loaded by floorbeam reaction is U(j)L(j) in panel (j – 1) – (j), Equation 4.3p for positive shear and Equation 4.3q for negative shear in panel (j) – (j + 1) provide

$$F^{+max}_{U(j)L(j)} = -\left(\frac{w((n-j+1)d)^2}{2d(n-1)}\right) \tag{4.7c}$$

and

$$F^{-max}_{U(j)L(j)} = \left(\frac{w(jd)^2}{2d(n-1)}\right) \tag{4.7d}$$

The force in end post member L0U1 from equilibrium at joint L0 is

$$F_{L0U1} = \frac{-R_0\sqrt{h^2+d^2}}{h} \tag{4.8}$$

Examples 4.6, 4.7, 4.8 and 4.9 illustrate the structural analyses of the statically determinate horizontal parallel-chord Pratt through-, Warren deck-, Warren through- and Howe through-trusses in Figure 4.1a, traversed by a uniformly distributed live load. The maximum and minimum live load forces in diagonal web members U(j – 1)L(j) and L(j – 1)U(j) are determined with Equations 4.5 and 4.6, respectively. The maximum and minimum live load forces in vertical web members U(j – 1) L(j – 1) or U(j)L(j), where the diagonal member in panel (j – 1) – (j) or (j) – (j + 1), respectively, is U(j – 1)L(j), are determined with Equation 4.7.

Example 4.6

The 150 ft long six-panel horizontal parallel-chord Pratt through-truss in Figure 4.1aa with h = 35 ft is subjected to a uniform dead load of 1500 lb/ft and uniform live load of 5500 lb/ft. Find the maximum and minimum forces in the truss members.

The maximum chord member and end post forces due to dead load and live load were determined in Example 4.5, with a uniformly distributed live load across the entire span. The web member forces due to dead load were also determined in Example 4.5. The maximum live load forces in diagonal web members from positive and negative panel shear forces are determined from Equations 4.5k and 4.5l as shown in Table E4.6a.

TABLE E4.6a
Maximum Diagonal Web Member Forces in a Horizontal Parallel-Chord Pratt Through Truss from Uniformly Distributed Live Load

Panel	Diagonal Web Member	j	F^{+max} (kips)	F^{-max} (kips)
1-2	U1L2	2	270.38	−16.90
2-3	U2L3	3	152.09	−67.60

Maximum live load forces in vertical web members U2L2 (j = 3) are determined from Equations 4.7a and 4.7b for positive and negative panel shear, respectively, as

$$F_{U2L2}^{+max} = -V_{2-3}^{+max} = -123.75 \text{ kips}$$

$$F_{U2L2}^{-max} = V_{2-3}^{-max} = 55.0 \text{ kips}$$

Forces in members U1L1 and U3L3 are determined by considering static equilibrium of free body diagrams at joints L1 and U3, respectively. A summary of web member forces is shown in Table E4.6b.

TABLE E4.6b
Dead Load and Maximum Live Load Web Member Forces in Horizontal Parallel-Chord Pratt Through Truss from Uniformly Distributed Live Load

Web Member		Dead Load Force (kips)	Maximum Live Load Force (kips)	Maximum Live Load Force due to Positive Shear F^{+max} (kips)	Maximum Live Load Force due to Negative Shear F^{-max} (kips)
U1L1	Vertical hanger	37.50	137.50	-	-
U2L2	Vertical	−18.75	-	−123.75	55.00
U3L3	Vertical	0	0	-	-
U1L2	Diagonal	69.13	-	270.38	−16.90
U2L3	Diagonal	23.04	-	152.09	−67.60

The maximum positive live load force in U1L2 is 6.7% greater than that calculated considering the span entirely loaded. The maximum positive live load force in U2L3 is 80.0% greater than that calculated considering the span entirely loaded.

Table E4.6c is a summary of the dead load and maximum live load member forces due to a uniformly distributed live load traversing the horizontal parallel-chord Pratt through truss in Figure 4.1a. Table E4.6d is a summary of the maximum and minimum member forces.

TABLE E4.6c
Dead Load and Maximum Live Load Member Forces in Horizontal Parallel-Chord Pratt Through Truss from Uniformly Distributed Live Load

Pratt Through Truss Member		Dead Load Force (kips)	Maximum Live Load Force (kips)	Maximum Live Load Force due to Positive Shear F^{+max} (kips)	Maximum Live Load Force due to Negative Shear F^{-max} (kips)
L0L1	Bottom chord	66.96	245.54	-	-
L1L2	Bottom chord	66.96	245.54	-	-
L2L3	Bottom chord	107.14	392.86	-	-
L0U1	End post	−115.17	−422.33	-	-
U1U2	Top chord	−107.14	−392.86	-	-
U2U3	Top chord	−120.54	−441.96	-	-
U1L1	Vertical hanger	37.50	137.50	-	-
U2L2	Vertical Post	−18.75	-	−123.75	55.00
U3L3	Vertical post	0	0	-	-
U1L2	Diagonal	69.13	-	270.38	−16.90
U2L3	Diagonal	23.04	-	152.09	−67.60

TABLE E4.6d
Maximum and Minimum Member Forces due to Dead Load and a Uniformly Distributed Live Load Traversing a Horizontal Parallel-Chord Pratt Through-Truss Span

Pratt Through-Truss Member		Maximum Dead and Live Load Force in Member (kips)	Minimum Dead and Live Load Force in Member (kips)
L0L1	Bottom chord	312.50	66.96
L1L2	Bottom chord	312.50	66.96
L2L3	Bottom chord	500.00	107.14
L0U1	End post	−537.50	−115.17
U1U2	Top chord	−500.00	−107.14
U2U3	Top chord	−562.50	−120.54
U1L1	Vertical hanger	175.00	37.50
U2L2	Vertical post	−142.50	36.25
U3L3	Vertical post	0	0
U1L2	Diagonal	339.51	52.23
U2L3	Diagonal	175.13	−44.56

Vertical post member U2L2 and diagonal member U2L3 experience stress reversal from live load (live load impact not included).

Example 4.7

The 150 ft long six-panel horizontal parallel chord Warren deck truss in Figure 4.1ac with h = 25 ft is subjected to a uniform dead load of 1500 lb/ft and uniform live load of 5500 lb/ft. Find the maximum and minimum magnitude forces in the truss members.

The Warren deck truss may be analyzed by using the method of sections and method of joints to determine truss member forces. Sections through panels of the Warren deck truss are the same as the sections through corresponding panels in the Pratt through-truss with exception of the orientation of the diagonal members in the end panel and even-numbered panels. Therefore, the live load forces in Warren deck truss members in panels 0-1 and 1-2 are of the same magnitude but opposite attitude as the live load forces in the corresponding Pratt truss members. Correspondingly, the live load forces in Warren deck-truss members in panel 2-3 are of the same magnitude and attitude as the live load forces in the corresponding Pratt truss members.

Investigation of static equilibrium of free body diagrams at joints U1 and U3 reveals that the force in the Warren deck-truss members U1L1 and U3L3 are in compression and of equal magnitude to the load applied at the joint (floorbeam reaction). Investigation at joint L2 reveals that the force in the Warren deck-truss member U2L2 is null and the force in Warren deck truss member L2L3 is equal in magnitude and attitude to the force in Warren deck-truss member L1L2.

The member forces in a 150 ft long six-panel horizontal parallel chord Pratt through truss in Figure 4.1aa with h = 25 ft subjected to a uniform dead load of 1500 lb/ft and uniform live load of 5500 lb/ft are as follows in Table E4.7a.

TABLE E4.7a
Maximum and Minimum Member Forces due to a Uniformly Distributed Live Load Traversing a Horizontal Parallel Chord Pratt Through Truss Span

Pratt Through-Truss Member		Maximum Dead and Live Load Force in Member (kips)	Minimum Dead and Live Load Force in Member (kips)
L0L1	Bottom chord	312.50(35/25) = 437.50	66.96(35/25) = 93.75
L1L2	Bottom chord	312.50(35/25) = 437.50	66.96(35/25) = 93.75
L2L3	Bottom chord	500(35/25) = 700.00	107.14(35/25) = 150.00
L0U1	End post	−537.50(1.414/1.229) = −618.75	−115.17(1.414/1.229) = −132.56
U1U2	Top chord	−500(35/25) =−700.00	−107.14(35/25) = −150.00
U2U3	Top chord	−562.50(35/25) = −787.50	−120.54(35/25) = −168.75
U1L1	Vertical hanger	175.00	37.50
U2L2	Vertical post	−142.50	36.25
U3L3	Vertical post	0	0
U1L2	Diagonal	339.51(1.414/1.229) = 390.68	52.23(1.414/1.229) = 60.10
U2L3	Diagonal	175.13(1.414/1.229) = 201.52	−44.56(1.414/1.229) = −51.26

Table E4.7b summarizes the dead load and maximum live load forces for a uniformly distributed live load traversing the Warren deck-truss span.

TABLE E4.7b
Dead Load and Maximum Live Load Member Forces in a Horizontal Parallel-Chord Warren Deck-Truss Span for a Uniformly Distributed Live Load

Member	Dead Load Force (kips)	Maximum Live Load Force (kips)	Maximum Live Load Force due to Positive Shear F^{+max} (kips)	Maximum Live Load Force due to Negative Shear F^{-max} (kips)
U0L1	132.56	486.06	-	-
L1L2	150.00	550.00	-	-
L2L3	150.00	550.00	-	-

TABLE E4.7b (*Continued*)
Dead Load and Maximum Live Load Member Forces in a Horizontal Parallel-Chord Warren Deck-Truss Span for a Uniformly Distributed Live Load

Member	Dead Load Force (kips)	Maximum Live Load Force (kips)	Maximum Live Load Force due to Positive Shear F^{+max} (kips)	Maximum Live Load Force due to Negative Shear F^{-max} (kips)
U0U1	−93.75	−343.75	-	-
U1U2	−93.75	−343.75	-	-
U2U3	−168.75	−618.75	-	-
U1L1	−37.50	−137.50	-	-
U2L2	0	0	-	-
U3L3	−37.50	−137.50	-	-
L1U2	−79.55	-	−390.68	−60.10
U2L3	26.52	-	201.52	−51.26

Table E4.7c is a summary of the resulting combined maximum and minimum member forces from the horizontal parallel-chord Warren deck truss in Figure 4.1ac.

TABLE E4.7c
Maximum and Minimum Member Forces due to Dead Load and a Uniformly Distributed Live Load Traversing a Horizontal Parallel Chord Warren Deck Truss Span

Warren Deck-Truss Member		Maximum Force in the Member (kips)	Minimum Force in the Member (kips)
U0L1	Bottom chord (end member)	618.75	132.56
L1L2	Bottom chord	700.00	150.00
L2L3	Bottom chord	700.00	150.00
U0U1	Top chord	−437.50	−93.75
U1U2	Top chord	−437.50	−93.75
U2U3	Top chord	−787.50	−168.8
U1L1	Vertical	−175.00	−37.50
U2L2	Vertical	0	0
U3L3	Vertical	−175.00	−37.50
L1U2	Diagonal	−390.68	−60.10
U2L3	Diagonal	201.52	−51.26

Diagonal member U2L3 experiences stress reversal from live load (not including live load impact).

Example 4.8

The 150 ft long six-panel horizontal parallel-chord Warren through-truss in Figure 4.1ab with h = 35 ft is subjected to a uniform dead load of 1500 lb/ft and uniform live load of 5500 lb/ft. Find the maximum and minimum magnitude forces in the truss members.

The Warren through truss may be analyzed by using the Method of Sections and Method of Joints to determine truss member forces. A review of the Warren through truss geometry reveals that it may be analyzed by referencing the Pratt through truss analysis of Example 4.6. Nevertheless, since Warren through truss railway spans are typically as prevalent as Pratt through truss railway spans, the analysis of the Warren through truss will be carried out in detail as for the Pratt through truss of Example 4.6.

The maximum chord member forces are determined using Equations 4.4c and 4.4d for chord members in panel (j − 1) − (j) where the diagonal web members are U(j − 1)L(j):

$$F_{L1L2}^{max} = \frac{M_{(1)}^{max}}{35} = \frac{2343.8 + 8593.8}{35} = 66.96 + 245.54 = 312.50 \text{ kips}$$

$$F_{U1U2}^{max} = \frac{M_{(2)}^{max}}{35} = \frac{-3750.0 - 13750.0}{35} = -107.14 - 392.86 = -500.0 \text{ kips}$$

The maximum chord member forces are determined using Equations 4.4e and 4.4f for chord members in panel (j − 1) − (j) where the diagonal web members are L(j-1)U(j).

$$F_{L2L3}^{max} = \frac{M_{(3)}^{max}}{35} = \frac{4218.8 + 15468.8}{35} = 120.54 + 441.96 = 562.50 \text{ kips}$$

$$F_{U2U3}^{max} = \frac{M_{(2)}^{max}}{35} = \frac{-3750.0 - 13750.0}{35} = -107.14 - 392.86 = -500.0 \text{ kips}$$

The maximum live load forces in diagonal web members from maximum positive and maximum negative panel shear forces are determined from Equations 4.5k and 4.5l for member U(j − 1)L(j) and Equations 4.6a and 4.6b for member L(j − 1)U(j) and are given in Table E4.8a.

TABLE E4.8a
Maximum Panel Shear Forces and Diagonal Web Member Forces in a Horizontal Parallel-Chord Warren Through Truss from a Uniformly Distributed Live Load

Panel	Diagonal Web Member	j	F^{+max} (kips)	F^{-max} (kips)
1-2	U1L2	2	270.40	−16.90
2-3	L2U3	3	−152.09	67.60

Equation 4.8 provides the force in end post member L0U1, which is a maximum for the uniformly distributed load entirely on the span:

$$F_{L0U1}^{DL} = \frac{-5(37.5)}{2}(1.229) = -115.17 \text{ kips}$$

$$F_{L0U1}^{LL} = \frac{-5(137.5)}{2}(1.229) = -422.46 \text{ kips}$$

Maximum live load forces in vertical web members are determined from equilibrium of forces at free body diagrams at joints L1, U2 and L3 as,

$$F^{max}_{U1L1} = F^{max}_{U3L3} = 137.50 \text{ kips}$$

$$F^{max}_{U2L2} = 0 \text{ kips}$$

Table E4.8b is a summary of the dead load and maximum live load member forces in the horizontal-chord Warren through-truss in Figure 4.1ab.

TABLE E4.8b
Dead Load and Maximum Live Load Member Forces in a Horizontal Parallel-Chord Warren Through-Truss Span for a Uniformly Distributed Live Load

Warren Through Truss Member		Dead Load Force (kips)	Maximum Live Load Force (kips)	Maximum Live Load Force due to Positive Shear F^{+max} (kips)	Maximum Live Load Force due to Negative Shear F^{-max} (kips)
L0U1	End post	−115.17	−422.36	-	-
U1U2	Top chord	−107.14	−392.86	-	-
U2U3	Top chord	−107.14	−392.86	-	-
L0L1	Bottom chord	66.96	245.54	-	-
L1L2	Bottom chord	66.96	245.54	-	-
L2L3	Bottom chord	120.54	441.96	-	-
U1L1	Vertical	37.50	137.50	-	-
U2L2	Vertical	0	0	-	-
U3L3	Vertical	37.50	137.50	-	-
U1L2	Diagonal	69.13	-	270.40	−16.90
L2U3	Diagonal	−23.04	-	−152.09	67.60

Table E4.8c is a summary of the resulting combined maximum and minimum member forces due to a uniformly distributed live load traversing the Warren through-truss in Figure 4.1ab.

TABLE E4.8c
Maximum and Minimum Member Forces due to Dead Load and a Uniformly Distributed Live Load Traversing a Horizontal Parallel-Chord Warren Through-Truss Span

Warren Through Truss Member		Maximum Force in the Member (kips)	Minimum Force in the Member (kips)
U0L1	End Post	−537.50	−115.17
L1L2	Bottom chord	312.50	66.96
L2L3	Bottom chord	562.50	120.54
U1U2	Top chord	−500.00	−107.14
U2U3	Top chord	−500.00	−107.14
U1L1	Vertical	137.50	37.50
U2L2	Vertical	0	0
U3L3	Vertical	137.50	37.50
U1L2	Diagonal	339.51	52.23
L2U3	Diagonal	−175.14	44.56

Diagonal member L2U3 experience stress reversal from live load (not including live load impact).

Example 4.9

The 150 ft long six-panel horizontal parallel top chord Howe through-truss in Figure 4.1ad with h = 35 ft is subjected to a uniform dead load of 1500 lb/ft and uniform live load of 5500 lb/ft. Find the maximum and minimum forces in the truss members.

The Howe through truss may be analyzed by using the method of sections and method of joints to determine truss member forces. A review of the Howe through truss geometry reveals that it may be analyzed by referencing the Pratt through-truss analysis of Example 4.6.

A section through panel 0-1 of the Howe through truss is the same as a section through panel 0-1 of the Pratt through truss. Sections through panel 1-2 and panel 2-3 of the Howe through truss are the same as the corresponding sections of the Pratt through truss with exception of the orientation of the diagonal. Therefore, live load forces in members of panel 0-1 of the Howe through truss are of the same magnitude and attitude as the corresponding members in panel 0-1 of the Pratt through truss with exception of the vertical member U1L1, which, due to participation in truss action, has forces of higher magnitude than the corresponding member U1L1 in the Pratt through truss. Chord member forces from live load in panels 1-2 and 2-3 of the Howe through-truss are of opposite magnitude but same attitude as the corresponding forces in members of the Pratt through-truss. Web member forces from live load in panels 1-2 and 2-3 of the Howe through-truss are of same magnitude but opposite attitude as the corresponding forces in members of the Pratt through-truss with exception that, similarly to vertical member U1L1, vertical web members in the Howe through-truss have forces of greater magnitude than the corresponding members in the Pratt through-truss.

Investigation of static equilibrium at joint L3 reveals that the force in the Howe through truss member U3L3 is of equal magnitude to the load applied at the joint (floorbeam reaction). A section through the bottom chord in panel 0-1 and the top chord in panel 1-2 of the Howe through truss indicates that member U1L1 will be of equal magnitude but opposite attitude than the vertical reaction at L0. A section through the bottom chord in panel 1-2 and the top chord in panel 2-3 of the Howe through truss indicates that live load forces in member U2L2 will be of equal magnitude but opposite attitude than the difference between the vertical reaction at L0 and the joint load at L1, which is the live load shear force in truss panels 1-2.

Table E4.9a is a summary of the dead load and maximum live load forces for a uniformly distributed live load traversing the Howe through truss span.

TABLE E4.9a

Dead Load and Maximum Live Load Member Forces in a Horizontal Parallel-Chord Howe Through-Truss Span for a Uniformly Distributed Live Load

Member	Dead Load Force (kips)	Maximum Live Load Force (kips)	Maximum Live Load Force due to Positive Shear F^{+max} (kips)	Maximum Live Load Force due to Negative Shear F^{-max} (kips)
L0L1	66.96	245.54	-	-
L1L2	107.14	392.86	-	-
L2L3	120.54	441.96	-	-
L0U1	−115.17	−422.46	-	-
U1U2	−66.96	−245.54	-	-
U2U3	−107.14	−392.86	-	-
U1L1	66.96	245.54	-	-
U2L2	56.25	-	220.00	−13.75
U3L3	37.50	137.50	-	-
L1U2	−69.13	-	−270.40	16.90
L2U3	−23.04	-	−152.10	67.60

Table E4.9b is a summary of the resulting combined maximum and minimum member forces due to a uniformly distributed live load traversing the horizontal-chord Howe through-truss in Figure 4.1ad.

TABLE E4.9b
Maximum and Minimum Member Forces due to Dead Load and a Uniformly Distributed Live Load Traversing a Horizontal Parallel-Chord Howe Through-Truss Span

Howe Through-Truss Member		Maximum Force in the Member (kips)	Minimum Force in the Member (kips)
L0L1	Bottom chord	245.54	66.96
L1L2	Bottom chord	392.86	107.14
L2L3	Bottom chord	441.96	120.54
L0U1	End Post	−422.46	−115.17
U1U2	Top chord	−245.54	−66.96
U2U3	Top chord	−392.86	−107.14
U1L1	Vertical	245.54	66.96
U2L2	Vertical	276.25	42.50
U3L3	Vertical	137.50	37.50
L1U2	Diagonal	−339.53	−52.23
L2U3	Diagonal	−175.14	44.56

Diagonal member L2U3 experience stress reversal from live load (not including live load impact).

It is generally considered that Howe truss spans are less efficient at resisting live loads due to the large magnitude of the forces in vertical web members in contrast to Pratt and Warren truss span vertical web members, as illustrated in Example 4.6 to Example 4.9. Pratt truss spans have smaller magnitude tensile force ranges in the bottom chord but will experience greater magnitude compressive forces in the top chord. Pratt and Warren truss forms are typically preferred for modern railway truss span design, fabrication, and erection.

Example 4.6 to Example 4.9 established maximum and minimum web member forces based on the locations of uniformly distributed live loads determined from Equation 4.5c for maximum positive shear and Equation 4.5g for maximum negative shear in panel (j – 1) – (j). However, it is expedient, and often sufficiently accurate, to approximate the positive and negative panel shear forces due to uniformly distributed live loads.

4.4.2.1.1 Approximation of the Maximum Panel Shear Forces for Uniformly Distributed Live Loads

Equations 4.5d and 4.5h, used to calculate maximum positive and negative panel shear forces due to a uniformly distributed live load, respectively, are based on the location, $x^{+}_{(j-1)-(j)}$, of Equation 4.5c and $x^{-}_{(j-1)-(j)}$, of Equation 4.5g. Alternatively, an adequately conservative approximation may be made to simplify[71] the calculation of the maximum and minimum live load forces in web members. Allocating the uniformly distributed live loads as concentrated loads at panel points, as shown in Figure 4.5a and Figure 4.5b, for positive and negative shear in panel (j-1)-(j), respectively, is mathematically inadmissible for a uniformly distributed load, but proves to be reasonably accurate and expedient.[72]

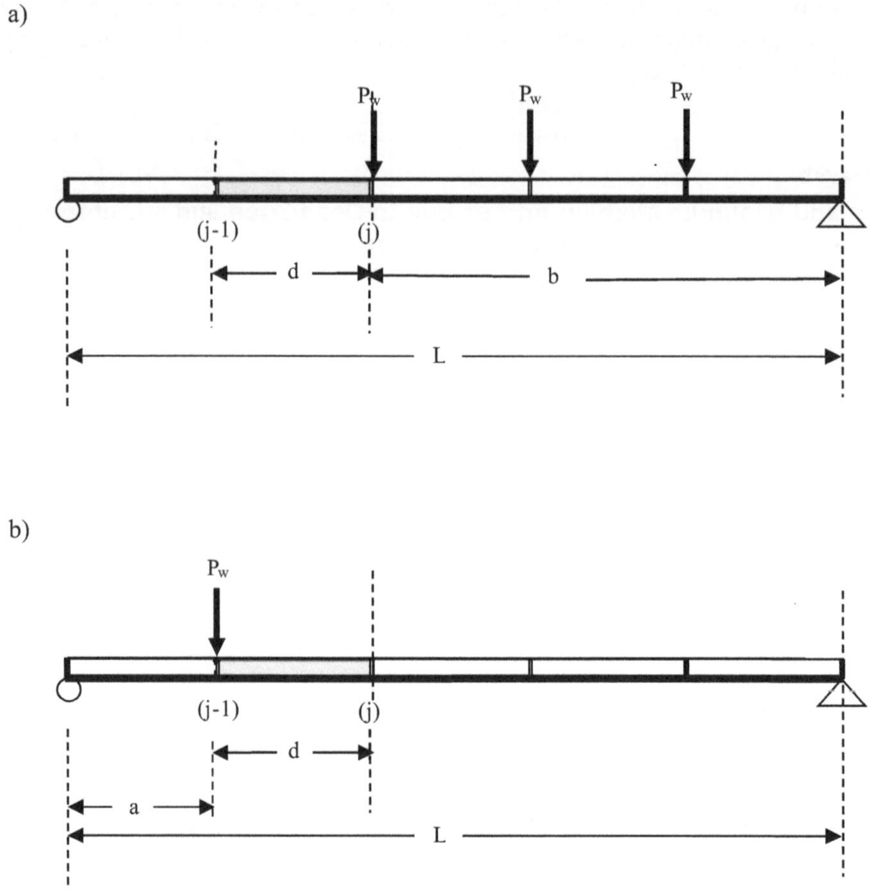

FIGURE 4.5 Allocation of uniformly distributed live load for approximate analysis of positive and negative panel shear.

From Figure 4.5a, the maximum positive panel shear, $V^{+max}_{(j-1)-(j)}$, in panel (j-1)-(j), where $j \le n/2$, is,

$$V^{+max}_{(j-1)-(j)} = \frac{P_W}{L}\sum_{n=0}^{\left(b_j/d\right)} nd = \frac{wd^2}{L}\sum_{n=0}^{\left(b_j/d\right)} n \tag{4.9a}$$

It is noted that Equation 4.9a is an arithmetic series[73] that can be expressed as,

$$V^{+max}_{(j-1)-(j)} = \frac{wd}{2n}(n-j)(n-j+1) \tag{4.9b}$$

and from Figure 4.5b, the maximum negative panel shear, $V^{-max}_{(j-1)-(j)}$, in panel (j-1)-(j), where $j \le n/2$, is,

$$V^{-max}_{(j-1)-(j)} = \frac{-P_W}{L}\sum_{n=0}^{\left(a_j/d\right)} nd = \frac{-wd^2}{L}\sum_{n=0}^{\left(a_j/d\right)} n \tag{4.9c}$$

which can also be expressed as

$$V^{-max}_{(j-1)-(j)} = \frac{-wd}{2n}(j)(j-1) \tag{4.9d}$$

Equations 4.9b and 4.9d are easily developed for any truss form without incurring excessively conservative errors in the maximum positive and maximum negative live load forces for web members. The magnitude of the error incurred by using the approximate method for positive shear may be determined by examining the difference, Δ_{V+}, between Equations 4.9b and 4.5d:

$$\Delta_{V+} = \frac{wd}{2n}(n-j)(n-j+1) - \frac{w((n-j)d)^2}{2d(n-1)} = \frac{wd^2}{2L}\left(\frac{a_{(j)}}{d}\right)\left(\frac{a_{(j)}}{d}+1\right) - \frac{wa_{(j)}^2}{2(L-d)} \tag{4.10a}$$

where $a_{(j)} = (n-j)d$ and $n = L/d$.

The maximum $a_{(j)}/d$ occurs where

$$\frac{d\Delta_{V+}}{d\left(\frac{a_{(j)}}{d}\right)} = \frac{wd}{2}\left(2\left(\frac{a_{(j)}}{d}\right)\left(\frac{1}{n} - \frac{1}{n-1}\right) + \frac{1}{n}\right) = 0 \tag{4.10b}$$

which yields,

$$\frac{a_{(j)}}{d} = \frac{n-1}{2} \tag{4.10c}$$

Substituting Equation 4.10c into Equation 4.10a gives

$$\Delta_{V+} = \frac{wd}{8}\left(\frac{n-1}{n}\right) \tag{4.10d}$$

Equation 4.10d indicates that the difference between approximate and exact values of maximum positive shear in panel (j-1)-(j) is greatest at central panels and is ((n-1)/n) less than 12.5% of the panel load, wd.

The magnitude of the error incurred by using the approximate method for negative shear may be determined by examining the difference, Δ_{V-}, between Equations 4.9d and 4.5h. A similar investigation of the difference reveals that Equation 4.10d is also valid for maximum negative shear in panel (j-1)-(j).

Example 4.10 illustrates the difference between the maximum positive shear forces in panels (j-1)-(j) determined by the approximate method (Equations 4.9b and 4.9d) and the exact method (Equations 4.5d and 4.5h).

Example 4.10

The 150 ft long six-panel horizontal-chord Pratt through-truss in Figure 4.1aa with h = 35 ft is subjected to a uniform dead load of 1500 lb/ft and uniform live load of 5500 lb/ft. Approximate the maximum live load positive panel shear forces and compare them with the actual values determined in Example 4.6.

$$W_{LL} = 5.50 \text{ kip/ft}$$

$$d = 25 \text{ ft}$$

$$L = 150 \text{ ft}$$

$$V^{+max}_{(j-1)-(j)} = 22.92 \sum_{n=0}^{\left(b_j/25\right)} n = 11.46(6-j)(7-j)$$

$$V^{-max}_{(j-1)-(j)} = 22.92 \sum_{n=0}^{\left(a_j/d\right)} n = 11.46(j)(j-1)$$

Table E4.10 illustrates that the approximate analysis results in only a small increase in member axial forces due to panel shear forces, particularly for members with larger axial forces.

TABLE E4.10
Approximate Live Load Positive Panel Shear Forces for a Uniformly Distributed Live Load

Maximum Positive Live Load Panel Shear					
Panel (j-1)-(j)	j	b_j (ft)	$b_j/25$	$V^{+max}_{(j-1)-(j)}$ (kips)	% Increase with Approximate Method (see Example 4.6)
0-1	1	125	5	343.75	0
1-2	2	100	4	229.20	4.2
2-3	3	75	3	137.50	11.1

Examples 4.6 to 4.10 illustrate the effects of uniformly distributed moving loads on horizontal parallel chord truss member forces. However, truss member forces due to concentrated moving loads must also be considered in the analysis of railway truss spans.

4.4.2.2 Concentrated Live Load Analysis

Maximum forces in chord members may be determined from a design live load[74] uniquely positioned, with a concentrated load at panel point (k),[75] to create maximum moments at truss panel point (k). Maximum and minimum forces in web members may be determined from a design live load specifically located, with a concentrated load at panel point (j) or (j – 1),[76] to create maximum positive and negative shear forces in truss panel (j-1)-(j). After establishing the position of the live load for maximum panel point moments, $M_{(k)}$, at panel point (k) and for maximum shear forces, $V_{(j-1)-(j)}$, in panel (j-1)-(j) using mathematical criteria,[77] the maximum and minimum member forces may be evaluated by static equilibrium or by constructing influence lines.

The locations of design live loads that create maximum panel point moments can be difficult to ascertain by observation. However, for a series of concentrated live loads in a constant pattern,[78] criteria may be established for the position of the live load creating maximum panel point moments and, consequently, the maximum forces in the chord members of horizontal parallel chord truss spans.

The moment, $M_{(k)}$, at panel point (k) of the floor system[79] in a truss of length, L, traversed by a series of concentrated loads with their resultant force, P_T at x_T, as shown in Figure 4.6a is,

$$M_{(k)} = R_0\left(a_k\right) - P_F\left(x_F\right) = \left(P_T \frac{x_T}{L}\right)\left(a_k\right) - P_F\left(x_F\right) = \left(P_T \frac{x_T}{n}\right)\left(k\right) - P_F\left(x_F\right) \quad (4.11a)$$

where a_k = kd, L = nd, d = the length of equal length panels in the truss span, k = the number of equal length panels between panel point (0) and panel point (k), n = the number of equal length panels in the truss span, P_T = the resultant of the concentrated loads on the truss span, and P_F = the resultant of the concentrated loads between panel point (0) panel point (k) with live loads moving from right to left (from panel point (n) to panel point (0)).

The maximum moment occurs when

$$\frac{dM_{(k)}}{dx} = 0 = \left(P_T \frac{k}{n}\right) - P_F \quad (4.11b)$$

or

$$\frac{P_T}{n} = \frac{P_F}{k} \quad (4.11c)$$

Equation 4.11c can be used to determine the location of the live load for maximum moment at panel point (k) within the span. Therefore, the criterion for the location of live loads for maximum panel point moment is when $\frac{P_T}{n} - \frac{P_F}{k}$ changes sign (positive to negative with loads moving in the direction shown in Figure 4.6a). In other words, this criterion states that, for live loads moving from right to left, the maximum force in chord members occurs when the average load between panel point (0) and panel point (k), $P_F/(kd)$, equals the average load on the entire span, $P_T/(nd)$.

The criterion of Equation 4.11c for maximum panel point moments is evidently met for uniformly distributed live loads because $\frac{P_T}{L} = \frac{P_F}{a_k}$. However, for a live load consisting of concentrated loads,[80] the criterion of Equation 4.11c for maximum panel point moments must be reviewed for various locations of the live load by a proficient 'trial and error' approach.

The locations of live loads that create maximum positive and negative shear forces in panels are also difficult to establish by observation. However, for concentrated live loads, criteria may also be established for the position of the live load for maximum positive and negative shear forces in panels.

The positive shear force, $V^+_{(j-1)-(j)}$, in panel (j − 1) − (j) in the truss span of length, L = nd, traversed by a live load, moving from panel point (n) to panel point (0) (right to left) over the truss span, with resultant force, P^+_T, at a distance, x^+_T, from panel point (n), as shown in Figure 4.6b, is,

$$V^+_{(j-1)-(j)} = R_0 - \sum_{i=0}^{j-1} W_i = P^+_T \frac{x^+_T}{nd} - P^+_F - \frac{P^+_{(j-1)-(j)} x^+_{(j-1)-(j)}}{d} = P^+_T \frac{x^+_T}{nd} - P^+_F - \frac{P^+_B x^+_B}{d} \quad (4.12a)$$

where

W_i = the load at panel point (i)

$R_0 = P^+_T \frac{x^+_T}{nd}$ = the reaction at panel point (0)

P^+_F = the resultant of the concentrated loads between panel point (0) and panel point (j-1)

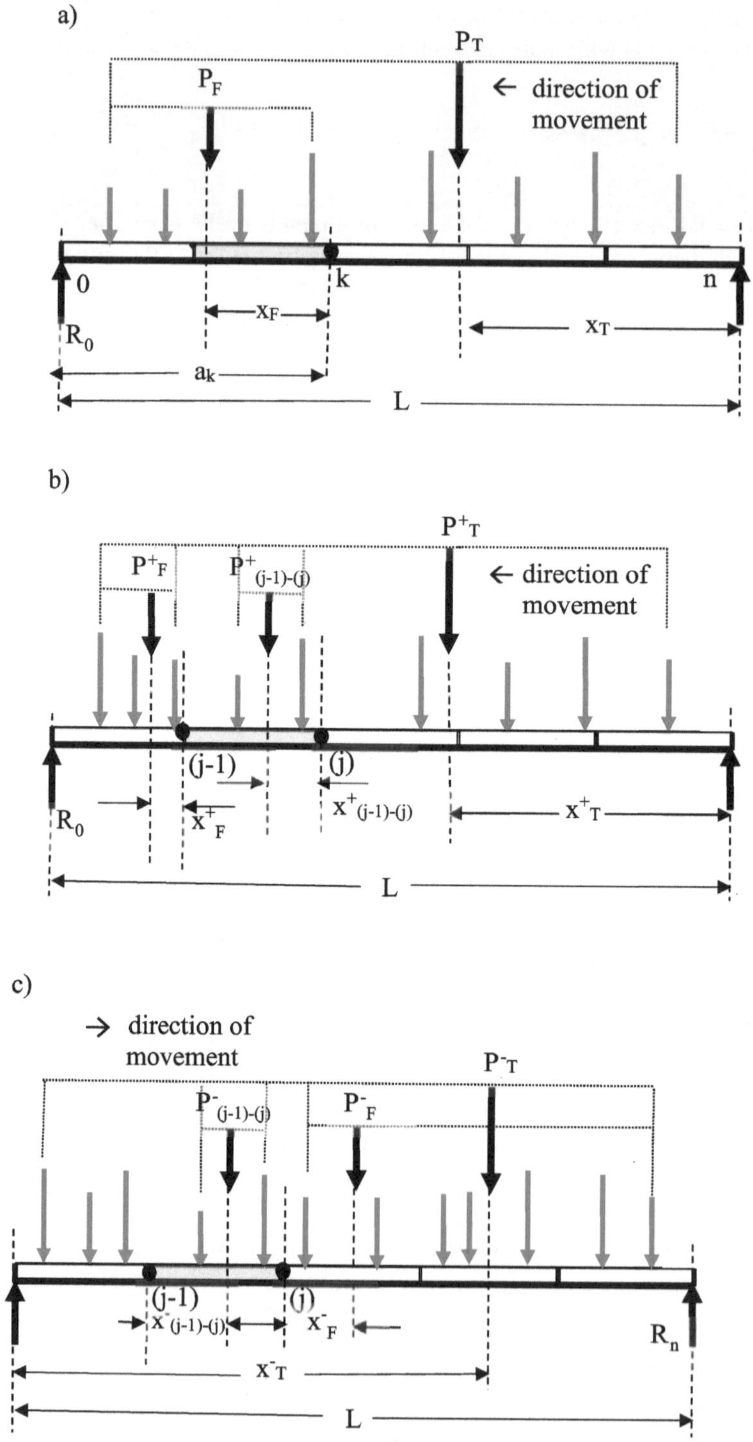

FIGURE 4.6 Concentrated live loads traversing a truss floor system.

x_F^+ = the distance between P_F^+ and panel point (j-1)
$P^+_{(j-1)-(j)} = P^+_B$ = the resultant of the concentrated loads between panel point (j-1) and panel point (j)[81]
$x^+_{(j-1)-(j)} = x^+_B$ = the distance between $P_{(j-1)-(j)} = P^+_B$ and panel point (j)

The maximum positive shear force, $V^{+max}_{(j-1)-(j)}$, in panel (j-1)-(j) may be determined by considering,

$$\frac{dV^+_{(j-1)-(j)}}{dx} = 0 = \frac{P_T^+}{nd} - \frac{P^+_{(j-1)-(j)}}{d} = \frac{P_T^+}{n} - P^+_{(j-1)-(j)} = \frac{P_T^+}{n} - P_B^+ \tag{4.12b}$$

or

$$\frac{P_T^+}{n} = P^+_{(j-1)-(j)} = P_B^+ \tag{4.12c}$$

Therefore, the maximum positive shear force in panel (j-1)-(j) occurs when the total load on the span divided by the number of panels in the span, $\frac{P_T^+}{n}$, equals the load, $P^+_{(j-1)-(j)} = P_B^+$, in panel (j-1)-(j).

The negative shear force, $V^-_{(j-1)-(j)}$, in panel (j-1)-(j) in the truss span of length, L = nd, traversed by a live load, moving from panel point (0) to panel point (n) (left to right) over the truss span, with resultant force, P^-_T, at a distance x_T^-, as shown in Figure 4.6c, is,

$$V^-_{(j-1)-(j)} = R_n - \sum_{i=j}^{n} W_i = P_T^- \frac{x_T^-}{nd} - P_F^- - \frac{P_B^- x_B^-}{d} \tag{4.13a}$$

where
$R_n = \frac{P_T^- x_T^-}{nd}$ = the reaction at panel point (n)
P_F^- = the resultant of the concentrated loads between panel point (j) and panel point (n)
x_F^- = the distance between P_F^- and panel point (j)
$P^-_{(j-1)-(j)} = P^-_B$ = the resultant of the concentrated loads between panel point (j-1) and panel point (j)
$x^-_{(j-1)-(j)} = x^-_B$ = distance from $P^-_{(j-1)-(j)} = P^-_B$ to panel point (j – 1)

The maximum negative shear force, $V^{-max}_{(j-1)-(j)}$, in panel (j – 1) – (j) may be determined by considering,

$$\frac{dV^-_{(j-1)-(j)}}{dx} = 0 = \frac{P_T^-}{nd} - \frac{P^-_{(j-1)-(j)}}{d} = \frac{P_T^-}{n} - P^-_{(j-1)-(j)} = \frac{P_T^-}{n} - P_B^- \tag{4.13b}$$

or

$$\frac{P_T^-}{n} = P^-_{(j-1)-(j)} = P_B^- \tag{4.13c}$$

Equation 4.13c indicates that the maximum negative shear force in panel (j – 1) – (j) occurs when the total load on the span divided by the number of panels in the span, $\frac{P_T^-}{n}$, equals the load, $P^-_{(j-1)-(j)} = P_B^-$, in panel (j – 1) – (j).

The criteria of Equations 4.12c and 4.13c determine the location of live loads for maximum positive and negative panel shear, from which maximum and minimum forces in web members are determined. For a live load consisting of concentrated loads, the criteria of Equations 4.12c and 4.13c must be refined for various locations of the live load by an informed trial-and-error approach. The criteria for the position of live loads for the maximum and minimum forces in vertical web members at panel point (j − 1) or (j) that are directly loaded by floorbeam reactions[82] is the live load position creating the maximum moment at panel point (j − 1) or (j), respectively, in the floor system.[83]

Figure 4.7 is an elevation of the floor system in a truss with concentrated live loads transferred to the truss panel points as floorbeam reactions, $P_{f(i)}$. The moments about panel points (k), (k − 1) and (k + 1) are

$$M_{(k)} = \sum_{i=0}^{i=k} P_{f(i)} x(i) \tag{4.14a}$$

$$M_{(k-1)} = \sum_{i=0}^{i=k} P_{f(i)} \left(x(i) - s_{(k)} \right) = M_{(k)} - s_{(k)} \sum_{i=0}^{i=k} P_{f(i)} \tag{4.14b}$$

$$M_{(k+1)} = M_{(k)} + s_{(k+1)} \sum_{i=0}^{i=k} P_{f(i)} + P_{f(k)} s_{(k+1)} \tag{4.14c}$$

Solving Equations 4.14a, 4.14b and 4.14c, for $P_{f(k)}$ yields,

$$P_{f(k)} = \frac{M_{(k-1)} - M_{(k)}}{s_{(k)}} + \frac{M_{(k+1)} - M_{(k)}}{s_{(k+1)}} \tag{4.14d}$$

which for equal panel lengths[84] $d = s_i$ is

$$P_{f(k)} = \frac{M_{(k-1)} - 2M_{(k)} + M_{(k+1)}}{d} \tag{4.14e}$$

Since the floorbeam reaction or panel point force is applied between two panels of equal length, d, Equation 4.14e reduces to,

$$P_{f(k)} = \frac{2M_F}{d} \tag{4.14f}$$

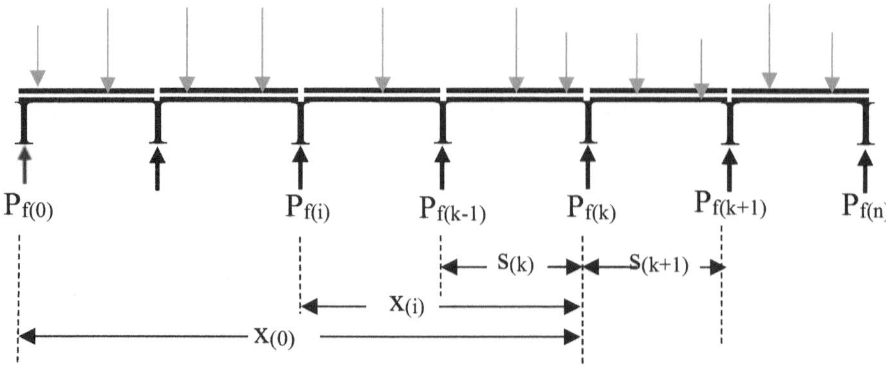

FIGURE 4.7 Concentrated live loads traversing a truss floor system

where $M_F = M_{(k+1)}/2 - M_{(k)}$ = the moment at the floorbeam in a simply supported beam with span length of 2d.

The location of the live load for maximum force in vertical web members directly loaded by floorbeams is the criteria for position for maximum moment at panel point (k). Therefore, Equation 4.14f, in conjunction with the location criteria for maximum moment of Equation 4.11c, can be used to determine the forces in truss vertical web members directly loaded by floorbeams.

4.4.2.2.1 Cooper's E Series Live Load Analysis

As shown in Chapter 3, the Cooper's E design live load recommended by AREMA (2024) consists of a series of concentrated live loads followed by a uniformly distributed live load.[85] The design live load positions are required for maximum panel point moments (for maximum chord member forces and maximum forces in vertical web members that are directly loaded by floorbeam reactions), and maximum positive and negative panel shear forces (for maximum and minimum web member forces).

Table 4.1 provides general guidance on axle or wheel location for maximum panel point moments with Cooper's E live load moving from right to left over typical railway truss spans. The criterion of Equation 4.11c may then be used to further confirm the axle or wheel load locations for maximum panel point moments at panel point (k).

The maximum positive and negative panel shear forces created by Cooper's E live load on typical railway truss spans will occur with the load partially on the span and the leading axles or wheels near the front of the Cooper's E design live load series positioned adjacent to the truss panel under consideration. Therefore, for Cooper's E design live loads, maximum shear forces in panel (j − 1) − (j), where $j \le n/2$, typically occur with wheels 2, 3 or 4 at panel point (j) for positive shear and at panel point (j − 1) for negative shear. Consequently, for typical panel lengths,[86] P^+_F and P^-_F, in Figure 4.6b and Figure 4.6c, respectively, are typically zero.

Since truss panels are relatively short, the position of the Cooper's E live load for maximum panel moment, and the associated maximum forces in vertical web members directly loaded by floorbeams, can be established by the criterion of Equation 4.11c or by inspection.[87]

Once the locations of the live load for maximum panel point moment (Equation 4.11c) and panel shear forces (Equations 4.12c and 4.13c) are established, maximum and minimum member forces may be determined from an investigation of static equilibrium. When the live load consists of a specified series of concentrated forces combined with uniformly distributed loads, as shown in Figure 4.8, the calculation of panel point moments and panel shear forces may be facilitated through a tabular presentation of distances, forces, and moment summations.

From Figure 4.8, the moment, M_C, at location, C, due to moving concentrated and uniformly distributed loads is as follows:

TABLE 4.1
Location of Cooper's E Design Live Load Axles or Wheels for Maximum Moment at Panel Points

Location of Panel Point from Left End of Truss Span (m)	Location of Panel Point from Left End of Truss Span (ft)	Cooper's E Axle or Wheel at Panel Point with Load Moving to Left
4.5 to 9.1	15 to 30	3 or 4
10.7 to 13.7	35 to 45	5 or 6
15.2 to 16.8	50 to 55	7 or 8
18.3 to 19.8	60 to 65	8 or 9
21.2 to 27.4	70 to 90	10, 11 or 12
30.5 to 36.6	100 to 120	13, 14 or 15
36.6 to 42.7	120 to 140	17 or 18

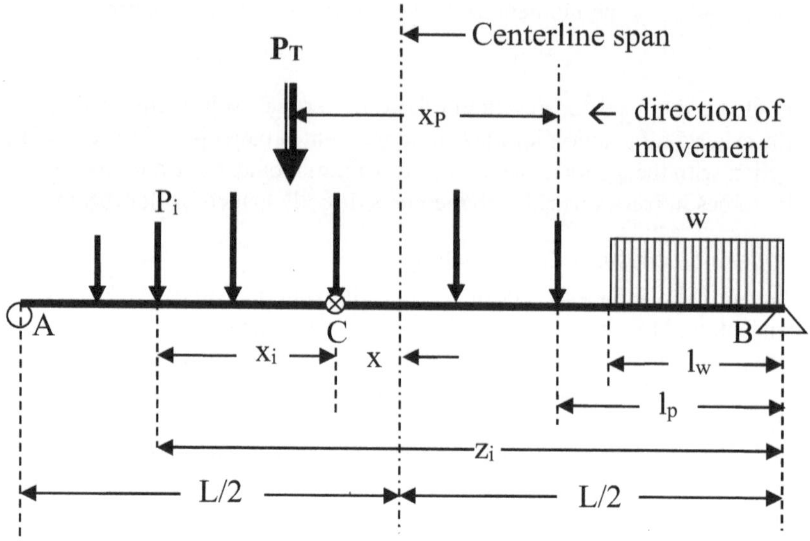

FIGURE 4.8 Live loads traversing a span.

$$M_C = R_A\left(\frac{L}{2} - x\right) - \Sigma P_i x_i \quad (4.15a)$$

where $\Sigma P_i x_i$ = the sum of moments due to live loads to the left of C.

The left reaction, R_A, is,

$$R_A = \frac{\Sigma P_i z_i + \frac{w l_w^2}{2}}{L} \quad (4.15b)$$

Substituting Equation 4.15b into Equation 4.15a yields

$$M_C = \left(\frac{\Sigma P_i z_i + \frac{w l_w^2}{2}}{L}\right)\left(\frac{L}{2} - x\right) - \Sigma P_i x_i \quad (4.15c)$$

The sum of the moments of concentrated loads about B is,

$$\Sigma P_i z_i = P_T\left(x_P\right) + (\Sigma P_i) l_P \quad (4.15d)$$

Substituting Equation 4.15d into Equation 4.15b and Equation 4.15c yields the following:

$$V_A = R_A = \frac{P_T\left(x_P\right) + (\Sigma P_i) l_P + \frac{w l_w^2}{2}}{L} \quad (4.15e)$$

and

TABLE 4.2
Moment Table for Cooper's E80 Wheel Loads

N	1	2	3	4	5	6	7	8	9	10	11	12	13	14	15	16	17	18
P	20	40	40	40	40	26	26	26	26	20	40	40	40	40	26	26	26	26
S_1	0	8	13	18	23	32	37	43	48	56	64	69	74	79	88	93	99	104
S_W	109	101	96	91	86	77	72	66	61	53	45	40	35	30	21	16	10	5
SP_1	20	60	100	140	180	206	232	258	284	304	344	384	424	464	490	516	542	568
SP_{18}	568	548	508	468	428	388	362	336	310	284	264	224	184	144	104	78	52	26
SM	0	320	840	1560	2480	3313	4273	5387	6640	7760	10320	13080	16040	19200	21467	23907	26467	29200
SM_{18}	32728	30548	26508	22668	19028	15588	13586	11714	9998	8412	7352	5552	3952	2552	1352	806	390	130
SM_{17}	29888	27808	23968	20328	16888	13648	11776	10034	8448	6992	6032	4432	3032	1832	832	416	130	
SM_{16}	27178	25198	21558	18118	14878	11838	10096	8484	7028	5702	4842	3442	2242	1242	442	156		
SM_{15}	24082	22222	18822	15622	12622	9822	8236	6780	5480	4310	3570	2410	1450	690	130			
SM_{14}	21632	19872	16672	13672	10872	8272	6816	5490	4320	3280	2640	1680	920	360				
SM_{13}	17456	15876	13036	10396	7956	5716	4494	3402	2466	1660	1200	600	200					
SM_{12}	15336	13856	11216	8776	6536	4496	3404	2442	1636	960	600	200						
SM_{11}	13416	12036	9596	7356	5316	3476	2514	1682	1006	460	200							
SM_{10}	11696	10416	8176	6136	4296	2656	1824	1122	576	160								
SM_9	9264	8144	6224	4504	2984	1664	1040	546	208									
SM_8	6992	6032	4432	3032	1832	832	416	130										
SM_7	5702	4842	3442	2242	1242	442	156											
SM_6	4310	3570	2410	1450	690	130												
SM_5	3280	2640	1680	920	360													
SM_4	1660	1200	600	200														
SM_3	960	600	200															
SM_2	460	200																
SM_1	160																	

LEGEND

N = wheel number
P = Cooper's E80 wheel load, kips, (1/2 axle load)
S_1 = distance, ft, from wheel N to wheel 1
S_W = distance, ft, from wheel N to uniform train load of 4,000 lb/ft
$S_1 + S_w = 109$ ft
SP_1 = sum of wheel loads, kips, between and including wheel 1to wheel N
SP_{18} = sum of wheel loads, kips, between and including wheel N to wheel 18
SM = sum of the moments, kips-ft, about wheel 1 of wheel loads between and including wheel 2 to wheel N
SM_{18}= sum of the moments, kips-ft, about the beginning of the uniform load of wheel loads between and including wheel N to wheel 18
SM_k = sum of the moments, kips-ft, about wheel load (k+1) of wheel loads between and including wheel N to wheel k where $k \leq 17$

TABLE 4.3
Moment Table for Cooper's EM360 Wheel Load

N	1	2	3	4	5	6	7	8	9	10	11	12	13	14	15	16	17	18
P	90	180	180	180	180	115	115	115	115	90	180	180	180	180	115	115	116	115
S_1	0.0	2.4	3.9	5.4	6.9	9.6	11.1	12.9	14.4	16.8	19.2	20.7	22.2	23.7	26.4	27.9	29.7	31.2
S_W	32.7	30.3	28.8	27.3	25.8	23.1	21.6	19.8	18.3	15.9	13.5	12.0	10.5	9.0	6.3	4.8	3.0	1.5
SP_1	90	270	450	630	810	925	1040	1155	1270	1360	1540	1720	1900	2080	2195	2310	2426	2541
SP_{18}	2541	2451	2271	2091	1911	1731	1616	1501	1386	1271	1181	1001	821	641	461	346	231	115
SM	0	439	1152	2139	3401	4543	5859	7387	9105	10641	14151	17936	21995	26328	29436	32782	36293	40040
SM_{18}	44878	41889	36349	31083	26092	21375	18630	16063	13710	11535	10081	7613	5419	3499	1854	1105	535	178
SM_{17}	40984	38131	32866	27874	23157	18715	16148	13759	11584	9588	8271	6077	4158	2512	1141	570	178	
SM_{16}	37267	34552	29561	24844	20401	16233	13844	11634	9637	7819	6640	4720	3074	1703	606	214		
SM_{15}	33022	30472	25809	21421	17308	13468	11293	9297	7514	5910	4895	3305	1988	946	178			
SM_{14}	29663	27249	22861	18748	14908	11343	9346	7528	5924	4498	3620	2304	1262	494				
SM_{13}	23936	21770	17875	14255	10910	7838	6162	4665	3381	2276	1645	823	274					
SM_{12}	21029	19000	15380	12034	8962	6165	4668	3349	2243	1316	823	274						
SM_{11}	18397	16504	13158	10087	7289	4766	3447	2306	1379	631	274							
SM_{10}	16038	14283	11211	8414	5891	3642	2501	1539	790	219								
SM_9	12703	11167	8535	6176	4092	2282	1426	749	285									
SM_8	9588	8271	6077	4158	2512	1141	570	178										
SM_7	7819	6640	4720	3074	1703	606	214											
SM_6	5910	4895	3305	1988	946	178												
SM_5	4498	3620	2304	1262	494													
SM_4	2276	1645	823	274														
SM_3	1316	823	274															
SM_2	631	274																
SM_1	219																	

LEGEND

N = wheel number
P = Cooper's EM360 wheel load, kN, (1/2 axle load)
S_1 = distance, m, from wheel N to wheel 1
S_W = distance, m, from wheel N to uniform train load of 60 kN/m
$S_1 + S_W = 32.7$ m
SP_1 = sum of wheel loads, kN, between and including wheel 1 to wheel N
SP_{18} = sum of wheel loads, kN, between and including wheel N to wheel 18
SM = sum of the moments, kN-m, about wheel 1 of wheel loads between and including wheel 2 to wheel N
SM_{18}= sum of the moments, kN-m, about the beginning of the uniform load of wheel loads between and including wheel N to wheel 18
SM_k = sum of the moments, kN-m, about wheel load (k+1) of wheel loads between and including wheel N to wheel k

$$M_C = \left(\frac{P_T\left(x_P\right) + (\Sigma P_i) l_P + \dfrac{w l_w^2}{2}}{L} \right) \left(\frac{L}{2} - x \right) - \Sigma P_i x_i \tag{4.15f}$$

Equations 4.15e and 4.15f illustrate that calculating the moment and shear force by static equilibrium at any location in the span due to concentrated and uniformly distributed live loads requires the following: the sum of the moments of all concentrated loads in front of, and about, the last concentrated load (at l_p from B in Figure 4.8) on the span, $P_T\ (x_P)$; the sum of all concentrated loads on the span, ΣP_i; and the sum of the moments about C of all concentrated loads in front of C, $\Sigma P_i x_i$.

Since the Cooper's E live load pattern is constant, it is efficient to develop a table to assist with the determination of the shear forces and moments using Equations 4.15e and 4.15f, respectively. Table 4.2 is developed for the wheel load (1/2 of axle load) of the Cooper's E80 live load and may be used to avoid protracted calculations of shear forces and moments associated with railway truss span analysis using US customary or imperial units. Table 4.3 is developed for the wheel load (1/2 of axle load) of the Cooper's EM360 live load and may be used to avoid lengthy calculations of shear forces and moments associated with railway truss span analysis using SI units.

Example 4.11 illustrates the use of mathematical criteria to determine the position of the Cooper's E80 live load for maximum panel point moment (Equation 4.11c) and maximum panel shear force (Equations 4.12c and Equation 4.13c) in the truss in Figure 4.1aa (a horizontal parallel chord Pratt through truss). Once these positions are established, Table 4.2 is used to determine the panel point moments and panel shear forces to establish the maximum and minimum member forces.[88] Equation 4.14f, in conjunction with the location criterion for maximum panel moment of Equation 4.11c, is used to determine the forces from the Cooper's E80 live load in truss vertical web members directly loaded by floorbeams.

Example 4.11

The 150 ft long six-panel horizontal parallel top chord Pratt through-truss in Figure 4.1aa with h = 35 ft is subjected to a uniform dead load of 1500 lb/ft and Cooper's E80 design live load. Calculate the maximum and minimum forces in the truss members. The Pratt truss member dead load forces are shown in Table E4.11a (see Examples 4.5 and 4.6).

TABLE E4.11a
Pratt Truss Member Dead Load Forces (Dead Load Apportioned to Bottom Chord)

Member		Dead Load Force (kips)
L0L1	Bottom chord	66.96
L1L2	Bottom chord	66.96
L2L3	Bottom chord	107.14
L0U1	End Post	−115.17
U1U2	Top chord	−107.14
U2U3	Top chord	−120.54
U1L1	Hanger	37.50
U2L2	Vertical	−18.75
U3L3	Vertical	0
U1L2	Diagonal	69.13
U2L3	Diagonal	23.04

All dead load is assumed to be apportioned to, and applied at, bottom chord panel points in Table E4.11a. When dead load is apportioned to both the bottom and top chord panel points, the dead load forces in vertical web members are affected. Truss vertical web members in compression will increase by the dead load apportioned to the top chord panel points and, conversely, the dead load forces in truss vertical web members in tension will decrease by the dead load apportioned to the top chord panel points. For example, if 1/3 of the dead load is apportioned to the top chord panel points, the truss member dead load forces are shown in Table E4.11b.

TABLE E4.11b
Pratt Truss Member Dead Load Forces (2/3 Dead Load Apportioned to Bottom and 1/3 to Top Chord Panel Points)

Member		Dead Load Force (kips)
L0L1	Bottom chord	66.96
L1L2	Bottom chord	66.96
L2L3	Bottom chord	107.14
L0U1	End post (top chord)	−115.17
U1U2	Top chord	−107.14
U2U3	Top chord	−120.54
U1L1	Hanger	25.00
U2L2	Vertical	−31.25
U3L3	Vertical	−12.50
U1L2	Diagonal	69.13
U2L3	Diagonal	23.04

Live load analysis of Pratt truss chord members:

Maximum live load forces in chord members U1U2, U2U3, L0L1, L1L2 and L2L3 are determined by considering moments about the opposite chord panel point that eliminate moments from the opposite chord and diagonal members in the panel and using Table 4.2 with the load positioned in accordance with the criterion of Equation 4.11c (live load location for maximum panel point moment).

From Table 4.1, the estimated location of Cooper's live load for maximum panel point moments at panel points L1, L2 and L3 are wheel 3 or 4 at panel point L1 (25 ft from left end of truss); wheel 7 or 8 at panel point L2 (50 ft from left end of truss) or wheel 10, 11 or 12 at panel point L3 (75 ft from left end of truss).

Member U1U2 (top chord in panel 1-2): Table 4.2 is used to determine moments about opposite panel point L2 (k = 2) with Cooper's E live load located in accordance with the criterion of Equation 4.11c as shown in Table E4.11c. If the value of $P_T(k/n)$ is between the sum of the wheel loads ahead of panel point L2, P_{Fa}, and the sum of the wheels loads from panel point L2 to L0, P_{Fb}, the criterion of Equation 4.11c is established and the wheel at L2 may be the position that creates the maximum moment at L2.

TABLE E4.11c
Criterion for Position of Live Load for Maximum Moment at Panel Point L2

	Axle at L2			
L2: k/n = 2/6 = 0.33	**6**	**7**	**8**	**9**
Distance to uniform load (ft)	77	72	66	61
Distance to axle no. 18 (ft)	72	67	61	56

TABLE E4.11c (*Continued*)
Criterion for Position of Live Load for Maximum Moment at Panel Point L2

	Axle at L2			
L2: k/n = 2/6 = 0.33	**6**	**7**	**8**	**9**
Last axle on span	18	18	18	18
Distance to axle no. 1 (ft)	32	37	43	48
First axle on span	1	1	1	1
Length of uniform load (ft)	100 − 77 = 23	100 − 72 = 28	100 − 66 = 34	100 − 61 = 39
Total load on span, P_T (kips)	568 + 23(4) = 660	568 + 28(4) = 680	568 + 34(4) = 704	568 + 39(4) = 724
P_T(k/n) (kips)	220	227	235	241
Sum of loads ahead of panel point L2, P_{Fa} (kips)	206 − 26 = 180	232 − 26 = 206	232	258
Sum of loads ahead of and including panel point L2, P_{Fb} (kips)	206	232	258	284
Criterion established	No	Yes	Yes	No

The criterion of Equation 4.11c indicates axles 7 or 8 at L2 for maximum panel point moment at L2, as shown in Table E4.11d. Panel point moments are determined from Table 4.2.

TABLE E4.11d
Moments at Panel Point L2 and Force in Member U1U2

	Axle at L2	
Moments at L2	**7**	**8**
Sum moments to uniform load (kip-ft)	32,728	32,728
Sum wheel loads to uniform load (kips)	568	568
Length of uniform load (ft)	28	34
Sum moments to L6 (kip-ft)	32,728+568(28) +4(28)²/2 = 50,200	32,728+568(34) +4(34)²/2 = 54,352
Reaction R_{y0} (kips)	50,200/150 = 334.67	54,352/150 = 362.35
Sum moments from first axle to axle at L2 (kip-ft)	4310	5702
Sum moments about L2 (kip-ft)	334.67(2)(25) − 4310 = 12,423	362.35(50) − 5702 = 12,415
Force in member U1U2 = F_{U1U2} = $-(\sum M_{L2})/h$ (kips)	−354.96	−354.73

$$F^{max}_{U1U2} = -354.96 \text{ kips}$$

Member U2U3 (top chord in panel 2-3): Table 4.2 is used to determine moments about opposite panel point L3 (k = 3) with Cooper's E live load located in accordance with Equation 4.11c as shown in Table E4.11e.

TABLE E4.11e
Criterion for Position of Live Load for Maximum Moment at Panel Point L3

	Axle at L3			
L3: k/n = 3/6 = 0.50	**10**	**11**	**12**	**13**
Distance to uniform load (ft)	53	45	40	35
Distance to axle no. 18 (ft)	48	40	35	30
Last axle on span	18	18	18	18
Distance to axle no. 1 (ft)	104 − 48 = 56	104 − 40 = 64	104 − 35 = 69	104 − 30 = 74
First axle on span	1	1	1	1
Length of uniform load on span (ft)	75 − 53 = 22	75 − 45 = 30	75 − 40 = 35	75 − 35 = 40
Total load on span, P_T (kips)	568 + 22(4) = 656	568 + 30(4) = 688	568 + 35(4) = 708	568 + 40(4) = 728
$P_T(k/n)$ (kips)	328	344	354	364
Sum of loads ahead of panel point L3, P_{Fa}(kips)	304 − 20 = 284	344 − 40 = 304	384 − 40 = 344	424 − 40 = 384
Sum of loads ahead of and including panel point L3, P_{Fb}(kips)	304	344	384	424
Criterion established	No	Yes	Yes	No

The criteria of Equation 4.11c indicates axles 11 or 12 at L3 for maximum panel point moment as shown in Table E4.11f. Panel point moments are determined from Table 4.2.

TABLE E4.11f
Moment at Panel Point L3 and Force in Member U2U3

	Axle at L3	
Moments at L3	**11**	**12**
Sum moments to uniform load (kip-ft)	32,728	32,728
Sum wheel loads to uniform load (kips)	568	568
Length of uniform load (ft)	30	35
Sum moments to L6 (kip-ft)	$32{,}728+568(30)+4(30)^2/2 = 51{,}568$	$32{,}728+568(35)+4(35)^2/2 = 53{,}858$
Reaction R_{y0} (kips)	51,568/150 = 343.79	53,858/150 = 359.05
Sum moments from first axle to axle at L3 (kip-ft)	11,696	13,416
Sum moments about L3 (kip-ft)	343.79(75)−11,696 = 14,088	359.05(75)−13,416 = 13,517
Force in member U2U3 (kips) = F_{U2U3} = $-(\sum M_{L3})/h$ (kips)	−402.52	−386.09

$$F^{max}_{U2U3} = -402.52 \text{ kips}$$

Members L0L1 and L1L2 (bottom chord in panels 0-1 and 1-2): Table 4.2 is used to determine moments about opposite panel point U1 (k = 1) with Cooper's E design live load located in accordance with Equation 4.11c as shown in Table E4.11g.

$$F_{L0L1} = F_{L1L2} \text{ (horizontal equilibrium at joint L1).}$$

TABLE E4.11g
Criterion for Position of Live Load for Maximum Moment at Panel Point U1

	Axle at L1		
U1: k/n = 1/6 = 0.167	**3**	**4**	**5**
Distance to uniform load (ft)	96	91	86
Distance to axle no. 18 (ft)	91	91–5=86	81
Last axle on span	18	18	18
Distance to axle no. 1 (ft)	104–91=13	104–86=18	104–81=23
First axle on span	1	1	1
Length of uniform load on span (ft)	125–96=29	125–91=34	125–86=39
Total load on span, P_T (kips)	568+29(4)=684	568+34(4)=704	568+39(4)=724
$P_T(k/n)$ (kips)	114	117	121
Sum of loads ahead of panel point L1, P_{Fa} (kips)	100–40=60	140–40=100	180–40=140
Sum of loads ahead of and including panel point L1, P_{Fb} (kips)	100	140	180
Criterion established	No	Yes	No

The criterion of Equation 4.11c indicates axle 4 at L1 for maximum panel point moment as shown in Table E4.11h. Panel point moments are calculated from Table 4.2.

TABLE E4.11h
Moment at Panel Point U1 and Force in Member L1L2

	Axle at L1
Moments at U1	**4**
Sum moments to uniform load (kip-ft)	32,728
Sum wheel loads to uniform load (kips)	568
Length of uniform load (ft)	34
Sum moments to L6 (kip-ft)	$32{,}728 + 568(34) + 4(34)^2/2 = 54{,}352$
Reaction R_{y0} (kips)	54,352/150 = 362.35
Sum moments from first axle to axle at L1 (kip-ft)	960
Sum moments about U1 (kip-ft)	362.35(25) – 960 = 8098
Force in member L1L2 (kips) = $F_{L1L2} = 8098/h = F_{L0L1}$	231.39

$$F^{max}_{L0L1} = F^{max}_{L1L2} = 231.39 \text{ kips}$$

Member L2L3 (bottom chord in panel 2-3): Table 4.2 is used to determine moments about opposite panel point U2 (k = 2) with Cooper's E design live load located in accordance with Equation 4.11c.

Maximum panel point moments at joint U2 are the same as that for L2 and, therefore, $F_{L2L3} = -F_{U1U2} = 354.96$ kips.

$$F^{max}_{L2L3} = 354.96 \text{ kips}$$

Maximum live load forces for the Pratt truss chord members are summarized in Table E4.11i.

TABLE E4.11i
Pratt Truss Chord Member Live Load Forces

Chord Member	Moment Center	Moment, $M_{(k)}$ (kip-ft)	Force (kips)
L1L2 = L0L1	U1 (k = 1)	8098	231.39 (T)
L2L3	U2 (k = 2)	12,423	354.96 (T)
U1U2	L2 (k = 2)	−12,423	−354.96 (C)
U2U3	L3 (k = 3)	−14,088	−402.52 (C)

Pratt truss end post member analysis (member L0U1):

Equilibrium at joint L0 provides that $F^{max}_{L0U1} = -1.720(F^{max}_{L0L1}) = -1.720(231.39) = -398.10$ kips

or $F^{max}_{L0U1} = -1.229(R_{y0} - W_{(1)}) = -1.229(362.35 - (960/25)) = -398.10$ kips

Live load analysis Pratt truss web members:

Maximum and minimum live load forces in web members U1L2, U2L3 and U2L2 that are not directly loaded by floorbeam reactions or are not zero-force members, in the Pratt truss are determined by considering positive and negative shear forces in panel (j − 1) − (j) with the load positioned in accordance with the criterion of Equations 4.12c and 4.13c, respectively (live load location for maximum panel shear forces). Vertical web members directly loaded by floorbeam reactions at panel point (k) use the criterion of Equation 4.11c (location for concentrated live loads for maximum moment) to determine maximum axial forces.

Member U1L2 (diagonal member in panel 1-2): Table 4.2 is used to determine positive and negative shear forces in panel 1-2 with Cooper's E live load located in accordance with Equation 4.12c for positive shear and Equation 4.13c for negative shear. If the value of P^{+}_{T}/n is between the sum of the wheel loads ahead of panel point L2, P^{+}_{Aa}, and the sum of the wheels loads from panel point L2 to L0, P^{+}_{Ab}, the criterion of Equation 4.12c is established and the wheel at L2 may be the position that creates the maximum positive shear in panel 1-2. Considering positive shear in Table E4.11j with the criterion of Equation 4.12c.

TABLE E4.11j
Criterion for Position of Live Load for Maximum Positive Shear in Panel 1-2

	Panel 1-2: Axle at L2			
L2, jd = 50 ft	2	3	4	5
Distance to axle no. 1 (ft)	8	13	18	23
First axle on span	1	1	1	1
Distance to axle no. 18 (ft)	96	91	86	81
Last axle on span	18	18	18	18
Distance of uniform load (ft)	100 − 96 − 5 = −1 (off span)	100 − 91 − 5 = 4	100 − 86 − 5 = 9	100 − 81 − 5 = 14
Total load on span, P^{+}_{T} (kips)	568	568 + 4(4) = 584	568 + 4(9) = 604	568 + 4(14) = 624
$P^{+}_{T}/n = P^{+}_{T}/6$ (kips)	95	97	101	104
Sum of loads ahead of panel point L2, P^{+}_{Aa} (kips)	20	60	100	140
Sum of loads ahead of and including panel point L2, P^{+}_{Ab} (kips)	60	100	140	180
Criterion established	No	Yes	Yes	No

The criterion of Equation 4.12c indicates axles 3 or 4 at L2 for maximum positive panel shear as shown in Table E4.11k. Therefore, there are no loads between panel point (0) and panel point (1). Table 4.2 and Equation 4.3j are used to determine panel shear forces.

TABLE E4.11k
Positive Shear Force in Panel 1-2 and Force in Member U1L2

	Axle at L2	
Panel 1-2: U1L2	**3**	**4**
Sum moments to L6 (kip-ft)	32,728 + 568(4) + 4(4)²/2 = 35,032	32,728 + 568(9) + 4(9)²/2 = 38,002
R_{y0} (kips)	35,032/150 = 233.55	38,002/150 = 253.35
Sum moments ahead of panel point L2 (kip-ft)	460	960
Panel load at L1 (kips)	460/25 = 18.40	960/25 = 38.40
Positive shear in panel 1-2 = $V^{+}_{(1)\text{-}(2)}$ (kips)	233.55 − 18.40 = 215.15	253.35 − 38.40 = 214.95
Maximum force in diagonal U1L2 = $F^{+}_{U1L2} = (V_{(1)\text{-}(2)})1.229$ (kips)	264.40	264.17

$$F^{+max}_{U1L2} = 264.40 \text{ kips}$$

The maximum negative live load shear in panel 1-2 is considered with the criterion of Equation 4.13c as shown in Table E4.11l.

TABLE E4.11l
Criterion for Position of Live Load for Maximum Negative Shear in Panel 1-2

	Panel 1-2: Axle at L1	
L1, jd= 25 ft	**2**	**3**
Distance to axle no. 1 (ft)	8	13
First axle on span	1	1
Distance to axle no. 18 (ft)	96	91
Last axle on span	6	7
Total load on span, P^{-}_{T} (kips)	206	232
$P^{-}_{T}/n = P^{-}_{T}/6$ (kips)	34	39
Sum of loads ahead of panel point L1, P^{-}_{Aa} (kips)	20	60
Sum of loads ahead of and including panel point L1, P^{-}_{Ab} (kips)	60	100
Criterion established	Yes	No

The criterion of Equation 4.13c indicates axle 2 at L1 for maximum negative shear in panel 1-2 as shown in Table E4.11m. Therefore, there are no loads between panel point

(2) and panel point (6). Table 4.2 and Equation 4.3k are used to determine panel shear forces.

TABLE E4.11m
Negative Shear Force in Panel 1-2 and Force in Member U1L2

Panel 1-2: U1L2	Axle at L1
	2
Sum moments to L0 (kip-ft)	3280 + 206(25 − (32 − 8)) = 3486
R_{y6} (kips)	3486/150 = 23.24
Sum moments ahead of panel point L1 (kip-ft)	160
Panel load at L2 (kips)	160/25 = 6.40
Negative shear in panel 1-2 = $V^{-}_{(1)-(2)}$ (kips)	23.24 − 6.40 = 16.84
Minimum force in diagonal U1L2 = F^{-max}_{U1L2} = $-(V^{-}_{(1)-(2)})1.229$ (kips)	−20.69

$$F^{-max}_{U1L2} = -20.69 \text{ kips}$$

Member U2L3 (diagonal member in panel 2-3): Table 4.2 is used to determine positive and negative shear forces in panel 2-3 with Cooper's E live load located in accordance with Equation 4.12c (Table E4.11n).

TABLE E4.11n
Criterion for Position of Live Load for Maximum Positive Shear in Panel 2-3

	Panel 2-3: Axle at L3		
L3: jd = 75 ft	**2**	**3**	**4**
Distance to axle no. 1 (ft)	8	13	18
First axle on span	1	1	1
Distance to axle no. 18 (ft)	96	91	86
Last axle on span	14	14 (15 at L6)	15 (16 at L6)
Distance of uniform load (ft)	75 – 96 – 5= −26 off span	off span	off span
Total load on span, P^{+}_{T} (kips)	464	464	490
P^{+}_{T}/n (kips)	77	77	82
Sum of loads ahead of panel point L3, P^{+}_{Aa} (kips)	20	60	100
Sum of loads ahead of and including panel point L3, P^{+}_{Ab} (kips)	60	100	140
Criterion established	No	Yes	No

The criterion of Equation 4.12c indicates axle 3 at L3 for maximum positive panel shear as shown in Table E4.11o. Therefore, there are no loads between panel point (0) and panel point (2). Table 4.2 and Equation 4.3j are used to determine panel shear forces.

TABLE E4.11o
Positive Shear in Panel 2-3 and Force In Member U2L3

	Axle at L3
Panel 2-3: U2L3	**3**
Sum moments to L6 (kip-ft)	17,456 + 464(9) = 21,632
R_{y0} (kips)	21,632/150 = 144.21
Sum moments ahead of panel point L3 (kip-ft)	460
Panel load at L2 (kips)	460/25 = 18.40
Positive shear in panel 2–3 = $V^{+}_{(2)\text{-}(3)}$ (kips)	144.21 − 18.40 = 125.81
Maximum force in diagonal U2L3 = F^{+}_{U2L3} = $(V^{+}_{(2)\text{-}(3)})1.229$ (kips)	154.62

$$F^{+max}_{U2L3} = 154.62 \text{ kips}$$

The maximum negative live load shear in panel 2-3 is considered as shown in Table E4.11p.

TABLE E4.11p
Criterion for Position of Live Load for Maximum Negative Shear in Panel 2-3

	Panel 2-3: Axle at L2	
L2, jd = 50 ft	**2**	**3**
Distance to axle no. 1 (ft)	8	13
First axle on span	1	1
Distance to axle no. 18 (ft)	96	91
Last axle on span	10	10
Distance to uniform load (ft)	50−96−5=−51 off span	off span
Total load on span, P^{-}_{T} (kips)	304	304
P^{-}_{T}/n (kips)	51	51
Sum of loads ahead of panel point L2, P^{-}_{Aa} (kips)	20	60
Sum of loads ahead of and including panel point L4, P^{-}_{Ab} (kips)	60	100
Criterion established	Yes	No

The criterion of Equation 4.13c indicates axle 2 at L2 for maximum negative shear in panel 2-3 as shown in Table E4.11q. Therefore, there are no loads between panel point (3) and panel point (6). Table 4.2 and Equation 4.3k are used to determine panel shear forces.

TABLE E4.11q
Negative Shear in Panel 2-3 and Force in Member U2L3

	Axle at L2
Panel 2-3: U2L3	**2**
Sum moments to L0 (kip-ft)	9264 + 304(50 − (56 − 8)) = 9872
R_{yn} (kips)	9872/150 = 65.81

(Continued)

TABLE E4.11q (*Continued*)
Negative Shear in Panel 2-3 and Force in Member U2L3

	Axle at L2
Panel 2-3: U2L3	**2**
Sum moments ahead of panel point L2 (kip-ft)	160
Panel load at L3 (kips)	160/25 = 6.40
Negative shear in panel 2–3 = $V^{-}_{(2)-(3)}$ (kips)	65.81 − 6.40 = 59.41
Minimum force in diagonal U2L3 = $F^{-max}_{U2L3} = -(V^{-}_{(2)-(3)})1.229$ (kips)	−73.01

$$F^{-max}_{U2L3} = -73.01 \text{ kips}$$

Member U1L1 (vertical hanger member at joint 1): Table 4.2 is used to determine moments about panel point L1 with Cooper's E live load located in accordance with Equation 4.11c with k = 1.

The force in member U1L1 is equal to the joint load at L1, which is the floorbeam reaction from the two adjacent 25 ft spans. The maximum floorbeam reaction at L1 may by computed by Equation 4.14f with maximum $M_{(1)}$ from loads on the 50 ft span from L0 to L2.

Inspection of the Cooper's E live load wheel configuration reveals it is likely that wheels 3, 4 or 5 and wheels 12, 13 or 14 create maximum moment in a 50 ft span.

The location of the load for maximum $M_{(1)}$ is, as shown in Table E4.11r.

TABLE E4.11r
Criterion for Position of Live Load for Maximum Moment at Panel Point L1

	Axle at L1			
L1: k/n = 1/2	**3**	**4**	**5**	**6**
Last axle on span of length 2d = 50 ft	7	7 (8 at L2)	8 (9 at L2)	10
Distance to axle no. 1 (ft)	13	18	23	32
First axle on span of length 2d	1	1	1	2
Total load on span, P_T (kips)	232	232	258	284
$P_T/2$ (kips)	116	116	129	142
Sum of loads ahead of panel point L1, P_{Fa} (kips)	60	100	140	160
Sum of loads ahead of and including panel point L1, P_{Fb} (kips)	100	140	180	186
Criterion established	No	Yes	No	No

The criterion of Equation 4.11c indicates axle 4 at L1 for maximum moment and floorbeam reaction as shown in Table E4.11r. The location of the load for maximum $M_{(1)}$ is as shown in Table E4.11s.

TABLE E4.11s
Criterion for Position of Live Load for Maximum Moment at Panel Point L1

	Axle at L1		
L1: 1/2	**12**	**13**	**14**
Last axle on span of length 2d = 50 ft	16	16 (17 at L2)	17 (18 at L2)
Distance to axle no. 1 (ft)	69	74	79

TABLE E4.11s (*Continued*)
Criterion for Position of Live Load for Maximum Moment at Panel Point L1

	Axle at L1		
L1: 1/2	**12**	**13**	**14**
First axle on span of length 2d	9	10	10
Total load on span, P_T (kips)	516−258=258	516−284=232	542−284=258
$P_T(0.50)$ (kips)	129	116	129
Sum of loads ahead of panel point L1, P_{Fa} (kips)	112−40=86	140−40=100	180−40=140
Sum of loads ahead of and including panel point L1, P_{Fb} (kips)	384−258=126	424−284=140	464−284=180
Criterion established	No	Yes	No

The criteria of Equation 4.11c indicates axle 13 at L1 for maximum moment and floorbeam reaction as shown in Table E4.11s.

It should be noted that, on a 50 ft span, axle 4 and axle 13 create the same live load pattern. Equation 4.14f and Table 4.2 are used to determine panel point moments and floorbeam reaction at joint L1 as shown in Table E4.11t.

TABLE E4.11t
Moment at Panel Point L1 and Floorbeam Reaction

	Axle at L1	
Moments at L1	**4**	**13**
Sum of moments to L2 = $M_{(2)}$ (kip-ft)	5702	5702
Reaction R_{y0} for span of length 2d = 50 ft (kips)	5702/50 = 114.04	114.04
Sum of moments from first axle to axle at L1 (kip-ft)	−960	−960
Sum of moments about L1 = $M_{(1)}$ (kip-ft)	114.04(25) − 960 = 1891	114.04(25) − 960 = 1891
Floorbeam reaction at L1 = $F_{U1L1} = P_{f(1)} = \frac{2M_{(1)}}{25}$ (kips)	2(1891)/(25) = 151.28	2(1891)/(25) = 151.28

Wheel configurations of wheel 4 and wheel 13 at L1 in the 50 ft long span L0-L2 are the same and, consequently, F^{+max}_{U1L1} = 151.28 kips.

If the Alternate Live Load recommended by AREMA (2024) is considered by placing one of the central 50 kip wheel loads at L1,

F_{U1L1} = (50/25)(14 + 19 + 25 + 20) = 156 kips, which indicates that the Alternate Live Load is equal to the effects of Cooper's E(156/151.28) = E82.5 live load in bending on a 50 ft span.

Member U2L2 (vertical member at joint 2): Table 4.2 is used to determine positive and negative shear forces in panel 2-3 with Cooper's E design live load located in accordance with Equations 4.12c and 4.13c, respectively.

From Equations 4.3n and 4.3o, Table E4.11o and Table E4.11q,

$$F^{+max}_{U2L2} = -V^{+max}_{(2)-(3)} = -125.81 \text{ kips}$$

$$F^{-max}_{U2L2} = V^{-max}_{(2)-(3)} = 59.41 \text{ kips}$$

Member U3L3 (vertical member at joint 3): Considering equilibrium of joint U3,

$$F_{U3L3} = 0 \text{ kips}$$

Web member maximum positive and maximum negative live load forces are shown in Table E4.11u.

TABLE E4.11u
Truss Web Member Live Load Forces

		Live Load Panel Shear (kips)		Live Load Force (kips)	
Web Member	**Panel**	**Positive**	**Negative**	**Positive**	**Negative**
L0U1 (end post)	0-1	323.95	0	−398.10	0
U1L2 (diagonal)	1-2	214.95	−16.84	264.40	−20.69
U2L3 (diagonal)	2-3	125.81	−59.41	154.62	−73.01
U1L1 (hanger)				156.00	-
U2L2 (vertical)	2-3	125.81	−59.41	−125.81	59.41
U3L3 (vertical)				0	-

Table E4.11v shows the dead load and maximum live load member forces for the Pratt truss in Figure 4.1a.

TABLE E4.11v
Dead Load and Maximum Cooper's E80 Live Load Member Forces

Through Pratt Truss Member		Dead Load Force (kips)	Maximum Live Load Force (kips)	Maximum Live Load Force due to Positive Shear F^{+max} (kips)	Maximum Live Load Force due to Negative Shear F^{-max} (kips)
L0L1	Bottom chord	66.96	231.39	-	-
L1L2	Bottom chord	66.96	231.39	-	-
L2L3	Bottom chord	107.14	354.96	-	-
L0U1	End Post	−115.17	−398.10	-	-
U1U2	Top chord	−107.14	−354.96	-	-
U2U3	Top chord	−120.54	−402.52	-	-
U1L1	Vertical	37.50	156.00	-	-
U2L2	Vertical	−18.75	-	−125.81	59.41
U3L3	Vertical	0	0	-	-
U1L2	Diagonal	69.13	-	264.40	−20.69
U2L3	Diagonal	23.04	-	154.64	−73.01

Table E4.11w shows the maximum and minimum member forces from Cooper's E80 live load.

TABLE E4.11w
Maximum and Minimum Member Forces from Cooper's E80 live load

Through Pratt Truss Member		Maximum Force (kips)	Minimum Force (kips)
L0L1	Bottom chord	298.35	66.96
L1L2	Bottom chord	298.35	66.96

TABLE E4.11w (*Continued*)
Maximum and Minimum Member Forces from Cooper's E80 live load

Through Pratt Truss Member		Maximum Force (kips)	Minimum Force (kips)
L2L3	Bottom chord	462.10	107.14
L0U1	End Post	−513.27	−115.17
U1U2	Top chord	−462.10	−107.14
U2U3	Top chord	−523.06	−120.54
U1L1	Vertical hanger	193.50	37.50
U2L2	Vertical	−144.56	40.66
U3L3	Vertical	0	0
U1L2	Diagonal	333.53	48.44
U2L3	Diagonal	177.68	−49.97

Vertical member U2L2 and diagonal member U2L3 experience stress reversal from live load (excluding live load impact).

Example 4.12

The 150 ft long six-panel horizontal parallel top chord Warren through truss in Figure 4.1ab with h = 35 ft is subjected to a uniform dead load of 1500 lb/ft and Cooper's E80 design live load. Determine the maximum and minimum live load forces in the truss members U2U3, L2L3 and L2U3 using the criteria for load locations of Equation 4.11c (for chord members U2U3 and L2L3), and Equations 4.12c and 4.13c (for diagonal web member L2U3). Table 4.2 can be used to determine panel moments and panel shear forces.

For the live load analysis of Warren truss chord members, maximum live load forces in chord members U2U3, and L2L3 are calculated by considering moments about the opposite chord panel point that eliminates moments from the opposite chord and diagonal members in the panel and using Table 4.2 with the load positioned in accordance with the criteria of Equation 4.11c.

Member U2U3 (top chord in panel 2-3): Table 4.2 is used to determine moments about opposite panel point L2 with Cooper's E live load located in accordance with Equation 4.11c with k = 2.

$$\text{From Table E4.11d, } M^{max}_{L2} = 12{,}423 \text{ kip-ft}$$

$$F^{max}_{U2U3} = -12{,}423/35 = -354.96 \text{ kips}$$

Member L2L3 (bottom chord in panel 2-3): Table 4.2 is used to determine moments about opposite panel point U3 with Cooper's E live load located in accordance with Equation 4.11c 11c with k = 3. Maximum panel point moments at joint U3 are the same as that for L3 in Example 4.11 for a Pratt truss.

$$\text{From Table E4.11f, } M^{max}_{U3} = 14{,}088 \text{ kip-ft}$$

$$F^{max}_{L2L3} = 14{,}088/35 = 402.52 \text{ kips}$$

Member L2U3 (diagonal member in panel 2-3): Table 4.2 is used with Equations 4.3l and 4.3m to determine positive and negative shear forces in panel 2-3 with Cooper's E live load located in accordance with Equations 4.12c and 4.13c, respectively. From Example 4.11,

$$V^{+max}_{(2)-(3)} = 125.81 \text{ kips (Table E4.11o)}$$

$$F^{+max}_{L2U3} = -125.81\ (1.229) = -154.64 \text{ kips (Equation 4.3l)}$$

$$V^{-max}_{(2)-(3)} = 59.41 \text{ kips (Table E4.11q)}$$

$$F^{-max}_{L2U3} = 59.41\ (1.229) = 73.01 \text{ kips (Equation 4.3m)}$$

Mathematical criteria, as shown in Examples 4.6 to 4.9, 4.11, and 4.12, are useful for establishing the positions of concentrated live loads to determine the maximum and minimum forces in truss members. Once the live load locations are established, calculating the maximum and minimum member forces for concentrated loads such as Cooper's E series live load is straightforward. Nevertheless, for more general loadings, influence lines provide a means of efficiently determining the maximum and minimum live load forces in truss members.

4.4.2.3 Analysis by Influence Lines

Influence lines for axial forces in truss members illustrate the variation in the axial force in the member as a unit load traverses the truss span.[89] Influence lines are constructed for moving load analysis of statically determinate trusses[90] by moving a unit concentrated load across the truss span and determining the value of the axial force in the truss member under consideration at each location of the unit load. Fortunately, constructing the influence lines is facilitated by determining the value of the member axial force at locations where changes in the influence line will obviously occur[91] and joining the member axial forces determined at those locations with straight lines.[92] Consequently, since trusses are loaded at the panel points, influence lines for truss members can be readily constructed by the judicious placement of unit loads at relevant panel points. The influence lines for the moment within panel (j – 1) – (j) at distance x_m from location (j – 1) and at panel point location (j) are shown in Figure 4.9.

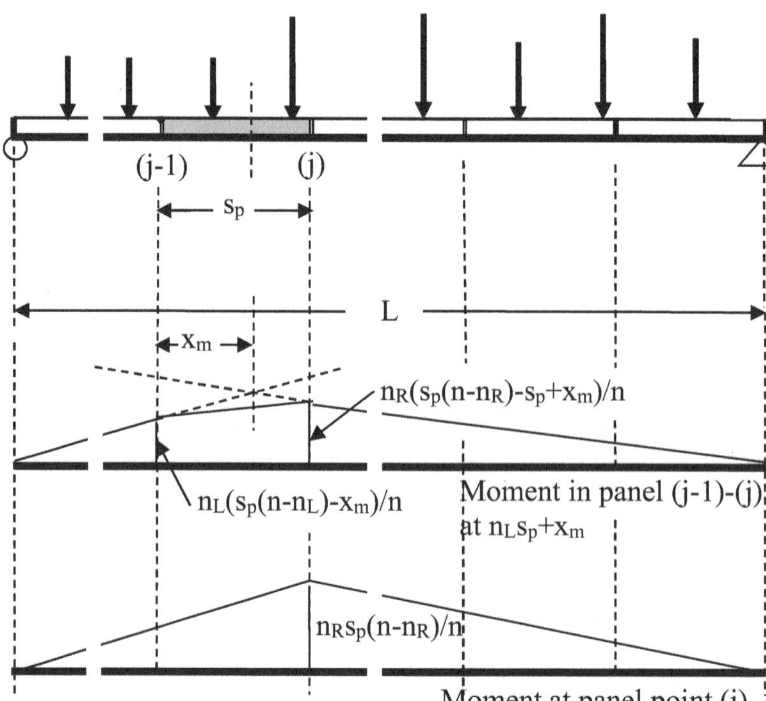

FIGURE 4.9 Influence lines for moment in panels and at panel points.

The influence lines are developed by calculating the bending moments at locations (j – 1) and (j) with a unit load placed successively at locations (j – 1) and (j) where

L = the length of the span = $n(s_p) = (n_L + n_R + 1)(s_p)$
n = the number of equal length panels in the span
n_L = the number of panels left of panel point (j – 1)
n_R = the number of panels right of panel point (j)
s_p = the length of the equal length panels in the span
x_m = the distance from panel point (j – 1) to the location where moment in the panel is required

The influence line for shear force in panel (j – 1) – (j) is shown in Figure 4.10.

The influence line is developed by determining the shear force at locations (j – 1) and (j) with a unit load placed successively at locations (j – 1) and (j) where

x_V = the distance from panel point (j – 1) to the location where shear in the panel is null $= \dfrac{s_p}{\left(1+\left(\dfrac{n_R}{n_L}\right)\right)}$

The influence line for floorbeam reaction at location (j), assuming simply supported stringer spans each side of joint (j), is shown in Figure 4.11.

The influence line is developed by determining the shear forces at locations (j – 1), (j) and (j + 1) with a unit load placed successively at locations (j – 1), (j) and (j + 1). The influence lines developed for panel point moment and panel shear force are useful in constructing influence lines for axial forces in truss chord and web members, respectively.

Influence lines for truss chord members are constructed by considering free body diagrams and equilibrium of moments with unit loads placed at pertinent panel points. Influence lines for truss web members are enabled by considering free body diagrams and equilibrium of horizontal and vertical forces with unit loads placed at significant panel points. Examples 4.13 and 4.14 outline the development of influence lines for chord and web members using the influence lines for panel point moment and panel shear force; respectively, they calculate influence lines for 150 ft long and 25 ft high Pratt and Warren through-truss spans. Practical Pratt and Warren through-truss spans can be taller (see Examples 4.4 to 4.12) to allow for lateral bracing in the plane of the top chord member. Nevertheless, the influence line calculations in Examples 4.13 and 4.14 illustrate the technique and method for establishing influence lines for truss spans.

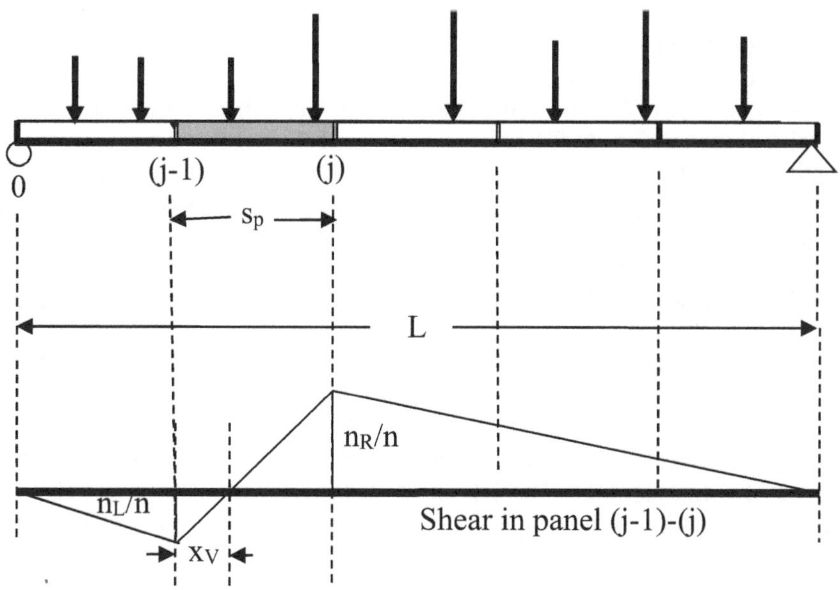

FIGURE 4.10 Influence lines for shear force in panels.

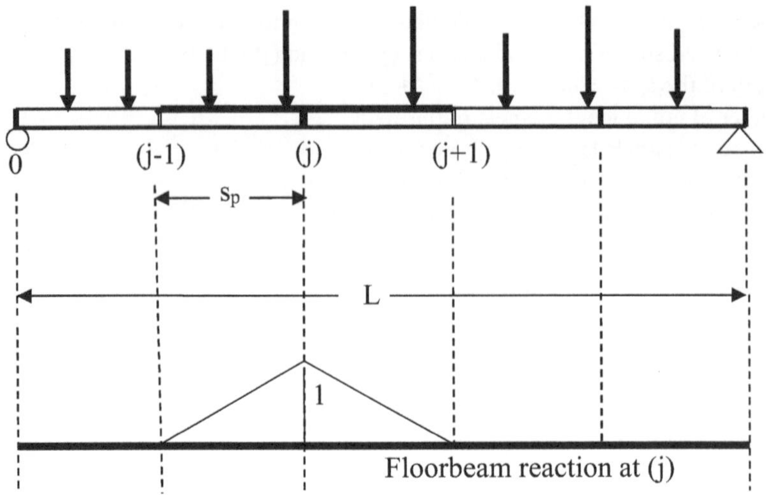

FIGURE 4.11 Influence line for floorbeam reaction.

Example 4.13

Calculate the influence lines for members of the Pratt through-truss span in Figure E4.13a.

Constructing the influence line for axial forces in member U1U2 of the Pratt truss in Figure E4.13a requires a section through panel 1-2 and equilibrium of moments about L2 with unit loads at L1 and L2. The influence line is shown in Figure E4.13b.

Constructing the influence line for axial forces in member L1L2 requires a section through panel 1-2 and equilibrium of moments about U1 with unit loads at L1 and L2. The influence line is shown in Figure E4.13c.

Constructing the influence line for axial forces in member U1L2 requires a section through panel 1-2 and horizontal force equilibrium with unit loads at L1 and L2. The influence line is shown in Figure E4.13d.

Constructing the influence line for axial forces in member U2U3 requires a section through panel 2-3 and equilibrium of moments about L3 with unit loads at L2 and L3. The influence line is shown in Figure E4.13e.

Constructing the influence line for axial forces in member L2L3 requires a section through panel 2-3 and equilibrium of moments about U2 with unit loads at L2 and L3. The influence line is shown in Figure E4.13f.

Constructing the influence line for axial forces in member U2L3 requires a section through panel 2-3 and horizontal force equilibrium with unit loads at L2 and L3. The influence line is shown in Figure E4.13g.

Constructing the influence line for axial forces in member U2L2 requires a section through U2L2 and adjacent chords with vertical force equilibrium in panel 2-3 with unit loads at L2 and L3. The influence line is shown in Figure E4.13h.

Constructing the influence line for axial forces in member L0U1 requires a section through panel 0-1 and horizontal force equilibrium with unit loads at L0 and L1. The influence line is shown in Figure E4.13i.

Constructing the influence line for axial forces in member U1L1 requires consideration of the panel point load at joint L1. The influence line is shown in Figure E4.13j.

Constructing the influence line for U3L3 is unnecessary as the axial force is zero due to the truss geometry at joint U3.

Constructing the influence line for L0L1 is unnecessary as the axial force is equal to that of L1L2 due to the truss geometry at joint L1.

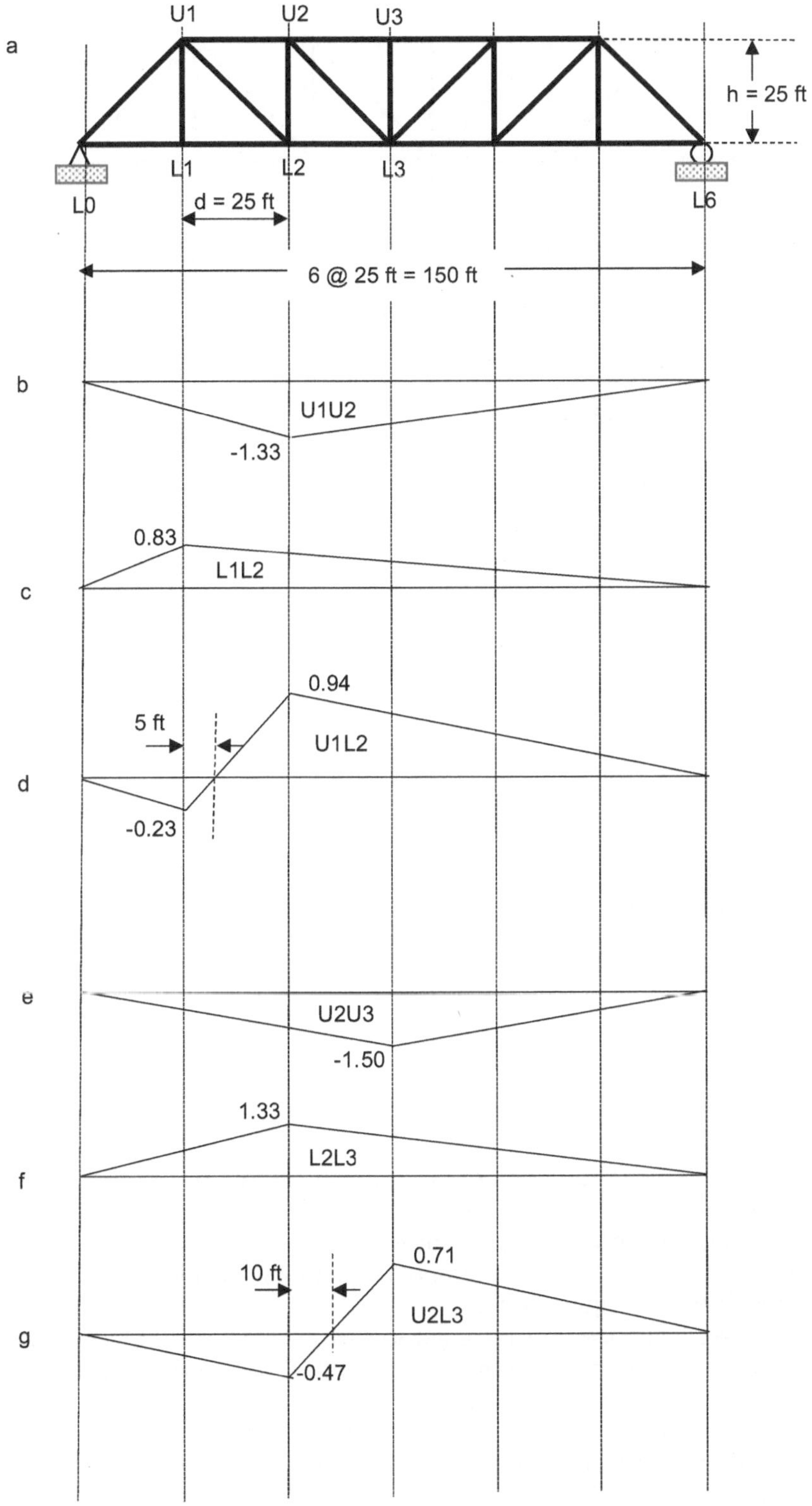

FIGURE E4.13 Influence lines for members of a Pratt truss span.

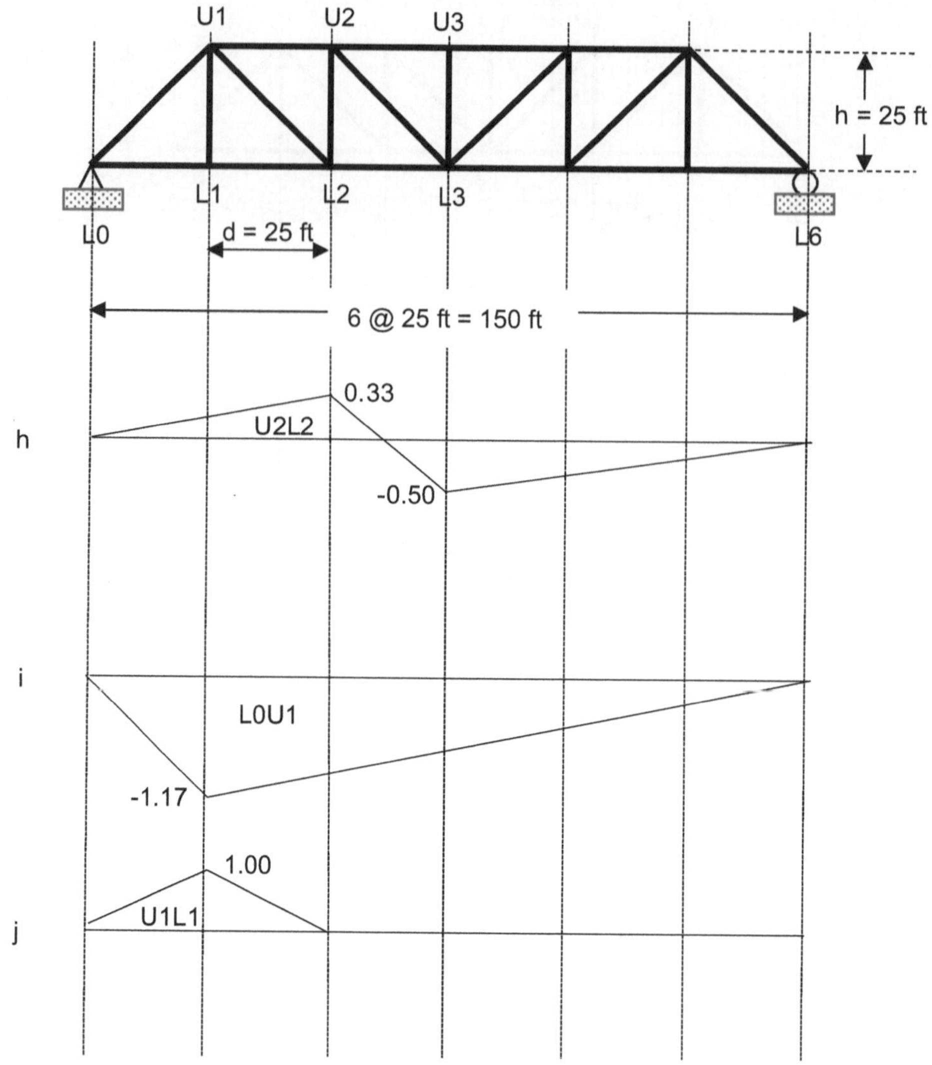

FIGURE E4.13 (Continued)

Example 4.14

Develop the influence lines for members U2U3, L2L3 and L2U3 of the Warren through-truss span in Figure E4.14a.

Constructing the influence line for axial forces in member U2U3 requires a section through panel 2-3 and equilibrium of moments about L2 with unit loads at L2 and L3. The influence line is shown in Figure E4.14b.

Constructing the influence line for axial forces in member L2L3 requires a section through panel 2-3 and equilibrium of moments about U3 with unit loads at L2 and L3. The influence line is shown in Figure E4.14c.

Constructing the influence line for axial forces in member U2L3 requires a section through panel 2-3 and horizontal force equilibrium with unit loads at L2 and L3. The influence line is shown in Figure E4.14d.

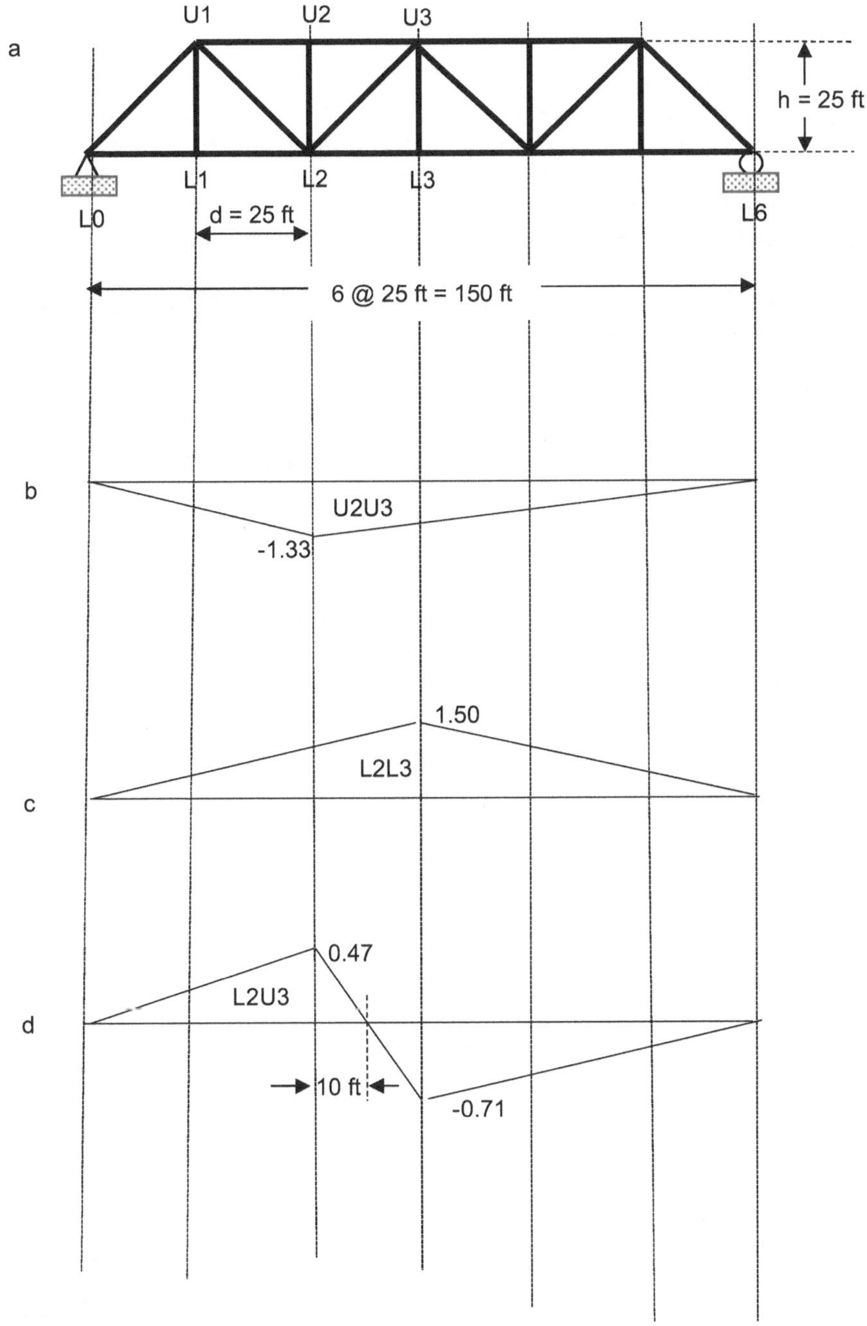

FIGURE E4.14 Influence lines for members of a Warren through-truss span.

The influence lines for members of the Pratt and Warren through-truss spans developed in Examples 4.13 and 4.14 can be used to determine the member forces resisting dead loads, uniformly distributed live loads and concentrated live loads.

The location of a uniformly distributed load for maximum member forces is visually evident from influence line diagrams. Member forces from uniformly distributed loads are determined from the area under the influence line as outlined in Examples 4.15 and 4.16 for 150 ft long and 25 ft high Pratt and Warren through-truss spans, respectively.[93]

Example 4.15

The 150 ft long six-panel horizontal parallel chord Pratt through-truss with h = 25 ft (Figure E4.13a) is subjected to a uniform dead load of 1500 lb/ft and uniform live load of 4000 lb/ft. Find the maximum and minimum forces in the truss members. Chord member forces are greatest when the span is fully loaded by the live load, and web member forces will be greatest with a partial live load on the span.

Combined dead load and live load force in member U1U2:

From Figure E4.13b, F_{U1U2} = 1.5(−1.33)(50/2 + 100/2) + 4.0(−1.33)(50/2 + 100/2) = −150.00 − 400.00 = −550.00 kips.

Combined dead load and live load force in member L1L2:

From Figure E4.13c, F_{L1L2} = 1.5(0.83)(25/2 + 125/2) + 4.0(0.83)(25/2 + 125/2) = 93.75 + 250.00 = 343.75 kips.

Combined dead load and live load force in member U1L2:

$x_V = \dfrac{25}{\left(1+\left(\frac{4}{1}\right)\right)}$ = 5 ft. From Figure E4.13d, Positive live load shear determines that

F^{+max}_{U1L2} = 1.5(−0.23)(25/2 + 5/2) +1.5(0.94)(20/2 + 100/2) + 4.0(0.94)(20/2 + 100/2) = 79.43 + 225.60 = 305.03 kips. Negative live load shear determines that F^{-max}_{U1L2} = 79.41 + 4.0(−0.23)(25/2 + 5/2) = 79.41 − 13.75 = 65.66 kips.

Combined dead load and live load force in member U2U3:

From Figure E4.13e, F_{U2U3} = 1.5(−1.50)(75) + 4.0(−1.50)(75) = −168.75 − 450.00 = −618.75 kips.

Combined dead load and live load force in member L2L3:

From Figure E4.13f, F_{L2L3} = 1.5(1.33)(50/2 + 100/2) + 4.0(1.33)(50/2 + 100/2) = 150.00 + 400.00 = 550.00 kips.

Combined dead load and live load force in member U2L3:

$x_V = \dfrac{25}{\left(1+\left(\frac{3}{2}\right)\right)}$ = 10 ft. From Figure E4.13g, positive live load shear indicates that

F^{+max}_{U2L3} = 1.5(−0.47)(50/2 + 10/2) +1.5(0.71)(15/2 + 75/2) + 4.0(0.71) (15/2 + 75/2) = 26.78 + 127.80 = 154.58 kips. Negative live load shear indicates that F^{-max}_{U2L3} = 26.78 + 4.0(−0.47)(50/2 + 10/2) = 26.78 − 56.40 = −29.62 kips.

Combined dead load and live load force in member U2L2:

From Figure E4.13h, positive live load shear indicates that F^{+max}_{U2L2} = 1.5(0.33) (50/2 + 10/2) +1.5(−0.50)(15/2 + 75/2) + 4.0(−0.50)(15/2 + 75/2) = −18.75 − 90.00 = −108.75 kips. Negative live load shear indicates that F^{-max}_{U2L2} = −18.75 + 4.0(0.33)(50/2 + 10/2) = −18.75 + 40.00 = 21.25 kips.

Combined dead load and live load force in member L0U1:

From Figure E4.13i, $F_{L0U1} = 1.5(-1.17)(25/2 + 125/2) + 4.0(-1.17)(25/2 + 125/2) = -131.63 - 351.00 = -482.63$ kips.

Combined dead load and live load force in member U1L1:

From Figure E4.13j, $F_{U1L1} = 1.5(1.0)(25/2)(2) + 4.0(1.0)(25/2)(2) = 37.50 + 100.00 = 137.50$.

Table E4.15a is a summary of the maximum member forces in the Pratt truss in Figure E4.13a loaded by dead load and a uniformly distributed live load.

TABLE E4.15a
Dead Load and Maximum Live Load Member Forces for a Uniformly Distributed Live Load Using Influence Lines

Pratt Through-Truss Member		Dead Load Force (kips)	Maximum Live Load Force (kips)	Maximum Live Load Force due to Positive Shear, F^{+max} (kips)	Maximum Live Load Force due to Negative Shear, F^{-max} (kips)
L0L1	Bottom chord	93.75	250.00	-	-
L1L2	Bottom chord	93.75	250.00	-	-
L2L3	Bottom chord	150.00	400.00	-	-
L0U1	End post	−131.63	−351.00	-	-
U1U2	Top chord	−150.00	−400.00	-	-
U2U3	Top chord	−168.75	−450.00	-	-
U1L1	Vertical hanger	37.50	100.00	-	-
U2L2	Vertical	−18.75	-	−90.00	40.00
U3L3	Vertical	0	0	-	-
U1L2	Diagonal	79.41	-	225.60	−13.75
U2L3	Diagonal	26.78	-	127.80	−56.40

Table E4.15b is a summary of the maximum and minimum member forces in the Pratt truss in Figure E4.13a.

TABLE E4.15b
Maximum and Minimum Member Forces in the Pratt Truss in Figure E4.13a Loaded by Dead Load and a Uniformly Distributed Live Load

Pratt Through-Truss Member		Minimum Force (kips)	Maximum Force (kips)
L0L1	Bottom chord	93.75	343.75
L1L2	Bottom chord	93.75	343.75
L2L3	Bottom chord	150.00	550.00
L0U1	End post	-131.63	-482.63
U1U2	Top chord	-150.00	-550.00
U2U3	Top chord	-168.75	-618.75
U1L1	Vertical hanger	37.50	137.50
U2L2	Vertical	21.25	-108.75
U3L3	Vertical	0	0
U1L2	Diagonal	65.66	305.01
U2L3	Diagonal	-29.62	154.58

Example 4.16

The 150 ft long six-panel horizontal parallel-chord Warren through-truss with h = 25 ft (Figure E4.14a) is subjected to a uniform dead load of 1500 lb/ft and uniform live load of 4000 lb/ft. Find the maximum and minimum forces in the truss members U2U3, L2L3 and L2U3. Chord member forces are greatest when the span is fully loaded by the live load and web member forces will be greatest with a partial live load on the span.

Combined dead load and live load force in member U2U3:

From Figure E4.14b, $F_{U2U3} = 1.5(-1.33)(75) + 4.0(-1.33)(75) = -150.00 - 400.00 = -550.00$ kips.

Combined dead load and live load force in member L2L3:

From Figure E4.14c, $F_{L2L3} = 1.5(1.50)(75) + 4.0(1.50)(75) = 168.75 + 450.00 = 618.75$ kips.

Combined dead load and live load force in member L2U3:

$x_V = \dfrac{25}{\left(1+\left(\dfrac{3}{2}\right)\right)} = 10$ ft. From Figure E4.14d, Positive live load shear indicates that

$F^{+max}_{U2L3} = 1.5(0.47)(50/2 + 10/2) + 1.5(-0.71)(15/2 + 75/2) + 4.0(-0.71)(15/2 + 75/2) = -26.78 - 127.80 = -154.58$ kips. Negative live load shear indicates that $F^{-max}_{U2L3} = -26.78 + 4.0(0.47)(50/2 + 10/2) = -26.78 + 56.40 = 29.62$ kips.

Table E4.16a is a summary of the maximum member forces in the Warren truss in Figure E4.14a loaded by dead load and a uniformly distributed live load.

TABLE E4.16A
Dead Load and Maximum Live Load Member Forces in Panel 2-3 in the Warren Truss in Figure E4.14a

Warren Through-Truss Member		Dead Load Force (kips)	Maximum Live Load Force (kips)	Maximum Live Load Force due to Positive Shear F^{+max} (kips)	Maximum Live Load Force due to Negative Shear F^{-max} (kips)
L2L3	Bottom chord	168.75	450.00	-	-
U2U3	Top chord	−150.00	−400.00	-	-
L2U3	Diagonal	−26.78	-	−127.80	56.40

Table E4.16b is a summary of the minimum and maximum member forces in panel 2-3 in the Warren truss in Figure E4.14a.

TABLE E4.16B
Maximum and Minimum Member Forces in the Warren Truss in Figure E4.14a Loaded by Dead Load and a Uniformly Distributed Live Load

Warren Through-Truss Member		Minimum Force (kips)	Maximum Force (kips)
L2L3	Bottom chord	168.75	618.75
U2U3	Top chord	−150.00	-550.00
L2U3	Diagonal	−154.58	29.62

The precise location of the Cooper's E live load for maximum member forces is not readily observable from influence line diagrams, and mathematical criteria for the location of the live load for maximum and minimum member forces may be required. Member forces from Cooper's E live load are determined from the ordinates of the influence lines at the locations of the concentrated loads and the area under the influence lines for uniformly distributed loads as outlined in Examples 4.17 and 4.18 for 150 ft long and 25 ft high Pratt and Warren through-truss spans, respectively.[94]

Example 4.17

The 150 ft long six-panel horizontal parallel-chord Pratt through-truss with h = 25 ft (Figure 4.13a) is subjected to a design live load of Cooper's E80. Determine the maximum forces in the members due to live load using the criterion for live load location of Equation 4.11c (for chord members) and Equations 4.12c and 4.13c (for web members) in conjunction with influence lines to determine truss member forces.

For the live load force in member U1U2, the criterion of Equation 4.11c with k = 2 indicates axle 7 at L2 for maximum moment. From Figure E4.13b,

$$F_{U1U2} = [20(0.26) + 40(0.42 + 0.52 + 0.62 + 0.72) + 26(0.90 + 1.00) + 26(0.94 + 0.89) + 20(0.81) + 40(0.73 + 0.68 + 0.63 + 0.58) + 26(0.49 + 0.44 + 0.38 + 0.33) + 4(0.28)(28/2)](-1.33) = -495.7 \text{ kips.}$$

For the live load force in L1L2, the criterion of Equation 4.11c with k = 1 indicates axle 4 at L1 for maximum moment. From Figure E4.13c,

$$F_{L1L2} = [20(0.28) + 40(0.60 + 0.80 + 1.00) + 40(0.96) + 26(0.89 + 0.84 + 0.80 + 0.76) + 20(0.70) + 40(0.63 + 0.59 + 0.55 + 0.51) + 26(0.44 + 0.40 + 0.35 + 0.31) + 4(0.27)(34/2)](0.83) = 322.2 \text{ kips.}$$

For the positive live load force in U1L2, the criterion of Equation 4.12c indicates axle 3 at L2 for maximum shear. From Figure E4.13d,

Positive shear indicates that $F^{+max}_{U1L2} = [20(0.35) + 40(0.75 + 1.00 + 0.95 + 0.90) + 26(0.81 + 0.76 + 0.70 + 0.65) + 20(0.57) + 40(0.49 + 0.44 + 0.39 + 0.34) + 26(0.25 + 0.20 + 0.14 + 0.09) + 4(0.4)(2/2)](0.94) = 303.4$ kips.

For the negative live load force in U1L2, the criterion of Equation 4.13c indicates axle 2 at L1 for maximum shear. From Figure E4.13d,

Negative shear indicates that $F^{-max}_{U1L2} = [40(1.00 + 0.80 + 0.60 + 0.40) + 26(0.04)](-0.23) + [20(0.15)](0.94) = -26.0 + 2.8 = -23.2$ kips.

For the live load force in U2U3, the criterion of Equation 4.11c with k = 3 indicates axle 11 at L3 for maximum moment. From Figure E4.13e,

$$F_{U2U3} = [20(0.15) + 40(0.25 + 0.32 + 0.39 + 0.45) + 26(0.57 + 0.64 + 0.72 + 0.79) + 20(0.89) + 40(1.00 + 0.93 + 0.87 + 0.80) + 26(0.68 + 0.61 + 0.53 + 0.47) + 4(0.40)(30/2)](-1.5) = -563.2 \text{ kips.}$$

For the live load force in L2L3, from Figures E4.13b and E4.13f,

$$F_{L2L3} = -F_{U1U2} = 495.7 \text{ kips.}$$

For the positive live load force in U2L3, the criterion of Equation 4.12c indicates axle 3 at L3 for maximum shear. From Figure E4.13g,

Positive shear indicates that $F^{+max}_{U2L3} = [20(0.14) + 40(0.67 + 1.00 + 0.93 + 0.87) + 26(0.75 + 0.68 + 0.60 + 0.53) + 20(0.43) + 40(0.32 + 0.25 + 0.19 + 0.12) + 26(0)](0.71) = 178.9$ kips.

For the negative live load force in U2L3, the criterion of Equation 4.13c indicates axle 2 at L2 for maximum shear. From Figure E4.13g,

Negative shear indicates that $F^{-max}_{U2L3} = [20(0.20) + 40(1.00 + 0.90 + 0.80 + 0.70) + 26(0.52 + 0.42 + 0.30 + 0.20) + 20(0.04)](-0.47) = -83.8$ kips.

For the live load force in U2L2, from Figures E4.13g and E4.13h,

Positive shear indicates that $F^{+max}_{U2L2} = -F^{+max}_{U2L3}(0.33/0.47) = -F^{+max}_{U2L3}(0.50/0.71) = -178.9(0.70) = -125.2$ kips.
Negative shear indicates that $F^{-max}_{U2L2} = -F^{-max}_{U2L3}(0.33/0.47) = -F^{-max}_{U2L3}(0.50/0.71) = 83.8(0.70) = 58.8$ kips.

For the live load force in L0U1, the criterion of Equation 4.12c indicates axle 4 at L1 for maximum shear. From Figure E4.13i,

$F_{L0U1} = [20(0.28) + 40(0.60 + 0.80 + 1.00 + 0.96) + 26(0.89 + 0.85 + 0.80 + 0.76) + 20(0.70) + 40(0.63 + 0.59 + 0.55 + 0.51) + 26(0.44 + 0.40 + 0.35 + 0.31) + 4(0.27)(34/2)](-1.17) = -454.4$ kips.

For the live load force in U1L1, from Figure E4.13j,

$F_{U1L1} = [20(0.28) + 40(0.60 + 0.80 + 1.00 + 0.80) + 26(0.44 + 0.24 + 0)](1.00) = 151.3$ kips, not including forces due to alternate live load

TABLE E4.17
Maximum Live Load Member Forces for Cooper's E80 Live Load on Pratt Through-Truss Using Influence Lines

Pratt Through-Truss Member		Maximum Live Load Force (kips)	Maximum Live Load Force due to Positive Shear F^{+max} (kips)	Maximum Live Load Force due to Negative Shear F^{-max} (kips)
L0L1	Bottom chord	322.2	-	-
L1L2	Bottom chord	322.2	-	-
L2L3	Bottom chord	495.7	-	-
L0U1	End post	−454.4	-	-
U1U2	Top chord	−495.7	-	-
U2U3	Top chord	−563.2	-	-
U1L1	Vertical hanger (not including alternate live load)	151.3	-	-
U2L2	Vertical	-	−125.2	58.8
U3L3	Vertical	0	-	-
U1L2	Diagonal	-	303.4	−23.2
U2L3	Diagonal	—	178.9	−83.8

Table E4.17 shows the maximum member forces from Cooper's E80 live load for the Pratt through-truss in Figure 4.13a.

Example 4.18

The 150 ft long six-panel horizontal parallel-chord Warren through truss with h = 25 ft (Figure 4.14a) is subjected to a design live load of Cooper's E80. Determine the maximum forces in members U2U3, L2L3 and L2U3 due to live load using the criteria for live load location of Equation 4.11c (for chord members) and Equations 4.12c and 4.13c (for web members) in conjunction with influence lines.

For the live load force in member U2U3, the criterion of Equation 4.11c with k = 2 indicates axle 7 at L2 for maximum moment. From Figure E4.14b,

$$F_{U2U3} = [20(0.26) + 40(0.42 + 0.52 + 0.62 + 0.72) + 26(0.90 + 1.00) + 26(0.94 + 0.89) + 20(0.81) + 40(0.73 + 0.68 + 0.63 + 0.58) + 26(0.49 + 0.44 + 0.38 + 0.33) + 4(0.28)(28/2)](-1.33) = -495.7 \text{ kips.}$$

For the live load force in L2L3, the criterion of Equation 4.11c with k = 3 indicates axle 11 at L3 for maximum moment. From Figure E4.14c,

$$F_{L2L3} = [20(0.15) + 40(0.25 + 0.32 + 0.39 + 0.45) + 26(0.57 + 0.64 + 0.72 + 0.79) + 20(0.89) + 40(1.00 + 0.93 + 0.87 + 0.80) + 26(0.68 + 0.61 + 0.53 + 0.47) + 4(0.40)(30/2)](1.5) = 563.2 \text{ kips.}$$

For the positive live load force in L2U3, the criterion of Equation 4.12c indicates axle 3 at L3 for maximum shear. From Figure E4.14d,

Positive shear indicates that $F^{+max}{}_{U2L3} = [20(0.14) + 40(0.67 + 1.00 + 0.93 + 0.87) - 26(0.75 + 0.68 + 0.60 + 0.53) - 20(0.43) - 40(0.32 + 0.25 + 0.19 + 0.12) - 26(0)](-0.71) = -178.9$ kips.

For the negative live load force in L2U3, the criterion of Equation 4.13c indicates axle 2 at L2 for maximum shear. From Figure E4.14d,

Negative shear indicates that $F^{-max}{}_{U2L3} = [20(0.20) + 40(1.00 + 0.90 + 0.80 + 0.70) + 26(0.52 + 0.42 + 0.30 + 0.20) + 20(0.04)](0.47) = 83.8$ kips.

Table E4.18 shows the maximum member forces from Cooper's E80 live load for the Warren through-truss in Figure 4.14a.

TABLE E4.18
Maximum Live Load Member Forces for Cooper's E80 Live Load on a Warren Through-Truss Using Influence Lines

Warren Through-Truss Member		Maximum Live Load Force (kips)	Maximum Live Load Force due to Positive Shear F^{+max} (kips)	Maximum Live Load Force due to Negative Shear F^{-max} (kips)
L2L3	Bottom chord	563.2	-	-
U2U3	Top chord	−495.7	-	-
L2U3	Diagonal	—	−178.9	+83.8

Examples 4.15 and 4.16 illustrate the effectiveness of influence line analysis for calculating truss member forces that resist uniformly distributed live loads. Examples 4.17 and 4.18 demonstrate the benefits of influence line analysis for calculating truss member forces that resist concentrated live loads. It is because of this efficacy that influence lines are used in algorithms for structural analysis in computer software.

4.5 STATICALLY DETERMINATE INCLINED CHORD TRUSS RAILWAY SPANS

Trusses with inclined chords enable longer spans. The fabrication geometry is more difficult, but even for moderate spans, reducing the chord and web member sizes[95] can decrease the overall fabrication and erection costs. Consequently, inclined-chord Pratt[96] and Warren trusses with an even number of panels[97] are often used for relatively long[98] simple-span railway truss superstructures.

The method of joints and method of sections are also useful for the structural analysis of statically determinate inclined chord trusses with single diagonal members in each panel and an even number of panels and joints.[99] However, as shown in Figure 4.12, the method of sections is particularly useful in establishing maximum and minimum forces in chord and web members of inclined-chord truss spans.

Taking a section through panel (j – 1) – (j)[100] of an inclined-chord Pratt (Parker) truss, as shown in Figure 4.12a, and taking moments about opposite chord panel point U(j – 1) yields the force in the bottom chord member L(j – 1)L(j) as

$$F_{L(j-1)L(j)} = \frac{\Sigma M_{U(j-1)}}{h_{(j-1)}} = \left((j-1)R_{y0} - \sum_{i=0}^{j-1}(j-1-i)W_i\right)\frac{d}{h_{(j-1)}} \tag{4.16a}$$

and taking moments about opposite chord panel point L(j) provides the force in the top chord member U (j – 1)U(j) as

$$F_{U(j-1)U(j)} = \frac{-\Sigma M_{L(j)}}{h_{(j)}} \frac{\sqrt{\left(h_{(j)} - h_{(j-1)}\right)^2 + d^2}}{d}$$

$$= -\left((j)R_{y0} - \sum_{i=0}^{j}(j-i)W_i\right)\left(\frac{\sqrt{\left(h_{(j)} - h_{(j-1)}\right)^2 + d^2}}{h_{(j)}}\right) \tag{4.16b}$$

where

$$R_{y0} = \sum_{i=0}^{n} \frac{(n-i)d}{nd} W_i = \sum_{i=0}^{n} \frac{(n-i)}{n} W_i$$

n = number of equal length panels between panel point (j = 0) and panel point (j = n), W_i = external load applied at joint (i) and d = panel length for trusses with equal length panels.

The free body diagrams in Figure 4.12 also indicate that the shear forces in truss panels (j – 1) – (j) are resisted by the vertical components of diagonal web members U(j – 1)L(j) or L(j – 1) U(j) and inclined chord member U(j – 1)U(j). Therefore, calculating the magnitude of the web

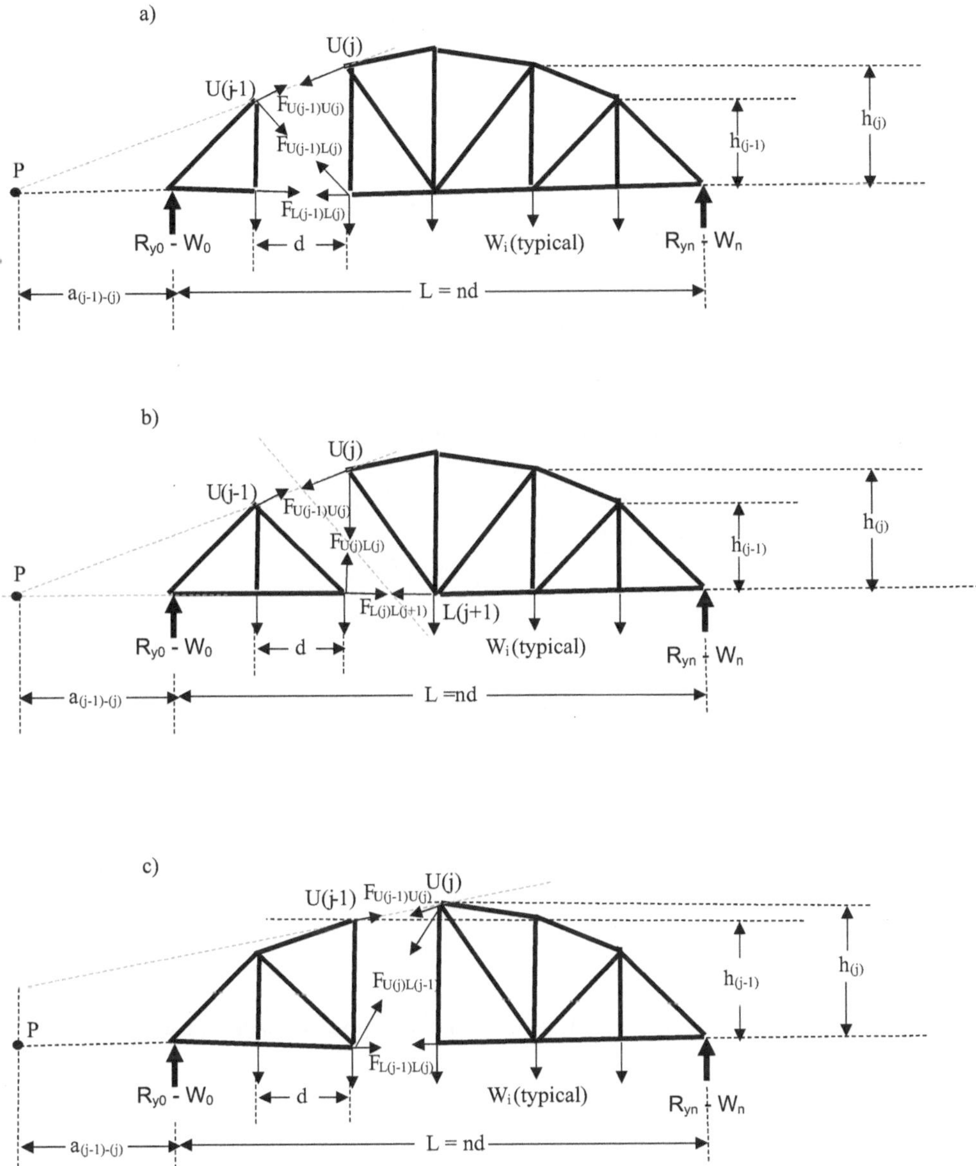

FIGURE 4.12 Free body diagrams of sections through inclined-chord Pratt (Parker) and Warren through-truss spans.

member forces from the panel shear forces requires knowledge of the simultaneous vertical component of the inclined chord force, $F^{V}_{U(j-1)U(j)}$, in the panel. Nevertheless, to obviate this requirement, a location can be established where at a section throughtruss panel (j – 1) – (j), there are no moments due to the inclined top chord member forces, allowing the forces in diagonal web members to be readily determined by the method of sections. This location, where there are no moments due to the inclined top chord member force, $F_{U(j-1)U(j)}$, is the point P intersected by lines coincident with the inclination of the top chord member and the bottom chord member at a section in panel (j – 1) – (j), as shown in Figure 4.12.

Taking a section through panel (j – 1) – (j) of an inclined chord Pratt (Parker) truss, as shown in Figure 4.12a and considering equilibrium of forces for positive panel shear, yields

$$\begin{aligned}\sum M_P = 0 &= -R_{y0}\left(a_{(j-1)-(j)}\right) + \sum_{i=0}^{j-1} W_i\left(a_{(j-1)-(j)} + id\right)\\ &\quad + F^{V+}_{U(j-1)L(j)}\left(a_{(j-1)-(j)} + (j-1)d\right) + F^{H+}_{U(j-1)L(j)}\left(h_{(j-1)}\right)\\ &= -R_{y0}\left(a_{(j-1)-(j)}\right) + \sum_{i=0}^{j-1} W_i\left(a_{(j-1)-(j)} + id\right)\\ &\quad + F^{V+}_{U(j-1)L(j)}\left(a_{(j-1)-(j)} + (j-1)d\right) + F^{V+}_{U(j-1)L(j)}(d)\ ,\end{aligned}$$

and the vertical component of the force in diagonal member U(j-1)L(j) due to positive panel shear is[101]

$$F^{V+}_{U(j-1)L(j)} = \frac{V^{+}_{(j-1)-(j)}\left(a_{(j-1)-(j)}\right) - M^{+0}_{(j-1)-(j)}}{\left(a_{(j-1)-(j)} + jd\right)} \tag{4.17a}$$

where

$V^{+}_{(j-1)-(j)}$ = the positive shear force in panel (j – 1) – (j) = $R_{y0} - \sum_{i=0}^{j-1} W_i$

$M^{+0}_{(j-1)-(j)}$ = the moment about L0 from joint loads R_{y0} and W_i between L0 and panel point (j-1) $= \left(\sum_{i=0}^{j-1} iW_i\right)d$

$a_{(j-1)-(j)}$ = the distance from panel point L0 to P at the location intersected by lines coincident with the inclination of U(j – 1)U(j) and L(j – 1)(j) = $\left(\left(\dfrac{h_{(j)}}{h_{(j)} - h_{(j-1)}}\right) - j\right)d.$

Therefore, the force in diagonal member U(j – 1)L(j) due to positive panel shear is

$$F^{+}_{U(j-1)L(j)} = \left(\frac{V^{+}_{(j-1)-(j)}\left(a_{(j-1)-(j)}\right) - M^{+0}_{(j-1)-(j)}}{\left(a_{(j-1)-(j)} + jd\right)}\right)\alpha_{(j-1)-(j)} \tag{4.17b}$$

where

$$\propto_{(j-1)-(j)} = \frac{\sqrt{h^2_{(j-1)} + d^2}}{h_{(j-1)}}$$

Taking a section through panel (j – 1) – (j) of an inclined-chord Pratt (Parker) truss, as shown in Figure 4.12a and considering equilibrium of forces for negative panel shear, yields

$$\sum M_P = 0 = -R_{yn}\left(a_{(j-1)-(j)} + nd\right) + \sum_{i=j}^{n} W_i\left(a_{(j-1)-(j)} + id\right) - F^{V-}_{U(j-1)L(j)}\left(a_{(j-1)-(j)} + jd\right),$$

and the vertical component of the force in diagonal member U(j-1)L(j) due to negative panel shear is

$$F^{V-}_{U(j-1)L(j)} = \frac{-V^{-}_{(j-1)-(j)}\left(a_{(j-1)-(j)}\right) + M^{-0}_{(j-1)-(j)}}{\left(a_{(j-1)-(j)} + jd\right)} \tag{4.17c}$$

where

$R_{yn} = \sum_{i=0}^{n} \frac{iW_i}{n}$

$V^{-}_{(j-1)-(j)}$ = the negative shear force in panel (j − 1) − (j) = $R_{yn} - \sum_{i=j}^{n} W_i$

$M^{-0}_{(j-1)-(j)}$ = the moment about L0 from loads R_{yn} and W_i between Ln and panel point (j)

$= -R_{yn}(nd) + \left(\sum_{i=j}^{n} iW_i\right)d = -\left(nR_{yn} - \sum_{i=j}^{n} iW_i\right)d$

Therefore, the force in diagonal member U(j − 1)L(j) due to negative panel shear is

$$F^{-}_{U(j-1)L(j)} = \left(\frac{-V^{-}_{(j-1)-(j)}\left(a_{(j-1)-(j)}\right) + M^{-0}_{(j-1)-(j)}}{\left(a_{(j-1)-(j)} + jd\right)}\right)\alpha_{(j-1)-(j)} \tag{4.17d}$$

Taking a section through top chord member U(j − 1)U(j) and bottom chord member L(j)L(j + 1) such that the vertical member U(j)L(j), where j < n/2, of an inclined-chord Pratt (Parker) truss is intersected as shown in Figure 4.12b, and considering equilibrium of forces for positive panel shear, yields[102] $\sum M_P = 0 = -F^{+}_{U(j)L(j)}\left(a_{(j-1)-(j)} + jd\right) - V^{+}_{(j)-(j+1)}\left(a_{(j-1)-(j)}\right) + M^{+0}_{(j)-(j+1)}$, and the force in vertical member U(j)L(j) due to positive panel shear is

$$F^{+}_{U(j)L(j)} = \left(\frac{-V^{+}_{(j)-(j+1)}\left(a_{(j-1)-(j)}\right) + M^{+0}_{(j)-(j+1)}}{\left(a_{(j-1)-(j)} + jd\right)}\right) \tag{4.18a}$$

where

$V^{+}_{(j)-(j+1)}$ = the positive shear force in panel (j) − (j + 1) = $R_{y0} - \sum_{i=0}^{j} W_i$

$M^{+0}_{(j)-(j+1)}$ = the moment about L0 from loads R_{y0} and W_i between L0 and joint (j)

$= \left(\sum_{i=0}^{j} iW_i\right)d.$

Taking a section through top chord member U(j − 1)U(j) and bottom chord member L(j)L(j + 1) such that the vertical member U(j)L(j), where j < n/2, of an inclined chord Pratt (Parker) truss is intersected as shown in Figure 4.12b, and considering equilibrium of forces for negative panel shear, yields

$$\sum M_P = 0 = F^{-}_{U(j)L(j)}\left(a_{(j-1)-(j)} + jd\right) - V^{-}_{(j)-(j+1)}\left(a_{(j-1)-(j)}\right) - M^{-0}_{(j)-(j+1)}$$

The force in vertical member U(j)L(j) due to negative panel shear is

$$F^{-}_{U(j)L(j)} = \left(\frac{V^{-}_{(j)-(j+1)}\left(a_{(j-1)-(j)}\right) - M^{-0}_{(j)-(j+1)}}{\left(a_{(j-1)-(j)} + jd\right)}\right) \tag{4.18b}$$

where

$V^{-}_{(j)-(j+1)}$ = the negative shear force in panel (j) – (j + 1) = $R_{yn} - \sum_{i=j+1}^{n} W_i$

$M^{-0}_{(j)-(j+1)}$ = the moment about L0 from loads R_{yn} and W_i between L(n) and joint (j + 1)

$= -\left(nR_{yn} - \sum_{i=j+1}^{n} iW_i\right)d$

An inclined-chord Warren truss has alternating panels in which the diagonal member in even-numbered panels, where j ≤ n/2, is oriented such that it is U(j – 1)L(j) and the diagonal in odd-numbered panels, where j ≤ n/2, is oriented such that it is L(j – 1)U(j). The forces in top chord, bottom chord and diagonal web members in panels of an inclined-chord Warren truss where the diagonal members are oriented as in a Parker truss will have the same magnitude and attitude (axial tension or compression) as those members in a Parker truss. However, the forces in top chord, bottom chord and diagonal web members in panels of an inclined-chord Warren truss where the diagonal members are opposite to the orientation of the diagonal members in a Parker truss will not have the same magnitude or, in the case of diagonal web members, the same attitude as those members in a Parker truss.

Taking a section through panel (j – 1) – (j) of an inclined-chord Warren truss, as shown in Figure 4.12c, and taking moments about opposite chord panel point U(j) yields the force in bottom chord member L(j – 1)L(j) as

$$F_{L(j-1)L(j)} = \left((j)R_{y0} - \sum_{i=0}^{j}(j-i)W_i\right)\left(\frac{d}{h_{(j)}}\right) = \left((j)\sum_{i=0}^{n}\frac{(n-i)}{n}W_i - \sum_{i=0}^{j}(j-i)W_i\right)\left(\frac{d}{h_{(j)}}\right) \quad (4.19a),$$

and taking moments about opposite panel point L(j-1), the force in the top chord member U(j – 1)U(j) is

$$F_{U(j-1)U(j)} = -\left((j-1)R_{y0} - \sum_{i=0}^{j-1}(j-1-i)W_i\right)\left(\frac{\sqrt{\left(h_{(j)} - h_{(j-1)}\right)^2 + d^2}}{h_{(j-1)}}\right)$$

$$= -\left((j-1)\sum_{i=0}^{n}\frac{(n-i)}{n}W_i - \sum_{i=0}^{j-1}(j-1-i)W_i\right)\left(\frac{\sqrt{\left(h_{(j)} - h_{(j-1)}\right)^2 + d^2}}{h_{(j-1)}}\right) \quad (4.19b)$$

Taking a section through panel (j – 1) – (j) of an inclined-chord Warren truss, as shown in Figure 4.12c, and considering equilibrium of forces yields the force in diagonal member L(j – 1) U(j) due to positive panel shear as

$$F^{+}_{L(j-1)U(j)} = \left(\frac{-V^{+}_{(j-1)-(j)}\left(a_{(j-1)-(j)}\right) + M^{+0}_{(j-1)-(j)}}{\left(a_{(j-1)-(j)} + (j-1)d\right)}\right)\alpha_{(j-1)-(j)} = F^{-}_{U(j-1)L(j)} \quad (4.20a)$$

where

$V^{+}_{(j-1)-(j)} = R_{y0} - \sum_{i=0}^{j-1} W_i$

$$M^{+0}_{(j-1)-(j)} = \left(\sum_{i=0}^{j-1} iW_i\right)d$$

$a_{(j-1)-(j)}$ = the distance from panel point L0 to P at the location intersected by lines coincident with the inclination of U(j-1)U(j) and L(j-1)(j) = $\left(\left(\frac{h_{(j)}}{h_{(j)} - h_{(j-1)}}\right) - (j)\right)d$

$$\propto_{(j-1)-(j)} = \frac{\sqrt{h^2_{(j)} + d^2}}{h_{(j)}}$$

Taking a section through panel (j – 1) – (j) of an inclined-chord Warren truss, as shown in Figure 4.12c, and considering equilibrium of forces yields the force in diagonal member L(j – 1) U(j) due to negative panel shear as

$$F^-_{L(j-1)U(j)} = \left(\frac{V^-_{(j-1)-(j)}\left(a_{(j-1)-(j)}\right) - M^{-0}_{(j-1)-(j)}}{\left(a_{(j-1)-(j)} + (j-1)d\right)}\right)\alpha_{(j-1)-(j)} = F^+_{U(j-1)L(j)} \tag{4.20b}$$

where

$$V^-_{(j-1)-(j)} = R_{yn} - \sum_{i=j}^{n} W_i$$

$$M^{-0}_{(j-1)-(j)} = -\left(nR_{yn} - \sum_{i=j}^{n} iW_i\right)d$$

The vertical web members of inclined chord Warren through trusses are either loaded directly by floorbeam reactions at the bottom chord[103] or join top chord members where there are no diagonal web members. Therefore, axial forces in the vertical web members of inclined chord Warren trusses can be readily determined considering only free body diagrams at bottom chord or top chord joints. At top chord joints, where there are no diagonal web members, non-central vertical members are loaded by the adjacent top chord member forces such that where the non-central vertical member is considered at joint j,

$$F_{U(j)L(j)} = \frac{F_{U(j-1)U(j)}\left(h_{(j)} - h_{(j-1)}\right)}{\sqrt{\left(h_{(j)} - h_{(j-1)}\right)^2 + d^2}} - \frac{F_{U(j)U(j+1)}\left(h_{(j+1)} - h_{(j)}\right)}{\sqrt{\left(h_{(j+1)} - h_{(j)}\right)^2 + d^2}} \tag{4.21a}$$

or where the non-central vertical member is deemed to be at joint j – 1,

$$F_{U(j-1)L(j-1)} = \frac{F_{U(j-2)U(j-1)}\left(h_{(j-1)} - h_{(j-2)}\right)}{\sqrt{\left(h_{(j-1)} - h_{(j-2)}\right)^2 + d^2}} - \frac{F_{U(j-1)U(j)}\left(h_{(j)} - h_{(j-1)}\right)}{\sqrt{\left(h_{(j)} - h_{(j-1)}\right)^2 + d^2}} \tag{4.21b}$$

At top chord joints, where there are no diagonal web members, central vertical members are loaded by the adjacent top chord member forces such that where the central vertical member is considered at joint (j = n/2)j,

$$F_{U\left(\frac{n}{2}\right)L\left(\frac{n}{2}\right)} = -\frac{F_{U\left(\frac{n}{2}-1\right)U(j)}\left(h_{\left(\frac{n}{2}\right)} - h_{\left(\frac{n}{2}-1\right)}\right)}{\sqrt{\left(h_{\left(\frac{n}{2}\right)} - h_{\left(\frac{n}{2}-1\right)}\right)^2 + d^2}} - \frac{F_{U\left(\frac{n}{2}\right)U\left(\frac{n}{2}+1\right)}\left(h_{\left(\frac{n}{2}\right)} - h_{\left(\frac{n}{2}+1\right)}\right)}{\sqrt{\left(h_{\left(\frac{n}{2}\right)} - h_{\left(\frac{n}{2}+1\right)}\right)^2 + d^2}} \qquad (4.21c)$$

Equations 4.16 and 4.19 are used to determine maximum forces $F^{max}_{L(j-1)L(j)}$ and $F^{max}_{U(j-1)U(j)}$, in the bottom and top chord members, respectively, of inclined-chord Pratt and Warren trusses. $F^{max}_{L(j-1)L(j)}$ and $F^{max}_{U(j-1)U(j)}$ occur with the span fully loaded for trusses traversed by uniformly distributed live loads. However, due to concentrated live loads, both occur with the live load at explicit locations on the span that create maximum panel point moments.

Equations 4.17 and 4.20 are used to determine maximum forces $F^{max}_{U(j-1)L(j)}$ and $F^{max}_{U(j)L(j-1)}$ in diagonal members of inclined-chord Pratt trusses and Warren trusses, respectively. Equation 4.18 is used to determine $F^{max}_{U(j(L(j)}$ in vertical members of inclined-chord Pratt trusses. For live loads, $F^{max}_{U(j-1)L(j)}$, $F^{max}_{U(j)L(j-1)}$ and $F^{max}_{U(j)L(j)}$ in diagonal and vertical members occur with the span partially loaded at specific locations on the span.

4.5.1 Dead Load Analysis

In the same manner as for horizontal parallel-chord truss spans, dead load of inclined-chord truss spans is typically modeled as a static uniformly distributed load allocated equally to panel points and proportioned between the truss top and bottom chord panel points. The dead load is often apportioned entirely to the loaded chord of the truss span.

4.5.2 Live Load Analysis

Uniformly distributed live loads across the entire span create maximum chord member forces.[104] However, the locations of uniformly distributed live loads that create maximum and minimum forces in web members must be established. Maximum and minimum forces in web members will occur with the uniformly distributed live loads partially within a truss panel (j – 1) – (j) as shown in Figures 4.13 and 4.14 for Pratt and Warren truss spans, respectively.

The locations of concentrated live loads that create maximum member forces in inclined chord truss spans can be ascertained by trial. However, the trial effort can be substantially reduced using mathematical criteria based on establishing live load locations on the span where the change in the member forces is zero. Once the locations of uniformly distributed and concentrated live loads that create maximum member forces are established, the maximum and minimum member forces can be determined using the method of joints, the method of sections and influence lines.

4.5.2.1 Uniformly Distributed Live Load Analysis

The locations of the uniformly distributed live loads for maximum and minimum web member forces in a truss panel are established by locating the positions of the uniformly distributed live loads that create maximum positive and maximum negative shear forces in the panel. However, since the inclined-chord members in truss panels resist panel shear forces,[105] they must be considered in establishing the locations of the live loads for maximum and minimum forces in the diagonal and vertical web members of inclined-chord truss spans.

Distances X^D and X^V in Figure 4.13 and X^{DW} in Figure 4.14 are the locations where the change in the member forces is zero.[106] Locations x^{D+}, x^{D-}, x^{V+} and x^{V-} of the uniformly distributed live loads in panel (j – 1) – (j) that result in maximum forces $F^{+max}_{U(j-1)L(j)}$, $F^{-max}_{U(j-1)L(j)}$, $F^{+max}_{U(j)L(j)}$ and $F^{-max}_{U(j)L(j)}$, in diagonal and vertical web members due to positive and negative panel shear forces in Parker[107] trusses are shown in Figure 4.13. Locations x^{DW+} and x^{DW-} of the uniformly distributed live loads in panel (j – 1) – (j) that result in $F^{+max}_{L(j-1)U(j)}$ and $F^{-max}_{L(j-1)U(j)}$ in diagonal web members due to positive and negative panel shear forces in inclined-chord Warren[108] trusses are shown in Figure 4.14.

Force $F^{+}_{U(j-1)L(j)}$ in diagonal member U(j – 1)L(j),[109] due to positive shear force in panel (j – 1) – (j), is from Equation 4.17b:

$$F^{+}_{U(j-1)L(j)} = \left(\frac{V^{+}_{(j-1)-(j)}\left(a_{(j-1)-(j)}\right) - M^{+0}_{(j-1)-(j)}}{\left(a_{(j-1)-(j)} + jd\right)} \right) \propto_{(j-1)-(j)}$$

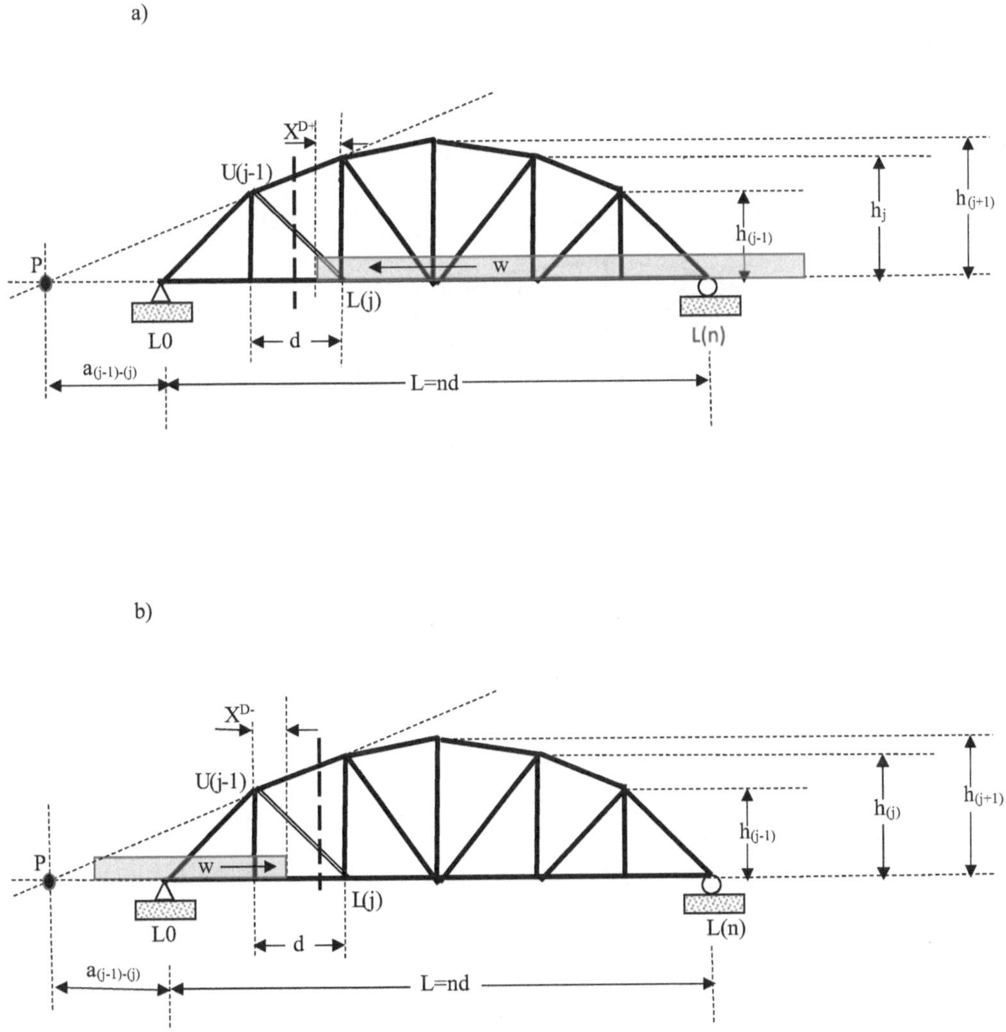

FIGURE 4.13 Uniformly distributed live loads partially on an inclined-chord Pratt truss span for diagonal web members.

c)

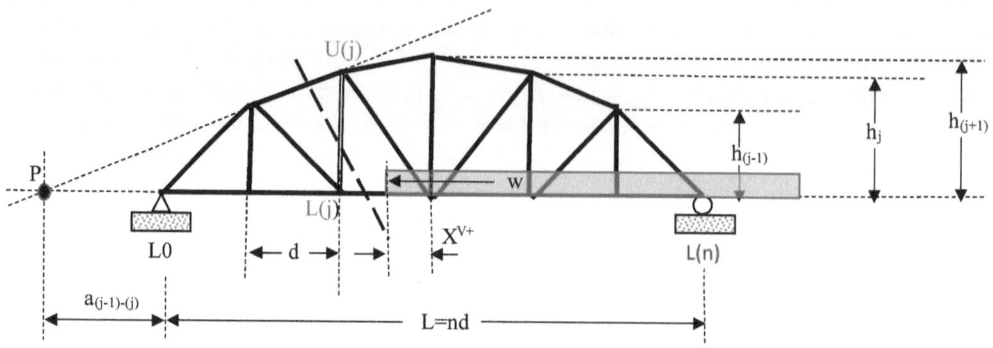

d)

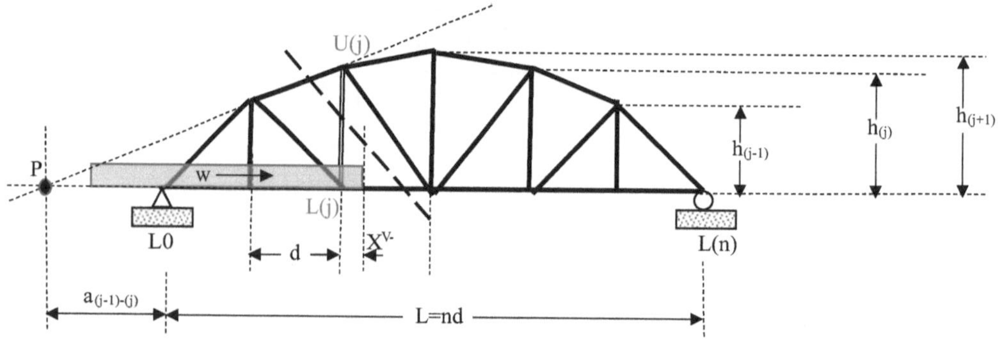

FIGURE 4.13 (Continued)

where in accordance with Figure 4.13a,

$$V^{+}{}_{(j-1)-(j)} = R_{y0} - \sum_{i=0}^{j-1} W_i$$

$$R_{y0} = \frac{w}{2}\left(\frac{\left((n-j)d + x^{D+}\right)^2}{nd}\right)$$

$$\sum_{i=0}^{j-1} W_i = W_{(j-1)} = \frac{w(x^{D+})^2}{2d}$$

$$V^{+}{}_{(j-1)-(j)}(a_{(j-1)-(j)}) = \frac{wa_{(j-1)-(j)}}{2}\left(\left(\frac{\left((n-j)d + x^{D+}\right)^2}{nd}\right) - \frac{(x^{D+})^2}{d}\right)$$

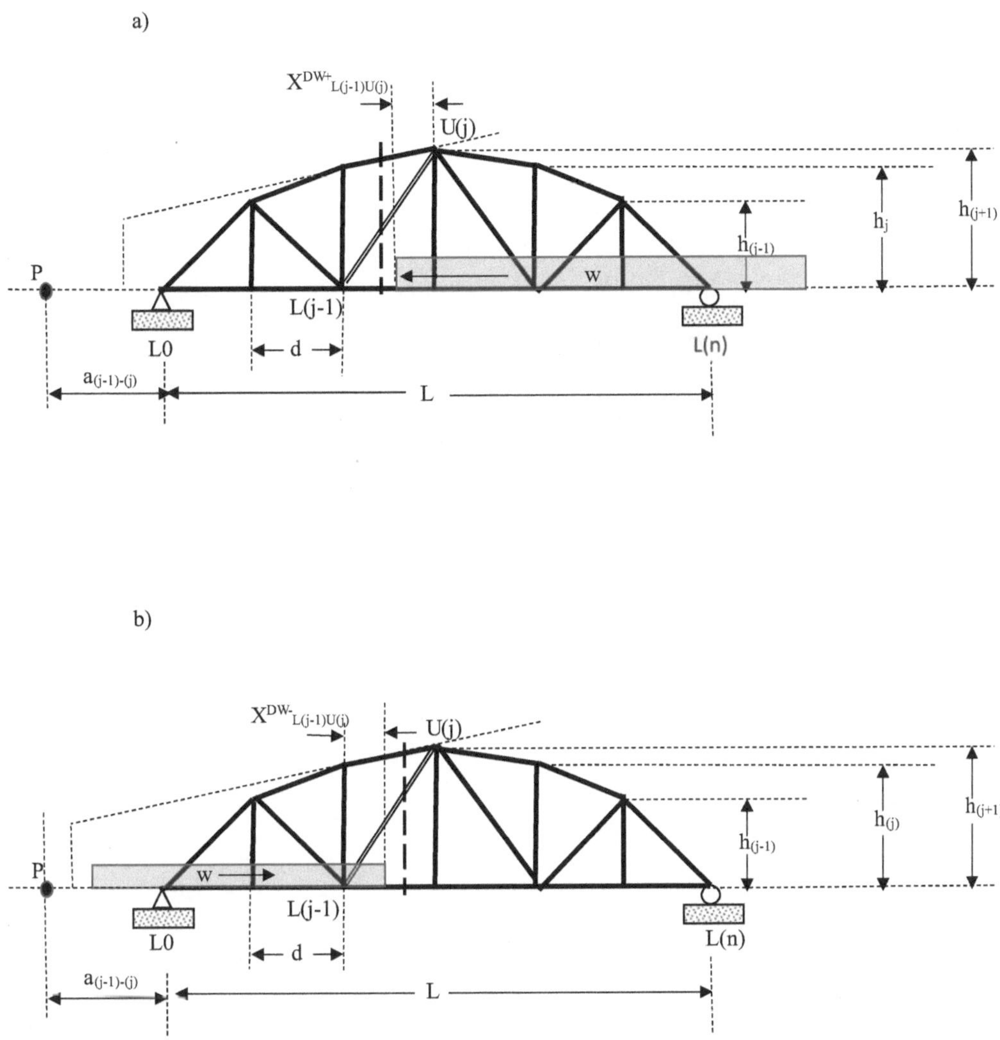

FIGURE 4.14 Uniformly distributed live load partially on an inclined-chord Warren truss span for vertical web members.

$$M^{+0}_{(j-1)-(j)} = \sum_{i=0}^{j-1} W_i(i) = W_{(j-1)}(j-1)d = \frac{w(j-1)\left(x^{D+}\right)^2}{2}$$

such that

$$F^{+}_{U(j-1)L(j)} = \left(\frac{\frac{wa_{(j-1)-(j)}}{2}\left(\left(\frac{\left((n-j)d + x^{D+}\right)^2}{nd}\right) - \frac{(x^{D+})^2}{d}\right) - \frac{w(j-1)(x^{D+})^2}{2}}{\left(a_{(j-1)-(j)} + jd\right)} \right) \propto_{(j-1)-(j)} \quad (4.22a)$$

Calculating $\frac{d\left(F^{+}_{U(j-1)L(j)}\right)}{dx}=0$ to find $x^{D+}_{(j-1)-(j)}$ of the uniformly distributed live load corresponding to maximum positive diagonal web member force $F^{+max}_{U(j-1)L(j)}$ provides

$$x^{D+}_{(j-1)-(j)}=\frac{(n-j)d}{n\left(1+\frac{(j-1)d}{a_{(j-1)-(j)}}\right)-1} \tag{4.22b}$$

$F^{-}_{U(j-1)L(j)}$ in diagonal member U(j-1)L(j) due to negative shear force in panel (j – 1) – (j) is, from Equation 4.17d,

$$F^{-}_{U(j-1)L(j)}=\left(\frac{-V^{-}_{(j-1)-(j)}\left(a_{(j-1)-(j)}\right)+M^{-0}_{(j-1)-(j)}}{a_{(j-1)-(j)}+jd}\right)\alpha_{(j-1)-(j)}$$

where in accordance with Figure 4.13b,

$$V^{-}_{(j-1)-(j)}=R_{yn}-\sum_{i=j}^{n}W_i$$

$$R_{yn}=\frac{w\left((j-1)d+x^{D-}\right)^2}{2nd}$$

$$\sum_{i=j}^{n}W_i=W_{(j)}=\frac{w\left(x^{D-}\right)^2}{2d}$$

$$V^{-}_{(j-1)-(j)}(a_{(j-1)-(j)})=\frac{wa_{(j-1)-(j)}}{2}\left(\left(\frac{\left((n-j)d+x^{D-}\right)^2}{nd}\right)-\frac{(x^{D-})^2}{d}\right)$$

$$M^{-0}_{(j-1)-(j)}=-ndR_{yn}+(jd)W_{(j)}=\frac{-w\left(\left((j-1)d+x^{D-}\right)^2-j\left(x^{D-}\right)^2\right)}{2}$$

$$F^{-}_{U(j-1)L(j)}=\left(\frac{-\left(\frac{wa_{(j-1)-(j)}}{2}\left(\left(\frac{\left((n-j)d+x^{D-}\right)^2}{nd}\right)-\frac{(x^{D-})^2}{d}\right)\right)-\left(\frac{w\left(\left((j-1)d+x^{D-}\right)^2-j\left(x^{D-}\right)^2\right)}{2}\right)}{a_{(j-1)-(j)}+jd}\right)\alpha_{(j-1)-(j)} \tag{4.23a}$$

Calculating $\frac{d\left(F^{-}_{U(j-1)L(j)}\right)}{dx}=0$ to find $x^{D-}_{(j-1)-(j)}$ of the uniformly distributed live load corresponding to the maximum negative diagonal web member force $F^{-max}_{U(j-1)L(j)}$ provides

$$x^{D-}_{(j-1)-(j)}=\frac{a_{(j-1)-(j)}+nd}{n+\frac{(n-1)a_{(j-1)-(j)}}{(j-1)d}}=d-x^{D+}_{(j-1)-(j)} \tag{4.23b}$$

Equation 4.23b is expected due to the shape of the uniformly distributed load shear force diagram in panel (j − 1) − (j).

From Equation 4.20a, force $F^{+}_{L(j-1)U(j)}$ in diagonal member L(j-1) U(j)[110] due to positive shear force in panel (j − 1) − (j) is

$$F^{+}_{L(j-1)U(j)} = \left(\frac{-V^{+}_{(j-1)-(j)}\left(a_{(j-1)-(j)}\right) + M^{+0}_{(j-1)-(j)}}{\left(a_{(j-1)-(j)} + (j-1)d\right)} \right) \alpha_{(j-1)-(j)}$$

Calculating $\frac{d\left(F^{+}_{L(j-1)U(j)}\right)}{dx} = 0$ provides $x^{DW+}_{(j-1)-(j)}$, corresponding to the maximum diagonal web member live load force $F^{+max}_{L(j-1)U(j)}$ as

$$x^{DW+}_{(j-1)-(j)} = \frac{(n-j)d}{n\left(1 + \frac{(j-1)d}{a_{(j-1)-(j)}}\right) - 1} = x^{D+}_{(j-1)-(j)} \tag{4.24}$$

The location of the uniformly distributed load for maximum positive panel shear is, as expected,[111] the same as for positive panel shear with the diagonal member oriented in the opposite direction ($x^{DW+}_{(j-1)-(j)} = x^{D+}_{(j-1)-(j)}$ of Equation 4.22b). This is also evident from a comparison of Equations 4.17b and 4.20a. Similarly, the location of the uniformly distributed load for maximum negative panel shear is the same as for negative panel shear with the diagonal member oriented in the opposite direction ($x^{DW-}_{(j-1)-(j)} = x^{D-}_{(j-1)-(j)}$ of Equation 4.23b).

Locations x^{V+} and x^{V-}, shown, respectively, in Figures 4.13c and Figure 4.13d in panel (j) − (j + 1) can be established by investigating the change in $F_{U(j)L(j)}$ as the uniformly distributed load traverses the span. Force $F^{+}_{U(j)L(j)}$ in vertical member U(j)L(j) in a Parker truss due to positive shear force in panel (j) − (j + 1) is, from Equation 4.18a,

$$F^{+}_{U(j)L(j)} = \left(\frac{-V^{+}_{(j)-(j+1)}\left(a_{(j-1)-(j)}\right) + M^{+0}_{(j)-(j+1)}}{\left(a_{(j-1)-(j)} + jd\right)} \right)$$

where

$$V^{+}_{(j)-(j+1)} = R_{y0} - \sum_{i=0}^{j} W_i$$

$$R_{y0} = \frac{w}{2}\left(\frac{\left((n-j-1)d + x^{V+}\right)^2}{nd} \right)$$

$$\sum_{i=0}^{j} W_i = W_{(j)} = \frac{w\left(x^{V+}\right)^2}{2d}$$

$$V^{+}_{(j)-(j+1)} = \frac{w}{2}\left(\frac{\left((n-j-1)d + x^{V+}\right)^2}{nd} \right) - \frac{w\left(x^{V+}\right)^2}{2d}$$

$$M^{+0}_{(j)-(j+1)} = \left(\sum_{i=0}^{j} iW_i\right)d = W_j\left(jd\right) = \frac{jw\left(x^{V+}\right)^2}{2}$$

$$F^{+}_{U(j)L(j)} = \frac{\frac{w}{2}\left(\left(-\left(\frac{\left((n-j-1)d+x^{V+}\right)^2}{nd}\right)+\frac{\left(x^{V+}\right)^2}{d}\right)\left(a_{(j-1)-(j)}\right)+\left(j\left(x^{V+}\right)^2\right)\right)}{\left(a_{(j-1)-(j)}+jd\right)} \tag{4.25a}$$

Calculating $\frac{d\left(F^{+}_{U(j)L(j)}\right)}{dx} = 0$ to find $x^{V+}_{(j)-(j+1)}$ of the live load corresponding to maximum vertical web member force due to positive panel shear $F^{+max}_{U(j)L(j)}$ provides

$$x^{V+}_{(j)-(j+1)} = \left(1 - \frac{a_{(j-1)-(j)} + nd}{2a_{(j-1)-(j)} + nd + \frac{a_{(j-1)-(j)}}{j}}\right)d \tag{4.25b}$$

Calculating $\frac{d\left(F^{-}_{U(j)L(j)}\right)}{dx} = 0$ to find $x^{V-}_{(j)-(j+1)}$ of the live load corresponding to the maximum vertical web member force due to negative panel shear $F^{-max}_{U(j)L(j)}$ provides

$$x^{V-}_{(j)-(j+1)} = d - x^{V+}_{(j)-(j+1)} \tag{4.25c}$$

Equation 4.25c is expected due to the shape of the uniformly distributed load shear force diagram in panel (j) – (j + 1).

Maximum forces in chord, diagonal and vertical members of a Parker truss traversed by a uniformly distributed live load are examined in Example 4.19. The maximum forces in the chord and diagonal members of an inclined-chord Warren truss traversed by a uniformly distributed live load are outlined in Example 4.20. Examples 4.19 and 4.20 outline the analysis of inclined-chord Pratt (Parker) and Warren through-truss spans with vertical post lengths of 25 ft, 35 ft and 40 ft to illustrate the methodology to determine member forces.[112]

Maximum force $F^{max}_{U(j)L(j)}$ in vertical member U(j)L(j) of an inclined-chord Warren truss that is not directly loaded by floorbeam reaction can be determined from Equation 4.21. However, the location of the uniformly distributed live load that creates the maximum force in member U(j)L(j) must be determined. This location is the one that creates $F^{max}_{U(j-1)U(j)}$ and $F^{max}_{U(j)U(j+1)}$ in top chord members U(j – 1)U(j) and U(j)U(j + 1), respectively. $F^{max}_{U(j-1)U(j)}$ and $F^{max}_{U(j)U(j+1)}$ are maximum when the span is fully loaded by the uniformly distributed live load as illustrated in Example 4.20.

Example 4.19

The six-panel inclined top chord Pratt (Parker) truss in Figure 4.12 is 150 ft long with h_1 = 25 ft, h_2 = 35 ft and h_3 = 40 ft. It is subjected to a uniformly distributed dead load of 1500 lb/ft and a uniformly distributed live load of 5500 lb/ft. Maximum and minimum forces in the truss members are required:

$$W_{DL} = (1.5)(25) = 37.5 \text{ kips}$$
$$W_{LL} = (5.5)(25) = 137.5 \text{ kips}$$

The bottom and top chord member forces in panels 1-2 and 2-3 are determined from Equations 4.16a and 4.16b, respectively, with j = 2 and j = 3:

$$R_{y0} = \sum_{i=0}^{n} \frac{(n-i)}{n} W_i = \sum_{i=0}^{6} \frac{(6-i)}{6} W_i = W_0 + \frac{W}{6}(5+4+3+2+1) = \frac{W}{2} + \frac{15W}{6} = \frac{18W}{6} = 3.00W$$

For j = 2,

$$F_{L1L2} = \left(\sum_{i=0}^{6} \frac{(6-i)}{6} W_i - \sum_{i=0}^{1} (1-i) W_i \right) \frac{25}{h_1} = 3.00W - \frac{W}{2} = 2.500W$$

$$F_{U1U2} = -\left(2\sum_{i=0}^{6} \frac{(6-i)}{6} W_i - \sum_{i=0}^{2} (2-i) W_i \right) \left(\frac{\sqrt{(h_2 - h_1)^2 + d^2}}{h_2} \right)$$

$$= -\left(\frac{2(18)W}{6} - \frac{2W}{2} - W \right) \left(\frac{\sqrt{725}}{35} \right) = -3.077W$$

For j = 3,

$$F_{L2L3} = \left((2)\sum_{i=0}^{6} \frac{(6-i)}{6} W_i - \sum_{i=0}^{2} (2-i) W_i \right) \frac{25}{h_2} = \left(\frac{2(18)W}{6} - W - W \right) \frac{25}{35} = 2.857W$$

$$F_{U2U3} = -\left((3)\sum_{i=0}^{6} \frac{(6-i)}{6} W_i - \sum_{i=0}^{3} (3-i) W_i \right) \left(\frac{\sqrt{(h_3 - h_2)^2 + d^2}}{h_3} \right)$$

$$= -\left(\frac{3(18)}{6} - \frac{3}{2} - 2 - 1 \right) W \left(\frac{\sqrt{650}}{40} \right) = -2.868W$$

The maximum force in end post L0U1, F^{max}_{L0U1} = −132.58 − 486.06 = −618.64 kips in accordance with Equation 4.8 (the section through panel 0-1 is identical for horizontal- and inclined-chord trusses), as shown in Table E4.19a.

TABLE E4.19A
Top and Bottom Chord Member Forces from a Uniformly Distributed Live Load Traversing a Parker Truss Span

Chord Member	j	$h_{(j)}$ (feet)	Dead Load Force (kips) (W=37.5 kips)	Maximum Live Load Force (kips) (W=137.5 kips)
L0L1	1	25	93.75 (T)	343.75 (T)
L1L2	2	35	93.75 (T)	343.75 (T)
L2L3	3	40	107.13 (T)	392.84 (T)
U1U2	2	35	−115.38 (C)	−423.09 (C)
U2U3	3	40	−107.55 (C)	−394.35 (C)

For web members,

$$a_{(1)-(2)} = \left(\left(\frac{h_{(2)}}{h_{(2)} - h_{(1)}}\right) - (j)\right) d = \left(\left(\frac{35}{35 - 25}\right) - (2)\right) 25 = 37.5\,\text{ft}$$

$$a_{(2)-(3)} = \left(\left(\frac{h_{(3)}}{h_{(3)} - h_{(2)}}\right) - (j)\right) d = \left(\left(\frac{40}{40 - 35}\right) - (3)\right) 25 = 125.0\,\text{ft}$$

$$\propto_{(1)-(2)} = \frac{\sqrt{h_{(1)}^2 + d^2}}{h_{(1)}} = \frac{\sqrt{25^2 + 25^2}}{25} = 1.414$$

$$\propto_{(2)-(3)} = \frac{\sqrt{h_{(2)}^2 + d^2}}{h_{(2)}} = \frac{\sqrt{35^2 + 25^2}}{25} = 1.229$$

Positive shear force in panel (j − 1) − (j) with the uniformly distributed live load traversing from L6 to L0 (right to left) is calculated from Equations 4.17b and 4.22b and Figure 4.13a as shown in Table E4.19b

TABLE E4.19B
Diagonal Web Member Forces due to Positive Panel Shear from Uniformly Distributed Live Load Traversing a Parker Truss Span

Diagonal Member U(j − 1)L(j): Positive Shear in Panel	j = 2	j = 3
$x^{D+}_{(j-1)-(j)} = \dfrac{(n-j)d}{n\left(1+\dfrac{(j-1)d}{a_{(j-1)-(j)}}\right)-1}\,(\text{ft})$	$\dfrac{(6-2)25}{6\left(1+\dfrac{(2-1)25}{37.5}\right)-1} = 11.11$	$\dfrac{(6-3)25}{6\left(1+\dfrac{(3-1)25}{125}\right)-1} = 10.14$
$R_{y0} = \dfrac{w}{2}\left(\dfrac{\left((n-j)d + x^{D+}_{(j-1)-(j)}\right)^2}{nd}\right)(\text{kips})$	$\dfrac{5.5}{2}\left(\dfrac{\left((6-2)25+11.11\right)^2}{6(25)}\right) = 226.3$	$\dfrac{5.5}{2}\left(\dfrac{\left((6-3)25+10.14\right)^2}{6(25)}\right) = 132.9$
$\sum^{j-1} W_i = W_{(j-1)} = \dfrac{w(x^{D+}_{(j-1)-(j)})^2}{2d}(\text{kips})$	$\dfrac{5.5(11.11)^2}{2(25)} = 13.58$	$\dfrac{5.5(10.14)^2}{2(25)} = 11.31$
$V^{+}_{(j-1)-(j)} = R_{y0} - \sum_{i=0}^{j-1} W_i(\text{kips})$	226.3 − 13.58 = 212.8	132.9 − 11.31 = 121.6
$M^{+0}_{(j-1)-(j)} = W_{(j-1)}(j-1)d = \dfrac{w(j-1)\left(x^{D+}_{(j-1)-(j)}\right)^2}{2}(\text{kip-ft})$	$\dfrac{5.5(2-1)(11.11)^2}{2} = 339.50$	$\dfrac{5.5(3-1)(10.14)^2}{2} = 565.50$
$V^{+}_{(j-1)-(j)}\left(a_{(j-1)-(j)}\right) - M^{+0}_{(j-1)-(j)}(\text{kips})$	212.8(37.5) − 339.50 = 7639.0	121.6(125) − 565.50 = 14632.6
$F^{+max}_{U(j-1)L(j)}(\text{kips})$	(7639.0/(50 + 37.5))(1.414) = 123.5 (T)	(14632.6/(75 + 125))(1.229) = 89.9 (T)

Negative shear force in panel (j − 1) − (j) with the uniformly distributed live load traversing from L0 to L6 (left to right) is calculated from Equations 4.17d and 4.23b and Figure 4.13b as in Table E4.19c.

TABLE E4.19C
Diagonal Web Member Forces due to Negative Panel Shear from Uniformly Distributed Live Load Traversing a Parker Truss Span

Diagonal member U(j − 1)L(j): negative shear in panel	j = 2	j = 3
$x^{D-}_{(j-1)-(j)} = \dfrac{a_{(j-1)-(j)} + nd}{n + \dfrac{(n-1)a_{(j-1)-(j)}}{(j-1)d}}$ (ft)	$\dfrac{37.5+150}{6+\dfrac{(6-1)37.5}{(2-1)25}} = 13.89$	$\dfrac{125+150}{6+\dfrac{(6-1)125}{(3-1)25}} = 14.86$
$R_{y6} = \dfrac{w\left((j-1)d + x^{D-}_{(j-1)-(j)}\right)^2}{2nd}$ (kips)	$\dfrac{5.5((2-1)25+13.89)^2}{2(6)(25)} = 27.73$	$\dfrac{5.5((3-1)25+14.86)^2}{2(6)(25)}$ $= 77.13$
$\sum_{i=j}^{n} W_i = W_{(j)} = \dfrac{w\left(x^{D-}_{(j-1)-(j)}\right)^2}{2d}$ (kips)	$\dfrac{5.5(13.89)^2}{2(25)} = 21.22$	$\dfrac{5.5(14.86)^2}{2(25)} = 24.29$
$V^{-}_{(j-1)-(j)} = R_{yn} - \sum_{i=j}^{n} W_i == R_{yn} - W_{(j)}$ (kips)	27.73 − 21.22 = 6.51	77.13 − 24.29 = 52.84
$M^{-0}_{(j-1)-(j)} = \left(-ndR_{yn} + \dfrac{j(w)\left(x^{D-}_{(j-1)-(j)}\right)^2}{2}\right)$ (kip-ft)	$-6(25)(27.73) + \dfrac{2(5.5)(13.89)^2}{2}$ $= -3098.4$	$-6(25)(77.13) +$ $\dfrac{3(5.5)(14.86)^2}{2} = -9747.7$
$-V^{-}_{(j-1)-(j)}\left(a_{(j-1)-(j)}\right) + M^{-0}_{(j-1)-(j)}$ (kip - ft)	−6.51(37.5) − 3098= −3342.2	−52.84(125) − 9747 = −16352
$F^{-max}_{U(j-1)L(j)}$ (kips)	−(3342.2/(37.5 + 2(25))(1.414) = −54.1 (C)	−(16352.0/(125 + 3(25)) (1.229) = −100.50 (C)

In the vertical web members, positive shear force in panel (j) − (j + 1) from Equations 4.18a and 4.25b and Figure 4.13c gives Table E4.19d.

TABLE E4.19d
Vertical Web Member Forces due to Positive Panel Shear from Uniformly Distributed Live Load Traversing a Parker Truss Span

Vertical member U(j)L(j): positive shear in panel	j = 2
$x^{V+}_{(j)-(j+1)} = d\left(1 - \dfrac{\left(a_{(j-1)-(j)} + nd\right)}{2a_{(j-1)-(j)} + nd + \dfrac{a_{(j-1)-(j)}}{j}}\right)$ (ft)	$25\left(1 - \dfrac{(37.5+6(25))}{2(37.5)+6(25)+\dfrac{37.5}{2}}\right) = 5.77$

(Continued)

TABLE E4.19d (*Continued*)
Vertical Web Member Forces due to Positive Panel Shear from Uniformly Distributed Live Load Traversing a Parker Truss Span

Vertical member U(j)L(j): positive shear in panel	j = 2
$R_{y0} = \frac{w}{2}\left(\frac{\left((n-j-1)d + x^{V+}_{(j)-(j+1)}\right)^2}{nd}\right)$(kips)	$\frac{5.5}{2}\left(\frac{\left((6-2-1)25+5.77\right)^2}{6(25)}\right) = 119.60$
$\sum_{i=0}^{j} W_i = W_{(j)} = \frac{w\left(x^{V+}_{(j)-(j+1)}\right)^2}{2d}$(kips)	$\frac{5.5(5.77)^2}{2(25)} = 3.66$
$V^{+}_{(j)-(j+1)} = R_{y0} - \sum_{i=0}^{j} W_i$(kips)	119.6 − 3.66 = 115.94
$M^{+0}_{(j)-(j+1)} = \left(\sum_{i=1}^{j} iW_i\right)d = W_j(jd) = \frac{j(w)\left(x^{V+}_{(j)-(j+1)}\right)^2}{2}$(kip - ft)	$\frac{5.5(2)(5.77)^2}{2} = 183.11$
$-V^{+}_{(j)-(j+1)}\left(a_{(j-1)-(j)}\right) + M^{+0}_{(j)-(j+1)}$(kip - ft)	−115.94(37.5) + 183.11 = −4164.6
$F^{+max}_{U(j)L(j)}$(kips)	−4164.6/(37.5 + 2(25)) = −47.60 (C)

Negative shear force in panel (j) − (j + 1) from Equations 4.18b and 4.25c and Figure 4.13d give Table E4.19e.

TABLE E4.19e
Vertical Web Member Forces due to Negative Panel Shear from Uniformly Distributed Live Load Traversing a Parker Truss Span

Vertical member U(j)L(j): negative shear in panel	j = 2
$x^{V-}_{(j)-(j+1)} = d - x^{V+}_{(j)-(j+1)}$(ft)	$25 - 5.77 = 19.23$
$R_{yn} = \frac{w\left(jd + x^{V-}_{(j)-(j+1)}\right)^2}{2(n)d}$(kips)	$\frac{5.5\left(2(25)+19.23\right)^2}{2(6)(25)} = 87.87$
$\sum_{i=j+1}^{6} W_i = W_{(j+1)} = \frac{w\left(x^{V-}_{(j)-(j+1)}\right)^2}{2d}$(kips)	$\frac{5.5(19.23)^2}{2(25)} = 40.68$
$V^{-}_{(j)-(j+1)} = R_{yn} - \sum_{i=j+1}^{6} W_i$ (kips)	87.87 − 40.68 = 47.19

TABLE E4.19e (*Continued*)
Vertical Web Member Forces due to Negative Panel Shear from Uniformly Distributed Live Load Traversing a Parker Truss Span

Vertical member U(j)L(j): negative shear in panel	j = 2
$M^{-0}_{(j)-(j+1)} = -\left(nR_{yn} - (j+1)W_{(j+1)}\right)d$ (kip-ft)	$-\left(6(87.87 - (2+1)40.68\right)25 = -10129.5$
$V^{-}_{(j)-(j+1)}\left(a_{(j-1)-(j)}\right) - M^{-0}_{(j)-(j+1)}$ (kip-ft)	47.19(37.5) + 10129.5 = 11899.1
$F^{-max}_{U(j)L(j)}$ (kips)	11899.1/(37.5 + 2(25)) = 136.0 (T)

The forces in vertical web members U1L1 and U3L3 are obtained using the method of joints at L1 and U3, respectively (Table E4.19f):

$$F_{U1L1} = W$$

$$F_{U3L3} = -2F_{U2U3}\left(\frac{h_3 - h_2}{\sqrt{(h_3 - h_2)^2 + d^2}}\right) = -2F_{U2U3}\left(\frac{40 - 35}{\sqrt{(40-35)^2 + 25^2}}\right) = -0.392F_{U2U3}$$

TABLE E4.19f
Vertical Web Member Forces due to Uniformly Distributed Live Load Traversing a Parker Truss Span

Vertical Web Member	$F_{U(j)L(j)}$ Dead Load Force (W = 37.5 kips)	$F_{U(j)L(j)}$ Live Load Force (W = 137.5 kips)
U1L1	37.50 (T)	137.50 (T)
U3L3	42.16 (T)	154.59 (T)

Table E4.19g gives the dead load and maximum live load forces.

TABLE E4.19g
Dead Load and Maximum Live Load Forces in Parker Truss Members

Parker Truss Member	Dead Load Force (kips) W = 37.50 kips	Maximum Live Load Force (kips) W = 137.5 kips
LOL1	93.75	343.75
L1L2	93.75	343.75
L2L3	107.13	392.84
LOU1	−132.60	−486.20
U1U2	−115.38	−423.09
U2U3	−107.55	−394.35
U1L1	37.50	137.50
U2L2	24.11	−47.60 (positive shear)
		136.00 (negative shear)
U3L3	42.16	154.59
U1L2	18.93	123.45 (positive shear)
		−54.10 (negative shear)
U2L3	−2.88	89.92 (positive shear)
		−100.50 (negative shear)

The maximum and minimum member forces in the Parker truss are given in Table E4.19h.

TABLE E4.19h
Maximum and Minimum Member Forces in the Parker Truss

Parker Truss Member	Minimum Force (kips)	Maximum Force (kips)
L0L1	93.75	437.50
L1L2	93.75	437.50
L2L3	107.13	500.00
L0U1	−132.60	−618.75
U1U2	−115.38	−538.50
U2U3	−107.55	−501.90
U1L1	37.50	175.00
U2L2	−23.49	160.11
U3L3	42.16	196.75
U1L2	−35.17	142.38
U2L3	87.04	−103.38

Members U2L2, U1L2 and U2L3 experience stress reversal due to live load not including live load impact.

Example 4.20

The six-panel inclined top chord Warren truss in Figure 4.12c is 150 ft long with h_1 = 25 ft, h_2 = 35 ft and h_3 = 40 ft. It is subjected to a uniformly distributed dead load of 1500 lb/ft and a uniformly distributed live load of 5500 lb/ft. Maximum and minimum forces in the truss members are required:

$W_{DL} = (1.5)(25) = 37.5$ kips
$W_{LL} = (5.5)(25) = 137.5$ kips
$n = 6$
$j = 3$ (panel 2-3 of Warren truss)

From Equations 4.19a and 4.19b, the chord member live load forces are calculated as

$$F_{L2L3} = \left((3)\sum_{i=0}^{6}\frac{(6-i)}{6}W_i - \sum_{i=0}^{3}(3-i)W_i\right)\left(\frac{25}{40}\right)$$

$$= \left(3(3) - \frac{3}{2} - 2 - 1\right)(137.5)(0.625) = 386.7 \text{ kips}$$

$$F_{U2U3} = -\left((3-1)\sum_{i=0}^{6}\frac{(6-i)}{6}W_i - \sum_{i=0}^{2}(3-1-i)W_i\right)\left(\frac{\sqrt{(40-35)^2+25^2}}{35}\right)$$

$$= -(2(3)-1-1)137.5(0.728) = -400.6 \text{ kips}$$

$$a_{(2)\text{-}(3)} = \left(\left(\frac{40}{40-35}\right) - (3)\right)25 = 125.0 \text{ ft}$$

$$\propto_{(2)\text{-}(3)} = \frac{\sqrt{40^2+625}}{40} = 1.179$$

Equation 4.20a, Equation 4.24, and Figure 4.14a give Table E4.20a.

TABLE E4.20a
Diagonal Web Member Forces due to Positive Panel Shear from Uniformly Distributed Live Load Traversing an Inclined-Chord Warren Truss Span

Diagonal Member L2U3: Positive Panel Shear

$x^{DW+}_{(j-1)-(j)} = x^{D+}_{(j-1)-(j)} = \dfrac{(n-j)d}{n-1+\left(\dfrac{(j-1)nd}{a_{(j-1)-(j)}}\right)}$ (ft)	$\dfrac{(6-3)25}{6+\left(\dfrac{(3-1)150}{125.0}\right)-1} = 10.14$
$R_{y0} = \dfrac{w}{2}\left(\dfrac{\left((n-j)d + x^{DW+}_{(j-1)-(j)}\right)^2}{nd}\right)$ (kips)	$\dfrac{5.5}{2}\left(\dfrac{\left((6-3)25+10.14\right)^2}{6(25)}\right) = 132.9$
$\sum_{i=0}^{j-1} W_i = W_{(j-1)} = \dfrac{w(x^{DW+}_{(j-1)-(j)})^2}{2d}$ (kips)	$\dfrac{5.5(10.14)^2}{2(25)} = 11.31$
$V^{+}_{(j1)-(j)} = R_{y0} - \sum_{i=0}^{j-1} W_i$ (kips)	132.90 − 11.31 = 121.58
$M^{+0}_{(j-1)-(j)} = \left(\sum_{i=0}^{j-1} iW_i\right)d = (j-1)W_{(j-1)}$ (kip-ft)	(3−1)(25)11.31 = 565.5
$-V^{+}_{(j-1)-(j)}\left(a_{(j-1)-(j)}\right) + M^{+0}_{(j-1)-(j)}$ (kip-ft)	−121.58(125) + 565.50 = −14632.0
$F^{+}_{L2U3} = \left(\dfrac{-V^{+}_{(j-1)-(j)}\left(a_{(j-1)-(j)}\right) + M^{+0}_{(j-1)-(j)}}{\left(a_{(j-1)-(j)} + (j-1)d\right)}\right)\alpha_{(j-1)-(j)}$ (kips)	-(14632.0/(50 + 125))(1.179) = −98.58 (C)

Equation 4.20b and Figure 4.14b give Table E4.20b.

TABLE E4.20b
Diagonal Web Member Forces due to Negative Panel Shear from Uniformly Distributed Live Load Traversing an Inclined-Chord Warren Truss Span

Diagonal Member L2U3: Negative Panel Shear

$x^{DW-}_{(j-1)-(j)} = x^{D-}_{(j-1)-(j)} = d - x^{D+}_{(j-1)-(j)}$ (ft)	25 − 10.14 = 14.86
$R_{yn} = \dfrac{w\left((j-1)d + x^{DW-}_{(j-1)-(j)}\right)^2}{2nd}$ (kips)	$\dfrac{5.5\left((3-1)(25)+14.86\right)^2}{2(6)(25)} = 77.13$
$\sum_{i=j}^{n} W_i = W_{(j)} = \dfrac{w\left(x^{DW-}_{(j-1)-(j)}\right)^2}{2d}$ (kips)	$\dfrac{5.5(14.86)^2}{2(25)} = 24.29$
$V^{-}_{(j-1)-(j)} = R_{yn} - \sum_{i=j}^{n} W_i$ (kips)	77.13 − 24.29 = 52.84
$M^{-0}_{(j-1)-(j)} = -\left(nR_{yn} - \sum_{i=j}^{n} iW_i\right)d = -\left(nR_{yn} - jW_j\right)d$ (kip-ft)	$-\left(6(77.13) - 3(24.29)\right)(25) = -9747.8$
$V^{-}_{(j-1)-(j)}\left(a_{(j-1)-(j)}\right) - M^{-0}_{(j-1)-(j)}$ (kip-ft)	52.84(125) + 9747.0 = 16352.8
$F^{-}_{L2U3} = \left(\dfrac{V^{-}_{(j-1)-(j)}\left(a_{(j-1)-(j)}\right) - M^{-0}_{(j-1)-(j)}}{\left(a_{(j-1)-(j)} + (j-1)d\right)}\right)\alpha_{(j-1)-(j)}$ (kips)	(16352.0/(125 + 2(25))(1.179) = 110.17 (T)

For vertical member U(j)L(j), from Equation 4.21 with j = 2,

$$F_{U2L2} = \frac{F_{U2U3}\left(h_{(3)} - h_{(2)}\right)}{\sqrt{\left(h_{(3)} - h_{(2)}\right)^2 + d^2}} - \frac{F_{U1U(2)}\left(h_{(2)} - h_{(1)}\right)}{\sqrt{\left(h_{(2)} - h_{(1)}\right)^2 + d^2}} = F_{U2U3}(0.196) - F_{U1U2}(0.371)$$

$F^{max}{}_{U2L2}$ will typically occur with $F^{max}{}_{U2U3}$ and $F^{max}{}_{U1U2}$. However, a few locations of live load are investigated in Table E4.20c to confirm this.

From Equation 4.19b, top chord member U1U2 forces are

$$F_{U1U2} = -\left((2)\sum_{i=0}^{6}\frac{(6-i)}{6}W_i - \sum_{i=0}^{2}(2-i)W_i\right)\left(\frac{\sqrt{\left(h_{(2)} - h_{(1)}\right)^2 + d^2}}{h_{(2)}}\right) = \left((2)\sum_{i=0}^{6}\frac{(6-i)}{6}W_i - \sum_{i=0}^{2}(2-i)W_i\right)(0.769)$$

Forces in top chord member U2U3 from Equation 4.19b are calculated as

$$F_{U2U3} = -\left((2)\sum_{i=0}^{6}\frac{(6-i)}{6}W_i - \sum_{i=0}^{2}(2-i)W_i\right)\left(\frac{\sqrt{\left(h_{(3)} - h_{(2)}\right)^2 + d^2}}{h_{(2)}}\right)$$

$$= -\left((2)\sum_{i=0}^{6}\frac{(6-i)}{6}W_i - \sum_{i=0}^{2}(2-i)W_i\right)(0.728)$$

TABLE E4.20c

Vertical Web Member Forces due to Top Chord Forces from Uniformly Distributed Live Load Traversing an Inclined-Chord Warren Truss Span

	Uniformly Distributed Live Load at Location					
i	**L4**	**L4 + 12.5 ft**	**L5**	**L5 + 6.25 ft**	**L5 + 12.5 ft**	**L6**
0	68.75	68.75	68.75	68.75	68.75	68.75
1	137.5	137.5	137.5	137.5	137.5	137.5
2	137.5	137.5	137.5	137.5	137.5	137.5
3	137.5	137.5	137.5	137.5	137.5	137.5
4	68.75	120.3	137.5	137.5	137.5	137.5
5	0	17.18	68.75	98.83	120.3	137.5
6	0	0	0	4.30	17.18	68.75
Ry0 (kips)	366.7	386.7	401.0	406.1	409.6	412.5
Ry0—W0 (kips)	297.9	318.0	332.3	337.3	340.9	343.8
2(Ry0—W0) (kips)	595.8	635.9	664.6	674.6	681.8	687.5
V U2L2 (kips)	137.5	137.5	137.5	137.5	137.5	137.5
−(2 (Ry0—W0)—V U2L2) (kips)	−458.3	−498.4	−527.1	−537.1	−544.3	−550.0
F_{U1U2}=(-(2 (Ry0—W0)—V U2L2)) αU(j − 1)U(j) (kips)	−352.5	−383.3	−405.3	−413.0	−418.5	−423.1
F_{U2L2} from F_{U1U2} (kips)	130.8	142.2	150.4	153.2	155.3	157.0
F_{U2U3}= =(−(2 (Ry0—W0)—V U2L2)) αU(j)U(j + 1) (kips)	−333.7	−362.9	−383.7	−391.0	−396.2	−400.6
F_{U2L2} from F_{U2U3} (kips)	65.40	71.12	75.21	76.64	77.66	78.52
F_{U2L2} (kips)	65.36	71.08	75.17	76.60	77.62	78.48

F_{U2L2} from F_{U2U3} is 50% of F_{U2L2} from F_{U1U2} due to the angle to the horizontal of top chord member U2U3 being 50% of the angle to the horizontal of top chord member U1U2

F^{max}_{U2L2} = 78.48 kips (span fully loaded by uniformly distributed live load), and the maximum and minimum member forces in the inclined chord Warren truss are given in Table E4.20d.

TABLE E4.20d
Maximum and Minimum Member Forces in the Inclined Chord Warren Truss

Warren Truss Member	Minimum Force (kips)	Maximum Force (kips)
L0L1	93.75	437.5
L1L2	93.75	437.5
L2L3	105.5	492.2
L0U1	−132.6	−618.8
U1U2	−115.4	−538.5
U2U3	−109.3	−509.9
U1L1	37.50	175.0
U2L2	21.43	100.0
U3L3	37.5	175.0
U1L2	−35.2	142.5
L2U3	−95.4	113.4

Member U1L2 and member L2U3 experience stress reversal due to live load, not including live load impact.

Examples 4.19 and 4.20 illustrate the effects of moving uniformly distributed loads on inclined-chord Pratt (Parker) and Warren truss member forces. Statically determinate Howe and other inclined-chord spans traversed by a uniformly distributed live load can be analyzed in a similar manner. However, in addition to only uniformly distributed loads, truss member forces due to concentrated live loads must also be considered in the structural analysis of railway truss spans.

4.5.2.2 Concentrated Live Load Analysis

After establishing the position of the live load for maximum panel point moments and maximum panel shear forces using mathematical criteria, the maximum and minimum forces in chord and web members may be evaluated by static equilibrium or by constructing influence lines.

For calculating maximum chord member forces, Table 4.1 provides general guidance on axle or wheel location for maximum panel point moment with Cooper's E live load moving from right to left over typical railway truss spans.[113] Furthermore, the criterion of Equation 4.11c can be used to refine, through trial and error, the axle location of the Cooper's E load configuration for maximum panel point moment in inclined-chord trusses.[114]

The maximum shear forces in panels and the associated web member forces created by Cooper's E live load on typical railway truss spans will occur with the load partially on the span and axles near the front of the Cooper's E live load positioned adjacent to the truss panel under consideration. A review of the Cooper's E load configuration reveals that maximum web member forces due to positive and negative panel shear forces typically occur with wheels 2, 3 or 4 at the applicable panel point.[115] Mathematical criteria can be established to further locate the axle location of the Cooper's E live load for maximum forces in diagonal members due to positive and negative panel shear forces.

For positive shear[116] in panel (j − 1) − (j) in Figure 4.15b, reaction R_{y0} at panel point L0 in the truss span of L = nd, traversed by a series of concentrated loads with resultant force P^+_T at distance $x = x^+_T$ from panel point (n), is

$$R_{y0} = P^+_T\left(\frac{x^+_T}{nd}\right) \tag{4.26a}$$

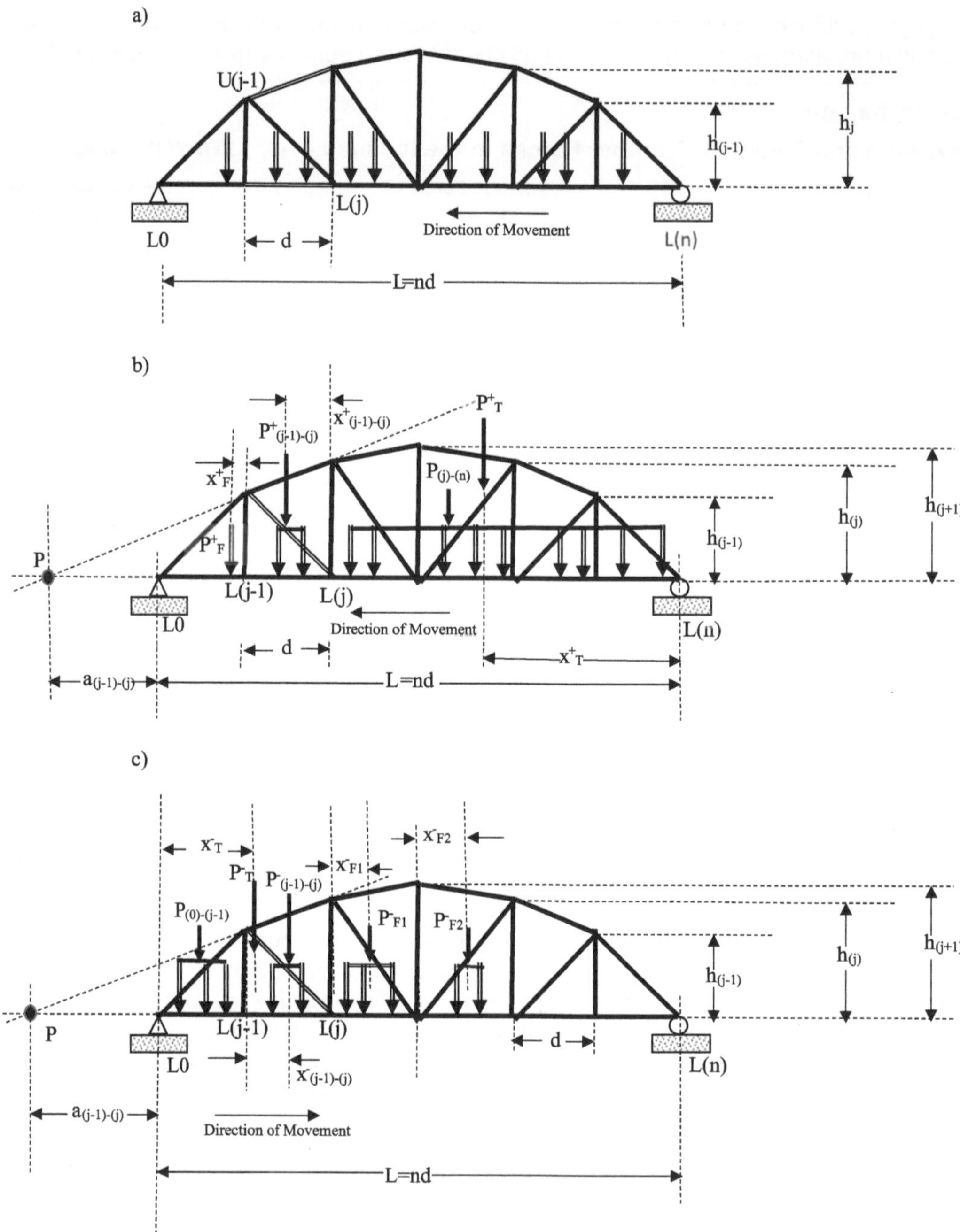

FIGURE 4.15 Concentrated live loads on a Parker truss span.

$F^{+V}_{U(j-1)L(j)}$ = the vertical component of the force in diagonal member U(j-1)L(j) due to positive panel shear forces, which from Equation 4.17a is

$$F^{+V}_{U(j-1)L(j)} = \left(\frac{V^{+}_{(j-1)-(j)}\left(a_{(j-1)-(j)}\right) - M^{+0}_{(j-1)-(j)}}{\left(a_{(j-1)-(j)} + jd\right)} \right)$$

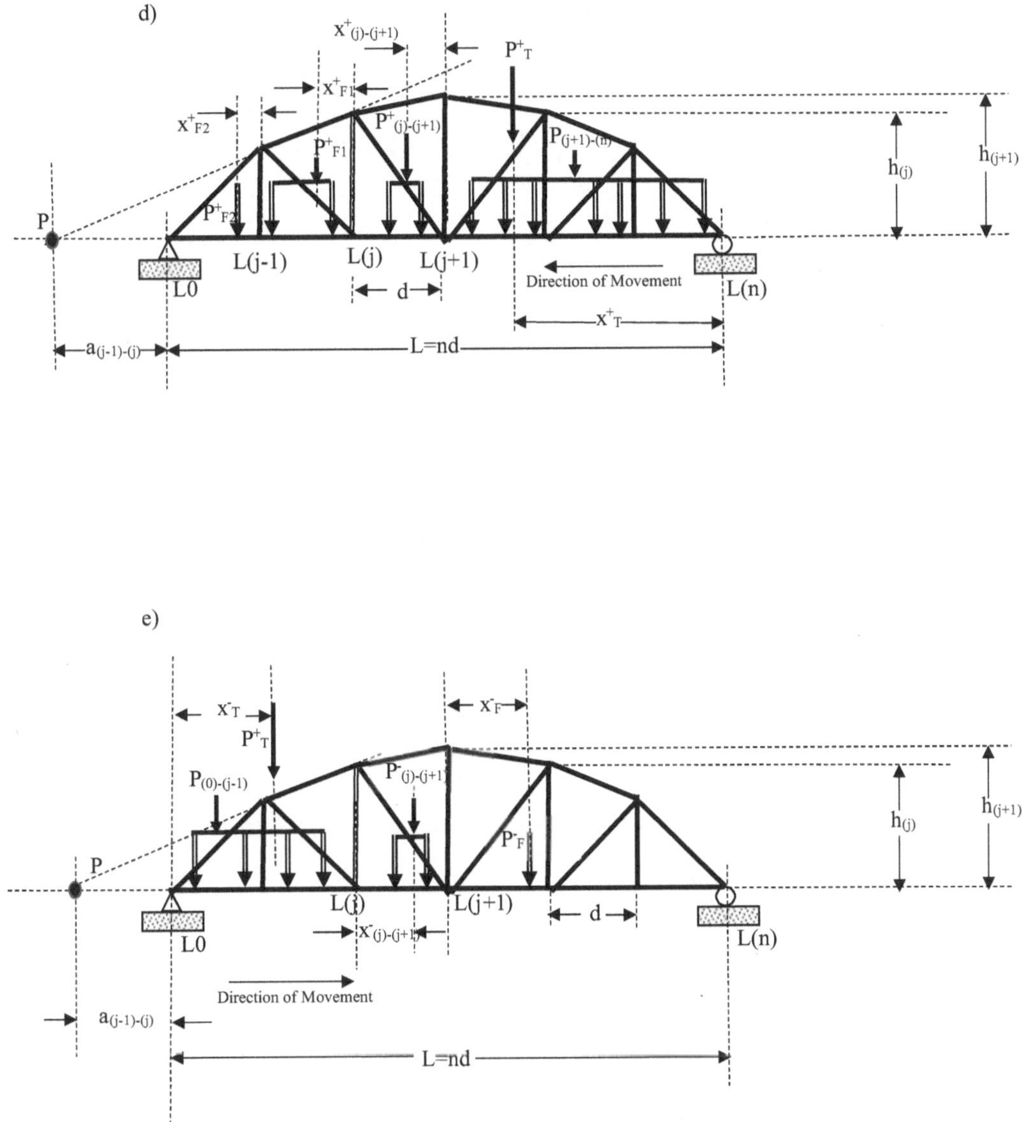

FIGURE 4.15 (Continued)

where in accordance with Figure 4.15b,

$P^{+}_{(j-1)-(j)} = P^{+}_{B}$ = the resultant of the concentrated live loads between panel point (j – 1) and panel point (j) at distance $x^{+}_{(j-1)-(j)} = x^{+}_{B}$ from panel point (j);

$P^{+}_{F} = \sum_{i=1}^{j-1} P^{+}_{Fi}$ = the resultant of concentrated live loads between panel point 0 and (j – 1);

P^{+}_{Fi} = the resultant of concentrated live loads between panel point (j – i – 1) and (j – i);

x^{+}_{F} = the distance from P^{+}_{F} to panel point (j – 1); and

$P_{(j)-(n)} = P^{+}_{R}$ = the resultant of concentrated live loads between panel point (j) and (n), which are not considered in the section in panel (j – 1) – (j) for positive shear:

$$V^{+}_{(j-1)-(j)} = \text{the shear force in panel } (j-1)-(j)$$

$$= R_{y0} - \sum_{i=0}^{j-1} W_i = R_{y0} - W_0 - \sum_{i=1}^{j-1} W_i = R_{y0} - \frac{P^{+}_{F} x^{+}_{F}}{(j-1)d} - \left(\frac{P^{+}_{F}\left((j-1)d - x^{+}_{F}\right)}{(j-1)d} + \frac{P^{+}_{(j-1)-(j)}\left(x^{+}_{(j-1)-(j)}\right)}{d} \right) \tag{4.26b}$$

$$V^{+}_{(j-1)-(j)} = P^{+}_{T}\left(\frac{x_T}{nd}\right) - P^{+}_{F} - \frac{P^{+}_{(j-1)-(j)}\left(x^{+}_{(j-1)-(j)}\right)}{d} = P^{+}_{T}\left(\frac{x_T}{nd}\right) - P^{+}_{F} - \frac{P^{+}_{B}\left(x^{+}_{B}\right)}{d} \tag{4.26c}$$

$$M^{+0}_{(j-1)-(j)} = \left(\sum_{i=0}^{j-1} iW_i\right)d = P^{+}_{F}\left((j-1)d - x^{+}_{F}\right) + (j-1)d\frac{P^{+}_{B}\left(x^{+}_{B}\right)}{d} \tag{4.26d}$$

$$F^{+V}_{U(j-1)L(j)} = \left(\frac{\left(\left(\frac{P^{+}_{T} x_T}{nd}\right) - P^{+}_{F} - \frac{P^{+}_{B}\left(x^{+}_{B}\right)}{d}\right)\left(a_{(j-1)-(j)}\right) - \left(P^{+}_{F}\left((j-1)d - x^{+}_{F}\right) + (j-1)d\frac{P^{+}_{B}\left(x^{+}_{B}\right)}{d}\right)}{\left(a_{(j-1)-(j)} + jd\right)} \right) \tag{4.26e}$$

The maximum force due to positive panel shear forces in diagonal member U(j – 1)L(j) is calculated by establishing $\frac{d\left(F^{+V}_{U(j-1)L(j)}\right)}{dx} = 0$, which yields

$$\frac{d\left(F^{+V}_{U(j-1)L(j)}\right)}{dx} = 0 = \left(\frac{P^{+}_{T}}{nd} - \left(\frac{P^{+}_{B}}{d}\right)\right)\left(a_{(j-1)-(j)}\right) - \left(-P^{+}_{F} + (j-1)P^{+}_{B}\right) \tag{4.26f}$$

or

$$\frac{P^{+}_{T}}{n} = P^{+}_{B}\left(1 + \frac{(j-1)d}{a_{(j-1)-(j)}}\right) - \frac{P^{+}_{F} d}{a_{(j-1)-(j)}} \tag{4.26g}$$

Equation 4.26g is the criterion to establish the position of the concentrated live loads for maximum force in diagonal member U(j – 1)L(j), where j ≤ 2, due to positive shear in panel (j – 1) – (j). Maximum web member forces due to positive shear forces in panel (j – 1) – (j) typically occur with Cooper E series wheels 2, 3 or 4 at panel point (j). Therefore, except for very short truss panels, there may be no live loads in panels between panel point 0 and (j – 1), and, consequently, $P^{+}_{F} = 0$, which indicates that maximum force $F^{+max}_{U(j-1)L(j)}$, in diagonal member U(j – 1)L(j) due to positive shear in panel (j – 1) – (j), occurs when the total load on the span, P^{+}_{T}, divided by the number of panels in the span, n, and the factor $\left(1 + \frac{(j-1)d}{a_{(j-1)-(j)}}\right)$ equals the load $P^{+}_{(j-1)-(j)} = P^{+}_{B}$ in panel (j – 1) – (j). If, due to horizontal parallel chord members, $a_{(j-1)-(j)} = \infty$, Equation 4.26g for a Parker truss is the same as Equation 4.12c for a horizontal-chord Pratt truss.

For negative shear[117] in panel (j – 1) – (j) in Figure 4.15c, R_{yn} at panel point L(n) in the truss span of L = nd, traversed by a series of concentrated loads with resultant force P_T at distance x_T^- from panel point (0), is

$$R_{yn} = P_T^-\left(\frac{x_T^-}{nd}\right) \quad (4.27a).$$

$F^{-V}_{U(j-1)L(j)}$ = the vertical component of the force in diagonal member U(j – 1)L(j) due to negative panel shear forces, which from Equation 4.17c is

$$F^{-V}_{U(j-1)L(j)} = \frac{-V^-_{(j-1)-(j)}\left(a_{(j-1)-(j)}\right) + M^{-0}_{(j-1)-(j)}}{\left(a_{(j-1)-(j)} + jd\right)}$$

where in accordance with Figure 4.15c, $P^-_{(j-1)-(j)} = P^-_B$ = the resultant of concentrated loads between (j – 1) and (j) at distance $x^-_{(j-1)-(j)} = x_B$ from (j – 1). Additionally,

$P_F^- = \sum_{i=1}^{n-j} P_{Fi}^-$ = the resultant of concentrated live loads between panel point (j) and (n); P_{Fi}^- = resultant of concentrated live loads between panel point (j + i – 1) and panel point (j + i); x_F^- = distance from P_F^- to panel point (j) and $P_{(0)-(j-1)} = P^-_R$ = the resultant of concentrated live loads between panel point 0 and (j – 1), which are not considered in the section in panel (j – 1) – (j) for negative shear:

$$V^-_{(j-1)-(j)} = R_{yn} - \sum_{i=j}^{n} W_i$$

$$V^-_{(j-1)-(j)} = P_T^-\frac{x_T^-}{nd} - \frac{P_B^-\left(x_B^-\right)}{d} - P_F^- \quad (4.27b)$$

$$M^{-0}_{(j-1)-(j)} = -\left(R_{yn}\left(nd\right) - \left(\sum\nolimits_{i=j}^{n} iW_i\right)d\right) = -P_T^-\,x_T^- + jP_B^-\left(x_B^-\right) + P_F^-\left(jd + x_F^-\right) \quad (4.27c)$$

$$F^{V-}_{U(j-1)L(j)} = \frac{-\left(P_T^-\frac{x_T^-}{nd} - \frac{P_B^-\left(x_B^-\right)}{d} - P_F^-\right)\left(a_{(j-1)-(j)}\right) + \left(-P_T^-\,x_T^- + jP_B^-\left(x_B^-\right) + P_F^-\left(jd + x_F^-\right)\right)}{\left(a_{(j-1)-(j)} + jd\right)}$$

(4.27d)

The maximum force due to negative panel shear forces in diagonal member U(j – 1)L(j) is determined by establishing $\frac{d\left(F^{V-}_{U(j-1)L(j)}\right)}{dx} = 0$, which yields

$$\frac{d\left(F^{-V}_{U(j-1)L(j)}\right)}{dx} = 0 = \left(\frac{-P_T^-}{nd} + \left(\frac{P_B^-}{d}\right)\right)\left(a_{(j)-(j+1)}\right) - P_T^- + j\left(P_B^-\right) + P_F^- \quad (4.27e)$$

or

$$\frac{P_T^-}{n} = P_B^-\left(\frac{1+\frac{jd}{a_{(j-1)-(j)}}}{1+\frac{nd}{a_{(j-1)-(j)}}}\right) + \frac{P_F^-}{n\left(1+\frac{a_{(j-1)-(j)}}{nd}\right)} \tag{4.27f}$$

Equation 4.27f is the criterion to establish the position of the concentrated live loads for maximum force in diagonal member U(j − 1)L(j), where j ≤ 2, due to negative shear in panel (j − 1) − (j). Maximum web member forces due to negative shear forces in panel (j − 1) − (j) typically occur with Cooper E series wheels 2, 3 or 4 at panel point (j − 1). As a result, except for very short truss panels, there may be no live loads in panels between panel point (n) and (j). Therefore, $P_F^- = 0$, which indicates that maximum force $F^{-max}_{U(j-1)L(j)}$ in diagonal member U(j − 1)L(j), due to negative shear in panel (j − 1) − (j), occurs when the total load on the span, P_T^-, divided by the number of panels in the span, n, and the factor $\left(\frac{1+\frac{jd}{a_{(j-1)-(j)}}}{1+\frac{nd}{a_{(j-1)-(j)}}}\right)$ equals $P_B^- = P^-_{(j-1)-(j)}$ in panel (j − 1) − (j). If, due to horizontal parallel chord members, $a_{(j-1)-(j)} = \infty$, Equation 4.27f for a Parker truss is the same as Equation 4.13c for a horizontal chord Pratt truss.

The magnitude of the panel shear force does not depend on the orientation of the diagonal member in the panel. Maximum force $F^{+max}_{L(j-1)U(j)}$ in diagonal member L(j-1)U(j),[118] due to positive shear in panel (j − 1) − (j), occurs in accordance with the criterion of Equation 4.26g. Accordingly, the criterion to establish the position of the concentrated live loads for maximum force in diagonal member L(j − 1)U(j), where j ≤ n/2, due to negative shear in panel (j − 1) − (j) is provided by Equation 4.27f.

$F^{+}_{U(j)L(j)}$, the force in vertical member U(j)L(j)[119] due to positive panel shear forces, as shown in Figure 4.15d, is determined at the section through members U(j − 1)U(j) and L(j)L(j + 1), and from Equation 4.18a is

$$F^+_{U(j)L(j)} = \left(\frac{-V^+_{(j)-(j+1)}\left(a_{(j-1)-(j)}\right) + M^0_{(j)-(j+1)}}{\left(a_{(j-1)-(j)} + jd\right)}\right)$$

where in accordance with Figure 4.15d,

$P^+_{(j)-(j+1)} = P^+_B$ = the resultant of concentrated loads between panel point (j) and (j + 1) at distance $x^+_{(j)-(j+1)} = x^+_B$ from (j + 1);

$P_F^+ = \sum_{i=1}^{j} P_{Fi}^+$ = the resultant of concentrated live loads between panel point (0) and (j);

P_{Fi}^+ = the resultant of concentrated live loads in panel i between panel point (j – i + 1) and (j – i);

x_F^+ = the distance from P_F^+ to panel point (j); and

$P_{(j+1)-(n)} = P^+_R$ = the resultant of concentrated live loads between panel point (j + 1) and (n), which are not considered in the section in panel (j − 1) − (j) for positive shear

$$V^+_{(j)-(j+1)} = \text{the shear force in panel}\left(j\right)-\left(j+1\right) = \frac{P_T^+ x_T^+}{nd} - \frac{P_B^+ x_B^+}{d} - P_F^+ \tag{4.28a}$$

$$M^{+0}_{(j)-(j+1)} = \left(\sum_{i=0}^{j} iW_i\right)d = j\left(P_B^+ x_B^+\right) + P_F^+\left(jd - x_F^+\right) \tag{4.28b}$$

$$F^{+}_{U(j)L(j)} = \left(\frac{-\left(\frac{P^{+}_{T} x^{+}_{T}}{nd} - \frac{P^{+}_{B} x^{+}_{B}}{d} - P^{+}_{F} \right)\left(a_{(j-1)-(j)}\right) + \left(j\left(P^{+}_{B} x^{+}_{B}\right) + P^{+}_{F}\left(jd - x^{+}_{F}\right)\right)}{\left(a_{(j-1)-(j)} + jd\right)} \right) \tag{4.28c}$$

The maximum force due to positive panel shear forces in vertical member U(j)L(j) is calculated by establishing $\frac{d\left(F^{+}_{U(j)L(j)}\right)}{dx} = 0$, which yields

$$\frac{d\left(F^{+}_{U(j)L(j)}\right)}{dx} = 0 = \left(-\frac{P^{+}_{T}}{nd} + \frac{P^{+}_{B}}{d} \right) a_{(j-1)-(j)} + (j) P^{+}_{B} - P^{+}_{F} \tag{4.28d}$$

or

$$\frac{P^{+}_{T}}{n} = P^{+}_{B}\left(1 + \frac{jd}{a_{(j-1)-(j)}}\right) - \frac{P^{+}_{F} d}{a_{(j-1)-(j)}} \tag{4.28e}$$

Equation 4.28e is the criterion to establish the position of the concentrated live loads for maximum force in vertical member U(j)L(j), where $j + 1 \le n/2$, due to positive shear at section through members U(j – 1)U(j) and L(j)L(j + 1). Maximum web member forces due to positive shear forces in panel (j) – (j + 1) typically occur with Cooper E series wheels 2, 3 or 4 at panel point (j + 1). Consequently, except for very short truss panels, there may be no live loads in panels between panel point 0 and (j), and therefore, $P^{+}_{F} = 0$, which indicates that $F^{+max}_{U(j)L(j)}$ in vertical member U(j)L(j), due to positive shear at section through members U(j – 1)U(j) and L(j)L(j + 1), occurs when the total load on the span, P^{+}_{T}, divided by the number of panels in the span, n and $\left(1 + \frac{jd}{a_{(j-1)-(j)}}\right)$ equals the load $P^{+}_{B} = P^{+}_{(j)-(j+1)}$. If due to horizontal parallel chord members, $a_{(j-1)-(j)} = \infty$, Equation 4.28e for a Parker truss is the same as Equation 4.12c for panel (j) – (j + 1) in a horizontal-chord Pratt truss.

For negative shear at section through members U(j – 1)U(j) and L(j)L(j + 1) in Figure 4.15e, $F^{-}_{U(j)L(j)}$ = the force in vertical member U(j)L(j)[120] due to negative panel shear forces, which from Equation 4.18b is

$$F^{-}_{U(j)L(j)} = \left(\frac{V^{-}_{(j)-(j+1)}\left(a_{(j-1)-(j)}\right) - M^{-0}_{(j)-(j+1)}}{\left(a_{(j-1)-(j)} + jd\right)} \right)$$

where in accordance with Figure 4.15e,

$P^{-}_{B} = P^{-}_{(j)-(j+1)}$ = the resultant of concentrated loads between panel point (j) and (j + 1) at distance $x^{-}_{(j)-(j+1)} = x^{-}_{B}$ from panel point (j);

$P^{-}_{F} = \sum_{i=1}^{n-j-1} P^{-}_{Fi}$ = the resultant of concentrated live loads between panel point (j + 1) and (n);

P^{-}_{Fi} = the resultant of concentrated live loads between panel point (j + 1) and (j + i + 1);

x^{-}_{F} = the distance from P^{-}_{F} to panel point (j + 1); and

$P_{(0)-(j)} = P^{-}_{R}$ = the resultant of concentrated live loads between panel point 0 and (j), which are not considered in the section in panel (j – 1) – (j) for negative shear:

$$V^{-}_{(j)-(j+1)} = \frac{P^{-}_{T}x^{-}_{T}}{nd} - \frac{P^{-}_{B}\left(x^{-}_{B}\right)}{d} - P^{-}_{F} \tag{4.29a}$$

$$M^{-0}_{(j)-(j+1)} = -\left(nR_{yn} - \sum_{i=j+1}^{n} iW_i\right)d = -P^{-}_{T}\left(x^{-}_{T}\right) + (j+1)P^{-}_{B}\left(x^{-}_{B}\right) + P^{-}_{F}\left((j+1)d + x^{-}_{F}\right) \tag{4.29b}$$

$$F^{+}_{U(j)L(j)} = \left(\frac{-\left(\frac{P^{-}_{T}x^{-}_{T}}{nd} - \frac{P^{-}_{B}x^{-}_{B}}{d} - P^{-}_{F}\right)\left(a_{(j-1)-(j)}\right) - \left(-P^{-}_{T}\left(x^{-}_{T}\right) + (j+1)P^{-}_{B}\left(x^{-}_{B}\right) + P^{-}_{F}\left((j+1)d + x^{-}_{F}\right)\right)}{\left(a_{(j-1)-(j)} + jd\right)}\right) \tag{4.29c}$$

The maximum force due to negative panel shear forces in vertical member U(j)L(j) is calculated by establishing $\frac{d\left(F^{-}_{U(j)L(j)}\right)}{dx} = 0$, which yields

$$\left(\frac{P^{-}_{T}}{nd} - \frac{P^{-}_{B}}{d}\right)a_{(j-1)-(j)} + P^{-}_{T} - (j+1)P^{-}_{B} - P^{-}_{F} = 0 \tag{4.29d}$$

or

$$\frac{P^{-}_{T}}{n} = \frac{P^{-}_{B}\left(1 + \frac{(j+1)d}{a_{(j)-(j+1)}}\right)}{\left(\frac{nd}{a_{(j)-(j+1)}} + 1\right)} + \frac{P^{-}_{F}}{\left(\frac{a_{(j)-(j+1)}}{d} + n\right)} \tag{4.29e}$$

Equation 4.29e is the criterion to establish the position of the concentrated live loads for maximum force in vertical member U(j)L(j), where $j + 1 \le n/2$, due to negative shear at the section through members U(j − 1)U(j) and L(j)L(j + 1). Where there are no loads between joint (j + 1) and joint (n), $P^{-}_{F} = 0$ and, accordingly, $F^{-max}_{U(j)L(j)}$ in vertical member U(j)L(j), due to negative shear as shown in Figure 4.15e, occurs when the total load on the span, P^{-}_{T}, divided by the number of panels in the span, n, and $\frac{\left(1 + \frac{(j+1)d}{a_{(j)-(j+1)}}\right)}{\left(\frac{nd}{a_{(j)-(j+1)}} + 1\right)}$ equals load $P^{-}_{B} = P^{-}_{(j)-(j+1)}$.

The criteria of Equations 4.26g and 4.27f are used to hone the axle or wheel location of the Cooper's E load configuration for maximum forces in diagonal members in inclined-chord Pratt (Parker) and Warren trusses. The criteria of Equations 4.28e and 4.29e are used to refine the axle or wheel location of the Cooper's E load configuration for maximum forces in vertical members in inclined chord Pratt (Parker) trusses.

Since truss panels are relatively short,[121] the position of the Cooper's E live load for maximum panel moment and associated forces in vertical web members directly loaded by floorbeams can often be established by inspection[122] and Equations 4.11c and 4.14f. Once the locations of the live loads for maximum panel point moment and maximum panel shear forces are established, the maximum chord and maximum positive and negative web member forces are determined from static equilibrium.

Example 4.21 illustrates the use of mathematical criteria to determine the position of the Cooper's E80 live load for maximum panel point moment (Equation 4.11c for chord members) and maximum positive and negative panel shear forces (Equations 4.26g and 4.27f for diagonal web members and Equations 4.28e and 4.29e for vertical web members) with the assistance of Table 4.2 to determine the member forces from Cooper's E80 live load traversing the Parker truss of Example 4.19. Example 4.21 outlines the analysis of an inclined-chord Pratt (Parker) through-truss span with post lengths of 25 ft, 35 ft and 40 ft to illustrate the methodology for calculating member forces.[123]

Example 4.21

The six-panel inclined top chord Pratt (Parker) truss in Figure 4.12 and Example 4.19 is 150 ft long with h_1 = 25 ft, h_2 = 35 ft and h_3 = 40 ft. It is subjected to a uniform dead load of 1500 lb/ft and Cooper's E80 design live load. Determine the maximum and minimum forces in the truss members: n = 6, and the truss member dead load forces are shown in Table E4.21a (see Example 4.19).

TABLE E4.21a
Parker Truss Member Dead Load Forces

Parker Truss Member		Dead Load Force (kips)
L0L1	Bottom chord	93.75
L1L2	Bottom chord	93.75
L2L3	Bottom chord	107.1
L0U1	End post	−132.6
U1U2	Top chord	−115.4
U2U3	Top chord	−107.6
U1L1	Hanger	37.50
U2L2	Vertical	24.11
U3L3	Vertical	42.16
U1L2	Diagonal	18.93
U2L3	Diagonal	−2.88

Note that all dead load is assumed to be applied at bottom chord panel points in Table E4.21a.

Maximum live load forces in chord members are determined by considering moments about the opposite chord panel point, using Table 4.2 with the load positioned in accordance with the criterion of Equation 4.11c (live load location for maximum panel point moment). Table 4.1 estimates that the Cooper's live load locations for maximum panel point moments are wheels 3 or 4 at panel point L1 (25 ft from left end of truss); wheels 7 or 8 at L2 (50 ft from left end of truss) or wheels 10, 11 or 12 at L3 (75 ft from left end of truss).

For chord members U1U2, U2U3, L0L1, L1L2 and L2L3, starting with top chord member U1U2, Example 4.11 shows that the criterion of Equation 4.11c with k = 2 indicates axle 7 at L2 for maximum panel point moment at L2 and

$$F_{U1U2} = \frac{-12,423}{35}\left(\frac{\sqrt{725}}{25}\right) = -382.3 \text{ kips.}$$

For top chord member U2U3, as shown in Example 4.11, the criterion of Equation 4.11c with k = 3 indicates axle 11 at L3 for maximum panel point moment at L3 and

$$F_{U2U3} = \frac{-14,008}{40}\left(\frac{\sqrt{650}}{25}\right) = -357.1 \text{ kips.}$$

For bottom chord members L0L1 and L1L2, as shown in Example 4.11, the criterion of Equation 4.11c with k = 1 indicates axle 4 at L1 for maximum panel point moment at U1 and

$$F_{L1L2} = \frac{8098}{25} = 323.9 \text{ kips} = F_{L0L1}.$$

For bottom chord member L2L3, as shown in Example 4.11, the criterion of Equation 4.11c with k = 2 indicates axle 7 at L2 for maximum panel point moment at U2 and

$$F_{L2L3} = \frac{12423}{35} = 354.9 \text{ kips}.$$

Chord member maximum live load forces are shown in Table E4.21b.

TABLE E4.21b
Truss Chord Member Live Load Forces

Chord Member	Moment Center	Maximum Live Load Panel Point Moment (kip-ft)	Maximum Live Load Force (kips)
L1L2 = L0L1	U1 (k = 1)	8098	323.9 (T)
L2L3	U2 (k = 2)	12,423	354.9 (T)
U1U2	L2 (k = 2)	−12,423	−382.3 (C)
U2U3	L3 (k = 3)	−14,088	−357.1 (C)

For end post member L0U1, since F_{L1L2} is known,

$$F_{L0U1} = \alpha_{(1)\text{-}(2)}(-F_{L1L2}) = -1.414(F_{L1L2}) = -1.414(323.90) = -458.0 \text{ kips (C)}$$

For diagonal web members U1L2 and U2L3, maximum diagonal web member live load forces are determined by considering positive and negative panel shear forces with the loads positioned in accordance with the criteria of Equations 4.26g and 4.27f, respectively.

For diagonal web member U1L2, use Table 4.2 to determine positive and negative shear forces in panel 1-2 with Cooper's E live load. Considering positive shear in panel 1-2, j = 2 and $P^+_A = 0$ with axles 1 to 5 at L2. The criterion of Equation 4.26g with $P^+_F = 0$ is (also as shown in Table E4.21c)

$$\frac{P_T^+}{6} = P_{(1)-(2)}^+\left(1+\frac{(2-1)25}{37.5}\right) = 1.67P_{(1)-(2)}^+$$

or

$$\frac{P_T^+}{10.0} = P_{(1)-(2)}^+$$

TABLE E4.21c
Criterion for Live Load Position for Maximum Positive Shear in Panel 1-2

	Panel 1-2: Axle at L2	
	2	**3**
Distance to axle no. 1 (ft)	8	13
First axle on span	1	1
Distance to axle no. 18 (ft)	96	91
Last axle on span	18	18
Distance to uniform load (ft)	4(25) – 96 – 5 = –1 (off span)	4(25) – 91 – 5 = 4
Total load on span, P^+_T (kips)	568	568 + 4(4) = 584
$P^+_T/10.0$ (kips)	56.8	58.4
$P^+_{a(1)\text{-}(2)}$ = Sum of loads ahead of panel point L2 (kips)	20	20 + 40 = 60
$P^+_{b(1)\text{-}(2)}$ = Sum of loads ahead of and including panel point L2 (kips)	20 + 40 = 60	60 + 40 = 100
Criterion established	Yes	No

The criterion indicates axle 2 at L2 for maximum positive shear force in member U1L2. Therefore, the assumption that $P^+_F = 0$ is valid. From Equation 4.17a, the maximum positive force $F^{+V}_{U(1)L(2)}$ in member U1L2 is as shown in Table E4.21d.

TABLE E4.21d
Maximum Force in Diagonal Web Member U1L2 from Positive Shear in Panel 1-2 with Cooper's E80 Live Load Traversing a Parker Truss Span

	Axle at L2
Positive Shear in Panel 1-2 (j = 2)	**2**
Sum moments to L(n) (kip-ft)	32,728 − 568(1) = 32,160
R_{y0} (kips)	32,160/150 = 214.4
P_T (kips)	568
Sum moments in panel 1-2 ahead of and about L2 (kip-ft)	160
$\sum_{i=0}^{j-1} W_i = W_{(j-1)}$ (kips)	160/25 = 6.4
$V^+_{(j-1)-(j)} = R_{y0} - \sum_{i=0}^{j-1} W_i = R_{y0} - W_{(j-1)}$ (kips)	214.4 − 6.4 = 208.0
$M^{+0}_{(j-1)-(j)} = \left(\sum_{i=0}^{j-1} iW_i\right) d = W_{(j-1)}(j-1)d$ (kip-ft)	6.4(1)(25) = 160
$a_{(j-1)-(j)} = \left(\left(\frac{h_{(j)}}{h_{(j)} - h_{(j-1)}}\right) - j\right) d$ (ft)	$\left(\left(\frac{35}{35-25}\right) - 2\right) 25 = 37.5$

(Continued)

TABLE E4.21d (*Continued*)
Maximum Force in Diagonal Web Member U1L2 from Positive Shear in Panel 1-2 with Cooper's E80 Live Load Traversing a Parker Truss Span

Positive Shear in Panel 1-2 (j = 2)	Axle at L2
	2
$F^{V+}_{U1L2}\left(\left(a_{(1)-(2)}+2d\right)\right)=V^{+}_{(1)-(2)}\left(a_{(1)-(2)}\right)-M^{+0}_{(1)-(2)}$ (kip-ft)	208.0(37.5) – 160 = 7640.0
F^{+V}_{U1L2} (kips)	7640.0/(37.5 + (2)(25)) = 7640.0/87.50 = 87.31
$\propto_{(1)-(2)}=\dfrac{\sqrt{h^2_{(2)}+d^2}}{h_{(2)}}$	$\dfrac{\sqrt{25^2+25^2}}{25}=1.414$
$F^{+}_{U1L2}=F^{V+}_{U1L2}\left(\alpha_{(1)-(2)}\right)$ (kips)	87.31(1.414) = 123.5 (T)

Considering negative shear, j = 2 and $P^-_A = 0$ with axles 1 to 5 at L1. The criterion of Equation 4.27f with $P^-_F = 0$ is as follows (see Table E4.21e):

$$\frac{P^-_T}{6}=P^-_{(1)-(2)}\left(\frac{1+\dfrac{2d}{a_{(1)-(2)}}}{1+\dfrac{6d}{a_{(1)-(2)}}}\right)=P^-_{(1)-(2)}\left(\frac{1+\dfrac{2(25)}{37.5}}{1+\dfrac{6(25)}{37.5}}\right)=0.467P_{(1)-(2)}$$

or

$$\frac{P^-_T}{2.80}=P^-_{(1)-(2)}$$

TABLE E4.21e
Criterion for Live Load Position for Maximum Negative Shear in Panel 1-2

	Panel 1-2: Axle at L1	
	3	4
Distance to axle no. 1 (ft)	13	18
First axle on span	1	1
Last axle on span	7	7 (8 at L6)
Total load on span, P^-_T (kips)	232	258 – 26 = 232
$P^-_T/2.8$ (kips)	83	83
$P^-_{a(1)-(2)}$ = Sum of loads ahead of panel point L1 (kips)	60	100
$P^-_{b(1)-(2)}$ = Sum of loads ahead of and including panel point L1 (kips)	100	140
Criterion established	Yes	No

The criterion indicates axle 3 at L1 for maximum negative shear force in member U1L2. Therefore, the assumption that $P^{-}_{F} = 0$ is valid. From Equation 4.17c, the maximum negative force, $F^{-V}_{U(1)L(2)}$, in member U1L2 is given in Table E4.21f.

TABLE E4.21f
Maximum Force in Diagonal Web Member U1L2 from Negative Shear in Panel 1-2 with Cooper's E80 Live Load Traversing a Parker Truss Span

	Axle at L1	
Negative Shear in Panel 1-2	**3**	**4**
Sum moments to L0 (kip-ft)	4310 + 232(25 − (37 − 13)) = 4542	5702
R_{y6} (kips)	4542/150 = 30.28	5702/150 = 38.01
Sum moments in panel 1-2 ahead of and about panel point L1 (kip-ft)	460	960
$W_{(2)}$ (kips)	460/25 = 18.40	38.40
$V^{-}_{(1)-(2)} = R_{y6} - \sum_{i=2}^{6} W_i = R_{y6} - W_{(2)}$ (kips)	30.28 − 18.40 = 11.88	38.01 − 38.40 = −0.39
$M^{-0}_{(1)-(2)} = -\left(6R_{y6} - 2W_{(2)}\right)d$ (kip-ft)	−(6(30.28) − 2(18.40))(25) = −3622	−(6(38.01) − 2(38.40))(25) = −3782
$-V^{-}_{(1)-(2)}\left(a_{(1)-(2)}\right) + M^{-0}_{(1)-(2)}$ (kip-ft)	−11.88(37.5) − 3622 = −4067.5	0.39(37.5) − 3782 = −3767
$F^{V-}_{U(1)L(2)}$ (kips)	−4067.5/(37.5 + 2(25)) = −46.49	−43.08
$F^{-}_{U1L2}\left(\alpha_{(1)-(2)}\right)$ (kips)	−46.49(1.414) = −65.7 (C)	−60.9 (C)

For diagonal web member U2L3, use Table 4.2 to determine positive and negative shear forces in panel 2-3 with Cooper's E live load. Considering positive shear and j = 3,

$$a_{(2)-(3)} = \left(\left(\frac{h_{(3)}}{h_{(3)} - h_{(2)}}\right) - 3\right)d = \left(\left(\frac{40}{40-35}\right) - 3\right)25 = 5.00(25) = 125.0 \text{ ft}$$

The criterion of Equation 4.26g with $P^{+}_{F} = 0$ is as follows, as shown in Table E4.21g:

$$\frac{P^{+}_{T}}{6} = P^{+}_{(2)-(3)}\left(1 + \frac{(3-1)d}{a_{(2)-(3)}}\right) = P^{+}_{(2)-(3)}\left(1 + \frac{(3-1)25}{125}\right) = 1.40P_{(2)-(3)}$$

or

$$\frac{P^{+}_{T}}{8.40} = P^{+}_{(2)-(3)}$$

TABLE E4.21g
Criterion for Live Load Position for Maximum Positive Shear in Panel 2-3

	Panel 2-3: Axle at L3	
	2	**3**
Distance to axle no. 1 (ft)	8	13
First axle on span	1	1
Distance to axle no. 18 (ft)	96	91
Last axle on span	14	14 (15 at L6)
Distance of uniform load (ft)	75 – 96 – 5 = –26 off span	off span
Total load on span, P^+_T (kips)	464	464
$P^+_T/8.4$ (kips)	55	55
$P^+_{a(2)\text{-}(3)}$ = Sum of loads ahead of panel point L3 (kips)	20	60
$P^+_{b(2)\text{-}(3)}$ = Sum of loads ahead of and including panel point L3 (kips)	60	100
Criterion established	Yes	No

The criterion indicates axle 2 at L3 for maximum positive force in U2L3. Therefore, the assumption that $P^+_F = 0$ is valid. From Equation 4.17a, maximum positive force $F^{+V}_{U(2)L(3)}$ in member U2L3 is as shown in Table E4.21h.

TABLE E4.21h
Maximum Force in Diagonal Web Member U2L3 from Positive Shear in Panel 2-3 with Cooper's E80 Live Load Traversing a Parker Truss Span

	Axle at L3
Positive Shear in Panel 2-3	**2**
Sum moments to L6 (kip-ft)	17,456 + 464(75 – (79 – 8)) = 17,456 + 464(4) = 19,312
R_{y0} (kips)	19,312/150 = 128.75
Sum moments in panel 2-3 ahead of and about L3 (kip-ft)	160
$W_{(2)}$ (kips)	160/25 = 6.40
$V^+_{(2)-(3)} = R_{y0} - \sum_{i=0}^{2} W_i = R_{y0} - W_{(2)}$ (kips)	128.75 – 6.40 = 122.35
$M^{+0}_{(2)-(3)} = d\sum_{i=0}^{2} iW_i = d(2)W_{(2)}$ (kip-ft)	6.40(2)(25) = 320
$F^{+V}_{U2L3}\left(a_{(2)-(3)} + 3d\right) = V^+_{(2)-(3)}a_{(2)-(3)} - M^0_{(2)-(3)}$ (kip-ft)	122.35(125) – 320 = 14,974
F^{+V}_{U2L3} (kips)	14,974/200 = 74.87
$\propto_{(2)-(3)} = \frac{\sqrt{h^2_{(3)} + d^2}}{h_{(3)}}$	$\frac{\sqrt{35^2 + 25^2}}{35} = 1.229$
$F^+_{U2L3}\left(\alpha_{(2)-(3)}\right)$ (kips)	74.87(1.229) = 92.02 (T)

Considering negative shear, j = 3, and the criterion of Equation 4.27f with $P^-_F = 0$ is as shown in Table E4.21i):

$$\frac{P_T^-}{6} = P_{(2)-(3)}^- \left(\frac{1 + \frac{3d}{a_{(2)-(3)}}}{1 + \frac{6d}{a_{(2)-(3)}}} \right) = P_{(2)-(3)}^- \left(\frac{1 + \frac{3(25)}{125}}{1 + \frac{6(25)}{125}} \right) = 0.727 P_{(2)-(3)}$$

or

$$\frac{P_T^-}{4.36} = P_{(2)-(3)}^-$$

TABLE E4.21i
Criterion for Live Load Position for Maximum Negative Shear in Panel 2-3

	Panel 2-3: Axle at L2		
	2	**3**	**4**
Distance to axle no. 1 (ft)	8	13	18
First axle on span	1	1	1
Last axle on span	10	10	11
Total load on span, P^-_T (kips)	304	304	344
$P^-_T/4.36$ (kips)	70	70	79
$P^-_{a(2)-(3)}$ = Sum of loads ahead of panel point L2 (kips)	20	60	100
$P^-_{b(2)-(3)}$ = Sum of loads ahead of and including panel point L2 (kips)	60	100	140
Criterion established	No	Yes	No

The criterion indicates axle 3 at L2 for maximum negative force in U2L3. Therefore, the assumption that $P^-_F = 0$ is valid. From Equation 4.17c, the maximum negative force $F^{-V}_{U(2)L(3)}$ in member U2L3 is as shown in Table E4.21j.

TABLE E4.21j
Maximum Force in Diagonal Web Member U2L3 from Negative Shear in Panel 2-3 with Cooper's E80 Live Load Traversing a Parker Truss Span

	Axle at L2
Negative Shear in Panel 2-3	**3**
Sum moments to L0 (kip-ft)	9264 + 304(50 − (56 − 13)) = 11,392.0
R_{y6} (kips)	11,392/150 = 75.95
Sum moments in panel 2-3 ahead of and about panel point L2	460
$W_{(3)}$ (kips)	460/25 = 18.40

(Continued)

TABLE E4.21j (*Continued*)

Maximum Force in Diagonal Web Member U2L3 from Negative Shear in Panel 2-3 with Cooper's E80 Live Load Traversing a Parker Truss Span

Negative Shear in Panel 2-3	Axle at L2 3
$V^-_{(2)-(3)} = R_{y6} - \sum_{i=3}^{6} W_i = R_{y6} - W_{(3)}$ (kips)	75.95 − 18.40 = 57.55
$M^{-0}_{(2)-(3)} = -\left(6R_{y6} - 3W_{(3)}\right)d$ (kip-ft)	−(6(75.95) − 3(18.40))25 = −10012.5
$-V^-_{(2)-(3)}\left(a_{(2)-(3)}\right) + M^{-0}_{(2)-(3)}$ (kip-ft)	−57.55(125.0) − 10012.5 = −17,205.3
F^{-V}_{U2L3} (kips)	−17,205.3/(125 + 3(25)) = −86.03 (C)
$F^-_{U2L3}\left(\alpha_{(2)-(3)}\right)$ (kips)	−86.03(1.229) = −105.73 (C)

Vertical web members U1L1, U2L2 and U3L3:

For vertical web member U1L1, maximum vertical member live load forces from loads applied directly as floorbeam reactions occur with the load positioned for maximum live load moment (Equation 4.11c). Use Table 4.2 to determine moments about panel point L1 with Cooper's E live load located in accordance with Equation 4.11c with k = 1 (see Example 4.11). Wheel 4 and wheel 13 at L1 in the span L0-L2 result in F_{U1L1} = 151.28 kips. If the alternate load is considered by placing one of the central 50 kip wheel loads at L1,

$$F_{U1L1} = (50/25)(14 + 19 + 25 + 20) = 156 \text{ kips}$$

For U2L2, the maximum vertical web member live load forces are determined by considering positive and negative panel shear forces with the load positioned in accordance with the criteria of Equations 4.28e and 4.29e, respectively, with Table 4.2 used to determine positive and negative shear forces with Cooper's E live load. The section through member U2L2 intersects the top chord in panel 1-2 and bottom chord in panel 2-3.

Considering positive shear, j = 2; $a_{(1)-(2)}$ = 37.5 ft (section through member U2L2 intersects the top chord in panel 1-2) and the criterion of Equation 4.28e with $P^+_F = 0$ is (also see Table E4.21k)

$$\frac{P^+_T}{6} = P^+_{(2)-(3)}\left(1 + \frac{2d}{a_{(1)-(2)}}\right) = 2.33P^+_{(2)-(3)}$$

or

$$\frac{P^+_T}{14.0} = P^+_{(2)-(3)}$$

TABLE E4.21k
Criterion for Cooper's E80 Live Load Position for Force in Member U2L2 from Positive Shear Force in Panel 2-3

	Panel 2-3: Axle at L3	
	2	3
Distance to axle no. 1 (ft)	8	13
First axle on span	1	1
Distance to axle no. 18 (ft)	96	91
Last axle on span	14	14 (15 at L6)
Distance of uniform load (ft)	75 – 96 – 5 = −26 off span	off span
Total load on span, P^{+}_{T} (kips)	464	464
$P^{+}_{T}/14.0$ (kips)	33	33
$P^{+}_{a(2)\text{-}(3)}$ = Sum of loads ahead of panel point L3 (kips)	20	60
$P^{+}_{b(2)\text{-}(3)}$ = Sum of loads ahead of and including panel point L3 (kips)	60	100
Criterion established	Yes	No

The criterion indicates axle 2 at L3 for maximum positive force in U2L2. Therefore, the assumption that $P^{+}_{F} = 0$ is valid. Positive shear at the section through U1U2 and L2L3 in Figure 4.15d provides Equation 4.18a for maximum positive force F^{+}_{U2L2} in vertical member U2L2, as shown in Table E4.21l.

TABLE E4.21l
Maximum Force in Vertical Web Member U2L2 from Positive Shear in Panel 2-3 with Cooper's E80 Live Load Traversing a Parker Truss Span

	Axle at L3
	2
Sum moments to L6 (kip-ft)	17,456 + 464(75 − (79−8)) = 17,456 + 464(4) = 19,312
R_{y0} (kips)	19,312/150 = 128.75
Sum moments ahead of panel point L3 (kip-ft)	160
$W_{(2)}$ (kips)	160/25 = 6.40
$V^{+}_{(2)-(3)} = R_{y0} - \sum_{i=0}^{2} W_i = R_{y0} - W_{(2)}$ (kips)	128.75 − 6.40 = 122.35
$M^{+0}_{(2)-(3)} = \left(\sum_{i=0}^{2} iW_i\right)d = W_{(2)}(2d)$ (kip-ft)	6.40(2)(25) = 320
$-V^{+}_{(2)-(3)}\left(a_{(1)-(2)}\right) + M^{+0}_{(2)-(3)}$ (kip-ft)	−122.35(37.5) + 320 = −4268.1
F^{+}_{U2L2} (kips)	−4268.1/87.5 = −48.8 (C)

Considering negative shear and the criterion of Equation 4.29e with $P^-_F = 0$, as shown in Table E4.21m:

$$\frac{P_T^-}{5} = \frac{P_{(2)-(3)}^-\left(1+\dfrac{(2+1)d}{a_{(2)-(3)}}\right)}{\left(\dfrac{6d}{a_{(2)-(3)}}+1\right)} = \frac{P_{(2)-(3)}^-\left(1+\dfrac{(2+1)25}{37.5}\right)}{\left(\dfrac{6(25)}{37.5}+1\right)} = 0.600P_{(2)-(3)}^-$$

or

$$\frac{P_T^-}{3.60} = P_{(2)-(3)}^-$$

TABLE E4.21m
Criterion for Cooper's E80 Live Load Position for Force in Member U2L2 from Negative Shear Force in Panel 2-3

	Panel 2-3: Axle at L2		
	2	**3**	**4**
Distance to axle no. 1 (ft)	8	13	18
First axle on span	1	1	1
Last axle on span	10	10	11
Total load on span, P^-_T (kips)	304	304	344
$P^-_T/3.60$ (kips)	84	84	96
$P^-_{a(2)-(3)}$ = Sum of loads ahead of panel point L2 (kips)	20	60	100
$P^-_{b(2)-(3)}$ = Sum of loads ahead of and including panel point L2 (kips)	60	100	140
Criterion established	No	Yes	No

The criterion indicates axle 3 at L2 for maximum negative force in member U2L2. Therefore, the assumption that $P^-_F = 0$ is valid. Maximum negative force in U2L2 from Equation 4.18b is as shown in Table E4.21n.

TABLE E4.21n
Maximum Force in Vertical Web Member U2L2 from Negative Shear in Panel 2-3 with Cooper's E80 Live Load Traversing a Parker Truss Span

	Axle at L2
	3
Sum moments to L0 (kip-ft)	9264 + 304(50 − (56 − 13)) = 11,392.0
R_{y6} (kips)	11,392/150 = 75.95
Sum moments ahead of panel point L2 (kip-ft)	460
$\sum_{i=0}^{3} W_i = W_{(3)}$ (kips)	460/25 = 18.40

TABLE E4.21n (*Continued*)
Maximum Force in Vertical Web Member U2L2 from Negative Shear in Panel 2-3 with Cooper's E80 Live Load Traversing a Parker Truss Span

	Axle at L2
	3
$V^{-}_{(2)-(3)} = R_{y6} - \sum_{i=2+1}^{6} W_i = R_{y6} - W_{(3)}$ (kips)	75.95 − 18.40 = 57.55
$M^{-0}_{(2)-(3)} = -\left(6R_{y6} - (2+1)W_{(2+1)}\right)d$ (kip-ft)	−(6(75.95) − (2 + 1)18.40)25 = −10,012.5
$V^{-}_{(2)-(3)}\left(a_{(1)-(2)}\right) - M^{-0}_{(2)-(3)}$ (kip-ft)	57.55(37.5) + 10,012.5 = 12,170.6
F^{-}_{U2L2} (kips)	12,170.6/(37.5 +2(25)) = 12,170.6/87.5 = 139.1 (T)

For vertical web member U3L3, considering the equilibrium of joint U3,

$$F_{U3L3} = 2(-F_{U2U3})\left(\frac{h_3 - h_2}{\sqrt{(h_3 - h_2)^2 + d^2}}\right) = 2(357.1)(0.196) = 140.0 \text{ kips.}$$

Web member live load forces are shown in Table E4.21o.

TABLE E4.21o
Truss Web Member Live Load Forces

	Maximum Live Load Force (kips)	
Web Member	Positive Panel Shear	Negative Panel Shear
U1L2 (diagonal)	123.5	−65.70
U2L3 (diagonal)	92.02	−105.7
U1L1 (hanger)	156.0	
U2L2 (vertical)	−48.8	139.1
U3L3 (vertical)	140.0	

Table E4.21p shows the member forces for the Parker truss in Figure 4.12a.

TABLE E4.21p
Dead Load and Maximum Cooper's E80 Live Load Forces for the Parker Truss in Figure 4.12a

Parker Truss Member	Dead Load Force (kips) W = 37.50 kips	Maximum Live Load Force (kips) Cooper's E80
L0L1	93.75	323.9
L1L2	93.75	323.9
L2L3	107.13	354.9
L0U1	−132.60	−458.0

(*Continued*)

TABLE E4.21p (*Continued*)
Dead Load and Maximum Cooper's E80 Live Load Forces for the Parker Truss in Figure 4.12a

Parker Truss Member	Dead Load Force (kips) W = 37.50 kips	Maximum Live Load Force (kips) Cooper's E80
U1U2	−115.38	−382.3
U2U3	−107.55	−357.1
U1L1	37.50	156.0
U2L2	24.11	−48.80 (positive shear)
		139.1 (negative shear)
U3L3	42.16	140.0
U1L2	18.93	123.5 (positive shear)
		−65.70 (negative shear)
U2L3	−2.88	92.02 (positive shear)
		−105.70 (negative shear)

Table E4.21q shows the maximum and minimum member forces for the Parker truss in Figure 4.12a.

TABLE E4.21q
Maximum and Minimum Member Forces for the Parker Truss in Figure 4.12a Traversed by Cooper's E80 Live Load

Parker Truss Member	Minimum Force (kips)	Maximum Force (kips)
L0L1	93.75	417.7
L1L2	93.75	417.7
L2L3	107.13	462.0
L0U1	−132.60	−590.6
U1U2	−115.38	−497.7
U2U3	−107.55	−464.7
U1L1	37.50	193.5
U2L2	−24.7	163.2
U3L3	42.16	182.2
U1L2	−46.8	142.4
U2L3	89.1	−108.6

Members U2L2, U1L2 and U2L3 experience stress reversal from live load (not considering live load impact).

Comparison of the maximum member forces in Examples 4.11 and 4.21 indicates that smaller member sizes are required for many members of the Parker truss. Smaller chord member forces in the Parker truss are due to the greater distance between chord members. Smaller web member forces, particularly for diagonal members, in the Parker truss are because the inclined top chord members resist part of the panel shear. However, the Parker truss requires longer members and more complex connection[124] geometry.

4.5.2.3 Analysis by Influence Lines

As is the case for horizontal parallel-chord trusses, influence lines are useful for establishing the maximum forces in inclined-chord truss members. Equations 4.16a and 4.16b for chord member forces and Equations 4.17b, 4.17d, 4.18a and 4.18b for web member forces can be utilized to construct influence lines for inclined-chord truss member forces. The influence lines are readily calculated with unit loads placed at appropriate panel points as shown in Example 4.22. Example 4.22 outlines the influence line analysis of inclined-chord Pratt (Parker) through-truss spans with post lengths of 25 ft, 35 ft and 40 ft to illustrate the methodology to determine member forces.[125] Once constructed, influence lines can be used to determine the maximum and minimum live load forces in inclined-chord truss chord and web members with the appropriate live load location.

Example 4.22

Construct the influence lines for the six-panel inclined top chord Pratt (Parker) truss in Figure 4.12. The Parker truss is 150 ft long with h_1 = 25 ft, h_2 = 35 ft and h_3 = 40 ft. The influence lines for bottom chord member forces using Equation 4.16a are calculated as follows and as shown in Figures 4.22a–i:

$$F_{L(j-1)L(j)} = \left((j-1)\sum_{i=0}^{n} \frac{(n-i)}{n} W_i - \sum_{i=0}^{j-1} (j-1-i) W_i \right) \frac{d}{h_{(j-1)}}$$

$$F_{L1L2} = \left(\frac{(2-1)}{6} \sum_{i=0}^{6} (6-i) W_i - \sum_{i=0}^{1} (1-i) W_i \right) \frac{25}{25} = \left(\frac{1}{6} \sum_{i=0}^{6} (6-i) W_i \right).$$

For $W_1 = 1$,

$$F_{L1L2} = \left(\frac{1}{6}(5)(1) \right) = 0.833 \text{ with unit load at L1.}$$

For $W_2 = 1$,

$$F_{L1L2} = \left(\frac{1}{6}(4)(1) \right) = 0.667 \text{ with unit load at L2.}$$

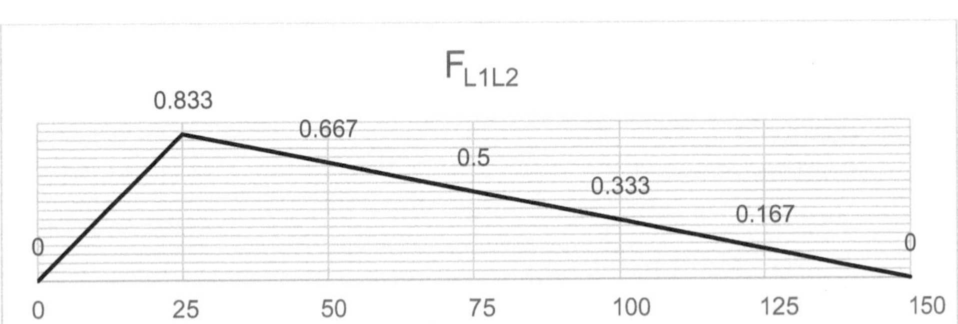

FIGURE E4.22a Influence line for member L1L2 of a Parker truss span.

$$F_{L2L3} = \left(\frac{(3-1)}{6}\sum_{i=0}^{6}(6-i)W_i - \sum_{i=0}^{2}(2-i)W_i\right)\frac{25}{35} = \left(\frac{2}{6}\sum_{i=0}^{6}(6-i)W_i - \sum_{i=0}^{2}(2-i)W_i\right)(0.714)$$

$W_2 = 1,$

$$F_{L2L3} = \left(\frac{2}{6}(4)(1) - 0\right)(0.714) = 0.952 \text{ with unit load at L2}$$

$W_3 = 1,$

$$F_{L2L3} = \left(\frac{2}{6}(3)(1) - 0\right)(0.714) = 0.714 \text{with unit load at L3}$$

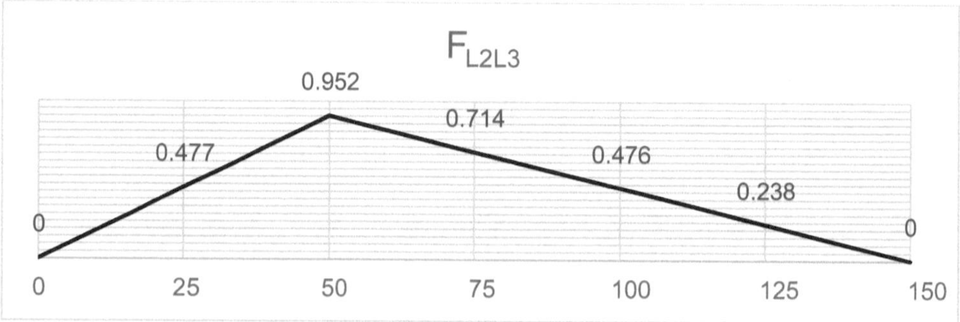

FIGURE E4.22b Influence line for L2L3 of a Parker truss span.

Influence lines for top chord member forces using Equation 4.16b are calculated as follows:

$$F_{U(j-1)U(j)} = -\left((j)\sum_{i=0}^{n}\frac{(n-i)}{n}W_i - \sum_{i=0}^{j}(j-i)W_i\right)\left(\frac{\sqrt{\left(h_{(j)} - h_{(j-1)}\right)^2 + d^2}}{h_{(j)}}\right)$$

$$F_{U1U2} = -\left(\frac{2}{6}\sum_{i=0}^{6}(6-i)W_i - \sum_{i=0}^{2}(2-i)W_i\right)\left(\frac{\sqrt{725}}{35}\right).$$

For $W_1 = 1,$

$$F_{U1U2} = -\left(\frac{2}{6}(5)(1) - (1)(1)\right)(0.769) = -0.513 \text{ with unit load at L1.}$$

For $W_2 = 1,$

$$F_{U1U2} = -\left(\frac{2}{6}(4)(1) - 0\right)(0.769) = -1.025 \text{ with unit load at L2.}$$

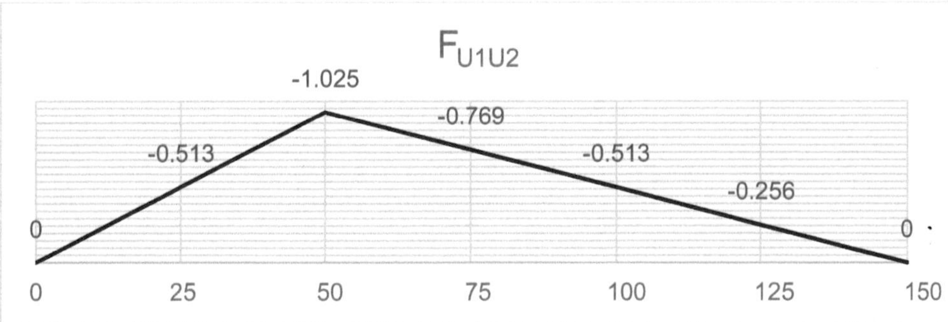

FIGURE E4.22c Influence line for U1U2 of a Parker truss span.

$$F_{U2U3} = -\left(\frac{3}{6}\sum_{i=0}^{6}(6-i)W_i - \sum_{i=0}^{3}(3-i)W_i\right)\left(\frac{\sqrt{650}}{40}\right)$$

For $W_2 = 1$,

$$F_{U2U3} = -\left(\frac{3}{6}(4)(1) - (1)(1)\right)(0.637) = -0.637 \text{ with unit load at L2.}$$

For $W_3 = 1$,

$$F_{U2U3} = -\left(\frac{3}{6}(3)(1) - 0\right)(0.637) = -0.956 \text{ with unit load at L3.}$$

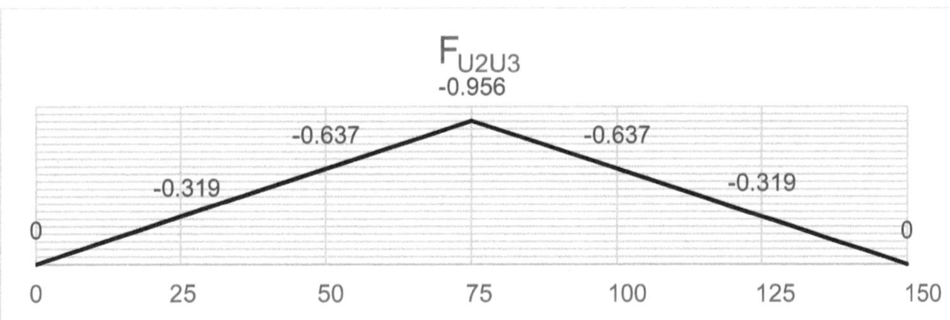

FIGURE E4.22d Influence line for U2U3 of a Parker truss span.

Influence lines for diagonal web member forces using Equations 4.17b and 4.17d for positive and negative panel shear forces, respectively, Panel 1-2 j = 2.

For unit load at L1 (i = 1) (negative panel shear),

$$W_1 = 1$$

$$W_2 = W_3 = W_4 = W_5 = 0$$

$$R_{y6} = \frac{1}{6}\sum_{i=0}^{6} W_i(i) = \frac{1}{6}(1)(1) = 0.167$$

$$V^{-}_{(1)\text{-}(2)} = R_{y6} - \sum_{i=2}^{6} W_i = 0.167$$

$$M^{-0}_{(1)-(2)} = -\left(6(0.167) - \sum_{i=2}^{6} iW_i\right)25 = -(1-0)25 = -25$$

$$F^{-}_{U1L2} = \left(\frac{-0.167(37.5) - 25}{87.5}\right)1.414 = -0.505 \text{ with unit load at i}$$

For unit load at L2 (i = 2) (positive panel shear),

$$W_2 = 1$$

$$W_1 = W_3 = W_4 = W_5 = 0$$

$$R_{y0} = \frac{1}{6}\sum_{i=0}^{6} W_i(6-i) = \frac{1}{6}(1)(4) = 0.667$$

$$V^{+}_{(1)\text{-}(2)} = R_{y0} - \sum_{i=0}^{1} W_i = 0.667 - 0 = 0.667$$

$$M^{+0}_{(1)\text{-}(2j)} = 25\sum_{i=0}^{1} iW_i = 25(0) = 0$$

$$F^{+}_{U1L2} = \left(\frac{0.667(37.5) - 0}{(87.5)}\right)1.414 = 0.286(1.414) = 0.404 \text{ with unit load at i} = 2$$

$$x^{+} = \frac{0.404(25)}{0.404 + 0.505} = 11.11\text{ft}$$

$$x^{-} = \frac{0.505(25)}{0.404 + 0.505} = 13.89\text{ft} \ .$$

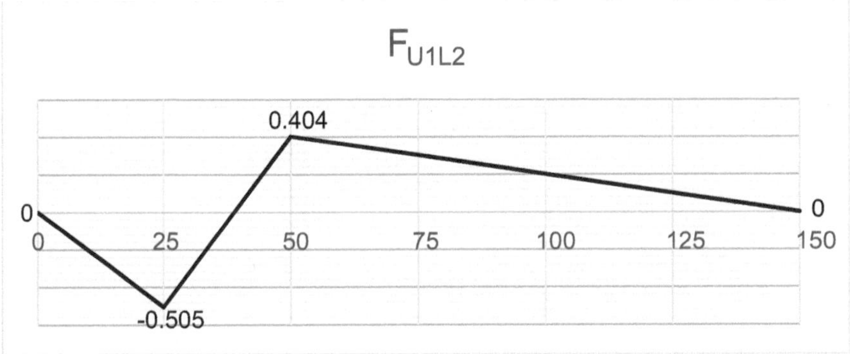

FIGURE E4.22e Influence line for U1L2 of a Parker truss span.

For panel 2-3, j = 3.
For unit load at L2 (i = 2) (negative panel shear),

$$W_2 = 1$$

$$W_1 = W_3 = W_4 = W_5 = 0$$

$$R_{y6} = \frac{1}{6}\sum_{i=0}^{6} W_i(i) = \frac{1}{6}(2)(1) = 0.333$$

$$V^-_{(2)-(3)} = 0.333 - \sum_{i=3}^{6} W_i = 0.333$$

$$M^{-0}_{(2)-(3)} = -\left(6(0.333) - \sum_{i=3}^{6} iW_i\right)25 = -(2-0)25 = -50$$

$$F^-_{U2L3} = \left(\frac{-0.333(125) - 50}{200}\right)1.229 = -0.458(1.229) = -0.563 \text{ with unit load at i = 2}$$

For unit load at L3 (i = 3) (positive panel shear),

$$W_3 = 1$$

$$W_1 = W_2 = W_4 = W_5 = 0$$

$$R_{y0} = \frac{1}{6}\sum_{i=0}^{6} W_i(6-i) = \frac{1}{6}(3)(1) = 0.500$$

$$V^+_{(2)-(3)} = 0.500 - \sum_{i=0}^{2} W_i = 0.500 - 0 = 0.500$$

$$M^{+0}_{(2)-(3)} = d\sum_{i=0}^{2} iW_i = 0$$

$$F^+_{U2L3} = \left(\frac{0.500(125) - 0}{200}\right)1.229 = -0.313(1.229) = 0.384 \text{ with unit load at i = 3}$$

$$x^{+} = \frac{0.384(25)}{0.384+0.563} = 10.14\,\text{ft}$$

$$x^{-} = \frac{0.563(25)}{0.384+0.563} = 14.86\text{ft}$$

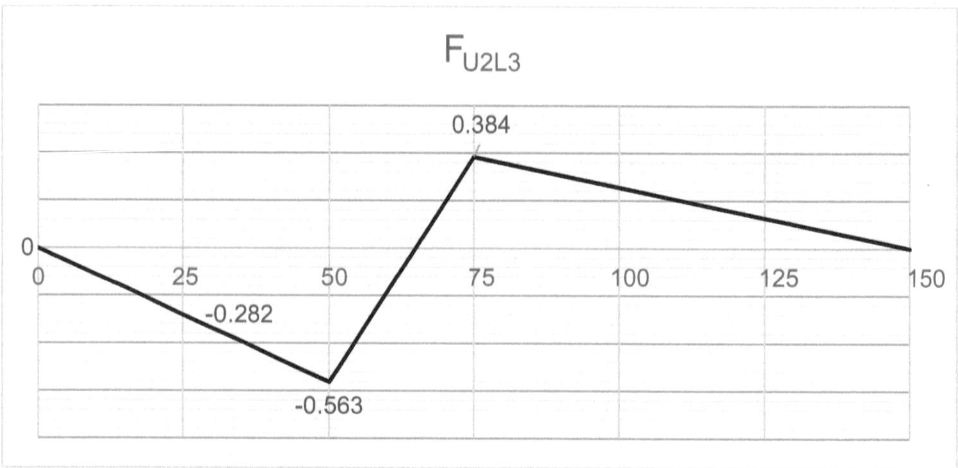

FIGURE E4.22f Influence line for U2L3 of a Parker truss span.

Influence lines for vertical web members not directly loaded by floorbeam reaction using Equations 4.18a and 4.18b are calculated as member U2L2 (panel 2-3)(j = 2).

For unit load at L2 (negative panel shear),

$$R_{y6} = \frac{1}{6}\sum_{i=0}^{6} W_i(i) = \frac{1}{6}(1)(2) = 0.333$$

$$V^{-}_{(2)\text{-}(3)} = R_{y6} - \sum_{i=3}^{6} W_i = 0.333$$

$$M^{-0}_{(2)\text{-}(3)} = -25(6R_{y6} - \sum_{i=3}^{6} iW_i) = -25\big(6(0.333)-0\big) = -50$$

$$F^{-}_{U(2)L(2)} = \frac{V^{-}_{(2)-(3)}\left(a_{(1)-(2)}\right) - M^{-0}_{(2)-(3)}}{\left(a_{(1)-(2)} + 2d\right)} = \frac{0.333(37.5)+50}{(87.5)} = 0.714 \quad \text{with unit load at i = 2}$$

For unit load at L3 (positive panel shear),

$$R_{y0} = \frac{1}{6}\sum_{i=0}^{6} W_i(6-i) = \frac{1}{6}(1)(3) = 0.500$$

$$V^{+}_{(2)\text{-}(3)} = R_{y0} - \sum_{i=0}^{2} W_i = 0.500 - 0 = 0.500$$

$$M^{+0}_{(2)\text{-}(3)} = d\sum_{i=0}^{2} iW_i = 25(0)(1) = 0$$

$$F^{+}_{U(2)L(2)} = \frac{-V^{+}_{(2)-(3)}\left(a_{(1)-(2)}\right) + M^{+0}_{(2)-(3)}}{\left(a_{(1)-(2)} + 2d\right)} = \frac{-0.500(37.5)+0}{(87.5)} = -0.214 \text{ with unit load at i = 3}$$

$$x^{+} = \frac{0.214(25)}{0.214+0.714} = 5.77\text{ft}$$

$$x^{-} = \frac{0.714(25)}{0.214+0.714} = 19.23\text{ft}$$

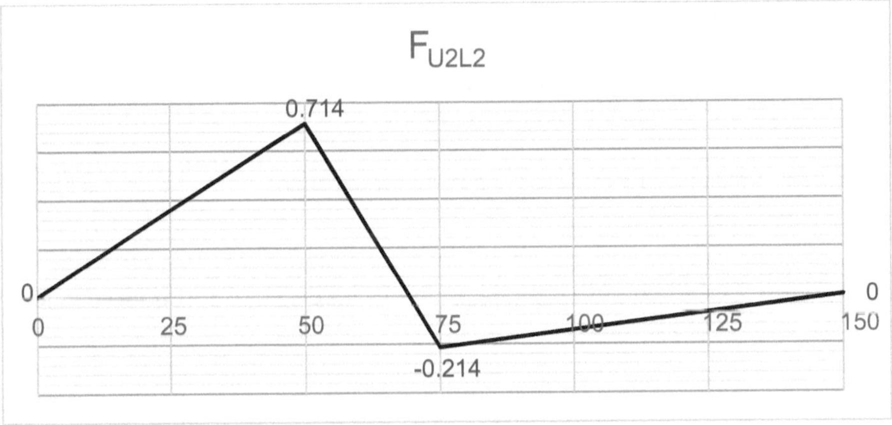

FIGURE E4.22g Influence line for U2L2 of a Parker truss span.

For member U3L3 (j = 3), unit load at L3 is

$$R_{y0} = R_{yn} = 0.500$$

$$V^{-} = V^{+} = 0.500 - 0.500 - 0$$

$$M^{-} = -d(nR_{yn} - \sum_{i=4}^{6} iW_i) = -25\left(6\left(0.500\right)\right) = -75$$

$$M^{+} = d\sum_{i=0}^{3} iW_i) = 25(3)(1) = 75$$

$$F^-_{U3L3} = \frac{V^-_{(3)-(4)}\left(a_{(2)-(3)}\right) - M^-_{(3)-(4)}}{\left(a_{(2)-(3)} + 3d\right)} = \frac{75}{(200.0)} = 0.375 \text{ with unit load at i = 3}$$

$$F^+_{U3L3} = \frac{-V^+_{(3)-(4)}\left(a_{(2)-(3)}\right) + M^+_{(3)-(4)}}{\left(a_{(2)-(3)} + 3d\right)} = \frac{75}{(200.0)} = 0.375 \text{ with unit load at i = 3}$$

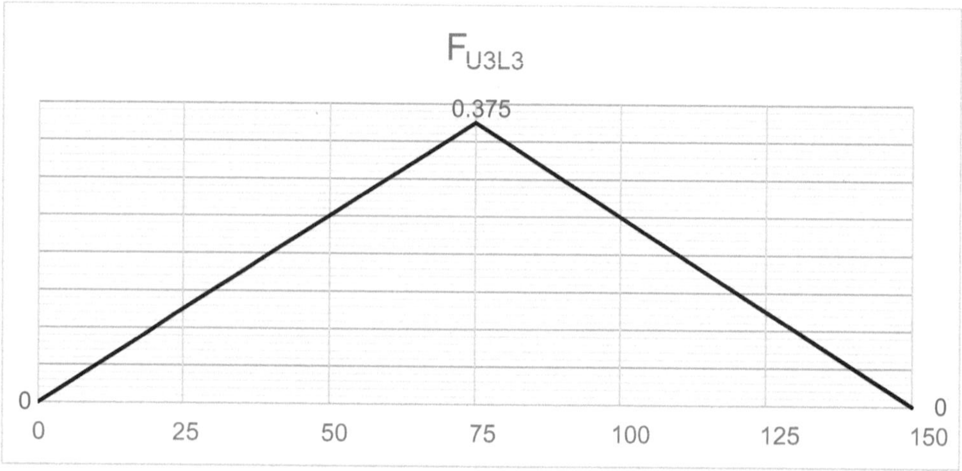

FIGURE E4.22h Influence line for U3L3 of a Parker truss span.

For influence lines for vertical web members directly loaded by floorbeam reaction, specifically for U1L1, see Figure E4.22i.

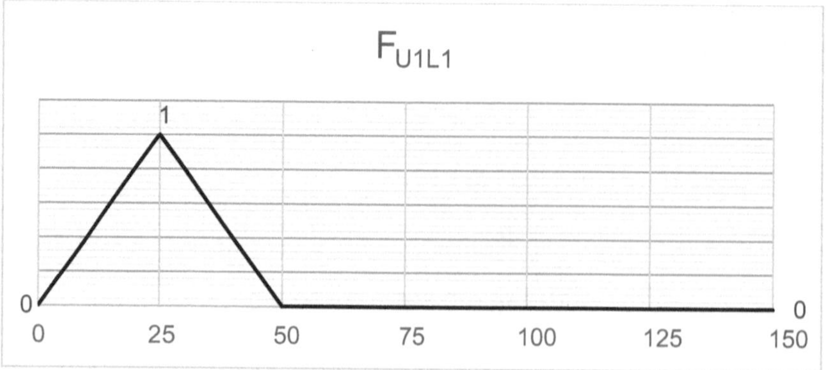

FIGURE E4.22i Influence line for member U1L1 of a Parker truss span.

The influence lines of Example 4.22 are used in Example 4.23 (uniformly distributed live load) and Example 4.24 (Cooper E80 concentrated live loads) to determine the maximum forces in the chord and web members of an inclined chord truss Parker truss.

Example 4.23

The six-panel inclined top chord Pratt (Parker) truss in Figure 4.12 is 150 ft long with h_1 = 25 ft, h_2= 35 ft and h_3 = 40 ft. It is subjected to a uniformly distributed dead load of 1500 lb/ft and a uniformly distributed live load of 5500 lb/ft. Use the influence lines from Example 4.22 to determine the maximum live load forces in the truss members (see Table E4.23).

TABLE E4.23
Member Forces by Influence Lines for Uniformly Distributed Live Load

Parker Truss Member	Influence Line Positive Area (ft)	Influence Line Negative Area (ft)	Positive Live Load Force in Member (kips)	Negative Live Load Force in Member (kips)
L1L2	62.475	-	343.6	-
L2L3	71.475	-	393.1	-
U1U2	-	−76.875	-	−422.8
U2U3	-	−71.70	-	−394.4
U1L2	22.444	−9.820	123.4	−54.0
U2L3	16.347	−18.258	89.9	−100.4
U1L1	25.0	-	137.5	-
U2L2	−8.642	24.715	−47.5	135.9
U3L3	28.125	-	154.7	-

The member forces due to a uniformly distributed live load determined with influence lines are essentially the same as those calculated in Example 4.19. Differences are attributable to rounding errors.

Example 4.24

The six-panel inclined top chord Pratt (Parker) truss in Figure 4.12 is 150 ft long with h_1 = 25 ft, h_2 = 35 ft and h_3 = 40 ft. It is subjected to a uniform dead load of 1500 lb/ft and Cooper's E80 live load. Use the influence lines from Example 4.22 to determine the maximum live load forces in truss members L2L3, U2U3, U1L2 and U2L3.

For L2L3, As shown in Example 4.21, the criterion of Equation 4.11c with k = 2 indicates axle 7 at L2:

$$F_{L2L3} = 20(0.248) + 40(0.400 + 0.496 + 0.591 + 0.686) + 26(0.858 + 0.953 + 0.896 + 0.848) + 20(0.772) + 40(0.696 + 0.648 + 0.600 + 0.553) + 26(0.467 + 0.419 + 0.362 + 0.314) + 4(28)(0.28)/2 = 340.24 + 15.68 = 355.9 \text{ kips}$$

For U2U3, as shown in Example 4.21, the criteria of Equation 4.11c with k = 3 indicates axle 11 at L3:

$$F_{U2U3} = 20(0.140) + 40(0.242 + 0.306 + 0.370 + 0.433) + 26(0.548 + 0.612 + 0.688 + 0.752) + 20(0.854) + 40(0.956 + 0.892 + 0.829 + 0.765) + 26(0.650 + 0.586 + 0.510 + 0.446) + 4(30)(0.400)/2 = 336.19 + 24.00 = 360.2 \text{ kips}$$

For U1L2, as shown in Example 4.21, for positive panel shear force the criterion of Equation 4.26g indicates axle 2 at L2:

$$F^{+}_{U1L2} = [20(0.280) + 40(1 + 0.950 + 0.900 + 0.850) + 26(0.760 + 0.710 + 0.650 + 0.600) + 20(0.520) + 40(0.440 + 0.390 + 0.340 + 0.290) + 26(0.200 + 0.150 + 0.090 + 0.040)]0.404 = 123.46 \text{ kips}$$

As shown in Example 4.21, for negative panel shear force, the criterion of Equation 4.27f indicates axle 3 at L1:

$$F^{-}_{U1L2} = [20(0.064)+40(0.640+1+0.800+0.600)+26(0.240+0.040)](-0.505) = -65.73 \text{ kips}$$

For U2L3, as shown in Example 4.21, for positive panel shear force, the criterion of Equation 4.26g indicates axle 2 at L3:

$$F^{+}_{U2L3} = 20(0.211) + 40(1 + 0.933 + 0.867 + 0.8) + 26(0.680 + 0.613 + 0.533 + 0.467) + 20(0.360) + 40(0.253 + 0.187 + 0.120 + 0.053) = 92.00 \text{ kips}$$

As shown in Example 4.21, for negative panel shear force, the criterion of Equation 4.27f indicates axle 3 at L2:

$$F^{-}_{U2L3} = 20(0.125) + 40(0.664 + 1 + 0.900 + 0.800) + 26(0.620 + 0.520 + 0.400 + 0.300) + 20(0.140) = 105.67 \text{ kips}$$

The member forces due to Cooper's E80 live load determined with influence lines are essentially the same as those calculated in Example 4.21. Differences are attributable to rounding errors. Influence lines for inclined-chord Warren truss chord members are determined in accordance with Equations 4.19a and 4.19b. Influence lines for inclined-chord Warren truss diagonal web members may be determined in accordance with Equations 4.20a and 4.20b.

4.6 STATICALLY DETERMINATE SUBDIVIDED PANEL TRUSS SPANS

Subdivided panel Pratt and Warren trusses are used for long[126] simple-span railway truss superstructures. These subdivided panel spans are horizontal- or inclined-chord trusses.[127] The subpanels are created with substruts, as shown in Figure 4.16, or with subties, as shown in Figure 4.17. Subdivided panels with substruts are most often used for modern railway spans as they typically provide for greater rigidity and, therefore, better serviceability[128] than that afforded by truss spans with subtie subdivided panels.

For substrut subdivided trusses with diagonal members U(j – 1)M(j) and M(j)L(j + 1),[129] panels subdivided by subvertical members M(j)L(j) have subpanels (j – 1) – (j) with upper diagonal member U(j – 1)M(j) and substrut member L(j – 1)M(j) and subpanels (j) – (j + 1) with lower diagonal member M(j)L(j + 1).

For substrut subdivided trusses with diagonal members L(j – 1)M(j) and M(j)U(j + 1),[130] panels subdivided by subvertical members M(j)L(j) have subpanels (j) – (j + 1) with upper diagonal member M(j)U(j + 1) and substrut member M(j)L(j + 1) and subpanels (j – 1) – (j) with lower diagonal member L(j – 1)M(j).

For subtie subdivided trusses with diagonal members U(j – 1)M(j) and M(j)L(j + 1) with panels subdivided by subvertical members M(j)L(j), subpanels (j) – (j + 1) have lower diagonal member M(j)L(j + 1), and subtie member M(j)U(j + 1) and subpanels (j – 1) – (j) have upper diagonal member U(j – 1)M(j).

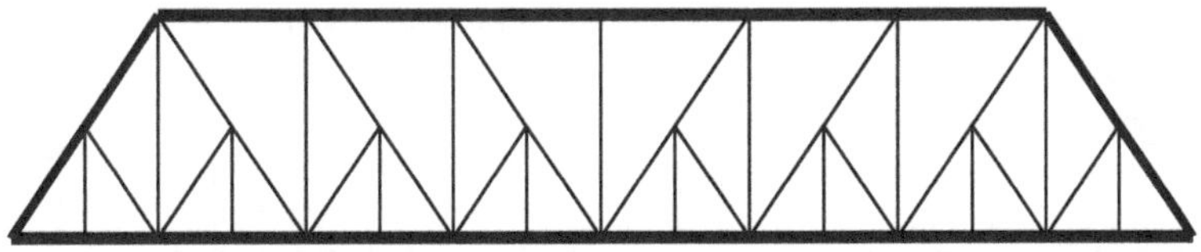

Subdivided horizontal chord Pratt (Baltimore) Truss with substruts

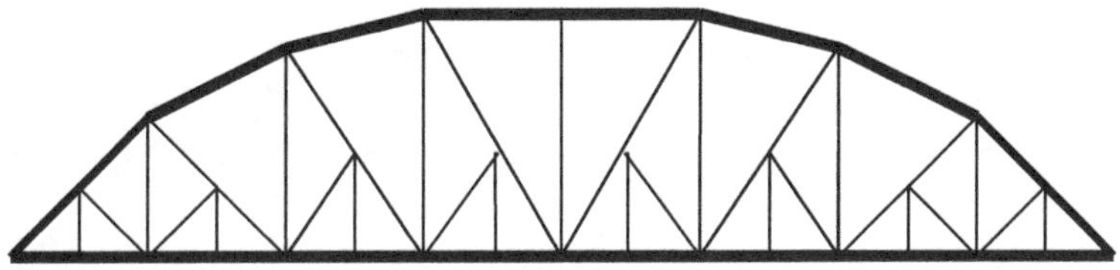

Subdivided inclined chord Pratt (Pennsylvania or Pettit) Truss with substruts

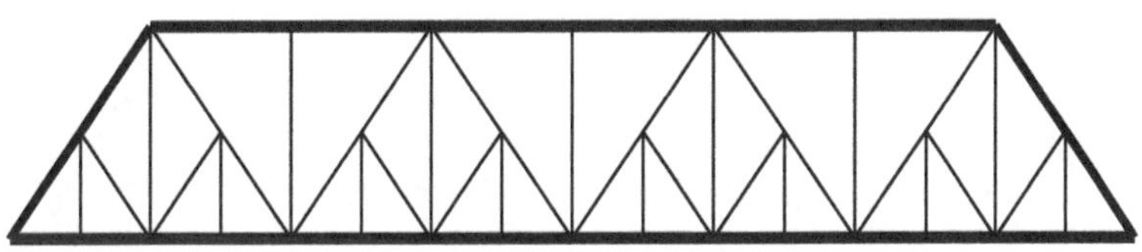

Subdivided horizontal chord Warren Truss with substruts

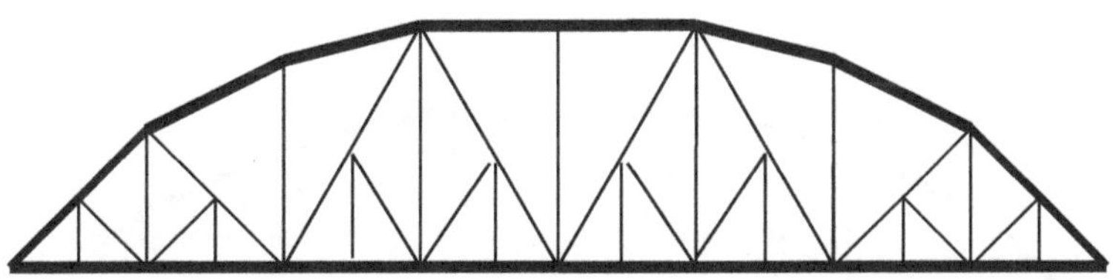

Subdivided inclined chord Warren Truss with substruts

FIGURE 4.16 Substrut subdivided Pratt and Warren truss spans.

For subtie subdivided trusses with diagonal members L(j – 1)M(j) and M(j)U(j + 1), panels subdivided by subvertical members M(j)L(j) have subpanels (j – 1) – (j) with lower diagonal member L(j – 1)M(j) and subtie member U(j – 1)M(j) and subpanels (j) – (j + 1) with upper diagonal member M(j)U(j + 1).

In subdivided truss spans, the length of subvertical members M(j)L(j) will determine whether the upper diagonal and lower diagonal members are aligned. For efficient design[131] and fabrication,

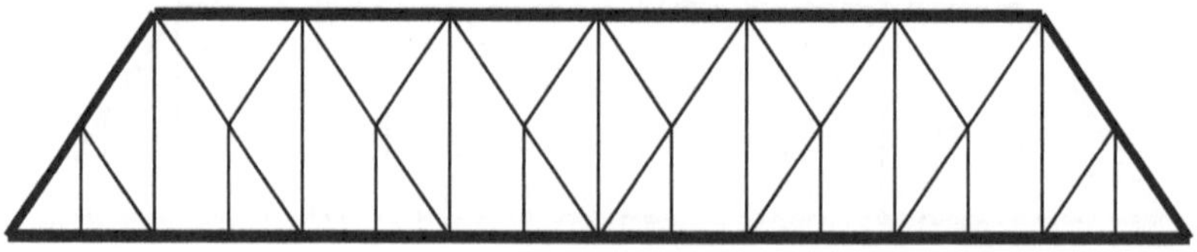

Subdivided horizontal chord Pratt (Baltimore) Truss with subties

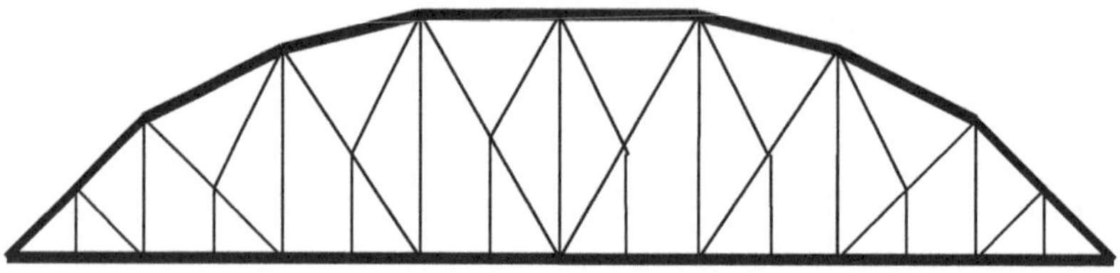

Subdivided inclined chord Pratt (Pennsylvania or Pettit) Truss with subties

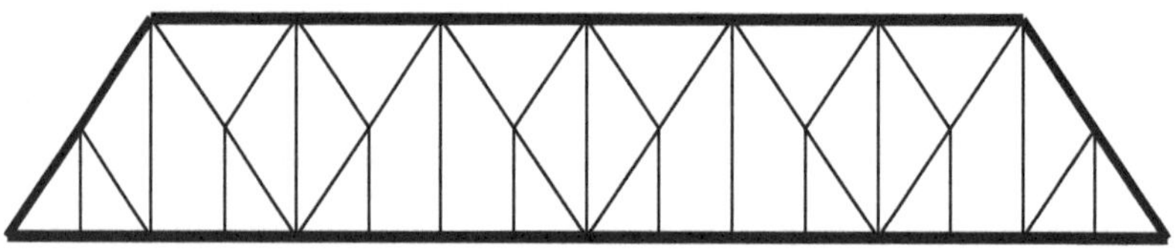

Subdivided horizontal chord Warren Truss with subties

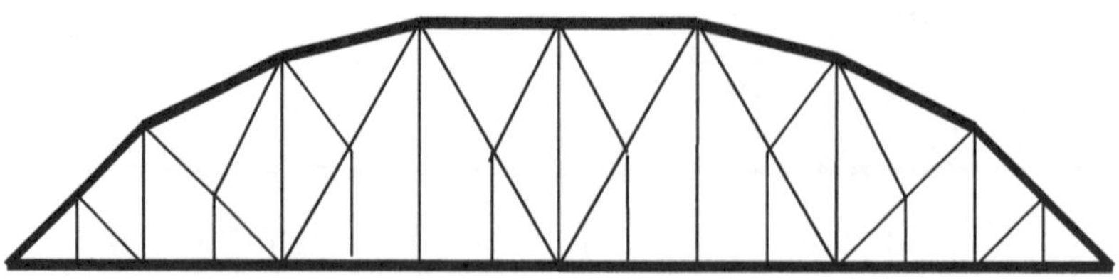

Subdivided inclined chord Warren Truss with subties

FIGURE 4.17 Subtie subdivided Pratt and Warren truss spans.

the upper and lower diagonal members are aligned. As with non-subdivided panel truss spans, maximum panel point moments and maximum chord member forces for uniformly distributed live loads occur with the span entirely loaded because a load positioned at any location on a beam creates a positive moment at every other location on the beam. Additionally, as with panels in non-subdivided truss spans, maximum positive and negative shear forces in subpanels occur with uniformly distributed live load from panel point (n) to distance x^+ and from panel point (0) to x^-, respectively. Distance x^+ for positive shear is the distance from the neutral point[132] to the right side of the subpanel under consideration and distance x^- for negative shear is the distance from the neutral point to the left side of the subpanel under consideration.

For concentrated live loads, the maximum panel point moments and maximum chord forces occur with a concentrated load at the panel point under consideration since moments at panel points are greater than moments within panels because shear forces at panel points become zero with a concentrated load at the panel point. Also, as with non-subdivided truss spans, mathematical criteria for the locations of concentrated live loads for maximum panel point moments can be established.

Furthermore, as with non-subdivided truss spans, mathematical criteria for the locations of concentrated live loads for positive and negative shear in subpanels can be established. If the subpanel under consideration contains only a lower or an upper diagonal member between chord members, a concentrated live load must be placed at the panel point on the right side of the subpanel under consideration for positive shear in the subpanel[133] and at the panel point on the left side of the subpanel under consideration for negative shear in the subpanel.[134] If the subpanel under consideration contains both a diagonal member and a substrut or subtie member, a concentrated live load must be placed at the panel point on the right side of the panel containing the subpanel under consideration for positive shear in the subpanel[135] and at the panel point on the left side of the panel containing the subpanel under consideration for negative shear in the subpanel.[136]

4.6.1 Subdivided Horizontal Parallel-Chord Pratt (Baltimore) Truss Spans with Substrut Members

As shown in Figure 4.18a, the forces in top chord member U(j – 1)U(j + 1) of a Baltimore truss with substrut members (BSS) can be determined by taking a section through subpanel (j) – (j + 1) and considering moments taken about opposite chord panel point L(j + 1), such that $F_{U(j-1)U(j+1)}$ is

$$F_{U(j-1)U(j+1)} = \frac{-\sum M_{L(j+1)}}{h} = -\left((j+1)R_{y0} - \sum_{i=0}^{j+1}(j+1-i)W_i\right)\left(\frac{d}{h}\right) \tag{4.30a}$$

where

R_{y0} = the reaction at panel point (0) due to total load P_T on the BSS span;
W_i = the load at panel point (i) due to load P_F on the BSS span;
d = equal subpanel length of the BSS span;
h = height of the BSS truss;
P_T = the resultant of the concentrated loads between panel point (0) and (n) (total load on the BSS truss span) at distance X_T from (n); and
P_F = the resultant of the concentrated loads between panel point (0) and (k = j + 1) at distance X_F from (j + 1) with live loads moving from right to left (from (n) to (0)) across the BSS span.

Maximum force $F^{max}_{U(j-1)U(j+1)}$ in top chord member U(j – 1)U(j + 1) of the BSS span due to a uniformly distributed live load occur with the span entirely loaded.

$F^{max}_{U(j-1)U(j+1)}$ in top chord member U(j – 1)U(j + 1) of the BSS span due to concentrated live loads is calculated with Equation 4.30a such that

$$F_{U(j-1)U(j+1)} = \frac{-\sum M_{L(j+1)}}{h} = -\left((j+1)\frac{P_T X_T}{nd} - \frac{P_F X_F}{d}\right)\left(\frac{d}{h}\right) \tag{4.30b}$$

where

$$R_{y0} = \frac{P_T X_T}{nd}$$

n = the number of subpanels in the BSS span

$$\sum_{i=0}^{j}(j+1-i)W_i = \frac{P_F X_F}{d}$$

$F^{max}_{U(j-1)U(j+1)}$ in top chord member U(j-1)U(j+1) of the BSS span occurs where

$$\frac{d\left(F_{U(j-1)U(j+1)}\right)}{dx} = 0 = \frac{-(j+1)\frac{P_T}{n} + P_F}{h}$$

or

$$\frac{P_T}{n} = \frac{P_F}{j+1} \quad (4.30c).$$

Therefore, $F^{max}_{U(j-1)U(j+1)}$ in top chord member U(j − 1)U(j + 1) of the BSS span due to concentrated live loads, such as the Cooper's E design load, are determined with the live load positioned on the span in accordance with the criterion of Equation 4.30c, which is, as expected, equal to Equation 4.11c with k = j + 1 (Figure 4.18).

The presence of subvertical members M(j)L(j) subdividing panel (j − 1) − (j + 1) in the BSS span in Figure 4.18a will affect the forces in bottom chord members L(j)L(j + 1). $F_{L(j)L(j+1)}$ in bottom chord members L(j)L(j + 1) of the BSS span are determined by taking a section through subpanel (j) − (j + 1), as shown in Figure 4.18a, and considering rotational equilibrium at joint U(j − 1). Accordingly, the forces in the bottom chord members L(j)L(j + 1)[137] of the BSS span are determined from

$$R_{y0}(j-1)d - d\sum_{i=0}^{j-1}(j-1-i)W_i + W_{(j)}d - F_{L(j)L(j+1)}(h) = 0 \quad (4.31a),$$

which yields

$$F_{L(j)L(j+1)} = \frac{M^{Wx}_{(j-1)} + 2M^{P}_{(j)}}{h} = F_{L(j-1)L(j)} \quad (4.31b)$$

where

$M^{Wx}_{(j-1)} = \left((j-1)R_{y0} - \sum_{i=0}^{j-1}(j-1-i)W_i\right)d$ = the moment about joint U(j − 1) in the BSS span from end reaction R_{y0} and applied joint loads W_i;

$M^{P}_{(j)} = \frac{W_{(j)}(d)}{2}$ = the moment at panel point (j) in the BSS span due to loads in equal length subpanels (j − 1) − (j) and (j) − (j + 1); and

$W_{(j)}$ = the floorbeam reaction or panel point force between two subpanels of equal length, d (see Equation 4.14f).

$F^{max}_{L(j-1)L(j)}$ and $F^{max}_{L(j)L(j+1)}$ in bottom chord members L(j − 1)L(j) and L(j)L(j + 1), respectively, in the BSS span due to a uniformly distributed live load occur with the span entirely loaded. Therefore, for a uniformly distributed live load w on subpanels of equal length d, the maximum moment at (j) occurs with the panel fully loaded such that $M^{P}_{(j)} = \frac{w(2d)^2}{8} = \frac{wd^2}{2} = \frac{W_{(j)}(d)}{2}$. Furthermore, if the BSS span is fully loaded by a uniformly distributed live load, the forces in members M1L2 and L2M3 in Figure 4.18a are equal,[138] which requires that the forces in the bottom chord members between joint (0) and joint (4) of the BSS span in Figure 4.18a also be equal.

$F^{max}_{L(j-1)L(j)}$ and $F^{max}_{L(j)L(j+1)}$ in bottom chord members L(j − 1)L(j) and L(j)L(j + 1), respectively, of the BSS span due to concentrated live loads are determined with the live load positioned on the

a) Section to determine chord and lower diagonal member forces in subdivided horizontal chord Pratt truss (Baltimore truss).

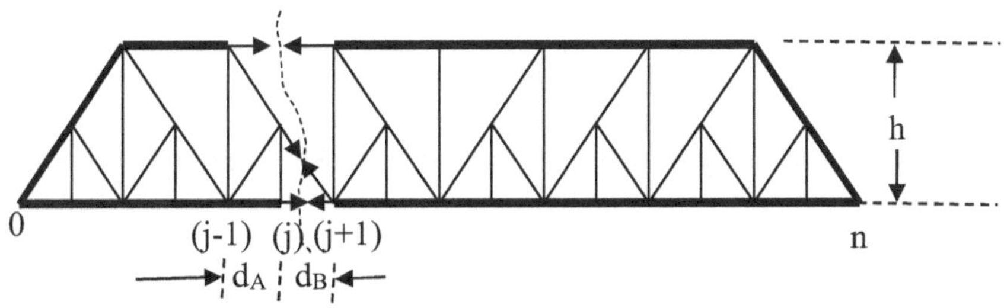

b) Section to determine chord and lower diagonal member forces in subdivided inclined chord Pratt truss (Pennsylvania truss).

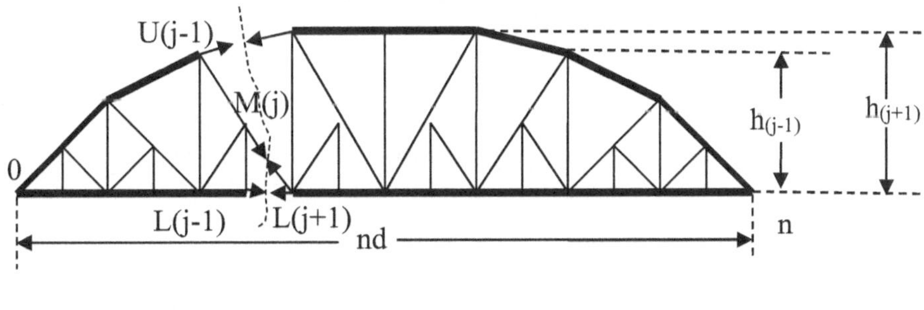

c) Section to determine upper diagonal member forces in subdivided horizontal chord Pratt truss (Baltimore truss).

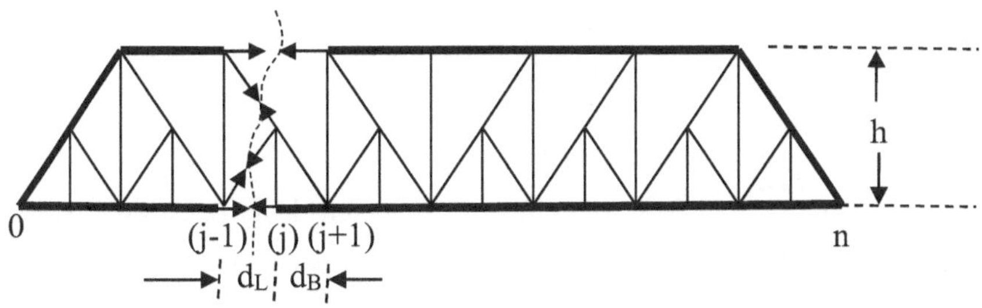

FIGURE 4.18 Sections through substrut subdivided Pratt truss spans to determine chord and web member forces.

d) Section to determine upper diagonal member forces in subdivided inclined chord Pratt truss (Pennsylvania truss).

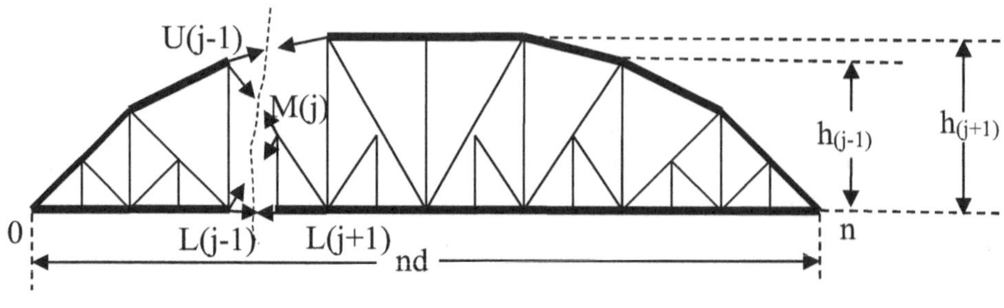

e) Section to determine upper end post member forces in horizontal chord Pratt truss.

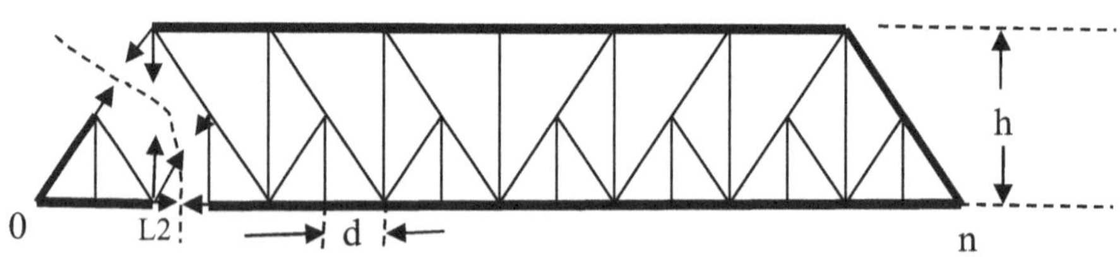

FIGURE 4.18 (Continued)

span in accordance with a criterion established by $\frac{d\left(F_{L(j)L(j+1)}\right)}{dx}=0$. Consequently, from Equation 4.31b,

$$\frac{d\left(F_{L(j)L(j+1)}\right)}{dx}=0=\frac{d\left(\frac{M_{(j-1)}^{Wx}+2M_P}{h}\right)}{dx}=\frac{d\left(\frac{M_{(j-1)}^{Wx}+W_{(j)}d}{h}\right)}{dx} \quad (4.31c).$$

Taking a section through subpanel (j) – (j + 1) of the BSS span reveals that the moment about joint U(j – 1) from end reaction R_{y0} and applied live loads $W_{(j)}$, P_F, P_A and P_B is calculated as[139]

$$M_{(j-1)}^{Wx}+2M_P=\frac{P_T X_T}{nd}(j-1)d-P_F X_F+W_{(j)}d \quad (4.31d)$$

where

$W_{(j)}=\frac{P_A(d-X_A)}{d_A}+\frac{P_B X_B}{d_B}=\frac{P_A(d-X_A)+P_B X_B}{d}$, for equal subpanel lengths, $d_A=d_B=d$;

P_F = the resultant load on the BSS span between panel point (0) and panel point (k = j – 1) at distance X_F from panel point (j – 1);

P_A = the resultant load on the BSS span between (j – 1) and (j) at X_A from (j);

P_B = the resultant load on the BSS span between (j) and (j + 1) at X_B from (j + 1); and
P_R = the resultant load on the BSS span between (j + 1) and (n) at distance X_R from (n).

Substituting $W_{(j)}$ into Equation 4.31d yields

$$M_{(j-1)}^{Wx} + 2M_P = \frac{P_T X_T}{n}(j-1) - P_F X_F + \left(P_A\left(d - X_A\right) + P_B X_B\right) \tag{4.31e}$$

The maximum force, $F_{L(j)L(j+1)}$, in the BSS span is determined with $\frac{d\left(F_{L(j)L(j+1)}\right)}{dx} = \frac{d\left(M_{(j-1)}^{Wx} + 2M_P\right)}{dx} = 0$ such that

$$\frac{d\left(M_{(j-1)}^{Wx} + 2M_P\right)}{dx} = \frac{P_T(j-1)}{n} - P_F - P_A + P_B = 0$$

or

$$\frac{P_T}{n} = \frac{P_F + P_A - P_B}{(j-1)} \tag{4.31f}$$

The criterion of Equation 4.31f can be used to determine the location of the concentrated live loads creating maximum bottom chord member force at any subpanel within a BSS span. Therefore, the criterion for the location of a series of concentrated loads for maximum bottom chord member force is when $\frac{P_T(j-1)}{n} - \left(P_F + P_A - P_B\right)$ changes sign as the live load traverses a BSS span.

Example 4.25

A 300 ft long and 50 ft high BSS span with 12 subpanels is loaded with a live load of 5.5 kips/ft. Determine the maximum forces due to live load in the top and bottom chord members:

$$R_{yo} = (5.5)(300/2) = 825 \text{ kips}$$

Forces due to live load in the top chord members are provided by Equation 4.30a:

$$F_{U2U4} = \frac{-\sum M_{L(4)}}{25} = -\left((3+1)(825) - \sum_{i=0}^{3+1}(4-i)W_i\right)\left(\frac{25}{50}\right)$$

$$= -\left(3300 - 4(68.75) - (3+2+1)137.5\right)\left(\frac{25}{50}\right) = -1100 \text{ kips where j = 3}$$

$$F_{U4U6} = \frac{-\sum M_{L(6)}}{25} = -\left((5+1)(825) - \sum_{i=0}^{5+1}(6-i)W_i\right)\left(\frac{25}{50}\right)$$

$$= -\left(4950.0 - 6(68.75) - (5+4+3+2+1)137.5\right)\left(\frac{25}{50}\right) = -1237.5 \text{ kips where j = 5}$$

Forces due to live load in the bottom chord members are provided by Equation 4.31a:

$$F_{L3L4} = \frac{M_{(2)}^{W} + 2M_{(3)}^{P}}{h} \text{ where j = 3}$$

$$M_{(2)}^{W} = \left(2(825) - \sum_{i=0}^{2}(2-i)W_i\right)25 = \left(1650 - 2(68.75) - 137.5\right)(25) = 34{,}375 \text{ kips} - \text{ft}$$

$$M_{(3)}^{P} = \frac{W_{(3)}(d)}{2} = \frac{137.5(25)}{2} = 1718.75 \text{ kips} - \text{ft}$$

$$F_{L3L4} = \frac{34{,}375 + 2(1718.75)}{50} = 756.25 \text{ kips} = F_{L2L3} = F_{L1L2} = F_{L0L1} \text{ as expected for a}$$

uniformly distributed live load entirely on the span

$$F_{L5L6} = \frac{M_{(4)}^{W} + 2M_{(5)}^{P}}{h} \text{ where } j = 5$$

$$M_{(4)}^{W} = \left(4(825) - \sum_{i=0}^{4}(4-i)W_i\right)25 = \left(3300 - 4(68.75) - (3+2+1)137.5\right)(25) = 55000 \text{ kips} - \text{ft}$$

$$M_{(5)}^{P} = \frac{W_{(5)}(d)}{2} = \frac{137.5(25)}{2} = 1718.75 \text{ kips} - \text{ft}$$

$$F_{L5L6} = \frac{55000 + 2(1718.75)}{50} = 1168.75 \text{ kips} = F_{L4L5}$$

Example 4.26

A 300 ft long and 50 ft high BSS span with 12 subpanels is loaded with Cooper's E80 live load. Determine the maximum forces due to live load in the top chord member U2U4 and bottom chord member L5L6.

For top chord member U2U4 (j = 3), the section shown in Figure 4.18a illustrates that the forces in U(j − 1)U(j + 1) can be determined by taking moments about joint L(j + 1) with the live load positioned in accordance with the criterion of Equation 4.30c (4.11c with k = j + 1):

$$\frac{P_T}{n} = \frac{P_F}{(j+1)}$$

For top chord member U2U4,

$$\frac{P_T}{3} = P_F$$

From inspection, the estimated location of Cooper's live load for maximum panel point moment is wheels 12, 13 or 14 at panel point L4 (100 ft from left end of truss).

Table 4.2 is used to determine moments about panel point L4 with Cooper's E live load located in accordance with the criterion of Equation 4.30c with k = 4 as shown in Table E4.26a.

TABLE E4.26a
Criterion for Maximum Force in Top Chord U2U4 in a BSS Span

	Axle at L4		
Location of Live Load	**12**	**13**	**14**
Distance to uniform load (ft)	40	35	30
Distance to axle no. 18 (ft)	40 – 5 = 35	30	25
Last axle on span	18	18	18
Distance to axle no. 1 (ft)	69	74	79
First axle on span	1	1	1
Length of uniform load (ft)	(12 – 4)(25) – 40 = 160	200 – 35 = 165	200 – 30 = 170
Total load on span, P_T (kips)	568 +160(4) = 1208	568 + 165(4) = 1228	568 + 170(4) = 1248
$P_T/3$ (kips)	402.7	409.3	416.0
Sum of loads ahead of panel point L4, P_{F1} (kips)	(384 – 40) = 344	(424 – 40) = 384	(464 – 40) = 424
Sum of loads ahead of and including panel point L4, P_{F2} (kips)	384	424	464
Criterion established	No	Yes	No

The criterion of Equation 4.30c indicates axle 13 at L4 for maximum panel point moment at L4 and, accordingly, the maximum force for the top chord member U2U4, as shown in Table E4.26b. Panel point moments are determined from Table 4.2.

TABLE E4.26b
Maximum Force in Member U2U4 for Cooper's E80 Live Load Traversing a BSS Span

	Axle at L4
Maximum Force in U2U4	**13**
Sum of moments to L12 (kip-ft)	$32{,}728 + 568(165) + 4(165)^2/2 = 180{,}898$
Reaction R_{y0} (kips)	180,898/300 = 603.0
Sum of moments from first axle to L4 (kip-ft)	15,336
Sum of moments about L4 = $\sum M_{L4}$ (kip-ft)	603.3(4)(25) – 15,336 = 44,963
Force in member U2U4 = $F_{U2U4} = -(\sum M_{L4})/h$ (kips)	–44,963/50 = –899.3

$$F^{max}_{U2U4} = -899.3 \text{ kips}$$

For bottom chord member L5L6 (j = 5), the section shown in Figure 4.18a illustrates that the forces in bottom chord members L(j)L(j+1) can be determined by taking moments about joint U(j-1) with the load positioned in accordance with the criterion of Equation 4.31f. From inspection, the estimated location of Cooper's live load for maximum panel point moment occurs with wheel 13 to 16 at panel point L5 (125 ft from left end of truss).

Table 4.2 is used to determine moments about panel point U4 with Cooper's E live load located in accordance with the criterion of Equation 4.31f as shown in Table E4.26c.

$$\frac{P_T}{3} = P_F + P_A - P_B$$

TABLE E4.26c
Criterion for Maximum Force in Bottom Chord Member L5L6 of a BSS Span

Location of Live Load	Axle at L5				
	14	**15**	**16**	**17**	**18**
Distance to uniform load (ft)	30	21	16	10	5
Distance to axle no. 18 (ft)	25	16	11	5	0
Last axle on span	18	18	18	18	18
Distance to axle no. 1 (ft)	79	88	93	99	104
First axle on span	1	1	1	1	1
Length of uniform load (ft)	175 – 30 = 145	175 – 21 = 154	175 – 16 = 159	175 – 10 = 165	175 – 5 = 170
Total load on span, P_T (kips)	568 +145(4) = 1148	568 + 154(4) = 1184	568 + 159(4) = 1204	568 + 165(4) = 1228	568 +170(4) = 1248
$(P_F + P_A)_L$ wheel to left of L5 (kips)	$\sum_{i=1}^{14} P_i = 464$	$\sum_{i=1}^{15} P_i = 490$	$\sum_{i=1}^{16} P_i = 516$	$\sum_{i=1}^{17} P_i = 542$	$\sum_{i=1}^{18} P_i = 568$
$(P_F + P_A)_R$ wheel to right of L5 (kips)	$\sum_{i=1}^{13} P_i = 424$	$\sum_{i=1}^{14} P_i = 464$	$\sum_{i=1}^{15} P_i = 490$	$\sum_{i=1}^{16} P_i = 516$	$\sum_{i=1}^{17} P_i = 542$
$(P_B)_L$ wheel to left of L5 (kips)	$\sum_{i=15}^{18} P_i = 104$	$\sum_{i=16}^{18+} P_i = 78 + 4(4) = 94$	$\sum_{i=17}^{18+} P_i = 52 + 9(4) = 88$	$\sum_{i=18}^{18+} P_i = 26 + 15(4) = 86$	20(4) = 80
$(P_B)_R$ wheel to right of L5 (kips)	$\sum_{i=14}^{17} P_i = 118$	$\sum_{i=15}^{18+} P_i = 104 + 4(4) = 120$	$\sum_{i=16}^{18+} P_i = 78 + 9(4) = 114$	$\sum_{i=17}^{18+} P_i = 52 + 15(4) = 112$	26 + 20(4) = 106
$P_T/3$ (kips)	382.7	394.8	401.2	409.2	416.0
$(P_F + P_A - P_B)_L$ wheel to left of L5 (kips)	(464 – 104) = 360	(490 – 94) = 396	(516 – 88) = 428	(542 – 86) = 456	(568 – 80) = 488
$(P_F + P_A - P_B)_R$ wheel to right of L5 (kips)	(424 – 118) = 306	(464 – 120) = 344	(490 – 114) = 376	(516 – 112) = 404	(542 – 106) = 436
Satisfies criterion	No	Yes (marginal)	Yes	Yes (marginal)	No

As shown in Table E4.26c, the criterion of Equation 4.31f indicates axles 15, 16 or 17 at L5 for maximum panel point moment at U4 and, accordingly, the maximum force of the bottom chord member L5L6. A review of the criterion indicates that it is likely that F^{max}_{L5L6} occurs with wheel 16 at L5. Using Table 4.2 with Equation 4.31b provides F^{max}_{L5L6} as shown in Table E4.26d.

TABLE E4.26d
Maximum Force in Bottom Chord Member L5L6 for Cooper's E80 Live Load Traversing a BSS Span

Maximum Force in Member L5L6	Axle at L5		
	15	16	17
Sum of moments to L12 (kip-ft)	32,728 + 568(154) + 4(154)²/2 = 167,632	32,728 + 568(159) + 4(159)²/2 = 173,602	32,728 + 568(165) + 4(165)²/2 = 180,898
Reaction R_{y0} (kips)	167,632/300 = 558.77	173,602/300 = 578.67	180,898/300 = 603.0
$4(25)R_{y0}$ (kips)	4(25)(558.77) = 55,877	4(25)(578.67) = 57,867	60,300
P_{FL} = wheel of P_F loads left of joint 4 (kips)	10	11	12
Distance P_{FL} to joint 4 (ft)	7	4	5
$25\sum_{i=0}^{4}(4-i)W_i = \sum_{i=0}^{P_{FL}} M_i$ (kips)	9264 + (7)(304) = 11392	11696 + 4(344) = 13072	13416 + 5(384) = 15336
$M_{(4)}^{Wx} = \left(4R_{y0} - \sum_{i=0}^{P_{FL}} M_i\right)$ (kip-ft)	55,877 − 11,392 = 44,485	57,867 − 13,072 = 44,795	60,300 − 15,336 = 44,964
Axle to right of L4	11	12	14 (13 at L4)
Axle to left of L6	18	18	18
Load at L5 = P_{L5} (kips)	26	26	26
Length of uniform load to left of L6 (ft)	154 − (12 − 6)(25) = 4	159 − (12 − 6)(25) = 9	165 − (6)(25) = 15
Sum of moments in subpanel 4-5 about L4 = $\sum M_{L4}$ (kip-ft)	2640	2410	3442
Force at L5 from subpanel 4-5 (kips)	(464 − 304) − 2640/25 = 160 − 105.60 = 54.40	(490 − 344) − 2410/25 = 146 − 96.4 = 49.60	(516 − 384) − 2242/25 = 132 − 89.68 = 42.32
Sum of moments in subpanel 5-6 about L6 = $\sum M_{L6}$ (kip-ft)	806 + 4(78) + (4.0) (4)²/2 = 1118 + 32 = 1150	390 + 9(52) +(4.0) (9)²/2 = 858 + 162 = 1020	130 + 15(26) +(4.0) (15)²/2 = 520 + 450 = 970
Force at L5 from subpanel 5-6	1150/25 = 46.00	1020/25 = 40.80	970/25 = 38.80
$W_{(5)}$ (kips)	26 + 54.40 + 46.00 = 126.40	26 + 49.60 + 40.80 = 116.40	26 + 42.32 + 38.80 = 107.12
$M_{(5)}^{P} = \frac{W_{(5)}(d)}{2}$ (kip-ft)	126.4(25)/2 = 1580	116.40(25)/2 = 1455	107.12(25)/2 = 1339
$F_{L5L6} = \frac{M_{(4)}^{Wx} + 2M_{(5)}^{P}}{h}$ (kips)	(44,485.3 + 2(1580))/50 = 47645.3/50 = 952.9	(44,795 + 2(1455))/50 = 47,705/50 = 954.1	(44,964+2(1339.0))/50 = 47,642/50 = 952.8

F^{max}_{L5L6} in bottom chord member L5L6 (and L4L5) due to Cooper's E80 live load is 954.1 kips with wheel 16 at joint L5.

As shown in Figure 4.18a, web members in a BSS span consist of upper diagonal U(j − 1) M(j), lower diagonal M(j)L(j + 1), substrut L(j − 1)M(j), subvertical M(j)L(j), interior vertical

U(j - 1)L(j - 1) and U(j + 1)L(j + 1) and end vertical members U2L2, where j ≤ n/2. The subdivided panels also result in upper end post M1U2 and lower end post members L0M1, where j ≤ n/2.

Figure 4.18a also reveals that the forces in lower diagonal member M(j)L(j + 1) of the BSS span may be determined in the same manner as the diagonal member U(j − 1)L(j) in non-subdivided horizontal chord Pratt trusses considering subpanel (j) − (j + 1).

For uniformly distributed live loads on BSS spans, the criteria of Equations 4.5c and 4.5g for maximum positive and negative live load shear in members U(j − 1)L(j) of non-subdivided Pratt truss spans, respectively, can be used in subpanel (j) − (j + 1) to determine maximum and minimum live load forces $F^{max}_{M(j)L(j+1)}$ and $F^{min}_{M(j)L(j+1)}$ in lower diagonal member M(j)L(j + 1) of the BSS span, such that

$$x^{+}_{(j)-(j+1)} = \frac{(n-j+1)d}{(n-1)} \tag{4.32a}$$

$$x^{-}_{(j)-(j+1)} = \frac{(j)d}{(n-1)} = d - x^{+}_{(j)-(j+1)} \tag{4.32b}$$

For concentrated live loads on BSS spans, the maximum and minimum forces in lower diagonal member M(j)L(j + 1) are determined, respectively, from the positive and negative shear force in subpanel (j) − (j + 1). For positive shear force in subpanel (j) − (j + 1) of the BSS span,

$$\begin{aligned} F^{+}_{M(j)L(j+1)} &= V^{+}_{(j)-(j+1)}\left(\frac{2\sqrt{\left(\frac{h}{2}\right)^2 + d^2}}{h}\right) = \left(R_{yo} - \sum_{i=0}^{j} W_i\right)\left(\frac{2\sqrt{\left(\frac{h}{2}\right)^2 + d^2}}{h}\right) \\ &= \left(\frac{P^{+}_{T}X^{+}_{T}}{nd} - P^{+}_{F} - P^{+}_{A} - \frac{P^{+}_{B}X^{+}_{B}}{d}\right)\left(\frac{2\sqrt{\left(\frac{h}{2}\right)^2 + d^2}}{h}\right) \end{aligned} \tag{4.33a}$$

$F^{+max}_{M(j)L(j+1)}$ occurs when $\frac{dF^{+}_{M(j)L(j+1)}}{dx}$, which yields,

$$\frac{P^{+}_{T}}{n} = P^{+}_{B} = P^{+}_{(j)-(j+1)} \tag{4.33b}$$

where for the live load traversing the BSS span from panel point (n) to panel point (0),

P^{+}_{T} = the total live load on the BSS span at distance X^{+}_{T} from panel point (n);
P^{+}_{A} = the live load between (j − 1) and (j) of the BSS span at distance X^{+}_{A} from (j);
P^{+}_{B} = the live load between (j) and (j + 1) of the BSS span at distance X^{+}_{B} from (j + 1); and
P^{+}_{F} = the live load between (0) and (j − 1) of the BSS span at distance X^{+}_{F} from (j − 1).

The criterion of Equation 4.33b can only occur with a concentrated load at panel point (j + 1) due to the shape of the influence diagram for lower diagonal member M(j)L(j + 1) (see Example 4.54).

For negative shear force in subpanel (j) − (j + 1) of the BSS span,

$$F^{-}_{M(j)L(j+1)} = V^{-}_{(j)-(j+1)}\left(\frac{2\sqrt{\left(\frac{h}{2}\right)^2 + d^2}}{h}\right) = \left(R_{yn} - \sum_{i=j+1}^{n} W_i\right)\left(\frac{2\sqrt{\left(\frac{h}{2}\right)^2 + d^2}}{h}\right)$$

$$= \left(\frac{P^{-}_T X^{-}_T}{nd} - P^{-}_F - \frac{P^{-}_B X^{-}_B}{d}\right)\left(\frac{2\sqrt{\left(\frac{h}{2}\right)^2 + d^2}}{h}\right) \quad (4.33c)$$

$F^{-max}_{M(j)L(j+1)}$ occurs when $\frac{dF^{-}_{M(j)L(j+1)}}{dx} = 0$, which yields

$$\frac{P^{-}_T}{n} = P^{-}_B = P^{-}_{(j)-(j+1)} \quad (4.33d)$$

where with the live load traversing the span from panel point (0) to panel point (n) of the BSS span,

P^{-}_T = the total live load on the BSS span at distance X^{-}_T from (0);
P^{-}_A = the live load between (j − 1) and (j) of the BSS span at distance X^{-}_A from (j − 1);
P^{-}_B = the live load between (j) and (j + 1) of the BSS span at distance X^{-}_B from (j); and
P^{-}_F = the live load between (j + 1) and panel point (n) of the BSS span at distance X^{-}_F from (j + 1).

The criterion of Equation 4.33d can only occur with a concentrated load at panel point (j) due to the shape of the influence diagram for lower diagonal member M(j)L(j + 1) (see Example 4.54).

Therefore, as foreseeable, the criteria of Equations 4.33b and 4.33d for the lower diagonal in subpanel (j) − (j + 1) of a BSS span are identical, respectively, to Equation 4.12c for maximum positive shear and Equation 4.13c for maximum negative shear in the diagonal member in panel (j − 1) − (j) of non-subdivided Pratt truss spans. The use of these criteria to determine $F^{+max}_{M(j)L(j+1)}$ and $F^{-max}_{M(j)L(j+1)}$ in a BSS span is outlined in Example 4.27.

Example 4.27

A 300 ft long and 50 ft high BSS span with 12 subpanels is loaded with Cooper's E80 live load. Calculate the maximum positive and negative shear force due to live load resisted by lower diagonal member M3L4 (in subpanel 3-4). Table 4.2 is used to calculate positive shear forces in subpanel 3-4 with Cooper's E live load located in accordance with the criterion of Equation 4.33b. Considering positive shear in Table E4.27a, j = 3, and

$$\frac{P^{+}_T}{12} = P^{+}_{(3)-(4)} = P^{+}_B$$

The criterion of Equation 4.33b indicates axle 3 at L4 for maximum positive shear in subpanel 3-4 as shown in Table E4.27b. Table 4.2 is used to determine positive shear forces in subpanel 3-4 with axles 3 and 4 at L4. The subpanel shear forces with axle 4 at L4 are also calculated as verification.

$$\alpha_{(3)-(4)} = \left(\frac{\sqrt{h^2_{(3)-(4)} + d^2}}{h_{(3)-(4)}}\right) = \left(\frac{\sqrt{(25)^2 + (25)^2}}{25}\right) = 1.414$$

TABLE E4.27a
Criterion for Position of Live Load for Positive Shear in Subpanel 3-4 in a BSS Span

Location of Live Load for Positive Shear in Subpanel 3-4	Axle at L4		
	2	3	4
Distance to axle no. 1 (ft)	8	13	18
First axle on span	1	1	1
Distance to axle no. 18 (ft)	96	91	86
Last axle on span	18	18	18
Length of uniform load (ft)	200 − 96 − 5 = 99	200 − 91 − 5 = 104	109
Total load on span, P^+_T (kips)	568 + 4(99) = 964	568 + 104(4) = 984	568 + 4(109) = 1004
$P^+_T/12$ (kips)	80.3	82.0	83.7
Sum of loads ahead of panel point L4 = $P^+_{a(3)\text{-}(4)}$ (kips). Wheel at L4 to right of L4	20	60	100
Sum of loads ahead of and including panel point L4 = $P^+_{b(3)\text{-}(4)}$ (kips). Wheel at L4 to left of L4	60	100	140
Criterion established	No	Yes	No

TABLE E4.27b
Maximum Force in Member M3L4 due to Positive Shear in Subpanel 3-4 for Cooper's E80 Live Load Traversing a BSS Span

Maximum Force in Member M3L4 due to Positive Shear in Subpanel 3-4	Axle at L4	
	3	4
Sum of live load moments to L12 (kip-ft)	32,728 + 568(104) + 4(104)²/2 = 113,432	32,728 + 568(109) + 4(109)²/2 = 118,402
R_{y0} = Reaction at L0 (kips)	113,432/300 = 378.1	118,402/300 = 394.7
Sum of live load moments in subpanel 3-4 about panel point L4 (kip-ft)	460	960
$W_{(3)}$ = panel load at L3 (kips)	460/25 = 18.4	960/25 = 38.4
$V^{+max}_{(3)\text{-}(4)}$ = Positive shear in subpanel 3–4 = $R_{y0} - W_{(j)} = R_{y0} - W_{(3)}$ (kips)	378.1 − 18.4 = 359.7	394.7 − 38.4 = 356.3
$F^{+max}_{M3L4} = V^{+max}_{(3)-(4)}\left(\alpha_{(3)-(4)}\right)$ = Maximum force from positive subpanel shear in diagonal M3L4 (kips)	508.6	503.8

$$F^{+max}_{M3L4} = 508.6 \text{ kips}$$

Table 4.2 is used to determine negative shear forces in subpanel 3-4 with Cooper's E live load located in accordance with Equation 4.33d considering negative shear in Table E4.27c.

$$\frac{P_T^-}{12} = P_{(3)-(4)}^- = P_B^-$$

TABLE E4.27c
Criterion for Position of Live Load for Negative Shear in Subpanel 3-4 in a BSS Span

Location of Live Load for Negative Shear in Subpanel 3-4	Axle at L3		
	2	3	4
Distance to axle no. 1 (ft)	8	13	18
First axle on span	1	1	1
Distance to axle no. 18 (ft)	96	91	86
Last axle on span	14 (4 ft from L0)	15 (at L0)	16 (at L0)
Total load on span, P^-_T (kips)	464	490	516
$P^-_T/12$ (kips)	38.7	40.8	43.0
Sum of loads ahead of panel point L3 = $P^-_{a(3)-(4)}$ (kips). Wheel at L3 to right of L3	20	60	100
Sum of loads ahead of and including panel point L3 = $P^-_{b(3)-(4)}$ (kips). Wheel at L3 to left of L3.	60	100	140
Criterion established	Yes	No	No

The criterion of Equation 4.33d indicates axle 2 at L3 for maximum negative shear in subpanel 3-4 as shown in Table E4.27d. The subpanel shear forces with axle 3 at L3 are also calculated as verification. Table 4.2 is used to determine negative shear forces in subpanel 3-4 with axle 2 and axle 3 at L3.

TABLE E4.27d
Maximum Force in Member M3L4 due to Negative Shear in Subpanel 3-4 for Cooper's E80 Live Load Traversing BSS Span

Maximum Force in Member M3L4 due to Negative Shear in Subpanel 3-4	Axle at L3	
	2	3
Sum of live load moments to L0 (kip-ft)	17,456 + 4(464) = 19,312	21,632
R_{y12} (kips)	19,312/300 = 64.37	21,632/300 = 72.11
Sum of live load moments in subpanel 3-4 about panel point L3 (kip-ft)	160	460
$W_{(4)}$ = panel load at L4 (kips)	160/25 = 6.40	460/25 = 18.40
$V^{-max}_{(3)-(4)}$ = Negative shear in subpanel 3-4 = $R_{y12} - W_{(4)}$ (kips)	64.37 − 6.40 = 57.97	72.11 − 18.40 = 53.71
$F^{-max}_{M3L4} = -V^{-max}_{(3)-(4)}\left(\alpha_{(3)-(4)}\right)$ (kips)	−81.97	−75.95

$$F^{-max}_{M3L4} = -81.97 \text{ kips}$$

The forces in substrut members L(j-1)M(j) of the BSS span are determined by considering the sum of the moments of the forces in panel (j − 1) − (j + 1) at joint M(j) about joint L(j + 1), as shown in Figure 4.19.

Resolving $F_{L(j-1)M(j)}$ at joint L(j-1) yields

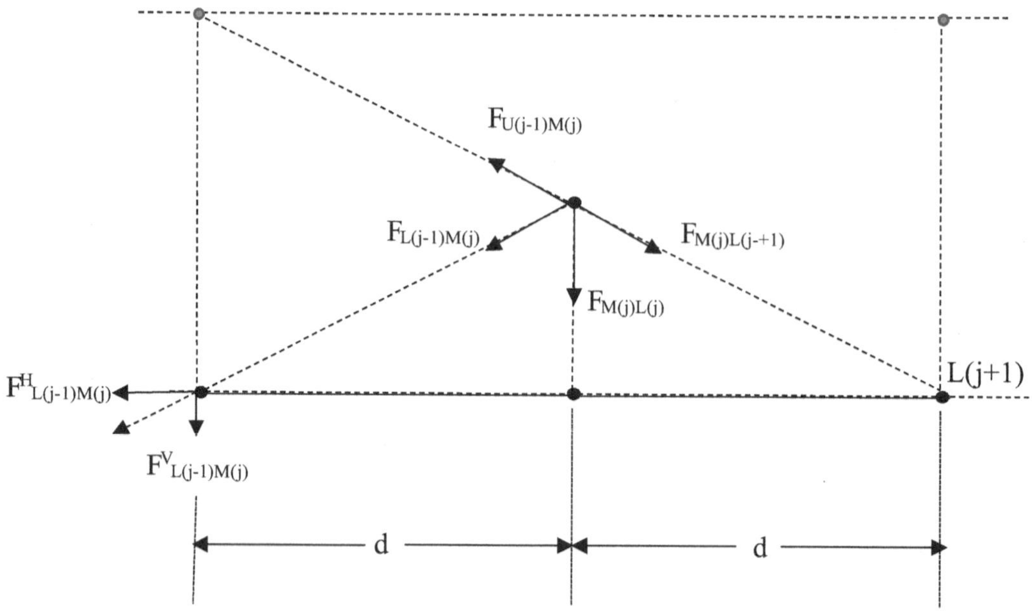

FIGURE 4.19 Panel in substrut subdivided Pratt (Baltimore) truss.

$$F_{L(j-1)M(j)} = \frac{-F_{M(j)L(j)}}{2}\left(\frac{\sqrt{\left(\frac{h}{2}\right)^2 + d^2}}{\left(\frac{h}{2}\right)}\right) \tag{4.34a}$$

where in accordance with Equation 4.14f,

$$F_{M(j)L(j)} = W_{(j)} = \frac{2M^P_{(j)}}{d}$$

Equation 4.34a shows that force $F_{L(j-1)M(j)}$ in substrut member L(j – 1)M(j) of the BSS span depends only on the magnitude of $F_{M(j)L(j)}$ in subvertical member M(j)L(j) and truss geometry. Therefore, $F^{max}{}_{L(j-1)M(j)}$ in L(j – 1)M(j) of the BSS span occurs when $F^{max}{}_{M(j)L(j)} = W^{max}{}_{(j)}$.

For a uniformly distributed live load w in subpanels (j – 1) – (j) and (j) – (j + 1), the force in substrut member L(j – 1)M(j) of the BSS span is maximum when equal-length subpanels (j – 1) – (j) and (j) – (j + 1) are fully loaded such that

$$M^P_{(j)} = \frac{w(2d)^2}{8} = \frac{wd^2}{2} = \frac{W_{(j)}d}{2} \tag{4.34b}$$

$$F^{max}_{M(j)L(j)} = W^{max}_{(j)} = wd \tag{4.34c}$$

$$F^{max}_{L(j-1)M(j)} = \frac{-wd}{2}\left(\frac{\sqrt{\left(\frac{h}{2}\right)^2 + d^2}}{\left(\frac{h}{2}\right)}\right) \tag{4.34d}$$

For concentrated live load forces in equal-length subpanels (j – 1) – (j) and (j) – (j + 1) of the BSS span, determine the maximum panel point moment $M^{Pmax}{}_{(j)}$ from Equation 4.14f using the criterion of Equation 4.11c to compute $F^{max}{}_{L(j-1)M(j)}$ in substrut member L(j-1)M(j) of the BSS span such that

$$F^{max}_{L(j-1)M(j)} = \frac{-M^{Pmax}_{(j)}}{d}\left(\frac{\sqrt{\left(\frac{h}{2}\right)^2 + d^2}}{\left(\frac{h}{2}\right)}\right) \tag{4.34e}$$

Example 4.28

Determine the maximum force in substrut member L2M3 of the Baltimore truss in Example 4.27 produced by a uniformly distributed live load of 4.0 kips/ft: h = 50 ft and j = 3. Equation 4.34d yields

$$F^{max}_{L2M3} = \frac{-4.0(25)}{2}\left(\frac{\sqrt{\left(\frac{50}{2}\right)^2 + 25^2}}{\left(\frac{50}{2}\right)}\right) = -50.0(1.414) = -70.71\,\text{kips}$$

Example 4.29

Determine the maximum force in substrut member L2M3 of the Baltimore truss in Example 4.27 produced by Cooper's E80 live load.

In accordance with Example 4.11, axle 4 or axle 13 at L3 creates $M_{(3)}{}^{Pmax}$ = 1891 kip-ft on two equal-length subpanels of d = 25 ft such that, from Equation 4.34e,

$$F^{max}_{L2M3} = \frac{-M^{Pmax}_{(3)}}{25}\left(\frac{\sqrt{\left(\frac{50}{2}\right)^2 + d^2}}{\left(\frac{50}{2}\right)}\right) = \frac{-1891}{25}(1.414) = -75.64(1.414) = -107.00\,\text{kips}$$

If the AREMA (2024) alternate load is considered by placing one of the central 50 kip wheel loads at L3,

F_{M3L3} = 156 kips, which provides $M^{Pmax}{}_{(3)}$ = 156(50)/4 = 1950 kip-ft and, hence, $F^{max}{}_{L2M3}$ = −1950(1.414)/25 = −110.29 kips.

In the BSS span in Figure 4.18c, the maximum and minimum forces in upper diagonal member U(j – 1) M(j) are determined by considering positive and negative shear forces in subpanel (j – 1) – (j).

The force in upper diagonal member U(j – 1)M(j) of the BSS span due to positive shear in subpanel (j – 1) – (j) is

$$F^{+}_{U(j-1)M(j)} = \left(V^{+}_{(j-1)-(j)} - F^{V}_{L(j-1)M(j)}\right)\frac{\sqrt{h^2+(2d)^2}}{h} \tag{4.35a}$$

where

$$V^{+}_{(j-1)-(j)} = R_{y0} - \sum_{i=0}^{j-1} W_i$$

$F^{V}_{L(j-1)M(j)} = \dfrac{F_{M(j)L(j)}}{2} = \dfrac{W_{(j)}}{2}$ = the vertical component of the force in substrut member L(j – 1)M(j) intersected by the section through subpanel (j – 1) – (j) as shown by Equation 4.34a

For a uniformly distributed live load on the BSS span,

$$R_{y0} = \frac{w\left((n-j)d + x^{+}_{(j-1)-(j)}\right)^2}{2nd}$$

$$\sum_{i=0}^{j-1} W_i = W_{(j-1)} = \frac{w\left(x^{+}_{(j-1)-(j)}\right)^2}{2d}$$

$$V^{+}_{(j-1)-(j)} = \frac{w\left((n-j)d + x^{+}_{(j-1)-(j)}\right)^2}{2nd} - \frac{w\left(x^{+}_{(j-1)-(j)}\right)^2}{2d}$$

$$W_{(j)} = \frac{w\left(x^{+}_{(j-1)-(j)}\right)\left(d - \dfrac{x^{+}_{(j-1)-(j)}}{2}\right)}{d} + \frac{wd}{2}$$

$$F^{+}_{U(j-1)M(j)} = \left(\left(\frac{w\left((n-j)d + x^{+}_{(j-1)-(j)}\right)^2}{2nd} - \frac{w\left(x^{+}_{(j-1)-(j)}\right)^2}{2d}\right) - \left(\frac{w\left(x^{+}_{(j-1)-(j)}\right)\left(d - \dfrac{x^{+}_{(j-1)-(j)}}{2}\right)}{2d} + \frac{wd}{4}\right)\right)\frac{\sqrt{h^2+(2d)^2}}{h}$$

$F^{+max}_{U(j-1)M(j)}$ occurs when $\dfrac{d\left(F^{+}_{U(j-1)M(j)}\right)}{dx} = 0$, which provides

$$x^{+}_{(j-1)-(j)} = \left(\frac{\frac{n}{2}-j}{\frac{n}{2}-1}\right)d = \frac{\left(1-\frac{2j}{n}\right)d}{\left(1-\frac{2}{n}\right)} \tag{4.35b}$$

The force in upper diagonal member U(j – 1)M(j) of the BSS span due to negative shear in subpanel (j – 1) – (j) is

$$F^{-}_{U(j-1)M(j)} = \left(-V^{-}_{(j-1)-(j)} - F^{V}_{L(j-1)M(j)}\right)\frac{\sqrt{h^2+(2d)^2}}{h} \tag{4.36a}$$

where for a uniformly distributed live load,

$$V^{-}_{(j-1)-(j)} = R_{y12} - \sum_{i=j}^{n} W_i$$

$$\sum_{i=j}^{n} W_i = W_{(j)}$$

$$R_{y12} = \frac{w\left((j-1)d + x^{-}_{(j-1)-(j)}\right)^2}{2nd}$$

$$W_{(j)} = \frac{w\left(x^{-}_{(j-1)-(j)}\right)^2}{2d}$$

$$V^{-}_{(j-1)-(j)} = \frac{w\left((j-1)d + x^{-}_{(j-1)-(j)}\right)^2}{2nd} - \frac{w\left(x^{-}_{(j-1)-(j)}\right)^2}{2d}$$

$$W_{(j-1)} = \frac{w\left(x^{-}_{(j-1)-(j)}\right)\left(d - \frac{x^{-}_{(j-1)-(j)}}{2}\right)}{d} + \frac{wd}{2}$$

$$F^{-}_{U(j-1)M(j)} = \frac{-w\left((j-1)d + x^{-}_{(j-1)-(j)}\right)^2}{2nd} + \frac{w\left(x^{-}_{(j-1)-(j)}\right)^2}{2d} - \frac{w\left(x^{-}_{(j-1)-(j)}\right)^2}{4d}$$

$F^{-max}_{U(j-1)M(j)}$ occurs when $\dfrac{d\left(F^{-}_{U(j-1)M(j)}\right)}{dx} = 0$, which provides

$$x^{-}_{(j-1)-(j)} = d - x^{+}_{(j-1)-(j)} \tag{4.36b}$$

Example 4.30

Determine the maximum forces due to positive and negative shear in upper diagonal member U2M3 produced by a uniformly distributed live load of 4.0 kips/ft on the Baltimore truss in Example 4.27 (j = 3). For positive subpanel shear from Equation 4.35b,

$$x^{+}_{(2)-(3)} = \left(\frac{6-3}{6-1}\right)25 = 15.0 \text{ ft}$$

$$R_{y0} = \frac{4.0\left((12-3)25+15\right)^2}{2(12)(25)} = 384 \text{ kips}$$

$$W_{(2)} = \frac{4.0(15.0)^2}{2(25)} = 18.0 \text{ kips}$$

$$W_{(3)} = \frac{4.0(15)\left(25-\frac{15}{2}\right)}{25} + \frac{4.0(25)}{2} = 42.0 + 50.0 = 92.0 \text{ kips}$$

$$V^{+max}_{(2)-(3)} = 384.0 - 18.0 = 366.0 \text{ kips}$$

$$F^{V}_{L2M3} = \frac{92.0}{2} = 46.0 \text{ kips}$$

$$F^{+max}_{U2M3} = (366.0-46.0)(1.414) = 320(1.414) = 452.5 \text{ kips}$$

For negative subpanel shear from Equation 4.36b,

$$x^{-}_{(2)-(3)} = 25 - 15.0 = 10.0 \text{ ft}$$

$$R_{y12} = \frac{4.0\left((2)25+10\right)^2}{2(12)(25)} = 24.0 \text{ kips}$$

$$W_{(3)} = \frac{4.0(10.0)^2}{2(25)} = 8.0 \text{ kips}$$

$$W_{(2)} = \frac{4.0(10)\left(25 - \frac{10}{2}\right)}{25} + \frac{4.0(25)}{2} = 32.0 + 50.0 = 82.0\text{kips}$$

$$V_{(2)-(3)}^{-max} = 24.0 - 8.0 = 16.0\text{kips}$$

$$F_{L2M3}^{V} = \frac{8.0}{2} = 4.0\text{kips}$$

$$F_{U2M3}^{-max} = (-16.0 - 4.0)(1.414) = -20.0(1.414) = -28.3\text{kips}$$

For concentrated live loads on the BSS span, the positive shear in subpanel (j – 1) – (j) to determine force $F_{U(j-1)M(j)}^{+}$ in upper diagonal member U(j – 1)M(j), as shown by Equation 4.35a, is

$$F_{U(j-1)M(j)}^{+} = \left(\left(\frac{P_T^+ x_T^+}{nd} - \frac{P_B^+\left(X_B^+\right)}{d} - P_F^+\right) - \left(\frac{P_B^+\left(d - X_B^+\right) + P_A^+\left(X_A^+\right)}{2d}\right)\right)\frac{\sqrt{h^2 + (2d)^2}}{h}$$

$$= \left(\frac{P_T^+ x_T^+}{nd} - \frac{P_B^+\left(X_B^+\right)}{2d} - P_F^+ - \frac{P_B^+}{2} - \frac{P_A^+\left(X_A^+\right)}{2d}\right)\frac{\sqrt{h^2 + (2d)^2}}{h} \tag{4.37a}$$

where

$$R_{y0} = \frac{P_T^+ x_T^+}{nd}$$

$$W_{(j-1)} = \frac{P_B^+\left(X_B^+\right)}{d} + \frac{P_F^+\left((j-1)d - X_F^+\right)}{(j-1)d} = \frac{P_B^+\left(X_B^+\right)}{d} + P_F^+ - \frac{P_F^+ X_F^+}{(j-1)d}$$

$$V_{(j-1)-(j)}^{+} = R_{y0} - \sum_{i=0}^{j-1} W_i = \left(\frac{P_T^+ x_T^+}{nd}\right) - \left(\frac{P_B^+\left(X_B^+\right)}{d} + P_F\right)$$

$$W_{(j)} = \frac{P_B^+\left(d - X_B^+\right)}{d} + \frac{P_A^+(A)}{d} = \frac{P_B^+\left(d - X_B^+\right) + P_A^+\left(X_A^+\right))}{d}$$

$P^+_T = P^+_F + P^+_A + P^+_B + P^+_R$ = the total load at distance X^+_T from panel point (n) with live load moving from joint (n) to joint (0) on the BSS span;

P^+_F = the live load on the truss between panel point (0) and panel point (j – 1) of the BSS span at distance, X^+_F, from (j – 1). P^+_F is typically 0 for positive shear in subpanel (j – 1) – (j) with Cooper's E live load on characteristic panel lengths;

P^+_A = the live load on the truss between (j) and (j + 1) of the BSS span at distance X^+_A from (j + 1);

P^+_B = load on the truss between panel points (j – 1) and (j) of the BSS span at X^+_B from (j). P^+_B is often zero when for positive shear in subpanel (j – 1) – (j) with Cooper's E live load on typical panel lengths;

$P^+_A + P^+_B$ = load on the truss between (j – 1) and (j + 1) of the BSS span; and
P^+_R = load on the truss between (j + 1) and (n) of the BSS span at distance X^+_R from joint (n).
$F^+_{U(j-1)M(j)}$ is maximum when,

$$\frac{d\left(F^+_{U(j-1)M(j)}\right)}{dx} = 0 = \left(\frac{P^+_T}{n} - \left(\frac{P^+_A + P^+_B}{2}\right)\right)$$

or

$$\frac{P^+_T}{n} = \left(\frac{P^+_A + P^+_B}{2}\right) \tag{4.37b}$$

The criterion of Equation 4.37b can only occur with a concentrated load at panel point (j + 1) due to the shape of the influence diagram for upper diagonal member U(j – 1)M(j) (see Example 4.57).

Therefore, the criterion for the location of a series of concentrated loads for maximum upper diagonal member positive shear force is when $\frac{P^+_T}{n} - \frac{\left(P^+_A + P^+_B\right)}{2}$ changes sign as the live load traverses a BSS span with loads moving from joint L(n) to joint L0. Considering negative shear in subpanel (j – 1) – (j) of the BSS span for concentrated live loads, $F^-_{U(j-1)M(j)}$, as shown by Equation 4.36a, is

$$F^-_{U(j-1)M(j)} = \left(-\frac{P^-_T x^-_T}{nd} + \frac{P^-_B\left(X^-_B\right)}{d} + P^-_F + P^-_A - \left(\frac{P^-_A\left(d - X^-_A\right) + P^-_B\left(X^-_B\right)}{2d}\right)\right)\frac{\sqrt{h^2 + (2d)^2}}{h}$$

$$= \left(-\frac{P^-_T x^-_T}{nd} + \frac{P^-_B\left(X^-_B\right)}{2d} + P^-_F + \frac{P_A}{2} + \frac{P^-_A X^-_A}{2d}\right)\frac{\sqrt{h^2 + (2d)^2}}{h} \tag{4.38a}$$

where

$$R_{y12} = \frac{P^-_T x^-_T}{nd}$$

$$W_{(j+1)} = \frac{P^-_A\left(X^-_A\right)}{d} + \frac{P^-_F\left((n-j-1)d - X^-_F\right)}{(n-j-1)d} = \frac{P^-_A\left(X^-_A\right)}{d} + P^-_F - \frac{P^-_F X^-_F}{(n-j-1)d}$$

$$V^-_{(j-1)-(j)} = R_{yn} - \sum_{i=j}^{n} W_i = \left(\frac{P^-_T x^-_T}{nd}\right) - P^-_F - P^-_A - \frac{P^-_B\left(X^-_B\right)}{d}$$

$$W_{(j)} = \frac{P^-_A\left(d - X^-_A\right)}{d} + \frac{P^-_B\left(X^-_B\right)}{d} = \frac{P^-_A\left(d - X^-_A\right) + P^-_B\left(X^-_B\right)}{d}$$

$P^-_T = P^-_F + P^-_A + P^-_B + P^-_R$ = total load on the truss at distance X^-_T from joint (0) (left end of truss span) with live load moving from joint (0) to joint (n) over the BSS span;
P^-_F = load on the truss between panel point (n) and panel point (j + 1) of the BSS span at X^-_F from (j + 1). P^-_F is typically 0 for negative shear due to Cooper's E live load in subpanel (j – 1) – (j);
P^-_A = load on the truss between (j) and (j + 1) of the BSS span at X^-_A from (j). P^-_A is often zero for negative shear due to Cooper's E live load in subpanel (j – 1) – (j);

P^-_B = load on the truss between (j − 1) and (j) of the BSS span at X^-_B from (j − 1); and
P^-_R = load on the truss between (j − 1) and panel point 0 of the BSS span at X^-_L from joint 0.
$F^-_{U(j-1)M(j)}$ is maximum when

$$\frac{d\left(F^-_{U(j-1)M(j)}\right)}{dx} = 0 = \left(-\frac{P^-_T}{nd} + \frac{P^-_A + P^-_B}{2d}\right)$$

or

$$\frac{P^-_T}{n} = \left(\frac{P^-_A + P^-_B}{2}\right) \tag{4.38b}$$

The criterion of Equation 4.38b can only occur with a concentrated load at panel point (j − 1) due to the shape of the influence diagram for upper diagonal member U(j − 1)M(j) (see Example 4.57). Therefore, as expected, the criterion of Equation 4.38b for the location of a series of concentrated live loads for minimum upper diagonal member force as the live load traverses a BSS span with loads moving from joint L0 to joint L(n) is equivalent to the criterion of Equation 4.37b.

Example 4.31

Determine the maximum forces produced by Cooper's E80 live load in upper diagonal members U2M3 and U4M5 of the Baltimore truss in Example 4.27.

For the positive shear in upper diagonal member U2M3, try wheels between wheel 4 and wheel 6 at panel point L4 for positive shear in subpanel 2-3; j = 3. The criterion of Equation 4.37b for maximum force in upper diagonal member U2M3 is as follows and as given in Table E4.31a.

$$\frac{P^+_T}{6} = P^+_A + P^+_B$$

P^+_B = live loads in subpanel 2-3
P^+_A = live loads in subpanel 3-4
$P^+_A + P^+_B$ = live loads in panel 2-4
Distance to L(n) from panel point L4 = (12 − 4)(25) = 200 ft
Distance to L0 from panel point L4 = (4)(25) = 100 ft

TABLE E4.31a
Criterion for Maximum Force due to Positive Shear in Member U2M3 of a BSS Span

Positive Shear in Subpanel 2-3 (j = 3)	Wheel at Panel Point 4		
Wheel at panel point L4	4	5	6
Distance to uniformly distributed load from wheel at panel point L4 (ft)	91	86	77
Distance to wheel 1 (ft)	109 − 91 = 18	109 − 86 = 23	109 − 77 = 32
Distance to L0 from wheel 1 (ft)	100 − 18 = 82	100 − 23 = 77	100 − 32 = 68
Length of uniformly distributed load (ft)	200 − 91 = 109	200 − 86 = 114	200 − 77 = 123

(Continued)

TABLE E4.31a (*Continued*)

Criterion for Maximum Force due to Positive Shear in Member U2M3 of a BSS Span

Positive Shear in Subpanel 2-3 (j = 3)	Wheel at Panel Point 4		
Total load on span = P^+_T (kips)	568 + 109(4.0) = 1004	568 + 114(4.0) = 1024	568 + 123(4.0) = 1060
$P^+_T/6$ (kips)	167.3	170.7	176.7
P^+_{Ba} (kips)	0	0	20
P^+_{Bb} (kips)	0	0	20
P^+_{Aa} (kips)	100	140	160
P^+_{Ab} (kips)	140	180	186
Load in panel 2-4 ahead of L4 = $(P^+_A + P^+_B)_a$ (kips)	100	140	180
Load in panel 2-4 including load at L4 = $(P^+_A + P^+_B)_b$ (kips)	140	180	206
Criterion met	No	Yes	No

The criterion of Equation 4.37b indicates maximum force due to positive shear in upper diagonal member U2M3 with wheel 5 at panel point L4. It is noted that $P^+_B = 0$ for wheels 4 and 5 at L4. However, $P^+_B \neq 0$ with wheel 6 at L4. Equation 4.35a provides the maximum force due to positive shear in upper diagonal member U2M3 as shown in Table E4.31b.

TABLE E4.31b

Maximum Force due to Positive Shear in Member U2M3 for Cooper's E80 Live Load Traversing a BSS Span

	U2M3 Positive Shear	
Axle at L4	5	6
Sum of moments about R_{y12} (kip-ft)	32,728 + 568(114) + 4.0(114)(114/2) = 123,472	32,728 + 568(123) + 4.0(123)²/2 = 132,850
R_{y0} (kips)	123,472/300 = 411.57	132,850/300 = 442.83
Wheel 1 to L2 (ft)	25 − (23 − 25) = 27 (in panel 3-4)	25 − (32 − 25) = 18 (in panel 2-3)
$W_{(2)}$ (kips)	0	20(25 − 18)/25 = 5.60
$V^+_{(2)-(3)} = R_{y12} - W_{(2)}$ (kips)	411.57 − 0 = 411.57	442.83 − 5.60 = 437.23
$W_{(3)}$ (kips)	1660/25 = 66.40	2640/25 + (20−5.60) = 105.60 +14.40 = 120.00
$F^V_{L2M3} = \frac{W_{(3)}}{2}$ (kips)	66.40/2 = 33.20	120.00/2 = 60.00
$F^+_{U2M3} = \left(V^+_{(2)-(3)} - F^V_{L2M3}\right)\frac{\sqrt{h^2 + (2d)^2}}{h}$ (kips)	(411.57 − 33.20)(1.414) = 535.0	(437.23 − 60.00)(1.414) = 533.4

$$F^{+max}_{U2M3} = 535.0 \text{ kips}$$

For the negative shear in upper diagonal member U2M3, try wheels between wheel 1 and wheel 3 at panel point L2 for negative shear in subpanel 2-3; j = 3.

P^-_A = live loads in subpanel 3-4

P^{-}_{B} = live loads in subpanel 2-3
$P^{-}_{A} + P^{-}_{B}$ = live loads in panel 2-4

The criterion of Equation 4.38b for minimum force in upper diagonal member U2M3 is given in Table E4.31c.

TABLE E4.31c
Criterion for Maximum Force due to Negative Shear in Member U2M3 of a BSS Span

Negative shear in subpanel 2-3 (j = 3)	Wheel at Panel Point 2		
Wheel at panel point L2	1	2	3
Last wheel on span	9	10	10
Distance of last wheel on span to L0 (ft)	50 − 48 = 2	50 − (56 − 8) = 2	50 − (56 − 13) = 7
Total load on span = P^{-}_{T} (kips)	284	304	304
$P^{-}_{T}/6$ (kips)	47.3	50.7	50.7
Load in panel 2-4 ahead of L2 = $(P^{-}_{A} + P^{-}_{B})_a$ (kips)	0	20	60
Load in panel 2-4 including load at L2 = $(P^{-}_{A} + P^{-}_{B})_b$ (kips)	20	60	100
Criterion met	No	Yes	No

The criterion of Equation 4.38b indicates maximum force due to negative shear in upper diagonal member U2M3 with wheel 2 at panel point L2. Therefore, $P^{-}_{A} = 0$. Equation 4.36a provides the maximum force due to negative shear in upper diagonal member U2M3 as given in Table E4.31d.

TABLE E4.31d
Maximum Force due to Negative Shear in Member U2M3 for Cooper's E80 Live Load Traversing a BSS Span

U2M3 Negative Shear		
Axle at L2	**2**	**3**
Last axle on span	10	10
Distance last axle on span to L0 (ft)	2	7
Sum of moments about R_{y0} (kip-ft)	9264 + 304(2) = 9,872	9264 + 304(7) = 11,392
R_{y12}(kips)	9,872/300 = 32.91	11,392/300 = 37.97
$W_{(3)}$ (kips)	160/25 = 6.40	460/25 = 18.40
$V^{+}_{(2)-(3)} = R_{y12} - W_{(2)}$ (kips)	32.91 − 6.40 = 26.51	37.97 − 18.40 = 19.57
$F^{V}_{L2M3} = \frac{W_{(3)}}{2}$ (kips)	6.40/2 = 3.20	18.40/2 = 9.20
$F^{-}_{U2M3} = \left(-V^{-}_{(2)-(3)} - F^{V}_{L2M3}\right)\frac{\sqrt{h^2+(2d)^2}}{h}$ (kips)	(−26.51 − 3.20) (1.414) = −42.0	(−19.57 − 9.20) (1.414) = −40.7

$$F^{-max}_{U2M3} = -42.0 \text{ kips}$$

For positive shear in upper diagonal member U4M5, try wheels between wheel 3 and wheel 5 at panel point L6 for positive shear in subpanel 4-5; j = 5.

P^+_B = live loads in subpanel 4-5
P^+_A = live loads in subpanel 5-6
$P^+_A + P^+_B$ = live loads in panel 4-6

The criterion of Equation 4.37b for maximum force in upper diagonal member U4M5 is (see Table E4.31e)

$$\frac{2P_T^+}{n} = P_A^+ + P_B^+$$

Distance to L(n) from panel point L6 = (12-6)(25) = 150 ft
Distance to L0 from panel point L6 = (6)(25) = 150 ft

TABLE E4.31e

Criterion for Maximum Force due to Positive Shear in Member U4M5 of a BSS Span

Positive Shear in Subpanel 4-5 (j=5)	Wheel at Panel Point 6		
Wheel at panel point L6	3	4	5
Distance to uniformly distributed load from wheel at panel point L6 (ft)	96	91	86
Distance to wheel 1 (ft)	109 − 96 = 13	109 − 91 = 18	109 − 86 = 23
Distance to L0 from wheel 1 (ft)	150 − 13 = 137	150 − 18 = 132	150 − 23 = 127
Length of uniformly distributed load (ft)	150 − 96 = 54	150 − 91 = 59	150 − 86 = 64
Total load on span = P^+_T (kips)	568 + 54(4.0) = 784	568 + 59(4.0) = 804	568 + 64(4.0) = 824
$P^+_T/6$ (kips)	130.7	134.0	137.3
Load in panel 4-6 ahead of L6 = $(P^+_A + P^+_B)_a$ (kips)	60	100	140
Load in panel 4-6 including load at L6 = $(P^+_A + P^+_B)_b$ (kips)	100	140	180
Criterion met	No	Yes	No

The criterion of Equation 4.37b indicates maximum force due to positive shear in upper diagonal member U4M5 with wheel 4 at panel point L6. Therefore, $P^+_B = 0$. Equation 4.35a provides the maximum force due to positive shear in upper diagonal member U4M5 (Table E4.31f).

TABLE E4.31f

Maximum Force due to Positive Shear in Member U4M5 for Cooper's E80 Live Load Traversing a BSS Span

	U4M5 Positive Shear (Axle 4 at L6)
Sum of moments about R_{y12} (kip-ft)	32,728 + 568(59) + 4.0(59)(59/2) = 73,202
R_{y0} (kips)	73,202/300 = 244.01
$W_{(4)}$ (kips)	0
$V_{+(4)-(5)} = R_{y0} - W_4$ (kips)	244.01 − 0 = 244.01
$W_{(5)}$ (kips)	960/25 = 38.40
$F_{L4M5} = W_{(5)}/2$ (kips)	38.40/2 = 19.20
$F^+_{U4M5} = \left(V^+_{(4)-(5)} - F^V_{L4M5}\right)\dfrac{\sqrt{h^2+(2d)^2}}{h}$	(244.01 − 19.20)(1.414) = 317.9

$$F_{U4M5}^{+max} = 317.9 \text{ kips}$$

For negative shear in upper diagonal member U4M5, try wheels between wheel 2 and wheel 5 at panel point L4 for negative shear in subpanel 4-5; j = 5:

P^-_A = live loads in subpanel 5-6
P^-_B = live loads in subpanel 4-5
$P^-_A + P^-_B$ = live loads in panel 4-6

The criterion of Equation 4.38b for maximum force due to negative shear in upper diagonal member U4M5 is given in Table E4.31g.

TABLE E4.31g
Criterion for Maximum Force due to Negative Shear in Member U4M5 of a BSS Span

Negative Shear in Subpanel 4-5 (j = 5)	Wheel at Panel Point 4			
Wheel at panel point L4	2	3	4	5
Last wheel on span	18	18	18	18
Distance of last wheel on span to L0 (ft)	100 − 96 = 4	100 − 91 = 9	100 − 86 = 14	100 − 81 = 19
Length of uniformly distributed load on span (ft)	4 − 5 = −1 (1 ft off span)	9 − 5 = 4	14 − 5 = 9	19 − 5 = 14
Total load on span = P^-_T (kips)	568	568 + 4.0(4) = 584	568 + 4.0(9) = 604	568 + 4.0(14) = 624
$P^-_T/6$ (kips)	94.7	97.3	100.7	104
Load in panel 4-6 ahead of L4 = $(P^-_A + P^-_B)_a$ (kips)	20	60	100	140
Load in panel 4-6 including load at L4 = $(P^-_A + P^-_B)_b$ (kips)	60	100	140	180
Criterion met	No	Yes	Yes (marginal)	No

The criterion of Equation 4.38b indicates maximum force due to negative shear in upper diagonal member U4M5 with wheels 3 or 4 at panel point L4. Therefore, $P^-_A = 0$. Equation 4.36a provides the maximum force due to negative shear in upper diagonal member U4M5 (Table E4.31h).

TABLE E4.31h
Maximum Force due to Negative Shear in Member U4M5 for Cooper's E80 Live Load Traversing a BSS Span

U4M5 Negative Shear		
Wheel at L4	3	4
Sum of moments about R_{y0} (kip-ft)	29,888 + 9(568) = 35,000	29,888 + 14(568) = 37,840
R_{y12} (kips)	35,000/300 = 116.67	126.1
$W_{(5)}$ (kips)	460/25 = 18.40	960/25 = 38.40
$V_{(4)\text{-}(5)} = R_{y12} - W_{(5)}$ (kips)	116.67 − 18.40 = 98.27	126.1 − 38.40 = 87.73
$F_{L4M5}^{V} = \dfrac{W_{(5)}}{2}$ (kips)	18.40/2 = 9.20	38.40/2 = 19.20
$F_{U4M5}^{-} = \left(-V_{(4)-(5)}^{-} - F_{L4M5}^{V}\right)\dfrac{\sqrt{h^2 + (2d)^2}}{h}$ (kips)	(−98.27 − 9.20)(1.414) = −152.0	(−87.73 − 19.20)(1.414) = −151.2

$$F_{U4M5}^{-max} = -152.0\text{kips}$$

The forces in interior vertical members, such as U(k)L(k) of the BSS span, where k = even numbers greater than or equal to 4 and less than n/2, as shown in Figure 4.18c, are determined by investigation of the free body diagram at isolated truss joint U(k). Equilibrium at U(k) indicates that the forces in interior vertical members U(k)L(k) of the BSS span resist the vertical component of the force in upper diagonal members U(k)M(k + 1) and external loads, if any, at joint U(k). Therefore, if no external loads exist at joint U(k),

$$F_{U(k)L(k)} = -F_{U(k)M(k+1)}\left(\frac{h}{\sqrt{h^2+(2d)^2}}\right) \tag{4.39}$$

Equilibrium at joint U(n/2) of the BSS span indicates the force in interior vertical member U(k)L(k), where k = n/2, is $F_{U(n/2)L(n/2)} = 0$.

Example 4.32

Determine the maximum forces produced by Cooper's E80 live load in interior vertical member U4L4 of the Baltimore truss of Example 4.27.

k = 4

Free body diagram at U4 indicates

$F_{U4L4} = -0.707F_{U4M5}$

From Example 4.31, Table E4.32 gives the maximum force.

TABLE E4.32
Maximum Force due to Positive and Negative Shear in Member U4L4 for Cooper's E80 Live Load Traversing a BSS Span

j	$F^{+}_{U(k)M(k+1)} = F^{+}_{U4M5}$ (kips)	$F^{-}_{U(k)M(k+1)} = F^{-}_{U4M5}$ (kips)	$F^{+}_{U(k)L(k)} = F^{+}_{U4L4}$ (kips)	$F^{-}_{U(k)L(k)} = F^{-}_{U4L4}$ (kips)
5	317.9	−152.0	−224.8	107.5

The forces in end vertical members U2L2 of the BSS span, shown in Figure 4.18a and Figure 4.18c, are determined by investigation of free body diagrams at isolated truss joints L2, which provides

$$F_{U2L2} = W_2 + \frac{W_1 + W_3}{2} \tag{4.40a}$$

From Equation 4.14e,

$$W_1 = \frac{-2M_{(1)} + M_{(2)}}{d} \tag{4.40b}$$

$$W_2 = \frac{M_{(1)} - 2M_{(2)} + M_{(3)}}{d} \tag{4.40c}$$

$$W_3 = \frac{M_{(2)} - 2M_{(3)} + M_{(4)}}{d} \qquad (4.40d)$$

For a uniformly distributed live load, $W_1 = W_2 = W_3 = W$, which when substituted into Equation 4.40a yields

$$F_{U2L2} = 2W \qquad (4.40e)$$

For concentrated live loads, substituting Equations 4.40b, 4.40c, and 4.40d into Equation 4.40a yields

$$F_{U2L2} = \frac{-2M_{(2)} + M_{(4)}}{2d} \qquad (4.40f)$$

However, as illustrated by Equation 4.14f, $\dfrac{-2M_{(2)} + M_{(4)}}{2}$ is the center moment M_c over a length $L_c = (j+2)d = 4d$ between panel point 0 and panel point (j+2) such that

$$F_{U2L2} = \frac{M_c}{d} \qquad (4.40g)$$

The maximum force in member U2L2 of the BSS span is determined with concentrated loads positioned for maximum moment at joint L2 in the span L_c.[140]

Example 4.33

Determine the maximum forces produced by a uniformly distributed live load of 4.0 kips/ft and Cooper's E80 live load in end vertical member U2L2 of the Baltimore truss in Example 4.27.

For a uniformly distributed live load, W = wd = 4.0(25) = 100 kips:

$$F_{U2L2} = 2(100) = 200 \text{ kips}$$

For Cooper's E80 live load, the maximum force in member U2L2 is determined using the criterion of Equation 4.11c with k = 2 (Table E4.33a):

$P_F = (k/n)P_T = (2/4)P_T = P_T/2$

TABLE E4.33a
Criterion for Maximum Force in Member U2L2 of a BSS Span

Location of Live Load	Wheel at L2			
	11	12	13	14
Wheel to right of L0	4	5	6	6
Wheel to left of L4	18	18	18	18
Length of uniformly distributed load (ft)	(64 + 50) − 109 = 5	(69 + 50) − 109 = 10	(74 + 50) − 109 = 15	(79 + 50) − 109 = 20
P_T (kips)	468 + (4.0)5 = 488	428 + 4.0(10) = 468	388 + 4.0(15) = 448	388 + 4.0(20) = 468
$P_T/2$ (kips)	244	234	224	234
P_{Fa} (kips)	304 − 100 = 204	344 − 140 = 204	384 − 180 = 204	424 − 180 = 244
P_{Fb} (kips)	244	244	204 + 40 = 244	244 + 40 = 284
Criterion met	Yes (marginal)	Yes	Yes	No

The criterion of Equation 4.11c indicates wheels 11, 12 or 13 at L2 for maximum panel point moment. It is likely that the maximum force in member U2L2 is created with axle 12 or axle 13 at joint L2. Using Table 4.2, Table E4.33b gives the maximum force.

TABLE E4.33b
Maximum Force in Member U2L2 for Cooper's E80 Live Load Traversing a BSS Span

Maximum Force in Member U2L2	Axle at L2		
	11	12	13
Sum of moments to L4 = $M_{(4)}$ (kip-ft)	20,328 + 468(5 + 5) + 4.0(5)(5)/2 = 20,328 + 4680 + 50 = 25,058	16,888 + 428(5 + 10) + 4.0(10)(10)/2 = 16,888 + 6420 + 200 = 23,508	13,648 + 388(5 + 15) + 4.0(15)15)/2 = 21,858
Reaction R_{y0} for span of length 4d = 100 ft (kips)	25,058/100 = 250.6	23,508/100 = 235.1	218.6
Sum of moments from first axle to axle at L2 (kip-ft)	6136	5316	4496
Sum of moments about L2 = $M_{(2)}$ (kip-ft)	250.6(50) − 6136 = 6393	235.1(50) − 5316 = 6438	218.6(50) − 4496 = 6433
$F_{U2L2} = \frac{M_{(2)}}{25}$ (kips)	6393/25 = 255.7	6438/25 = 257.5	257.3

$$F_{U2L2}^{max} = 257.5 \text{ kips}$$

The force F_{M1U2} in upper end post member M1U2 of the BSS span is determined by taking moments about panel point L2 as shown in Figure 4.18e such that

$$F_{M1U2} = \frac{-\left(2(R_{y0} - W_0) - W_1\right)d}{\sqrt{\left(\frac{h}{2}\right)^2 + d^2}} \tag{4.41}$$

Example 4.34

Determine the maximum forces produced by a uniformly distributed live load of 4.0 kips/ft and Cooper's E80 live load in upper end post member M1U2 of the Baltimore truss in Example 4.27.

For a uniformly distributed live load of 4.0 kips/ft,
$R_{y0} = 4.0(300/2) = 600$ kips
$W_0 = 4.0(25)/2 = 50$ kips
$W_1 = 4.0(25) = 100$ kips

$$F_{M1U2}^{max} = \frac{-\left(2(25)(600-50) - 25(100)\right)}{\sqrt{\left(\frac{50}{2}\right)^2 + 25^2}} = \frac{-25{,}000}{35.36} = -707.1 \text{ kips}$$

For Cooper's E80 live load, in accordance with Table 4.1, the maximum panel point moment at L2 occurs with wheels 6, 7 or 8 at L2. The criterion of Equation 4.11c, with k = 2 for upper end post member M1U2 (j = 2), is

$$\frac{P_T}{6} = P_F$$

Moment center at joint L2 is 2(25) = 50 ft from L0 (k = 2).

Table 4.2 is used to determine moments about panel point L2 with Cooper's E80 live load located in accordance with the criterion of Equation 4.11c as shown in Table E4.34a.

TABLE E4.34a
Criterion for Maximum Force in Member M1U2 of a BSS Span

	Axle at L2		
Location of Live Load	**6**	**7**	**8**
Distance to uniform load (ft)	77	72	66
Distance to axle no. 18 (ft)	77 − 5 = 72	67	61
Last axle on span	18	18	18
Distance to axle no. 1 (ft)	32	37	43
First axle on span	1	1	1
Length of uniform load (ft)	250 − 77 = 173	250 − 72 = 178	250 − 66 = 184
Total load on span, P_T (kips)	568 +173(4) = 1260	568 + 178(4) = 1280	568 + 184(4) = 1304
$P_T/6$ (kips)	210.0	213.3	217.3
Sum of loads ahead of panel point L2 = P_{F1} (kips)	180	206	232
Sum of loads ahead of and including panel point L2 = P_{F2} (kips)	206	232	258
Criterion established	No (marginal)	Yes	No

The criterion of Equation 4.11c indicates axle 7 at L2 for maximum panel point moment at L2 and, accordingly, the maximum force for upper end post member M1U2, as shown in Table E4.34b. Panel point moments are determined from Table 4.2. The live load force with wheel 6 at L2 is also calculated as validation of the criterion.

TABLE E4.34b
Maximum Force in Member M1U2 for Cooper's E80 Live Load Traversing a BSS Span

	Axle at L2	
Maximum Force in Member M1U2	**6**	**7**
Sum moments to uniform load (kip-ft)	32728	32,728
Sum wheel loads to uniform load (kips)	568	568
Length of uniform load (ft)	173	178
Sum moments to L12 (kip-ft)	$32{,}728 + 568(173) + 4(173)^2/2 =$ 190,850	$32{,}728 + 568(178) + 4(178)^2/2 =$ 197,200
Reaction R_{y0} (kips)	190,850/300 = 636.2	197,200/300 = 657.3
Sum moments from first axle to axle at L2 (kip-ft)	3280	4310
Sum moments about L2 = $\sum M_{(2)}$ (kip-ft)	636.2(2)(25) − 3280 = 28,528.3	657.3(2)(25) − 4310 = 28,556.7
Force in member M1U2 = F_{M1U2} = $-(\sum M_{(2)})/35.36$ (kips)	−806.8	−807.7

$F_{M1U2}^{max} = -807.7 \text{ kips}$

Force F_{L0M1} in lower end post member L0M1 of the BSS span is calculated by considering force equilibrium of the free body diagram at L0 and L1 such that

$$F_{L0M1} = -(R_{y0} - W_0)\left(\frac{\sqrt{(2d)^2 + h^2}}{h}\right) = -F_{L0L1}\left(\frac{\sqrt{(2d)^2 + h^2}}{2d}\right)$$
$$= -F_{L1L2}\left(\frac{\sqrt{(2d)^2 + h^2}}{2d}\right) \quad (4.42a)$$

Example 4.35

Determine the maximum forces produced by a uniformly distributed live load of 4.0 kips/ft and Cooper's E80 live load in lower end post member L0M1 of the Baltimore truss in Example 4.27.

For a uniformly distributed live load of 4.0 kips/ft,

$R_{y0} = 4.0(300/2) = 600.0 \text{ kips}$

$$F_{L0M1} = -\left(600.0 - \frac{4.0(25)}{2}\right)\left(\frac{\sqrt{(50)^2 + 50^2}}{50}\right) = -(550.0)(1.414) = -777.7\text{kips}$$

For Cooper's E80 live load, Table 4.1 indicates wheels 3, 4 or 5 at L1 for maximum panel point moment at M1 (for maximum force in member L1L2 and, by equilibrium at joint L1, member L0L1) and the criterion of Equation 4.11c with k = 1 is given in Table E4.35a.

TABLE E4.35a
Criterion for Maximum Force in Member L1L2 to Determine Maximum Force in Member L0M1 of a BSS Span

	Axle at L1		
Location of Live Load	**3**	**4**	**5**
Distance to uniform load (ft)	96	91	86
Distance to axle no. 18 (ft)	91	91 – 5 = 86	81
Last axle on span	18	18	18
Distance to axle no. 1 (ft)	104 – 91 = 13	104 – 86 = 18	104 – 81 = 23
First axle on span	1	1	1
Length of uniform load on span (ft)	275 – 96 = 179	275 – 91 = 184	275 – 86 = 189
Total load on span, P_T (kips)	568 + 179(4) = 1284	568 + 184(4) = 1304	568 + 189(4) = 1324
$P_T(k/n) = P_T(1/12)$ (kips)	107.0	108.7	110.3
Sum of loads ahead of panel point L1 = PF_a (kips)	100 – 40 = 60	140 – 40 = 100	180 – 40 = 140
Sum of loads ahead of and including panel point L1 = PF_b (kips)	100	140	180
Criterion established	No	Yes	No

The criterion of Equation 4.11c indicates wheel 4 at L1 for maximum moment at panel point M1. Table 4.2 is used to develop the maximum force in member L1L2 and, therefore, L0M1 as in Table E4.35b.

TABLE E4.35b
Maximum Force in Member L1L2 to Determine Maximum Force in Member L0M1 for Cooper's E80 Live Load Traversing a BSS Span

	Axle at L1
Maximum Force in L1L2 and L0L1	**4**
Sum moments to uniform load (kip-ft)	32,728
Sum wheel loads to uniform load (kips)	568
Length of uniform load (ft)	275 – 91 = 184
Sum moments to L12 (kip-ft)	32,728 + 568(184) + 4(184)²/2 = 204,952
Reaction R_{y0} (kips)	204,952/300 = 683.2
Sum moments from first axle to axle at L1 (kip-ft)	960
Sum moments about U1 (kip-ft)	683.2(25) – 960 = 16,119.3
Maximum force in L1L2 = $F_{L1L2} = F_{L0L1}$ (kips)	16119.3/(50/2) = 644.77
$F_{L0M1}^{max} = -1.414F_{L0L1}^{max} = -1.414F_{L1L2}^{max}$ (kips)	$-1.414(644.77) = -911.7$

The maximum forces due to uniformly distributed and concentrated loads in member M1L2 of the BSS span are calculated by reviewing equilibrium at joint L2, which indicates that the force in M1L2 is equal to the force in substrut member L2M3 and therefore, in accordance with Equation 4.34a, is

$$F_{M1L2} = F_{L2M3} = \frac{F_{M3L3}}{2}\left(\frac{\sqrt{\left(\frac{h}{2}\right)^2 + d^2}}{\left(\frac{h}{2}\right)}\right) = \frac{-W_{(3)}}{2}\left(\frac{\sqrt{\left(\frac{h}{2}\right)^2 + d^2}}{\left(\frac{h}{2}\right)}\right) \tag{4.42b}$$

4.6.2 Subdivided Inclined Chord Pratt (Pennsylvania) Truss Spans with Substrut Members

As shown in Figure 4.18b[141] the forces in top chord member U(j – 1)U(j + 1) of a Pennsylvania truss span with substrut members (PSS) are calculated by taking a section through subpanel (j) – (j + 1) and considering moments taken about opposite chord panel point L(j + 1) such that $F_{U(j-1)U(j+1)}$ is

$$F_{U(j-1)U(j+1)} = \frac{-\Sigma M_{L(j+1)}}{h_{(j+1)}} \frac{\sqrt{\left(h_{(j+1)} - h_{(j-1)}\right)^2 + (2d)^2}}{2d}$$
$$= -\left((j+1)R_{y0} - \sum_{i=0}^{j+1}(j+1-i)W_i\right)\left(\frac{\sqrt{\left(h_{(j+1)} - h_{(j-1)}\right)^2 + (2d)^2}}{2h_{(j+1)}}\right) \tag{4.43a}$$

Maximum force $F^{max}_{U(j-1)U(j+1)}$ in top chord member U(j – 1)U(j + 1) of the PSS span due to a uniformly distributed live load occurs with the span entirely loaded.

$F^{max}_{U(j-1)U(j+1)}$ in top chord member U(j – 1)U(j + 1) of the PSS span due to concentrated live loads, such as the Cooper's E design load, are determined with the live load positioned on the span in accordance with the criterion of Equation 4.11c, which is $\frac{P_T}{n} = \frac{P_F}{k}$ where k = j + 1. As shown in Figure 4.18b, the forces in bottom chord member L(j)L(j + 1) of the PSS span are calculated by taking a section through subpanel (j) – (j + 1) and considering equilibrium at joint U(j – 1). Accordingly, the force in bottom chord members L(j – 1)L(j) and L(j)L(j + 1) for equal-length subpanels in the PSS span (see Equation 4.31b) is

$$F_{L(j)L(j+1)} = \frac{M^{Wx}_{(j-1)} + 2M^{P}_{(j)}}{h_{(j-1)}} \tag{4.43b}$$

where

$$M^{Wx}_{(j-1)} = \left((j-1)R_{y0} - \sum_{i=0}^{j-1}(j-1-i)W_i \right)d$$

$$M^{P}_{(j)} = \frac{W_{(j)}(d)}{2}$$

For a uniformly distributed live load, $F^{max}_{L(j-1)L(j)}$ and $F^{max}_{L(j)L(j+1)}$ of the PSS span are determined with

$$M^{P}_{(j)} = \frac{w(2d)^2}{8} = \frac{wd^2}{2}$$

For concentrated live loads, the criterion of Equation 4.31f for the position of live load creating $F^{max}_{L(j-1)L(j)}$ and $F^{max}_{L(j)L(j+1)}$ in the PSS span is

$$\frac{P_T}{n} = \frac{P_F + P_A - P_B}{(j-1)}$$

where $P_T = P_F + P_A + P_B + P_R$ = the total load at distance X_T from joint (n) at the right end of the truss span with live load moving from joint (n) to joint (0) of the PSS span;

P_F = the resultant load on the truss between panel point (0) and panel point (j – 1) of the PSS span at distance X_F from panel point (j – 1);

P_A = the resultant load on the truss between panel point (j – 1) and (j) of the PSS span at X_A from (j);

P_B = the resultant load on the truss between (j) and (j + 1) of the PSS span at X_B from (j + 1); and

P_R = load on the truss between panel points (j + 1) and (n) of the PSS span at X_R from joint (n).

The criterion of Equation 4.31f can be used to determine the location of the concentrated live loads creating maximum bottom chord member force at any panel within a PSS span.

Example 4.36

A 300 ft long PSS truss with twelve subpanels is loaded with a live load of 5.5 kips/ft. The top chord is inclined with truss height of 25 ft at joints 1 and 11, 38 ft at joints 2 and 10, 47.5 ft at joints 4 and 8 and 50 ft at joint 6. Subvertical member M1L1 is 25 ft long, subvertical member M3L3 is 19 ft long and subvertical member M5L5 is 23.75 ft long. Determine the maximum forces due to live load in the top and bottom chord members.

$R_{y0} = (5.5)(300/12) = 825$ kips
$h_1 = 25$ ft
$h_2 = 38$ ft
$h_4 = 47.5$ ft
$h_6 = 50$ ft

Forces due to live load in the top chord members are provided by Equation 4.43a:

$$F_{U2U4} = \frac{-\Sigma M_{L(4)}}{h_{(4)}} \frac{\sqrt{\left(h_{(4)} - h_{(2)}\right)^2 + (2d)^2}}{2d} = -\left((3+1)(825) - \sum_{i=0}^{3+1}(3+1-i)W_i\right)$$

$$\left(\frac{\sqrt{\left(h_{(4)} - h_{(2)}\right)^2 + (2(25))^2}}{2h_{(4)}}\right) = -\left(3300 - 2(137.5) - (3+2+1)(137.5)\right)$$

$$\left(\frac{\sqrt{(47.5-38.0)^2 + (2(25))^2}}{2(47.5)}\right) = -2200\left(\frac{50.89}{2(47.5)}\right) = -1178.6 \text{ kips}$$

where j = 3

$$F_{U4U6} = \frac{-\Sigma M_{L(6)}}{h_{(6)}} \frac{\sqrt{\left(h_{(6)} - h_{(4)}\right)^2 + (2d)^2}}{2d} = -\left((5+1)(825) - \sum_{i=0}^{5+1}(5+1-i)W_i\right)$$

$$\left(\frac{\sqrt{\left(h_{(6)} - h_{(6)}\right)^2 + (2(25))^2}}{2h_{(6)}}\right) = -\left(4950 - \left(\frac{6}{2} + 5 + 4 + 3 + 2 + 1\right)137.5\right)$$

$$\left(\frac{\sqrt{(50.0-47.5)^2 + (2(25))^2}}{2(50.0)}\right) = -2475.0\left(\frac{50.06}{2(50.0)}\right) = -1239.0 \text{ kips}$$

where j = 5

Forces due to live load in the bottom chord members are provided by Equation 4.43b where j = 3:

$$F_{L3L4} = \frac{M_{(2)}^W + 2M_{(3)}^P}{h_2}$$

$$M_{(2)}^W = \left(2(825) - \sum_{i=0}^{2}(2-i)137.5\right)25 = \left(1650 - 2(68.75) - 137.5\right)(25) = 34{,}375 \text{ kips} - \text{ft}$$

$$M^P_{(3)} = \frac{W_{(3)}(d)}{2} = \frac{137.5(25)}{2} = 1718.75 \text{ kips} - \text{ft}$$

$$F_{L3L4} = \frac{34,375 + 2(1718.5)}{38.0} = 995.1 \text{ kips} = F_{L2L3} = F_{L1L2} = F_{L0L1} \text{ as expected for a}$$

uniformly distributed live load entirely on the span

$$F_{L5L6} = \frac{M^W_{(4)} + 2M^P_{(5)}}{h_4} \text{ where j} = 5$$

$$M^W_{(4)} = \left(4(825) - \sum_{i=0}^{4}(4-i)W_i\right)25 = (3300 - (2+3+2+1)137.5)(25) = 55,000 \text{ kips} - \text{ft}$$

$$M^P_{(5)} = \frac{W_{(5)}(d)}{2} = \frac{137.5(25)}{2} = 1718.75 \text{ kips} - \text{ft}$$

$$F_{L5L6} = \frac{55,000 + 2(1718.75)}{47.5} = 1230.3 \text{ kips} = F_{L4L5}$$

Example 4.37

A 300 ft long PSS truss with twelve subpanels is loaded with Cooper's E80 live load. The top chord is inclined with truss height of 25 ft at joints 1 and 11, 38 ft at joints 2 and 10, 47.5 ft at joints 4 and 8 and 50 ft at joint 6. Subvertical member M1L1 is 25 ft long, subvertical member M3L3 is 19 ft long and subvertical member M5L5 is 23.75 ft long. Determine the maximum forces due to live load in the top chord member U2U4 and bottom chord member L5L6.

The section shown in Figure 4.18b illustrates that the forces in top chord members U(j − 1)U(j + 1) can be determined by taking moments about joint L(j + 1) with the live load positioned in accordance with the criterion of Equation 4.11c with k = j + 1. As shown by Example 4.26, the maximum panel point moment at joint L4, which creates the maximum force F^{max}_{U2U4}, due to Cooper's E80 live load, occurs with wheel 13 at joint L4 such that considering the inclination of member U2U4,

$$F^{max}_{U2U4} = \frac{-\Sigma M_{L(j+1)}}{h_{(j+1)}} \frac{\sqrt{\left(h_{(j+1)} - h_{(j-1)}\right)^2 + (2d)^2}}{2d} = \frac{-44,963}{h_4}\left(\frac{\sqrt{\left(h_{(4)} - h_{(2)}\right)^2 + (2(25))^2}}{2(25)}\right)$$

$$= \frac{-44,963}{47.5}\left(\frac{\sqrt{(47.5 - 38.0)^2 + (2(25))^2}}{2(25)}\right) = -946.6\left(\frac{50.894}{50}\right) = -963.5 \text{ kips}$$

The section shown in Figure 4.18b also shows that the forces in bottom chord members L(j)L(j + 1) can be determined by taking moments about joint U(j − 1) with the load positioned in accordance with the criterion of Equation 4.31f. As shown by Example 4.26,

F^{max}_{L5L6} in bottom chord member L5L6 due to Cooper's E80 live load occurs in conjunction with the maximum moment about joint U4 with wheel 16 at joint L5, which is 47,704 kip-ft. Therefore, F^{max}_{L5L6} in bottom chord member L5L6 is

$$F^{max}_{L5L6} = \frac{47,704}{47.5} = 1004.3\text{kips}$$

As shown in Figure 4.18b, web members in a PSS truss consist of upper diagonal U(j – 1)M(j), lower diagonal M(j)L(j + 1), substrut L(j – 1)M(j), subvertical M(j)L(j), interior vertical U(j – 1)L(j – 1) and end vertical members U2L2. The subdivided panels also result in upper end post M1U2 and lower end post L0M1.

Figure 4.18b also reveals that the forces in lower diagonal members M(j)L(j + 1) of the PSS span are calculated in the same way as the diagonal members in non-subdivided inclined-chord Pratt trusses. Taking a section through subpanel (j) – (j + 1) of a PSS span, as shown in Figure 4.18b, and considering equilibrium of forces for positive panel shear yields the vertical component of the force in lower diagonal member M(j)L(j + 1) due to positive panel shear as

$$F^{V+}_{M(j)L(j+1)} = \frac{V^{+}_{(j)-(j+1)}\left(a_{(j)-(j+1)}\right) - M^{+0}_{(j)-(j+1)}}{\left(a_{(j)-(j+1)} + (j+1)d\right)} \tag{4.44a}$$

where

$$R_{y0} = \sum_{i=0}^{n} \frac{(n-i)W_i}{n}$$

$V^{+}_{(j)-(j+1)}$ = the positive shear force in subpanel (j) – (j + 1) $= R_{y0} - \sum_{i=0}^{j} W_i$

$M^{+0}_{(j)-(j+1)}$ = the moment about L0 from joint loads R_{y0} and W_i between L0 and panel point (j) $= \left(\sum_{i=0}^{j} iW_i\right)d$

$a_{(j)-(j+1)}$ = the distance from panel point L0 to point P at the location intersected by lines coincident with the inclination of U(j – 1)U(j + 1) and L(j)(j + 1) $= \left(\left(\frac{2h_{(j+1)}}{h_{(j+1)} - h_{(j-1)}}\right) - j - 1\right)d.$

Therefore, the force in lower diagonal member M(j)L(j + 1) of the PSS span due to positive panel shear is

$$F^{+}_{M(j)L(j+1)} = \left(\frac{V^{+}_{(j)-(j+1)}\left(a_{(j)-(j+1)}\right) - M^{+0}_{(j)-(j+1)}}{a_{(j)-(j+1)} + (j+1)d}\right)\alpha_{(j)-(j+1)} \tag{4.44b}$$

where

$$\propto_{(j)-(j+1)} = 2\frac{\sqrt{\left(\frac{h_{(j-1)}}{2}\right)^2 + d^2}}{h_{(j-1)}}$$

Taking a section through subpanel (j) – (j + 1) of a PSS span, as shown in Figure 4.18b, and considering equilibrium of forces for negative panel shear, the vertical component of the force in lower diagonal member M(j)L(j + 1) due to negative panel shear is

$$F^{V-}_{M(j)L(j+1)} = \frac{-V^{-}_{(j)-(j+1)}\left(a_{(j)-(j+1)}\right) - M^{-0}_{(j)-(j+1)}}{\left(a_{(j)-(j+1)} + (j+1jd\right)} \tag{4.44c}$$

where

$$R_{yn} = \sum_{i=0}^{n} \frac{iW_i}{n}$$

$V^{-}_{(j)-(j+1)}$ = the negative shear force in subpanel (j)-(j+1) = $R_{yn} - \sum_{i=j+1}^{n} W_i$

$M^{-0}_{(j)-(j+1)}$ = the moment about L0 from loads R_{yn} and W_i between L(n) and panel point (j + 1) $= \left(nR_{yn} - \sum_{i=j+1}^{n} iW_i\right)d.$

Therefore, the force in lower diagonal member M(j)L(j + 1) of the PSS span due to negative panel shear is

$$F^{-}_{M(j)L(j+1)} = \left(\frac{-V^{-}_{(j)-(j+1)}\left(a_{(j-1)-(j+1)}\right) - M^{-0}_{(j)-(j+1)}}{a_{(j-1)-(j)+1} + (j+1)d}\right)\alpha_{(j)-(j+1)} \tag{4.44d}$$

For uniformly distributed live loads on PSS spans, the criteria of Equations 4.22b and 4.23b for maximum positive and negative live load shear, respectively, in subpanel (j) – (j + 1) are appropriate for calculating maximum and minimum forces $F^{max}_{M(j)L(j+1)}$ and $F^{min}_{M(j)L(j+1)}$, in lower diagonal members such that

$$x^{+}_{(j)-(j+1)} = \frac{(n-j+1)d}{n\left(1+\dfrac{(j)d}{a_{(j)-(j+1)}}\right)-1} \tag{4.45a}$$

$$x^{-}_{(j)-(j+1)} = \frac{a_{(j)-(j+1)} + nd}{n + \dfrac{(n-1)a_{(j)-(j+1)}}{(j)d}} = d - x^{+}_{(j)-(j+1)} \tag{4.45b}$$

For concentrated live loads on PSS spans, the criterion for maximum positive shear in subpanel (j) – (j + 1) to determine maximum forces in lower diagonal members M(j)L(j + 1) is determined from Equation 4.44a as

$$F^{V+}_{M(j)L(j+1)} = \frac{V^{+}_{(j)-(j+1)}\left(a_{(j)-(j+1)}\right) - M^{+0}_{(j)-(j+1)}}{\left(a_{(j)-(j+1)} + (j+1)d\right)}$$

where

$$R_{y0} = P_T^+\left(\frac{x_T^+}{nd}\right)$$

$$V_{(j)-(j+1)}^+ = \text{the shear force in subpanel (j) − (j + 1)} = R_{y0} - \sum_{i=0}^{j} W_i = P_T^+\left(\frac{x_T^+}{nd}\right) - P_F^+ - P_A^+ - \frac{P_B^+\left(x_B^+\right)}{d}$$

$$M_{(j)-(j+1)}^{+0} = \left(\sum_{i=0}^{j} iW_i\right)d = P_A^+\left(jd - x_A^+\right) + P_F^+\left((j-1)d - X_F^+\right) + jP_B^+X_B^+$$

$$F_{M(j)L(j+1)}^{+V} = \left(\frac{\left(P_T^+\left(\frac{x_T^+}{nd}\right) - P_F^+ - P_A^+ - \frac{P_B^+X_B^+}{d}\right)\left(a_{(j)-(j+1)}\right) - \left(P_A^+\left(jd - x_A^+\right) + P_F^+\left((j-1)d - X_F^+\right) + jP_B^+X_B^+\right)}{\left(a_{(j)-(j+1)} + (j+1)d\right)}\right)$$

The maximum force due to positive panel shear forces in diagonal member M(j)L(j + 1) of the PSS span is calculated by establishing $\frac{d\left(F_{M(j)L(j+1)}^{+V}\right)}{dx} = 0$, which yields

$$\frac{d\left(F_{U(j-1)L(j)}^{+V}\right)}{dx} = 0 = \left(\frac{P_T^+}{nd} - \frac{P_B^+}{d}\right)\left(a_{(j)-(j+1)}\right) + P_A^+ + P_F^+ - jP_B^+$$

or,

$$\frac{P_T^+}{n} = P_B^+\left(1 + \frac{jd}{a_{(j)-(j+1)}}\right) - \frac{(P_A^+ + P_F^+)d}{a_{(j)-(j+1)}} \tag{4.46a}$$

The criterion of Equation 4.46a can only occur with a concentrated load at panel point (j + 1).

Equation 4.46a is the criterion to establish the position of the concentrated live loads for maximum force in lower diagonal member M(j)L(j + 1) of the PSS span due to positive shear in subpanel (j) − (j + 1):

P_A^+ = the resultant of live loads between panel point (j − 1) and panel point (j) of the PSS span;
P_B^+ = the resultant of live loads between (j) and (j + 1) of the PSS span; and
P_F^+ = the resultant of live loads between (0) and (j − 1) of the PSS span.

When $P_F^+ = 0$, Equation 4.46a is equal to Equation 4.26g. When $P_A^+ = P_F^+ = 0$, maximum force $F^{+max}_{M(j)L(j+1)}$ in lower diagonal member M(j)L(j + 1) of the PSS span, due to positive shear in subpanel (j) − (j + 1), occurs when the total load on the span, P_T^+, divided by the number of subpanels in the span, n, and the factor $\left(1 + \frac{jd}{a_{(j)-(j+1)}}\right)$ equals the load $P_{(j)-(j+1)}^+ = P_B^+$ in subpanel (j) − (j + 1). If, due to horizontal parallel chord members, $a_{(j)-(j+1)} = \infty$, Equation 4.46a for the lower diagonal member of a PSS span is the same as Equation 4.33b for the lower diagonal member of a BSS span and Equation 4.12c for the diagonal member of a non-subdivided horizontal-chord Pratt truss.

For concentrated live loads on PSS spans, the criterion for maximum negative shear in subpanel (j) – (j + 1) to determine minimum forces in lower diagonal members M(j)L(j + 1) is determined from Equation 4.44c as

$$F^{V-}_{M(j)L(j+1)} = \frac{-V^{-}_{(j)-(j+1)}\left(a_{(j)-(j+1)}\right) - M^{-0}_{(j)-(j+1)}}{\left(a_{(j)-(j+1)} + (j+1)d\right)}$$

where

$$V^{-}_{(j)-(j+1)} = R_{yn} - \sum_{i=j+1}^{n} W_i = \frac{P^{-}_T X^{-}_T}{nd} - P^{-}_F - \frac{P^{-}_B X^{-}_B}{d}$$

$$M^{-0}_{(j)-(j+1)} = \sum_{i=j+1}^{n} W_i\,(id) = -ndR_{yn} + P^{-}_F\left((j+1)d + X^{-}_F\right) + \frac{P^{-}_B X^{-}_B}{d}(j+1)d$$

$$F^{V-}_{M(j)L(j+1)} = \frac{-\left(\frac{P^{-}_T X^{-}_T}{nd} - P^{-}_F - \frac{P^{-}_B X^{-}_B}{d}\right)\left(a_{(j)-(j+1)}\right) + \left(P^{-}_T X^{-}_T - P^{-}_F\left((j+1)d + X^{-}_F\right) - (j+1)P^{-}_B X^{-}_B\right)}{\left(a_{(j)-(j+1)} + (j+1)d\right)}$$

The maximum force due to negative panel shear forces in diagonal member M(j)L(j + 1) of the PSS span is calculating by establishing $\frac{d\left(F^{-V}_{M(j)L(j+1)}\right)}{dx} = 0$, which yields

$$\frac{P^{-}_T}{n} = P^{-}_{(j)-(j+1)}\left(\frac{1+\frac{(j+1)d}{a_{(j)-(j+1)}}}{1+\frac{nd}{a_{(j)-(j+1)}}}\right) + \frac{P^{-}_F}{1+\frac{nd}{a_{(j)-(j+1)}}} = P^{-}_B\left(\frac{1+\frac{(j+1)d}{a_{(j)-(j+1)}}}{1+\frac{nd}{a_{(j)-(j+1)}}}\right) + \frac{P^{-}_F d}{a_{(j)-(j+1)}\left(1+\frac{nd}{a_{(j)-(j+1)}}\right)} \qquad (4.46b)$$

The criterion of Equation 4.4 6b can only occur with a concentrated load at panel point (j).

Equation 4.46b is the criterion to establish the position of the concentrated live loads for maximum force in diagonal member M(j)L(j + 1) of the PSS span due to negative shear in subpanel (j) – (j + 1):

P^{-}_A = the resultant of live loads between panel point (j – 1) and panel point (j) of the PSS span;

P^{-}_B = the resultant of live loads between (j) and (j + 1) of the PSS span; and

P^{-}_F = the resultant of live loads between panel point (n) and (j + 1) of the PSS span.

When $P^{-}_F = 0$, Equation 4.46b is equal to Equation 4.27f, and $F^{-max}_{M(j)L(j+1)}$ in lower diagonal member M(j)L(j + 1) due to negative shear in subpanel (j) – (j + 1) occurs when the total load on the PSS span, P^{-}_T, divided by the number of subpanels in the span, n, and the factor $\left(\frac{1+\frac{(j+1)d}{a_{(j)-(j+1)}}}{1+\frac{nd}{a_{(j)-(j+1)}}}\right)$ equals the load $P^{-}_{(j)-(j+1)} = P^{-}_B$ in subpanel (j) – (j + 1). If, due to horizontal parallel chord members, $a_{(j-1)-(j)} = \infty$, Equation 4.46b for the lower diagonal member of a PSS span is the same as Equation

4.33d for the lower diagonal of a BSS span and Equation 4.13c for the diagonal member of a horizontal-chord Pratt truss.

Example 4.38

A 300 ft long and 50 ft high PSS truss and twelve subpanels is loaded with Cooper's E80 live load. The top chord is inclined with truss height of 25 ft at joints 1 and 11, 38 ft at joints 2 and 10, 47.5 ft at joints 4 and 8, and 50 ft at joint 6. Subvertical member M1L1 is 25 ft long, subvertical member M3L3 is 19 ft long and subvertical member M5L5 is 23.75 ft long. Determine the maximum positive and negative shear force due to live load resisted by the lower diagonal member M3L4:

$$a_{(j)\text{-}(j+1)} = \left(\left(\frac{2h_{(j+1)}}{h_{(j+1)} - h_{(j-1)}}\right) - (j+1)\right)d$$

$j = 3$

$$a_{(3)\text{-}(4)} = \left(\left(\frac{2h_{(4)}}{h_{(4)} - h_{(2)}}\right) - (3+1)\right)25 = \left(\left(\frac{2(47.5)}{47.5 - 38.0}\right) - (3+1)\right)25 = (10-4)25 = 150\text{ft}$$

$$\propto_{(3)-(4)} = \frac{\sqrt{h_{(2)}^2 + (50)^2}}{h_{(2)}} = \frac{\sqrt{(38)^2 + (50)^2}}{38} = 1.653$$

where $P^+_A = P^+_F = 0$. The maximum force due to positive panel shear forces in lower diagonal member M3L4 is calculated using the criterion of Equation 4.46a in panel 3-4 as

$$P^+_A = P^+_{(2)\text{-}(3)} = 0$$

$$P^+_B = P^+_{(3)\text{-}(4)}$$

$$\frac{P^+_T}{n} = P^+_{(3)-(4)}\left(1 + \frac{(3)(25)}{150}\right)$$

or

$$\frac{P^+_T}{18} = P^+_{(3)-(4)}$$

For lower diagonal web member M3L4, use Table 4.2 to determine positive shear forces in subpanel 3-4 with Cooper's E live load using the criterion in Table E4.38a.

The criterion of Equation 4.46a indicates axle 2 at L4 for maximum positive force in lower diagonal M3L4. Therefore, the assumption that $P^+_A = P^+_F = 0$ is valid. However, maximum positive forces for axle 1 and axle 3 at L4 are also calculated to verify the criterion. Maximum positive force in M3L4 from Equation 4.44b is given in Table E4.38b.

TABLE E4.38a

Criterion for Live Load Position for Maximum Force due to Positive Shear in Member M3L4 for Cooper's E80 Live Load Traversing a PSS Span

	Panel 3-4: Axle at L4		
Subpanel 3-4: M3L4	**1**	**2**	**3**
Distance to axle no. 1 (ft)	0	8	13
First axle on span	1	1	1
Distance to axle no. 18 (ft)	104	96	91
Last axle on span	18	18	18
Distance of uniform load (ft)	200 − 104 − 5 = 91	200 − 96 − 5 = 99	200 − 91 − 5 = 104
Total load on span, P^+_T (kips)	568 + 4(91) = 932	568 + 4(99) = 964	568 + 4(104) = 984
$P^+_T/18$	51.8	53.6	54.7
Loads in subpanel 3-4: Sum of loads ahead of panel point L4 = $P^+_{a(3)-(4)}$ (kips)	0	20	60
Loads in subpanel point 3-4: Sum of loads ahead of and including panel point L4 = $P^+_{b(3)-(4)}$ (kips)	20	60	100
Criterion established	No	Yes	No

TABLE E4.38b

Maximum Force in Member M3L4 due to Positive Shear Force due to Cooper's E80 Live Load Traversing a PSS Span

	Axle at L4		
Positive Shear in Subpanel 3-4	**1**	**2**	**3**
Sum of live load moments to L12 (kip-ft)	32,728 + 568(91) + 4(91)²/2 = 100,978	32,728 + 568(99) + 4(99)²/2 = 108,562	32,728 + 568(104) + 4(104)²/2 = 113,432
R_{y0} (kips)	100,978/300 = 336.59	108,562/300 = 361.87	113,432/300 = 378.11
P^+_T = total load on truss span (kips)	932	964	984
$R_{y12} = P^+_T - R_{y0}$ (kips)	932 − 336.59 = 595.41	964 − 361.87 = 602.13	984 − 378.11 = 605.89
Sum moments in subpanel 3–4 ahead of and about L4 (kip-ft)	0	160	460
$W_{(3)}$ (kips)	0	160/25 = 6.40	460/25 = 18.40
$V^{+max}_{(3)-(4)} = R_{y0} - \sum_{i=0}^{3} W_i = R_{y0} - W_{(3)}$ (kips)	336.59 − 0 = 336.59	361.87 − 6.40 = 355.47	378.11 − 18.40 = 359.71
$M^{+0}_{(3)-(4)} = d\sum_{i=0}^{3} iW_i = d(3)W_{(3)}$ (kip-ft)	0	6.40(3)(25) = 480	18.40(3)(25) = 1380
$F^{+Vmax}_{M3L4}\left(a_{(3)-(4)} + 4d\right) = V^{+max}_{(3)-(4)}a_{(3)-(4)} - M^{+0}_{(3)-(4)}$ (kip-ft)	336.59(150) = 50,489	355.47(150) − 480 = 52,841	359.71(150) − 1380 = 52,576.5
F^{+Vmax}_{M3L4} (kips)	50,489/(150 + 4(25)) = 201.95	52,841/(150 + 4(25)) = 211.36	52,576.5/(150 + 4(25)) = 210.31
$F^{+max}_{M3L4} = F^{+Vmax}_{M3L4}\left(\alpha_{(3)-(4)}\right)$ (kips)	201.95(1.653) = 333.83 (T)	211.36(1.653) = 349.31 (T)	210.31(1.653) = 347.64 (T)

$$F^{+max}_{M3L4} = 349.3 \text{ kips}$$

The maximum force due to negative panel shear force in lower diagonal member M3L4 from the criterion of Equation 4.46b is

$$P^-_F = 0$$
$$P^-_B = P^-_{(3\text{-}(4)}$$

$$\frac{P_T^-}{12} = P_{(3)-(4)}^- \left(\frac{1 + \frac{(4)(25)}{150}}{1 + \frac{300}{150}} \right) = P_{(3)-(4)}^- (0.556)$$

or

$$\frac{P_T^-}{6.67} = P_{(3)-(4)}^-$$

For lower diagonal web member M3L4, use Table 4.2 to determine negative shear forces in subpanel 3-4 with Cooper's E80 live load as shown in Table E4.38c.

TABLE E4.38c
Criterion for Live Load Position for Maximum Force due to Negative Shear in Member M3L4 for Cooper's E80 Live Load Traversing a PSS Span

	Axle at L3		
Subpanel 3-4: M3L4	**2**	**3**	**4**
Distance to axle no. 1 (ft)	8	13	18
First axle on span	1	1	1
Distance to axle no. 18 (ft)	96	91	86
Last axle on span	14	15	16
Distance from last axle on span to end of span (ft)	75 – 79 + 8 = 4	75 – 88 + 13 = 0	75 – 93 + 18 = 0
Total load on span, P^-_T (kips)	464	490	516
$P^-_T/6.67$ (kips)	69.6	73.5	77.4
Loads in subpanel 3-4: Sum of loads ahead of panel point L3 = $P^-_{a(3)-(4)}$ (kips)	20	60	100
Loads in subpanel 3-4: Sum of loads ahead of and including panel point L3 = $P^-_{b(3)-(4)}$ (kips)	60	100	140
Criterion established	No	Yes	No

The criterion indicates axle 3 at L3 for maximum negative shear force in M3L4. Therefore, the assumption that $P^-_A = P^-_F$ 0 = is valid. However, maximum positive forces for axle 2 and axle 4 at L3 are also calculated to verify the criterion. Maximum negative force in M3L4 from Equation 4.44d is given in Table E4.38d.

TABLE E4.38d
Maximum Force in Member M3L4 due to Negative Shear Force in Subpanel 3-4 due to Cooper's E80 Live Load Traversing a PSS Span

	Axle at L3		
Negative Shear in Subpanel 3-4	**2**	**3**	**4**
Sum moments to L0 (kip-ft)	17,456 + 464(4) = 19,312	21,632	24,082
R_{y12} (kips)	19312/300 = 64.37	21632/300 = 72.11	80.27
P^-_T (kips)	464	490	516
R_{yo} (kips)	464 − 64.37 = 399.63	490 − 72.11 = 417.89	435.73
Sum moments in subpanel 3-4 ahead of and about panel point L3	160	460	960
$W_{(4)}$ (kips)	160/25 = 6.40	460/25 = 18.40	38.40
$V^-_{(3)-(4)} = R_{y12} - W_{(4)}$ (kips)	64.37 − 6.40 = 57.97	72.11 − 18.40 = 53.71	41.87
$M^{-0}_{(3)-(4)} = 100\left(3R_{y12} - W_{(4)}\right)$ (kip-ft)	100(3(64.37) − 6.40) = 18,671.0	100(3(72.11) − 18.40) = 19,793.0	20,241.0
$F^{-V}_{M3L4}\left(a_{(3)-(4)} + 4d\right) = -V^-_{(3)-(4)}\left(a_{(3)-(4)}\right) - M^{-0}_{(3)-(4)}$ (kip-ft)	−57.97(150) − 18671.0 = −27,366.5	−53.71(150) − 19793.0 = −27,849.5	−41.87(150) − 20241.0 = 26,521.5
F^{-V}_{M3L4} (kips)	−27,366.5/(150 + 4(25)) = −109.5	−27,849.5/(150 + 4(25)) = −111.40	−26,521.5/(150 + 100) = −106.1
$F^-_{M3L4}\left(\alpha_{(3)-(4)}\right)$ (kips)	−109.5(1.653) = −180.1 (C)	−111.40(1.653) = −184.1 (C)	−175.4 (C)

$$F^{-max}_{M3L4} = -184.1 \text{ kips}$$

The forces in substrut member L(j − 1)M(j) of a PSS span are calculated by considering the sum of the moments of the forces in panel (j − 1) − (j + 1) at joint M(j) about joint L(j + 1). Resolving $F_{L(j-1)M(j)}$ at joint L(j-1), as shown in Figure 4.19, yields[142]

$$F_{L(j-1)M(j)} = \frac{-F_{M(j)L(j)}}{2}\left(\frac{\sqrt{\left(\frac{h_{(j-1)}}{2}\right)^2 + d^2}}{\left(\frac{h_{(j-1)}}{2}\right)}\right) \tag{4.47a}$$

where

$$F_{M(j)L(j)} = W_{(j)} = \frac{2M^P_{(j)}}{d}$$

For a uniformly distributed live load w in subpanels (j – 1) – (j) and (j) – (j + 1), the force in substrut member L(j – 1)M(j) of a PSS span is maximum when equal-length subpanels (j – 1) – (j) and (j) – (j + 1) are fully loaded, such that

$$F^{max}_{L(j-1)M(j)} = \frac{-wd}{2}\left(\frac{\sqrt{\left(\frac{h_{(j-1)}}{2}\right)^2 + d^2}}{\left(\frac{h_{(j-1)}}{2}\right)}\right) \tag{4.47b}$$

For concentrated live load forces in equal-length subpanels (j – 1) – (j) and (j) – (j + 1), determine the maximum panel point moment $M^{Pmax}_{(j)}$ from Equation 4.14f with the criterion of Equation 4.11c to compute $F^{max}_{L(j-1)M(j)}$ in substrut member L(j – 1)M(j) as follows:

$$F^{max}_{L(j-1)M(j)} = \frac{-M^{Pmax}_{(j)}}{d}\left(\frac{\sqrt{\left(\frac{h_{(j-1)}}{2}\right)^2 + d^2}}{\left(\frac{h_{(j-1)}}{2}\right)}\right) \tag{4.47c}$$

Example 4.39

Determine the maximum force in substrut members L2M3 and L4M5 of the Pennsylvania truss in Example 4.36 produced by a uniformly distributed live load of 4.0 kips/ft and Cooper's E80 live load.

For L2M3 with uniformly distributed live load, j = 3 and h_2 = 38.0 ft, and Equation 4.47b yields

$$F^{max}_{L2M3} = \frac{-4.0(25)}{2}\left(\frac{\sqrt{\left(\frac{38}{2}\right)^2 + 25^2}}{\left(\frac{380}{2}\right)}\right) = -50.0\left(\frac{31.40}{19}\right) = -50.0(1.653) = -82.63 \text{ kips}$$

For L4M5 with uniformly distributed live load, j = 5 and h_4 = 47.5 ft, and Equation 4.47b yields

$$F^{max}_{L4M5} = \frac{-4.0(25)}{2}\left(\frac{\sqrt{\left(\frac{47.5}{2}\right)^2 + 25^2}}{\left(\frac{47.5}{2}\right)}\right) = -50.0\left(\frac{34.48}{23.75}\right) = -50.0(1.452) = -72.60 \text{ kips}$$

For L2M3 with Cooper's E80 live load, in accordance with Example 4.11, axle 4 or axle 13 at L3 creates $M_{(3)}^{Pmax}$ = 1891 kip-ft on two equal-length subpanels of d = 25 ft, such that, from Equation 4.47c,

$$F_{L2M3}^{max} = \frac{-1891}{25}(1.653) = -125.0 \text{ kips}.$$

If the alternate load is considered by placing one of the central 50 kip wheel loads at L3,

F_{M3L3} = 156 kips, which provides $M^{Pmax}{}_{(3)}$ = 156(50)/4 = 1950 kip-ft and $F^{max}{}_{L2M3}$ = −1950(1.653)/25 =
−128.9 kips.

For L4M5 with Cooper's E80 live load, in accordance with Example 4.11, axle 4 or axle 13 at L5 creates $M_{(5)}{}^{Pmax}$ = 1891 kip-ft on two equal-length subpanels of d = 25 ft, such that, from Equation 4.47c,

$$F_{L4M5}^{max} = \frac{-1891}{25}(1.452) = -109.8 \text{ kips}.$$

If the alternate load is considered by placing one of the central 50 kip wheel loads at L5,

F_{M5L5} = 156 kips, which provides $M^{Pmax}{}_{(5)}$ = 156(50)/4 = 1950 kip-ft and $F^{max}{}_{L4M5}$ = −1950(1.452)/25
= −113.3 kips.

In the PSS span, the maximum and minimum forces in upper diagonal member U(j − 1)M(j) are determined by considering positive and negative shear forces in subpanel (j − 1) − (j) as shown in Figure 4.18d.

The force in upper diagonal member U(j − 1)M(j) of the PSS span due to positive shear in subpanel (j − 1) − (j) is

$$F^{+}_{U(j-1)M(j)} = \left(V^{+}_{(j-1)-j} - F^{+}_{L(j-1)M(j)} - F^{+}_{U(j-1)U(j+1)}\right)\frac{\sqrt{h^{2}_{(j-1)}+(2d)^{2}}}{h_{(j-1)}} \tag{4.48}$$

where

$$V^{+}_{(j-1)-(j)} = R_{y0} - \sum_{i=0}^{j-1} W_i$$

$F^{V}_{L(j-1)M(j)} = \dfrac{F_{M(j)L(j)}}{2} = \dfrac{W_{(j)}}{2}$ = the vertical component of the force in substrut member L(j − 1) M(j) intersected by the section through subpanel (j − 1) − (j).

$F^{V}_{U(j-1)U(j+1)} = F_{U(j-1)U(j+1)}\left(\dfrac{\sqrt{(h_{(j+1)}-h_{(j-1)})^{2}+(2d)^{2}}}{(h_{(j+1)}-h_{(j-1)})}\right)$ = the vertical component of the force in top chord member U(j − 1)U(j + 1) intersected by the section through subpanel (j − 1) − (j).

However, since $F^{V}_{U(j-1)U(j+1)}$ in a PSS span must be predetermined to determine $F^{+}_{U(j-1)M(j)}$, it is advantageous to take moments about a point P at the intersection of the lines extending from the top chord and bottom chord members in panel (j − 1) − (j + 1).[143]

Taking a section through subpanel (j − 1) − (j) of a PSS span where j ≤ n/2, and considering equilibrium of forces about point P for positive panel shear, yields

$$F^{+}_{U(j-1)M(j)} = \left(\frac{V^{+P}_{(j-1)-(j)}\left(a_{(j-1)-(j+1)}\right) - M^{+0P}_{(j-1)-(j)}}{a_{(j-1)-(j+1)} + (j+1)d} \right) \alpha_{(j-1)-(j)} \tag{4.49a}$$

where

$V^{+P}_{(j-1)-(j)}$ = the positive shear force in subpanel (j – 1) – (j) = $R_{y0} - \sum_{i=0}^{j-1} W_i - \frac{W_{(j)}}{2}$

$M^{+0P}_{(j-1)-(j)}$ = the moment about L0 from joint loads R_{y0} and W_i between L0 and panel point (j – 1)

$$= \left(\sum_{i=0}^{j-1} iW_i + (j-1)\frac{W_{(j)}}{2} \right) d$$

$a_{(j-1)-(j+1)}$ = the distance from panel point L0 to point P at the location intersected by lines coincident with the inclination of U(j – 1)U(j + 1) and L(j – 1)(j + 1) = $\left(\left(\frac{2h_{(j+1)}}{h_{(j+1)} - h_{(j-1)}} \right) - j - 1 \right) d$

$\propto_{(j-1)-(j)} = \frac{2\sqrt{\left(\frac{h_{(j-1)}}{2}\right)^2 + d^2}}{h_{(j-1)}}$ = factor due to inclination of the upper diagonal web member U(j – 1) M(j) in subpanel (j – 1) – (j).

Similarly, taking a section through subpanel (j – 1) – (j), where j ≤ n/2, of a PSS span and considering equilibrium of forces about point P for negative panel shear yields

$$F^{-}_{U(j-1)M(j)} = \left(\frac{-V^{-P}_{(j-1)-(j)}\left(a_{(j-1)-(j)}\right) + M^{-0P}_{(j-1)-(j)}}{a_{(j-1)-(j)} + (j+1)d} \right) \alpha_{(j-1)-(j)} \tag{4.49b}$$

where

$V^{-P}_{(j-1)-(j)}$ = the negative shear force in subpanel (j – 1) – (j) = $R_{yn} - \sum_{i=j}^{n} W_i + \frac{W_{(j)}}{2}$

$M^{-0P}_{(j-1)-(j)}$ = the moment about L0 from loads R_{yn} and W_i between L(n) and panel point (j) =

$$-\left(nR_{yn} - \sum_{i=j}^{n} iW_i + \left(\frac{W_{(j)}}{2}\right)(j-1) \right) d$$

For a uniformly distributed live load considering positive shear in subpanel (j – 1) – (j),

$$R_{y0} = \frac{w\left((n-j)d + x^{+}_{(j-1)-(j)}\right)^2}{2nd}$$

$$W_{(j-1)} = \frac{w\left(x^{+}_{(j-1)-(j)}\right)^2}{2d}$$

$$W_{(j)} = \frac{w\left(x^{+}_{(j-1)-(j)}\right)\left(d - \frac{x^{+}_{(j-1)-(j)}}{2}\right)}{d} + \frac{wd}{2}$$

$F^{+max}_{U(j-1)M(j)}$ occurs when $\dfrac{d\left(F^{+}_{U(j-1)M(j)}\right)}{dx} = 0$, which provides

$$x^{+}_{(j-1)-(j)} = \frac{\left((n-j)-\frac{n}{2}\left(1+\frac{(j-1)d}{a_{(j-1)-(j+1)}}\right)\right)d}{\frac{(j-1)nd}{2a_{(j-1)-(j+1)}}+\frac{n}{2}-1} \tag{4.50a}$$

Due to the shear diagram for uniformly distributed loads, the location, $x^{-}_{(j-1)-(j+1)}$ in subpanel (j – 1) – (j) is

$$x^{-}_{(j-1)-(j)} = d - x^{+}_{(j-1)-(j)} \tag{4.50b}$$

Example 4.40

Determine the maximum and minimum force in upper diagonal members U2M3 and U4M5 of the Pennsylvania truss in Example 4.36 produced by a uniformly distributed live load of 4.0 kips/ft:

j = 3 and $a_{(j-1)-(j+1)}$ = 150 ft.

For positive shear in subpanel 2-3 from Equations 4.49a and 4.50a,

$$x^{+}_{(2)-(3)} = \frac{\left((12-3)-\frac{12}{2}\left(1+\frac{(3-1)25}{150}\right)\right)25}{\frac{(3-1)12(25)}{2(150)}+\frac{12}{2}-1} = \frac{25}{7} = 3.57 \text{ ft}$$

$$R_{y0} = \frac{4.0\left((12-3)25+3.57\right)^2}{2(12)(25)} = 348.3 \text{ kips}$$

$$W_{(2)} = \frac{4.0(3.57)^2}{2(25)} = 1.02 \text{ kips}$$

$$W_{(3)} = \frac{4.0(3.57)\left(25-\frac{3.57}{2}\right)}{25} + \frac{4.0(25)}{2} = 13.26 + 50.0 = 63.26 \text{ kips}$$

$$V^{+Pmax}_{(2)-(3)} = 348.3 - 1.02 - \frac{63.26}{2} = 315.7 \text{ kips}$$

$$M^{+0P}_{(j-1)-(j)} = \left(\sum_{i=0}^{j-1} iW_i + (j-1)\frac{W_{(j)}}{2}\right)d = (j-1)d\left(W_{(j-1)} + \frac{W_{(j)}}{2}\right)$$

$$= (3-1)(25)\left(1.02+\frac{63.26}{2}\right) = 50(32.65) = 1632.5 \text{ kips} - \text{ft}$$

$$F_{U2M3}^{+max} = \left(\frac{315.7(150) - 1632.5}{150 + (3+1)(25)} \right) 1.653 = 302.3 \text{ kips}$$

For negative shear in subpanel 2-3 from Equations 4.49b and 4.50b,

$$x_{(2)-(3)}^{-} = 25 - 3.57 = 21.43 \text{ ft}$$

$$R_{y12} = \frac{4.0((2)25 + 21.43)^2}{2(12)(25)} = 34.01 \text{ kips}$$

$$W_{(3)} = \frac{4.0(21.43)^2}{2(25)} = 36.74 \text{ kips}$$

$$V_{(j-1)-(j)}^{-Pmax} = 34.01 - 36.74 + \frac{36.74}{2} = 15.64 \text{ kips}$$

$$M_{(j-1)-(j)}^{-0P} = -\left(12(34.01) - 3(36.74) + \frac{36.74}{2}(3-1) \right) 25 = -8366.0 \text{ kips} - \text{ft}$$

$$F_{U2M3}^{-max} = \left(\frac{-15.64(150) - 8366.0}{150 + (4)25} \right) 1.653 = -70.83 \text{ kips}$$

$j = 5$ and $a_{(j-1)-(j+1)} = 850$ ft.

For positive shear in subpanel 4-5 from Equations 4.49a and 4.50a,

$$x_{(4)-(5)}^{+} = \frac{\left((12-5) - \frac{12}{2}\left(1 + \frac{(5-1)25}{850} \right) \right) 25}{\frac{(5-1)12(25)}{2(850)} + \frac{12}{2} - 1} = \frac{7.35}{5.71} = 1.29 \text{ ft}$$

$$R_{y0} = \frac{4.0((12-5)25 + 1.29)^2}{2(12)(25)} = 207.2 \text{ kips}$$

$$W_{(4)} = \frac{4.0(1.29)^2}{2(25)} = 0.133 \text{ kips}$$

$$W_{(5)} = \frac{4.0(1.29)\left(25 - \frac{1.29}{2} \right)}{25} + \frac{4.0(25)}{2} = 5.02 + 50.0 = 55.02 \text{ kips}$$

$$V_{(4)-(5)}^{+Pmax} = 207.2 - 0.133 - \frac{55.02}{2} = 179.54 \text{ kips}$$

$$M_{(4)-(5)}^{+0P} = \left(\sum_{i=0}^{j-1} iW_i + (j-1)\frac{W_{(j)}}{2}\right)d = (j-1)d\left(W_{(j-1)} + \frac{W_{(j)}}{2}\right)$$

$$= (5-1)(25)\left(0.133 + \frac{55.02}{2}\right) = 100(27.64) = 2764.3 \text{ kips} - \text{ft}$$

$$F_{U4M5}^{+max} = \left(\frac{179.54(850) - 2764.3}{850 + (5+1)(25)}\right)1.452 = 217.6 \text{ kips}$$

For negative shear in subpanel 4-5 from Equations 4.49b and 4.50b,

$$x_{(4)-(5)}^{-} = 25 - 1.29 = 23.71 \text{ ft}$$

$$R_{y12} = \frac{4.0((4)25 + 23.71)^2}{2(12)(25)} = 102.03 \text{ kips}$$

$$W_{(5)} = \frac{4.0(23.71)^2}{2(25)} = 44.97 \text{ kips}$$

$$V_{(4)-(5)}^{-Pmax} = 102.03 - 44.97 + \frac{44.97}{2} = 79.55 \text{ kips}$$

$$M_{(4)-(5)}^{-0P} = -\left(12(102.03) - 5(44.97) + \frac{44.97}{2}(5-1)\right)25 = -27236.3 \text{ kips} - \text{ft}$$

$$F_{U4M5}^{-max} = \left(\frac{-79.55(850) - 27236.3}{850 + (5+1)25}\right)1.452 = -137.7 \text{ kips}$$

Considering positive shear in subpanel (j − 1) − (j) of the PSS span for concentrated live loads, $F_{U(j-1)M(j)}^{+}$, as shown by Equation 4.49a, is

$$\frac{F_{U(j-1)M(j)}^{+}}{{}^{-}{}_{(j-1)-(j)}} = \left(\frac{\left(\frac{P_T^+X_T^+}{nd} - P_F^+ - \frac{P_B^+}{2} - \frac{P_B^+X_B^+}{2d} - \frac{P_A^+X_A^+}{2d}\right)\left(a_{(j-1)-(j+1)}\right) - \left((j-1)\left(\frac{P_A^+(X_A^+)}{2} + \left(P_F^+ + \frac{P_B^+}{2}\right)d - \frac{P_F^+X_F^+}{(j-1)} + \frac{P_B^+(X_B^+)}{2}\right)\right)}{a_{(j-1)-(j+1)} + (j+1)d}\right) \quad (4.51a)$$

where

$$R_{y0} = \frac{P_T^+ X_T^+}{nd}$$

$$W_{(j-1)} = \frac{P_B^+\left(X_B^+\right)}{d} + \frac{P_F^+\left((j-1)d - X_F^+\right)}{(j-1)d} = \frac{P_B^+\left(X_B^+\right)}{d} + P_F^+ - \frac{P_F^+ X_F^+}{(j-1)d}$$

$$W_{(j)} = \frac{P_B^+\left(d - X_B^+\right) + P_A^+\left(X_A^+\right)}{d}$$

$$V_{(j-1)-(j)}^{+P} = R_{y0} - \sum_{i=0}^{j-1} W_i - \frac{W_{(j)}}{2} = \frac{P_T^+ X_T^+}{nd} - P_F^+ - \frac{P_B^+\left(X_B^+\right)}{d} - \left(\frac{P_B^+\left(d - X_B^+\right) + P_A^+\left(X_A^+\right)}{2d}\right)$$

$$= \frac{P_T^+ X_T^+}{nd} - P_F^+ - \frac{P_B^+}{2} - \frac{P_B^+ X_B^+}{2d} - \frac{P_A^+ X_A^+}{2d}$$

$$M_{(j-1)-(j)}^{+0P} = \left(\sum_{i=0}^{j-1} iW_i + (j-1)\frac{W_{(j)}}{2}\right)d = P_B^+\left(jd - X_B^+\right) + P_F^+\left((j-1)d - X_F^+\right) + (j-1)d$$

$$\left(\frac{P_B^+\left(d - X_B^+\right) + P_A^+\left(X_A^+\right)}{2d}\right) = (j-1)\left(\frac{P_B^+\left(X_B^+\right)}{2} + \left(P_F^+ + \frac{P_B^+}{2}\right)d - \frac{P_F^+ X_F^+}{(j-1)} + \frac{P_A^+\left(X_A^+\right)}{2}\right)$$

$P^+_T = P^+_F + P^+_A + P^+_B + P^+_R$ = the resultant total load on the truss at distance X^+_T from joint (n) (right end of truss span) with live load moving from (n) to joint (0) across the PSS span.

P^+_F = the resultant live load on the truss between panel point (0) and panel point (j – 1) of the PSS span at X^+_F from (j – 1). It should be noted that for the Cooper's E load configuration, P^+_F is typically zero when positioning live loads for positive subpanel shear.

P^+_B = the resultant live load on the truss between (j – 1) and (j) of the PSS span at X^+_B from (j). It should be noted that for the Cooper's E load configuration, P^+_B is often zero when positioning live loads for positive subpanel shear.

P^+_A = the resultant live load on the truss between (j) and (j + 1) of the PSS span at X^+_A from (j + 1).

P^+_R = the resultant live load on the truss between (j + 1) and (n) of the PSS span at X^+_R from (n).

$F^+_{U(j-1)M(j)}$ is maximum when

$$\frac{d\left(F^+_{U(j-1)M(j)}\right)}{dx} = 0 = \left(\frac{P_T^+}{nd} - \frac{P_A^+ + P_B^+}{2d}\right)\left(a_{(j-1)-(j+1)}\right) - \left((j-1)\left(\frac{P_A^+}{2} - \frac{P_F^+}{(j-1)} + \frac{P_B^+}{2}\right)\right)$$

or

$$\frac{P_T^+}{n} = \left(\frac{P_A^+ + P_B^+}{2}\right)\left(1 + \frac{(j-1)d}{a_{(j-1)-(j+1)}}\right) - \frac{P_F^+ d}{a_{(j-1)-(j+1)}} \tag{4.51b}$$

The criterion of Equation 4.51b can only occur with a concentrated load at panel point (j + 1).

Therefore, when $P^+_F = 0$, the criterion for the location of a series of concentrated live loads for maximum upper diagonal member force due to positive shear in subpanel (j – 1) – (j) is when $\frac{P_T^+}{n} - \left(\frac{P_A^+ + P_B^+}{2}\right)\left(1 + \frac{(j-1)d}{a_{(j-1)-(j+1)}}\right)$ changes sign as the live load traverses a PSS span with loads moving from joint L(n) to joint L0.

For concentrated live loads on PSS spans, the criterion for maximum negative shear in subpanel (j – 1) – (j) to determine minimum forces in upper diagonal members U(j – 1)M(j) is

$$\frac{P_T^-}{n} = \frac{\left(\frac{P_A^- + P_B^-}{2}\right)\left(1 + \frac{(j+1)d}{a_{(j-1)-(j+1)}}\right)}{\left(\frac{nd}{a_{(j-1)-(j+1)}} + 1\right)} + \frac{P_F^- d}{a_{(j-1)-(j+1)}} \tag{4.52}$$

where

$P^-_T = P^-_F + P^-_A + P^-_B + P^-_L$ = the resultant total load on the truss at distance X^-_T from joint (0) (left end of truss span) with live load moving from (0) to joint (n) across the PSS span.

P^-_F = the resultant load on the truss between panel point (n) and panel point (j + 1) of the PSS span at X^-_F from (j + 1) = 0 for negative shear in panel (j – 1) – (j). It should be noted that for the Cooper's E load configuration, P^-_F is typically zero when positioning live loads for negative subpanel shear.

P^-_A = the resultant load on the truss between (j) and (j + 1) of the PSS span at X^-_A from (j). It should be noted that for the Cooper's E load configuration, P^-_A is often zero when positioning live loads for negative subpanel shear.

P^-_B = the resultant load on the truss between (j – 1) and (j) of the PSS span at X^-_B from (j – 1).

P^-_R = the resultant load on the truss between (j – 1) and (0) of the PSS span at X^-_R from joint (0).

The criterion of Equation 4.52 can only occur with a concentrated load at panel point (j – 1). Therefore, where $P^-_F = 0$, the criterion for the location of a series of concentrated live loads for maximum upper diagonal member force due to negative shear in subpanel (j – 1) – (j) is when $\frac{P_T^-}{n} - \frac{\left(\frac{P_A^- + P_B^-}{2}\right)\left(1 + \frac{(j+1)d}{a_{(j-1)-(j+1)}}\right)}{\left(\frac{nd}{a_{(j-1)-(j+1)}} + 1\right)}$ changes sign as the live load traverses a PSS span with loads moving from joint L0 to joint L(n). When $a_{(j-1)-(j+1)} = \infty$, chord members are parallel and horizontal, and Equation 4.51b for the live load position criterion for maximum upper diagonal force due to positive shear and Equation 4.52 for the live load position criterion for maximum upper diagonal force due to negative shear in a PSS span are equal to Equations 4.37b and 4.38b, respectively, for the live load position criterion for maximum upper diagonal forces in a BSS span.

Example 4.41

Determine the maximum forces produced by Cooper's E80 live load in upper diagonal members U2M3 and U4M5 of the Pennsylvania truss in Example 4.37.

For the positive shear for U2M3, try wheels between wheel 3 and wheel 5 at panel point L4 with j = 3:

$$a_{(2)-(4)} = \left(\left(\frac{2(47.5)}{47.5 - 38.0}\right) - 3 - 1\right)25 = 150\,\text{ft}$$

$$\propto_{(2)-(3)} = \frac{2\sqrt{\left(\frac{38.0}{2}\right)^2 + 25^2}}{38.0} = 1.653$$

where $P^+_F = 0$, the criterion of Equation 4.51b for maximum force due to positive shear in upper diagonal member U2M3 is given in Table E4.41a and as follows:

$$P_T^+ = 12\left(\frac{P_A^+ + P_B^+}{2}\right)\left(1 + \frac{(3-1)25}{a_{(j-1)-(j+1)}}\right) = 8.0\left(P_A^+ + P_B^+\right)$$

P^+_B = live loads in subpanel 2-3
P^+_A = live loads in subpanel 3-4
Distance to L(n) from panel point L4 = (12 − 4)(25) = 200 ft
Distance to L0 from panel point L4 = (4)(25) = 100 ft

TABLE E4.41a
Criterion for Maximum Positive Force in Member U2M3 with Cooper's E80 Live Load Traversing a PSS Span

Positive Shear in Subpanel 2-3 (j = 3)		Wheel at Panel Point L4	
Wheel at panel point L4	3	4	5
Distance to uniformly distributed load from wheel at panel point L4 (ft)	96	91	86
Distance to wheel 1 (ft)	109 − 96 = 13	109 − 91 = 18	109 − 86 = 23
Distance to L0 from wheel 1 (ft)	100 − 13 = 87	100 − 18 = 82	100 − 23 = 77
Length of uniformly distributed load (ft)	200 − 96 = 104	200 − 91 = 109	200 − 86 = 114
Total load on span = P^+_T (kips)	568 + 104(4.0) = 984	568 + 109(4.0) = 1004	568 + 114(4.0) = 1024
$P^+_T/8$ (kips)	123.0	125.5	128.0
Load in panel 2-4 ahead of L4 = $(P^+_A + P^+_B)_a$ (kips)	60	100	140
Load in panel 2-4 including load at L4 = $(P^+_A + P^+_B)_b$ (kips)	100	140	180
Criterion met	No	Yes	No

The criterion of Equation 4.51b indicates maximum force due to positive shear in upper diagonal member U2M3 with wheel 4 at panel point L4. Therefore, $P^+_F = P^+_A = W_{(j-1)} = 0$. Equation 4.49a provides the maximum force due to positive shear in U2M3 as in Table E4.41b.

TABLE E4.41b
Maximum Positive Force in Member U2M3 due to Cooper's E80 Live Load Traversing a PSS Span

U2M3 Positive Shear, j = 3 (Axle 4 at L4)	
Sum of moments about R_{y12} (kip-ft)	32,728 + 568(109) + 4.0(109)(109/2) = 118402
R_{y0} (kips)	118,402/300 = 394.67
$W_{(j-1)}$ (kips)	0
$W_{(j)}$ (kips)	960/25 = 38.4

(Continued)

TABLE E4.41b *(Continued)*
Maximum Positive Force in Member U2M3 due to Cooper's E80 Live Load Traversing a PSS Span

U2M3 Positive Shear, j = 3 (Axle 4 at L4)

$V^{+P}_{(j-1)-(j)} = R_{y0} - W_{(j-1)} - \dfrac{W_{(j)}}{2}$ (kips)	394.67 − 0 − 38.4/2 = 375.47
$M^{+0P}_{(j-1)-(j)} = \left((j-1)W_{(j-1)} + (j-1)\dfrac{W_{(j)}}{2}\right)d$ (kip-ft)	(2(0) + 2(38.4/2))25 = 960
$F^{+}_{U(j-1)M(j)} = \left(\dfrac{V^{+P}_{(j-1)-(j)}\left(a_{(j-1)-(j+1)}\right) - M^{+0P}_{(j-1)-(j)}}{a_{(j-1)-(j+1)} + (j+1)d}\right)\alpha_{(j-1)-(j)}$ (kips)	((375.47)(150) − 960)/(150 + 4(25))1.653 = (55360.5/250)1.653 = 366.0

$$F^{+max}_{U2M3} = 366.0\,\text{kips}$$

For the negative shear in U2M3, try wheels between wheel 2 and wheel 5 at panel point L2. Where $P^{-}_{F} = 0$, the criterion of Equation 4.52 for maximum force due to negative shear in U2M3 is given in Table E4.41c and as follows:

$$P^{-}_{T} = \frac{12\left(\dfrac{P^{-}_{A} + P^{-}_{B}}{2}\right)\left(1 + \dfrac{(3+1)25}{150}\right)}{\left(\dfrac{12(25)}{150} + 1\right)} = 3.33\left(P^{-}_{A} + P^{-}_{B}\right)$$

P^{-}_{A} = live loads in subpanel 3-4
P^{-}_{B} = live loads in subpanel 2-3

TABLE E4.41c
Criterion for Maximum Negative Force in Member U2M3 with Cooper's E80 Live Load Traversing a PSS Span

Negative Shear in Subpanel 2-3 (j = 3)	Wheel at Panel Point 2			
Wheel at panel point L2	2	3	4	5
Last wheel on span	10	10	11	12
Distance of last wheel on span to L0 (ft)	2	7	4	4
Total load on span = P^{-}_{T} (kips)	304	304	344	384
$P^{-}_{T}/3.33$ (kips)	91.2	91.2	103.2	115.2
Load in panel 2-4 ahead of L2 = $(P^{-}_{A} + P^{-}_{B})_a$ (kips)	20	60	100	140
Load in panel 2-4 including load at L2 = $(P^{-}_{A} + P^{-}_{B})_b$ (kips)	60	100	140	180
Criterion met	No	Yes	Yes	No

The criterion of Equation 4.52 indicates maximum force due to negative shear in U2M3 with wheel 3 or 4 at panel point L2. Therefore, $P^-_F = P^-_A = 0$. Equation 4.49b provides the maximum force due to negative shear in U2M3 as shown in Table E4.41d.

TABLE E4.41d
Maximum Negative Force in Member U2M3 due to Cooper's E80 Live Load Traversing a PSS Span

U2M3 Negative Shear, j = 3

Wheel at L2	**3**	**4**
Sum of moments about R_{y0} (kip-ft)	9264 + 7(304) = 11392	11,696 + 344(4) = 13072
R_{y12}(kips)	11,392/300 = 37.97	13,072/300 = 43.57
$W_{(3)}$ (kips)	460/25 = 18.40	960/25 = 38.40
$V^{-P}_{(j-1)-(j)} = R_{yn} - \sum_{i=j}^{n} W_i + \frac{W_{(j)}}{2}$ (kips)	37.97 − 18.40 + (18.40/2) = 28.77	43.57 − 38.40 + (38.40/2) = 24.37
$M^{-0P}_{(j-1)-(j)} = -\left(nR_{yn} - \sum_{i=j}^{n} iW_i + \left(\frac{W_{(j)}}{2}\right)(j-1)\right)d$ (kip-ft)	−(300(37.97) − 3(25) (18.40) + (3 − 1)(25) (18.40/2)) = −(11,391 − 1380 + 460) = −10,471	−(300(43.57) − 3(25) (38.40) + 25(38.40) = −(13,071 − 2880 + 960) = −11,151
$F^-_{U(j-1)M(j)} = \left(\frac{-V^{-P}_{(j-1)-(j)}\left(a_{(j-1)-(j)}\right) + M^{-0P}_{(j-1)-(j)}}{a_{(j-1)-(j)} + (j+1)d}\right)\alpha_{(j-1)-(j)}$(kips)	((−28.77(150) − 10,471)/ (150 + (3 + 1) (25)))1.653 = ((−4315.5 −10471)/(250))1.653 = (−14786.5/250)1.653 = −59.15(1.653) = −97.77	((−24.37(150) −11151)/250)1.653 = −97.90

$$F^{-max}_{U2M3} = -97.90\,\text{kips}$$

For the positive shear for upper diagonal member U4M5, try wheels between wheel 3 and wheel 5 at panel point L6 with j = 5:

$$a_{(4)\text{-}(6)} = \left(\left(\frac{2(50)}{50-47.5}\right) - 5 - 1\right)25 = 850\,\text{ft}$$

$$\propto_{(4)-(5)} = \frac{2\sqrt{\left(\frac{47.5}{2}\right)^2 + 25^2}}{47.5} = 1.452$$

Where $P^+_F = 0$, the criterion of Equation 4.51b for maximum force due to positive shear in upper diagonal member U4M5 is given in Table E4.41e and as follows:

$$P^+_T = 12\left(\frac{P^+_A + P^+_B}{2}\right)\left(1 + \frac{(5-1)25}{a_{(j-1)-(j+1)}}\right) = 6.71\left(P^+_A + P^+_B\right)a$$

P^+_A = live loads in subpanel 5-6
P^+_B = live loads in subpanel 4-5
Distance to L(n) from panel point L6 = (12 − 6)(25) = 150 ft
Distance to L0 from panel point L6 = (6)(25) = 150 ft

TABLE E4.41e
Criterion for Maximum Positive Force in Member U4M5 with Cooper's E80 Live Load Traversing a PSS Span

Positive Shear in Subpanel 4-5 (j = 5)		Wheel at Panel Point 6	
Wheel at panel point L6	3	4	5
Distance to uniformly distributed load from wheel at panel point L6 (ft)	96	91	86
Distance to wheel 1 (ft)	109 − 96 = 13	109 − 91 = 18	109 − 86 = 23
Distance to L0 from wheel 1 (ft)	150 − 13 = 137	150 − 18 = 132	150 − 23 = 127
Length of uniformly distributed load (ft)	150 − 96 = 54	150 − 91 = 59	150 − 86 = 64
Total load on span = P^+_T (kips)	568 + 54(4.0) = 784	568 + 59(4.0) = 804	568 + 64(4.0) = 824
$P^+_T/6.71$ (kips)	116.8	119.8	122.8
Load in panel 4-6 ahead of L6 = $(P^+_A + P^+_B)_a$ (kips)	60	100	140
Load in panel 4-6 including load at L6 = $(P^+_A + P^+_B)_b$ (kips)	100	140	180
Criterion met	No	Yes	No

The criterion of Equation 4.51b indicates maximum force due to positive shear in U4M5 with wheel 4 at panel point L6. Therefore $P^+_F = P^+_A = 0$ and $W_{(j-1)} = 0$. Equation 4.49a provides the maximum force due to positive shear in U4M5 as shown in Table E4.41f.

TABLE E4.41f
Maximum Positive Force in Member U4M5 due to Cooper's E80 Live Load Traversing a PSS Span

U4M5 Positive Shear, j = 5 (Axle 4 at L6)	
Sum of moments about R_{y12} (kip-ft)	32,728 + 568(59) + 4.0(59)(59/2) = 73202
R_{y0} (kips)	73,202/300 = 244.0
$W_{(4)}$ (kips)	0
$W_{(5)}$ (kips)	960/25 = 38.4
$V^{+P}_{(j-1)-(j)} = R_{y0} - W_{(j-1)} - \frac{W_{(j)}}{2}$ (kips)	244.0 − 38.4/2 = 224.8
$M^{+0P}_{(j-1)-(j)} = \left((j-1)W_{(j-1)} + (j-1)\frac{W_{(j)}}{2}\right)d$ (kip-ft)	(4)(0) + 4(38.4/2))25 = 1920
$F^+_{U(j-1)M(j)} = \left(\frac{V^{+P}_{(j-1)-(j)}\left(a_{(j-1)-(j+1)}\right) - M^{+0P}_{(j-1)-(j)}}{a_{(j-1)-(j+1)} + (j+1)d}\right)\alpha_{(j-1)-(j)}$ (kips)	((224.8)(850) − 1920)/(850 + 6(25))1.452 = (189168.5/1000)1.452 = 274.67

$$F_{U4M5}^{+max} = 274.67\,\text{kips}$$

For the negative shear in U4M5, try wheels between wheel 2 and wheel 5 at panel point L4 with j = 5. Where $P^-_F = 0$, the criterion of Equation 4.52 for maximum force due to negative shear in U4M5 is given in Table E4.41g and as follows:

$$P_T^- = \frac{12\left(\dfrac{P_A^- + P_B^-}{2}\right)\left(1 + \dfrac{(5+1)25}{850}\right)}{\left(\dfrac{12(25)}{850} + 1\right)} = 5.22\left(P_A^- + P_B^-\right)$$

P^-_A = live loads in subpanel 5-6
P^-_B = live loads in subpanel 4-5

TABLE E4.41g
Criterion for Maximum Negative Force in Member U4M5 with Cooper's E80 Live Load Traversing a PSS Span

Negative Shear in Subpanel 4-5 (j = 5)	Wheel at Panel Point 4			
Wheel at panel point L4	2	3	4	5
Last wheel on span	18	18	18	18
Distance of last wheel on span to L0 (ft)	4	9	14	19
Length of uniformly distributed load on span (ft)	4 − 5 = −1 off span	9 − 5 = 4	14 − 5 = 9	19 − 5 = 14
Total load on span = P^-_T (kips)	568	568 + 4(4) = 584	568 + 4(9) = 604	568 + 14(4) = 624
$P^-_T/5.22$ (kips)	108.8	111.9	115.7	119.5
Load in panel 4-6 ahead of L4 = $(P^-_A + P^-_B)_a$ (kips)	20	60	100	140
Load in panel 4-6 including load at L4 = $(P^-_A + P^-_B)_b$ (kips)	60	100	140	180
Criterion met	No	No	Yes	No

The criterion of Equation 4.52 indicates maximum force due to negative shear in U4M5 with wheel 4 at panel point L4. Therefore, $P^-_F = P^-_A = 0$. Equation 4.49b provides the maximum force due to negative shear in U4M5 as shown in Table E4.41h.

TABLE E4.41h
Maximum Negative Force in Member U4M5 due to Cooper's E80 Live Load Traversing a PSS Span

U4M5 Negative Shear, j = 5	
Wheel at L4	4
Sum of moments about R_{y0} (kip-ft)	32,728 + 568(9) + 4(9)(9/2) = 38,002
R_{y12} (kips)	38,002/300 = 126.67
$W_{(5)}$ (kips)	960/25 = 38.40

(Continued)

TABLE E4.41h (*Continued*)
Maximum Negative Force in Member U4M5 due to Cooper's E80 Live Load Traversing a PSS Span

U4M5 Negative Shear, j = 5

$V^{-P}_{(j-1)-(j)} = R_{yn} - \sum_{i=j}^{n} W_i + \dfrac{W_{(j)}}{2}$ (kips)	126.67 − 38.40 +(38.40/2) = 107.47
$M^{-0P}_{(j-1)-(j)} = -\left(nR_{yn} - \sum_{i=j}^{n} iW_i + \left(\dfrac{W_{(j)}}{2}\right)(j-1)\right)d$ (kip-ft)	−(300(126.67) − 5(25)(38.40) + 25(4)(38.40/2) = −(38,002 − 4800 + 1920) = −35,122
$F^{-}_{U(j-1)M(j)} = \left(\dfrac{-V^{-P}_{(j-1)-(j)}\left(a_{(j-1)-(j)}\right) + M^{-0P}_{(j-1)-(j)}}{a_{(j-1)-(j)} + (j+1)d}\right)\alpha_{(j-1)-(j)}$ (kips)	((−107.47(850) − 35,122)/1000)1.452 = −183.6

$$F^{-max}_{U4M5} = -183.6\,\text{kips}$$

The forces in non-central vertical members U(k)L(k) of PSS spans, where k = even numbers greater than or equal to 4 and less than n/2, shown in Figures 4.18b and 4.18d, are determined by investigating the free body diagram at isolated truss joint U(k). Equilibrium at U(k) indicates that the forces in U(k)L(k) of PSS spans resist the vertical component of the force in upper diagonal member U(k) M(k + 1), top chord member U(k − 2)U(k), top chord member U(k)U(k + 2) and, if present, external loads at U(k). Therefore, if no external loads exist at U(k),

$$F_{U(k)L(k)} = -F_{U(k)M(k+1)}\left(\frac{h_{(k)}}{\sqrt{h^2_{(k)} + (2d)^2}}\right) - F_{U(k-2)U(k)}\left(\frac{h_{(k)} - h_{(k-2)}}{\sqrt{\left(h_{(k)} - h_{(k-2)}\right)^2 + (2d)^2}}\right) + F_{U(k)U(k+2)}\left(\frac{h_{(k+2)} - h_{(k)}}{\sqrt{\left(h_{(k+2)} - h_{(k)}\right)^2 + (2d)^2}}\right) \tag{4.53a}$$

Since the forces in top chord members U(k − 2)U(k) and U(k)U(k + 2) for typical inclined chords will tend to cancel, it is likely that the live load position for maximum and minimum forces in members U(k)L(k) are nearly coincident with the live load locations for maximum forces due to positive and negative shear in upper diagonal member U(k)M(k + 1).

Equilibrium at joint U(n/2) of a PSS span indicates that the forces in interior vertical members U(n/2)L(n/2) resist the vertical component of the force in top chord member U(n/2–2)U(n/2), top chord member U(n/2)U(n/2+2) and, if present, external loads at joint U(n/2). Therefore, if no external loads exist at joint U(n/2),

$$F_{U\left(\frac{n}{2}\right)L\left(\frac{n}{2}\right)} = -\frac{F_{U\left(\frac{n}{2}-2\right)U\left(\frac{n}{2}\right)}\left(h_{\left(\frac{n}{2}\right)} - h_{\left(\frac{n}{2}-2\right)}\right)}{\sqrt{\left(h_{\left(\frac{n}{2}\right)} - h_{\left(\frac{n}{2}-2\right)}\right)^2 + 4d^2}} - \frac{F_{U\left(\frac{n}{2}\right)U\left(\frac{n}{2}+2\right)}\left(h_{\left(\frac{n}{2}+2\right)} - h_{\left(\frac{n}{2}\right)}\right)}{\sqrt{\left(h_{\left(\frac{n}{2}+2\right)} - h_{\left(\frac{n}{2}\right)}\right)^2 + 4d^2}} \tag{4.53b}$$

Example 4.42

Determine the maximum forces produced by Cooper's E80 live load in interior vertical member U4L4 of the Pennsylvania truss of Example 4.37.

k = 4
j = k + 1 = 5

Free body diagram at U4 indicates,

$$F_{U4L4} = -F_{U4M5}\left(\frac{h_{(4)}}{\sqrt{h_{(4)}^2 + (2d)^2}}\right) - F_{U2U4}\left(\frac{h_{(4)} - h_{(2)}}{\sqrt{\left(h_{(4)} - h_{(2)}\right)^2 + (2d)^2}}\right)$$

$$+ F_{U4U6}\left(\frac{h_{(6)} - h_{(4)}}{\sqrt{\left(h_{(6)} - h_{(4)}\right)^2 + (2d)^2}}\right) = -F_{U4M5}(0.689) - F_{U2U4}(0.187) + F_{U4U6}(0.050)$$

In accordance with Example 4.41, Table E4.42a gives the maximum force in member U4M5.

TABLE E4.42a
Maximum Positive Force in Member U4M5 due to Cooper's E80 Live Load Traversing a PSS Span

U4M5 Positive Shear, j = 5				
Wheel at L6	**2**	**3**	**4**	**5**
Sum of moments about R_{y12} (kip-ft)	32,728 + 568(49)+ 4.0(49)(49/2) = 65,362	32,278 + 568(54) +4.0(54) (54)/2 = 69,232	32,728 + 568(59) + 4.0(59)(59/2) = 73,202	77,272
R_{y0} (kips)	217.9	230.77	73,202/300 = 244.0	257.57
$W_{(3)}$ (kips)	0	0	0	0
$W_{(5)}$ (kips)	160/25 = 6.40	460/25 = 18.40	960/25 = 38.4	1660/25= 66.40
$V^{+P}_{(j-1)-(j)} = R_{y0} - W_{(j-1)} - \frac{W_{(j)}}{2}$ (kips)	214.7	230.77 − 18.40/2 = 221.57	244.0 − 38.4/2 = 224.8	224.37
$M^{+0P}_{(j-1)-(j)} = \left((j-1)W_{(j-1)} + (j-1)\frac{W_{(j)}}{2}\right)d$ (kip-ft)	320	4(18.40/2)25 = 920	(4)(0) + 4(38.4/2)) 25 = 1920	3320
$F^{+}_{U(j-1)M(j)} = \left(\frac{V^{+P}_{(j-1)-(j)}\left(a_{(j-1)-(j+1)}\right) - M^{+0P}_{(j-1)-(j)}}{a_{(j-1)-(j+1)} + (j+1)d}\right)$ $\alpha_{(j-1)-(j)}$ (kips)	(214.7(850) − 320)/1000) 1.452 = 264.50	((221.57(850) − 920)/1000) 1.452= 272.1	((224.8)(850) − 1920)/(850 + 6(25)) 1.452 = (189168.5/1000) 1.452 = 274.67	272.1

$$F^{+}_{U4M5} = 274.67 \text{ kips}$$

$$F_{U4L4} = -F_{U4M5}(0.689) - F_{U2U4}(0.187) + F_{U4U6}(0.050)$$

From Equation 4.43a, top chord forces are as follows and as in Table E4.42b:

$$F_{U2U4} = -\left(4R_{y0} - \sum_{i=0}^{3}(4-i)W_i\right)(0.536) = -\left(4R_{y0} - 0\right)(0.536)$$

$$F_{U4U6} = -\left(6R_{y0} - \sum_{i=0}^{5}(6-i)W_i\right)(0.500) = -\left(6R_{y0} - W_{(5)}\right)(0.500)$$

TABLE E4.42b
Maximum Positive Force in Member U4L4 due to Cooper's E80 Live Load Traversing a PSS Span

	k	j	Wheel 2 at L6	Wheel 3 at L6	Wheel 4 at L6	Wheel 5 at L6
R_{y0} (kips)			217.9	230.77	244.0	257.57
$W_{(5)}$ (kips)			6.40	18.40	38.40	66.40
F^{+}_{U4M5} (kips)	4	5	264.5	272.1	274.67	272.1
F^{+}_{U2U4} (kips)	2	3	−467.2	−494.77	−523.1	−552.2
F^{+}_{U4U6} (kips)	4	5	−650.5	−683.1	−712.8	−739.5
F^{+}_{U4L4} (kips)	4	5	−127.5	−129.3	−127.2	−121.2

$F^{+max}_{U4L4} = -129.3 \text{ kips}$ with wheel 3 at L6 (close to wheel 4 at L6) for maximum force due to positive shear in member U4M5. Table E4.42c gives the forces in accordance with Example 4.41.

TABLE E4.42c
Maximum Negative Force in member U4M5 due to Cooper's E80 live load traversing PSS span.

U4M5 Negative Shear, j = 5				
Wheel at L4	3	4	5	6
Sum of moments about R_{y0} (kip-ft)	32,728 + 568(4) + 4(4)(4/2) = 35,032	32,728 + 568(9) + 4(9)(9/2) = 38,002	32,728 + 568(14) + 4(14)(14/2) = 41,072	32,728 + 568(23) + 4(23)(23)/2 = 46,850
R_{y12} (kips)	116.77	38,002/300 = 126.67	136.9	156.17

TABLE E4.42c (*Continued*)
Maximum Negative Force in member U4M5 due to Cooper's E80 live load traversing PSS span.

U4M5 Negative Shear, j = 5

Wheel at L4	**3**	**4**	**5**	**6**
$W_{(5)}$ (kips)	460/25 = 18.40	960/25 = 38.40	1660/25 = 66.40	3280/25 = 131.2
$V^{-P}_{(j-1)-(j)} = R_{yn} - \sum_{i=j}^{n} W_i + \frac{W_{(j)}}{2}$ (kips)	116.77 −18.40 + (18.40/2) = 107.57	126.67 − 38.40 + (38.40/2) = 107.47	103.70	156.17 − 131.2 + (131.2/2)= 90.57
$M^{-0P}_{(j-1)-(j)} = -\left(nR_{yn} - \sum_{i=j}^{n} iW_i + \left(\frac{W_{(j)}}{2}\right)(j-1)\right)$ d (kip-ft)	−(300(116.77) − 5(25)(18.40) + 25(4) (18.40/2)) = −33,651	−(300(126.67) − 5(25)(38.40) + 25(4)(38.40/2) = −(38,002 − 4800 + 1920)) = −35,122	−36,090	-(300(156.17)− 5(25)(131.2)+ 25(4)(131.2/2))= 37,011
$F^{-}_{U(j-1)M(j)} = \left(\frac{-V^{-P}_{(j-1)-(j)}\left(a_{(j-1)-(j)}\right) + M^{-0P}_{(j-1)-(j)}}{a_{(j-1)-(j)} + (j+1)d}\right)$ $\alpha_{(j-1)-(j)}$ (kips)	((−107.57(850)− 33,651)/1000 (1.452) = −181.6	((−107.47(850) −35,122)/1000) 1.452 = −183.6	−180.4	-(((90.57)(850)− 37,011)/ 1000)1.452 = −165.5

$$F^{-}_{U4M5} = -183.6 \text{ kips}$$

From Equation 4.43a (see Table E4.42d), reestablished for negative shear with load traversing from left to right,

$$F^{-}_{U(j-1)U(j+1)} = -\left(n-(j+1))R_{yn} + \sum_{i=j+1}^{n}(j+1-i)W_i\right)\left(\frac{\sqrt{\left(h_{(j+1)} - h_{(j-1)}\right)^2 + (2d)^2}}{2h_{(j+1)}}\right)$$

$$F^{-}_{U2U4} = -\left(8R_{yn} + \sum_{i=4}^{12}(4-i)W_i\right)(0.536) = -\left(8R_{y12} - W_{(5)}\right)(0.536)$$

$$F^{-}_{U4U6} = -\left(6R_{yn} + \sum_{i=6}^{12}(6-i)W_i\right)(0.500) = -\left(6R_{y12} + 0\right)(0.500)$$

TABLE E4.42d
Maximum Negative Force in Member U4L4 due to Cooper's E80 Live Load Traversing a PSS Span

	k	j	Wheel 3 at L4	Wheel 4 at L4	Wheel 5 at L4	Wheel 6 at L4
R_{y12} (kips)			116.77	126.67	136.9	156.17
$W_{(5)}$ (kips)			18.40	38.40	66.40	131.20
$F^{-}_{U(4)M(5)}$ (kips)	4	5	−181.6	−183.6	−180.4	−165.5
$F^{-}_{U(2)U(4)}$ (kips)	2	3	−490.85	−522.58	−551.44	−599.33
$F^{-}_{U(4)U(6)}$ (kips)	4	5	−350.31	−380.00	−410.70	−468.51
$F^{-}_{U(4)L(4)}$ (kips)	4	5	199.40	205.0	206.6	202.7

$F^{-max}_{U4L4} = 206.6\,\text{kips}$ with wheel 5 at L4 (close to wheel 4 at L4) for maximum force due to negative shear in member U4M5.

It is apparent that by considering equilibrium at joint L2, the forces in end vertical member U2L2 of PSS spans can be calculated in the same way as for U2L2 in BSS spans. Therefore, forces in U2L2 of PSS spans can be calculated in accordance with Equations 4.40e and 4.40g for uniformly distributed and concentrated live loads, respectively.

It is also evident that due to the configurations of BSS and PSS spans between panel point (0) and panel point (2) in Figure 4.18, the forces in upper end post member M1U2, lower end post member L0M1 and diagonal member M1L2 of PSS spans can be calculated in the same way both spans. Therefore, forces in M1U2 and L0M1 are calculated using Equations 4.41 and 4.42a, respectively, with $h = h_2$. The force in member M1L2 of a PSS truss can be calculated as for a BSS truss by considering equilibrium at joint L2 and Equation 4.47a.

4.6.3 Subdivided Horizontal Chord Pratt (Baltimore) Truss Spans with Subtie Members

Member forces in Baltimore trusses with subtie members (BST), as shown in Figure 4.17, are calculated in the same way as the forces for BSS trusses. Furthermore, since Figures 4.16 and 4.17 illustrate that the substrut and subtie subdivided truss forms are identical between panel point (0) and panel point (2), forces in members L0M1 (lower end post), M1U2 (upper end post) and M1L2 of BST spans are determined in the same manner for subdivided subtie and substrut spans.

Forces in bottom chord members of BST spans are calculated by equilibrium of forces about joint U(j-1) on a section through subpanel (j − 1) − (j), which yields

$$F_{L(j-1)L(j)} = F_{L(j)L(j+1)} = \left((j-1)R_{y0} - \sum_{i=0}^{j-1} W_i\,(j-1-i)\right)\frac{d}{h} \tag{4.54a}$$

Equilibrium at joint (j) indicates that force $F_{L(j-1)L(j)}$ in bottom chord member L(j − 1)L(j) equals $F_{L(j)L(j+1)}$ in bottom chord members L(j)L(j + 1). The maximum bottom chord forces in panel (j − 1) − (j + 1) for a uniformly distributed live load occur with span entirely loaded, and for concentrated live loads, they occur with the criterion of Equation 4.11c with k = j − 1 at panel point (j − 1) such that

$$\frac{P_T}{n} = \frac{P_F}{j-1} \tag{4.54b}$$

where n = the number of subpanels in the BST span; P_T = the resultant total load on the BST span at distance X_T from joint (n) at the right end of the truss span with live load moving from (n) to joint 0 and P_F = the resultant load on the BST span between (0) and (j − 1) at distance X_F from (j − 1).

The presence of subvertical members M(j)L(j) subdividing panel (j − 1) − (j + 1) will affect the forces in top chord members U(j − 1)U(j + 1). Therefore, panel point moments from $W_{(j)} = F_{M(j)L(j)}$, the joint load at (j) due to loads in panel (j − 1) − (j + 1), need to be considered.

The top chord member forces in BST spans are calculated by taking a section through subpanel (j − 1) − (j) and considering equilibrium at joint L(j + 1). Accordingly, the force in U(j − 1)U(j + 1) for equal-length panels 2d is determined, noting that force $W_{(j)}$ is on the right side of the section through subpanel (j − 1) − (j), from

$$R_{y0}(j+1)d - d\sum_{i=0}^{j+1}(j+1-i)W_i + W_{(j)}d + F_{U(j-1)U(j+1)}(h) = 0 \tag{4.55a}$$

which yields

$$F_{U(j-1)U(j+1)} = \frac{-R_{y0}(j+1)d + d\sum_{i=0}^{j+1}(j+1-i)W_i - W_{(j)}d}{h} = \frac{M^{Wx}_{(j+1)} - 2M^{P}_{(j)}}{h} \tag{4.55b}$$

where

$M^{Wx}_{(j+1)} = -\left((j+1)R_{y0} - \sum_{i=0}^{j+1}(j+1-i)W_i\right)d$ = the moment about joint L(j + 1) from end reaction R_{y0} and applied joint loads W_i;

$M^{P}_{(j)} = \frac{W_{(j)}(d)}{2}$ = the moment at panel point (j) due to loads in equal-length subpanels (j − 1) − (j) and (j) − (j + 1); and

$W_{(j)}$ = the floorbeam reaction or panel point force between two subpanels of equal length d (see Equation 4.14f).

Maximum force $F^{max}_{U(j-1)U(j+1)}$ in top chord member U(j − 1)U(j + 1) of the BST span due to a uniformly distributed live load occurs with the span entirely loaded. $F^{max}_{U(j-1)U(j+1)}$ in top chord members U(j − 1)U(j + 1) of the BST span due to concentrated live loads are determined with the live load positioned on the span in accordance with a criterion established by $\frac{d\left(F_{U(j-1)U(j+1)}\right)}{dx} = 0$.

Taking a section through subpanel (j − 1) − (j) reveals the force in member U(j − 1)U(j + 1) due to moments about joint L(j + 1) from end reaction R_{y0} and applied live loads P_F and P_B is

$$F_{U(j-1)U(j+1)} = \frac{-\frac{P_T X_T}{nd}(j+1)d + P_F(X_F + 2d) + 2P_B(X_B + d)}{h} \tag{4.55c}$$

where for the BST span,

P_T = the resultant total load on the BST span at distance X_T from joint (n) at the right end of the truss span with live load moving form joint (n) to joint (0);

P_F = the resultant load on the BST span between panel point (0) and panel point (j − 1) at X_F from (j − 1); and

P_B = the resultant load on the BST span between (j − 1) and (j) at X_B from (j).

$F^{max}_{U(j-1)U(j+1)}$ is determined with $\frac{d\left(F_{U(j-1)U(j+1)}\right)}{dx} = 0$ such that

$$\frac{d\left(M_{(j+1)}^{Wx}+2M_{(j)}^{P}\right)}{dx}=\frac{-P_T\left(j+1\right)}{n}+P_F+2P_B=0$$

or

$$\frac{P_T}{n}=\frac{P_F+2P_B}{\left(j+1\right)} \tag{4.55d}$$

The criterion of Equation 4.55d can be used to determine the location of the concentrated live loads creating maximum top chord member force at any panel within a BST span. Therefore, the criterion for the location of a series of concentrated loads for maximum top chord member force in a BST span is when $\frac{P_T(j+1)}{n}-\left(P_F+2P_B\right)$ changes sign as the live load traverses the span.

Example 4.43

Determine the maximum forces produced by a uniformly distributed live load of 4.0 kips/ft and Cooper's E80 live load in bottom chord member L5L6 and top chord member U4U6 of a 300 foot long twelve-subpanel BST truss with height of 50 ft.

Maximum force in L5L6 (j = 5) due to uniformly distributed live load of 4 kip/ft:

$$\begin{aligned}F_{L5L6}=F_{L4L5}&=((5-1)(4.0)(150)-\sum_{i=0}^{5-1}W_i(4-i))\frac{d}{h}\\&=\left(4(600)-\left(4(50)+3(100)+2(100)+100\right)\right)\frac{25}{50}\\&=800.0\text{ kips}\end{aligned}$$

Maximum force in U4U6 (j = 5) due to uniformly distributed live load of 4 kip/ft:

$$\begin{aligned}M_{(6)}^{Wx}&=\left(-(6)600+\sum_{i=0}^{6}(6-i)W_i\right)25\\&=\left(-3600+6(50)+(5+4+3+2+1)100\right)25\\&=-45{,}000\text{ kips}-\text{ft}\end{aligned}$$

$$M_{(5)}^{P}=\frac{100(25)}{2}=1250\,\text{kips}-\text{ft}$$

$$F_{U4U6}=\frac{-45{,}000-2(1250)}{50}=\frac{-47{,}500}{50}=-950.0\text{ kips}$$

Maximum force in L5L6 (j = 5) due to Cooper's E80 live load:

The criterion of Equation 4.54b (Equation 4.11c with k = j − 1) indicates

$$\frac{P_T}{12}=\frac{P_F}{(5-1)}$$

or

$$\frac{P_T}{3} = P_F$$

From Example 4.26, the criterion provides that the maximum force in L5L6 occurs with axle 13 at L4 such that

$$F^{max}_{L5L6} = 899.3 \text{ kips}$$

Maximum force in U4U6 (j = 5) due to Cooper's E80 live load:

The criterion of Equation 4.55d indicates $\frac{P_T}{2} = P_F + 2P_B$ as calculated in Table E4.43a.

TABLE E4.43a

Criterion for Maximum Force in Top Chord Member U4U6 in a BST Span

Location of Live Load	Axle at L5				
	12	**13**	**14**	**15**	**16**
Distance to uniform load (ft)	40	35	30	21	16
Distance to axle no. 18 (ft)	35	30	25	16	11
Last axle on span	18	18	18	18	18
Distance to axle no. 1 (ft)	69	74	79	88	93
First axle on span	1	1	1	1	1
Length of uniform load (ft)	175 – 40 = 135	175 – 35 = 140	175 – 30 = 145	175 – 21 = 154	175 – 16 = 159
Total load on span, P_T (kips)	568 + 135(4) = 1108	568 + 140(4) = 1128	568 + 145(4) = 1148	568 + 154(4) = 1184	568 + 159(4) = 1204
P_F (between L0 and L4) (kips)	$\sum_{i=1}^{8} P_i = 258$	$\sum_{i=1}^{9} P_i = 284$	$\sum_{i=1}^{9} P_i = 284$	$\sum_{i=1}^{10} P_i = 304$	$\sum_{i=1}^{11} P_i = 344$
$(P_B)_L$ (between L4 and L5) (kips)	$\sum_{i=9}^{11} P_i = 86$	$\sum_{i=10}^{12} P_i = 100$	$\sum_{i=10}^{13} P_i = 140$	$\sum_{i=11}^{14} P_i = 160$	$\sum_{i=12}^{15} P_i = 146$
$(P_B)_R$ (between L4 and L5 including load at L5) (kips)	$\sum_{i=9}^{12} P_i = 126$	$\sum_{i=10}^{13} P_i = 140$	$\sum_{i=10}^{14} P_i = 180$	$\sum_{i=11}^{15} P_i = 186$	$\sum_{i=12}^{16} P_i = 172$
$P_T/2$ (kips)	554	564	574	592	602
$(P_F + 2P_B)_L$ wheel to left of L6 (kips)	258 + 2(86) = 430	284 + 2(100) = 484	284 + 2(140) = 564	304 + 2(160) = 624	344 + 2(146) = 636
$(P_F + 2P_B)_R$ wheel to right of L6 (kips)	258 + 2(126) = 510	284 + 2(140) = 564	284 + 2(180) = 644	304 + 2(186) = 676	344 + 2(172) = 688
Satisfies criterion	No	Yes (marginal)	Yes	No	No

As shown in Table E4.43b, the criterion of Equation 4.55e indicates axles 13 or 14 at L5 for maximum panel point moment at L6 and, accordingly, the maximum force the top chord member U4U6. It is likely that axle 14 at L5 creates the maximum panel point moment at L6.

TABLE E4.43b
Force in Top Chord Member U4U6 due to Cooper's E80 Live Load Traversing a BST Span

Maximum Force in Member U4U6	Axle at L5		
	13	14	15
Sum of moments to L12 (kip-ft)	32,728 + 568(140) + 4(140)²/2 = 151,448	32,728 + 568(145) + 4(145)²/2 = 157,138	32,728 + 568(154) + 4(154)²/2 = 167,632
Reaction R_{y0} (kips)	151,448/300 = 504.8	157,138/300 = 523.8	167,632/300 = 558.8
$6(25)R_{y0}$ (kips)	75,724	78,569	83,816
$25\sum_{i=0}^{6}(6-i)W_i$ (kips)	27,178	29,888	29,888 + 568(9) + 4(4)(2) = 35,032
$M_{(6)}^{Wx} = -\left(6(R_{y0}) - \sum_{i=0}^{6}(6-i)W_i\right)25$ (kip-ft)	−(75,724 − 27,178) = −48,546	−(78,569 − 29,888) = −48,681	−(83,816 − 35,032) = −48,784
$W_{(5)}$ (kips)	151.28	146.88	126.40
$M_{(5)}^{P} = \dfrac{-W_{(5)}(d)}{2}$ (kip-ft)	1891	−146.88(25)/2 = 1836	−126.40(25)/2 = 1580
$F_{U4U6} = \dfrac{M_{(6)}^{Wx} - 2M_{(5)}^{P}}{h}$ (kips)	(−48,546 − 2(1891))/50 = −52,328/50 = −1046.6	−48,681 − 2(1836))50 = −52,353/50 = −1047.1	(−48,784 − 2(1580))/50 = −51,944/50 = −1038.9

The maximum force, F^{max}_{U4U6}, in top chord member U4U6 due to Cooper's E80 live load is −1047.1 kips with wheel 14 at joint L5.

The forces in subtie members M(j)U(j + 1) of the BST span are calculated by considering the sum of the moments of the forces in panel (j − 1) − (j + 1) at joint M(j) about joint U(j − 1), as shown in Figure 4.20

Resolving $F_{M(j)U(j+1)}$ at U(j + 1) yields that the vertical component of the force in subtie member M(j)U(j + 1) is equal to half the load at joint L(j), $W_{(j)}/2$, such that

$$F_{M(j)U(j+1)} = F_{M(j)L(j)}\left(\frac{\sqrt{\left(\frac{h}{2}\right)^2 + d^2}}{h}\right) = W_{(j)}\left(\frac{\sqrt{\left(\frac{h}{2}\right)^2 + d^2}}{h}\right) \tag{4.56a}$$

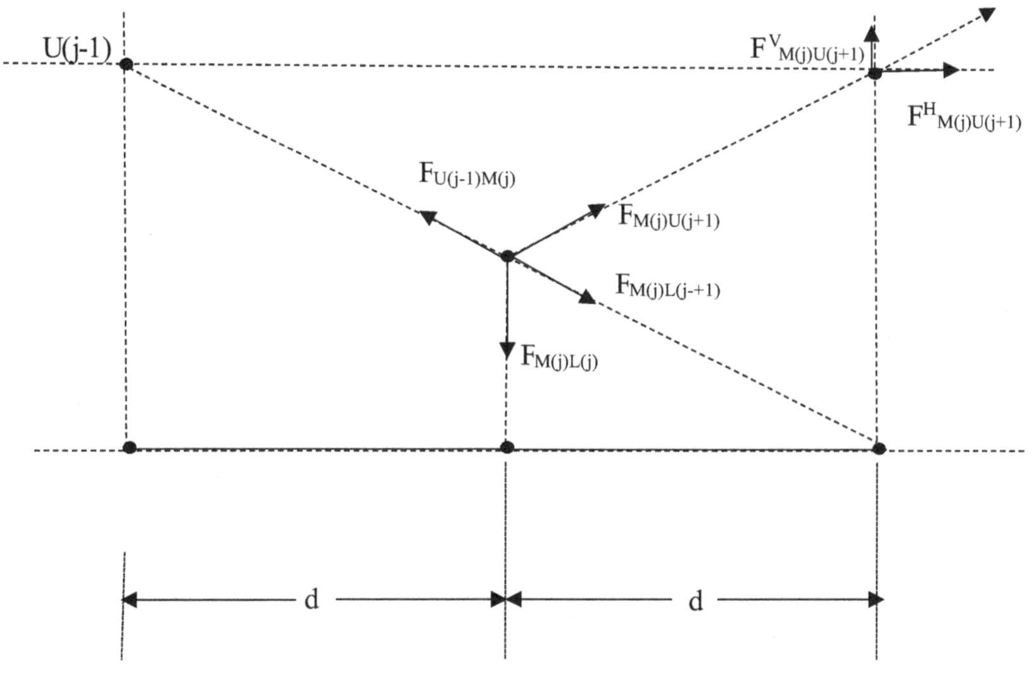

FIGURE 4.20 Panel in subtie subdivided Pratt (Baltimore) truss.

For a uniformly distributed live load, the entire panel (j – 1) – (j + 1) is loaded and for a series of concentrated loads, the criterion of Equation 4.11c with k = j is used to find the maximum value of $W_{(j)}$.

A section through subpanel (j – 1) – (j) reveals that the maximum panel shear forces in upper diagonal member U(j – 1)M(j) of the BST span due to concentrated live loads are calculated in the same was as for a horizontal chord Pratt truss with the criteria of Equations 4.12c and 4.13c or a BSS truss with the criteria of Equations 4.33a and 4.33b for positive shear and negative shear, respectively. These criteria can only occur with a concentrated load at panel point (j – 1) for positive shear and a concentrated load at panel point (j) for negative shear.

A section through subpanel (j) – (j + 1) reveals that the vertical component of the panel shear forces in lower diagonal member M(j)L(j + 1) is calculated as

$$F^{+V}_{M(j)L(j+1)} = V_{(j)-(j+1)} + F^{V}_{M(j)U(j+1)} = R_{y0} - \sum_{i=0}^{j} W_i + F^{V}_{M(j)U(j+1)} \tag{4.57a}$$

Substituting Equation 4.56a into Equation 4.57a provides

$$F^{+}_{M(j)L(j+1)} = \left(R_{y0} - \sum_{i=0}^{j} W_i + \frac{F_{M(j)L(j)}}{2}\right)\frac{\sqrt{\left(\frac{h}{2}\right)^2 + d^2}}{\frac{h}{2}} = \left(R_{y0} - \sum_{i=0}^{j} W_i + \frac{W_{(j)}}{2}\right)\frac{\sqrt{\left(\frac{h}{2}\right)^2 + d^2}}{\frac{h}{2}} \tag{4.57b}$$

For uniformly distributed live loads, Equation 4.57b provides that the maximum positive shear in subpanel (j) – (j + 1) occurs with the uniformly distributed live load at $x^{+}_{(j)\text{-}(j+1)}$ to the left of panel point (j + 1), where

$$x^{+}_{(j)-(j+1)} = \frac{(n-j-1)d}{(n-2)} \tag{4.57c}$$

Similarly, the maximum negative shear in subpanel (j) – (j + 1) for uniformly distributed live loads occurs with the uniformly distributed live load at $x^{-}_{(j)\text{-}(j+1)}$ to the right of panel point (j), where

$$x^{-}_{(j)-(j+1)} = d - x^{+}_{(j)-(j+1)} \tag{4.57d}$$

For concentrated live loads, Equation 4.57b delivers the equivalent criteria of Equation 4.37b for positive shear to locate the Cooper's E load wheel such that for positive shear in subpanel (j) – (j + 1),

$$\frac{P_T^+}{n} = \left(\frac{P_A^+ + P_B^+}{2}\right) \tag{4.57e}$$

and, in accordance with Equation 4.38b for negative shear in subpanel (j) – (j + 1),

$$\frac{P_T^-}{n} = \left(\frac{P_A^- + P_B^-}{2}\right) \tag{4.57f}$$

where for the BST truss,

n = the number of subpanels of equal length, d;

j = the panel point at vertical hanger member M(j)L(j) between panels (j – 1) – (j) and (j) – (j + 1);

P^{+}_{A} = the resultant concentrated loads in subpanel (j – 1) – (j) at distance X^{+}_{A} from (j) for positive shear;

P^{+}_{B} = resultant concentrated loads in subpanel (j) – (j + 1) at X^{+}_{B} from (j + 1) for positive shear;

P^{-}_{A} = the resultant concentrated loads in subpanel (j – 1) – (j) at X^{-}_{A} from (j – 1) for negative shear; and

P^{-}_{B} = resultant concentrated loads in subpanel (j) – (j + 1) at X^{-}_{B} from (j) for negative shear.

The criterion of Equation 4.57e can only occur with a concentrated load at panel point (j + 1), and the criterion of Equation 4.57f can only occur with a concentrated load at j – 1.

Example 4.44

Determine the maximum forces due to positive shear produced by a uniformly distributed live load of 4.0 kips/ft and Cooper's E80 live load in lower diagonal member M3L4 of the 300 ft long twelve-subpanel Baltimore truss with height of 50 ft: j = 3 and d = 300/12 = 25 ft. For a uniformly distributed live load,

$$x^{+}_{(3)-(4)} = \frac{(12-3-1)25}{(12-2)} = 20$$

$$F^{+}_{M3L4} = \left(R_{y0} - \sum_{i=0}^{3} W_i + \frac{W_{(3)}}{2}\right)\frac{\sqrt{\left(\frac{50}{2}\right)^2 + d^2}}{\frac{h}{2}} = \left(R_{y0} - \frac{W_{(3)}}{2}\right)(1.414)$$

$$R_{y0} = \frac{w\left(\left(n-(j+1)\right)d + x^{+}_{(j)-(j+1)}\right)^2}{2(nd)} = \frac{4.0\left((12-3-1)25+20\right)^2}{2(300)} = 322.67 \text{ kips}$$

$$W_{(j+1)} = \frac{w\left(x^{+}_{(j)-(j+1)}\right)\left(d - \frac{x^{+}_{(j)-(j+1)}}{2}\right)}{d} + \frac{wd}{2}$$
$$= W_{(4)} = \frac{4.0(20)(25-10)}{25} + \frac{4.0(25)}{2}$$
$$= 48.0 + 50.0 = 98.0 \text{ kips}$$

$$W_{(j)} = \frac{w\left(x^{+}_{(j)-(j+1)}\right)^2}{2d} = W_{(3)} = \frac{4.0(20)^2}{2(25)} = 32.0 \text{ kips}$$

$$F^{+max}_{M3L4} = \left(322.67 - \frac{32.0}{2}\right)(1.414) = 306.7(1.414) = 433.6 \text{ kips}$$

For Cooper's E80 live load, in accordance with the criterion of Equation 4.57e, the maximum force due to positive shear in lower diagonal member M3L4 occurs with the criterion that $\frac{P_T^+}{6} = P_A^+ + P_B^+$, which occurs with axle 5 at L4 as shown in Example 4.31. Therefore, $P^+_A = 0$ and $F^{+}_{M3L4} = \left(R_{y0} - \frac{W_{(3)}}{2}\right)(1.414)$ as shown in Table E4.44.

TABLE E4.44
Maximum Positive Shear in Member M3L4 due to Cooper's E80 Live Load Traversing a BST Span

M3L4 Positive Shear	
Wheel at L4	**5**
Sum of moments about R_{y12} (kip-ft)	32,728 + 568(114) + 4.0(114)(114/2) = 123,472
R_{y0} (kips)	123,472/300 = 411.57
$F_{L2M3} = W_{(3)}$ (kips)	1660/25 = 66.40
$F^{+max}_{M3L4} = \left(R_{y0} - \frac{W_{(3)}}{2}\right)\frac{\sqrt{h^2+(2d)^2}}{h}$ (kips)	(411.57 − 33.20)(1.414) = 535.0

$$F^{+max}_{M3L4} = 535.0 \text{ kips}$$

The forces in end vertical member U2L2 of the BST span can be developed as for the BSS truss such that for a uniformly distributed load, the maximum force in member U2L2 is

$$F_{U2L2} = \frac{3W}{2} \quad (4.58a),$$

and for a series of concentrated loads, the maximum force in U2L2 is

$$F_{U2L2} = \frac{3M_c}{2d} \quad (4.58b)$$

where M_c is the moment at joint L2 in a beam of span $L_c = (j + 1)d = 3d$ from joint L0 to joint L3 in the BST span. The maximum force in member U2L2 is determined with concentrated loads positioned in accordance with the criterion of Equation 4.11c for maximum moment at joint L2 in the span L_c.[144] The minimum force in U2L2 is due to dead load.

Equilibrium at joint U(k) indicates that the force in interior vertical member U(k)L(k) of the BST span, where k is an even number equal to or greater than 4 and less than n/2, depends on the vertical components of the force in subtie member M(k – 1)U(k) and upper diagonal member U(k)M(k + 1) and the applied force at joint U(k).[145] As shown in the development of Equation 4.56a, the vertical component of the force in subtie member M(k – 1)U(k) is $W_{(k-1)}/2$. The vertical component of the load in upper diagonal member U(k)M(k + 1) is $V_{(k)-(k+1)}$, and the shear in subpanel (k) – (k + 1). Therefore, the force in interior vertical member U(k)L(k) of the BST span due to positive shear is

$$F^{+}_{U(k)L(k)} = -\frac{W_{(k-1)}}{2} - V^{+}_{(k)-(k+1)} \quad (4.59a),$$

and the force in interior vertical member U(k)L(k) due to negative shear is

$$F^{-}_{U(k)L(k)} = -\frac{W_{(k-1)}}{2} + V^{-}_{(k)-(k+1)} \quad (4.59b).$$

For a uniformly distributed load, $F^{+max}_{U(k)L(k)}$ in the BST span occurs with the uniform load from L(n) to partially within subpanel (k) – (k + 1) or with the uniform load from L(n) to partially within subpanel (k – 2) – (k – 1).[146] For uniformly distributed load, $F^{-max}_{U(k)L(k)}$ in the BST span occurs with the uniform load from L0 to partially within subpanel (k) – (k + 1) or with the uniform load from L0 to partially within subpanel (k – 2) – (k – 1).[147]

For concentrated loads, $F^{+max}_{U(k)L(k)}$ in the BST span occurs with the concentrated loads from L(n) to within subpanel (k) – (k + 1) with the load position criteria of Equations 4.12c and 4.33b that $\frac{P^{+}_{T}}{n} = P^{+}_{(k)-(k+1)} = P^{+}_{B}$. The criteria of Equations 4.12c and 4.33b can only occur with a concentrated load at panel point (k + 1).

For concentrated loads, $F^{+max}_{U(k)L(k)}$ in the BST span can also occur with the concentrated loads from L(n) to within subpanel (k – 2) – (k – 1) with the load position criteria that

$$V^{+}_{(k)-(k+1)} = R_{y0} - \sum_{i=0}^{k} W_i = \frac{P^{+}_{T}X^{+}_{T}}{nd} - P^{+}_{F} - P^{+}_{A} - P^{+}_{C} - \frac{P^{+}_{B}X^{+}_{B}}{d}$$

$$W_{(k-1)} = \frac{P^{+}_{C}\left(d - X^{+}_{C}\right)}{d} + \frac{P^{+}_{A}X^{+}_{A}}{d}$$

$$F^{+}_{U(k)L(k)} = -\frac{W_{(k-1)}}{2} - V^{+}_{(k)-(k+1)}$$

$$= -\frac{P^{+}_{C}\left(d - X^{+}_{C}\right)}{2d} - \frac{P^{+}_{A}X^{+}_{A}}{2d} - \frac{P^{+}_{T}X^{+}_{T}}{nd} + P^{+}_{F} + P^{+}_{A} + P^{+}_{C} + \frac{P^{+}_{B}X^{+}_{B}}{d}$$

where for the BST span,

n = the total number of subpanels in the truss span;
P^{+}_{T} = the resultant live loads from panel point (0) to panel point (n) at distance X^{+}_{T} from panel point (n);
P^{+}_{A} = the resultant live loads in panel (k – 1) – (k) at X^{+}_{A} from panel point (k);
P^{+}_{B} = the resultant live loads in panel (k) – (k + 1) at X^{+}_{B} from (k + 1);
P^{+}_{C} = the resultant of the live loads in panel (k – 2) – (k – 1) at X^{+}_{C} from (k – 1); and
P^{+}_{F} = the resultant live loads from panel point (0) to panel point (k – 2) at X^{+}_{F} from (k – 2).

$F^{+max}_{U(k)L(k)}$ occurs when $\frac{d(F^{+}_{U(k)L(k)})}{dx} = 0$, which yields

$$\frac{P^{+}_{T}}{n} = \frac{P^{+}_{C}}{2} - \frac{P^{+}_{A}}{2} + P^{+}_{B} \tag{4.59c}$$

The criterion of Equation 4.59c can only occur with a concentrated load at panel point (k + 1) or (k – 1).

For concentrated loads, $F^{-max}_{U(k)L(k)}$ in the BST span occurs with the concentrated loads from L0 to within subpanel (k) – (k + 1) with the load position criteria of Equations 4.13c and 4.33d that $\frac{P^{-}_{T}}{n} = P^{-}_{(k)-(k+1)} = P^{-}_{B}$. For concentrated loads, $F^{-max}_{U(k)L(k)}$ in the BST span also occurs with concentrated loads from L0 to within subpanel (k – 2) – (k – 1) with the following load position criteria:

$$V^{-}_{(k)-(k+1)} = R_{yn} - \sum_{i=k+1}^{n} W_i = \frac{P^{-}_{T}X^{-}_{T}}{nd} - P^{-}_{F} - P^{-}_{A} - P^{-}_{C} - \frac{P^{-}_{B}X^{-}_{B}}{d}$$

$$F^{-}_{U(k)L(k)} = -\frac{W_{(k-1)}}{2} + V^{-}_{(k)-(k+1)} = V^{-}_{(k)-(k+1)} = \frac{P^{-}_{T}X^{-}_{T}}{nd} - P^{-}_{F} - P^{-}_{A} - P^{-}_{C} - \frac{P^{-}_{B}X^{-}_{B}}{d}$$

where for the BST span,

P^{-}_{T} = the resultant live loads from panel point (0) to panel point (n) at distance X^{-}_{T} from panel point (0)
P^{-}_{A} = the resultant live loads in panel (k + 1) – (k + 2) at X^{-}_{A} from panel point (k + 1);
P^{-}_{B} = the resultant live loads in panel (k) – (k + 1) at X^{-}_{B} from panel point (k);
P^{-}_{C} = the resultant live loads in panel (k + 2) – (k + 3) at X^{-}_{C} from panel point (k + 2); and
P^{-}_{F} = the resultant live loads from panel point (n) to panel point (k + 3) at X^{-}_{F} from panel point (k + 3).

$F^{-max}_{U(k)L(k)}$ occurs when $\frac{d(F^{-}_{U(k)L(k)})}{dx} = 0$, which yields

$$\frac{P^{-}_{T}}{n} = P^{-}_{B} \tag{4.59d}$$

which is the same as the criteria expressed in Equations 4.13c and 4.33d. The criterion of Equation 4.59d can only occur with a concentrated load at panel point (k).

The maximum force in member U(n/2)L(n/2) (central vertical member) in a BST truss are calculated the same way as for interior members U(k)L(k). For a uniformly distributed live load, the entire panels (n/2 – 2) – (n/2 + 2) are loaded, and for a series of concentrated loads, the criterion of Equation 4.11c with k = (n/2 – 1) is used to find the maximum $W_{(k-1)}$, such that from equilibrium at joint U(n/2),

$$F_{U\left(\frac{n}{2}\right)L\left(\frac{n}{2}\right)} = \frac{-W_{(k-1)} - W_{(k+1)}}{2} \quad (4.59e)$$

Example 4.45

Determine the maximum forces in vertical members U2L2, U4L4 and U6L6 due to a uniformly distributed live load of 4.0 kip/ft and Cooper's E80 live load on the 300 foot long twelve-subpanel BST truss with height of 50 ft.

For member U2L2 with uniformly distributed live load (Equation 4.58a):

$$F_{U2L2}^{max} = \frac{3(4.0)(25)}{2} = 150 \text{ kips}$$

For U2L2 with Cooper's E80 live load (Equation 4.58b), the location of the load for maximum $M_{(2)}$ is as shown in Table E4.45a.

TABLE E4.45a
Criterion for Maximum Moment at L2 in a BST Span from L0 to L3

	Axle at L2				
L2: k/n = 2/3	3	4	5	6	7
Last axle on length d = 25 ft	7	7 (8 at L3)	8 (9 at L3)	10	10
Distance from last axle on span to L3 (ft)	25 − (37 − 13) = 1	25 − (37 − 18) = 6	25 − (43 − 23) = 5	25 − (56 − 32) = 1	25 − (56 − 37) = 6
Distance to axle no. 1 (ft)	13	18	23	32	37
First axle on length 2d = 50 ft	1	1	1	1	1
Distance from first axle on span to L0 (ft)	50 − 13 = 37	50 − 18 = 32	50 − 23 = 27	50 − 32 = 18	50 − 37 = 13
Total load on span, P_T (kips)	232	232	258	304	304
P_T/1.5 (kips)	154.7	154.7	172	202.7	202.7
Sum of loads ahead of panel point L2, P_{Fa}	60	100	140	180	206
Sum of loads ahead of and including panel point L2, P_{Fb}	100	140	180	206	232
Criterion established	No	No	Yes	Yes	No (marginal)

Equation 4.58b and Table 4.2 are used to determine panel point moments and floorbeam reaction at joint L2 as shown in Table E4.45b.

TABLE E4.45b
Moment at Panel Point L2 of BST Span from L0 to L3

Moments at L2	Axle at L2: 4	5	6	7
Sum of moments to L3 = $M_{(3)}$ (kip-ft)	5702	6992	9264 + 304(1) = 9568	9264 + 304(6) = 11,088
Reaction R_{y0} for span of length 3d = 75 ft (kips)	5702/75 = 76.03	6992/75 = 93.23	9568/75 = 127.57	147.84
Sum of moments from first axle to axle at L2 (kip-ft)	−960	−1660	−3280	−4310
Sum of moments about L2 = $M_{(2)}$ (kip-ft)	76.03(50) − 960 = 2841.3	93.23(50) − 1660 = 3001.3	127.57(50) − 3280 = 3098.7	147.84(50) − 4310 = 3082.0
Floorbeam reaction at L2 = $F_{U2L2} = \frac{3M_{(2)}}{2(25)}$ (kips)	170.48	180.80	185.92	184.92

F^{max}_{U2L2} = 185.92 kips (exclusive of F^{max}_{U2L2} due to alternate live load)

For member U4L4 with uniformly distributed live load (Equations 4.59a and 4.59b), for positive shear, with the approximate method indicates live load W from panel point L5 to panel point L(n) is as follows:

$$R_{y0} = 4.0(25)((12\text{-}5)!)(25)/300 = 233.33 \text{ kips}$$

$$V^+_{(4)-(5)} = R_{y0} - \sum_{i=0}^{4} W_i = R_{y0} = 233.33 \text{ kips}$$

$$F^+_{U4L4} = -\frac{W_{(3)}}{2} - V^+_{(4)-(5)} = -V^+_{(4)-(5)} = -233.33 \text{ kips}$$

For positive shear, the approximate method indicates W from panel point L3 to panel point L(n) as follows:

$$R_{y0} = 4.0(25)((12\text{-}3)!)(25)/300 = 375.00 \text{ kips}$$

$$V^+_{(4)-(5)} = R_{y0} - \sum_{i=0}^{4} W_i = R_{y0} - W_{(3)} - W_{(4)}$$

$$= 375.00 - 2(4.0)(25) = 175.00 \text{ kips}$$

$$F^+_{U4L4} = -\frac{W_{(3)}}{2} - V^+_{(4)-(5)} = -\frac{4.0(25)}{2} - 175.00 = -225.00 \text{kips}$$

Therefore, .$F^{+max}_{U4L4} = -233.33 \text{kips}$

For negative shear, the approximate method indicates W from L0 to L4 as follows:

$$R_{y12} = 4.0(25)4!(25)/300 = 83.33 \text{ kips}$$

$$V^-_{(4)-(5)} = R_{y12} - \sum_{i=5}^{12} W_i = R_{y12} = 83.33 \text{ kips}$$

$$F^-_{U4L4} = V^-_{(4)-(5)} = \frac{-4.0(25)}{2} + 83.33 = 33.33 \text{ kips}$$

For negative shear, the approximate method indicates W from L0 to L2 as

$$R_{y12} = 4.0(25)2!(25)/300 = 25.00 \text{ kips}$$

$$V^-_{(4)-(5)} = R_{y12} - \sum_{i=5}^{12} W_i = R_{y12} = 25.00 \text{kips}$$

$W_{(3)} = 0$ for live load from panel point L0 to panel point L2

$$F^-_{U4L4} = V^-_{(4)-(5)} = 25.00 \text{kips}$$

Therefore, $F^{-max}_{U4L4} = 33.33 \text{kips}$.

For member U4L4 with Cooper's E80 live load (Equations 4.59a and 4.59b), the criterion for positive shear of Equation 4.12c or Equation 4.33b with load at k + 1 yields the following, as also shown in Table E4.45c:

$$\frac{P^+_T}{12} = P^+_{(k)-(k+1)} = P^+_B$$

TABLE E4.45c
Criterion for Positive Shear in Subpanel 4-5 of a BST Span

	Axle at L5		
Member U4L4 (k=4)	**2**	**3**	**4**
Distance to uniformly distributed load (ft)	(12 − 5)(25) − 101 = 74	175 − 96 = 79	175 − 91 = 84
P_T (kips)	568 + 4.0(74) = 864	568 + 4.0(79) = 884	568 + 4.0(84) = 904
$P_T/12$ (kips)	72.0	73.7	75.3
P_{Ba} (kips)	20	60	100
P_{Bb} (kips)	60	100	140
Criterion Met	No	Yes	No

The criterion indicates axle 3 at L5 for maximum positive shear in member U4L4, as shown in Table E4.45d.

TABLE E4.45d
Maximum Force in Member U4L4 of a BST Span for Positive Shear with the Criterion of Equation 4.33b with Axle 3 at L5

	Axle 3 at L5
R_{y0} (kips)	$(29,888 + 568(79+5) + 4.2(79)^2/2)/300 = 90,082/300 = 300.27$
$W_{(3)}$ (kips)	0
$W_{(4)}$ (kips)	460/25 = 18.40
$V^+_{(4)-(5)}$ (kips)	300.27 − 18.40 = 281.87
$F^+_{U4L4} = -\frac{W_{(3)}}{2} - V^+_{(4)-5)}$ (kips)	−281.87

The criterion for positive shear of Equation 4.59c with load at panel point k − 1 yields the following, as shown in Table E4.45e:

$$\frac{P_T^+}{12} = \frac{P_C^+}{2} - \frac{P_A^+}{2} + P_B^+$$

TABLE E4.45e
Criterion for Maximum Force in Member U4L4 of a BST Span for Positive Shear with Criterion of Equation 4.59c

	Axle at L3			
Member U4L4 (k = 4)	**2**	**3**	**4**	**5**
Distance of uniformly distributed load on span (ft)	225 − 101 = 124	225 − 96 = 129	225 − 91 = 134	225 − 86 = 139
P_T (kips)	568 + 4.0(124) = 1064	568 + 4.0(129) = 1084	568 + 4.0(134) = 1104	568 + 4.0(139) = 1124
$P_T/12$ (kips)	88.7	90.3	92.0	93.7
P_{Ca} (kips)	20	60	100	140
P_{Cb} (kips)	60	100	140	180
P_{Aa} (kips)	146	132	92	78
P_{Ab} (kips)	146	132	118	104
P_{Ba} (kips)	98	72	86	100
P_{Bb} (kips)	98	72	86	100
$\left(\frac{P_C^+}{2} - \frac{P_A^+}{2} + P_B^+\right)_a$ (kips)	35.0	36.0	90.0	131.0
$\left(\frac{P_C^+}{2} - \frac{P_A^+}{2} + P_B^+\right)_b$ (kips)	55.0	56.0	97.0	138.0
Criterion Met	No	No	Yes	No

The criterion of Equation 4.59c indicates axle 4 at L3 for maximum force in member U4L4, as shown in Table E4.45f.

TABLE E4.45f
Maximum Force in Member U4L4 of a BST Span for Positive Shear with Criterion of Equation 4.59c with Axle 4 at L3

Axle 4 at L3	
R_{y0} (kips)	(29,888 + 568(134+5) + 4.2(134)²/2)/300 = 144,752/300 = 482.51
$W_{(2)}$ (kips)	960/25 = 38.40
$W_{(3)}$ (kips)	(100 − 38.40) + 40 + 1242/25 = 151.28
$W_{(4)}$ (kips)	(92 − 1242/25) + 26 + 920/25 = 105.12
$V^{+}_{(4)-(5)}$ (kips)	482.51 − 38.40 − 151.28 − 105.12 = 187.71
$F^{+}_{U4L4} = -\frac{W_{(3)}}{2} - V^{+}_{(4)-5)}$ (kips)	−151.28/2 − 187.71 = −263.35

The criterion for positive shear of Equation 4.59c with load at panel point k + 1 = 5 yields the following with axle 3 at L5, $P^{+}_{A} = P^{+}_{C} = 0$, and Equation 4.59c is

$$\frac{P_T^+}{12} = P_B^+,$$

which provides that $F^{+}_{U4L4} = -281.87$ kips as shown in Table E4.45d. Therefore $F^{+max}_{U4L4} = -281.87$ kips (with axle 3 at L5).

The criterion for negative shear of Equation 4.59d with load at k = 4 yields the following as well as Table E4.45g:

$$\frac{P_T^-}{n} = P_B^-$$

TABLE E4.45g
Criterion for Maximum Force in Member U4L4 of a BST Span for Negative Shear with Criterion of Equation 4.59d

Member U4L4 (k = 4)	Axle at L4			
	1	2	3	4
Distance to uniformly distributed load (ft)	4(25) − 109 = −9 (off span)	4(25) −101 = −1 (off span)	4(25) − 96 = 4	4(25) − 91 = 9
P_T (kips)	568	568	568 + 4.0(4) = 584	568 + 4.0(9) = 604
$P_T/12$ (kips)	47.3	47.3	48.7	50.3
P_{Ba} (kips)	0	20	60	100
P_{Bb} (kips)	20	60	100	140
Criterion Met	No	Yes	No	No

The criteria of Equations 4.13c, 4.33d and 4.59d indicate axle 2 at panel point L4 for maximum force in U4L4 due to negative shear, as shown in Table E4.45h.

TABLE E4.45h
Maximum Force in Member U4L4 of a BST Span for Negative Shear with Criterion of Equation 4.59d with Axle 2 at L4

	Axle 2 at L4
R_{y12} (kips)	(29,888 + 568(4))/300 = 32,160/300 = 107.20
$W_{(3)}$ (kips)	((3(40) + 26) − 1826/25) + (1040 + ((26)(3) + 20)2)/25 = 72.96 + 49.44 = 122.40
$V^{-}_{(4)\text{-}(5)}$ (kips)	107.20 − 160/25 = 100.80
$F^{-}_{U4L4} = -\dfrac{W_{(3)}}{2} + V^{+}_{(4)-5)}$ (kips)	−122.40/2 + 100.80 = 39.60

$$F^{-max}_{U4L4} = 39.60 \text{ kips}$$

For member U6L6 with uniformly distributed live load of 4.0 kips/ft (Equation 4.59e),

$$F_{U6L6} = \frac{-W_{(5)} - W_{(7)}}{2} = \frac{-4.0(2)(25)}{2} - 100 \text{ kips}$$

For U6L6 with Cooper's E80 live load (Equation 4.59e), as shown in Example 4.11, axle 4 at L5 provides $W^{max}_{(5)}$ = 151.28 kips.

With axle 4 at L5, Table 4.2 indicates that

$W_{(7)}$ = (344 − 258) − ((576 + (344 − 258)(25 − (64 − 43)))/25 + 2410/25 = 86 − (576 + (86)(4))/25 + 96.40 = 86 − 36.80 + 96.40 = 49.20 + 96.40 = 145.60 kips

$$F_{U6L6} = \frac{-W_{(5)} - W_{(7)}}{2} = \frac{-151.28 - 145.60}{2} = -148.44 \text{kips}$$

TABLE E4.45i
Force in Member U6L6 of a BST Span

Axle at L5	$W_{(5)}$ (kips)	$W_{(7)}$ (kips)	F_{U6L6} (kips)
3	144.88	135.60	−140.24
4	151.28	145.60	−148.44
5	146.88	150.00	−148.44
6	125.12	138.56	−131.84

As a verification step, Table E4.45i shows the results when axles 3, 5 and 6 at L5 are also investigated.

Therefore, F^{max}_{U6L6} = −148.44 kips.

Due to the configuration of the end post of the BST truss in Figure 4.18, forces in upper end post member M1U2 and lower end post member L0M1 are determined in accordance with Equations 4.41 and 4.42a, respectively. Once the force in end vertical member U2L2 is determined by Equations 4.58a or 4.58b for uniformly distributed and concentrated live loads, respectively, the force in diagonal member M1L2 can be ascertained by considering vertical force equilibrium at joint L2.

4.6.4 Subdivided Inclined Chord Pratt (Pennsylvania) Truss Spans with Subtie Members

The maximum forces in members of a Pennsylvania truss with subties (PST)[148] are calculated in the same way as the PSS trusses outlined in Section 4.6.2 and the BST trusses in Section 4.6.3.

For example, the maximum force due to concentrated loads in bottom chord members of PST trusses are calculated with the criteria of Equation 4.54b (Equation 4.11c with $k = j - 1$) and Equation 4.54a with $h = h_{(j-1)}$. Likewise, the maximum force due to concentrated loads in top chord members of PST trusses are calculated with the criteria of Equations 4.55b and 4.11e with $h = h_{(j+1)}$, with due consideration of the top chord inclination such that[149]

$$F_{U(j-1)U(j+1)} = \frac{M^{Wx}_{(j+1)} + 2M^{P}_{(j)}}{h_{(j+1)}} \left(\frac{\sqrt{\left(h_{(j+1)} - h_{(j-1)}\right)^2 + (2d)^2}}{2d} \right) \tag{4.60}$$

Example 4.46

Determine the maximum forces produced by Cooper's E80 live load in bottom chord member L5L6 and top chord member U4U6 of the 300 foot long twelve-subpanel PST truss The top chord is inclined with truss height of 25 ft at joints 1 and 11, 38 ft at joints 2 and 10, 47.5 ft at joints 4 and 8 and 50 ft at joint 6. Subvertical member M1L1 is 25 ft long, subvertical member M3L3 is 19 ft long and subvertical member M5L5 is 23.75 ft long, indicating the upper and lower diagonal members in each panel are aligned.

The maximum force in member L5L6 (j= 5) due to Cooper's E80 live load occurs with the criterion of Equation 4.54b that $\frac{P_T}{3} = P_F$, which in accordance with Examples 4.26 and 4.37 occurs with axle 13 at L4. Therefore, from Example 4.37,

$$F^{max}_{L5L6} = \frac{\Sigma M_{(j-1)}}{h_{(j-1)}} = \frac{\Sigma M_4}{47.5} = \frac{44,963}{47.5} = 946.6 \text{ kips}$$

The maximum force in top chord member U4U6 (j=5) due to Cooper's E80 live load occurs with the criterion of Equation 4.55e that $\frac{P_T}{2} = P_F + 2P_B$, which in accordance with Example 4.43 occurs with axle 14 at L5. Therefore, from Example 4.43 and Equation 4.60,

$$F^{max}_{U4U6} = -1047.1 \left(\frac{\sqrt{(50 - 47.5)^2 + (50)^2}}{50} \right) = -1048.4 \text{ kips}.$$

The forces in subtie members M(j)U(j+1) of a PST span are calculated by considering the sum of the moments of the forces in panel (j – 1) – (j + 1) at joint M(j) about a point coincident with a horizontal line from joint U(j + 1) and a line coextensive with the upper and lower diagonal members U(j – 1)M(j) and M(j)L(j + 1) such that, as shown in Figure 4.21,[150]

$$d_{S1} = 2d \left(\frac{h_{(j+1)}}{h_{(j-1)}} - 1 \right) \tag{4.61a}$$

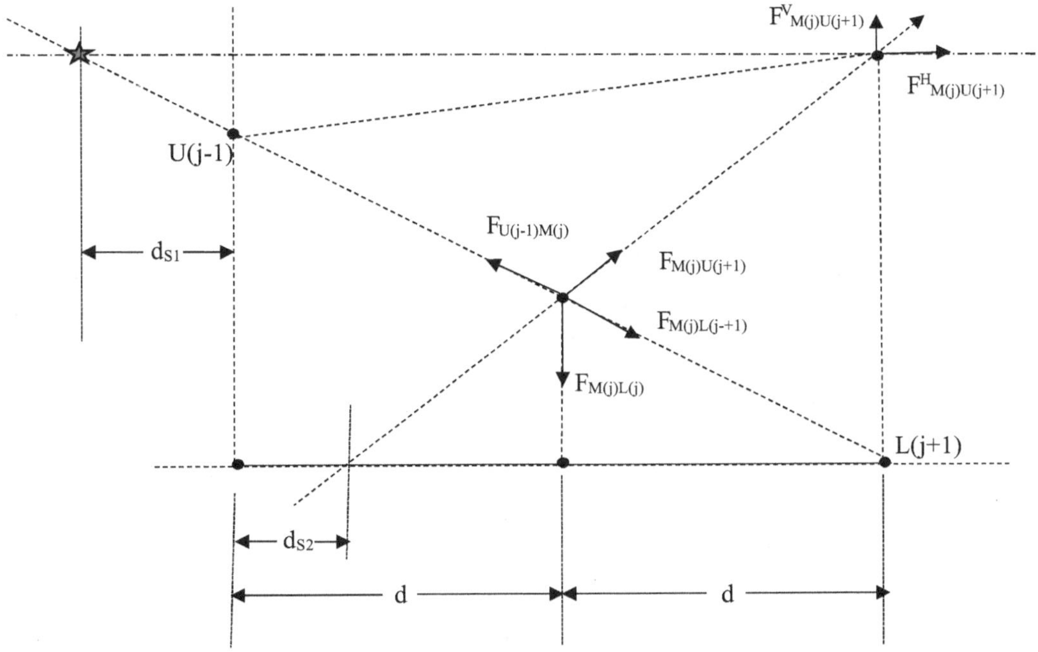

FIGURE 4.21 Panel in subtie subdivided inclined chord Pratt (Pennsylvania) truss.

$$d_{S2} = d\left(2 - \frac{h_{(j+1)}}{h_{(j+1)} - \frac{h_{(j-1)}}{2}}\right) \tag{4.61b}$$

Resolving $F_{M(j)U(j+1)}$ at joint U(j+1) yields

$$F_{M(j)U(j+1)} = F_{M(j)L(j)}\left(1 - \frac{h_{(j-1)}}{2h_{(j+1)}}\right)\left(\frac{h_{(j+1)}}{\sqrt{\left(h_{(j+1)}\right)^2 - \left(\frac{h_{(j-1)}}{2}\right)^2 - d^2}}\right)$$

$$= W_{(j)}\left(1 - \frac{h_{(j-1)}}{2h_{(j+1)}}\right)\left(\frac{h_{(j+1)}}{\sqrt{\left(h_{(j+1)}\right)^2 - \left(\frac{h_{(j-1)}}{2}\right)^2 - d^2}}\right) \tag{4.61c}$$

The maximum forces in subvertical members M(j)L(j) and subtie members M(j)U(j + 1) of the PST span due to Cooper's E loading are calculated with the criterion of Equation 4.11c with k = j to determine maximum floorbeam reaction $W_{(j)}$ in panel (j − 1) − (j + 1).

Example 4.47

Determine the maximum forces produced by Cooper's E80 live load in subtie member M5U6 of the 300 foot long twelve-subpanel PST truss with height of 50 ft. The top chord is inclined with truss height of 25 ft at joints 1 and 11, 38 ft at joints 2 and 10, 47.5 ft at joints 4 and 8 and 50 ft at joint 6. Subvertical member M1L1 is 25 ft long, subvertical member M3L3 is 19 ft long and subvertical member M5L5 is 23.75 ft long, indicating the upper and lower diagonal members in each panel are aligned.

From Equation 4.14f and Example 4.11, the maximum force in subvertical member M5L5 = $W_{(5)}$ = 2(1891)/25 = 151.28 kips (156.0 kips considering the alternate live load). The maximum force in subtie member M5U6 is

$$F_{M5U6} = \left(W_{(5)} \left(1 - \frac{h_{(4)}}{2h_{(6)}} \right) \right) \left(\frac{h_{(6)}}{\sqrt{\left(h_{(6)}\right)^2 - \left(\frac{h_{(4)}}{2}\right)^2 - d^2}} \right)$$

$$= \left(151.28 \left(1 - \frac{47.5}{2(50)} \right) \right) \left(\frac{50.0}{\sqrt{(50.0)^2 - \left(\frac{47.5}{2}\right)^2 - 25^2}} \right)$$

$$= 79.42(1.38) = 109.68 \text{ kips} (113.10 \text{ kips considering the alternate live load})$$

The forces in upper diagonal member U(j – 1)M(j) in subpanel (j – 1) – (j) of a PST truss are calculated by taking a section through subpanel (j – 1) – (j) and resolving $F_{U(j-1)M(j)}$ into horizontal and vertical components at joint L(j + 1) to preclude the moment about point P due to the horizontal component of $F_{U(j-1)M(j)}$. Therefore, the force in upper diagonal member U(j – 1)M(j) of a PST truss due to positive shear in subpanel (j – 1) – (j) is

$$F^{+}_{U(j-1)M(j)} = \left(\frac{V^{+}_{(j-1)-(j)} \left(a_{(j-1)-(j)} \right) - M^{+0}_{(j-1)-(j)}}{\left(a_{(j-1)-(j)} + (j+1)d \right)} \right) \propto_{(j-1)-(j)} \quad (4.62a),$$

and the force in upper diagonal member U(j – 1)M(j) of a PST truss due to negative shear in subpanel (j – 1) – (j) is

$$F^{-}_{U(j-1)M(j)} = \left(\frac{-V^{-}_{(j-1)-(j)} \left(a_{(j-1)-(j)} \right) + M^{-0}_{(j-1)-(j)}}{a_{(j-1)-(j)} + (j+1)d} \right) \alpha_{(j-1)-(j)} \quad (4.62b)$$

where

$$V^{+}_{(j-1)-(j)} = R_{y0} - \sum_{i=0}^{j-1} W_i \;;$$

$$M^{+0}_{(j-1)-(j)} = \left(\sum_{i=0}^{j-1} iW_i\right)d;$$

$$V^{-}_{(j-1)-(j)} = R_{yn} - \sum_{i=j}^{n} W_i; \text{ and}$$

$$M^{-0}_{(j-1)-(j)} = -\left(nR_{yn} - \sum_{i=j}^{n} iW_i\right)d.$$

The location of a uniformly distributed live load for positive shear in subpanel (j – 1) – (j) is given by Equation 4.22b and for negative shear in subpanel (j – 1) – (j) by Equation 4.23b. The location of concentrated live loads for positive shear in subpanel (j – 1) – (j) is given by Equation 4.26g and for negative shear in subpanel (j – 1) – (j) by Equation 4.27f such that

$$\frac{P_T^+}{n} = P_B^+\left(1 + \frac{(j-1)d}{a_{(j-1)-(j)}}\right) - \frac{P_F^+ d}{a_{(j-1)-(j)}} \tag{4.62c}$$

$$\frac{P_T^-}{n} = P_B^-\left(\frac{1 + \dfrac{jd}{a_{(j-1)-(j)}}}{1 + \dfrac{nd}{a_{(j-1)-(j)}}}\right) + \frac{P_F^-}{n\left(1 + \dfrac{a_{(j-1)-(j)}}{nd}\right)} \tag{4.62d}$$

The criterion of Equation 4.62c can only occur with a concentrated load at panel point (j), and the criterion of Equation 4.62d can only occur with a concentrated load at panel point (j – 1).

Example 4.48

Determine the maximum forces produced by Cooper's E80 live load in upper diagonal members U2M3 and U4M5 of the 300 foot long twelve-subpanel PST truss with height of 50 ft. The top chord is inclined with truss height of 25 ft at joints 1 and 11, 38 ft at joints 2 and 10, 47.5 ft at joints 4 and 8 and 50 ft at joint 6. Subvertical member M1L1 is 25 ft long, subvertical member M3L3 is 19 ft long and subvertical member M5L5 is 23.75 ft long.

For U2M3, use Table 4.2 to determine positive and negative shear forces in subpanel 2-3 with Cooper's E live load. Considering positive shear and j = 3,

$$a_{(2)-(3)} = a_{(3)-(4)} = \left(\left(\frac{2(47.5)}{47.5-38.0}\right) - (3+1)\right)25 = 150\,\text{ft}$$

The criterion of Equation 4.62c with $P^+_F = 0$ is as shown in Table E4.48a and as follows:

$$\frac{P_T^+}{12} = P^+_{(2)-(3)}\left(1 + \frac{(3-1)d}{a_{(2)-(3)}}\right) = P^+_{(2)-(3)}\left(1 + \frac{(3-1)25}{150}\right) = 1.33P^+_{(2)-(3)}$$

or

$$\frac{P_T^+}{16} = P^+_{(2)-(3)}$$

TABLE E4.48a

Criterion for Live Load Position for Maximum Positive Force in Member U2M3 due to Cooper's E80 Live Load Traversing a PST Span

	Subpanel 2-3		
Axle at L3	**2**	**3**	**4**
Distance to axle no. 1 (ft)	8	13	18
First axle on span	1	1	1
Distance to axle no. 18 (ft)	96	91	91 – 5 = 86
Last axle on span	18	18	18
Distance of uniform load (ft)	(12 – 3)(25) – 101 = 124	(12 – 3)(25) – 96 = 129	225 – 91 = 134
Total load on span, P^+_T (kips)	568 + 124(4) = 1064	568 + 129(4) = 1084	568 + 134(4) = 1104
$P^+_T/16$ (kips)	66.5	67.8	69.0
$P^+_{a(2)\text{-}(3)}$ = Sum of loads ahead of panel point L3 (kips)	20	60	100
$P^+_{b(2)\text{-}(3)}$ = Sum of loads ahead of and including panel point L3 (kips)	60	100	140
Criterion established	No	Yes	No

The criterion indicates axle 3 at L3 for maximum positive force in U2M3. Therefore, the assumption that $P^+_F = 0$ is valid.

From Equation 4.62a, the maximum positive force F^{+V}_{U2M3} in U2M3 is given as in Table E4.48b.

TABLE E4.48b

Maximum Force in Diagonal Web Member U2M3 from Positive Shear in Subpanel 2-3 due to Cooper's E80 Live Load Traversing a PST Span

Positive Shear in Subpanel 2-3	**Axle 3 at L3**
Sum moments to L12 (kip-ft)	32,728 + 568(129) + 4.0(129)(129/2) = 139,282
R_{y0} (kips)	139,282/300 = 464.27
Sum moments in subpanel 2-3 ahead of and about L3 (kip-ft)	460
$W_{(2)}$ (kips)	18.40
$V^+_{(2)-(3)} = R_{y0} - \sum_{i=0}^{2} W_i = R_{y0} - W_{(2)}$ (kips)	464.27 – 18.40 = 445.87
$M^{+0}_{(2)-(3)} = d\sum_{i=0}^{2} iW_i = d(2)W_{(2)}$ (kip-ft)	2(25)18.40 = 920
$F^{+V}_{U2M3}\left(a_{(2)-(3)} + 4d\right) = V^+_{(2)-(3)}a_{(2)-(3)} - M^{+0}_{(2)-(3)}$ (kip-ft)	445.87(150) – 920 = 65,961
F^{+V}_{U2M3} (kips)	65,961/(150 + 4(25)) = 263.84
$\propto_{(2)-(3)} = \dfrac{\sqrt{h^2_{(2)} + 4d^2}}{h_{(2)}}$	1.653
$F^+_{U2M3}\left(\alpha_{(2)-(3)}\right)$ (kips)	263.84(1.653) = 436.1

$$F_{U2M3}^{+max} = 436.1 \text{ kips}$$

Considering negative shear and j = 3, the criterion of Equation 4.62d with $P^-_F = 0$ is as given in Table E4.48c or as follows:

$$\frac{P_T^-}{12} = P_{(2)-(3)}^- \left(\frac{1 + \frac{3(25)}{150}}{1 + \frac{12(25)}{150}} \right) = 0.500 P_{(2)-(3)}^-$$

or

$$\frac{P_T^-}{6} = P_{(2)-(3)}^-$$

TABLE E4.48c
Criterion for Live Load Position for Maximum Negative Force in Member U2M3 due to Cooper's E80 Live Load Traversing a PST Span

	Subpanel 2-3			
Axle at L2	**1**	**2**	**3**	**4**
Distance to axle no. 1 (ft)	0	8	13	18
First axle on span	1	1	1	1
Last axle on span	9	10	10	11
Total load on span, P^-_T (kips)	284	304	304	344
$P^-_T/6.0$ (kips)	47.3	50.7	50.7	57.3
$P^-_{a(2)-(3)}$ = Sum of loads ahead of panel point L2 (kips)	0	20	60	100
$P^-_{b(2)-(3)}$ = Sum of loads ahead of and including panel point L2 (kips)	20	60	100	140
Criterion established	No	Yes	No	No

The criterion indicates axle 2 at L2 for maximum negative force in U2M3. Therefore, the assumption that $P^-_F = 0$ is valid.

From Equation 4.62b, the maximum negative force F^{-V}_{U2M3} in U2M3 is as given in Table E4.48d.

TABLE E4.48d
Maximum Force in Upper Diagonal Web Member U2M3 from Negative Shear in Subpanel 2-3 due to Cooper's E80 Live Load Traversing a PST Span

	Axle at L2	
Negative Shear in Subpanel 2-3	**2**	**3**
Sum moments to L0 (kip-ft)	9264 + 304(50 − (56 − 8) = 9872	9264 + 304(50 − (56 − 13)) = 11,392.0
R_{y12} (kips)	9872/300 = 32.91	11,392/300 = 37.97
Sum moments in subpanel 2-3 ahead of and about panel point L2 (kip-ft)	160	460
$W_{(3)}$ (kips)	160/25 = 6.40	460/25 = 18.40

(Continued)

TABLE E4.48d (*Continued*)
Maximum Force in Upper Diagonal Web Member U2M3 from Negative Shear in Subpanel 2-3 due to Cooper's E80 Live Load Traversing a PST Span

	Axle at L2	
Negative Shear in Subpanel 2-3	**2**	**3**
$V^-_{(2)-(3)} = R_{y12} - \sum_{i=3}^{12} W_i = R_{y12} - W_{(3)}$ (kips)	32.91 − 6.40 = 26.51	37.97 − 18.40 = 19.57
$M^{-0}_{(2)-(3)} = -\left(12R_{y12} - 3W_{(3)}\right)d$ (kip-ft)	−(12(32.91) − 3(6.40))25 = −9393	−(12(37.97) − 3(18.40))25 = −10,011
$-V^-_{(2)-(3)}\left(a_{(2)-(3)}\right) + M^{-0}_{(2)-(3)}$ (kip-ft)	−26.51(150) − 9393 = −13,369.5	−19.57(150.0) − 10,011 = −12,946.5
F^{-V}_{U2M3}(kips)	−13,369.5/(150 + 4(25)) = −53.48 (C)	−12,946.5/(150 + 4(25)) = −51.79 (C)
$F^-_{U2M3}\left(\alpha_{(2)-(3)}\right)$(kips)	−88.40 (C)	−85.60 (C)

$$F^{-max}_{U2M3} = -88.40 \text{ kips}$$

For upper diagonal web member U4M5, use Table 4.2 to determine positive and negative shear forces in subpanel 4-5 with Cooper's E live load. For considering positive shear, j = 5 and

$$a_{(4)\text{-}(5)} = a_{(4)\text{-}(6)} = \left(\left(\frac{2(50.0)}{50.0-47.5}\right) - (5+1)\right)25 = 850\,\text{ft}.$$

The criterion of Equation 4.62c with $P^+_F = 0$ is given in Table E4.48e and as follows:

$$\frac{P^+_T}{12} = P^+_{(4)-(5)}\left(1+\frac{(j-1)d}{a_{(2)-(3)}}\right) = P^+_{(4)-(5)}\left(1+\frac{(5-1)25}{850}\right) = 1.118P^+_{(4)-(5)}$$

or

$$\frac{P^+_T}{13.4} = P^+_{(4)-(5)}$$

TABLE E4.48e
Criterion for Live Load Position for Maximum Positive Force in Member U4M5 due to Cooper's E80 Live Load Traversing a PST Span

	Subpanel 4-5		
Axle at L5	**2**	**3**	**4**
Distance to axle no. 1 (ft)	8	13	18
First axle on span	1	1	1
Distance to axle no. 18 (ft)	96	91	91 − 5 = 86
Last axle on span	18	18	18

TABLE E4.48e (*Continued*)
Criterion for Live Load Position for Maximum Positive Force in Member U4M5 due to Cooper's E80 Live Load Traversing a PST Span

	Subpanel 4-5		
Axle at L5	**2**	**3**	**4**
Distance of uniform load (ft)	(12 − 5)(25) − 101 = 74	(12 − 5)(25) − 96 = 79	175 − 91 = 84
Total load on span, P^+_T (kips)	568 + 74(4) = 864	568 + 79(4) = 884	568 + 84(4) = 904
$P^+_T/13.4$ (kips)	64.5	66.0	67.5
$P^+_{a(4)-(5)}$ = Sum of loads ahead of panel point L5 (kips)	20	60	100
$P^+_{b(4)-(5)}$ = Sum of loads ahead of and including panel point L5 (kips)	60	100	140
Criterion established	No	Yes	No

The criterion indicates axle 3 at L5 for maximum positive force in U4M5. Therefore, the assumption that $P^+_F = 0$ is valid for d = 25 ft.

From Equation 4.62a, the maximum positive force F^{+V}_{U4M5} in U4M5 is given in Table E4.48f.

TABLE E4.48f
Maximum Force in Diagonal Web Member U4M5 from Positive Shear in Subpanel 4-5 due to Cooper's E80 Live Load Traversing a PST Span

Positive Shear in Subpanel 4-5	**Axle 3 at L5**
Sum moments to L12 (kip-ft)	32,728 + 568(79) + 4.0(79)(79/2) = 90,082
R_{y0} (kips)	90,082/300 = 300.27
Sum moments in subpanel 4-5 ahead of and about L5 (kip-ft)	460
$W_{(4)}$ (kips)	18.40
$V^+_{(4)-(5)} = R_{y0} - \sum_{i=0}^{4} W_i = R_{y0} - W_{(4)}$ (kips)	300.27 − 18.40 = 281.87
$M^{+0}_{(4)-(5)} = d\sum_{i=0}^{4} iW_i = d(4)W_{(4)}$ (kip-ft)	4(25)18.40 = 1840
$F^{+V}_{U4M5}\left(a_{(4)-(5)} + 6d\right) = V^+_{(4)-(5)}a_{(4)-(5)} - M^{+0}_{(4)-(5)}$ (kip-ft)	281.87(850) − 1840 = 237,750
F^{+V}_{U4M5} (kips)	237,750/(850 + 6(25)) = 237.75
$\propto_{(4)-(5)} = \dfrac{\sqrt{h^2_{(4)} + 4d^2}}{h_{(4)}}$	1.452
$F^+_{U4M5}\left(\alpha_{(4)-(5)}\right)$ (kips)	237.75(1.452) = 345.19

$$F^{+max}_{U4M5} = 345.2 \text{ kips}$$

Considering negative shear and j = 5, the criterion of Equation 4.62d with $P^-_F = 0$ is given in Table E4.48g and as follows:

$$\frac{P_T^-}{12} = P_{(4)-(5)}^- \left(\frac{1 + \dfrac{5(25)}{850}}{1 + \dfrac{12(25)}{850}} \right) = 0.848 P_{(4)-(5)}^-$$

or

$$\frac{P_T^-}{10.2} = P_{(4)-(5)}^-$$

TABLE E4.48g

Criterion for Live Load Position for Maximum Negative Force in Member U4M5 due to Cooper's E80 Live Load Traversing a PST Span

	Subpanel 4-5			
Axle at L4	**1**	**2**	**3**	**4**
Distance to axle no. 1 (ft)	0	8	13	18
First axle on span	1	1	1	1
Last axle on span	17	18	18	18
Length of uniform load (ft)	0	0	4	9
Total load on span, P^-_T (kips)	542	568	568 + 4(4.0) = 584	568 + 9(4.0) = 604
$P^-_T/10.2$ (kips)	53.1	55.7	57.3	59.2
$P^-_{a(4)-(5)}$ = Sum of loads ahead of panel point L4 (kips)	0	20	60	100
$P^-_{b(4)-(5)}$ = Sum of loads ahead of and including panel point L4 (kips)	20	60	100	140
Criterion established	No	Yes	No (marginal)	No

The criterion indicates axle 2 at L4 for maximum negative force in U4M5. Therefore, the assumption that $P^-_F = 0$ is valid for d = 25 ft.

From Equation 4.62b, the maximum negative force F^{-V}_{U4M5} in member U4M5 is given in Table E4.48h.

TABLE E4.48h

Maximum Force in Upper Diagonal Web Member U4M5 from Negative Shear in Subpanel 4-5 due to Cooper's E80 Live Load Traversing a PST Span

	Axle at L4	
Negative Shear in Subpanel 4-5	**2**	**3**
Sum moments to L0 (kip-ft)	29,888 + 568(4) = 32,160	29,888 + 568(9) + 4(4)(2) = 35,032
R_{y12} (kips)	32,160/300 = 107.20	35,032/300 = 116.77
Sum moments in subpanel 4-5 ahead of and about panel point L4 (kip-ft)	160	460
$W_{(5)}$ (kips)	160/25 = 6.40	460/25 = 18.40

TABLE E4.48h (*Continued*)
Maximum Force in Upper Diagonal Web Member U4M5 from Negative Shear in Subpanel 4-5 due to Cooper's E80 Live Load Traversing a PST Span

	Axle at L4	
Negative Shear in Subpanel 4-5	**2**	**3**
$V^{-}_{(4)-(5)} = R_{y12} - \sum_{i=5}^{12} W_i = R_{y12} - W_{(5)}$ (kips)	107.20 − 6.40 = 100.8	116.77 − 18.40 = 98.37
$M^{-0}_{(4)-(5)} = -\left(12R_{y12} - 5W_{(5)}\right)d$ (kip-ft)	−(12(107.20) − 5(6.40))25 = −31,360	−(12(116.77) − 5(18.40))25 = −32,731
$-V^{-}_{(4)-(5)}\left(a_{(4)-(5)}\right) + M^{-0}_{(4)-(5)}$ (kip-ft)	−100.8(850) − 31,360 = −117,040	−98.37(850.0) − 32,731 = −116,346
F^{-V}_{U4M5} (kips)	−117,040/(850 + 6(25)) = −117.04 (C)	−116,345/(850 + 6(25)) = −116.35 (C)
$F^{-}_{U4M5}\left(\alpha_{(4)-(5)}\right)$ (kips)	−169.94 (C)	−168.93 (C)

$$F^{-max}_{U4M5} = -169.9 \text{ kips}$$

The forces in lower diagonal member M(j)L(j + 1) due to shear in subpanel (j) – (j + 1) in a PST truss are calculated by taking a section through subpanel (j) – (j + 1) and considering equilibrium of forces about location P at the junction of the coextension of lines from bottom chord member L(j)L(j + 1) and top chord member U(j – 1)U(j + 1). Resolution of the force in subtie member M(j) U(j + 1) at location ds_2 from panel point (j – 1), as shown in Figure 4.21, and resolution of the force in lower diagonal member M(j)L(j + 1) at L(j + 1) to preclude moments at location P from the horizontal components of $F_{M(j)U(j+1)}$ and $F_{M(j)L(j+1)}$ for positive shear provide

$$\Sigma M_P = 0 = -R_{y0}\left(a_{(j-1)-(j+1)}\right) + \sum_{i=0}^{j} W_i\left(a_{(j-1)-(j+1)} + id\right) - F^{V}_{M(j)U(j+1)} \left(a_{(j-1)-(j+1)} + (j-1)d + d_{s2}\right) + F^{+V}_{M(j)L(j+1)}\left(a_{(j-1)-(j+1)} + (j+1)d\right)$$

which yields the vertical component of the force in lower diagonal member M(j)L(j+1) due to positive shear as

$$F^{+V}_{M(j)L(j+1)} = \frac{V^{+}_{(j)-(j+1)}\left(a_{(j-1)-(j+1)}\right) - M^{+0}_{(j)-(j+1)}}{\left(a_{(j-1)-(j+1)} + (j+1)d\right)} \tag{4.63a}$$

where

$$V^{+}_{(j)-(j+1)} = R_{y0} - \sum_{i=0}^{j} W_i + F^{V}_{M(j)U(j+1)}$$

$$M^{+0}_{(j)-(j+1)} = \sum_{i=0}^{j} W_i(id) - F^{V}_{M(j)U(j+1)}\left((j-1)d + d_{s2}\right)$$

From Equations 4.61b and 4.61c,[151]

$$d_{S2} = \left(2 - \frac{h_{(j+1)}}{h_{(j+1)} - \dfrac{h_{(j-1)}}{2}} \right) d$$

$$F^{V}_{M(j)U(j+1)} = W_{(j)} \left(1 - \frac{h_{(j-1)}}{2h_{(j+1)}} \right),$$

and taking a section through subpanel (j) – (j + 1) for negative shear yields

$$\Sigma M_P = 0 = -R_{yn}(a_{(j-1)-(j+1)} + nd) - F^{-V}_{M(j)L(j+1)}\left(a_{(j-1)-(j+1)} + (j+1)d\right)$$
$$+ \sum_{i=j+1}^{n} W_i \left(a_{(j-1)-(j+1)} + id\right) + F^{V}_{M(j)U(j+1)}\left(a_{(j-1)-(j+1)} + (j-1)d + d_{S2}\right)$$

such that the vertical component of the force in lower diagonal member M(j)L(j + 1) due to negative shear is

$$F^{-V}_{M(j)L(j+1)} = \frac{-V^{-}_{(j)-(j+1)}\left(a_{(j-1)-(j+1)}\right) + M^{-0}_{(j)-(j+1)}}{\left(a_{(j-1)-(j+1)} + (j+1)d\right)} \tag{4.63b}$$

where

$$V^{-}_{(j)-(j+1)} = R_{yn} - \sum_{i=j+1}^{n} W_i - F^{V}_{M(j)U(j+1)}$$

and

$$M^{-0}_{(j)-(j+1)} = (-nR_{yn} + \sum_{i=j+1}^{n} W_i\,(id) + F^{V}_{M(j)U(j+1)}\,(j-1))d + F^{V}_{M(j)U(j+1)}d_{s2}$$

For concentrated live loads, the criterion for the location of live load for maximum force in lower diagonal member M(j)L(j + 1) of a PST span due to positive shear is equivalent to that for the upper diagonal U(j – 1)M(j) in a PSS truss as stated by Equation 4.51b such that

$$\frac{P^{+}_{T}}{n} = \left(\frac{P^{+}_{A} + P^{+}_{B}}{2}\right)\left(1 + \frac{(j-1)d}{a_{(j-1)-(j+1)}}\right) - \frac{P^{+}_{F}d}{a_{(j-1)-(j+1)}} \tag{4.64a}$$

The criterion of Equation 4.64a can only occur with a concentrated load at panel point (j + 1).

When, as is typical, $P^{+}_{A} = P^{+}_{F} = 0$, Equation 4.64a is

$$\frac{P^{+}_{T}}{n} = \left(\frac{P^{+}_{B}}{2}\right)\left(1 + \frac{(j-1)d}{a_{(j-1)-(j+1)}}\right) \tag{4.64b}$$

For concentrated live loads, the criterion for the location of live load for maximum force in lower diagonal member M(j)L(j + 1) of the PST span due to negative shear is equivalent to that for the upper diagonal U(j-1)M(j) in a PSS truss as given by Equation 4.52 such that

$$\frac{P_T^-}{n} = \frac{\left(\frac{P_A^- + P_B^-}{2}\right)\left(1+\frac{(j+1)d}{a_{(j-1)-(j+1)}}\right)}{\left(\frac{nd}{a_{(j-1)-(j+1)}}+1\right)} + \frac{P_F^- d}{a_{(j-1)-(j+1)}} \tag{4.65a}$$

The criterion of Equation 4.65a can only occur with a concentrated load at panel point (j – 1).

When, as is typical, $P^-_A = P^-_F = 0$, Equation 4.65a is

$$\frac{P_T^-}{n} = \frac{\left(\frac{P_B^-}{2}\right)\left(1+\frac{(j+1)d}{a_{(j-1)-(j+1)}}\right)}{\left(\frac{nd}{a_{(j-1)-(j+1)}}+1\right)} \tag{4.65b}$$

where for a PST span,

P^+_A = sum of the concentrated loads in subpanel (j – 1) – (j);
P^+_B = sum of the concentrated loads in subpanel (j) – (j + 1);
P^-_A = sum of the concentrated loads in subpanel (j – 1) – (j); and
P^-_B = sum of the concentrated loads in subpanel (j) – (j + 1).

Equations 4.64b and 4.65b are used in Example 4.49 to determine the location of Cooper's E80 live load for maximum forces in lower diagonal member M(j)L(j + 1) of a PST span due to positive and negative shear in subpanel (j) – (j + 1), respectively.

Example 4.49

Determine the maximum forces produced by Cooper's E80 live load in lower diagonal member M3L4 of the 300 foot long twelve-subpanel PST truss with height of 50 ft. The top chord is inclined with truss height of 25 ft at joints 1 and 11, 38 ft at joints 2 and 10, 47.5 ft at joints 4 and 8 and 50 ft at joint 6. Subvertical member M1L1 is 25 ft long, subvertical member M3L3 is 19 ft long and subvertical member M5L5 is 23.75 ft long.

j = 3

$$a_{(2)-(4)} = \left(\left(\frac{2(47.5)}{47.5-38.0}\right) - (3+1)\right)25 = 150\,\text{ft}$$

Positive shear in subpanel 3-4 with criterion of Equation 4.64b (also see Table E4.49a):

$$\frac{P_T^+}{12} = \left(\frac{P_B^+}{2}\right)\left(1+\frac{(3-1)25}{150}\right)$$

or

$$\frac{P_T^+}{8} = P_B^+$$

TABLE E4.49a

Criterion for Live Load Position for Maximum Positive Shear in Subpanel 3-4 for Cooper's E80 Live Load Traversing a PST Span

	Subpanel 3-4			
Axle at L4	**2**	**3**	**4**	**5**
Distance to axle no. 1 (ft)	8	13	18	23
First axle on span	1	1	1	1
Last axle on span	18	18	18	18
Length of uniform load on span (ft)	(12 − 4)(25) − 101 = 99	(12 − 4)(25) − 96 = 104	(12 − 4)(25) − 91 = 109	(12 − 4)(25) − 86 = 114
Total load on span, P^+_T (kips)	568 + 99(4) = 964	568 + 104(4) = 984	568 + 109(4) = 1004	568 + 114(4) = 1024
$P^+_T/8.0$ (kips)	120.5	123.0	125.5	128.0
$P_B{}^+_a$ (kips)	20	60	100	140
$P_B{}^+_b$ (kips)	60	100	140	180
Criterion established	No	No	Yes	No

The criterion indicates axle 4 at L4 for maximum positive force in M3L4. Therefore, the assumption that $P^+_A = P^+_F = 0$ is valid.

The forces due to positive shear in lower diagonal member M3L4 of the PST truss are determined with Equation 4.63a and Table E4.49b:

$$F^{+V}_{M3L4} = \frac{V^+_{(3)-(4)}(150) - M^{+0}_{(3)-(4)}}{(150 + (3+1)25)} = \frac{V^+_{(3)-(4)}(150) - M^{+0}_{(3)-(4)}}{250}$$

$$V^+_{(3)-(4)} = R_{y0} - \sum_{i=0}^{3} W_i + F^{+V}_{M3U4} = R_{y0} - W_{(3)} + F^{+V}_{M3U4}$$

$$= R_{y0} - W_{(3)} + W_{(3)}\left(1 - \frac{h_{(2)}}{2h_{(4)}}\right) = R_{y0} - W_{(3)}\left(\frac{h_{(2)}}{2h_{(4)}}\right)$$

$$= R_{y0} - W_{(3)}\left(\frac{38}{2(47.5)}\right) = R_{y0} - 0.40W_{(3)}$$

Equation 4.61b,

$$d_{S2} = d\left(2 - \frac{h_{(j+1)}}{h_{(j+1)} - \frac{h_{(j-1)}}{2}}\right) = 25\left(2 - \frac{h_{(4)}}{h_{(4)} - \frac{h_{(2)}}{2}}\right) = 25\left(2 - \frac{47.5}{47.5 - \frac{38.0}{2}}\right)$$

$$= \frac{25}{3} = 8.33 \text{ ft}$$

and Equation 4.61c,

$$F^{+V}_{M(j)U(j+1)} = W_{(j)}\left(1 - \frac{h_{(j-1)}}{2h_{(j+1)}}\right) = W_{(3)}\left(1 - \frac{38.0}{2(47.5)}\right) = 0.60W_{(3)}$$

$$M^{+0}_{(3)-(4)} = \sum_{i=0}^{j} W_i(id) - F^{+V}_{M(j)U(j+1)}\left((j-1)d + d_{s2}\right)$$

$$= W_{(j)}\left(jd - \left(1 - \frac{h_{(j-1)}}{2h_{(j+1)}}\right)\left((j-1)d + d_{s2}\right)\right)$$

$$= W_{(3)}\left(3(25) - \left(1 - \frac{38}{2(47.5)}\right)\left((3-1)(25) + 8.33\right)\right)$$

$$= W_{(3)}\left(75 - (0.60)(58.33)\right)$$

$$= 40.0W_{(3)}$$

$$F^{+V}_{M3L4} == \frac{\left(R_{y0} - 0.40W_{(3)}\right)(150) - 40.0W_{(3)}}{250} = 0.60R_{y0} - 0.40W_{(3)}$$

TABLE E4.49b
Maximum Force in Member M3L4 due to Positive Shear in Subpanel 3-4 with Cooper's E80 Live Load Traversing a PST Span

	Axle at L4		
Positive Shear in Subpanel 3-4	**3**	**4**	**5**
Length of uniformly distributed load (ft)	(12 − 4)25 − 96 = 104	109	114
Sum moments to L12 (kip-ft)	32,728 + 568(104) + 4.0(104)(104/2) = 113,432	118,402	123,472
R_{y0} (kips)	113,432/300 = 378.11	394.67	411.57
Sum moments in subpanel 3-4 ahead of and about L4 (kip-ft)	460	960	1660
$W_{(3)}$ (kips)	460/25 = 18.40	38.40	66.40
$V^{+}_{(3)-(4)}$(kips)	370.75	379.31	385.01
$M^{+0}_{(3)-(4)}$ (kip-ft)	736	1536	2656

(Continued)

TABLE E4.49b (*Continued*)
Maximum Force in Member M3L4 due to Positive Shear in Subpanel 3-4 with Cooper's E80 Live Load Traversing a PST Span

	Axle at L4		
Positive Shear in Subpanel 3-4	**3**	**4**	**5**
F^{+V}_{M3L4} (kips)	219.51	221.44	220.38
$\propto_{(3)-(4)} = \frac{\sqrt{\frac{h^2_{(2)}}{4} + d^2}}{\frac{h_{(2)}}{2}}$	1.653	1.653	1.653
$F^{+}_{M3L4}\left(\alpha_{(3)-(4)}\right)$ (kips)	362.84	366.04	364.29

$$F^{+max}_{M3L4} = 366.0\,\text{kips}$$

For negative shear in subpanel 3-4 with criterion of Equation 4.65b and Table E4.49c:

$$\frac{P^-_T}{12} = \frac{\left(\frac{P^-_B}{2}\right)\left(1 + \frac{(3+1)25}{150}\right)}{\left(\frac{300}{150} + 1\right)}$$

or

$$\frac{P^-_T}{3.33} = P^-_B$$

TABLE E4.49c
Criterion for Live Load Position for Maximum Negative Shear in Subpanel 3-4 with Cooper's E80 Live Load Traversing a PST Span

	Subpanel 3-4			
Axle at L2	**2**	**3**	**4**	**5**
Distance to axle no. 1 (ft)	8	13	18	23
First axle on span	1	1	1	1
Last axle on span	10	10	11	12
Distance from last load on span to L0 (ft)	50 − (56 − 8) = 2	50 − (56 − 13) = 7	50 − (64 − 18) = 4	50 − (69 − 23) = 4
Total load on span, P^-_T (kips)	304	304	344	384
$P^-_T/3.33$ (kips)	91.3	91.3	103.3	115.3
$P_B{}^-{}_a$ (kips)	20	60	100	140
$P_B{}^-{}_b$ (kips)	60	100	140	180
Criterion established	No	Yes (marginal)	Yes (marginal)	No

The criterion indicates axle 3 or axle 4 at L2 for maximum negative force in M3L4. Therefore, the assumption that $P^{+}_{A} = P^{+}_{F} = 0$ is valid.

The force due to negative shear in lower diagonal member M3L4 of the PST truss is determined with Equation 4.63b. and Table E4.49d.

$$F^{-V}_{M3L4} = \frac{-V^{-}_{(3)-(4)}(150) + M^{-0}_{(3)-(4)}}{250}$$

where

$$V^{-}_{(3)-(4)} = R_{yn} - 0.60W_{(3)}$$

$$M^{-0}_{(3)-(4)} = \left(-12R_{yn} + 0.60W_{(3)}(3-1)\right)25 + 0.60W_{(3)}(8.33)$$

$$= \left(-12R_{yn} + 1.20W_{(3)}\right)25 + 5.0W_{(3)}$$

TABLE E4.49d
Maximum Force in Member M3L4 due to Negative Shear in Subpanel 3–4 with Cooper's E80 Live Load Traversing a PST Span

	Negative Shear in Subpanel 3-4	
Axle at L2	**3**	**4**
Sum of moments to L0 (kip-ft)	9264 + 304(50 − (56 − 13)) = 11,392.0	11,696 + 344(50 − (64 − 18)) = 13,072
R_{y12} (kips)	11392/300 = 37.97	13072/300 = 43.57
Sum of moments in subpanel 2-3 ahead of and about panel point L2 (kip-ft)	460	960
$W_{(3)}$ (kips)	460/25 = 18.40	38.40
$V^{-}_{(3)-(4)}$ (kips)	37.97 − 0.60(18.40) = 26.93	20.53
$M^{-0}_{(3)-(4)}$ (kip-ft)	((−12(37.97) + 1.20(18.40))25 + 5.0(18.40) = −10,747	−11,727
$-V^{-}_{(3)-(4)}\left(a_{(2)-(4)}\right) + M^{-0}_{(3)-(4)}$ (kip-ft)	−26.93(150) − 10,747 = −14,787	−14,807
F^{-V}_{M3L4} (kips)	−14,787/(150 + 4(25)) = −59.15	−14,807/250 = −59.23
$F^{-}_{M3L4}\left(\alpha_{(3)-(4)}\right)$ (kips)	−59.15(1.653) = −97.77	−97.90

As expected, the marginal nature of the criterion for axle 3 and axle 4 at panel point L2 provides similar results for the maximum force due to negative shear in member M3L4.

$$F^{-max}_{M3L4} = -97.90 \text{kips}$$

The forces in end vertical member U2L2 of a PST span can be developed as for the BST truss such that for a uniformly distributed load, the maximum force in U2L2 is

$$F_{U2L2} = \frac{3W}{2} \tag{4.66a}$$,

and for a series of concentrated loads, the maximum force in U2L2 is

$$F_{U2L2} = \frac{3M_c}{2d} \tag{4.66b}$$

where M_c is the moment at joint L2 in a beam of span $L_c = (j+1)d = 3d$ from joint L0 to joint L3 in the PST span. The maximum force in U2L2 is determined with concentrated loads positioned in accordance with the criterion of Equation 4.11c for maximum moment at joint L2 in the span L_c.[152] The minimum force in U2L2 is due to dead load.

The forces in interior vertical member U(k)L(k) of the PST span, where k is an even number greater than or equal to 4 and less than n/2, in a PST truss are calculated by taking a section through adjacent top chord member U(k − 2)U(k) and bottom chord member L(k)L(k + 1) intersecting subtie member M(k – 1)U(k) and vertical web member U(k)L(k) such that due to positive shear in subpanel (k) – (k + 1),

$$F^{+}_{U(k)L(k)} = \left(\frac{-V^{+}_{(k)-(k+1)}\left(a_{(k-2)-(k)}\right) + M^{+0}_{(k)-(k+1)} + M^{S}_{(k-2)-(k)}}{a_{(k-2)-(k)} + (k)d}\right) \tag{4.66c}$$

where

$$V^{+}_{(k)-(k+1)} = R_{y0} - \sum_{i=0}^{k} W_i$$

$$M^{+0}_{(k)-(k+1)} = \left(\sum_{i=0}^{k} iW_i\right)d$$

$$M^{S}_{(k-2)-(k)} = F^{V}_{M(k-1)U(k)}\left(d_{S2} + (k-2)d + a_{(k-2)-(k)}\right)$$

From Equation 4.61b,

$d_{S2} = d\left(2 - \dfrac{h_{(k)}}{h_{(k)} - \frac{h_{(k-2)}}{2}}\right)$ = horizontal distance from L(k-2) to line extended from subtie member M(k-1)U(k)[153]

and from Equation 4.61c,

$$F^{V}_{M(k-1)U(k)} = F_{M(k-1)L(k-1)}\left(1 - \frac{h_{(k-2)}}{2h_{(k)}}\right) = W_{(k-1)}\left(1 - \frac{h_{(k-2)}}{2h_{(k)}}\right).$$

The force in vertical member U(k)L(k) due to negative shear in subpanel (k) – (k + 1) is

$$F^{-}_{U(k)L(k)} = \left(\frac{V^{-}_{(k)-(k+1)}\left(a_{(k-2)-(k)}\right) - M^{-0}_{(k)-(k+1)} - M^{S}_{(k-2)-(k)}}{a_{(k-2)-(k)} + (k)d}\right) \tag{4.66d}$$

where

$$V^{-}_{(k)-(k+1)} = R_{yn} - \sum_{i=k+1}^{n} W_i$$

$$M^{-0}_{(k)-(k+1)} = -\left(nR_{yn} - \sum_{i=0}^{k+1} iW_i\right)d.$$

The locations of concentrated loads creating maximum forces in U(k)L(k) will resemble the positions of concentrated loads creating maximum forces in upper diagonal member U(k)M(k + 1).

Example 4.50

Determine the maximum forces produced by Cooper's E80 live load in vertical web member U4L4 of the 300 ft long twelve-subpanel PST truss with height of 50 ft. The top chord is inclined with truss height of 25 ft at joints 1 and 11, 38 ft at joints 2 and 10, 47.5 ft at joints 4 and 8 and 50 ft at joint 6. Subvertical member M1L1 is 25 ft long, subvertical member M3L3 is 19 ft long and subvertical member M5L5 is 23.75 ft long.

k = 4
j = 3

$$a_{(2)-(4)} = \left(\left(\frac{2(47.5)}{47.5-38.0}\right) - (3+1)\right)25 = 150\,\text{ft}$$

For compression in member U4L4 due to positive shear, Example 4.48 indicates that maximum force in upper diagonal U4M5 due to positive shear will occur for axle 3 at L5. Therefore, axle 2 at L5 and axle 3 at L5 will be investigated for maximum compression in member U4L4 as shown in Table E4.50a.

TABLE E4.50a
Maximum Force in Vertical Web Member U4L4 of PST Span from Positive Shear in Subpanel 4-5

	Positive Shear in Subpanel 4-5	
Axle at L5	2	3
Sum moments to L12 (kip-ft)	32,728 + 568(74) + 4.0(74)(74)/2 = 85,712	32,728 + 568(79) + 4.0(79)(79/2) = 90,082
R_{y0} (kips)	85,712/300 = 285.71	90,082/300 = 300.27
Sum moments in subpanel 4-5 ahead of and about L5 (kip-ft)	160	460
$W_{(4)}$ (kips)	160/25 = 6.40	18.40
Sum moments in subpanel 3-4 ahead of and about L4 (kip-ft)	0	0
$W_{(3)}$ (kips)	0	0
$F_{M3U4} = W_{(3)}\left(1 - \frac{38.0}{2(47.5)}\right)$ (kips)	0	0

(Continued)

TABLE E4.50a (*Continued*)
Maximum Force in Vertical Web Member U4L4 of PST Span from Positive Shear in Subpanel 4-5

Positive Shear in Subpanel 4-5		
Axle at L5	**2**	**3**
$d_{S2} = \dfrac{\left(1-\dfrac{h_{(2)}}{h_{(4)}}\right)d}{\left(1-\dfrac{h_{(2)}}{2h_{(4)}}\right)}$ (ft)	$\dfrac{\left(1-\dfrac{38.0}{47.5}\right)25}{\left(1-\dfrac{38.0}{2(47.5}\right)} = \dfrac{0.20(25)}{0.60} = 8.33$	8.33
$M^{S}_{(2)-(4)} = F^{V}_{M3U4}\left(d_{S2} + (3-1)d + a_{(2)-(4)}\right)$ (kip-ft)	0	0
$V^{+}_{(4)-(5)} = R_{y0} - \sum_{i=0}^{4} W_i = R_{y0} - W_{(4)}$ (kips)	285.71 − 6.40 = 279.31	300.27 − 18.40 = 281.87
$M^{+0}_{(4)-(5)} = d\sum_{i=0}^{4} iW_i = 4dW_{(4)}$(kip-ft)	6.40(4)(25) = 640	4(25)18.40 = 1840
$F^{+max}_{U4L4}\left(a_{(2)-(4)} + 4d\right) = -V^{+}_{(4)-(5)}a_{(2)-(4)} + M^{+0}_{(4)-(5)} + M^{S}_{(2)-(4)}$(kip-ft)	−279.31(150) + 640 + 0 = −41,257	−281.87(150) + 1840 + 0 = −40,441
F^{+max}_{U4L4} (kips)	−41,257/(150 + 4(25)) = −165.03	−40,441/(150 + 4(25)) = −161.76

$$F^{+max}_{U4L4} = -165.0 \text{ kips}$$

For tension in U4L4 due to negative shear, Example 4.48 indicates that maximum force in upper diagonal U4M5 due to negative shear will occur for axle 2 at L4. Therefore, as validation, axle 2 at L4, axle 3 at L4 and axle 4 at L4 will be investigated for maximum tension in U4L4 as shown in Table E4.50b.

TABLE E4.50b
Maximum Force in Vertical Web Member U4L4 of PST Span from Negative Shear in Subpanel 4-5

Negative Shear in Subpanel 4-5			
Axle at L4	**2**	**3**	**4**
Last axle on span	18	18	18
Distance to L0 (ft)	4(25) − (101 − 5)=4	4(25) − (96 − 5) = 9	4(25) − (91 − 5) = 14
Length of uniform load on span (ft)	4 − 5 = −1 (off span)	9 − 5 = 4	14 − 5 = 9
Sum moments to L0 (kip-ft)	29,888 + 568(4) = 32,160	29,888 + 568(9) + 4.0(4)(4)/2 = 35,032	29,888 + 568(14) + 4.0(9)(9)/2 = 38,002
R_{y12} (kips)	32,160/300 = 107.20	35,032/300 = 116.77	38,002/300 = 126.67
Sum moments in subpanel 4-5 ahead of and about L4 (kip-ft)	160	460	960
$W_{(5)}$ (kips)	160/25 = 6.40	460/25 = 18.40	960/25 = 38.40
$V^{-}_{(4)-(5)} = R_{y12} - \sum_{i=5}^{12} W_i = R_{y12} - W_{(5)}$(kips)	107.20 − 6.40 = 100.80	116.77 − 18.40 = 98.37	126.67 − 38.40 = 88.27

TABLE E4.50b (*Continued*)
Maximum Force in Vertical Web Member U4L4 of PST Span from Negative Shear in Subpanel 4-5

	Negative Shear in Subpanel 4-5		
Axle at L4	**2**	**3**	**4**
$V^-_{(4)-(5)}a_{(2)-(4)}$ (kips)	100.80(150) = 15,120	98.37(150) = 14,756	88.27(150) = 13,241
$M^{-0}_{(4)-(5)} = d\sum_{i=0}^{5} iW_i = -\left(12R_{y12} - 5W_{(5)}\right)25$ (kip-ft)	−(12(107.20) − 5(6.40))25 = −31,360	−(12(116.77) − 5(18.40))25 = −32,731	−(12(126.67) − 5(38.40))25 = −33,201
ΣM_{L2} (kip-ft)	$\sum_{axle=7}^{10} M_{L2}$ = 1040 + (304 − 206)(2) = 1236	$\sum_{axle=8}^{10} M_{L2}$ =546 + (304 − 232)(7) = 1050	$\sum_{axle=8}^{11} M_{L2}$ = 1122 + (344 − 232)(4) = 1570
ΣM_{L3} (kip-ft)	$\sum_{axle=3}^{6} M_{L3}$ = 1680 + (206 − 60)(1) = 1826	$\sum_{axle=4}^{7} M_{L3}$ =1450 + (232 − 100)1 = 1582	$\sum_{axle=5}^{7} M_{L3}$ = 690 + (232 − 140)(6) = 1242
$W_{(3)}$ (kips)	1236/25 + ((206 − 60) − 1826/25) = 49.44 + 73.00 = 122.44	1050/25 + ((232 − 100) − 1582/25) = 42.00 + 68.72 = 110.72	1570/25 + ((232 − 140) − 1242/25) = 62.80 + 42.32 = 105.12
$F^V_{M3U4} = W_{(3)}\left(1 - \frac{38.0}{2(47.5)}\right)$ (kips)	122.44(0.60)=73.46	110.72(0.60)=66.43	105.12(0.60) = 63.07
$d_{S2} = \frac{\left(1 - \frac{38.0}{47.5}\right)25}{\left(1 - \frac{38.0}{2(47.5}\right)}$ (ft)	8.33	8.33	8.33
$M^S_{(2)-(4)} = F^V_{M(3)U(4)}\left(8.33 + (3-1)25 + 150\right)$ (kip-ft)	73.46(8.33 + 2(25) + 150) = 15,305	66.43(8.33 + 2(25) + 150) = 13,840	63.07(8.33 + 2(25) + 150) = 13,140
$F^{-max}_{U4L4}\left(150 + 4(25)\right) = V^-_{(4)-(5)}(150) - M^{-0}_{(4)-(5)} - M^S_{(2)-(4)}$ (kip-ft)	100.80(150) + 31,360 − 15,305 = 31,175	98.37(150) + 32,731 − 13,840 = 33,647	88.27(150) + 33,201 − 13,140 = 33,302
F^{-max}_{U4L4} (kips)	31,175/(150 + 4(25)) = 124.72	33,647/(150 + 4(25)) = 134.59	33,302/(150 + 4(25)) = 133.21

$$F^{-max}_{U4L4} = 134.6 \text{ kips}$$

The maximum force in member U(n/2)L(n/2) (central vertical member) in a PST truss are calculated from equilibrium at joint U(n/2) by considering top chord and subtie forces. For a uniformly distributed live load, the entire panels (n/2 – 2) – (n/2 + 2) are loaded, and for a series of concentrated loads, the criterion of Equations 4.11c is used to find the maximum $W_{(n/2-1)}$.

Due to the configuration of the end post of the PST truss in Figure 4.18, forces in upper end post member M1U2 and lower end post member L0M1 are determined in accordance with Equations 4.41 and 4.42a, respectively, with $h = h_2$. Once the force in end vertical member U2L2 is determined by Equations 4.58a or 4.58b for uniformly distributed and concentrated live loads, respectively, the force in diagonal member M1L2 can be ascertained by considering vertical force equilibrium at joint L2.

4.6.5 Subdivided Warren Truss Spans

Panels (j – 1) – (j + 1) in horizontal- and inclined-chord Warren truss spans subdivided at joint (j) and with diagonal members oriented such that upper diagonal members are U(j-1)M(j) and lower diagonal members are M(j)L(j+1) are analyzed identically to panels in subdivided Pratt truss spans (Baltimore and Pennsylvania trusses) as outlined in Sections 4.6.1 to 4.6.4.

However, panels (j – 1) – (j + 1) in horizontal and inclined chord Warren truss spans subdivided at joint (j) and with diagonal members oriented such that upper diagonal members are M(j)U(j + 1) and lower diagonal members are L(j – 1)M(j) are analyzed similarly, but not identically, to panels in subdivided Pratt truss spans, as outlined in the subsequent Section 4.6.5.1 and Section 4.6.5.2 for inclined-chord Warren truss spans with substruts and subties, respectively.

4.6.5.1 Subdivided Inclined-Chord Warren Truss Spans with Substrut Members

The forces in top chord member U(j – 1)U(j + 1) of inclined-chord Warren truss spans with substrut members (WISS), where the upper diagonal member is M(j)U(j + 1), are calculated by taking a section through subpanel (j – 1) – (j) and moments about joint L(j – 1).

Forces in bottom chord member L(j – 1)L(j) where the upper diagonal member is M(j)U(j + 1) are calculated by taking a section through subpanel (j – 1) – (j) and moments about U(j + 1).

Forces in lower diagonal member L(j – 1)M(j) are calculated by taking a section through subpanel (j – 1) – (j) and moments about the location P that is the intersection between the extension of a line from horizontal bottom chord member L((j – 1)L(j) and a line from inclined top chord member U(j – 1)U(j + 1).

Forces in upper diagonal members M(j)U(j + 1) are determined by taking a section through subpanel (j) – (j + 1) and moments about location P.

To determine the forces in inclined top chord member U(j – 1)U(j + 1), taking a section through subpanel (j – 1) – (j) and moments about panel point L(j – 1) in a WISS truss with diagonal members oriented such that lower diagonal members are L(j – 1)M(j) provides

$$R_{y0}(j-1)d-\sum_{i=0}^{j-1}W_i(j-1-i)d+\left(F_{U(j-1)U(j+1)}h_{(j-1)}\right)\frac{2d}{\sqrt{\left(h_{(j+1)}-h_{(j-1)}\right)^2+4d^2}}=0$$

such that

$$F_{U(j-1)U(j+1)}=\left(\frac{\left(-R_{y0}(j-1)+\sum_{i=0}^{j-1}W_i(j-1-i)\right)}{h_{(j-1)}}\right)\frac{\sqrt{\left(h_{(j+1)}-h_{(j-1)}\right)^2+4d^2}}{2} \tag{4.67a}$$

For concentrated live loads,

$$R_{y0}=\frac{P_TX_T}{nd}$$

$$\sum_{i=0}^{j-1}W_i(j-1-i)d=P_FX_F$$

$$F_{U(j-1)U(j+1)}=\left(\frac{\left(-\frac{P_TX_T}{n}(j-1)+P_FX_F\right)}{h_{(j-1)}}\right)\frac{\sqrt{\left(h_{(j+1)}-h_{(j-1)}\right)^2+4d^2}}{2}$$

where for the WISS span, P_T is the resultant of all loads on the span at distance X_T from panel point (n) and P_F is the resultant loads between panel point (0) and panel point (j – 1) at distance X_F from (j – 1).

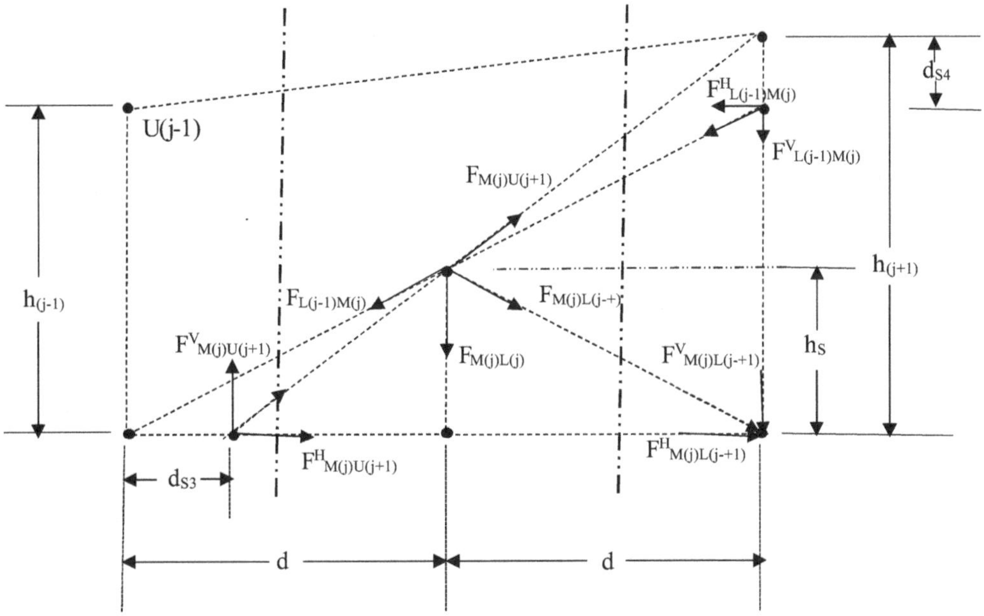

FIGURE 4.22 Panel in a substrut subdivided inclined chord Warren truss.

The maximum value of $F_{U(j-1)U(j+1)}$ is obtained by $\frac{d(F_{U(j-1)U(j+1)})}{dx} = 0$, which provides the criterion for the location of concentrated live loads for maximum force in top chord members as follows:

$$\frac{P_T}{n} = \frac{P_F}{(j-1)} \tag{4.67b}$$

which is the criterion of Equation 4.11c with k = j − 1.

To determine the forces in horizontal bottom chord member L(j − 1)L(j), consider a section through subpanel (j − 1) − (j) and moments about panel point U(j + 1) in a WISS truss. Diagonal members are oriented with the lower diagonal members L(j − 1)M(j). Note that the load at panel point (j) is to the right of the section through (j − 1) − (j) referring to the geometry in Figure 4.22.[154] The resolved $F_{L(j-1)M(j)}$ at (j + 1) is given as

$$R_{y0}(j+1)d - \sum_{i=0}^{j+1} W_i(j+1-i)d + W_{(j)}d - F_{L(j-1)L(j)}h_{(j+1)} - F_{L(j-1)M(j)}\left(\frac{d}{\sqrt{h_S^2+d^2}}\right)d_{S4} = 0$$

which results in

$$F_{L(j-1)L(j)} = \frac{\left(R_{y0}(j+1) - \sum_{i=0}^{j+1} W_i(j+1-i) + W_{(j)} - F_{L(j-1)M(j)}\left(\frac{d_{S4}}{\sqrt{h_S^2+d^2}}\right)\right)d}{h_{(j+1)}}$$

$$= \frac{M^{Wx}_{(j+1)} + 2M^P_{(j)}}{h_{(j+1)}} \tag{4.67c}$$

where

$$M_{(j+1)}^{Wx} = \left(R_{y0}(j+1) - \sum_{i=0}^{j+1} W_i (j+1-i) \right) d$$

$$M_{(j)}^{P} = \left(\frac{W_{(j)}}{2} - F_{L(j-1)M(j)} \left(\frac{d_{S4}}{\sqrt{h_S^2 + d^2}} \right) \right) d$$

$$d_{S4} = h_{(j+1)} - 2h_S$$

h_S = length of member $M_{(j)}L_{(j)}$
For concentrated live loads,

$$W_{(j)} = \frac{P_A X_A - P_B (d - X_{B)}}{d}$$

where for the WISS span, P_B = the resultant loads between panel point (j – 1) and panel point (j) at distance X_B from (j) and P_A = the resultant loads between (j) and (j + 1) at X_A from (j + 1):

$$F_{L(j-1)L(j)} = \frac{\left(\frac{P_T X_T}{n}(j+1) - P_F (X_F + 2d) - P_A X_A - P_B (X_B + d) \right)}{h_{(j+1)}}$$

The maximum value of $F_{L(j-1)L(j)}$ is obtained by $\frac{d\left(F_{L(j-1)L(j)}\right)}{dx} = 0$, which provides the criterion for the location of concentrated live loads for maximum force in bottom chord members as follows:

$$\frac{P_T}{n} = \frac{P_F + P_A + P_B}{j+1} \tag{4.67d}$$

Example 4.51

Determine the maximum forces in top chord member U4U6 and bottom chord member L4L5 of a 300 ft long subdivided twelve-subpanel WISS truss traversed by Cooper's E80 live load with h_1 = 25.0 ft, h_2 = 38.0 ft, h_4 = 47.5 ft and h_6 = 50.0 ft. Subvertical member M1L1 is 25 ft long, subvertical member M3L3 is 19 ft long and subvertical member M5L5 is 25 ft long. Therefore, diagonal member L4M5 and diagonal member M5U6 are aligned because, as shown by Figure 4.22, $h_{(5)} = \frac{h_{(6)}}{2}$.

n = 12
j = 5

For top chord member U4U6, Equation 4.66b yields

$$\frac{P_T}{3} = P_F$$

In accordance with Example 4.26 (BSS), Example 4.37 (PSS) and Example 4.46 (PST), the criterion indicates axle 13 at L4 for maximum force in WISS top chord member U4U6 as given in Table E4.51a.

TABLE E4.51a
Maximum Force in WISS Top Chord Member U4U6

	Axle at L4
Maximum Force in U4U6	**13**
Sum of moments to L12 (kip-ft)	$32{,}728 + 568(165) + 4(165)^2/2 = 180{,}898$
Reaction R_{y0} (kips)	$180{,}898/300 = 603.0$
Sum of moments from first axle to L4 (kip-ft)	15,336
Sum of moments about L4 = $\sum M_{L4}$ (kip-ft)	$603.3(4)(25) - 15{,}336 = 44{,}963$
$\propto_{(4)-(6)} = \dfrac{\sqrt{(50-47.5)^2 + 4(25)^2}}{2(25)}$	1.0012
Force in member U4U6 = $F_{U4U6}\,\alpha_{(4)-(6)} = -(\sum M_{L4}/h_{(4)})\,\alpha_{(4)-(6)}$ (kips)	$(-44{,}963/47.5)1.0012 = -947.8$

$$F_{U4U6}^{max} = -947.8 \text{ kips}$$

For bottom chord member L4L5, Equation 4.67b yields the criterion shown in Table E4.51b:

$$\frac{P_T}{2} = P_F + P_A + P_B$$

TABLE E4.51b
Criterion for Maximum Force in Member L4L5 due to Concentrated Live Loads Traversing a WISS Span

	Location of Live Load for L4L5		
Axle at L5	**13**	**14**	**15**
Distance to uniform load (ft)	35	30	21
Distance to axle no. 18 (ft)	30	25	16
Last axle on span	18	18	18
Distance to axle no. 1 (ft)	109 − 35 = 74	109 − 30 = 79	88
First axle on span	1	1	1
Length of uniform load (ft)	175 − 35 = 140	175 − 30 = 145	175 − 21 = 154
Total load on span, P_T (kips)	568 + 140(4) = 1128	568 + 145(4) = 1148	568 + 154(4) = 1184
$(P_F + P_A)_L$ wheel to left of L5 (kips)	$\sum_{i=1}^{13} P_i = 424$	$\sum_{i=1}^{14} P_i = 464$	$\sum_{i=1}^{15} P_i = 490$
$(P_F + P_A)_R$ wheel to right of L5 (kips)	$\sum_{i=1}^{12} P_i = 384$	$\sum_{i=1}^{13} P_i = 424$	$\sum_{i=1}^{14} P_i = 464$
$(P_B)_L$ wheel to left of L5 (kips)	$\sum_{i=13}^{18} P_i = 184$	$\sum_{i=15}^{18} P_i = 104$	$\sum_{i=16}^{18+} P_i = 78 + 4(4) = 94$
$(P_B)_R$ wheel to right of L5 (kips)	$\sum_{i=12}^{17} P_i = 198$	$\sum_{i=14}^{17} P_i = 118$	$\sum_{i=15}^{17} P_i = 78$
$P_T/2$ (kips)	564.0	574.0	592.0
$(P_F + P_A + P_B)_L$ wheel to left of L5 (kips)	424 + 198 = 622	(464 + 118) = 582	(490 + 78) = 568
$(P_F + P_A + P_B)_R$ wheel to right of L5 (kips)	384 + 184 = 568	(424 + 104) = 528	(464 + 94) = 558
Satisfies criterion	No (marginal)	Yes	No

The criterion indicates axle 14 at L5 for maximum force in bottom chord member L4L5. Axle 13 at L5 does not provide maximum force in L4L5, but it appears that this live load location will yield close to the maximum force. Therefore, forces in bottom chord member L4L5 are calculated for axles 13 to 15 to verify the criterion and review the changes in the force in L4L5 as Cooper's E80 live load traverses the span from right to left (joint L(n) to L(0)). From Equation 4.67c (also see Table E4.51c),

$$F_{L4L5} = \frac{\left(6R_{y0} - \sum_{i=0}^{6} W_i(6-i) + W_{(5)} - F_{L4M5}\left(\frac{d_{S4}}{\sqrt{h_S^2 + 25^2}}\right)\right)25}{h_{(6)}} = \frac{M_{(6)}^{Wx} + 2M_{(5)}^{P}}{h_{(6)}}$$

$h_S = 25$ ft
$d_{S4} = h_{(6)} - 2h_S = 50 - 2(25) = 0$ (the force in member L4M5 is resolved at U(j+1))

$$M_{(6)}^{Wx} = \left(6R_{y0} - \sum_{i=0}^{6} W_i(6-i)\right)25$$

$$M_{(5)}^{P} = \left(\frac{W_{(5)}}{2} - 0\right)d = \frac{W_{(5)}d}{2}$$

$$F_{L4L5}^{max} = F_{L5L6}^{max} = 1047.1\,\text{kips}$$

TABLE E4.51c
Maximum Force in Member L4L5 due to Concentrated Live Loads Traversing a WISS Span

	Maximum Force in Member L4L5		
Axle at L5	13	14	15
Sum of moments to L12 (kip-ft)	32,728 + 568(140) + 4(140)²/2 =151,448	32,728 + 568(145) + 4(145)²/2 = 157,138	32,728 + 568(154) + 4(154)²/2 = 167,632
Reaction R_{y0} (kips)	151,448/300 = 504.8	157,138/300 = 523.8	167,632/300 = 558.77
(j+1)(25)R_{y0} (kips)	6(25)(504.8) = 75,724	6(25)(523.8) = 78,569	6(25)(558.77) =83,816
P_{FL} = wheel of P_F loads left of joint (j+1) (kips)	17	18	18
Distance P_{FL} to joint (j+1) (ft)	0	0	9
$25\sum_{i=0}^{6}(6-i)W_i = \sum_{i=0}^{P_{FL}} M_i$ (kip-ft)	27,178	29,888	29,888 + 568(9) + 4(9 − 5)²/2 = 35,032
$M_{(6)}^{Wx} = \left((6(25))R_{y0} - \sum_{i=0}^{P_{FL}} M_i\right)$(kip-ft)	75,724 − 27,178 = 48,546	78,569 − 29,888 = 48,681	83,816 − 35032 = 48,784
Axle to right of L4	10	10	11
Axle at L5	13	14	15
Axle to left of L6	16 (17 at L6)	17 (18 at L6)	18
Load at L5 = P_{L5} (kips)	40	40	26
Length of uniform load to left of L6 (ft)	0	150 − (12 − 6)25 = 0	154 − (12 − 6)(25) = 4

TABLE E4.51c (*Continued*)
Maximum Force in Member L4L5 due to Concentrated Live Loads Traversing a WISS Span

	Maximum Force in Member L4L5		
Axle at L5	**13**	**14**	**15**
Sum of moments in subpanel 4-5 about L4 = ΣM_{L4} (kip-ft)	960	1660	2640
Force at L5 from subpanel 4-5 (kips)	(384 − 284) − 960/25 = 61.60	(424 − 284) − 1660/25 = 140 − 66.40 = 73.60	(464 − 304) − 2640/25 = 160 − 105.60 = 54.40
Sum of moments in subpanel 5-6 about L6 = ΣM_{L6} (kip-ft)	1242	832	806 + 4(78) + (4.0)(4)²/2 = 1118 + 32 = 1150
Force at L5 from subpanel 5-6	1242/25 = 49.68	832/25 = 33.28	1150/25 = 46.00
$W_{(5)}$ (kips)	61.60 + 49.68 + 40 = 151.28	40 + 73.60 + 33.28 = 146.88	26 + 54.40 + 46.00 = 126.40
$M^P_{(5)} = \frac{W_{(5)}(d)}{2}$ (kip-ft)	1891	146.88(25)/2 = 1836	126.4(25)/2 = 1580
$F_{L4L5} = \frac{M^{Wx}_{(6)} + 2M^P_{(5)}}{h_{(j+1)}}$ (kips)	(48,546 + 2(1891))/50 = 1046.6	(48,681 + 2(1836))/50 = 1047.1	(48,784 + 2(1580))/50 = 1038.9

Additionally, from Example 4.43, for a Pratt BST truss, taking a section through subpanel 4-5 and taking moments about joint L6 provides the maximum force in top chord member U4U6 as −1047.1 kips. Consequently, the maximum force in bottom chord member L4L5 of the WISS truss is 1047.1 kips.

Example 4.52

Determine the maximum forces in bottom chord member L4L5 of a 300 ft long subdivided twelve-subpanel WISS truss traversed by Cooper's E80 live load with $h_1 = 25.0$ ft, $h_2 = 38.0$ ft, $h_4 = 47.5$ ft and $h_6 = 50.0$ ft. Subvertical member M1L1 is 25 ft long, subvertical member M3L3 is 19 ft long and subvertical member M5L5 is 23.75 ft long. Therefore, diagonal member L5M5 and diagonal member M5U6 are not aligned because, as shown by Figure 4.22, $h_{(5)} \neq \frac{h_{(6)}}{2}$.

$$n = 12$$
$$j = 5$$

As shown in Example 4.51, the criterion indicates axle 14 at L5 for maximum force in bottom chord member L4L5. From Equation 4.67c,

$$F_{L4L5} = \frac{\left(6R_{y0} - \sum_{i=0}^{6} W_i(6-i) + W_{(5)} - F_{L4M5}\left(\frac{d_{S4}}{\sqrt{h_S^2 + 25^2}}\right)\right)25}{h_{(6)}} = \frac{M^{Wx}_{(6)} + 2M^P_{(5)}}{h_{(6)}}$$

$h_S = 23.75$ ft

$d_{S4} = h_{(6)} - 2h_S = 50 - 2(23.75) = 2.50$ ft (the force in member L4M5 creates a moment at joint U(j+1))

$$M_{(6)}^{Wx} = \left(6R_{y0} - \sum_{i=0}^{6} W_i(6-i)\right)25$$

$$M_{(5)}^{P} = \left(\frac{W_{(5)}}{2} - \frac{F_{L(j-1)M(j)}}{2}\left(\frac{d_{S4}}{\sqrt{h_S^2 + d^2}}\right)\right)d$$

To determine $M^P_{(5)}$ in Equation 4.67c, F_{L4M5} must be known for axle 14 at L5,

$$F_{L4M5}^{+} = \left(\frac{-V_{(4)-(5)}^{+}\left(a_{(4)-(6)}\right) + M_0^{+}}{a_{(4)-(6)} + (5-1)25}\right)\left(\frac{\sqrt{(23.75)^2 + 25^2}}{23.75}\right) = \left(\frac{-V_{(4)-(5)}^{+}(850) + M_0^{+}}{654.31}\right)$$

Table 4.2 gives the forces in Tables E4.52a and E4.52b.

TABLE E4.52a
Force in Member L4M5 due to Concentrated Live Loads Traversing a WISS Span with Axle 14 at Panel Point L5

	Positive Shear in Subpanel 4-5
Axle at L5	**14**
Distance axle 1 to panel point (0) (ft)	125 − 79 = 46
Length of uniform load on the span (ft)	175 − 30 = 145
Sum moments to L12 (kip-ft)	32,728 + 568(145) + 4.0(145)(145/2) = 157,138
R_{y0} (kips)	157,138/300 = 523.79
Axles in panel 4-5	10 to 14 (14 at L5)
Sum moments in panel 4-5 about panel point L5 (kip-ft)	1660
Load at L4 from loads in panel 4-5 (kips)	1660/25 = 66.40
Sum of loads W_i to panel point 4 (kips)	284 + 66.40 = 350.4
$V_{(4)-(5)}^{+}$ (kips)	523.79 − 350.4 = 173.39
Sum moments from panel point 4 to panel point (0) (kip-ft)	6640 + 284(46) = 19,704
M^{+0} = Sum moments from panel point 5 to panel point (0) (kip-ft)	19,704 + 66.40(4)(25) = 26,344
$F_{L4M5}^{+} = \left(\frac{-V_{(4)-(5)}^{+}(850) + M^{+0}}{654.31}\right)$ (kip)	(−173.39(850) + 26,344)/654.31 = −184.98

TABLE E4.52b
Maximum Force in Member L4L5 due to Concentrated Live Loads Traversing a WISS Span

	Maximum Force in Member L4L5
Axle at L5	**14**
Sum of moments to L12 (kip-ft)	32,728 + 568(145) + 4(145)²/2 = 157,138
Reaction R_{y0} (kips)	157,138/300 = 523.8

TABLE E4.52b (*Continued*)
Maximum Force in Member L4L5 due to Concentrated Live Loads Traversing a WISS Span

Maximum Force in Member L4L5	
Axle at L5	**14**
$(j+1)(25)R_{y0}$ (kips)	6(25)(523.8) = 78,569
P_{FL} = wheel of P_F loads left of joint (j + 1) (kips)	18
Distance P_{FL} to joint (j + 1) (ft)	0
$25\sum_{i=0}^{6}(6-i)W_i = \sum_{i=0}^{P_{FL}} M_i$ (kips-ft)	29,888
$M_{(6)}^{Wx} = \left((6(25))R_{y0} - \sum_{i=0}^{P_{FL}} M_i\right)$ (kip-ft)	78,569 − 29,888 = 48,681
Axle to right of L4	10
Axle at L5	14
Axle to left of L6	17 (18 at L6)
Load at L5 = P_{L5} (kips)	40
Length of uniform load to left of L6 (ft)	150 − (12 − 6)25 = 0
Sum of moments in subpanel 4-5 about L4 = ΣM_{L4} (kip-ft)	1660
Force at L5 from subpanel 4-5 (kips)	(424 − 284) −1660/25 = 140 − 66.40 = 73.60
Sum of moments in subpanel 5-6 about L6 = ΣM_{L6} (kip-ft)	832
Force at L5 from subpanel 5-6 (kips)	832/25 = 33.28
$W_{(5)}$ (kips)	40 + 73.60 + 33.28 = 146.88
$M_{(5)}^{P} = \left(\frac{146.88}{2} - \frac{F_{L(j-1)M(j)}}{2}\left(\frac{2.50}{34.48}\right)\right)25$ (kip-ft)	1836 − 0.906 $F_{L(j-1)M(j)}$ = 1836 − 0.906(−184.98) = 2003.5
$F_{L4L5} = \frac{M_{(6)}^{Wx} + 2M_{(5)}^{P}}{h_{(j+1)}}$ (kips)	(48,681 + 2(2003.5)/50 = 1053.8

$$F_{L4L5}^{max} = F_{L5L6}^{max} = 1053.8\,\text{kips}$$

The force in lower diagonal member L(j − 1)M(j) of the WISS span, where j = 5, 9, 13. . . ≤ (n/2 − 1) and (n) is an even number,[155] is calculated by taking a section through subpanel (j − 1) − (j) and determining the positive shear in subpanel (j − 1) − (j) by taking moments about location P such that

$$F_{L(j-1)M(j)}^{+} = \left(\frac{-V_{(j-1)-(j)}^{+}\left(a_{(j-1)-(j+1)}\right) + M_0^{+}}{\left(a_{(j-1)-(j+1)} + (j-1)d\right)}\right)\left(\frac{\sqrt{(h_S)^2 + d^2}}{h_S}\right) \tag{4.68a}$$

where

$$V_{(j-1)-(j)}^{+} = R_{y0} - \sum_{i=0}^{j-1} W_i$$

$$M_0^{+} = \sum_{i=0}^{j-1} W_i\,(id)$$

For concentrated live loads and positive shear in panel (j – 1) – (j),

$$R_{y0} = \frac{P_T^+ X_T^+}{nd}$$

$$W_0 = \frac{P_F^+ X_F^+}{(j-1)d}$$

$$\sum_{i=1}^{j-1} W_i = P_F^+ - \frac{P_F^+ X_F^+}{(j-1)d} + \frac{P_B^+ X_B^+}{d}$$

$$V_{(j-1)-(j)}^+ = \frac{P_T^+ X_T^+}{nd} - P_F^+ - \frac{P_B^+ X_B^+}{d}$$

$$M_0^+ = \sum_{i=0}^{j-1} W_i(id) = (j-1)dP_F^+ - P_F^+ X_F^+ + (j-1)P_B^+ X_B^+$$

where for the WISS span,

P^+_B = resultant of the concentrated loads in panel (j – 1) – (j) at distance X^+_B left of panel point (j);
P^+_A = resultant of the concentrated loads in panel (j) – (j + 1) at X^+_A left of (j + 1); and
P^+_F = resultant of the concentrated loads in panel (0) – (j – 1) at X^+_F left of (j – 1).
$F^{+max}_{L(j-1)M(j)}$ occurs when $\frac{d(F^+_{L(j-1)M(j)})}{dx} = 0$, which yields

$$\frac{P_T^+}{n} = P_B^+\left(1 + \frac{(j-1)d}{a_{(j-1)-(j+1)}}\right) - \frac{P_F^+ d}{a_{(j-1)-(j+1)}} \tag{4.68b}$$

The criterion of Equation 4.68b can only occur with an axle at panel point (j).

As expected, from Equation 4.68b, the criterion for member L(j – 1)M(j) in a WISS truss is the same criterion as Equation 4.26g for diagonal member U(j – 1)L(j) in non-subdivided inclined top chord Pratt and Warren truss spans and Equation 4.62c for upper diagonal member U(j – 1)M(j) in a PST span.

The force in lower diagonal member L(j – 1)M(j), where j = 5, 9, 13. . . ≤ (n/2 – 1) and (n) is an even number, of a WISS truss is calculated by taking a section through subpanel (j – 1) – (j) and determining the negative shear in subpanel (j – 1) – (j) by taking moments about location, P. Consequently, the criterion for the force in lower diagonal member L(j-1)M(j) of the WISS span due to negative shear in subpanel (j – 1) – (j) is equivalent to Equation 4.62d for upper diagonal member U(j – 1)M(j) in a PST span.

$$\frac{P_T^-}{n} = P_B^-\left(\frac{1 + \frac{jd}{a_{(j-1)-(j+1)}}}{1 + \frac{nd}{a_{(j-1)-(j+1)}}}\right) + \frac{P_F^-}{n\left(1 + \frac{a_{(j-1)-(j+1)}}{nd}\right)} \tag{4.68c}$$

The criterion of Equation 4.68c can only occur with an axle at panel point (j – 1).

Example 4.53

Determine the maximum force due to positive shear for Cooper's E80 live load in lower diagonal member L4M5 of a 300 ft long subdivided twelve-subpanel WISS truss with h_1 = 25.0 ft, h_2 = 38.0 ft, h_4 = 47.5 ft and h_6 = 50.0 ft. Subvertical member M1L1 is 25 ft long, subvertical member M3L3 is 19 ft long and subvertical member M5L5 is 25 ft long. Therefore, diagonal member L4M5 and diagonal member M5U6 are aligned because, as shown by Figure 4.22, $h_{(5)} = \frac{h_{(6)}}{2}$.

$$j = 5$$

$$\alpha_{(j-1)-(j)} = \alpha_{(4)-(5)} = \frac{\sqrt{\left(\frac{50}{2}\right)^2 + (25)^2}}{25} = 1.414$$

$$a_{(4)-(6)} = \left(\left(\frac{2(50.0)}{50.0-47.5}\right) - (5+1)\right)25 = 850\text{ ft}$$

Assuming P^+_F = 0, the criterion of Equation 4.68b indicates

$$\frac{P_T^+}{12} = P_B^+\left(1 + \frac{(5-1)25}{850}\right)$$

$$\frac{P_T^+}{13.4} = P_B^+$$

Example 4.48 shows that the criterion indicates axle 3 at L5. Therefore, the assumption that P^+_F = 0 is valid and Equation 4.68a yields

$$F^+_{L4M5} = \left(\frac{-V^+_{(4)-(5)}\left(a_{(4)-(6)}\right) + M^+_0}{a_{(4)-(6)} + (5-1)25}\right)\left(\frac{\sqrt{25^2+25^2}}{25}\right) = \left(\frac{-V^+_{(4)-(5)}(850) + M^+_0}{671.75}\right)$$

$$V^+_{(4)-(5)} = R_{y0} - W_{(4)}$$

$$M^+_0 = 4(25)W_{(4)} = 100W_{(4)}$$

The force in member L4M5 will also be determined for axle 2 and axle 4 at panel point L5 as validation in the following Table E4.53.

TABLE E4.53
Maximum Force in Member L4M5 due to Positive Shear from Cooper's E80 Live Load on a WISS Span

	Positive Shear in Subpanel 4-5		
Axle at L5	**2**	**3**	**4**
Length of uniform load on the span (ft)	(12 − 5)(25) − 101 = 74	(12 − 5)(25) − 96 = 79	(12 − 5)(25) − 91 = 84
Sum moments to L12 (kip-ft)	32,728 + 568(74) + 4.0(74)(74/2) = 85,712	32,728 + 568(79) + 4.0(79)(79/2) = 90,082	94,552

(Continued)

TABLE E4.53 (*Continued*)
Maximum Force in Member L4M5 due to Positive Shear from Cooper's E80 Live Load on a WISS Span

	Positive Shear in Subpanel 4-5		
Axle at L5	**2**	**3**	**4**
R_{y0} (kips)	85,712/300 = 285.71	90,082/300 = 300.27	315.17
Sum moments in subpanel 4-5 ahead of and about L5 (kip-ft)	160	460	960
$W_{(4)}$ (kips)	160/25 = 6.40	18.40	38.40
$V^{+}_{(4)-(5)} = R_{y0} - W_{(4)}$(kips)	285.71 − 6.40 = 279.31	300.27 − 18.40 = 281.87	276.77
$M^{+0}_{(4)-(5)} = 4dW_{(4)}$ (kip-ft)	4(25)(6.40) = 640	4(25)18.40 = 1840	3840
$F^{+}_{L4M5} = \left(\dfrac{-V^{+}_{(4)-(5)}(850) + M^{+0}}{671.75}\right)$(kip)	(−279.31(850) + 640)/671.75 = −352.47	(−281.87(850) + 1840)/671.75 = −237,750/671.75 = −353.92	(−276.77(850) + 3840)/671.75 = −344.49

$$F^{+max}_{L4M5} = -353.92\,\text{kips}$$

Example 4.54

Determine the maximum force due to positive shear for Cooper's E80 live load in lower diagonal member L4M5 of a 300 ft long subdivided twelve-subpanel WISS truss with h_1 = 25.0 ft, h_2 = 38.0 ft, h_4 = 47.5 ft and h_6 = 50.0 ft. Subvertical member M1L1 is 25 ft long, subvertical member M3L3 is 19 ft long and subvertical member M5L5 is 23.75 ft long. Therefore, diagonal member L4M5 and diagonal member M5U6 are not aligned because, as shown by Figure 4.22, $h_{(5)} \neq \frac{h_{(6)}}{2}$.

$$j = 5$$

$$\alpha_{(j-1)-(j)} = \alpha_{(4)-(5)} = \frac{\sqrt{\left(\frac{50}{2}\right)^2 + (23.75)^2}}{23.75} = 1.452$$

$$a_{(4)-(6)} = \left(\left(\frac{2(50.0)}{50.0 - 47.5}\right) - (5+1)\right)25 = 850\text{ ft}$$

$$F^{+}_{L4M5} = \left(\frac{-V^{+}_{(4)-(5)}\left(a_{(4)-(6)}\right) + M^{+}_{0}}{a_{(4)-(6)} + (5-1)25}\right)\left(\frac{\sqrt{23.75^2 + 25^2}}{23.75}\right) = \left(\frac{-V^{+}_{(4)-(5)}(850) + M^{+}_{0}}{654.31}\right)$$

From Example 4.53,

$$V^{+}_{(4)-(5)} = 281.87\text{ kips}$$
$$M^{+0}_{(4)-(5)} = 1840\text{ kips} - \text{ft}$$

$$F_{L4M5}^{+max} = \left(\frac{-281.87(850)+1840}{654.31} \right) = -363.36 \text{ kips}$$

The forces in upper diagonal member M(j)U(j+1) of a WISS truss are calculated by taking a section through panel (j) – (j + 1), as shown in Figure 4.22, and determining the shear in panel (j) – (j + 1) by taking moments about location P for positive shear:

$$\Sigma M_P = 0 = -R_{y0}(a_{(j-1)-(j+1)}) + \sum_{i=0}^{j} W_i \left(a_{(j-1)-(j+1)} + id\right) - F_{M(j)U(j+1)}^{+V}$$
$$(a_{(j-1)-(j+1)} + (j-1)d + d_{S3}) + F_{M(j)L(j+1)}^{+V}\left(a_{(j-1)-(j+1)} + (j+1)d\right)$$

which yields

$$F_{M(j)U(j+1)}^{+V} = \frac{-V_{(j)-(j+1)}^{+} a_{(j-1)-(j+1)} + M_{(j)-(j+1)}^{+0}}{(a_{(j-1)-(j+1)} + (j-1)d + d_{S3}) - \dfrac{(a_{(j-1)-(j+1)} + (j+1)d)d_{S3}}{2d}} \tag{4.69a}$$

where

$$V_{(j)-(j+1)}^{+} = R_{y0} - \sum_{i=0}^{j} W_i - F_{M(j)L(j+1)}^{+V}$$

$$M_{(j)-(j+1)}^{+0} = d\sum_{i=0}^{j} iW_i + (j+1)dF_{M(j)L(j+1)}^{+V}$$

Taking moments about L(j-1) in Figure 4.22 indicates

$$F_{M(j)L(j+1)}^{+V} = \frac{F_{M(i)U(i+1)}^{+V} d_{S3}}{2d} - \frac{W_{(j)}}{2}$$

$$V_{(j)-(j+1)}^{+} = R_{y0} - \sum_{i=0}^{j} W_i + \frac{W_{(j)}}{2}$$

$$M_{(j)-(j+1)}^{+0} = d\sum_{i=0}^{j} iW_i - (j+1)d\left(\frac{W_{(j)}}{2}\right)$$

$$d_{S3} = \left(2 - \frac{h_{(j+1)}}{h_{(j+1)} - h_S}\right)d$$

For concentrated live loads and positive shear in panel (j) – (j + 1),

$$R_{y0} = \frac{P_T^+ X_T^+}{nd}$$

$$W_0 = \frac{P_F^+ X_F^+}{(j-1)d}$$

$$W_{(j)} = \frac{P_A^+\left(d - X_A^+\right)}{d} + \frac{P_B^+ X_B^+}{d}$$

$$\sum_{i=1}^{j} W_i = P_F^+ + P_A^+ + \frac{P_B^+ X_B^+}{d}$$

$$V_{(j)-(j+1)}^+ = \frac{P_T^+ X_T^+}{nd} - P_F^+ - P_A^+ - \frac{P_B^+ X_B^+}{d} + \frac{P_A^+\left(d - X_A^+\right)}{2d} + \frac{P_B^+ X_B^+}{2d}$$

$$M_0^+ = \sum_{i=0}^{j} W_i\left(id\right) = \left(j-1\right)dP_F^+ - P_F^+ X_F^+ + P_A^+\left(jd - X_A^+\right) + jP_B^+ X_B^+ - \left(j+1\right)\left(\frac{P_A^+\left(d - X_A^+\right)}{2} + \frac{P_B^+ X_B^+}{2}\right)$$

where for the WISS span,

P^+_B = resultant of the concentrated loads in panel (j) – (j + 1) at distance X^+_B left of panel point (j + 1);

P^+_A = resultant of the concentrated loads in panel (j – 1) – (j) at X^+_A left of (j); and

P^+_F = resultant of the concentrated loads in panels (0) – (j – 1) at X^+_F left of (j – 1).

$F_{M(j)U(j+1)}^{+max}$ occurs when $\frac{d(F^+_{M(j)U(j+1)})}{dx} = 0$, which yields

$$\frac{P_T^+}{n} = \frac{P_A^+ + P_B^+}{2}\left(1 + \frac{\left(j-1\right)d}{a_{(j-1)-(j+1)}}\right) - \frac{P_F^+ d}{a_{(j-1)-(j+1)}} \tag{4.69b}$$

which, as expected, is the same criterion for positive shear in panel (j – 1) – (j) creating the force in upper diagonal member U(j – 1)M(j) of a PSS truss as provided by Equation 4.61b.

The forces in upper diagonal member M(j)U(j + 1) of a WISS truss are calculated by taking a section through panel (j) – (j + 1), as shown in Figure 4.22, and determining the negative shear in panel (j) – (j + 1) by taking moments about location P such that

$$F_{M(j)U(j+1)}^{-V} = \frac{V_{(j)-(j+1)}^- a_{(j-1)-(j+1)} + M_{(j)-(j+1)}^{-0}}{\left(a_{(j-1)-(j+1)} + \left(j-1\right)d + d_{S3}\right) - \dfrac{\left(a_{(j-1)-(j+1)} + \left(j+1\right)d\right)d_{S3}}{2d}} \tag{4.69c}$$

where

$$V_{(j)-(j+1)}^- = nR_{yn} - \sum_{i=j+1}^{n} W_i - \frac{W_{(j)}}{2}$$

$$M_{(j)-(j+1)}^{-0} = \left(nR_{yn} - \sum_{i=j+1}^{n} iW_i + \frac{\left(j+1\right)W_{(j)}}{2}\right)d$$

Accordingly, the criterion for negative shear in panel (j) – (j + 1) creating the force in upper diagonal member M(j)U(j + 1) of WISS truss is Equation 4.52.

Example 4.55

Determine the maximum force due to positive and negative panel shear for Cooper's E80 live load in upper diagonal member M5U6 of a 300 ft long sub strut subdivided twelve subpanel inclined top chord Warren truss with h_1 = 25.0 ft, h_2 = 38.0 ft, h_4 = 47.5 ft and

h_6 = 50.0 ft. Subvertical member M1L1 is 25 ft long, subvertical member M3L3 is 19 ft long and subvertical member M5L5 is 25 ft long. Therefore, diagonal member L4M5 and diagonal member M5U6 are aligned because, as shown by Figure 4.22, $h_{(5)} = \frac{h_{(6)}}{2}$.

j = 5

$$\alpha_{M5U6} = \frac{\sqrt{(25)^2 + (25)^2}}{25} = 1.414$$

$$a_{(4)-(6)} = \left(\left(\frac{2(50.0)}{50.0 - 47.5}\right) - (5+1)\right)25 = 850 \text{ ft}$$

For $P^+_A = P^+_F = 0$, Equation 4.69b yields the following as shown in Table E4.55a:

$$\frac{P_T^+}{12} = \frac{P_B^+}{2}\left(1 + \frac{(j-1)d}{a_{(j-1)-(j+1)}}\right) = \frac{P_T^+}{6.71} = P_B^+$$

TABLE E4.55a
Criterion for Maximum Positive Force in Upper Diagonal Member M5U6 of a WISS Truss Traversed by Cooper's E80 Live Load

	Subpanel 5-6		
Axle at L6	**3**	**4**	**5**
Distance to axle no. 1 (ft)	13	18	18
First axle on span	1	1	1
Distance to axle no. 18 (ft)	91	91 – 5 = 86	81
Last axle on span	18	18	18
Distance of uniform load (ft)	(12 – 6)(25) – 96 = 54	150 – 91 = 59	150 – 86 = 64
Total load on span, P^+_T (kips)	568 + 54(4) = 784	568 + 59(4) = 804	568 + 64(4) = 824
$P^+_T/6.71$ (kips)	116.8	119.8	122.8
$P^+_{a(5)-(6)}$ = Sum of loads ahead of panel point L6 (kips)	60	100	140
$P^+_{b(5)-(6)}$ = Sum of loads ahead of and including panel point L6 (kips)	100	140	180
Criterion established	No	Yes	No

The criterion indicates axle 4 at L6 for maximum force due to positive shear in upper diagonal member M5U6. Therefore, the assumption that $P^+_A = P^+_F = 0$ is valid, and Equation 4.69a yields the following as shown in Table E4.55b:

h_S = 25 ft (length of member M5L5)

$$d_{S3} = \left(2 - \frac{50.0}{50.0 - 25}\right)25 = 0$$

$$F^{+V}_{M5U6} = \frac{-V^{+}_{(5)-(6)}850 + M^{+0}_{(5)-(6)}}{\left(850+(5-1)(25)+0\right) - \dfrac{\left(850+(5+1)25\right)0}{2(25)}} = \frac{-V^{+}_{(5)-(6)}850 + M^{+0}_{(5)-(6)}}{950.0}$$

Where $P^{+}_{A} = P^{+}_{F} = 0$,

$$V^{+}_{(5)-(6)} = R_{y0} - \sum_{i=0}^{5} W_i + \frac{W_{(5)}}{2} = R_{y0} - \frac{W_{(5)}}{2}$$

$$M^{+0}_{(j)-(j+1)} = d\sum_{i=0}^{j} iW_i - (j+1)d\left(\frac{W_{(j)}}{2}\right) = \frac{dW_{(j)}}{2}(j-1) = \frac{25W_{(j)}}{2}(5-1) = 50.0W_{(5)}$$

TABLE E4.55b
Maximum Positive Force in Upper Diagonal Member M5U6 of a WISS Truss Traversed by Cooper's E80 Live Load

	Positive Shear in Panel 5-6		
Axle at L6	**3**	**4**	**5**
Length of uniform load on the span (ft)	(12 − 6)(25) − 96 = 54	(12 − 6)(25) − 91=59	64
Sum moments to L12 (kip-ft)	32,728 + 568(54) + 4.0(54)(54/2) = 69,232	73,202	77,272
R_{y0} (kips)	69,232/300 = 230.77	244.0	257.57
Sum moments in panel 5-6 ahead of and about L6 (kip-ft)	460	960	1660
$W_{(5)}$ (kips)	18.40	38.40	66.40
$V^{+}_{(5)-(6)}$(kips)	230.77 − 18.40/2 = 221.57	224.81	224.37
$M^{+0}_{(5)-(6)}$(kip-ft)	920	1920	3320
$F^{+}_{M5U6} = \left(\dfrac{-V^{+}_{(5)-(6)}(850) + M^{+0}}{950.0}\right)(1.414)$ (kip-ft)	((−221.57(850) + 920)950)1.414 = −278.95	−281.56	−278.92

$F^{+max}_{M5U6} = -281.56$ kips

For $P^{-}_{A} = P^{-}_{F} = 0$, Equation 4.52 yields the following as given in Table E4.55c:

$$\frac{P^{-}_{T}}{12} = \frac{\left(\dfrac{P^{-}_{B}}{2}\right)\left(1+\dfrac{(5+1)25}{850}\right)}{\left(\dfrac{300}{850}+1\right)}$$

or

$$\frac{P_T^-}{5.22} = P_B^-$$

TABLE E4.55c
Criterion for Maximum Negative Force in Upper Diagonal Member M5U6 of a WISS Truss Traversed by Cooper's E80 Live Load

	Subpanel 5-6		
Axle at L4	**3**	**4**	**5**
Distance to axle no. 1 (ft)	13	18	18
First axle on span	1	1	1
Distance to axle no. 18 (ft)	91	91 – 5 = 86	81
Last axle on span	18	18	18
Distance of uniform load (ft)	(4)(25) – 96 = 4	100 – 91 = 9	100 – 86 = 14
Total load on span, P^+_T (kips)	568 + 4(4) = 584	568 + 9(4) = 604	568 + 14(4) = 624
$P^-_T/5.22$ (kips)	111.9	115.7	119.5
$P^-_{a(5)\text{-}(6)}$ = Sum of loads ahead of panel point L6 (kips)	60	100	140
$P^+_{b(5)\text{-}(6)}$ = Sum of loads ahead of and including panel point L6 (kips)	100	140	180
Criterion established	No	Yes	No

The criterion indicates axle 4 at L4 for maximum force due to negative shear in upper diagonal member M5U6. Therefore, the assumption that $P^-_A = P^-_F = 0$ is valid, and Equation 4.69c yields the following as also given in Table E4.55d:

$$F_{M5U6}^{-V} = \frac{V^-_{(5)-(6)}850 + M^{-0}_{(5)-(6)}}{950.0}$$

Where $P^-_A = P^-_F = 0$,

$$V^-_{(5)-(6)} = R_{yn} - W_{(6)} - \frac{W_{(5)}}{2} = R_{yn} - \frac{W_{(5)}}{2}$$

$$M^{-0}_{(5)-(6)} = \left(12R_{yn} - 6W_{(6)} - \frac{6W_{5)}}{2}\right)25 = \left(300R_{yn} - 75W_{(5)}\right)$$

TABLE E4.55d
Maximum Negative Force in Upper Diagonal Member M5U6 of a WISS Truss Traversed by Cooper's E80 Live Load

	Negative Shear in Panel 5-6		
Axle at L4	**3**	**4**	**5**
Length of uniform load on the span (ft)	4	9	14
Sum moments to L0 (kip-ft)	32,728 + 568(4) + 4.0(4)(4/2) = 35,032	38,002	41,072

(Continued)

TABLE E4.55d (*Continued*)

Maximum Negative Force in Upper Diagonal Member M5U6 of a WISS Truss Traversed by Cooper's E80 Live Load

Negative Shear in Panel 5-6			
Axle at L4	**3**	**4**	**5**
R_{y12} (kips)	35,032/300 = 116.77	126.67	136.91
Sum moments in panel 4-5 ahead of and about L4 (kip-ft)	460	960	1660
$W_{(5)}$ (kips)	18.40	38.40	66.40
$V^{-}_{(5)-(6)}$(kips)	(116.77) − (18.40/2) = 107.57	107.47	103.71
$M^{-0}_{(5)-(6)}$(kip-ft)	300(116.77) − 75(18.4) =33,652	35,121	36,093
$F^{-}_{M5U6} = \left(\dfrac{V^{-}_{(5)-(6)}(850)+M^{-0}}{950.0}\right)(1.414)$ (kip-ft)	((107.57(850) + 33,652)/950)1.414 =186.18	188.24	184.93

$F^{-max}_{M5U6} = 188.24$ kips

Example 4.56

Determine the maximum force due to positive and negative panel shear for Cooper's E80 live load in upper diagonal member M5U6 of a 300 ft long subdivided twelve-subpanel WISS truss with $h_1 = 25.0$ ft, $h_2 = 38.0$ ft, $h_4 = 47.5$ ft and $h_6 = 50.0$ ft. Subvertical member M1L1 is 25 ft long, subvertical member M3L3 is 19 ft long and subvertical member M5L5 is 23.75 ft long. Therefore, diagonal member L4M5 and diagonal member M5U6 are not aligned because, as shown by Figure 4.22, $h_{(5)} \neq \frac{h_{(6)}}{2}$.

$j = 5$

$h_S = 23.75$ ft (length of member M5L5)

$h_{(j+1)} - h_S = 50 - 23.75 = 26.25$ ft

$$\alpha_{M5U6} = \frac{\sqrt{(25)^2 + (26.25)^2}}{26.25} = 1.381$$

$$a_{(4)-(6)} = \left(\left(\frac{2(50.0)}{50.0-47.5}\right) - (5+1)\right)25 = 850\text{ft}$$

For $P^+_A = P^+_F = 0$, Equation 4.69b yields

$$\frac{P_T^+}{12} = \frac{P_B^+}{2}\left(1 + \frac{(j-1)d}{a_{(j-1)-(j+1)}}\right) = \frac{P_T^+}{6.71} = P_B^+$$

As shown in Example 4.55, the criterion indicates axle 4 at L6 for maximum force due to positive shear in upper diagonal member M5U6. Therefore, the assumption that $P^+_A = P^+_F = 0$ is valid, and Equation 4.69a yields the following as well as Table E4.56a:

$$d_{S3} = \left(2 - \frac{50.0}{50.0-23.75}\right)25 = 2.38$$

$$F^{+V}_{M5U6} = \frac{-V^{+}_{(5)-(6)}850 + M^{+0}_{(5)-(6)}}{\left(850+(5-1)(25)+2.38\right) - \dfrac{\left(850+(5+1)25\right)2.38}{2(25)}} = \frac{-V^{+}_{(5)-(6)}850 + M^{+0}_{(5)-(6)}}{904.78}$$

Where $P^{+}_{A} = P^{+}_{F} = 0$,

$$V^{+}_{(5)-(6)} = R_{y0} - \sum_{i=0}^{5} W_i + \frac{W_{(5)}}{2} = R_{y0} - \frac{W_{(5)}}{2}$$

$$M^{+0}_{(j)-(j+1)} = d\sum_{i=0}^{j} iW_i - (j+1)d\left(\frac{W_{(j)}}{2}\right) = \frac{dW_{(j)}}{2}(j-1) = \frac{25W_{(j)}}{2}(5-1) = 50.0W_{(5)}$$

TABLE E4.56a
Maximum Positive Force in Upper Diagonal Member M5U6 of a WISS Truss Traversed by Cooper's E80 Live Load

Positive Shear in Panel 5-6			
Axle at L6	**3**	**4**	**5**
Length of uniform load on the span (ft)	(12 − 6)(25) − 96 = 54	(12 − 6)(25) – 91 = 59	64
Sum moments to L12 (kip-ft)	32728 + 568(54) + 4.0(54)(54/2) = 69,232	73,202	77,272
R_{y0} (kips)	69,232/300 = 230.77	244.0	257.57
Sum moments in panel 5-6 ahead of and about L6 (kip-ft)	460	960	1660
$W_{(5)}$ (kips)	18.40	38.40	66.40
$V^{+}_{(5)-(6)}$ (kips)	230.77 − 18.40/2 = 221.57	224.81	224.37
$M^{+0}_{(5)-(6)}$ (kip-ft)	920	1920	3320
$F^{+}_{M5U6} = \left(\dfrac{-V^{+}_{(5)-(6)}(850)+M^{+0}}{904.78}\right)(1.381)$ (kip-ft)	((−221.57(850) + 920)904.78)1.381 = −286.06	−288.74	−286.03

$F^{+max}_{M5U6} = -288.74$ kips

For $P^{-}_{A} = P^{-}_{F} = 0$, Equation 4.52 yields

$$\frac{P^{-}_{T}}{12} = \frac{\left(\dfrac{P^{-}_{B}}{2}\right)\left(1+\dfrac{(5+1)25}{850}\right)}{\left(\dfrac{300}{850}+1\right)}$$

or

$$\frac{P^{-}_{T}}{5.22} = P^{-}_{B}$$

As shown in Example 4.55, the criterion indicates axle 4 at L4 for maximum force due to negative shear in upper diagonal member M5U6. Therefore, the assumption that $P^{-}_{A} = P^{-}_{F} = 0$ is valid and Equation 4.69c yields the following as given in Table E4.56b:

$$F^{-V}_{M5U6} = \frac{V^{-}_{(5)-(6)}850 + M^{-0}_{(5)-(6)}}{904.78}$$

Where $P^{-}_{A} = P^{-}_{F} = 0$,

$$V^{-}_{(5)-(6)} = R_{yn} - W_{(6)} - \frac{W_{(5)}}{2} = R_{yn} - \frac{W_{(5)}}{2}$$

$$M^{-0}_{(5)-(6)} = \left(12R_{yn} - 6W_{(6)} - \frac{6W_{(5)}}{2}\right)25 = \left(300R_{yn} - 75W_{(5)}\right)$$

TABLE E4.56b
Minimum Negative Force in Upper Diagonal Member M5U6 of a WISS Truss Traversed by Cooper's E80 Live Load

	Negative Shear in Panel 5-6		
Axle at L4	**3**	**4**	**5**
Length of uniform load on the span (ft)	4	9	14
Sum moments to L0 (kip-ft)	32,728 + 568(4) + 4.0(4)(4/2) = 35,032	38,002	41,072
R_{y12} (kips)	35,032/300 = 116.77	126.67	136.91
Sum moments in panel 4-5 ahead of and about L4 (kip-ft)	460	960	1660
$W_{(5)}$ (kips)	18.40	38.40	66.40
$V^{-}_{(5)-(6)}$ (kips)	(116.77) − (18.40/2) = 107.57	107.47	103.71
$M^{-0}_{(5)-(6)}$ (kip-ft)	300(116.77) − 75(18.4) = 33,652	35,121	36,093
$F^{-}_{M5U6} = \left(\frac{V^{-}_{(5)-(6)}(850) + M^{-0}}{904.78}\right)(1.381)$ (kip-ft)	((107.57(850) + 33,652)/904.78)1.381 = 190.92	193.04	189.64

$$F^{-max}_{M5U6} = 193.04 \text{ kips}$$

Vertical member U(k)L(k) of the WISS span, where k is an even number greater than 2 and less than (n/2), can be analyzed like member U(k)L(k) in non-subdivided Warren truss spans by considering equilibrium at joint U(k). Vertical members U(n/2)L(n/2) in WISS trusses can be analyzed like members U(n/2)L(n/2) in PST trusses by considering equilibrium at joint U(n/2) and the forces in upper diagonal members M(n/2-1)U(n/2) and U(n/2)M(n/2+1).[156] Vertical member U2L2 in WISS trusses are calculated using Equation 4.40e for uniformly distributed live loads and Equation 4.40g for concentrated live loads.

Due to the configuration of the end post of the subdivided WISS span, forces in upper end post member M1U2 and lower end post member L0M1 are determined in accordance with Equations 4.41 and 4.42a, respectively, with $h = h_2$. Once the force in end vertical member U2L2 is determined by Equations 4.40e or 4.40g for uniformly distributed and concentrated live loads, respectively, the force in diagonal member M1L2 can be ascertained by considering vertical force equilibrium at joint L2.

4.6.5.2 Subdivided Inclined Chord Warren Truss Spans with Subtie Members

The forces in top chord member U(j − 1)U(j + 1) of inclined-chord Warren truss spans with subtie members (WIST), where upper diagonal members are M(j)U(j + 1), are calculated by taking a section through subpanel (j) − (j + 1) and moments about joint L(j − 1). The forces in bottom chord member L(j − 1)L(j) where upper diagonal members are M(j)U(j + 1) are calculated by taking a section through subpanel (j) − (j + 1) and moments about U(j + 1). Forces in lower diagonal member L(j − 1)M(j) are calculated by taking a section through subpanel (j − 1) − (j) and moments about location P, that is, the intersection between the extension of a line from horizontal bottom chord member L((j − 1)L(j) and a line from inclined top chord member U(j − 1) U(j + 1). Forces in upper diagonal members M(j)U(j + 1) are determined by taking a section through subpanel (j) − (j + 1) and moments about P.

Taking a section through subpanel (j) − (j + 1) and moments about panel point L(j − 1) in a subtie subdivided Warren truss with diagonal members oriented such that lower diagonal members are L(j − 1)M(j) to determine the forces in inclined top chord member U(j − 1)U(j + 1) provides

$$F_{U(j-1)U(j+1)} = \frac{\left(-R_{y0}(j-1)+\sum_{i=0}^{j-1}W_i(j-1-i)\right)-W_{(j)}+F_{M(j)U(j+1)}\left(\dfrac{h_{(j+1)}-2h_S}{\sqrt{\left(h_{(j+1)}-h_S\right)^2+d^2}}\right)}{h_{(j-1)}}\left(\frac{\sqrt{\left(h_{(j+1)}-h_{(j-1)}\right)^2+4d^2}}{2}\right) \quad (4.70a)$$

where

h_S = the length of member M(j)L(j), which is typically equal to $h_{(j+1)}/2$ to provide aligned upper and lower diagonal members and

$F_{M(j)U(j+1)}$ = the force in upper diagonal member M(j)U(j+1), which must be determined for the live load position of concern (typically, the live load position that creates the maximum force in inclined top chord member U(j − 1)U(j + 1)).

In the usual case, $h_S = h_{(j+1)}/2$, and Equation 4.70a is

$$F_{U(j-1)U(j+1)} = \left(\frac{\left(-R_{y0}(j-1)+\sum_{i=0}^{j-1}W_i(j-1-i)\right)-W_{(j)}}{h_{(j-1)}}\right)\left(\frac{\sqrt{\left(h_{(j+1)}-h_{(j-1)}\right)^2+4d^2}}{2}\right)$$

Taking a section through subpanel (j) − (j + 1) and moments about panel point U(j + 1) in a subtie subdivided Warren truss with diagonal members oriented such that lower diagonal members are L(j − 1)M(j), to determine the forces in horizontal bottom chord member L(j − 1)L(j),

$$F_{L(j)L(j+1)} = F_{L(j-1)L(j)} = \frac{\left(R_{y0}(j+1)-\sum_{i=0}^{j+1}W_i(j+1-i)\right)d}{h_{(j+1)}} \quad (4.70b)$$

The force in lower diagonal member L(j − 1)M(j), where j = 5, 9, 13. . . ≤ (n/2 − 1) and (n) is an even number[157] of a WIST truss is calculated by taking a section through subpanel (j − 1) − (j) and determining the positive shear in subpanel (j − 1) − (j) by taking moments about location P such that

$$F^{+}_{L(j-1)M(j)} = \frac{-R_{y0}\left(a_{(j-1)-(j+1)}\right) + \sum_{i=0}^{j-1} W_i\left(a_{(j-1)-(j+1)} + id\right) + F^{+V}_{U(j-1)(M(j)}\left(a_{(j-1)-(j+1)} + (j+1)d + \dfrac{\left(2h_s - h_{(j-1)}\right)d}{h_{(j-1)} - h_s}\right)}{a_{(j-1)-(j+1)} + (j-1)d} \left(\frac{\sqrt{h_S^2 + d^2}}{h_S}\right) \quad (4.70c)$$

where

$F^{+V}_{U(j-1)(M(j)} = \left(W_{(j)} - F^{+V}_{M(j)U(j+1)}\left(\frac{h_{(j+1)} - 2h_S}{h_{(j+1)} - h_S}\right)\right)\left(\frac{h_{(j-1)} - h_S}{h_{(j-1)}}\right)$ = the vertical component of the force in subtie member U(j-1)M(j) due to positive shear, as indicated in Equation 4.70f, which must be determined for the live load position of concern (typically, the live load position that creates the maximum force in lower diagonal member L(j-1)M(j)):

$F^{+V}_{M(j)U(j+1)} = \left(\dfrac{-R_{y0}\left(a_{(j-1)-(j+1)}\right) + \sum_{i=0}^{j} W_i\left(a_{(j-1)-(j+1)} + id\right)}{a_{(j-1)-(j+1)} + (j-1)d + \left(\frac{h_{(j+1)} - 2h_S}{h_{(j+1)} - h_S}\right)d}\right)$ = the vertical component of the force in upper diagonal member M(j)U(j+1) due to positive shear, as provided in Equation 4.70d, which must be determined for the live load position of concern (typically, the live load position that creates the maximum force in lower diagonal member L(j-1)M(j)).

In the usual case, $h_S = h_{(j+1)}/2$, and Equation 4.70c is

$$F^{+}_{L(j-1)M(j)} = \frac{-R_{y0}\left(a_{(j-1)-(j+1)}\right) + \sum_{i=0}^{j-1} W_i\left(a_{(j-1)-(j+1)} + id\right) + W_{(j)}\left(\dfrac{h_{(j-1)} - h_S}{h_{(j-1)}}\right)\left(a_{(j-1)-(j+1)} + (j+1)d + \dfrac{\left(2h_s - h_{(j-1)}\right)d}{h_{(j-1)} - h_s}\right)}{a_{(j-1)-(j+1)} + (j-1)d} \left(\frac{\sqrt{h_S^2 + d^2}}{h_S}\right)$$

The forces in upper diagonal member M(j)U(j + 1) of a WIST truss are calculated by taking a section through panel (j) − (j + 1) and determining the shear in panel (j) − (j + 1) by taking moments about P such that for positive shear,

$$F^{+V}_{M(j)U(j+1)} = \left(\frac{-R_{y0}\left(a_{(j-1)-(j+1)}\right) + \sum_{i=0}^{j} W_i\left(a_{(j-1)-(j+1)} + id\right)}{a_{(j-1)-(j+1)} + (j-1)d + \left(\dfrac{h_{(j+1)} - 2h_S}{h_{(j+1)} - h_S}\right)d}\right) \quad (4.70d)$$

$$F^{+}_{M(j)U(j+1)} = \left(\frac{-R_{y0}\left(a_{(j-1)-(j+1)}\right) + \sum_{i=0}^{j} W_i \left(a_{(j-1)-(j+1)} + id\right)}{a_{(j-1)-(j+1)} + (j-1)d + \left(\frac{h_{(j+1)} - 2h_S}{h_{(j+1)} - h_S} \right) d} \right) \left(\frac{\sqrt{\left(h_{(j+1)} - h_S\right)^2 + d^2}}{h_{(j+1)} - h_S} \right) \quad (4.70e)$$

In the usual case, $h_S = h_{(j+1)}/2$, and Equation 4.70e is

$$F^{+}_{M(j)U(j+1)} = \left(\frac{-R_{y0}\left(a_{(j-1)-(j+1)}\right) + \sum_{i=0}^{j} W_i \left(a_{(j-1)-(j+1)} + id\right)}{a_{(j-1)-(j+1)} + (j-1)d} \right) \left(\frac{\sqrt{\left(h_{(j+1)} - h_S\right)^2 + d^2}}{h_{(j+1)} - h_S} \right)$$

Taking moments about L(j − 1) of forces at M(j) indicates that the forces in subtie member U(j − 1)M(j) are

$$F^{+}_{U(j-1)(M(j)} = \left(W_{(j)} - F^{+V}_{M(j)U(j+1)} \left(\frac{h_{(j+1)} - 2h_S}{h_{(j+1)} - h_S} \right) \right) \left(\frac{\sqrt{\left(h_{(j-1)} - h_S\right)^2 + d^2}}{h_{(j-1)}} \right) \quad (4.70f).$$

In the usual case, $h_S = h_{(j+1)}/2$, and Equation 4.70f is

$$F^{+}_{U(j-1)(M(j)} = W_{(j)} \left(\frac{\sqrt{\left(h_{(j-1)} - h_S\right)^2 + d^2}}{h_{(j-1)}} \right) \quad (4.70g).$$

The forces in lower diagonal members and upper diagonal members due to negative shear are calculated as in Equations 4.70c and 4.70e at sections (j − 1) − (j) and (j) − (j + 1), respectively. Forces in vertical members are calculated by isolating joints as free-body diagrams.

4.6.6 Influence Lines for Subdivided Truss Spans

As is the case for non-subdivided horizontal and inclined chord trusses, influence lines are useful for establishing the maximum member forces in subdivided truss spans. Equations for chord and web member forces in subdivided truss spans can be utilized to construct influence lines for subdivided truss member forces. The influence lines are readily calculated with unit loads placed at appropriate panel points as shown in Example 4.57 for a PST span. Once constructed, influence lines can be used to determine the maximum and minimum live load forces in subdivided truss chord and web members with the appropriate live load location.

Example 4.57

Develop the influence lines for bottom chord member L5L6, top chord member U4U6, upper diagonal member U2M3 and lower diagonal member M3U4 of PST truss in Examples 4.46, 4.48 and 4.49.

The influence line for bottom chord member L5L6 can be developed with Equation 4.54a applied to unit loads placed at panel points, which results in Figure E4.57a. For example, with a unit load located at panel point L4,

$R_{y0} = 200/300 = 0.667$
$F_{L5L6} = 4(0.667 - 0)(25/47.5) = 1.4035$

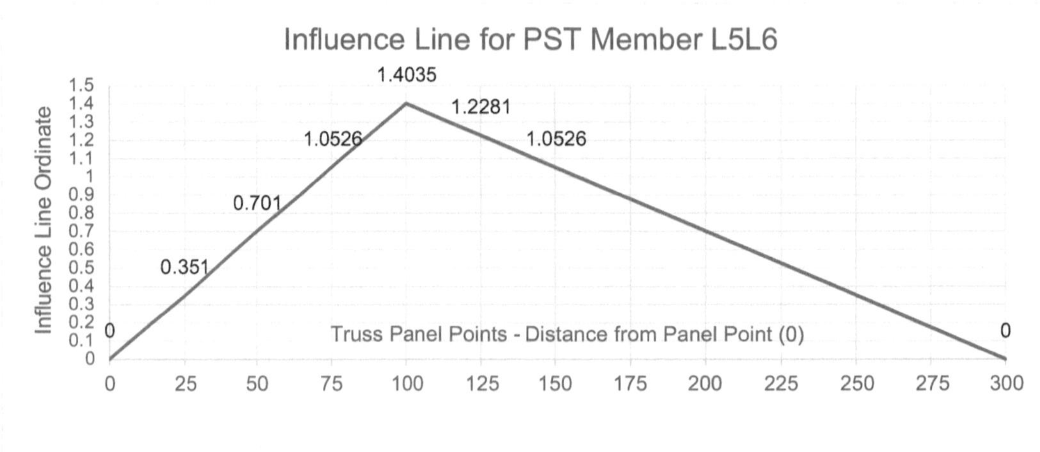

FIGURE E4.57a Influence line for bottom chord member L5L6 of a PST span.

The influence line for top chord member U4U6 can be developed with Equation 4.55b applied to unit loads placed at panel points, which results in Figure E4.57b.

$$\propto = \frac{\sqrt{(50-47.5)^2 + 50^2}}{50} = 1.0013$$

With a unit load located at panel point L4,

$R_{y0} = 200/300 = 0.667$
$F_{U4U6} = (-6(0.667) + 2 - 0)(25/50)(1.0012) = -1.0013$

With a unit load located at panel point L5,

$R_{y0} = 175/300 = 0.5833$
$F_{U4U6} = (-6(0.5833) + 1 - 1)(25/50)(1.0012) = -1.7522$

With a unit load located at panel point L6,

$R_{y0} = 150/300 = 0.500$
$F_{U4U6} = (-6(0.500) + 0 - 0)(25/50)(1.0012) = -1.5019$

Note that Figure E4.57b indicates a peak value in the influence line for U4U6 at panel point 5. This occurs because the load at panel point L5 is to the right of the section through

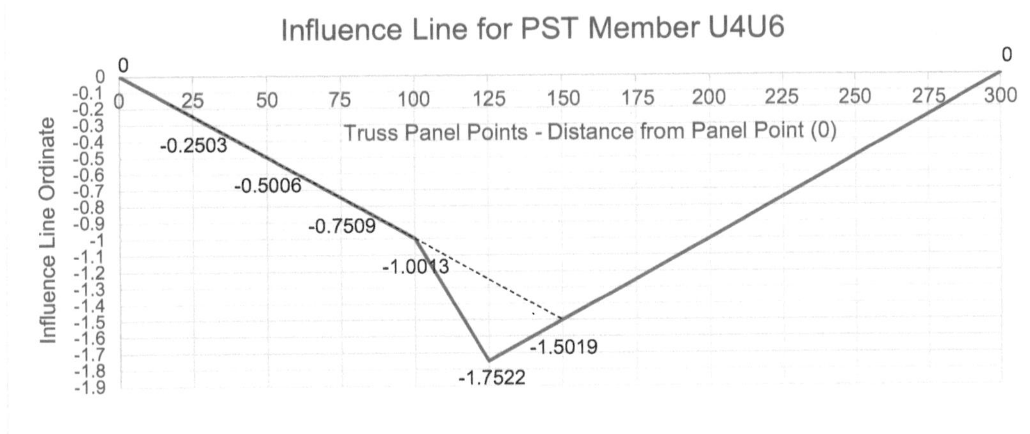

FIGURE E4.57b Influence line for top chord member U4U6 of a PST span.

subpanel 4-5 and to the left of the moment center at L6. Therefore, the influence line in panel 5-6 is an extension of the influence line between panel point (0) and panel point (4). Such peaks in top chord members U(j − 1)U(j + 1) of subdivided truss spans will occur when the load at panel point (j) is to the right of the section through subpanel (j − 1) − (j) and to the left of the moment center at L(j + 1). The influence line in subpanel (j) − (j + 1) is an extension of the influence line between panel point (0) and panel point (j − 1).

The influence line for upper diagonal member U2M3 can be developed with Equations 4.62a and 4.62b for positive and negative panel shear, respectively, applied to unit loads placed at panel points, which results in Figure E4.57c:

$$a_{(2)-(3)} = 150 \text{ ft}$$
$$a_{(2)-(3)} + (j + 1)d = 150 + 4(25) = 250 \text{ ft}$$
$$\alpha = 1.653 \text{ for } j = 3$$

With a unit load located at panel point L2,

$$R_{y0} = 250/300 = 0.8333$$
$$F_{U2M3} = (((0.8333 - 1)(150) - 25(2))/250)(1.653) = -0.496$$

With a unit load at L3,

$$R_{y0} = 225/300 = 0.750$$
$$F_{U2M3} = (((0.750 - 0)(150) - 25(0))/250)(1.653) = 0.744$$

Note that the influence line for member U2M3 passes from negative to positive through subpanel 2-3. Therefore, loads must be placed at panel points 2 and 3 for maximum forces in U2M3 due to positive and negative shear. Therefore, the influence line for diagonal members in subpanel sections that contain only top chord, diagonal and bottom chord members requires that loads be placed at the panel points to the left and right side of the subpanel under consideration. The influence line crosses the x-axis (zero shear) 15 ft to the left of panel point 3, as predicted by Equation 4.22b.

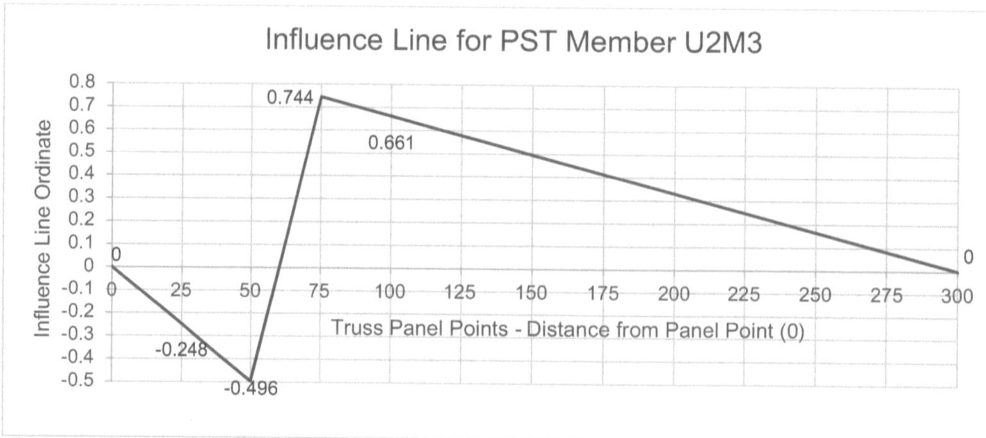

FIGURE E4.57c Influence line for upper diagonal member U2M3 of a PST span.

The influence line for lower diagonal member M3L4 can be developed with Equations 4.63a and 4.63b for positive and negative panel shear, respectively, applied to unit loads placed at panel points, which results in Figure E4.57d:

$$V = R_{y0} - \sum_{i=0}^{j} W_i + 0.60W_{(j)}$$

$$M^0 = d\sum_{i=0}^{j} iW_i - 35W_{(j)}$$

$a_{(2)\text{-}(3)} = 150$ ft

$a_{(2)\text{-}(3)} + (j+1)d = 150 + 4(25) = 250$ ft

$\alpha = 1.653$ for $j = 3$

With a unit load located at panel point L2,

$R_{y0} = 250/300 = 0.8333$

$V = 0.833 - 1 + 0 - 0.167$

$M^0 = (25)(2)(1) - 0 = 50$

$F_{U2M3} = (-0.167(150) - 50)/250)(1.653) = -0.496$

With a unit load at panel point L3,

$R_{y0} = 225/300 = 0.750$

$V = 0.750 - 1 + 0.60(1) = 350$

$M^0 = (25)(3)(1) - 35(1) = 40$

$F_{U2M3} = (0.35(150) - 40)/250)(1.653) = 0.0827$

With a unit load at panel point L4,

$R_{y0} = 225/300 = 0.667$

$V = 0.667 - 0 + 0 = 0.667$

$M^0 = (25)(0) - (0) = 0$

$F_{U2M3} = (0.667(150) - 0)/250)(1.653) = 0.661$

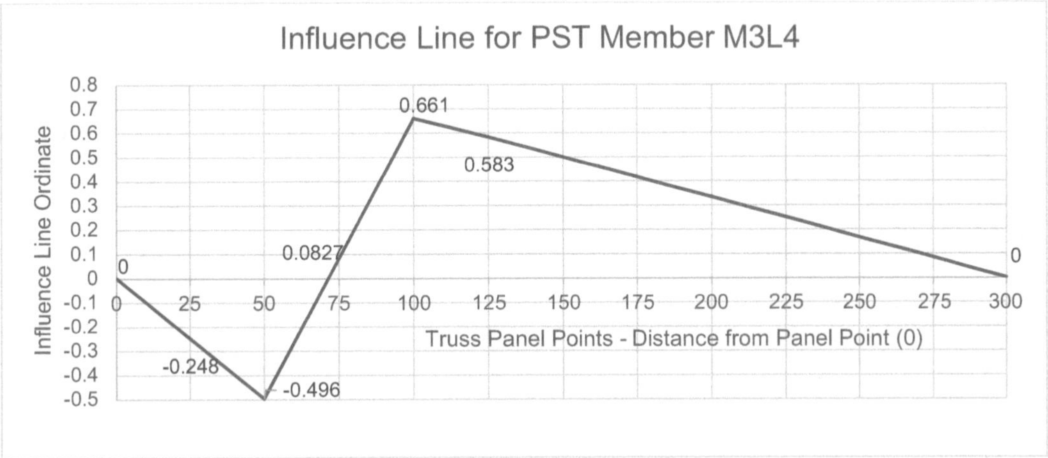

FIGURE E4.57d Influence line for lower diagonal member M3L4 of a PST span.

Note that the influence line for member M3L4 passes from negative to positive through panel 2-4. Therefore, loads must be placed at panel points 2 and 4 for maximum forces in M3L4 due to positive and negative shear. Therefore, the influence line for diagonal members in subpanel sections that contain top chord, diagonal, substrut or subtie and bottom chord members required that loads be placed at the panel points to the left and right side of the panel under consideration. The influence line crosses the x-axis (zero shear) 3.57 feet to the left of panel point 3.

4.6.7 Influence Lines for Compound Truss Spans

Very long spans are typically continuous spans or of cantilever construction. Continuous spans are externally statically indeterminate and for that reason are only rarely constructed where conditions such as unyielding foundations exist and where member fit-up and temperature effects can be accommodated. Cantilever construction allows for very long spans that can be efficiently designed and constructed as statically determinate structures.

Long-span steel railway bridges can be economically constructed as cantilever bridges (see Chapter 1). The economical relative lengths of the cantilever arms, L_c; anchor spans, L_a and suspended spans, L_s, will vary with live to dead load bending moment ratio. For the relatively high ratios of steel railway superstructures, L_a/L_c of 1 to 2 are typically used depending on the suspended span length, L_s. In steel railway superstructures, L_c/L_s typically range from 0.4 to 2. The relative lengths of the cantilever arm, anchor and suspended spans are also governed by site conditions that dictate the location of piers at a crossing (Unsworth, 2018). Influence lines for cantilever superstructures are constructed by considering the unit loads traversing the bridge. The ordinates of the influence lines are readily determined by calculating the reaction, bending moment and shear due to unit loads at locations where the influence lines change direction. Example 4.58 outlines the construction of influence lines for members in a cantilever truss bridge.

Example 4.58

Construct influence lines for the members in panel 2-3 in the anchor arm of the cantilever truss bridge in Figure E4.58. Following inspection and placement of unit loads at L0, L2, L3, L6 and considering the hinge at the end of the cantilever and suspended spans, the influence lines for axial forces in L2-L3, U2-U3 and U2-L3 are shown in Figure E4.58. Influence lines for axial force in other members of the trusses are constructed in a similar manner.

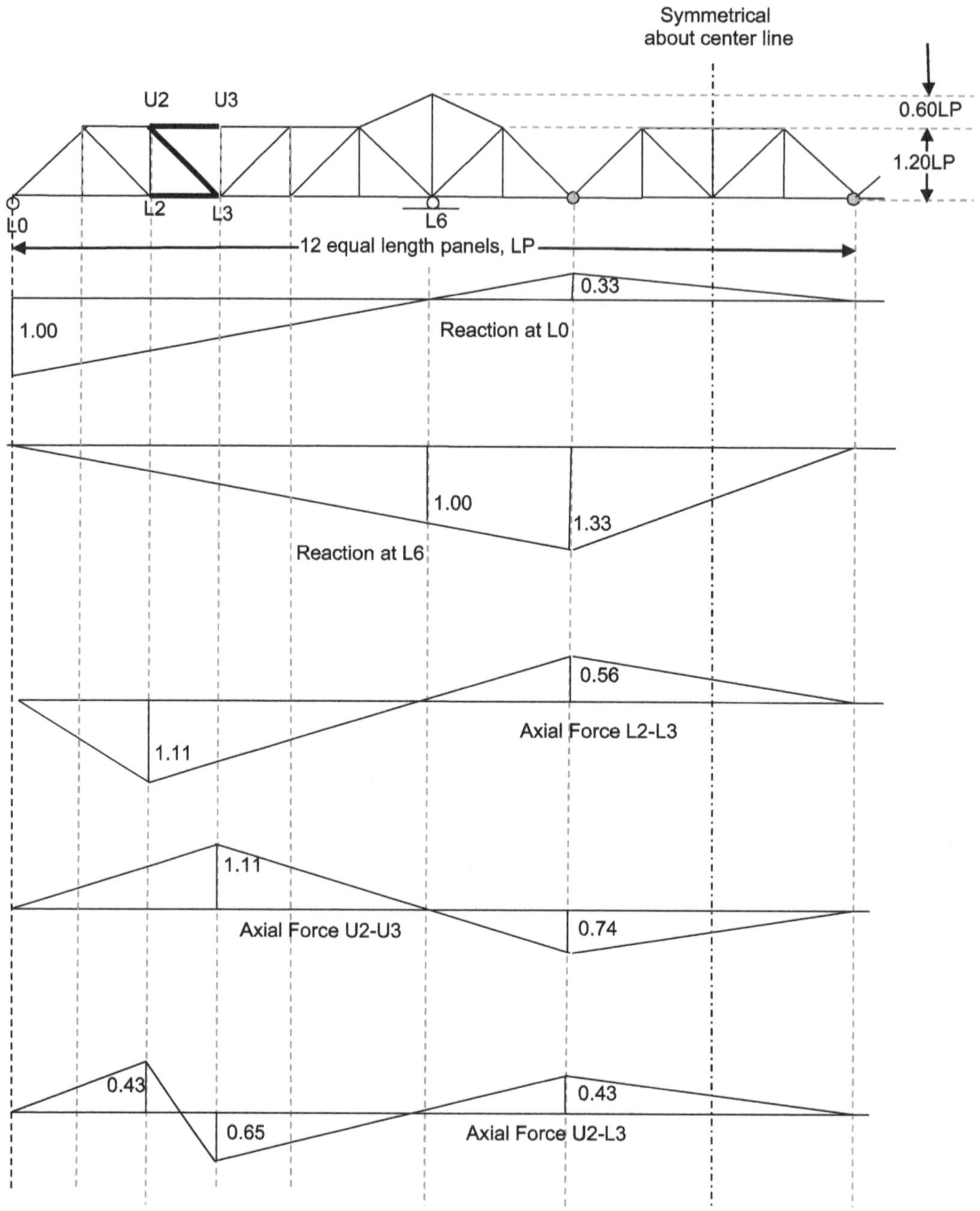

FIGURE E4.58 Influence lines for reactions and member forces in panel 2-3 of Cantilever truss bridge.

NOTES

1 See Chapter 2 for a discussion of simple, compound and complex truss types.
2 Typically for vehicle roadways, railways or rivers with high water levels.
3 For example, some high-level crossings use deck trusses with vertical end posts to reduce the height of substructures.
4 For example regulatory, hydraulic, clearance and/or geotechnical conditions.
5 Economical truss designs typically have diagonal members at orientations of 40° to 60° angles with the horizontal.

6 Typically horizontal parallel- and inclined-chord Warren and Pratt forms for modern steel truss spans of moderate length. For long modern steel truss spans, subdivided panel truss forms are generally used.
7 Long panels can be uneconomical due to floor system weight.
8 Typically a minimum of 7.0 m (23 ft) from the top of the rail to the underside of structural steel portal or lateral sway bracing members.
9 A Warren truss without vertical members is often referred to as a triangular truss. These are inefficient for moderate-length and long railway spans due to the distances between panel points. Since trusses are not often used for short spans, Warren trusses without vertical members are typically not used in railway bridges.
10 The vertical web members must support floorbeams. Vertical members in Warren trusses are either members directly loaded by floorbeam reactions (vertical members in tension in Warren through truss spans and compression in Warren deck truss spans) or, in horizontal chord Warren trusses, zero-force members. Vertical members in inclined chord Warren trusses are members either directly loaded by floorbeam reactions or loaded by forces from the inclined chord members at the joint between the vertical member and the inclined chord members.
11 For trusses with single diagonal web members (only one diagonal member in each truss panel), chord member forces are calculated from opposite panel point moments, and web member forces are calculated by considering panel shear forces. These forces in horizontal-chord trusses are resisted solely by web members; in inclined-chord trusses, they are resisted by both web members and the inclined chords.
12 The diagonal web members of Howe through-trusses are in compression under balanced loading, and the vertical members experience tensile forces of considerable magnitude. Early railroad bridges, with relatively light live loads, that used timber chords, timber diagonals and steel rod vertical members were economical to construct in the Howe through-truss form. For example, wood diagonals and wood chords were often connected with cast-iron bearing blocks. The use of metal truss members in the latter part of the of the 19th century precipitated the widespread use of Pratt and Warren truss forms, with diagonal web members in tension or compression under balanced loading since metals performed well in both tension and compression. However, the early use of tension-only members (such as eyebars) for diagonal web members precipitated the need for "counters" or diagonal web members of the opposite inclination or orientation, particularly in central panels of the truss, to counter the reversal of stress as live loads traversed the span. When active, these counters change the truss panel geometry creating an unsymmetrical planar truss that must be considered in moving load structural analyses.
13 In comparison with the vertical members in Pratt and Warren truss forms.
14 See Chapter 7 for a detailed review of secondary stresses in truss members. Warren truss chord member secondary flexural stresses are caused by elongation of the vertical web members in tension. These members in tension create relatively large bottom chord rotations at adjacent panel points, and the vertical web members at those panel points connected at the top chord (even if zero-force members) can create top chord secondary flexural stresses. Similar effects are manifested in subdivided panel trusses used for long spans. The "K" truss form in Figure 4.1e can be used to alleviate this effect. The K truss form was used in the design of the immense cantilever trusses of the Quebec bridge to mitigate secondary stresses (see Chapter 1).
15 The Baltimore truss is a subdivided panel horizontal chord Pratt truss. The Baltimore truss shown in Figure 4.1b is subdivided by substruts.
16 Petit and Pennsylvania trusses are inclined-chord Baltimore trusses. The Petit and Pennsylvania trusses shown in Figure 4.1b are subdivided by substruts.
17 For example, cantilever truss bridges are compound truss spans. The indeterminacy of compound truss cantilever bridges can be reduced by introducing construction devices such as links and hinges that provide additional equations of condition. In this manner, cantilever truss bridges can be designed and constructed as statically determinate compound span bridges.
18 Type and configuration depend on site characteristics as well as economic and in some cases aesthetic considerations.
19 Static and live (moving) primary loads must be considered.
20 Wind can be considered as a secondary force.
21 Typically temperature changes, member length variations and/or support settlements.
22 Good estimates of initial member properties will often lead to only a single analytical iteration being required in the analysis of statically indeterminate truss forms.
23 In particular the method of sections.
24 In addition, the classical methods of structural analysis are particularly useful when only a single or a few truss members require evaluation, such as during truss retrofit and rehabilitation projects.

25 Castigliano's Second Theorem and the method of virtual forces (method of consistent displacements or method of redundant forces) are used to calculate member forces in statically indeterminate trusses. Castigliano's Second Theorem is valid when only external loads are applied to the truss joints; The method of virtual work is more general and valid for the determination of truss joint displacements due to external static and live loads, temperature changes, support settlements and changes in fabricated member lengths. The method of virtual work is typically used for analyzing joint displacements in both statically determinate and indeterminate trusses.
26 Truss member sizes must be known to determine the displacement compatibility equations.
27 When NM + NR < 2NJ, there are not enough unknowns, NM + NR, to support the simultaneous solution of 2NJ equations of equilibrium, and the truss is necessarily statically unstable.
28 A free body diagram is an illustration of part of the truss span including all external loads, reactions and internal member forces at members intersected by the part of the truss being considered. Free body diagrams are useful for investigating static equilibrium at truss joints or sections of the truss.
29 Influence lines for axial forces in truss members illustrate the variations in the axial force in the member as a unit load traverses the truss span.
30 Such as hinges and hangers in cantilever truss spans that enable a statically determinate analysis. These provide condition equations in addition to the equilibrium equations, thereby providing a statically determinate truss.
31 Including the entire truss as a free body diagram to determine external (support) reactions.
32 Positive shear at a section is considered when the resultant forces to the right side of the section are downward and the resultant forces to the left are upward. Negative shear at a section is considered when the resultant forces to the right side of the section are upward and the resultant forces to the left are downward.
33 Attitude is the vector direction of the axial force, causing either compression or tension in the member.
34 For a horizontal parallel-chord Pratt truss, the vertical web member is loaded by the diagonal web member in the adjacent panel.
35 Taking moments about the opposite panel point eliminates moments created by the opposite chord and diagonal member in the section.
36 Typically spans between 45 m (150 ft) and 75 m (250 ft) in length.
37 In addition to the analytical and performance benefits of statically determinate trusses, secondary stresses due to member fit-up of intersecting members in central panels are avoided.
38 Inclined end post members are common in through-truss spans. Vertical end post members are typically not used with the exception of deck-truss spans where reducing substructure height is beneficial. Therefore, analytical methods will focus on horizontal parallel-chord, inclined-chord and subdivided panel Pratt and Warren through-truss spans with inclined end post members. Other truss forms are analyzed in a similar manner.
39 Dead loads are commonly modeled as uniformly distributed loads across the entire span. Uniformly distributed live loads can be only partially on the span as the live load traverses the span.
40 Howe through-trusses with wood chord members and wood compression diagonals (using simple cast-iron bearing block connections to the chord members) with iron rod vertical web members were economical and easy to construct to resist the relatively light live loads of the mid- to late-19th century.
41 The vertical web member forces in Howe truss spans are typically considerably greater than those of Pratt and Warren truss spans under most load conditions.
42 The panel point where the moments due to the diagonal member forces in the panel and the opposite chord member forces are null.
43 With the first panel point (joint at left end of the span) at j = 0 and the last panel point (joint at right end of the span) at j = n. The panel point at the center of the statically determinate truss with an even number of panels is j = n/2.
44 Dead loads are typically symmetrical about the centerline of the truss span, and live loads traversing from each end of the truss spans provide symmetrical loading conditions in truss panels.
45 U(j − 1)L(j) is the orientation of diagonal web members in Pratt truss panels and alternate panels of Warren trusses.
46 L(j − 1)U(j) is the orientation of diagonal web members in alternate Warren truss panels.
47 Positive shear at a section is considered when the resultant forces to the right side of the section are downward and the resultant forces to the left are upward, which occurs with live loads traversing from right to left over the span (live load moving from L(n) to L(0)).
48 Negative shear at a section is considered when the resultant forces to the right side of the section are upward and the resultant forces to the left are downward, which occurs with live loads traversing from left to right over the span (live load moving from L(0) to L(n)).

49 Where k = j − 1, the vertical member U(k)L(k) is loaded by the diagonal member U(j − 1)L(j) in panel (j − 1) − (j).

50 Where k = j, the vertical member U(k)L(k) is loaded by the diagonal member U(j)L(j + 1) in panel (j) − (j + 1).

51 Typically includes track, floor system, bracing and truss self-weight.

52 Concentrated dead loads (such as dead loads due to appurtenances on the truss span) can be properly added to the truss members and/or panel points.

53 Track, floor system and truss span bracing weights are typically applied directly at panel points. Truss self-weight can also be apportioned to the panel points. The truss and bracing weights are apportioned between the top and bottom chord panel points. However, truss member self-weight must be transferred through beam–column behaviour to the panel points. Nevertheless, bending due to truss member self-weight typically needs to be considered only as a secondary stress for unusual, complex or long-span truss design.

54 The top chord of deck-truss spans and bottom chord of through-truss spans.

55 Representing locomotive axles or wheels.

56 Representing trailing freight cars.

57 Nevertheless, in some deck-truss spans without a floor system, live loads are transferred directly to the top chord. In such cases, bending of the top chord must be considered in conjunction with axial forces from truss behaviour, thereby producing combined axial and bending stresses.

58 Particularly web members in panels near the center of truss spans.

59 For example during retrofit design or quality assurance design reviews.

60 For example, influence line analysis is often the analytical foundation of algorithms used in moving load analysis software.

61 Equivalent uniformly distributed load charts for shear forces and bending moments can be prepared that depend on span length and location on the span. Equivalent uniformly distributed loads for bending moment w_{eM}, can be prepared considering $w_{eM} = \frac{2M}{x(L-x)}$ where M = bending moment at location x due to concentrated loads on a span of length L. Equivalent uniformly distributed loads for shear force w_{eV} can be prepared considering $w_{eV} = \frac{2VL}{(L-x)^2}$ where V = shear force at location x due to concentrated loads on a span of L.

62 Intuitively, the uniformly distributed live load entirely on the span creates maximum chord member forces in panel (j − 1) − (j).

63 Intuitively, the uniformly distributed live load adjacent to either side of panel (j − 1) − (j) creates the maximum web member forces in panel (j − 1) − (j).

64 It should be noted that uniformly distributed loads across the entire span provide maximum and minimum chord member forces but not maximum and minimum web member forces.

65 Panel (j − 1) − (j) for $j \leq n/2$ is sufficient because of the symmetry of truss geometry and loading.

66 If the vertical web member in a Pratt truss panel (j − 1) − (j) is member U(j)L(j), then $F_{U(j)L(j)} = -V_{(j)-(j+1)} = -w(n/2 - j + 2)$.

67 In the middle of the 19th century, European engineers such as Schwedler and Winkler (see Chapter 1) were using partial uniformly distributed loads to determine the maximum shear forces in railway bridge spans (Kurrer, 2008).

68 $x^{+}_{(j-1)-(j)}$ and $x^{-}_{(j-1)-(j)}$ are the neutral points in panel (j − 1) − (j).

69 The vertical web member experiences truss action.

70 Vertical member U(j − 1)L(j − 1) in a Pratt truss is loaded by the diagonal member U(j − 1)L(j) in panel (j − 1) − (j).

71 The approximate method (often referred to as the conventional method) precludes the requirement to locate the position of the uniformly distributed live load within panel (j − 1) − (j) using Equations 4.5c for positive shear and 4.5g for negative shear.

72 Load P_w at joint (j) in Figure 4.5a or at joint (j − 1) in Figure 4.5b requires that panel (j − 1) − (j) be fully loaded, which implies that $P_w = wd/2$ at joint (j) or (j − 1) (where w is the magnitude of the uniformly distributed live load). Therefore, assuming $P_w = 0$ at joint (j) for negative shear or (j − 1) for positive shear results in a slightly conservative but convenient calculation of the magnitude of the maximum positive and negative live load shear in panel (j − 1) − (j).

73 The sum of an arithmetic series is equal to the sum of the first term in the series and the last term in the series multiplied by half the number of terms in the series.

74 Typically a design load consisting of concentrated and uniformly distributed live loads. The uniformly distributed live loads can be resolved to equivalent concentred live loads so the analysis can proceed based on considering only concentrated live loads.

75 The maximum moment at a panel point (k) will occur with a concentrated load at (k) since the panel point moment is a maximum where the shear force becomes zero, which occurs with a concentrated load at (k).

76 Since the panel point moment is a maximum where the shear force becomes zero, the shear force must change from positive to negative at panel point (j) for positive shear and from negative to positive at (j – 1) for negative shear, which will occur as the concentrated load at the panel point moves from left to right at (j) for positive shear and from right to left at (j – 1) for negative shear. Therefore, the maximum shear in panel (j) – (j – 1) will occur with a concentrated load at (j) for positive shear and at (j – 1) for negative shear.

77 Based on analyses of moving load effects on panel point moments and panel shear forces, criteria can be calculated to establish the location of a series of concentrated loads that create maximum effects. For short spans, the position of live load for maximum effects may be readily observed and determined by only a few cycles of trial-and-error calculations.

78 Typical of specified or recommended design live loads. For constant load patterns, it is efficient to develop a load table to assist with calculating the panel point moments and panel shear forces.

79 The live load moves on longitudinal stringers between transverse floorbeams connected to the truss panel points.

80 Such as the Cooper's E series load recommended by Chapter 15 of the 2024 AREMA MRE.

81 The reason for defining the load in panel (j – 1) – (j) as P_B will become apparent when analyzing subdivided panel trusses where the subdivided panel loads are defined as P_A for loads in the subdivided panel that does not contain the member being considered and P_B for loads in the subdivided panel containing the member being considered.

82 Typically called hangers, suspenders or hip verticals for tension members.

83 Typically longitudinal stringers between transverse floorbeams.

84 Typically, through- and deck-trusses have equal panel lengths.

85 The Cooper's E series design live load has eighteen concentrated forces (locomotive axles) followed by a uniformly distributed load (trailing freight cars). The distance from the first axle or wheel of the Cooper's E series concentrated load to the start of the uniformly distributed load is 33.2 m (109 ft).

86 Typically, truss panels (stringers) are between 4.5 m (15 ft) and 9 m (30 ft) long.

87 For example, axle or wheel numbers 3, 4, 5,12, 13 or 14 will create the maximum moment in panels of about 7.5 m (25 ft) long.

88 Excluding the forces in vertical web members directly loaded by floorbeam reactions that do not participate in truss action.

89 Typically, the live load traverses the truss span via a floor system of longitudinal stringers and transverse floorbeams framing into the truss span panel points.

90 Influence lines are also constructed for moving load analysis of statically indeterminate trusses using the Muller–Breslau principle.

91 For example at supports, panel points and hinges (a construction condition installed in compound truss spans to provide a statically determinant structure).

92 For axial forces in statically determinate trusses, influence lines consist of straight-line segments.

93 The member forces determined by influence lines for a 150 ft long and 25 ft high Pratt through-truss span subjected to uniformly distributed loads will be used in subsequent chapters containing examples for calculating statically indeterminate member forces, joint deflections and secondary member forces.

94 The member forces determined by influence lines for a 150 ft long and 25 ft high Pratt through-truss span subjected to Cooper's E80 live load will be used in subsequent chapters containing examples for calculating statically indeterminate member forces, joint deflections and secondary member forces.

95 Midspan chord member forces are reduced due to the increased distance between the chords, and the inclined top chord members will resist some panel shear due to their orientation.

96 An inclined-chord Pratt truss is often referred to as a Parker truss.

97 Inclined-chord trusses with an odd number of panels will require at least one double-diagonal panel at the center panel, thereby rendering the truss statically indeterminate.

98 Typically truss spans between about 60 m (200 ft) and 90 m (300 ft) long.

99 The number of joints is equal to the number of panels with end joint L0 for through-trusses or U0 for deck-truss spans.

100 Where $j \leq n/2$ with the first panel point (joint at left end of the span) at $j = 0$ and the last panel point (joint at right end of the span) at $j = n$.

101 This result can also be obtained by resolving the force in member U(j – 1)L(j) at panel point L(j).

102 Vertical member U(j)L(j) is loaded by diagonal member U(j)L(j + 1) in panel (j) – (j + 1), and the section in Figure 4.12b cuts the top chord in panel (j – 1) – (j).

103 These vertical members are hangers.
104 Minimum chord member forces are dead load forces since the attitude (axial compression or tension) of the chord forces is the same for dead and live loads.
105 Since horizontal chord members do not resist panel shear, the web member forces can be determined directly in terms of only the panel shear force and orientation of the web member.
106 For example, assuming the live loads are moving from right to left (from Ln to L0) across the span, a location exists where the member forces change from increasing to decreasing in magnitude.
107 Parker trusses have diagonal members U(j – 1)L(j) in panels (j – 1) – (j) where j ≤ n/2.
108 Warren trusses have diagonal members U(j – 1)L(j) in even numbered panels (j – 1) – (j) and diagonal members L(j – 1)U(j) in odd numbered panels (j – 1) – (j), where j ≤ n/2.
109 Diagonals with the orientation U(j – 1)L(j) are found in every panel of Pratt trusses and every even-numbered panel of Warren trusses where j ≤ n/2.
110 Diagonals with the orientation L(j – 1)U(j) are found in every panel of Howe trusses and every odd-numbered panel of Warren trusses where j ≤ n/2.
111 This expected because for uniformly distributed live loads, the magnitude of the panel shear forces does not depend on the orientation of the diagonal member in the panel.
112 A post height of 25 ft does not necessarily allow for typical lateral bracing in the plane of the top chord of the through-truss span. However, Examples 4.19 and 4.20 indicate a methodology for calculating member forces in inclined-chord through-truss spans.
113 Railway live loads traversing a truss span from right to left are typically used for structural analysis as the Cooper's E live load configuration is typically illustrated with the leading axle (axle no. 1) on the left.
114 The criterion of Equation 4.11c is independent of the section taken and therefore of the truss form.
115 For loads moving from right to left over the span, axles 2, 3 or 4 at panel point (j) typically create maximum positive shear forces in panel (j – 1) – (j) where j ≤ n/2. Therefore, except for very short truss panels, there will be no live loads in panels between panel point (0) and panel point (j – 1). For loads moving from left to right over the span, axles 2, 3 or 4 at (j – 1) typically create maximum negative shear forces in panel (j – 1) – (j). Therefore, except for very short truss panels, there will be no live loads in panels between panel point (j) and panel point (n).
116 The concentrated loads move to the left from panel point Ln to panel point L0.
117 The concentrated loads move to the right from L0 to L(n).
118 Diagonal members with orientation L(j – 1)U(j) are found in Howe truss panels and alternate panels of Warren truss spans where j ≤ 2.
119 If the vertical member is taken as U(j – 1)L(j-1), the force in the member is calculated by taking a section through members U(j – 2)U(j – 1) and L(j – 1)L(j) such that $F^{+}_{U(j-1)L(j-1)} = \left(\dfrac{-V^{+}_{(j-1)-(j)}\left(a_{(j-2)-(j-1)}\right) + M^{+0}_{(j-1)-(j)}}{\left(a_{(j-2)-(j-1)} + (j-1)d\right)} \right)$.
120 If the vertical member is taken as U(j – 1)L(j – 1), the force in the member is calculated by taking a section through members U(j – 2)U(j – 1) and L(j – 1)L(j) such that $F^{-}_{U(j-1)L(j-1)} = \left(\dfrac{V^{-}_{(j-1)-(j)}\left(a_{(j-2)-(j-1)}\right) + M^{-0}_{(j-1)-(j)}}{\left(a_{(j-2)-(j-1)} + (j-1)d\right)} \right)$.
121 Typically, truss panels (stringers) are between about 6 m (20 ft) and 9 m (30 ft) long.
122 For example, axle or wheel numbers 3, 4 or 5 and 12, 13 and 14 will create the maximum moment in panels of about 25 ft long.
123 A post height of 25 ft does not necessarily allow for typical lateral bracing in the plane of the top chord of a Parker truss span. However, Example 4.21 indicates a methodology for calculating member forces in inclined-chord Pratt through-truss spans.
124 In modern truss spans connections between members are made with bolted gusset plates (see Chapter 12).
125 A post height of 25 ft does not necessarily allow for typical lateral bracing in the plane of the top chord of a through-truss span. However, Example 4.22 indicates a methodology for calculating member forces in inclined-chord through-truss spans.
126 Typically truss spans longer than about 90 m (300 ft).
127 Subdivided horizontal-chord Pratt trusses are often referred to as Baltimore trusses. Subdivided inclined-chord Pratt trusses are commonly referred to as Pennsylvania or Petit trusses.
128 Improved deflection and vibration control.

129 Pratt type truss spans.
130 Warren type truss spans.
131 If, due to the length of subvertical member M(j)L(j), the upper and lower diagonal members are not aligned, the eccentricities will create larger forces and secondary bending moments (see Chapter 7) in other truss members in panel (j – 1) – (j + 1).
132 The neutral point is the point of zero shear in the subpanel under consideration. Loads located to the right of the neutral point create positive shear in the subpanel, and loads located to the left of the neutral point create negative shear in the subpanel.
133 If the subpanel under consideration is (j – 1) – (j), a concentrated live load must be located at panel point (j) for positive shear in the subpanel. If the subpanel under consideration is (j) – (j + 1), a concentrated live load must be located at panel point (j + 1) for positive shear in the subpanel. This situation is similar to diagonal members in non-subdivided truss spans.
134 If the subpanel under consideration is (j – 1) – (j), a concentrated live load must be located at panel point (j – 1) for negative shear in the subpanel. If the subpanel is (j) – (j + 1), a concentrated live load must be located at panel point (j) for negative shear in the subpanel. This situation is similar to diagonal members in non-subdivided truss spans.
135 If the subpanel under consideration is (j – 1) – (j) or (j) – (j + 1), a concentrated live load must be located at panel point (j + 1) for positive shear in the subpanel. The presence of a substrut or subtie member in the subpanel represents a different situation from that of non-subdivided truss spans. The distribution of shear across panel (j – 1) – (j + 1) is outlined in the subsequent section concerning the influence lines of subdivided truss spans.
136 If the subpanel under consideration is (j – 1) – (j) or (j) – (j + 1), a concentrated live load must be located at panel point (j – 1) for negative shear in the subpanel. The presence of a substrut or subtie member in the subpanel represents a different situation from that of non-subdivided truss spans. The distribution of shear across panel (j – 1) – (j + 1) is outlined in the subsequent section concerning the influence lines of subdivided truss spans.
137 The forces in member L(j – 1)L(j) are equal to the forces in member L(j)L(j + 1) by translational equilibrium at joint L(j).
138 As shown by equilibrium at joint L2.
139 Loads P_R, between panel point (j + 1) and panel point (n), are not directly considered in a free body diagram of the truss to the left of the section through (j) – (j + 1) but are a component of the total force on the span P_T.
140 L2 is at the 1/2 point along the span L_c.
141 In Figure 4.18b, the location of joint M(j) is assumed to be at $h_{(j-1)}/2$ such that the upper diagonal member U(j – 1)M(j) and the lower diagonal member M(j)L(j + 1) are aligned. If the location of joint M(j) is not at $h_{(j-1)}/2$, force eccentricities will exist that must be accounted for when taking moments about joints at (j – 1) and (j + 1). For example, when taking moments about panel point L(j + 1), an eccentric force from upper diagonal member U(j – 1)M(j) must be considered. Similarly, when taking moments about panel point U(j – 1), an eccentric force from lower diagonal member M(j)L(j + 1) must be considered. As information, such eccentricities are considered for inclined-chord Warren truss spans with non-aligned upper and lower diagonal members in Examples 4.52, 4.54 and 4.56, which illustrate the effect of non-aligned diagonal member eccentricities in bottom chord, lower diagonal and upper diagonal members, respectively. In typical inclined-chord truss designs, the upper and lower diagonal members are aligned for efficient design (lower member forces) and fabrication.
142 If the length of subvertical M(j)L(j) is not equal to $h_{(j-1)}/2$, then Equation 4.47a is $F_{L(j-1)M(j)} = \frac{-F_{M(j)L(j)}}{2}\left(\frac{\sqrt{h_s^2+d^2}}{h_s}\right)$ where h_S = the length of subvertical M(j)L(j).
143 Since substrut member L(j – 1)M(j) frames into the bottom chord member at L(j – 1), there is no eccentricity of the force in L(j – 1)M(j). However, the force in L(j – 1)M(j) will be affected by the length of subvertical member M(j)L(j) as shown in Equation 4.47a.
144 L2 is at the 2/3 point along the span L_c.
145 Typically, the applied load at joint U(j – 1) is zero unless some of the dead load is distributed to the upper chord of the truss.
146 The approximate method (see Section 4.4.2.1.1) with loads from joint (n) to joint (k + 1) or joint (n) to joint (k – 1) is typically used for uniformly distributed loads.
147 The approximate method (see Section 4.4.2.1.1) with loads from joint (0) to joint (k) or joint (0) to joint (k – 2) is typically used for uniformly distributed loads.

148 As with PSS truss spans, the location of joint M(j) is assumed to be $h_{(j-1)}/2$ such that the upper diagonal member U(j – 1)M(j) and the lower diagonal member M(j)L(j + 1) are aligned. If the location of joint M(j) is not $h_{(j-1)}/2$, force eccentricities will exist that must be accounted for when taking moments about joints at (j – 1) and (j + 1). As information, such eccentricities are considered for inclined-chord Warren truss spans with non-aligned upper and lower diagonal members in Examples 4.52, 4.54 and 4.56, which illustrate the effects of non-aligned diagonal member eccentricities in bottom chord, lower diagonal and upper diagonal members, respectively. In typical inclined-chord truss designs, the upper and lower diagonal members are aligned for efficient design (lower member forces) and fabrication.

149 The force in bottom and top chord members of PST truss spans is determined by taking a section through subpanel (j – 1) – (j). The force in top chord member U(j – 1)U(j + 1) is resolved at U(j + 1) to preclude the vertical force components in top chord member U(j – 1)U(j + 1) for moments taken about L(j + 1).

150 In Equation 4.61b, the length of subvertical member M(j)L(j) is assumed to be equal to $h_{(j-1)}/2$ such that upper diagonal member U(j – 1)M(j) is aligned with lower diagonal member M(j)L(j + 1). If U(j – 1)M(j) and M(j)L(j + 1) are not aligned, Equation 4.61b is $d_{S2} = d\left(2 - \frac{h_{(j+1)}}{h_{(j+1)} - h_S}\right)$ where h_S is the length of subvertical. member M(j)L(j).

151 In Equations 4.61b and 4.61c, the length of subvertical member M(j)L(j) is assumed to be equal to $h_{(j-1)}/2$, which is typical, such that upper diagonal member U(j – 1)M(j) is aligned with lower diagonal member M(j)L(j + 1).

152 L2 is at the 2/3 point along the span L_c.

153 In a horizontal chord (Baltimore) truss, $d_{S2} = 0$ as the subtie member M(k)U(k + 1) is the same length as the upper diagonal member U(k – 1)M(k).

154 The force in member L(j – 1)M(j) must be considered when taking moments about joint U(j + 1). If the length of member M(j)L(j) = $h_S = h_{(j+1)}/2$, the force in member L(j – 1)M(j) is resolved at U(j + 1) and $d_{S4} = 0$. Figure 4.22 shows the general case where $h_S \neq h_{(j+1)}/2$.

155 The subdivided Warren truss with single diagonal members in each panel is statically determinate with an even number of panels. An odd number of panels will require a center panel with two diagonals, thereby rendering the truss statically indeterminate.

156 In PST trusses, the force in member U(n/2)L(n/2) depends on the forces in the subtie members. In WISS trusses, the force in member U(n/2)L(n/2) depends on the forces in upper diagonal members.

157 The subdivided Warren truss with single diagonal members in each panel is statically determinate with an even number of panels. An odd number of panels will require a center panel with two diagonals, thereby rendering the truss statically indeterminate.

REFERENCES

Eisenbeis, K.R., 2010, *Railroad Relocation –500' Truss Span*, Presentation at AREMA Technical Conference, Orlando, FL, August 31.

Kurrer, K.E., 2008, *The History of the Theory of Structures*, Ernst and Sohn (Wiley), Berlin, Germany.

Unsworth, J.F., 2018, *Design and Construction of Modern Steel Railway Bridges*, 2nd edition, CRC Press, Taylor & Francis, Boca Raton, FL.

BIBLIOGRAPHY

Armenakas, A.E., 1988, *Classical Structural Analysis—A Modern Approach*, McGraw-Hill, New York.

Hool, G.A., and Kinnie, W.S., 1923, *Stresses in Framed Structures*, McGraw-Hill, New York.

Utku, S., Norris, C.H., and Wilbur, J.B., 1991, *Elementary Structural Analysis*, 4th edition, McGraw-Hill, New York.

5 Statically Indeterminate[1] Railway Truss Span Analysis

5.1 INTRODUCTION

Truss spans may be internally statically indeterminate with more members than those required to satisfy the conditions of equilibrium and/or externally statically indeterminate with redundant support conditions. Therefore, the structural analysis of internally and externally statically indeterminate truss spans requires additional conditions, equal in number to the degree of indeterminacy, to determine internal member forces. Classical energy methods of statically indeterminate truss analysis use Castigliano's Second Theorem[2] or the principle of virtual work to develop the additional equations, based on displacement compatibility,[3] necessary to determine the member forces.

Castigliano's Second Theorem indicates that if a truss subjected to external forces is in equilibrium, the displacement of a joint in the direction of the applied external force at the joint is equal to the partial derivative, with respect to the applied external force at the joint, of the strain energy of the truss due to the external forces. Since only external forces are independent forces, non-independent reaction forces[4] and forces due to changes in temperature and/or member length cannot be considered in the analysis of statically indeterminate trusses by Castigliano's Second Theorem. Castigliano's Second Theorem is used in conjunction with the theorem of least work to develop equations that define the conditions of displacement compatibility. Forces are defined in terms of the displacements and are independent variables as components of the equations of equilibrium to determine displacements and, subsequently, the forces.

The principle of virtual work[5] indicates that for a truss in equilibrium that remains in equilibrium when subjected to an infinitesimal displacement,[6] the external virtual work performed by virtual forces is equal to the internal virtual work of displacement performed by internal member stresses due to the virtual forces. Since the principle of virtual work involves only virtual forces and does not include the work performed by external forces on the truss, the method is not restricted to the analysis of external forces on statically indeterminate trusses and is useful for calculating member forces in statically indeterminate trusses due to displacements associated with temperature change, deviations in member length and/or support settlements. Therefore, the virtual work method uses redundant forces or reactions as the primary variables to derive the additional equations that are the necessary conditions of truss displacement, which are used in conjunction with the equations of displacement superposition to determine the internal member forces.

Modern techniques of statically indeterminate truss analysis are the flexibility (or force) and stiffness (or displacement) methods.[7] Nevertheless, this chapter will focus on the classical energy methods of statically indeterminate truss analysis to gain a percipient understanding of structural behavior in response to external loads, support settlements, temperature changes and variations in member lengths due to mis-fabrication. These classical energy methods are amenable to manual calculations that can be performed efficiently with modern computer spreadsheets.

5.2 CASTIGLIANO'S SECOND THEOREM FOR PIN-CONNECTED ELASTIC TRUSS SPAN MEMBER FORCES

In 1879, Castigliano expressed his second theorem as, "If the internal work of a framed structure is expressed as a function of the external forces the resulting expression is such that its differential coefficients give the relative displacements of their points of application."

Mathematically, this translates as the first partial derivative of the internal work or strain energy with respect to any external load P_i is equal to the displacement of the point of application of that

DOI: 10.1201/9781003469971-6

force in the direction of its line of action, Δ_{Pi}. Accordingly, for an elastic truss without rigid-body displacements[8] or internal member forces related to temperature change and/or variations in member lengths,[9] Δ_{Pi} in the direction of P_i is

$$\frac{\partial U}{\partial P_i} = \Delta_{Pi} \tag{5.1}$$

where U = the strain energy (internal work) as a function of the external forces.

The strain energy U_j of an elastic pin-connected truss member j is

$$U_j = \frac{F_j^2 L_j}{2A_j E} \tag{5.2}$$

where

F_j = the internal force in member j of the statically indeterminate truss due to external loads;
L_j= the length of truss member j;
A_j = the gross cross-sectional area of truss member j; and
E = the modulus of elasticity.

Therefore, the U of an elastic pin-connected truss is

$$U = \Sigma U_j = \frac{1}{2}\Sigma\frac{F_j^2 L_j}{EA_j} \tag{5.3}$$

Equation 5.1 yields a displacement of

$$\frac{\partial U}{\partial P_i} = \Delta_{Pi} = \frac{\partial}{\partial P_i}\Sigma\frac{F_j^2 L_j}{2EA_j} = \Sigma F_j \frac{\partial F_j}{\partial P_i}\frac{L_j}{EA_j} \tag{5.4}$$

$\frac{\partial F_j}{\partial P_i}$ is the rate of change of the force F_j in member j with respect to P_i and can be determined by considering a unit load at redundant members[10] of a statically indeterminate truss. If u_j is the force produced in truss member j due to the unit load, $\frac{\partial F_j}{\partial P_i} = \frac{u_j}{1}$, Equation 5.4 becomes

$$\Delta_{Pi} = \Sigma F_j \frac{L_j}{EA_j} u_j \tag{5.5}$$

The use of Castigliano's Second Theorem for the analysis of a statically indeterminate truss is shown in Example 5.1.

Example 5.1

The 150 ft long six-panel horizontal parallel-chord Pratt through-truss in Figure E5.1a with height of 25 ft is subjected to a uniform dead load of 1500 lb/ft and uniform live load of 5500 lb/ft. Find the forces in the truss members for the live load location producing the internal force in member U1L2 due to maximum positive shear force in panel 1-2 using Castigliano's Second Theorem.

The truss in Figure E5.1a has three independent reactions, NR, twenty-seven members, NM, and fourteen joints, NJ. Therefore, the degree of indeterminacy, ID_c = NM + NR − 2NJ = 27 + 3 − 2(14) = 2. Therefore, two members need to be removed to make the truss statically determinate. Remove L2U3 and U3L4, and the truss is a statically

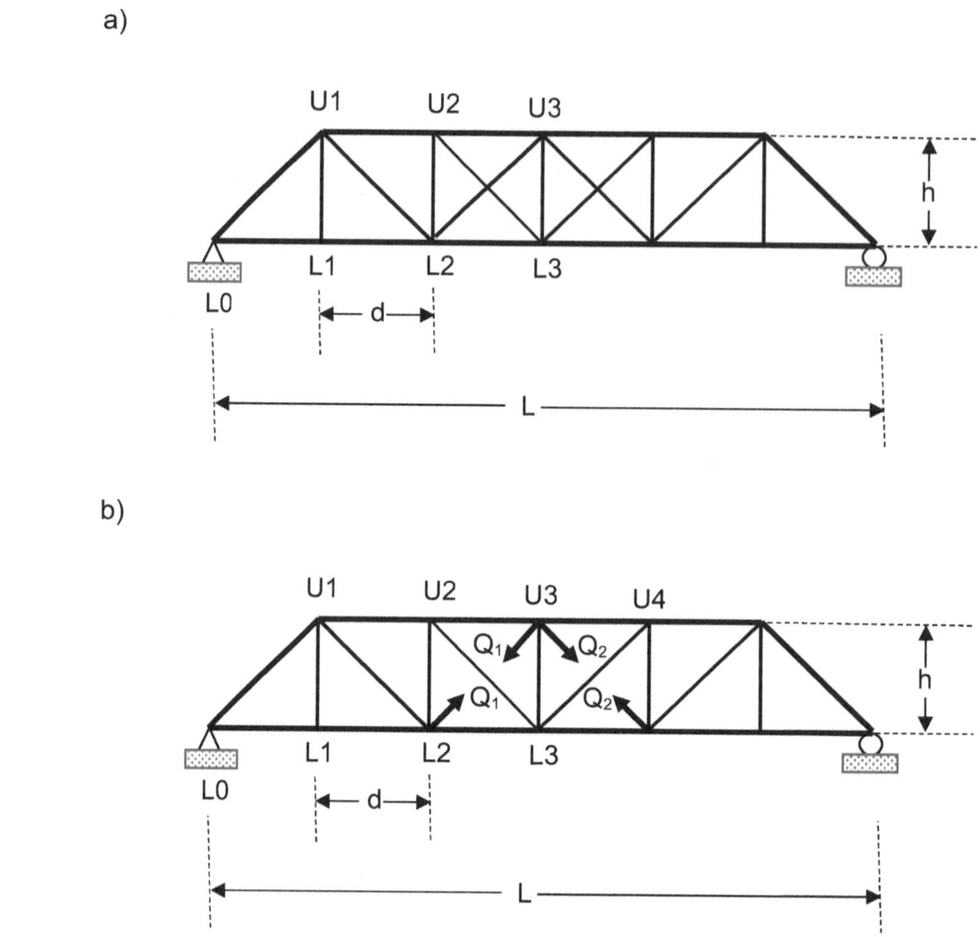

FIGURE E5.1 Pratt truss with redundant members L2U3 and U3L4.

TABLE E5.1a
Joint Loads from Uniformly Distributed Live Load at Location for Maximum Positive Shear in Panel 1-2

Joint	Joint Load with Uniformly Distributed Live Load 30 ft from L0
L0	0
L1	44.00
L2	134.75
L3	137.50
L4	137.50
L5	137.50
L6	68.75

determinate Pratt truss as shown in Figure E5.1b. Q_1 is the force at removed member L2U3, and Q_2 is the force at U3L4. From Equation 4.5c, the maximum positive live load shear force in member U1L2 occurs at $x^+ = (n - j)d/(n - 1) = (6 - 2)(25)/(6 - 1) = 20.0$ ft (also see Table E5.1a).

From Equation 4.5k, the maximum positive live load shear force in member U1L2 =

$$F^{+max}_{U(j-1)L(j)} = \left(\frac{w((n-j)d)^2}{2d(n-1)}\right)\left(\frac{\sqrt{h^2+d^2}}{h}\right) = \left(\frac{5.5((6-2)25)^2}{2(25)(6-1)}\right)\left(\frac{\sqrt{25^2+25^2}}{25}\right) = 311.1\ \text{kips}$$

Member forces determined from structural analysis of the statically determinate truss for the live load location producing maximum positive shear force in member U1L2 with redundant member loads Q_1 and Q_2 yields Table E5.1b.

Ensuring compatibility of displacements at members L2U3 and U3L4,

$$\frac{\partial U}{\partial Q_1} = \frac{\partial U}{\partial Q_2} = 0$$

and Equation 5.4 gives the following as well as Table E5.1c:

$$\frac{\partial U}{\partial Q_1} = \sum \frac{F_j L_j}{EA_j}\frac{\partial F_j}{\partial Q_1} = 0$$

$$\frac{\partial U}{\partial Q_2} = \sum \frac{F_j L_j}{EA_j}\frac{\partial F_j}{\partial Q_2} = 0$$

TABLE E5.1b
Member Forces Determined from Structural Analysis of a Statically Determinate Truss

Member		Dead Load Force (kips)	Live Load Force (kips)	Total Force, F_j (kips)	L_j (in)	A_j (in²)	L_j / A_j (in⁻¹)
L0L1	Bottom chord	93.8	264.0	357.8	300.0	20	15.00
L1L2	Bottom chord	93.8	264.0	357.8	300.0	20	15.00
L2L3	Bottom chord	150.0	484.0	634.0-0.707Q_1	300.0	35	8.57
L3L4	Bottom chord	150.0	517.0	667.0-0.707Q_2	300.0	35	8.57
L4L5	Bottom chord	93.8	327.5	421.3	300.0	20	15.00
L5L6	Bottom chord	93.8	327.5	421.3	300.0	20	15.00
L0U1	End post	−132.6	−373.4	−506.0	424.3	25	16.97
U1U2	Top chord	−150.0	−484.0	−634.0	300.0	25	12.00
U2U3	Top chord	−168.8	−569.3	−738.1-0.707Q_1	300.0	30	10.00
U3U4	Top chord	−168.8	−569.3	−738.1-0.707Q_2	300.0	30	10.00
U4U5	Top chord	−150.0	−517.0	−667.0	300.0	25	12.00
U5L6	End post	−132.6	−462.8	−595.4	424.3	25	16.97
U1L1	Vertical	37.5	44.0	81.5	300.0	10	30.00
U2L2	Vertical	−18.8	−85.3	−104.1-0.707Q_1	300.0	5	60.00
U3L3	Vertical	0	0	−0.707Q_1-0.707Q_2	300.0	5	60.00
U4L4	Vertical	−18.8	−52.3	−71.1-0.707Q_2	300.0	5	60.00
U5L5	Vertical	37.5	137.5	175.0	300.0	10	30.00
U1L2	Diagonal	79.6	311.1	390.7	424.3	10	42.43
U2L3	Diagonal	26.5	120.6	147.1 + Q_1	424.3	5	84.86
L2U3	Diagonal removed	0	0	Q_1	424.3	5	84.86
L3U4	Diagonal	26.5	73.9	100.4 + Q_2	424.3	5	84.86
U3L4	Diagonal removed	0	0	Q_2	424.3	5	84.86
L4U5	Diagonal	79.6	268.3	347.9	424.3	10	42.43

TABLE E5.1c

Displacement Compatibility in the Direction of Forces in Removed Members

Member		$\frac{F_jL_j}{A_j}$ (kips/in)	$\frac{\partial F_j}{\partial Q_1}$	$\frac{F_jL_j}{A_j}\frac{\partial F_j}{\partial Q_1}$ (kips/in)	$\frac{\partial F_j}{\partial Q_2}$	$\frac{F_jL_j}{A_j}\frac{\partial F_j}{\partial Q_2}$ (kips/in)
L0L1	Bottom chord	5367.0	0	0	0	0
L1L2	Bottom chord	5367.0	0	0	0	0
L2L3	Bottom chord	$5433.4\text{-}6.06Q_1$	−0.707	$-3842.0 + 4.29Q_1$	0	0
L3L4	Bottom chord	$5716.2\text{-}6.06Q_2$	0	0	−0.707	$-4041.9 + 4.29Q_2$
L4L5	Bottom chord	6319.5	0	0	0	0
L5L6	Bottom chord	6319.5	0	0	0	0
L0U1	End post	−8586.8	0	0	0	0
U1U2	Top chord	−7608.0	0	0	0	0
U2U3	Top chord	$-7381.0-7.07Q_1$	−0.707	$5219.1 + 5.0Q_1$	0	0
U3U4	Top chord	$-7381.0-7.07Q_2$	0	0	−0.707	$5219.1 + 5.0Q_2$
U4U5	Top chord	−8004.0	0	0	0	0
U5L6	End post	−10103.9	0	0	0	0
U1L1	Vertical	2445.0	0	0	0	0
U2L2	Vertical	$-6246.0\text{-}42.43Q_1$	−0.707	$4416.6 + 30.0Q_1$	0	0
U3L3	Vertical	$-42.43Q_1\text{-}42.43Q_2$	−0.707	$30.0Q_1 + 30.0Q_2$	−0.707	$30.0Q_1 + 30.0Q_2$
U4L4	Vertical	$-4266.0\text{-}42.43Q_2$	0	0	−0.707	$3016.5 + 30.0Q_2$
U5L5	Vertical	5250.0	0	0	0	0
U1L2	Diagonal	16,577.4	0	0	0	0
U2L3	Diagonal	$12{,}482.9 + 84.86Q_1$	1.00	$12{,}482.9 + 84.86Q_1$	0	0
U3L2	Diagonal removed	$84.86Q_1$	1.00	$84.86Q_1$	0	0
L3U4	Diagonal	$8519.9 + 84.86Q_2$	0	0	1.00	$8519.9 + 84.86Q_2$
U3L4	Diagonal removed	$84.86Q_2$	0	0	1.00	$84.86Q_2$
L4U5	Diagonal	14761.4	0	0	0	0
Σ				$18{,}276.6 + 239.0Q_1 + 30.0Q_2$		$12{,}713.6 + 30.0Q1 + 239.0Q_2$

$$\sum \frac{F_jL_j}{A_j}\frac{\partial F_j}{\partial Q_1} = 18{,}276.6 + 239.0(Q_1) + 30.0(Q_2) = 0$$

$$\sum \frac{F_jL_j}{A_j}\frac{\partial F_j}{\partial Q_2} = 12{,}713.6 + 30.0(Q_1) + 239.0(Q_2) = 0$$

The solution of these equations provides the following as well as Table E5.1d.

$Q_1 = -70.9$ kips (force in removed member L2U3)

$Q_2 = -44.3$ kips (force in removed member U3L4)

TABLE E5.1d

Member Forces Determined from Structural Analysis of a Statically Indeterminate Truss

Member		F_{Pj} (kips)	
L0L1	Bottom chord	357.8	357.8
L1L2	Bottom chord	357.8	357.8
L2L3	Bottom chord	$634.0\text{-}0.707Q_1$	684.1
L3L4	Bottom chord	$667.0\text{-}0.707Q_2$	698.3

TABLE E5.1d (*Continued*)
Member Forces Determined from Structural Analysis of a Statically Indeterminate Truss

Member		F_{Pj} (kips)	
L4L5	Bottom chord	421.3	421.3
L5L6	Bottom chord	421.3	421.3
L0U1	End post	−506.0	−506.0
U1U2	Top chord	−634.0	−634.0
U2U3	Top chord	$-738.1-0.707Q_1$	−688.0
U3U4	Top chord	$-738.1-0.707Q_2$	−706.8
U4U5	Top chord	−667.0	−667.0
U5L6	End post	−595.4	−595.4
U1L1	Vertical	81.5	81.5
U2L2	Vertical	$-104.1-0.707Q_1$	−54.0
U3L3	Vertical	$-0.707Q_1-0.707Q_2$	81.4
U4L4	Vertical	$-71.1-0.707Q_2$	−39.8
U5L5	Vertical	175.0	175.0
U1L2	Diagonal	390.7	390.7
U2L3	Diagonal	$147.1 + Q_1$	76.2
L2U3	F_1 Redundant member	Q_1	−70.9
L3U4	Diagonal	$100.4 + Q_2$	56.1
U3L4	F_2 Redundant member	Q_2	−44.3
L4U5	Diagonal	347.9	347.9

Example 5.1 illustrates the use of Castigliano's Second Theorem for the analysis of member forces in a statically indeterminate truss. However, Castigliano's Second Theorem is only applicable for displacements caused by external loads. Alternatively, and with greater generality, the virtual force method can be used to determine statically indeterminate pin-connected elastic truss member forces F_j due to external loads, support movements, temperature changes and/or variations in member lengths.

5.3 VIRTUAL WORK ANALYSIS OF PIN-CONNECTED ELASTIC TRUSS SPANS

Redundant members can be removed from a statically indeterminate truss to form a statically determinate truss under the compatibility condition that the displacement of a redundant member is equal to the displacement of the redundant member in the statically determinate truss. Therefore, the redundant members (internally redundant) and/or support reactions (externally redundant) can be replaced by a virtual system of forces that maintain static equilibrium of the statically indeterminate truss. Enforcement of displacement compatibility at the redundant members and/or support reactions enables determination of the virtual forces X_i and, subsequently, the forces in all members of the statically indeterminate truss.

Equation 5.3 provides the strain energy U of an elastic pin-connected truss as,

$$U = \frac{1}{2}\sum_{j=1}^{NM} \frac{F_j^2 L_j}{EA_j} \tag{5.6}$$

where NM = the number of members in the statically indeterminate truss and F_j = the internal force in member j due to external loads, support movements, temperature changes and/or variations in member lengths applied to the statically indeterminate truss.

Considering superposition and compatibility of displacements of the statically determinate and indeterminate truss members gives the following:

$$F_j = f_j + \sum_{i=1}^{n} f_{ji} X_i \tag{5.7}$$

where

f_j = the internal force in member j due to external loads, support movements, temperature changes and/or variations in member lengths applied to the statically determinate truss;

n = number of redundant members and/or reactions in the statically indeterminate truss span;

X_i = the ratio of the force in member j in the statically determinate truss to the load at redundant member and/or support reaction that causes the force;

f_{ji} = the internal force in a redundant member and/or support reaction due to external loads, support movements, temperature changes and/or variations in member lengths applied to the statically determinate truss; and

$f_{ji}X_i$ = internal force in redundant member and/or support reaction due to external loads, support movements, temperature changes and/or variations in member lengths applied to the statically indeterminate truss.

Substituting Equation 5.7 in Equation 5.6 yields

$$U = \frac{1}{2}\sum_{j=1}^{NM} \frac{\left(f_j + \sum_{i=1}^{n} f_{ji} X_i\right)^2 L_j}{EA_j} \tag{5.8}$$

The change in strain energy with respect to the redundant forces must be zero[11] and differentiating Equation 5.8 with respect to X_k yields

$$\frac{\partial U}{\partial X_k} = 0 = \sum_{j=1}^{NM} \frac{\left(f_j + \sum_{i=1}^{n} f_{ji} X_i\right) X_k L_j}{EA_j} \tag{5.9}$$

where X_k = internal force in redundant member and/or support reaction k due to external loads, support movements temperature changes and/or variations in member lengths applied to the statically determinate truss.

Therefore, for member j, the change in strain energy with respect to the redundant force X_k is

$$\frac{\partial U}{\partial X_k} = \frac{\left(f_j + \sum_{i=1}^{n} f_{ji} X_i\right) X_k L_j}{EA_j} = \left(f_j + \sum_{i=1}^{n} f_{ji} X_i\right) X_k \left(\frac{L_j}{A_j}\right) = 0 \tag{5.10}$$

For example, Equation 5.10 provides the three simultaneous Equation 5.11 for the redundant forces in a truss with three redundant members and/or support reactions. Consequently, for member j with n = 3,

$$\frac{\partial U}{\partial X_1} = 0 = \left(f_j X_1 + f_{j1} X_1^2 + f_{j2} X_2 X_1 + f_{j3} X_3 X_1\right)\left(\frac{L_j}{A_j}\right) \tag{5.11a}$$

$$\frac{\partial U}{\partial X_2} = 0 = \left(f_j X_2 + f_{j1} X_1 X_2 + f_{j2} X_2^2 + f_{j3} X_3 X_2\right)\left(\frac{L_j}{A_j}\right) \tag{5.11b}$$

$$\frac{\partial U}{\partial X_3} = 0 = \left(f_j X_3 + f_{j1} X_1 X_3 + f_{j2} X_2 X_3 + f_{j3} X_3^2\right)\left(\frac{L_j}{A_j}\right) \tag{5.11c}$$

Redundant truss members or reactions should be selected to create a stable and readily analyzed statically determinate truss. Examples 5.2 and 5.3 outline the use of Equation 5.10 for calculating member forces in the statically indeterminate truss span of Example 5.1.

Example 5.2

The 150 ft long six-panel horizontal parallel chord Pratt through-truss in Figure E5.1a with height of 25 ft is subjected to a uniform dead load of 1500 lb/ft and uniform live load of 5500 lb/ft. Find the forces in the truss members for the live load location producing maximum internal force in member U1L2 due to positive panel shear using the principle of virtual work.

As shown in Example 5.1, the degree of indeterminacy is 2. Since the truss is internally statically indeterminate, choose L2U3 and U3L4 as redundant members because with those members removed, the truss is a statically determinate Pratt truss. Member forces determined from structural analysis of the statically determinate truss (members L2U3 and U3L4 removed) for the live load location producing maximum positive shear force in member U1L2 (see Example 5.1) yields Table E5.2a.

For the statically determinate Pratt truss with L2U3 and U3L4 as redundant members, unit loads at joints L2, U3 and L4 only affect members in the central panels of the truss (L2L3, L3L4, U2U3, U3U4, U2L2, U3L3, U4L4, U2L3, L2U3, L3U4 and U3L4). However, for completeness and generality, the statically determinate analysis with unit loads at joints

TABLE E5.2a
Member Forces from Structural Analysis of the External Forces on the Statically Determinate Truss

Member		Dead Load (DL) Force (kips)	Live Load (LL) Force (kips)	DL + LL Force, F_j (kips)
L0L1	Bottom chord	93.8	264.0	357.8
L1L2	Bottom chord	93.8	264.0	357.8
L2L3	Bottom chord	150.0	484.0	634.0
L3L4	Bottom chord	150.0	517.0	667.0
L4L5	Bottom chord	93.8	327.5	421.3
L5L6	Bottom chord	93.8	327.5	421.3
L0U1	End post	−132.6	−373.4	−506.0
U1U2	Top chord	−150.0	−484.0	−634.0
U2U3	Top chord	−168.8	−569.3	−738.1
U3U4	Top chord	−168.8	−569.3	−738.1
U4U5	Top chord	−150.0	−517.0	−667.0
U5L6	End post	−132.6	−462.8	−595.4
U1L1	Vertical	37.5	44.0	81.5
U2L2	Vertical	−18.8	−85.3	−104.1
U3L3	Vertical	0	0	0
U4L4	Vertical	−18.8	−52.3	−71.1
U5L5	Vertical	37.5	137.5	175.0
U1L2	Diagonal	79.6	311.1	390.7
U2L3	Diagonal	26.5	120.6	147.1
U3L2	Diagonal redundant	0	0	0
L3U4	Diagonal	26.5	73.9	100.4
U3L4	Diagonal redundant	0	0	0
L4U5	Diagonal	79.6	268.3	347.9

L2, U3 and L4 includes all members of the truss. Structural analysis of the statically determinate truss with unit loads X1 at redundant member L2U3 and X2 at redundant member U3L4 provides Table E5.2b.

Equation 5.10 for member j with n = 2 gives the following as well as Tables E5.2c through E5.2e:

$$\frac{\partial U}{\partial X_1} = 0 = \left(f_j X_1 + f_{j1} X_1^2 + f_{j2} X_2 X_1\right)\left(\frac{L_j}{A_j}\right)$$

$$\frac{\partial U}{\partial X_2} = 0 = \left(f_j X_2 + f_{j1} X_1 X_2 + f_{j2} X_2^2\right)\left(\frac{L_j}{A_j}\right)$$

$$\frac{\partial U}{\partial X_1} = 0 = \left(f_j X_1 + f_{j1} X_1^2 + f_{j2} X_2 X_1\right)\left(\frac{L_j}{A_j}\right) = 18{,}275.2 + 239.0(f_{j1}) + 30.0(f_{j2}) = 0$$

$$\frac{\partial U}{\partial X_2} = 0 = \left(f_j X_2 + f_{j1} X_1 X_2 + f_{j2} X_2^2\right)\left(\frac{L_j}{A_j}\right) = 12{,}712.4 + 30.0(f_{j1}) + 239.0(f_{j2}) = 0$$

TABLE E5.2b

Member Forces due to Unit Load X_1 at Redundant Member L2U3 and Unit Load X_2 at Redundant Member U3L4

Member		X_1	X_2
L0L1	Bottom chord	0	0
L1L2	Bottom chord	0	0
L2L3	Bottom chord	−0.707	0
L3L4	Bottom chord	0	−0.707
L4L5	Bottom chord	0	0
L5L6	Bottom chord	0	0
L0U1	End post	0	0
U1U2	Top chord	0	0
U2U3	Top chord	−0.707	0
U3U4	Top chord	0	−0.707
U4U5	Top chord	0	0
U5L6	End post	0	0
U1L1	Vertical	0	0
U2L2	Vertical	−0.707	0
U3L3	Vertical	−0.707	−0.707
U4L4	Vertical	0	−0.707
U5L5	Vertical	0	0
U1L2	Diagonal	0	0
U2L3	Diagonal	1.00	0
L2U3	F_1 Redundant	1.00	0
L3U4	Diagonal	0	1.00
U3L4	F_2 Redundant	0	1.00
L4U5	Diagonal	0	0

TABLE E5.2c
Change in Strain Energy with Respect to Unit Load X_1

Member		L_j (in)	A_j (in^2)	L_j / A_j (in^{-1})	X_1	$X_1(L_j/A_j)$ (in^{-1})	$X_1^2(L_j/A_j)$ (in^{-1})	$X_1X_2(L_j/A_j)$ (in^{-1})
L0L1	Bottom chord	300.0	20	15.00	0	0	0	0
L1L2	Bottom chord	300.0	20	15.00	0	0	0	0
L2L3	Bottom chord	300.0	35	8.57	−0.707	−6.06	4.28	0
L3L4	Bottom chord	300.0	35	8.57	0	0	0	0
L4L5	Bottom chord	300.0	20	15.00	0	0	0	0
L5L6	Bottom chord	300.0	20	15.00	0	0	0	0
L0U1	End post	424.3	25	16.97	0	0	0	0
U1U2	Top chord	300.0	25	12.00	0	0	0	0
U2U3	Top chord	300.0	30	10.00	−0.707	−7.07	5.0	0
U3U4	Top chord	300.0	30	10.00	0	0	0	0
U4U5	Top chord	300.0	25	12.00	0	0	0	0
U5L6	End post	424.3	25	16.97	0	0	0	0
U1L1	Vertical	300.0	10	30.00	0	0	0	0
U2L2	Vertical	300.0	5	60.00	−0.707	−42.42	30.0	0
U3L3	Vertical	300.0	5	60.00	−0.707	−42.42	30.0	30.0
U4L4	Vertical	300.0	5	60.00	0	0	0	0
U5L5	Vertical	300.0	10	30.00	0	0	0	0
U1L2	Diagonal	424.3	10	42.43	0	0	0	0
U2L3	Diagonal	424.3	5	84.86	1.00	84.86	84.86	0
L2U3	F_1 Redundant	424.3	5	84.86	1.00	84.86	84.86	0
L3U4	Diagonal	424.3	5	84.86	0	0	0	0
U3L4	F_2 Redundant	424.3	5	84.86	0	0	0	0
L4U5	Diagonal	424.3	10	42.43	0	0	0	0
Σ							239.0	30.0

TABLE E5.2d
Change in Strain Energy with Respect to Unit Load X_2

Member		L_j (in)	A_j (in^2)	L_j / A_j (in^{-1})	X_2	$X_2(L_j/A_j)$ (in^{-1})	$X_2^2(L_j/A_j)$ (in^{-1})	$X_2X_1(L_j/A_j)$ (in^{-1})
L0L1	Bottom chord	300.0	20	15.00	0	0	0	0
L1L2	Bottom chord	300.0	20	15.00	0	0	0	0
L2L3	Bottom chord	300.0	35	8.57	0	0	0	0
L3L4	Bottom chord	300.0	35	8.57	−0.707	−6.06	4.28	0
L4L5	Bottom chord	300.0	20	15.00	0	0	0	0
L5L6	Bottom chord	300.0	20	15.00	0	0	0	0
L0U1	End post	424.3	25	16.97	0	0	0	0
U1U2	Top chord	300.0	25	12.00	0	0	0	0
U2U3	Top chord	300.0	30	10.00	0	0	0	0
U3U4	Top chord	300.0	30	10.00	−0.707	−7.07	5.0	0
U4U5	Top chord	300.0	25	12.00	0	0	0	0
U5L6	End post	424.3	25	16.97	0	0	0	0
U1L1	Vertical	300.0	10	30.00	0	0	0	0

(Continued)

TABLE E5.2d (*Continued*)
Change in Strain Energy with Respect to Unit Load X_2

Member		L_j (in)	A_j (in²)	L_j / A_j (in⁻¹)	X_2	$X_2(L_j/A_j)$ (in⁻¹)	$X_2^2(L_j/A_j)$ (in⁻¹)	$X_2X_1(L_j/A_j)$ (in⁻¹)
U2L2	Vertical	300.0	5	60.00	0	0	0	0
U3L3	Vertical	300.0	5	60.00	−0.707	−42.42	30.0	30.0
U4L4	Vertical	300.0	5	60.00	−0.707	−42.42	30.0	0
U5L5	Vertical	300.0	10	30.00	0	0	0	0
U1L2	Diagonal	424.3	10	42.43	0	0	0	0
U2L3	Diagonal	424.3	5	84.86	0	0	0	0
L2U3	F_1 Redundant	424.3	5	84.86	0	0	0	0
L3U4	Diagonal	424.3	5	84.86	1.00	84.86	84.86	0
U3L4	F_2 Redundant	424.3	5	84.86	1.00	84.86	84.86	0
L4U5	Diagonal	424.3	10	42.43	0	0	0	0
Σ							239.0	30.0

TABLE E5.2e
Change in Strain Energy with Respect to Statically Determinate Member Forces

Member		f_j (kips)	$f_jX_1(L_j/A_j)$ (kips/in)	$f_jX_2(L_j/A_j)$ (kips/in)	$f_jX_1X_2(L_j/A_j)$ (kips/in)
L0L1	Bottom chord	357.8	0	0	0
L1L2	Bottom chord	357.8	0	0	0
L2L3	Bottom chord	634.0	−3842.0	0	0
L3L4	Bottom chord	667.0	0	−4042.0	0
L4L5	Bottom chord	421.3	0	0	0
L5L6	Bottom chord	421.3	0	0	0
L0U1	End post	−506.0	0	0	0
U1U2	Top chord	−634.0	0	0	0
U2U3	Top chord	−738.1	5218.4	0	0
U3U4	Top chord	−738.1	0	5218.4	0
U4U5	Top chord	−667.0	0	0	0
U5L6	End post	−595.4	0	0	0
U1L1	Vertical	81.5	0	0	0
U2L2	Vertical	−104.1	4415.9	0	0
U3L3	Vertical	0	0	0	0
U4L4	Vertical	−71.1	0	3016.1	0
U5L5	Vertical	175.0	0	0	0
U1L2	Diagonal	390.7	0	0	0
U2L3	Diagonal	147.1	12,482.9	0	0
L2U3	F_1 redundant member	0	0	0	
L3U4	Diagonal	100.4	0	8519.9	0
U3L4	F_2 redundant member	0	0	0	0
L4U5	Diagonal	347.9	0	0	0
Σ			18,275.2	12,712.4	0

TABLE E5.2f
Member Forces in the Statically Indeterminate Truss

Member		f_j (kips)	$f_{j1}X_1$ (kips)	$f_{j2}X_2$ (kips)	$f_{j1}X_1 + f_{j2}X_2$ (kips)	F_j (kips)
L0L1	Bottom chord	357.8	0	0	0	357.8
L1L2	Bottom chord	357.8	0	0	0	357.8
L2L3	Bottom chord	634.0	50.1	0	50.1	684.1
L3L4	Bottom chord	667.0	0	31.3	31.3	698.3
L4L5	Bottom chord	421.3	0	0	0	421.3
L5L6	Bottom chord	421.3	0	0	0	421.3
L0U1	End post	−506.0	0	0	0	−506.0
U1U2	Top chord	−634.0	0	0	0	−634.0
U2U3	Top chord	−738.1	50.1	0	50.1	−688.0
U3U4	Top chord	−738.1	0	31.3	31.3	−706.8
U4U5	Top chord	−667.0	0	0	0	−667.0
U5L6	End post	−595.4	0	0	0	−595.4
U1L1	Vertical	81.5	0	0	0	81.5
U2L2	Vertical	−104.1	50.1	0	50.1	−54.0
U3L3	Vertical	0	50.1	31.3	81.4	81.4
U4L4	Vertical	−71.1	0	31.3	31.3	−39.8
U5L5	Vertical	175.0	0	0	0	175.0
U1L2	Diagonal	390.7	0	0	0	390.7
U2L3	Diagonal	147.1	−70.9	0	−70.9	76.2
L2U3	F_1 redundant member	0	−70.9	0	−70.9	−70.9
L3U4	Diagonal	100.4	0	−44.3	−44.3	56.1
U3L4	F_2 redundant member	0	0	−44.3	−44.3	−44.3
L4U5	Diagonal	347.9	0	0	0	347.9

The solution of these equations provides the following as well as Table E5.2f:

$f_{j1} = -70.9$ kips (force in redundant member L2U3)
$f_{fj2} = -44.3$ kips (force in redundant member U3L4)

When only external forces are considered, the method of virtual work is essentially the same as Castigliano's Second Theorem approach outlined in Example 5.1.

Example 5.3

The 150 ft long six-panel horizontal parallel-chord Pratt through-truss in Figure E5.1a with h = 25 ft is subjected to a uniform dead load of 1500 lb/ft and Cooper's E80 live load. Find the forces in the truss members for the live load location with axle 7 at L2 (Table E5.3a).

As shown in Examples 5.1 and 5.2, the degree of indeterminacy is 2. Since the truss is internally statically indeterminate, choose L2U3 and U3L4 as redundant members because with those members removed, the truss is a statically determinate Pratt truss. The member forces in the statically determinate truss with Cooper's E80 axle 7 at joint L2 with the train proceeding from right to left are given in Table E5.3b.

TABLE E5.3a
Joint Loads due to Cooper's E80 Live Load with Axle 7 and L2

Joint	Joint Live Load with Cooper's E80 Live Load Axle 7 at L2
L0	16.00
L1	140.40
L2	114.72
L3	132.08
L4	128.08
L5	98.72
L6	50.00

TABLE E5.3b
Dead and Live Load Member Forces due to External Loads in the Statically Determinate Truss

Member		Dead Load (DL) Force (kips)	E80 Live Load (LL) Force (kips)	DL + LL Force, F_j (kips)
L0L1	Bottom chord	93.8	318.7	412.5
L1L2	Bottom chord	93.8	318.7	412.5
L2L3	Bottom chord	150.0	496.9	646.9
L3L4	Bottom chord	150.0	491.9	641.9
L4L5	Bottom chord	93.8	295.3	389.1
L5L6	Bottom chord	93.8	295.3	389.1
L0U1	End post	−132.6	−450.7	−583.3
U1U2	Top chord	−150.0	−496.9	−646.9
U2U3	Top chord	−168.8	−560.5	−729.3
U3U4	Top chord	−168.8	−560.5	−729.3
U4U5	Top chord	−150.0	−491.9	−641.9
U5L6	End post	−132.6	−417.7	−550.3
U1L1	Vertical	37.5	140.4	177.9
U2L2	Vertical	−18.8	−63.6	−82.4
U3L3	Vertical	0	0	0
U4L4	Vertical	−18.8	−68.5	−87.3
U5L5	Vertical	37.5	98.7	136.2
U1L2	Diagonal	79.6	252.1	331.7
U2L3	Diagonal	26.5	89.9	116.4
U3L2	Diagonal redundant	0	0	0
L3U4	Diagonal	26.5	96.9	123.4
U3L4	Diagonal redundant	0	0	0
L4U5	Diagonal	79.6	278.1	357.7

For the statically determinate Pratt truss with L2U3 and U3L4 as redundant members, unit loads at joints L2, U3 and L4 only affect members in the central panels of the truss (L2L3, L3L4, U2U3, U3U4, U2L2, U3L3, U4L4, U2L3, L2U3, L3U4 and U3L4). However, for completeness and generality,[12] the statically determinate analysis with unit loads at joints L2, U3 and L4 includes all members of the truss.

From Example 5.2 (same redundant members selected) the changes in strain energy with respect to unit loads X_1 and X_2 are as follows and as in Table E5.3c:

$$\left(f_{j1}X_1^2 + f_{j2}X_2X_1\right) = 239.0(f_{j1}) + 30.0(f_{j2})$$

$$\left(f_{j1}X_1X_2 + f_{j2}X_2^2\right) = 30.0(f_{j1}) + 239.0(f_{j2})$$

TABLE E5.3c
Change in Strain Energy with Respect to f_j (Member Forces) and Unit Loads X_1 and X_2

Member		f_j (kips)	X_1	$f_jX_1(L_j/A_j)$ (kips/in)	X_2	$f_jX_2(L_j/A_j)$ (kips/in)	$f_jX_1X_2(L_j/A_j)$ (kips/in)
L0L1	Bottom chord	412.5	0	0	0	0	0
L1L2	Bottom chord	412.5	0	0.0	0	0.0	0
L2L3	Bottom chord	646.9	−0.707	−3919.6	0	0.0	0
L3L4	Bottom chord	641.9	0	0.0	−0.707	−3889.3	0
L4L5	Bottom chord	389.1	0	0.0	0	0.0	0
L5L6	Bottom chord	389.1	0	0.0	0	0.0	0
L0U1	End post	−583.3	0	0.0	0	0.0	0
U1U2	Top chord	−646.9	0	0.0	0	0.0	0
U2U3	Top chord	−729.3	−0.707	5156.2	0	0.0	0
U3U4	Top chord	−729.3	0	0.0	−0.707	5156.2	0
U4U5	Top chord	−641.9	0	0.0	0	0.0	0
U5L6	End post	−550.3	0	0.0	0	0.0	0
U1L1	Vertical	177.9	0	0.0	0	0.0	0
U2L2	Vertical	−82.4	−0.707	3495.4	0	0.0	0
U3L3	Vertical	0	−0.707	0.0	−0.707	0.0	0
U4L4	Vertical	−87.3	0	0.0	−0.707	3703.3	0
U5L5	Vertical	136.2	0	0.0	0	0.0	0
U1L2	Diagonal	331.7	0	0.0	0	0.0	0
U2L3	Diagonal	116.4	1.00	9877.7	0	0.0	0
L2U3	F_1 redundant member	0	1.00	0.0	0	0.0	0
L3U4	Diagonal	123.4	0	0.0	1.00	10471.7	0
U3L4	F_2 redundant member	0	0	0.0	1.00	0.0	0
L4U5	Diagonal	357.7	0	0.0	0	0.0	0
Σ				14,609.7		15,441.9	0

From Equation 5.10 for member j with n = 2,

$$\frac{\partial U}{\partial X_1} = 0 = \left(f_jX_1 + f_{j1}X_1^2 + f_{j2}X_2X_1\right)\left(\frac{L_j}{A_j}\right) = 14{,}609.7 + 239.0(f_{j1}) + 30.0(f_{j2}) = 0$$

$$\frac{\partial U}{\partial X_2} = 0 = \left(f_jX_2 + f_{j1}X_1X_2 + f_{j2}X_2^2\right)\left(\frac{L_j}{A_j}\right) = 15{,}441.9 + 30.0(f_{j1}) + 239.0(f_{j2}) = 0$$

The solution of these equations provides the following and Table E5.3d:

$f_{j1} = -53.87$ kips (force in redundant member L2U3)
$f_{j2} = -57.85$ kips (force in redundant member U3L4)

TABLE E5.3d
Member Forces in the Statically Indeterminate Truss

Member		f_j (kips)	$f_{j1}X_1$ (kips)	$f_{j2}X_2$ (kips)	$f_{j1}X_1 + f_{j2}X_2$ (kips)	F_j (kips)
L0L1	Bottom chord	412.5	0	0	0	412.5
L1L2	Bottom chord	412.5	0.0	0.0	0	412.5
L2L3	Bottom chord	646.9	38.1	0.0	38.1	685.0
L3L4	Bottom chord	641.9	0.0	40.9	40.9	682.8
L4L5	Bottom chord	389.1	0.0	0.0	0	389.1
L5L6	Bottom chord	389.1	0.0	0.0	0	389.1
L0U1	End post	−583.3	0.0	0.0	0	−583.3
U1U2	Top chord	−646.9	0.0	0.0	0	−646.9
U2U3	Top chord	−729.3	38.1	0.0	38.1	−691.2
U3U4	Top chord	−729.3	0.0	40.9	40.9	−688.4
U4U5	Top chord	−641.9	0.0	0.0	0	−641.9
U5L6	End post	−550.3	0.0	0.0	0	−550.3
U1L1	Vertical	177.9	0.0	0.0	0	177.9
U2L2	Vertical	−82.4	38.1	0.0	38.1	−44.3
U3L3	Vertical	0	38.1	40.9	79.0	79.0
U4L4	Vertical	−87.3	0.0	40.9	40.9	−46.4
U5L5	Vertical	136.2	0.0	0.0	0	136.2
U1L2	Diagonal	331.7	0.0	0.0	0	331.7
U2L3	Diagonal	116.4	−53.9	0.0	−53.9	62.5
L2U3	F_1 redundant member	0	−53.9	0.0	−53.9	−53.9
L3U4	Diagonal	123.4	0.0	−57.9	−57.9	65.6
U3L4	F_2 redundant member	0	0.0	−57.9	−57.9	−57.9
L4U5	Diagonal	357.7	0.0	0.0	0	357.7

In addition to the determination of member forces in internally statically indeterminate trusses due to external loads, virtual force methods can be used to determine member forces due to support settlements of externally statically indeterminate truss spans.[13] Member forces due to temperature changes and/or variations of member length[14] in statically indeterminate truss spans are also determined by virtual force methods. Examples 5.4 to 5.7 outline the determination of member forces in statically indeterminate trusses due to support movements, temperature change and variation of member lengths.

Example 5.4

The 150 ft long six-panel horizontal parallel-chord Pratt through-truss in Figure E5.4a with h = 25 ft is supported at joints L0, L2 and L6. Support L0 settles vertically 0.55 in. Find the forces in the truss members.

$ID_E = NR - 3 = NS + NS_F - 3$
$NS = 3$
$NS_F = 1$
$NR = 4$
$ID_E = 4 - 3 = 1$

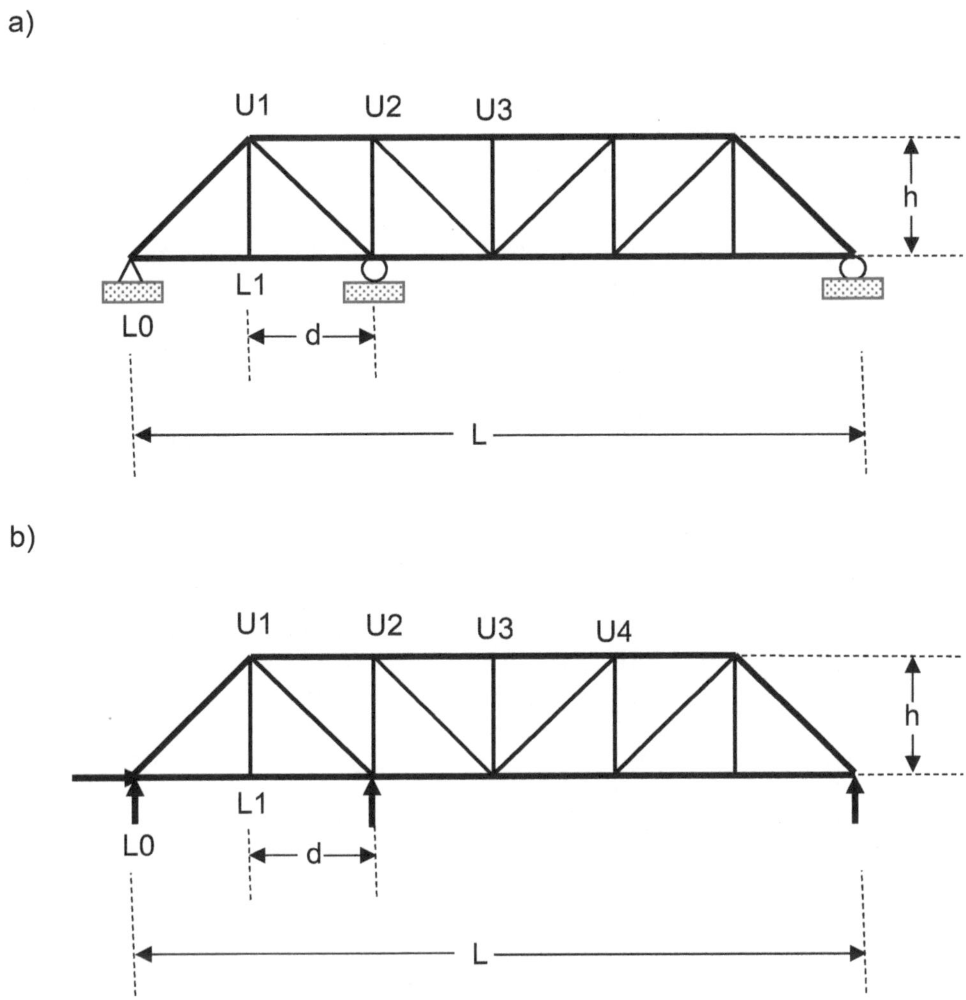

FIGURE E5.4 Pratt truss with redundant reaction at L2.

The truss is externally statically indeterminate to the first degree. As shown in Figure E5.4b, choose the vertical reaction at joint L2 as the redundant reaction. From Equation 5.10 for member j with n = 1,

$$\frac{\partial U}{\partial X_1} = 0 = \left(f_j X_1 + f_{j1} X_1^2\right)\left(\frac{L_j}{A_j}\right)$$

With a positive unit load replacing the redundant support L2, the reaction at support L0 = −0.667 and, therefore, the virtual displacement at redundant support L2 = Δ_{vL2}= −0.667(0.55) = −0.367 in:

$$f_j X_1 = \Delta_{vL2}(EA_j/L_j)$$

$$f_j X_1(L_j/A_j) = \Delta_{vL2} E = -0.367(29000) = -10{,}633.3 \text{ kip/in}$$

Structural analysis of the statically determinate truss in Figure E5.4b with unit load at L2 yields the findings in Table E5.4a.

TABLE E5.4a
Member Forces due to Unit Load X_1 at L2

Member		L_j (in)	A_j (in²)	L_j/A_j (in⁻¹)	X_1	$X_1^2(L_j/A_j)$ (in⁻¹)
L0L1	Bottom chord	300.0	20	15.00	−0.667	6.67
L1L2	Bottom chord	300.0	20	15.00	−0.667	6.67
L2L3	Bottom chord	300.0	35	8.57	−1.333	15.23
L3L4	Bottom chord	300.0	35	8.57	−0.667	3.81
L4L5	Bottom chord	300.0	20	15.00	−0.333	1.66
L5L6	Bottom chord	300.0	20	15.00	−0.333	1.66
L0U1	End post	424.3	25	16.97	0.943	15.09
U1U2	Top chord	300.0	25	12.00	1.333	21.32
U2U3	Top chord	300.0	30	10.00	1.000	10.00
U3U4	Top chord	300.0	30	10.00	1.000	10.00
U4U5	Top chord	300.0	25	12.00	0.667	5.34
U5L6	End post	424.3	25	16.97	0.471	3.76
U1L1	Vertical	300.0	10	30.00	0	0.00
U2L2	Vertical	300.0	5	60.00	−0.333	6.65
U3L3	Vertical	300.0	5	60.00	0	0.00
U4L4	Vertical	300.0	5	60.00	0.333	6.65
U5L5	Vertical	300.0	10	30.00	0	0.00
U1L2	Diagonal	424.3	10	42.43	−0.943	37.73
U2L3	Diagonal	424.3	5	84.86	0.471	18.83
L3U4	Diagonal	424.3	5	84.86	−0.471	18.83
L4U5	Diagonal	424.3	10	42.43	−0.471	9.41
Σ						199.33

Since support L2 did not settle,

$$-10{,}633.3+(f_{j1})199.33 = 0$$

$$f_{j1} = 10{,}633.3/199.33 = 53.35 \text{ kips (reaction at L2)}$$

The member forces due to the settlement of support L0 are shown in Table E5.4b.

TABLE E5.4b
Member Forces due to Settlement of Support L0

Member		X_1	$F_j = f_{j1}X_1$ (kips)
L0L1	Bottom chord	−0.667	−35.58
L1L2	Bottom chord	−0.667	−35.58
L2L3	Bottom chord	−1.333	−71.12

TABLE E5.4b (*Continued*)
Member Forces due to Settlement of Support L0

Member		X_1	$F_j = f_{j1}X_1$ (kips)
L3L4	Bottom chord	−0.667	−35.58
L4L5	Bottom chord	−0.333	−17.77
L5L6	Bottom chord	−0.333	−17.77
L0U1	End post	0.943	50.31
U1U2	Top chord	1.333	71.12
U2U3	Top chord	1.000	53.35
U3U4	Top chord	1.000	53.35
U4U5	Top chord	0.667	35.58
U5L6	End post	0.471	25.13
U1L1	Vertical	0	0.00
U2L2	Vertical	−0.333	−17.77
U3L3	Vertical	0	0.00
U4L4	Vertical	0.333	17.77
U5L5	Vertical	0	0.00
U1L2	Diagonal	−0.943	−50.31
U2L3	Diagonal	0.471	25.13
L3U4	Diagonal	−0.471	−25.13
L4U5	Diagonal	−0.471	−25.13

Example 5.5

The 150 ft long six-panel horizontal parallel-chord Pratt through-truss in Figure 5.4a with h = 25 ft is supported at joints L0, L2 and L6. Support L0 settles vertically 0.55 in and support L2 settles vertically 0.75 in. Find the forces in the truss members.

The truss is externally statically indeterminate to the first degree. Choose the vertical reaction at joint L2 as the redundant reaction. From Equation 5.10 for member j with n = 1,

$$\frac{\partial U}{\partial X_1} = 0 = \left(f_j X_1 + f_{j1} X_1^2\right)\left(\frac{L_j}{A_j}\right)$$

With a positive unit load at redundant support L2, the reaction at support L0 = −0.667 and the displacement at support L2 = −0.667(0.55) = −0.367 in:

$$f_j = -0.367(29000) = -10{,}633.3 \text{ kip/in (see Example 5.4)}$$

Since support L2 settled 0.75 in (also see Table E5.5),

$$-10{,}633.3 + (f_{j1})199.3 = -0.75(29000) = -21{,}750 \text{ kip/in}$$

$$f_{j1} = -11{,}116.7/199.3 = -55.78 \text{ kips (reaction at L2)}$$

TABLE E5.5
Member Forces due to Settlement of Supports L0 and L2

Member		X_1	$F_j = f_{j1}X_1$ (kips)
L0L1	Bottom chord	−0.667	37.21
L1L2	Bottom chord	−0.667	37.21
L2L3	Bottom chord	−1.333	74.35
L3L4	Bottom chord	−0.667	37.21
L4L5	Bottom chord	−0.333	18.57
L5L6	Bottom chord	−0.333	18.57
L0U1	End post	0.943	−52.60
U1U2	Top chord	1.333	−74.35
U2U3	Top chord	1.000	−55.78
U3U4	Top chord	1.000	−55.78
U4U5	Top chord	0.667	−37.21
U5L6	End post	0.471	−26.27
U1L1	Vertical	0	0.00
U2L2	Vertical	−0.333	18.57
U3L3	Vertical	0	0.00
U4L4	Vertical	0.333	−18.57
U5L5	Vertical	0	0.00
U1L2	Diagonal	−0.943	52.60
U2L3	Diagonal	0.471	−26.27
L3U4	Diagonal	−0.471	26.27
L4U5	Diagonal	−0.471	26.27

Example 5.6

The 150 ft long six-panel horizontal parallel-chord Pratt through-truss in Figure E5.1a with $h = 25$ ft is subjected to a change in temperature, $\Delta T = 80$ °F at the bottom chord members. Find the forces in the truss members. The coefficient of thermal expansion of steel, α_T, is (6.5×10^{-6}) °F^{-1}.

Choose L2U3 and U3L4 as redundant members. From Equation 5.10 for member j with n = 2,

$$\frac{\partial U}{\partial X_1} = 0 = \left(f_jX_1 + f_{j1}X_1^2 + f_{j2}X_2X_1\right)\left(\frac{L_j}{A_j}\right) = f_jX_1 + f_{j1}(239) + f_{j2}(30)$$

$$\frac{\partial U}{\partial X_2} = 0 = \left(f_jX_2 + f_{j1}X_1X_2 + f_{j2}X_2^2\right)\left(\frac{L_j}{A_j}\right) = f_jX_2 + f_{j1}(30) + f_{j2}(239)$$ (see Example 5.2 and Example 5.3 with same redundant members selected):

where
$X_1 = X_2 = -0.707$ (angle between bottom chord and redundant members L2U3 and U3L4)
$f_jX_1 = f_jX_2 = \alpha_T E(\Delta T)A_j$
$f_jX_1(L_j/A_j) = f_jX_2(L_j/A_j) = \alpha_T EX_1(\Delta T)L_j = \alpha_T EX_2(\Delta T)L_j = 0.189(-0.707)(\Delta T)(L_j)$ kips/in $= 0.189(-0.707)(80)(300) = -3198.5$ kips/in for bottom chord members

From Equation 5.10,

$$-3198.5 + 239.0(f_{j1}) + 30.0(f_{j2}) = 0$$

$$-3198.5 + 30.0(f_{j1}) + 239.0(f_{j2}) = 0$$

The solution of these equations provides

f_{j1} = 11.89 kips (force in redundant member L2U3)
f_{j2} = 11.89 kips (force in redundant member U3L4)

Members of the statically indeterminate truss are subjected to the forces (in Table E5.6) from a temperature increase of 80 °F at the bottom chord members.

TABLE E5.6
Member Forces due to Temperature Change

Member		X_1	X_2	$f_{j1}X_1$ (kips)	$f_{j2}X_2$ (kips)	$F_j = f_{j1}X_1 + f_{j2}X_2$ (kips)
L0L1	Bottom chord	0	0	0	0	0
L1L2	Bottom chord	0	0	0	0	0
L2L3	Bottom chord	−0.707	0	−8.41	0	−8.41
L3L4	Bottom chord	0	−0.707	0	−8.41	−8.41
L4L5	Bottom chord	0	0	0	0	0
L5L6	Bottom chord	0	0	0	0	0
L0U1	End post	0	0	0	0	0
U1U2	Top chord	0	0	0	0	0
U2U3	Top chord	−0.707	0	−8.41	0	−8.41
U3U4	Top chord	0	−0.707	0	−8.41	−8.41
U4U5	Top chord	0	0	0	0	0
U5L6	End post	0	0	0	0	0
U1L1	Vertical	0	0	0	0	0
U2L2	Vertical	−0.707	0	−8.41	0	−8.41
U3L3	Vertical	−0.707	−0.707	−8.41	−8.41	−16.82
U4L4	Vertical	0	−0.707	0	−8.41	−8.41
U5L5	Vertical	0	0	0	0	0
U1L2	Diagonal	0	0	0	0	0
U2L3	Diagonal	1.00	0	11.89	0	11.89
L2U3	F_1 Redundant	1.00	0	11.89	0	11.89
L3U4	Diagonal	0	1.00	0	11.89	11.89
U3L4	F_2 Redundant	0	1.00	0	11.89	11.89
L4U5	Diagonal	0	0	0		0

Example 5.7

The bottom chord members of the 150 ft long six-panel horizontal parallel-chord Pratt through-truss in Figure E5.1 with h = 25 ft are fabricated 0.25 in too short. Find the forces in the truss members with the bottom chord members “forced” into place.

Choose L2U3 and U3L4 as redundant members. From Equation 5.10 for member j with n = 2,

$$\frac{\partial U}{\partial X_1} = 0 = \left(f_j X_1 + f_{j1} X_1^2 + f_{j2} X_2 X_1\right)\left(\frac{L_j}{A_j}\right) = f_j X_1 + f_{j1}(239) + f_{j2}(30)$$

$$\frac{\partial U}{\partial X_2} = 0 = \left(f_j X_2 + f_{j1} X_1 X_2 + f_{j2} X_2^2\right)\left(\frac{L_j}{A_j}\right) = f_j X_2 + f_{j1}(30) + f_{j2}(239)$$ (see Examples 5.2, 5.3 and 5.6 with same redundant members selected):

where

$f_j = (\Delta L_j)E = -0.25(29000) = -7250$ kips/in for bottom chord members
$f_j X_1(L_j/A_j) = f_j X_2(L_j/A_j) = (\Delta L_j)X_1E(L_j/A_j) = (\Delta L_j)X_2E(L_j/A_j) = (-0.25)(-0.707)(29000)(L_j/A_j) = 5126.5(L_j/A_j)$ kips/in for redundant members L2U3 and U3L4

From Equation 5.10,

$$5126.5 + 239.0(f_{j1}) + 30.0(f_{j2}) = 0$$

$$5126.5 + 239.0(f_{j1}) + 30.0(f_{j2}) = 0$$

The solution of these equations provides

$f_{j1} = -19.05$ kips (force in redundant member L2U3)
$f_{j2} = -19.05$ kips (force in redundant member U3L4)

Members of the statically indeterminate truss are subjected to the forces in Table E5.7 from a variation of member length of −0.25 in of the bottom chord members.

TABLE E5.7
Member Forces due to Bottom Chord Shortening

Member		$f_{j1}X_1$ (kips)	$f_{j2}X_2$ (kips)	$F_j = f_{j1}X_1 + f_{j2}X_2$ (kips)
L0L1	Bottom chord	0	0	0
L1L2	Bottom chord	0	0	0
L2L3	Bottom chord	13.47	0	13.47
L3L4	Bottom chord	0	13.47	13.47
L4L5	Bottom chord	0	0	0
L5L6	Bottom chord	0	0	0
L0U1	End post	0	0	0
U1U2	Top chord	0	0	0
U2U3	Top chord	13.47	0	13.47
U3U4	Top chord	0	13.47	13.47
U4U5	Top chord	0	0	0
U5L6	End post	0	0	0
U1L1	Vertical	0	0	0
U2L2	Vertical	13.47	0	13.47
U3L3	Vertical	13.47	13.47	26.95
U4L4	Vertical	0	13.47	13.47

TABLE E5.7 (*Continued*)
Member Forces due to Bottom Chord Shortening

Member		$f_{j1}X_1$ (kips)	$f_{j2}X_2$ (kips)	$F_j = f_{j1}X_1 + f_{j2}X_2$ (kips)
U5L5	Vertical	0	0	0
U1L2	Diagonal	0	0	0
U2L3	Diagonal	−19.05	0	−19.05
L2U3	F_1 redundant	−19.05	0	−19.05
L3U4	Diagonal	0	−19.05	−19.05
U3L4	F_2 redundant	0	−19.05	−19.05
L4U5	Diagonal	0	0	0

5.4 INFLUENCE LINES FOR PIN-CONNECTED ELASTIC TRUSS SPANS

Equation 5.7, $F_j = f_j + \sum_{i=1}^{n} f_{ji}X_i$, provides a basis for calculating influence lines for statically indeterminate truss spans where f_j = member forces due to unit load at all possible load locations (panel points and joints) and X_i = force in member j due to unit loads at redundant members. Example 5.8 is an example of calculating the ordinates of influence lines for statically indeterminate truss spans.

Example 5.8

The influence lines for members of the statically indeterminate 150 ft long six-panel horizontal parallel-chord Pratt through-truss in Figure E5.1 with h = 25 ft are required for a live load analysis.

With L2U3 and U3L4 as redundant members, Equation 5.10 for member j with n = 2 provides

$$\frac{\partial U}{\partial X_1} = 0 = \left(f_jX_1 + f_{j1}X_1^2 + f_{j2}X_2X_1\right)\left(\frac{L_j}{A_j}\right) = f_jX_1 + f_{j1}(239) + f_{j2}(30)$$

$$\frac{\partial U}{\partial X_2} = 0 = \left(f_jX_2 + f_{j1}X_1X_2 + uf_{j2}X_2^2\right)\left(\frac{L_j}{A_j}\right) = f_jX_2 + f_{j1}(30) + f_{j2}(239)$$ (see Examples 5.2, 5.3 and 5.6 with the same redundant members selected).

X_1 = force in j due to unit loads at redundant member L2U3
X_2 = force in j due to unit loads at redundant member U3L4

With unit loads at joints L1, L2, L3, L4 and L5, the influence coefficients for the statically determinate member forces f_j can be used to determine the influence coefficients for the statically indeterminate member forces F_j (Table E5.8a).

$$-27.62 + f_{j1}(239) + f_{j2}(30) = 0$$

$$28.63 + f_{j1}(30) + f_{j2}(239) = 0$$

TABLE E5.8a
Rate of Change of Strain Energy due to Unit Loads X_1 and X_2 and Unit Load at L1

UNIT LOAD AT L1

Member		L_j/A_j	f_j (kips)	X_1	$X_1f_j(L_j/A_j)$ (kips/in)	X_2	$X_2f_j(L_j/A_j)$ (kips/in)
L0L1	Bottom chord	15.00	0.833	0	0.00	0	0.00
L1L2	Bottom chord	15.00	0.833	0	0.00	0	0.00
L2L3	Bottom chord	8.57	0.667	−0.707	−4.04	0	0.00
L3L4	Bottom chord	8.57	0.333	0	0.00	−0.707	-2.02
L4L5	Bottom chord	15.00	0.167	0	0.00	0	0.00
L5L6	Bottom chord	15.00	0.167	0	0.00	0	0.00
L0U1	End post	16.97	−1.179	0	0.00	0	0.00
U1U2	Top chord	12.00	−0.667	0	0.00	0	0.00
U2U3	Top chord	10.00	−0.500	−0.707	3.54	0	0.00
U3U4	Top chord	10.00	−0.500	0	0.00	−0.707	3.54
U4U5	Top chord	12.00	−0.333	0	0.00	0	0.00
U5L6	End post	16.97	−0.236	0	0.00	0	0.00
U1L1	Vertical	30.00	1.000	0	0.00	0	0.00
U2L2	Vertical	60.00	0.167	−0.707	−7.08	0	0.00
U3L3	Vertical	60.00	0	−0.707	0.00	−0.707	0.00
U4L4	Vertical	60.00	−0.167	0	0.00	−0.707	7.08
U5L5	Vertical	30.00	0	0	0.00	0	0.00
U1L2	Diagonal	42.43	−0.236	0	0.00	0	0.00
U2L3	Diagonal	84.86	−0.236	1.00	−20.03	0	0.00
L2U3	F_1 redundant	84.86	0	1.00	0.00	0	0.00
L3U4	Diagonal	84.86	0.236	0	0.00	1.00	20.03
U3L4	F_2 redundant	84.86	0	0	0.00	1.00	0.00
L4U5	Diagonal	42.43	0.236	0	0.00	0	0.00
Σ					−27.62		28.63

The solution of these equations provides the following as well as Tables E5.8b and E5.8c:

f_{j1} = 0.133 kips (force in redundant member L2U3)
f_{j2} = −0.136 kips (force in redundant member U3L4)

$$F_j = f_j + \sum_{i=1}^{n} f_{ji}X_i = f_j + X_1f_{j1} + X_2f_{j2}$$

$$-55.10 + f_{j1}(239) + f_{j2}(30) = 0$$

$$57.12 + f_{j1}(30) + f_{j2}(239) = 0$$

TABLE E5.8b
Influence Coefficients for Unit Load at L1 in a Statically Indeterminate Truss Span

		UNIT LOAD AT L1					
Member		**f_j (kips)**	**X_1**	**X_1f_j (kips)**	**X_2**	**X_2f_j (kips)**	**$F_j = f_j + X_1f_{j1} + X_2f_{j2}$ (kips)**
L0L1	Bottom chord	0.833	0	0	0	0	0.833
L1L2	Bottom chord	0.833	0	0	0	0	0.833
L2L3	Bottom chord	0.667	−0.707	−0.094	0	0	0.573
L3L4	Bottom chord	0.333	0	0	−0.707	0.096	0.429
L4L5	Bottom chord	0.167	0	0	0	0	0.167
L5L6	Bottom chord	0.167	0	0	0	0	0.167
L0U1	End post	−1.179	0	0	0	0	−1.179
U1U2	Top chord	−0.667	0	0	0	0	−0.667
U2U3	Top chord	−0.500	−0.707	−0.094	0	0	−0.594
U3U4	Top chord	−0.500	0	0	−0.707	0.096	−0.404
U4U5	Top chord	−0.333	0	0	0	0	−0.333
U5L6	End post	−0.236	0	0	0	0	−0.236
U1L1	Vertical	1.000	0	0	0	0	1
U2L2	Vertical	0.167	−0.707	−0.094	0	0	0.073
U3L3	Vertical	0	−0.707	−0.094	−0.707	0.096	0.002
U4L4	Vertical	−0.167	0	0	−0.707	0.096	−0.071
U5L5	Vertical	0	0	0	0	0	0
U1L2	Diagonal	−0.236	0	0	0	0	−0.236
U2L3	Diagonal	−0.236	1.00	0.133	0	0	−0.103
L2U3	F_1 redundant	0	1.00	0.133	0	0	0.133
L3U4	Diagonal	0.236	0	0	1.00	−0.136	0.1
U3L4	F_2 redundant	0	0	0	1.00	−0.136	−0.136
L4U5	Diagonal	0.236	0	0	0	0	0.236

TABLE E5.8c
Rate of Change of Strain Energy due to Unit Loads X_1 and X_2 and Unit Load at L2

		UNIT LOAD AT L2					
Member		**L_j/A_j**	**f_j (kips)**	**X_1**	**$X_1f_j(L_j/A_j)$ (kips/in)**	**X_2**	**$X_2f_j(L_j/A_j)$ (kips/in)**
L0L1	Bottom chord	15.00	0.667	0	0	0	0
L1L2	Bottom chord	15.00	0.667	0	0.00	0	0.00
L2L3	Bottom chord	8.57	1.333	−0.707	−8.08	0	0.00
L3L4	Bottom chord	8.57	0.667	0	0.00	−0.707	−4.04
L4L5	Bottom chord	15.00	0.333	0	0.00	0	0.00
L5L6	Bottom chord	15.00	0.333	0	0.00	0	0.00
L0U1	End post	16.97	−0.943	0	0.00	0	0.00
U1U2	Top chord	12.00	−1.333	0	0.00	0	0.00
U2U3	Top chord	10.00	−1.000	−0.707	7.07	0	0.00
U3U4	Top chord	10.00	−1.000	0	0.00	−0.707	7.07

(Continued)

TABLE E5.8c (*Continued*)
Rate of Change of Strain Energy due to Unit Loads X_1 and X_2 and Unit Load at L2

UNIT LOAD AT L2

Member		L_j/A_j	f_j (kips)	X_1	$X_1f_j(L_j/A_j)$ (kips/in)	X_2	$X_2f_j(L_j/A_j)$ (kips/in)
U4U5	Top chord	12.00	−0.667	0	0.00	0	0.00
U5L6	End post	16.97	−0.471	0	0.00	0	0.00
U1L1	Vertical	30.00	0	0	0.00	0	0.00
U2L2	Vertical	60.00	0.333	−0.707	−14.13	0	0.00
U3L3	Vertical	60.00	0	−0.707	0.00	−0.707	0.00
U4L4	Vertical	60.00	−0.333	0	0.00	−0.707	14.13
U5L5	Vertical	30.00	0	0	0.00	0	0.00
U1L2	Diagonal	42.43	0.943	0	0.00	0	0.00
U2L3	Diagonal	84.86	−0.471	1.00	−39.97	0	0.00
L2U3	F_1 redundant	84.86	0	1.00	0.00	0	0.00
L3U4	Diagonal	84.86	0.471	0	0.00	1.00	39.97
U3L4	F_2 redundant	84.86	0	0	0.00	1.00	0.00
L4U5	Diagonal	42.43	0.471	0	0.00	0	0.00
Σ					−55.10		57.12

The solution of these equations provides the following as well as Tables E5.8d and E5.8e:

f_{j1} = 0.265 kips (force in redundant member L2U3)
f_{j2} = −0.272 kips (force in redundant member U3L4)

$$85.75 + f_{j1}(239) + f_{j2}(30) = 0$$

$$85.75 + f_{j1}(30) + f_{j2}(239) = 0$$

TABLE E5.8d
Influence Coefficients for Unit Load at L2 in a Statically Indeterminate Truss Span

UNIT LOAD AT L2

Member		f_j (kips)	X_1	X_1f_j (kips)	X_2	X_2f_j (kips)	$F_j = f_j + X_1f_{j1} + X_2f_{j2}$ (kips)
L0L1	Bottom chord	0.667	0	0	0	0	0.667
L1L2	Bottom chord	0.667	0	0	0	0	0.667
L2L3	Bottom chord	1.333	−0.707	−0.187	0	0	1.146
L3L4	Bottom chord	0.667	0	0	−0.707	0.192	0.859
L4L5	Bottom chord	0.333	0	0	0	0	0.333
L5L6	Bottom chord	0.333	0	0	0	0	0.333
L0U1	End post	−0.943	0	0	0	0	−0.943
U1U2	Top chord	−1.333	0	0	0	0	−1.333
U2U3	Top chord	−1.000	−0.707	−0.187	0	0	−1.187

TABLE E5.8d (*Continued*)
Influence Coefficients for Unit Load at L2 in a Statically Indeterminate Truss Span

UNIT LOAD AT L2

Member		f_j (kips)	X_1	X_1f_j (kips)	X_2	X_2f_j (kips)	$F_j = f_j + X_1f_{j1} + X_2f_{j2}$ (kips)
U3U4	Top chord	−1.000	0	0	−0.707	0.192	−0.808
U4U5	Top chord	−0.667	0	0	0	0	−0.667
U5L6	End post	−0.471	0	0	0	0	−0.471
U1L1	Vertical	0	0	0	0	0	0
U2L2	Vertical	0.333	−0.707	−0.187	0	0	0.146
U3L3	Vertical	0	−0.707	−0.187	−0.707	0.192	0.005
U4L4	Vertical	−0.333	0	0	−0.707	0.192	−0.141
U5L5	Vertical	0	0	0	0	0	0
U1L2	Diagonal	0.943	0	0	0	0	0.943
U2L3	Diagonal	−0.471	1.00	0.265	0	0	−0.206
L2U3	F_1 redundant	0	1.00	0.265	0	0	0.265
L3U4	Diagonal	0.471	0	0	1.00	−0.272	0.199
U3L4	F_2 redundant	0	0	0	1.00	−0.272	−0.272
L4U5	Diagonal	0.471	0	0	0	0	0.471

TABLE E5.8e
Rate of Change of Strain Energy due to Unit Loads X_1 and X_2 and Unit Load at L3

UNIT LOAD AT L3

Member		L_j/A_j	f_j (kips)	X_1	$X_1f_j(L_j/A_j)$ (kips/in)	X_2	$X_2f_j(L_j/A_j)$ (kips/in)
L0L1	Bottom chord	15.00	0.500	0	0	0	0
L1L2	Bottom chord	15.00	0.500	0	0.00	0	0.00
L2L3	Bottom chord	8.57	1.000	−0.707	−6.06	0	0.00
L3L4	Bottom chord	8.57	1.000	0	0.00	−0.707	−6.06
L4L5	Bottom chord	15.00	0.500	0	0.00	0	0.00
L5L6	Bottom chord	15.00	0.500	0	0.00	0	0.00
L0U1	End post	16.97	−0.707	0	0.00	0	0.00
U1U2	Top chord	12.00	−1.000	0	0.00	0	0.00
U2U3	Top chord	10.00	−1.500	−0.707	10.61	0	0.00
U3U4	Top chord	10.00	−1.500	0	0.00	−0.707	10.61
U4U5	Top chord	12.00	−1.000	0	0.00	0	0.00
U5L6	End post	16.97	−0.707	0	0.00	0	0.00
U1L1	Vertical	30.00	0	0	0.00	0	0.00
U2L2	Vertical	60.00	−0.500	−0.707	21.21	0	0.00
U3L3	Vertical	60.00	0	−0.707	0.00	−0.707	0.00
U4L4	Vertical	60.00	−0.500	0	0.00	−0.707	21.21
U5L5	Vertical	30.00	0	0	0.00	0	0.00
U1L2	Diagonal	42.43	0.707	0	0.00	0	0.00
U2L3	Diagonal	84.86	0.707	1.00	60.00	0	0.00
L2U3	F_1 redundant	84.86	0	1.00	0.00	0	0.00

(*Continued*)

TABLE E5.8e (*Continued*)
Rate of Change of Strain Energy due to Unit Loads X_1 and X_2 and Unit Load at L3

UNIT LOAD AT L3

Member		L_j/A_j	f_j (kips)	X_1	$X_1 f_j(L_j/A_j)$ (kips/in)	X_2	$X_2 f_j(L_j/A_j)$ (kips/in)
L3U4	Diagonal	84.86	0.707	0	0.00	1.00	60.00
U3L4	F_2 redundant	84.86	0	0	0.00	1.00	0.00
L4U5	Diagonal	42.43	0.707	0	0.00	0	0.00
Σ					85.75		85.75

The solution of these equations provides the following as well as Table E5.8f:

f_{j1} = 0.319 kips (force in redundant member L2U3)
f_{j2} = 0.319 kips (force in redundant member U3L4)

TABLE E5.8f
Influence Coefficients for Unit Load at L3 in a Statically Indeterminate Truss Span

UNIT LOAD AT L3

Member		f_j (kips)	X_1	$X_1 f_j$ (kips)	X_2	$X_2 f_j$ (kips)	$F_j = f_j + X_1 f_{j1} + X_2 f_{j2}$ (kips)
L0L1	Bottom chord	0.500	0	0	0	0	0.5
L1L2	Bottom chord	0.500	0	0	0	0	0.5
L2L3	Bottom chord	1.000	−0.707	−0.226	0	0	0.774
L3L4	Bottom chord	1.000	0	0	−0.707	−0.226	0.774
L4L5	Bottom chord	0.500	0	0	0	0	0.5
L5L6	Bottom chord	0.500	0	0	0	0	0.5
L0U1	End post	−0.707	0	0	0	0	−0.707
U1U2	Top chord	−1.000	0	0	0	0	−1
U2U3	Top chord	−1.500	−0.707	−0.226	0	0	−1.726
U3U4	Top chord	−1.500	0	0	−0.707	−0.226	−1.726
U4U5	Top chord	−1.000	0	0	0	0	−1
U5L6	End post	−0.707	0	0	0	0	−0.707
U1L1	Vertical	0	0	0	0	0	0
U2L2	Vertical	−0.500	−0.707	−0.226	0	0	−0.726
U3L3	Vertical	0	−0.707	−0.226	−0.707	−0.226	−0.452
U4L4	Vertical	−0.500	0	0	−0.707	−0.226	−0.726
U5L5	Vertical	0	0	0	0	0	0
U1L2	Diagonal	0.707	0	0	0	0	0.707
U2L3	Diagonal	0.707	1.00	0.319	0	0	1.026
L2U3	F_1 redundant	0	1.00	0.319	0	0	0.319
L3U4	Diagonal	0.707	0	0	1.00	0.319	1.026
U3L4	F_2 redundant	0	0	0	1.00	0.319	0.319
L4U5	Diagonal	0.707	0	0	0	0	0.707

The ordinates for the influence lines are given in Table E5.8g.

TABLE E5.8g
Ordinates for the Influence Lines for Members of a Statically Indeterminate Truss Span

Member		F_j (kips)				
		Unit Load at				
		L1	L2	L3	L4	L5
L0L1	Bottom chord	0.833	0.667	0.5	0.333	0.167
L1L2	Bottom chord	0.833	0.667	0.5	0.333	0.167
L2L3	Bottom chord	0.573	1.146	0.774	0.859	0.429
L3L4	Bottom chord	0.429	0.859	0.774	1.146	0.573
L4L5	Bottom chord	0.167	0.333	0.5	0.667	0.833
L5L6	Bottom chord	0.167	0.333	0.5	0.667	0.833
L0U1	End post	−1.179	−0.943	−0.707	−0.471	−0.236
U1U2	Top chord	−0.667	−1.333	−1	−0.667	−0.333
U2U3	Top chord	−0.594	−1.187	−1.726	−0.808	−0.404
U3U4	Top chord	−0.404	−0.808	−1.726	−1.187	−0.594
U4U5	Top chord	−0.333	−0.667	−1	−1.333	−0.667
U5L6	End post	−0.236	−0.471	−0.707	−0.943	−1.179
U1L1	Vertical	1	0	0	0	0
U2L2	Vertical	0.073	0.146	−0.726	−0.141	−0.071
U3L3	Vertical	0.002	0.005	−0.452	0.005	0.002
U4L4	Vertical	−0.071	−0.141	−0.726	0.146	0.073
U5L5	Vertical	0	0	0	0	1
U1L2	Diagonal	−0.236	0.943	0.707	0.471	0.236
U2L3	Diagonal	−0.103	−0.206	1.026	0.199	0.1
L2U3	Diagonal	0.133	0.265	0.319	−0.272	−0.136
L3U4	Diagonal	0.1	0.199	1.026	−0.206	−0.103
U3L4	Diagonal	−0.136	−0.272	0.319	0.265	0.133
L4U5	Diagonal	0.236	0.471	0.707	0.943	−0.236

The influence lines for members L0L1, L3L4, U2U3, U2L3 and L2U3, are shown in Figure E5.8 on the following pages.

The Muller–Breslau principle is also used with virtual work methods to construct the influence lines for members of statically indeterminate trusses. The principle states that the influence line for any restraint in a truss is the elastic curve created by a corresponding unit virtual displacement. The corresponding unit virtual displacement is applied at the location of the restraint with the same attitude (direction) as the restraint.

5.5 APPROXIMATE ANALYSIS OF PIN-CONNECTED ELASTIC TRUSS SPANS

Energy methods of statically indeterminate truss span analysis are rigorous but approximate in the sense that only axial force members with frictionless pins are considered in the truss span. Nevertheless, there are approximate methods of statically indeterminate truss span analysis based

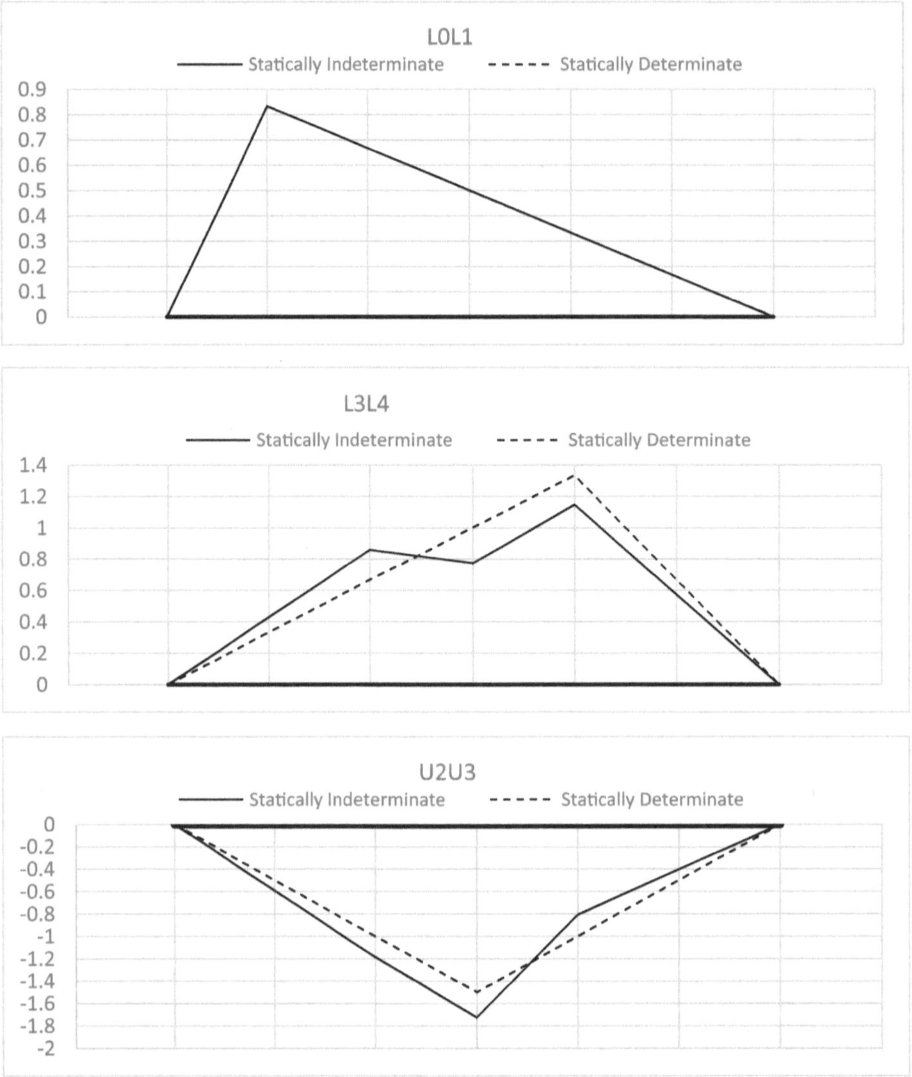

FIGURE E5.8 Influence lines for statically determinate and statically indeterminate truss members.

on experience and reviews of accuracy that are often sufficient for initial, conceptual and preliminary truss span design. Rational approximate methods make assumptions equal to the number of redundant members and/or supports to render the statically indeterminate truss statically determinate.

Many railroad truss spans are of moderate length with horizontal parallel chords and single-intersection web members for ease of design,[15] fabrication and erection. Horizontal parallel-chord truss spans with two diagonals in the central panels are used to obtain a stiffer structure,[16] and, inevitably, trusses with an odd number of panels require two diagonals in at least the center panel. Statically indeterminate trusses are rendered as statically determinate by assuming that diagonal members in panels with two diagonal members equally resist the shear force in the panel in tension and compression.[17] Alternatively, diagonal members in panels with two diagonal members can be assumed to act in tension only to create a statically determinate truss.[18] Examples 5.9 and 5.10 illustrate approximate methods for analyzing the truss span in Example 5.1.

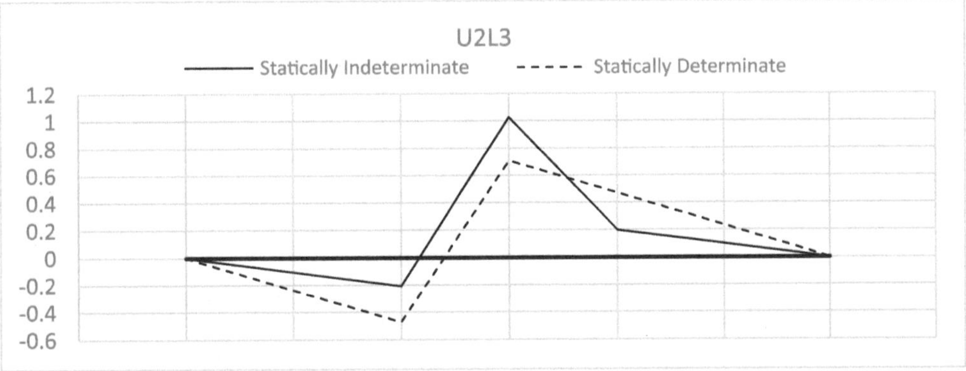

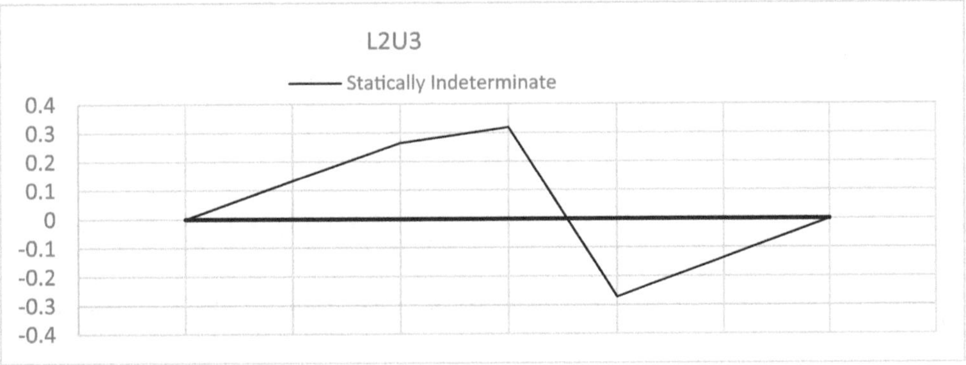

FIGURE E5.8 *(Continued)*

Example 5.9

The 150 ft long six-panel horizontal parallel-chord Pratt through-truss in Figure E5.1 with h = 25 ft is subjected to a uniform dead load of 1500 lb/ft and uniform live load of 5500 lb/ft. Find the maximum forces in the truss members for the live load location producing maximum positive shear force in member U1L2 (live load on 120 ft of the span from the right hand side). The joint loads for this uniformly distributed live load position are given in Tables E5.9a and E5.9b.

TABLE E5.9a
Dead and Live Loads at Truss Joints due to the External Live Load Location Producing Maximum Positive Shear Force in Panel 1-2

Joint	Dead Load (kips)	Live Load (kips)	Dead + Live Load (kips)
L0	18.75	0	18.75
L1	37.50	44.00	81.50
L2	37.50	134.75	172.25
L3	37.50	137.5	175.00
L4	37.50	137.5	175.00
L5	37.50	137.5	175.00
L6	18.75	68.75	87.50
Σ	225.00	660.00	885.00

TABLE E5.9b

Dead and Live Load Reactions due to the External Live Load Location Producing Maximum Positive Shear Force in Panel 1-2

Reaction	Dead Load	Live Load	Dead + Live Load (kips)
L0	112.5	264.0	376.50
L6	112.5	396.0	508.50
Σ	225.00	660.00	885.00

The degree of indeterminacy is 2. Since the truss is internally statically indeterminate, choose two experience-based approximation assumptions to create a statically determinate Pratt truss. Assume that the diagonals in the two center panels resist panel shear with equal but opposite (one diagonal in tension, the other in compression) internal forces (Table E5.9c).

TABLE E5.9c

Dead and Live Panel Shear Forces in Panels 2-3 and 3-4 due to the External Live Load Location Producing Maximum Positive Shear Force in Panel 1-2

Panel	Live Load Panel Shear (kips)	Dead Load Panel Shear (kips)	Compression Diagonal Member (kips)	Tension Diagonal Member (kips)
2-3	264.0 – 0 – 44.0 – 134.75 = 85.25	112.50 – 18.75 – 37.50 – 37.50 = 18.75	−73.5	73.5
3-4	85.25 – 137.5 = −52.25	18.75 – 37.5 = −18.75	−50.2	50.2

Using the method of sections with the free body diagram of the left side of the truss sectioned at the center of panel 2-3 provides the forces in members L2L3 and U2U3. Using the method of sections with the free body diagram of the left side of the truss sectioned at the center of panel 3-4 provides the forces in members L3L4 and U3U4. Investigating the equilibrium of joints U2, L2, U1, U4, L4 and U5 provides the remaining member forces as in Table E5.9d.

TABLE E5.9d

Exact and Approximate Member Forces in a Statically Indeterminate Truss due to the External Live Load Location Producing Maximum Positive Shear Force in Panel 1-2

Member		Exact Member Force, F_j (kips)	Approximate Member Force (kips)	% Difference
L0L1	Bottom chord	357.8	359.7	0.53
L1L2	Bottom chord	357.8	359.7	0.53
L2L3	Bottom chord	684.1	686.0	0.28
L3L4	Bottom chord	698.3	701.8	0.50
L4L5	Bottom chord	421.3	421.3	0.00

TABLE E5.9d (*Continued*)
Exact and Approximate Member Forces in a Statically Indeterminate Truss due to the External Live Load Location Producing Maximum Positive Shear Force in Panel 1-2

Member		Exact Member Force, F_j (kips)	Approximate Member Force (kips)	% Difference
L5L6	Bottom chord	421.3	421.3	0.00
L0U1	End post	−506.0	−508.6	0.51
U1U2	Top chord	−634.0	−634.0	0.00
U2U3	Top chord	−688.0	−686.0	0.29
U3U4	Top chord	−706.8	−701.8	0.71
U4U5	Top chord	−667.0	−666.3	0.10
U5L6	End post	−595.4	−595.4	0.00
U1L1	Vertical	81.5	81.5	0.00
U2L2	Vertical	−54.0	−51.5	4.63
U3L3	Vertical	81.4	87.5	7.49
U4L4	Vertical	−39.8	−35.5	10.80
U5L5	Vertical	175.0	175.0	0.00
U1L2	Diagonal	390.7	388.0	0.69
U2L3	Diagonal	76.2	73.5	3.54
L2U3	F_1 redundant member	−70.9	−73.5	3.67
L3U4	Diagonal	56.1	50.2	10.52
U3L4	F_2 redundant Member	−44.3	−50.2	13.32
L4U5	Diagonal	347.9	346.5	0.40

Table E5.9d shows very close agreement, with differences between the exact and approximate member forces of less than 1% for all members except U2L2, U3, L3, U4L4, U2L3, L2U3, L3U4 and U3L4. Moreover, the differences between the exact and approximate analyses for these web members in the two central panels, each with two diagonals, is relatively small. Furthermore, these members are relatively lightly loaded and typically designed for serviceability (slenderness).

Example 5.10

The 150 ft long six-panel horizontal parallel-chord Pratt through-truss in Figure E5.1 with h = 25 ft is subjected to a uniform dead load of 1500 lb/ft and Cooper's E80 live load. Find the forces in the truss members for the live load location with axle 7 at L2. The joint loads for this Cooper's E80 live load position are given in Tables E5.10a and E5.10b.
The degree of indeterminacy is 2. Since the truss is internally statically indeterminate, choose two experience-based approximation assumptions to create a statically determinate Pratt truss. Assume that the diagonals in the two center panels resist panel shear with equal but opposite (one diagonal in tension, the other in compression) internal forces (see Table E5.10c).

Using the method of sections and method of joints provides the remaining member forces as in Table 5.10d.

Table E5.10d shows very close agreement, with differences between the exact and approximate member forces of less than 1% for all members except U2L2, U3, L3, U4L4,

TABLE E5.10a
Dead and Live Loads at Truss Joints due to the External Live Load Location with Axle 7 at L2

Joint	Dead Load (kips)	Live Load (kips)	Dead + Live Load (kips)
L0	18.75	16.00	34.75
L1	37.50	140.40	177.9
L2	37.50	114.72	152.22
L3	37.50	132.08	169.58
L4	37.50	128.08	165.58
L5	37.50	98.72	136.22
L6	18.75	50.00	68.75
Σ	225.00	680.00	905.00

TABLE E5.10b
Dead and Live Load Reactions due to the External Live Load Location with Axle 7 at L2

Reaction	Dead Load	Live Load	Dead + Live Load (kips)
L0	112.5	334.67	447.17
L6	112.5	345.33	457.83
Σ	225.00	680.00	905.00

TABLE E5.10c
Dead and Live Panel Shear Forces in Panels 2-3 and 3-4 due to the External Live Load Location with Axle 7 at L2

Panel	Live Load Panel Shear (kips)	Dead Load Panel Shear (kips)	Compression Diagonal Member (kips)	Tension Diagonal Member (kips)
2-3	334.67 − 16.00 − 140.40 − 114.72 = 63.55	112.50 − 18.75 − 37.50 − 37.50 = 18.75	−58.2	58.2
3-4	63.55 − 132.08 = −68.53	18.75 − 37.5 = −18.75	−61.7	61.7

TABLE E5.10d
Exact and Approximate Member Forces in the Statically Indeterminate Truss due to the External Live Load Location with Axle 7 at L2

	Member	Exact Member Force, F_j (kips)	Approximate Member Force (kips)	% Difference
L0L1	Bottom chord	412.5	411.1	0.34
L1L2	Bottom chord	412.5	411.1	0.34
L2L3	Bottom chord	685.0	688.1	0.45
L3L4	Bottom chord	682.8	685.6	0.41

TABLE E5.10d (*Continued*)
Exact and Approximate Member Forces in the Statically Indeterminate Truss due to the External Live Load Location with Axle 7 at L2

	Member	Exact Member Force, F_j (kips)	Approximate Member Force (kips)	% Difference
L4L5	Bottom chord	389.1	390.0	0.23
L5L6	Bottom chord	389.1	390.0	0.23
L0U1	End post	−583.3	−582.8	0.09
U1U2	Top chord	−646.9	−646.9	0.00
U2U3	Top chord	−691.2	−688.1	0.45
U3U4	Top chord	−688.4	−685.6	0.41
U4U5	Top chord	−641.9	−642.0	0.02
U5L6	End post	−550.3	−548.1	0.40
U1L1	Vertical	177.9	177.9	0.00
U2L2	Vertical	−44.3	−41.3	6.77
U3L3	Vertical	79.0	84.8	7.34
U4L4	Vertical	−46.4	−43.6	6.03
U5L5	Vertical	136.2	136.2	0.00
U1L2	Diagonal	331.7	332.2	0.15
U2L3	Diagonal	62.5	58.2	6.88
L2U3	F_1 redundant member	−53.9	−58.2	7.98
L3U4	Diagonal	65.6	61.7	5.95
U3L4	F_2 redundant member	−57.9	−61.7	6.56
L4U5	Diagonal	357.7	356.1	0.45

U2L3, L2U3, L3U4 and U3L4. Moreover, the differences between the exact and approximate analyses for these web members in the two central panels, each with two diagonals, is relatively small. Furthermore, these members are relatively lightly loaded and typically designed for serviceability (slenderness).

NOTES

1 Also known as redundant or hyperstatic truss spans.
2 Theorem of the Differential Coefficients of the Internal Work—Part 2 in Castigliano's book "The Theory of Equilibrium of Elastic Systems and Its Applications" (Castigliano, 1879). Castigliano, a railway engineer, first proposed the theorem in an 1873 thesis. Proof of Castigliano's Second Theorem can be found in many books on structural mechanics.
3 The analysis of pin-connected truss displacements requires that member gross cross-sectional area be known. Approximate analysis and access to similar truss span design documents will assist with a preliminary estimate of member gross cross-sectional area.
4 Non-independent reaction forces are due to support settlements.
5 The principle of virtual work is applied through the method of virtual work, otherwise known as the dummy load method or method of virtual forces.
6 The infinitesimal displacements must be compatible but are not necessarily dependent on the external forces acting on the truss.
7 These methods are outlined by modern authors such as Armenakas (1988), Mau (2012) and Ghali, Neville and Brown (2007). They are typically presented as matrix formulations.
8 Typical of trusses with unyielding foundations.
9 Variations in member length are typically the result of mis-fabrication of members and/or connections.
10 Redundant members are those in excess of the number of members required for static equilibrium.

11 Since the redundant forces prevent displacements.
12 Even though the unit loads at joints L2, U3 and L4 only affect members in the central panels, all truss members are indicated in the tables to aid in the systematic analysis of the statically indeterminate truss span.
13 Typically continuous spans that may experience support settlements.
14 Typically due to mis-fabrication of members and/or gusset plate connections.
15 For single-intersection web member trusses, chord member forces are calculated from panel point moments and web member forces from consideration of panel shear forces.
16 Stiffer trusses will perform better from a live load serviceability (deflection and vibration) perspective.
17 For unsymmetrical live load, the diagonal members are not equally stressed. However, there is typically only a small difference in the magnitude of the diagonal member forces, and therefore, equal distribution of the panel shear force to each diagonal member in the panel is generally sufficiently accurate.
18 To attain a tension-only member condition, the slenderness of the diagonal members must be such that the diagonal members in the panel will elastically buckle and only carry a marginal compression force. The panel shear is then resisted by the other diagonal member in tension. Many old pin-connected trusses were designed with very slender counter rods to attain a statically determinate truss. However, depending on live load position, an unsymmetrical statically determinate truss will result for trusses with an odd number of panels.

REFERENCES

Armenakas, A.E., 1988, *Classical Structural Analysis—A Modern Approach*, McGraw-Hill, New York.

Castigliano, C.A.P., 1879, *The Theory of Equilibrium of Elastic Systems and Its Application*, Dover Publications, New York, 1966.

Ghali, A., Neville, A.M., and Brown, T.G., 2007, *Structural Analysis—A Unified Classical and Matrix Approach*, 5th edition, Spon, London.

Mau, S.T., 2012, *Introduction to Structural Analysis—Displacement and Force Methods*, CRC Press, Taylor and Francis, Boca Raton, FL.

BIBLIOGRAPHY

Fife, W.M., and Wilbur, J.B., 1937, *Theory of Statically Indeterminate Structures*, McGraw-Hill, New York.

Hool, G.A., and Kinnie, W.S., 1923, *Stresses in Framed Structures*, McGraw-Hill, New York.

Wang, C.-K., 1953, *Statically Indeterminate Structures*, McGraw-Hill, New York.

Williams, A., 2009, *Structural Analysis in Theory and Practice*, Butterworth-Heinemann, Burlington, MA.

6 Displacements of Truss Spans

6.1 INTRODUCTION

Pin-connected[1] elastic truss displacements are the result of changes in member length[2] caused by member internal forces. The member internal forces are the result of external loads, changes in temperature, changes in member length from fabrication errors and/or settlements of the truss supports. Truss displacements may be required to determine member secondary stresses,[3] serviceability[4] and camber[5] design. Truss members without perforated cover plates use the gross cross-sectional area for displacement calculations. However, the effective cross-sectional area of members with perforated cover plates is the gross area reduced by the area determined by dividing the volume of a perforation by the distance between the centers of perforations. The displacements of truss spans can be determined graphically or analytically.

Graphical analysis, such as the Williot–Mohr method, calculates the displacements of all truss joints, but since truss member changes in length and displacements are smaller than truss member lengths may be inaccurate. Therefore, classical energy methods such as Castigliano's Second Theorem[6] and the method of virtual work or modern matrix[7] analysis methods are typically used to determine the displacements of pin-connected elastic truss spans.

Classical methods of truss displacement analysis provide clear insight into truss behavior and are useful tools with which to verify displacements obtained by modern matrix methods. An ideally pin-connected truss with external loads applied to just joints creates constant axial forces in members and joint displacements due only to member axial deformations. Therefore, Castigliano's Second Theorem or the method of virtual work are effective analytical methods for the determination of truss joint displacements. Castigliano's Second Theorem, used to determine elastic truss joint displacements due to external loads only, is essentially the same as applying the method of virtual work. However, the method of virtual work also allows for considering displacements due to other external effects such as temperature changes, member length variations and support settlements.

6.2 CASTIGLIANO'S SECOND THEOREM FOR PIN-CONNECTED ELASTIC TRUSS SPAN JOINT DISPLACEMENTS

For an elastic truss without rigid-body displacements or internal member forces related to temperature change and/or variations in member lengths, the displacement component Δ_{Pi} in the direction of the force P_i is

$$\frac{\partial U}{\partial P_i} = \Delta_{Pi} \tag{6.1}$$

where U = the strain energy (internal work) as a function of the external forces.

The strain energy U_j of an elastic pin-connected truss member j, is calculated as

$$U_j = \frac{F_j^2 L_j}{2A_j E} \tag{6.2}$$

where

F_j = internal force in member j of the truss due to external loads;
L_j= length of j of the truss;

DOI: 10.1201/9781003469971-7

A_j = gross cross-sectional area of j; and
E = modulus of elasticity.

Therefore, U for an elastic ideally pin-connected truss is calculated as

$$U = \Sigma U_j = \frac{1}{2}\Sigma \frac{F_j^2 L_j}{EA_j} \tag{6.3}$$

and since only F_j depends on P_i, Equation 6.3 gives

$$\Delta_{pi} = \frac{\partial U}{\partial P_i} = \frac{\partial}{\partial P_i}\Sigma \frac{F_{pj}^2 L_j}{2EA_j} = \Sigma\left(\frac{F_{pj} L_j}{A_j E}\right)\frac{\partial F_{pj}}{\partial P_i} \tag{6.4}$$

$\frac{\partial F_{pj}}{\partial P_i}$ is the rate of change of the force F_{pj} in a truss member j with respect to external load P_i, and $\frac{F_{pj} L_j}{EA_j}$ is the displacement of truss member j from P_i. The rate of change of F_{pj} with respect to P_i, $\frac{\partial F_{pj}}{\partial P_i}$, is equal to the rate of change of F_{1j} with respect to external load $P_i = 1$, and Equation 6.4 becomes

$$\Delta_{pi} = \Sigma\left(\frac{F_{pj} L_j}{A_j E}\right) F_{1j} \tag{6.5}$$

Equation 6.5 indicates that truss member displacements, Δ_{pi}, are calculated from the sum of the member displacements due to external loads multiplied by the force in the member due to a unit load coincident in direction to the required horizontal or vertical displacement. As shown in Example 6.1, the angular rotations of members are also calculated using unit load couples.

Example 6.1

The statically determinate 150 ft long six-panel horizontal parallel top chord Pratt truss with height of 25 ft shown in Figure E6.1 is subjected to a uniform dead load of 1500 lb/ft and uniform live load of 5500 lb/ft distributed across the entire span. Use Castigliano's Second Theorem to determine truss joint displacements and member angular rotations due to external loads. Specifically, determine the vertical displacement and camber required at

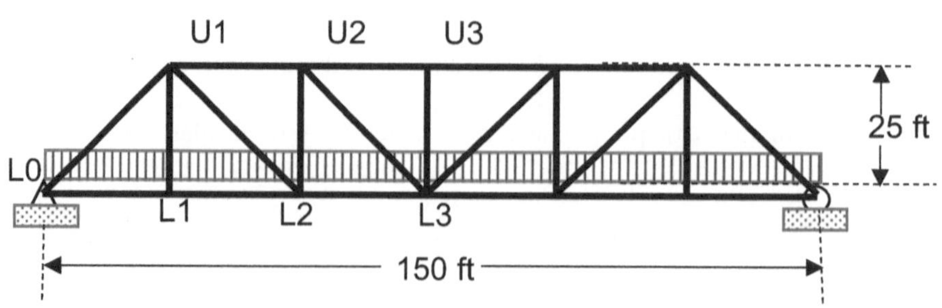

FIGURE E6.1 Pratt truss with distributed live load.

joints L1, L2 and L3; horizontal displacement of joints L3 and L6 (expansion bearing) and rotation of member U2U3.

Castigliano's second theorem indicates

$$\Delta_{pi} = \frac{\partial U}{\partial P_i} = \frac{\partial}{\partial P_i}\Sigma\frac{F_{pj}^2 L_j}{2EA_j} = \Sigma\left(\frac{F_{pj}L_j}{A_j E}\right)\frac{\partial F_{pj}}{\partial P_i} = \Sigma\left(\frac{F_{pj}L_j}{A_j E}\right)F_{1j}$$

The forces in member j, F_{1j}, with respect to external load P_i =1 are given in Table E6.1a.

Truss members without perforated cover plates use the gross cross-sectional area for deflection calculations. However, the effective cross-sectional area of perforated cover plate truss members is the gross area reduced by the area determined by dividing the volume of a perforation by the distance center to center of perforations. Vertical displacement at joint L1 is given in Table E6.1b.

The camber of trusses must be equal to the deflection produced by the dead load plus a live load of 3,000 lb per foot of track:

Deflection produced by dead load = (1.5/(1.5 + 5.5))1.91 = 0.41 in

Deflection produced by live load of 3000 lb/ft = (3.0/(1.5 + 5.5))1.91 =0.82 in

TABLE E6.1a
Member Loads for the Rate of Change of the Force with Respect to P_i = 1

$$\frac{\partial F_{pj}}{\partial P_i} = F_{1j}$$

Member	P Vertical at L1	P Vertical at L2	P Vertical at L3	P Horizontal at L3	P Horizontal at L6	P Couple at U2 and U3
L0L1	0.8333	0.6667	0.5000	1.0000	1.0000	−0.1667
L1L2	0.8333	0.6667	0.5000	1.0000	1.0000	−0.1667
L2L3	0.6667	1.3333	1.0000	1.0000	1.0000	−0.3333
L3L4	0.3333	0.6667	1.0000	0	1.0000	0.3333
L4L5	0.1667	0.3333	0.5000	0	1.0000	0.1667
L5L6	0.1667	0.3333	0.5000	0	1.0000	0.1667
L0U1	−1.1785	−0.9428	−0.7071	0	0	0.2357
U1U2	−0.6667	−1.3333	−1.0000	0	0	0.3333
U2U3	−0.5000	−1.0000	−1.5000	0	0	−0.5000
U3U4	−0.5000	−1.0000	−1.5000	0	0	−0.5000
U4U5	−0.3333	−0.6667	−1.0000	0	0	−0.3333
U5L6	−0.2357	−0.4714	−0.7071	0	0	−0.2357
U1L1	1.0000	0	0	0	0	0
U2L2	0.1667	0.3333	−0.5000	0	0	0.1667
U3L3	0	0	0	0	0	−1.0000
U4L4	−0.1667	−0.3333	−0.5000	0	0	−0.1667
U5L5	0	0	0	0	0	0
U1L2	−0.2357	0.9248	0.7071	0	0	−0.2357
U2L3	−0.2357	−0.4714	0.7071	0	0	1.1785
L3U4	0.2357	0.4714	0.7071	0	0	0.2357
L4U5	0.2357	0.4714	0.7071	0	0	0.2357

TABLE E6.1b
Vertical Displacement at Joint L1 due to Uniformly Distributed Load on an Entire Truss Span

Member	F_{pj} (kips)	A_j (in²)	L_j (in)	$\frac{F_{pj}L_j}{A_jE}$ (in)	$\frac{\partial F_{pj}}{\partial P_i}$	$\left(\frac{F_{pj}L_j}{A_jE}\right)\frac{\partial F_{pj}}{\partial P_i}$ (in)
L0L1	437.50	20	300	0.0485	0.8333	0.1885
L1L2	437.50	20	300	0.0485	0.8333	0.1885
L2L3	700.00	35	300	0.0443	0.6667	0.1377
L3L4	700.00	35	300	0.0443	0.3333	0.0691
L4L5	437.50	20	300	0.0485	0.1667	0.0378
L5L6	437.50	20	300	0.0485	0.1667	0.0378
L0U1	−618.61	25	424	−0.0775	−1.1785	0.4261
U1U2	−700.00	25	300	−0.0621	−0.6667	0.1932
U2U3	−787.50	30	300	−0.0582	−0.5000	0.1358
U3U4	−787.50	30	300	−0.0582	−0.5000	0.1358
U4U5	−700.00	25	300	−0.0621	−0.3333	0.0966
U5L6	−618.61	25	424	−0.0775	−0.2357	0.0854
U1L1	175.00	10	300	0.0388	1.0000	0.1811
U2L2	−87.50	5	300	−0.0388	0.1667	−0.0303
U3L3	0.00	5	300	0	0	0.0000
U4L4	−87.50	5	300	−0.0388	−0.1667	0.0303
U5L5	175.00	10	300	0.0388	0	0.0000
U1L2	371.23	10	424	0.1163	−0.2357	−0.1279
U2L3	123.76	5	424	0.0775	−0.2357	−0.0854
L3U4	123.76	5	424	0.0775	0.2357	0.0854
L4U5	371.23	10	424	0.1163	0.2357	0.1279
Σ						1.9133

Camber at L1 = 0.82 + 0.41 = 1.23 in

The vertical displacement at joint L2 is given in Table E6.1c.

The camber of trusses must be equal to the deflection produced by the dead load plus a live load of 3,000 lb per foot of track:

Deflection produced by dead load = (1.5/(1.5 + 5.5))3.26 = 0.70 in

Deflection produced by live load of 3000 lb/ft = (3.0/(1.5 + 5.5))3.26 =1.40 in

Camber at L2 = 0.70 + 1.40 = 2.10 in

The vertical displacement at joint L3 is given in Table E6.1d.

The camber of trusses must be equal to the deflection produced by the dead load plus a live load of 3,000 lb per foot of track:

Deflection produced by dead load = (1.5/(1.5+5.5))4.24 = 0.91 in

Deflection produced by live load of 3000 lb/ft = (3.0/(1.5 + 5.5))4.24 = 1.82 in

Camber at L3 = 1.82 + 0.91 = 2.72 in

Assuming impact of 26%, the live load plus impact vertical displacement at L3 = 1.26(4.23) = 5.33 in, which is span length divided by 338. The AREMA Chapter 15 allowable deflection of span/640 for live load and impact load is exceeded, indicating that larger member sizes are required for serviceability and camber determination. Horizontal displacement of joint L3 is given in Table E6.1e.

Horizontal displacement of joint L6 is given in Table E6.1f.

Angular rotation of member U2U3 is given in Table E6.1g.

TABLE E6.1c
Vertical Displacement at Joint L2 due to Uniformly Distributed Load on an Entire Truss Span

Member	F_{pj} (kips)	A_j (in²)	L_j (in)	$\frac{F_{pj}L_j}{A_jE}$ (in)	$\frac{\partial F_{pj}}{\partial P_i}$	$\left(\frac{F_{pj}L_j}{A_jE}\right)\frac{\partial F_{pj}}{\partial P_i}$ (in)
L0L1	437.50	20	300	0.0485	0.6667	0.1507
L1L2	437.50	20	300	0.0485	0.6667	0.1507
L2L3	700.00	35	300	0.0443	1.3333	0.2758
L3L4	700.00	35	300	0.0443	0.6667	0.1377
L4L5	437.50	20	300	0.0485	0.3333	0.0756
L5L6	437.50	20	300	0.0485	0.3333	0.0756
L0U1	−618.61	25	424	−0.0775	−0.9428	0.3411
U1U2	−700.00	25	300	−0.0621	−1.3333	0.3864
U2U3	−787.50	30	300	−0.0582	−1.0000	0.2716
U3U4	−787.50	30	300	−0.0582	−1.0000	0.2716
U4U5	−700.00	25	300	−0.0621	−0.6667	0.1932
U5L6	−618.61	25	424	−0.0775	−0.4714	0.1703
U1L1	175.00	10	300	0.0388	0	0.0000
U2L2	−87.50	5	300	−0.0388	0.3333	−0.0602
U3L3	0.00	5	300	0	0	0.0000
U4L4	−87.50	5	300	−0.0388	−0.3333	0.0602
U5L5	175.00	10	300	0.0388	0	0.0000
U1L2	371.23	10	424	0.1163	0.9248	0.5021
U2L3	123.76	5	424	0.0775	−0.4714	−0.1703
L3U4	123.76	5	424	0.0775	0.4714	0.1703
L4U5	371.23	10	424	0.1163	0.4714	0.2557
Σ						3.2583

TABLE E6.1d
Vertical Displacement at Joint L3 due to Uniformly Distributed Load on Entire Truss Span

Member	F_{pj} (kips)	A_j (in²)	L_j (in)	$\frac{F_{pj}L_j}{A_jE}$ (in)	$\frac{\partial F_{pj}}{\partial P_i}$	$\left(\frac{F_{pj}L_j}{A_jE}\right)\frac{\partial F_{pj}}{\partial P_i}$ (in)
L0L1	437.50	20	300	0.0485	0.5000	0.1134
L1L2	437.50	20	300	0.0485	0.5000	0.1134
L2L3	700.00	35	300	0.0443	1.0000	0.2067
L3L4	700.00	35	300	0.0443	1.0000	0.2067
L4L5	437.50	20	300	0.0485	0.5000	0.1134
L5L6	437.50	20	300	0.0485	0.5000	0.1134
L0U1	−618.61	25	424	−0.0775	−0.7071	0.2557
U1U2	−700.00	25	300	−0.0621	−1.0000	0.2898
U2U3	−787.50	30	300	−0.0582	−1.5000	0.4074
U3U4	−787.50	30	300	−0.0582	−1.5000	0.4074

(Continued)

TABLE E6.1d (*Continued*)
Vertical Displacement at Joint L3 due to Uniformly Distributed Load on Entire Truss Span

Member	F_{pj} (kips)	A_j (in²)	L_j (in)	$\frac{F_{pj}L_j}{A_jE}$ (in)	$\frac{\partial F_{pj}}{\partial P_i}$	$\left(\frac{F_{pj}L_j}{A_jE}\right)\frac{\partial F_{pj}}{\partial P_i}$ (in)
U4U5	−700.00	25	300	−0.0621	−1.0000	0.2898
U5L6	−618.61	25	424	−0.0775	−0.7071	0.2557
U1L1	175.00	10	300	0.0388	0	0.0000
U2L2	−87.50	5	300	−0.0388	−0.5000	0.0905
U3L3	0.00	5	300	0	0	0.0000
U4L4	−87.50	5	300	−0.0388	−0.5000	0.0905
U5L5	175.00	10	300	0.0388	0	0.0000
U1L2	371.23	10	424	0.1163	0.7071	0.3836
U2L3	123.76	5	424	0.0775	0.7071	0.2557
L3U4	123.76	5	424	0.0775	0.7071	0.2557
L4U5	371.23	10	424	0.1163	0.7071	0.3836
Σ						4.2327

TABLE E6.1e
Horizontal Displacement at Joint L3 due to Uniformly Distributed Load on Entire Truss Span

Member	F_{pj} (kips)	A_j (in²)	L_j (in)	$\frac{F_{pj}L_j}{A_jE}$ (in)	$\frac{\partial F_{pj}}{\partial P_i}$	$\left(\frac{F_{pj}L_j}{A_jE}\right)\frac{\partial F_{pj}}{\partial P_i}$ (in)
L0L1	437.50	20	300	0.0485	1.0000	0.2263
L1L2	437.50	20	300	0.0485	1.0000	0.2263
L2L3	700.00	35	300	0.0443	1.0000	0.2067
Σ						0.6593

TABLE E6.1f
Horizontal Displacement at Joint L6 due to Uniformly Distributed Load on Entire Truss Span

Member	F_{pj} (kips)	A_j (in²)	L_j (in)	$\frac{F_{pj}L_j}{A_jE}$ (in)	$\frac{\partial F_{pj}}{\partial P_i}$	$\left(\frac{F_{pj}L_j}{A_jE}\right)\frac{\partial F_{pj}}{\partial P_i}$ (in)
L0L1	437.50	20	300	0.0485	1.0000	0.2263
L1L2	437.50	20	300	0.0485	1.0000	0.2263
L2L3	700.00	35	300	0.0443	1.0000	0.2067
L3L4	700.00	35	300	0.0443	1.0000	0.2067
L4L5	437.50	20	300	0.0485	1.0000	0.2263
L5L6	437.50	20	300	0.0485	1.0000	0.2263
Σ						1.3186

TABLE E6.1g
Angular Rotation of Member U2U3 due to Uniformly Distributed Load on Entire Truss Span

Member	F_{pj} (kips)	A_j (in^2)	L_j (in)	$\frac{F_{pj}L_j}{A_jE}$ (in)	$\frac{\partial F_{pj}}{\partial P_i}$	$\left(\frac{F_{pj}L_j}{A_jE}\right)\frac{\partial F_{pj}}{\partial P_i}$ (in)
L0L1	437.50	20	300	0.0485	−0.1667	−0.0378
L1L2	437.50	20	300	0.0485	−0.1667	−0.0378
L2L3	700.00	35	300	0.0443	−0.3333	−0.0691
L3L4	700.00	35	300	0.0443	0.3333	0.0691
L4L5	437.50	20	300	0.0485	0.1667	0.0378
L5L6	437.50	20	300	0.0485	0.1667	0.0378
L0U1	−618.61	25	424	−0.0775	0.2357	−0.0854
U1U2	−700.00	25	300	−0.0621	0.3333	−0.0966
U2U3	−787.50	30	300	−0.0582	−0.5000	0.1358
U3U4	−787.50	30	300	−0.0582	−0.5000	0.1358
U4U5	−700.00	25	300	−0.0621	−0.3333	0.0966
U5L6	−618.61	25	424	−0.0775	−0.2357	0.0854
U1L1	175.00	10	300	0.0388	0	0.0000
U2L2	−87.50	5	300	−0.0388	0.1667	−0.0303
U3L3	0.00	5	300	0	−1.0000	0.0000
U4L4	−87.50	5	300	−0.0388	−0.1667	0.0303
U5L5	175.00	10	300	0.0388	0	0.0000
U1L2	371.23	10	424	0.1163	−0.2357	−0.1279
U2L3	123.76	5	424	0.0775	1.1785	0.4261
L3U4	123.76	5	424	0.0775	0.2357	0.0854
L4U5	371.23	10	424	0.1163	0.2357	0.1279
Σ						0.7831
Angular rotation						0.7831/300 = 2.610 × 10^{-3}

Castigliano's Second Theorem is useful for calculating pin-connected elastic truss span displacements when the displacements, Δ_{pi}, are linear functions of the external loads, P_i, and the external loads are independent.[8] Therefore, Castigliano's Second Theorem is not applicable for the analysis of displacements caused by temperature change, variations in member length[9] and support settlements. However, the method of virtual work is used to consider pin-connected elastic truss span displacements due to external loads, temperature changes, variations in member lengths and/or support settlements.

6.3 METHOD OF VIRTUAL WORK FOR PIN-CONNECTED ELASTIC TRUSS SPAN JOINT DISPLACEMENTS

The internal virtual work, $W_{\Delta j}$, associated with truss joint displacements for a truss member j is

$$W_{\Delta j} = \int_0^{L_j} F_{Qj}\varepsilon_j dl = F_{Qj}\Delta L_j \tag{6.6}$$

where L_j = length of member j; F_{Qj} = internal force in member j due to virtual loads Q_i, ε_j = axial strain in member j and ΔL_j = axial change in length of member j.

Equation 6.7 is used for an ideal pin-connected truss with P_i and Q_i loads applied only at joints and creating axial forces of F_{PJ} and F_{QJ} in member j:

$$W_\Delta = \Sigma W_{\Delta j} = \Sigma F_{Qj} \Delta L_j \tag{6.7}$$

Equating the work performed by P_i through Δ_{Pi} and that performed by virtual forces Q_i gives the following:

$$\Sigma Q_i \Delta_{Pi} = \Sigma F_{Qj} \Delta L_j \tag{6.8}$$

If displacements due to external loads, $\frac{F_{Pj} L_j}{A_j E}$, temperature change, $\alpha_t (\Delta T_j) L_j$ and member length change, Δl_j, are considered,

$$\Delta L_j = \left(\alpha_t (\Delta T_j) + \frac{F_{Pj}}{A_j E} \right) L_j + \Delta l_j \tag{6.9}$$

and Equation 6.8 becomes

$$\Sigma Q_i \Delta_{Pi} = \Sigma F_{Qj} \left(\alpha_t (\Delta T_j) L_j + \frac{F_{Pj} L_j}{A_j E} + \Delta l_j \right) \tag{6.10}$$

where

α_t = coefficient of thermal expansion of steel;[10]
ΔT_j = change in temperature of member j;
F_{pj} = internal force in j due to external loads P_i;
A_j = gross cross-sectional area of j;
E = modulus of elasticity of steel;[11] and
Δl_j = change in length of j due to mechanical devices or fabrication error.

If displacements due to external loads, temperature change, member length change and support settlements are considered, Equation 6.10 becomes

$$\Sigma Q_i \Delta_{Pi} = \Sigma F_{Qj} \left(\alpha_t (\Delta T_j) L_j + \frac{F_{Pj} L_j}{A_j E} + \Delta l_j \right) - W_{QR} \tag{6.11}$$

where W_{QR} is the work performed by the virtual reactions from Q_i through the support movements Δ_r.[12]

If only displacements due to external loads are considered, Equation 6.11 becomes

$$\Sigma Q_i \Delta_{Pi} = \Sigma F_{Qj} \left(\frac{F_{Pj} L_j}{A_j E} \right) \tag{6.12}$$

which with virtual load Q_i = 1 is

$$\Delta_{Pi} = \Sigma F_{Qj}\left(\frac{F_{Pj}L_j}{A_jE}\right) \tag{6.13}$$

Equation 6.13 is equivalent to Equation 6.5 indicating that Castigliano's Second Theorem to determine the displacements of ideally pin-connected elastic truss spans from external loads is essentially the same as applying the method of virtual work.

For each member, j, the displacements of Equation 6.13 can be determined from a structural analysis for P_i. Additionally, for each member and each displacement to be determined, a truss analysis is required for $Q_i = 1$ associated with the required displacement location and direction. Furthermore, for each displacement to be determined, the maximum corresponding displacement from the structural analysis of the external moving load must be established. Nevertheless, for static loads and uniformly distributed moving (live) loads, only a few[13] structural analyses of the truss are typically required for the external effects (e.g., external loads, temperature changes) to obtain the required joint displacements.

It is evident that except for relatively simple P_i, and where only a few Δ_{Pi} are to be determined, the use of Castigliano's Second Theorem and the method of virtual work can be laborious. Nevertheless, truss displacements determined by Castigliano's Second Theorem or the method of virtual work are most useful when only a few displacements are under consideration for only a few load positions.

The method of virtual work is most effective for calculating truss joint displacements when they are systematically organized. Examples 6.2 to 6.5 outline the use of the method of virtual work to determine truss joint displacements due to dead load, a uniformly distributed live load and Cooper's E80 live load. Examples 6.6 to 6.9 indicate how the method of virtual work can be used to determine truss joint displacements due to temperature change, member length variation and support settlement.

Example 6.2

The statically determinate 150 ft long six-panel horizontal parallel top chord Pratt truss from Example 6.1 with h = 25 ft is subjected to a uniform dead load of 1500 lb/ft and uniform live load of 4000 lb/ft distributed across the entire span. Using the method of virtual work, determine truss joint displacements and member angular rotations due to external loads. Specifically, determine the vertical displacement of joints L1, L2 and L3; horizontal displacement of joints L3 and L6 (expansion bearing) and rotation of member U2U3.

To determine the displacement of a joint j due to only external loads, Equation 6.13 is,

$$\Delta_{Pi} = \Sigma F_{Qj}\left(\frac{F_{Pj}L_j}{A_jE}\right)$$

and statically determinate truss analyses are required to determine F_{Pj} and F_{Qj} for each member j of the truss with the member cross-sectional area shown in Table E6.2a. $Q_i = 1$ is applied vertically at joints L1, L2 and L3 to obtain F_{Qj} for each member. The results of these analyses for the vertical displacement of joints L1, L2 and L3 from external dead load and live load are shown in Table E6.2a.

A summary of joint (panel point) vertical deflections due to a uniform dead load of 1500 lb/ft and uniform live load of 4000 lb/ft distributed across the entire span is shown in Table E6.2b.

TABLE E6.2a
Vertical Displacements at Joints L1, L2 and L3

Dead Load Vertical Displacement at L3

Member		F_{Pj} (kips)	L_j (in)	A_j (in²)	F_{Qj} (kips)	$F_{Pj}L_jF_{Qj}/A_j$ (kip/in)
L0L1	Bottom chord	93.8	300	20	0.5	703.5
L1L2	Bottom chord	93.8	300	20	0.5	703.5
L2L3	Bottom chord	150	300	35	1	1285.7
L3L4	Bottom chord	150	300	35	1	1285.7
L4L5	Bottom chord	93.8	300	20	0.5	703.5
L5L6	Bottom chord	93.8	300	20	0.5	703.5
L0U1	End post (top chord)	−132.6	424	25	−0.71	1596.7
U1U2	Top chord	−150	300	25	−1	1800.0
U2U3	Top chord	−168.8	300	30	−1.5	2532.0
U3U4	Top chord	−168.8	300	30	−1.5	2532.0
U4U5	Top chord	−150	300	25	−1	1800.0
U5L6	End post (top chord)	−132.6	424	25	−0.71	1596.7
U1L1	Hanger	37.5	300	10	0	0.0
U2L2	Vertical	−18.8	300	5	−0.5	564.0
U3L3	Vertical	0	300	5	0	0.0
U4L4	Vertical	−18.8	300	5	−0.5	564.0
U5L5	Hanger	37.5	300	10	0	0.0
U1L2	Diagonal	79.6	424	10	0.71	2396.3
U2L3	Diagonal	26.5	424	5	0.71	1595.5
L3U4	Diagonal	26.5	424	5	0.71	1595.5
L4U5	Diagonal	79.6	424	10	0.71	2396.3
$\sum$ (kip/in)						26354.4
Δ (in)						0.91

Live Load Vertical Displacement at L3

Member		F_{Pj} (kips)	L_j (in)	A_j (in²)	F_{Qj} (kips)	$F_{Pj}L_jF_{Qj}/A_j$ (kip/in)
L0L1	Bottom chord	250	300	20	0.5	1875.0
L1L2	Bottom chord	250	300	20	0.5	1875.0
L2L3	Bottom chord	400	300	35	1	3428.6
L3L4	Bottom chord	400	300	35	1	3428.6
L4L5	Bottom chord	250	300	20	0.5	1875.0
L5L6	Bottom chord	250	300	20	0.5	1875.0
L0U1	End post (top chord)	−353.7	424	25	−0.71	4259.1
U1U2	Top chord	−400	300	25	−1	4800.0
U2U3	Top chord	−450	300	30	−1.5	6750.0
U3U4	Top chord	−450	300	30	−1.5	6750.0
U4U5	Top chord	−400	300	25	−1	4800.0
U5L6	End Post (top chord)	−353.7	424	25	−0.71	4259.1
U1L1	Hanger	100	300	10	0	0.0
U2L2	Vertical	−50	300	5	−0.5	1500.0
U3L3	Vertical	0	300	5	0	0.0
U4L4	Vertical	−50	300	5	−0.5	1500.0
U5L5	Hanger	100	300	10	0	0.0
U1L2	Diagonal	212.1	424	10	0.71	6385.1
U2L3	Diagonal	70.7	424	5	0.71	4256.7

TABLE E6.2a (*Continued*)
Vertical Displacements at Joints L1, L2 and L3

Live Load Vertical Displacement at L3

Member		F_{Pj} (kips)	L_j (in)	A_j (in²)	F_{Qj} (kips)	$F_{Pj}L_jF_{Qj}/A_j$ (kip/in)
L3U4	Diagonal	70.7	424	5	0.71	4256.7
L4U5	Diagonal	212.1	424	10	0.71	6385.1
Σ (kip/in)						70258.9
Δ (in)						2.42

Dead Load Vertical Displacement at L1

Member		F_{Pj} (kips)	L_j (in)	A_j (in²)	F_{Qj} (kips)	$F_{Pj}L_jF_{Qj}/A_j$ (kip/in)
L0L1	Bottom chord	93.8	300	20	0.833	1172.0
L1L2	Bottom chord	93.8	300	20	0.833	1172.0
L2L3	Bottom chord	150	300	35	0.667	857.6
L3L4	Bottom chord	150	300	35	0.333	428.1
L4L5	Bottom chord	93.8	300	20	0.167	235.0
L5L6	Bottom chord	93.8	300	20	0.167	235.0
L0U1	End post (top chord)	−132.6	424	25	−1.18	2653.7
U1U2	Top chord	−150	300	25	−0.667	1200.6
U2U3	Top chord	−168.8	300	30	−0.5	844.0
U3U4	Top chord	−168.8	300	30	−0.5	844.0
U4U5	Top chord	−150	300	25	−0.333	599.4
U5L6	End post (top chord)	−132.6	424	25	−0.236	530.7
U1L1	Hanger	37.5	300	10	1	1125.0
U2L2	Vertical	−18.8	300	5	0.167	−188.4
U3L3	Vertical	0	300	5	0	0.0
U4L4	Vertical	−18.8	300	5	−0.167	188.4
U5L5	Hanger	37.5	300	10	0	0.0
U1L2	Diagonal	79.6	424	10	−0.236	−796.5
U2L3	Diagonal	26.5	424	5	−0.236	−530.3
L3U4	Diagonal	26.5	424	5	0.236	530.3
L4U5	Diagonal	79.6	424	10	0.236	796.5
Σ (kip/in)						11897.2
Δ (in)						0.41

Live Load Vertical Displacement at L1

Member		F_{Pj} (kips)	L_j (in)	A_j (in²)	F_{Qj} (kips)	$F_{Pj}L_jF_{Qj}/A_j$ (kip/in)
L0L1	Bottom chord	250	300	20	0.833	3123.8
L1L2	Bottom chord	250	300	20	0.833	3123.8
L2L3	Bottom chord	400	300	35	0.667	2286.9
L3L4	Bottom chord	400	300	35	0.333	1141.7
L4L5	Bottom chord	250	300	20	0.167	626.3
L5L6	Bottom chord	250	300	20	0.167	626.3
L0U1	End post (top chord)	−353.7	424	25	−1.18	7078.5
U1U2	Top chord	−400	300	25	−0.667	3201.6
U2U3	Top chord	−450	300	30	−0.5	2250.0
U3U4	Top chord	−450	300	30	−0.5	2250.0
U4U5	Top chord	−400	300	25	−0.333	1598.4

(*Continued*)

TABLE E6.2a (*Continued*)
Vertical Displacements at Joints L1, L2 and L3

Live Load Vertical Displacement at L1

Member		F_{Pj} (kips)	L_j (in)	A_j (in²)	F_{Qj} (kips)	$F_{Pj}L_jF_{Qj}/A_j$ (kip/in)
U5L6	End post (top chord)	−353.7	424	25	−0.236	1415.7
U1L1	Hanger	100	300	10	1	3000.0
U2L2	Vertical	−50	300	5	0.167	−501.0
U3L3	Vertical	0	300	5	0	0.0
U4L4	Vertical	−50	300	5	−0.167	501.0
U5L5	Hanger	100	300	10	0	0.0
U1L2	Diagonal	212.1	424	10	−0.236	−2122.4
U2L3	Diagonal	70.7	424	5	−0.236	−1414.9
L3U4	Diagonal	70.7	424	5	0.236	1414.9
L4U5	Diagonal	212.1	424	10	0.236	2122.4
Σ (kip/in)						31722.8
Δ (in)						1.09

Dead Load Vertical Deflection at L2

Member		F_{Pj} (kips)	L_j (in)	A_j (in²)	F_{Qj} (kips)	$F_{Pj}L_jF_{Qj}/A_j$ (kip/in)
L0L1	Bottom chord	93.8	300	20	0.667	938.5
L1L2	Bottom chord	93.8	300	20	0.667	938.5
L2L3	Bottom chord	150	300	35	1.333	1713.9
L3L4	Bottom chord	150	300	35	0.667	857.6
L4L5	Bottom chord	93.8	300	20	0.333	468.5
L5L6	Bottom chord	93.8	300	20	0.333	468.5
L0U1	End post (top chord)	−132.6	424	25	−0.943	2120.7
U1U2	Top chord	−150	300	25	−1.33	2394.0
U2U3	Top chord	−168.8	300	30	−1	1688.0
U3U4	Top chord	−168.8	300	30	−1	1688.0
U4U5	Top chord	−150	300	25	−0.667	1200.6
U5L6	End post (top chord)	−132.6	424	25	−0.471	1059.2
U1L1	Hanger	37.5	300	10	0	0.0
U2L2	Vertical	−18.8	300	5	0.333	−375.6
U3L3	Vertical	0	300	5	0	0.0
U4L4	Vertical	−18.8	300	5	−0.333	375.6
U5L5	Hanger	37.5	300	10	0	0.0
U1L2	Diagonal	79.6	424	10	0.943	3182.7
U2L3	Diagonal	26.5	424	5	−0.471	−1058.4
L3U4	Diagonal	26.5	424	5	0.471	1058.4
L4U5	Diagonal	79.6	424	10	0.471	1589.6
Σ (kip/in)						20308.3
Δ (in)						0.70

Live Load Vertical Deflection at L2

Member		F_{Pj} (kips)	L_j (in)	A_j (in²)	F_{Qj} (kips)	$F_{Pj}L_jF_{Qj}/A_j$ (kip/in)
L0L1	Bottom chord	250	300	20	0.667	2501.3
L1L2	Bottom chord	250	300	20	0.667	2501.3
L2L3	Bottom chord	400	300	35	1.333	4570.3

TABLE E6.2a (*Continued*)
Vertical Displacements at Joints L1, L2 and L3

		Live Load Vertical Deflection at L2				
Member		**F_{Pj} (kips)**	**L_j (in)**	**A_j (in²)**	**F_{Qj} (kips)**	**$F_{Pj}L_jF_{Qj}/A_j$ (kip/in)**
L3L4	Bottom chord	400	300	35	0.667	2286.9
L4L5	Bottom chord	250	300	20	0.333	1248.8
L5L6	Bottom chord	250	300	20	0.333	1248.8
L0U1	End post (top chord)	−353.7	424	25	−0.943	5656.8
U1U2	Top chord	−400	300	25	−1.33	6384.0
U2U3	Top chord	−450	300	30	−1	4500.0
U3U4	Top chord	−450	300	30	−1	4500.0
U4U5	Top chord	−400	300	25	−0.667	3201.6
U5L6	End post (top chord)	−353.7	424	25	−0.471	2825.4
U1L1	Hanger	100	300	10	0	0.0
U2L2	Vertical	−50	300	5	0.333	−999.0
U3L3	Vertical	0	300	5	0	0.0
U4L4	Vertical	−50	300	5	−0.333	999.0
U5L5	Hanger	100	300	10	0	0.0
U1L2	Diagonal	212.1	424	10	0.943	8480.4
U2L3	Diagonal	70.7	424	5	−0.471	−2823.8
L3U4	Diagonal	70.7	424	5	0.471	2823.8
L4U5	Diagonal	212.1	424	10	0.471	4235.7
$\sum$ (kip/in)						54141.1
Δ (in)						1.87

$Q_i = 1$ is applied horizontally at joints L3 and L6 to obtain F_{Qj} for each member. The results of the analysis for the horizontal displacement of joints L3 and L6 from external dead load and live load are shown in Table E6.2c.

A summary of joint (panel point) horizontal deflections due to a uniform dead load of 1500 lb/ft and uniform live load of 4000 lb/ft distributed across the entire span is shown in Table E6.2d.

Virtual load couple $Q_i = 1/L_{U2U3}$ is applied vertically and in opposite directions at joints U2 and U3 to obtain F_{Qj} for each member. The results of the analysis for the angular rotation of member U2U3 from external dead load and live load are shown in Table E6.2e.

TABLE E6.2b
Vertical Displacements at Joints L1, L2 and L3

Panel Point	Dead Load Vertical Deflection (in.)	Live Load Vertical Deflection (in.)	Dead + Live Load Vertical Deflection (in.)
L1	0.41	1.09	1.50
L2	0.70	1.87	2.57
L3	0.91	2.42	3.33

TABLE E6.2c
Horizontal Displacements at Joints L1, L2 and L3

Dead Load Horizontal Displacement at L3						
Member		F_{Pj} (kips)	L_j (in)	A_j (in²)	F_{Qj} (kips)	$F_{Pj}L_jF_{Qj}/A_j$ (kip/in)
L0L1	Bottom chord	93.8	300	20	1	1407.0
L1L2	Bottom chord	93.8	300	20	1	1407.0
L2L3	Bottom chord	150	300	35	1	1285.7
Σ (kip/in)						4099.7
Δ (in)						0.14

Live Load Horizontal Displacement at L3						
Member		F_{Pj} (kips)	L_j (in)	A_j (in²)	F_{Qj} (kips)	$F_{Pj}L_jF_{Qj}/A_j$ (kip/in)
L0L1	Bottom chord	250	300	20	1	3750.0
L1L2	Bottom chord	250	300	20	1	3750.0
L2L3	Bottom chord	400	300	35	1	3428.6
Σ (kip/in)						10928.6
Δ (in)						0.38

Dead Load Horizontal Displacement of L6						
Member		F_{Pj} (kips)	Length (in)	A (in²)	F_{Qj} (kips)	$F_{Pj}L_jF_{Qj}/A_j$ (kip/in)
L0L1	Bottom chord	93.8	300	20	1.00	1407.0
L1L2	Bottom chord	93.8	300	20	1.00	1407.0
L2L3	Bottom chord	150	300	35	1.00	1285.7
L3L4	Bottom chord	150	300	35	1.00	1285.7
L4L5	Bottom chord	93.8	300	20	1.00	1407.0
L5L6	Bottom chord	93.8	300	20	1.00	1407.0
Σ (kip/in)						8199.4
Δ (in)						0.28

Live Load Horizontal Displacement of L6						
Member		F_{Pj} (kips)	Length (in)	A (in²)	F_{Qj} (kips)	$F_{Pj}L_jF_{Qj}/A_j$ (kip/in)
L0L1	Bottom chord	250	300	20	1.00	3750.0
L1L2	Bottom chord	250	300	20	1.00	3750.0
L2L3	Bottom chord	400	300	35	1.00	3428.6
L3L4	Bottom chord	400	300	35	1.00	3428.6
L4L5	Bottom chord	250	300	20	1.00	3750.0
L5L6	Bottom chord	250	300	20	1.00	3750.0
Σ (kip/in)						21857.1
Δ (in)						0.75

TABLE E6.2d
Horizontal Displacements at Joints L3 and L6

Panel Point	Dead Load Horizontal Deflection (in.)	Live Load Horizontal Deflection (in.)	Dead + Live Load Horizontal Deflection (in.)
L3	0.14	0.38	0.52
L6	0.28	0.75	1.03

TABLE E6.2e
Angular Rotation of Member U2U3

Dead Load Rotation of Member U2U3

Member		F_{Pj} (kips)	L_j (in)	A_j (in²)	F_{Qj} (kips)	$F_{Pj}L_jF_{Qj}/A_j$ (kip/in)
L0L1	Bottom chord	93.8	300	20	−0.167	−0.78
L1L2	Bottom chord	93.8	300	20	−0.167	−0.78
L2L3	Bottom chord	150	300	35	−0.333	−1.43
L3L4	Bottom chord	150	300	35	0.333	1.43
L4L5	Bottom chord	93.8	300	20	0.167	0.78
L5L6	Bottom chord	93.8	300	20	0.167	0.78
L0U1	End post (top chord)	−132.6	424	25	0.236	−1.77
U1U2	Top chord	−150	300	25	0.333	−2.00
U2U3	Top chord	−168.8	300	30	−0.5	2.81
U3U4	Top chord	−168.8	300	30	−0.5	2.81
U4U5	Top chord	−150	300	25	−0.333	2.00
U5L6	End post (top chord)	−132.6	424	25	−0.236	1.77
U1L1	Hanger	37.5	300	10	0	0.00
U2L2	Vertical	−18.8	300	5	0.167	−0.63
U3L3	Vertical	0	300	5	−1	0.00
U4L4	Vertical	−18.8	300	5	−0.167	0.63
U5L5	Hanger	37.5	300	10	0	0.00
U1L2	Diagonal	79.6	424	10	−0.236	−2.66
U2L3	Diagonal	26.5	424	5	1.18	8.84
L3U4	Diagonal	26.5	424	5	0.236	1.77
L4U5	Diagonal	79.6	424	10	0.236	2.66
∑ (kip/in)						16.23
θ (rad)						0.56E−03

Live Load Rotation of Member U2U3

Member		F_{Pj} (kips)	L_j (in)	A_j (in²)	F_{Qj} (kips)	$F_{Pj}L_jF_{Qj}/A_j$ (kip/in)
L0L1	Bottom chord	250	300	20	−0.167	−2.09
L1L2	Bottom chord	250	300	20	−0.167	−2.09
L2L3	Bottom chord	400	300	35	−0.333	−3.81
L3L4	Bottom chord	400	300	35	0.333	3.81
L4L5	Bottom chord	250	300	20	0.167	2.09
L5L6	Bottom chord	250	300	20	0.167	2.09
L0U1	End post (top chord)	−353.7	424	25	0.236	−4.72
U1U2	Top chord	−400	300	25	0.333	−5.33
U2U3	Top chord	−450	300	30	−0.5	7.50
U3U4	Top chord	−450	300	30	−0.5	7.50
U4U5	Top chord	−400	300	25	−0.333	5.33
U5L6	End post (top chord)	−353.7	424	25	−0.236	4.72
U1L1	Hanger	100	300	10	0	0.00
U2L2	Vertical	−50	300	5	0.167	−1.67
U3L3	Vertical	0	300	5	−1	0.00
U4L4	Vertical	−50	300	5	−0.167	1.67
U5L5	Hanger	100	300	10	0	0.00
U1L2	Diagonal	212.1	424	10	−0.236	−7.07

(Continued)

TABLE E6.2e (*Continued*)
Angular Rotation of Member U2U3

Live Load Rotation of Member U2U3						
Member		F_{Pj} (kips)	L_j (in)	A_j (in²)	F_{Qj} (kips)	$F_{Pj}L_jF_{Qj}/A_j$ (kip/in)
U2L3	Diagonal	70.7	424	5	1.18	23.58
L3U4	Diagonal	70.7	424	5	0.236	4.72
L4U5	Diagonal	212.1	424	10	0.236	7.07
Σ (kip/in)						43.30
θ (rad)						1.49E−03

Example 6.3

The 150 ft long six-panel horizontal parallel chord Pratt truss shown in Figure E6.3 with h = 25 ft is subjected to a uniform live load of 4000 lb/ft distributed across 80% of the span (120 ft from the right end of the truss span), which is the location for maximum shear in panel 1-2 (maximum member force in U1L2). Using the method of virtual work, determine truss joint displacements due to external loads. Specifically, determine the vertical and horizontal displacement of joint L3 due to live load.

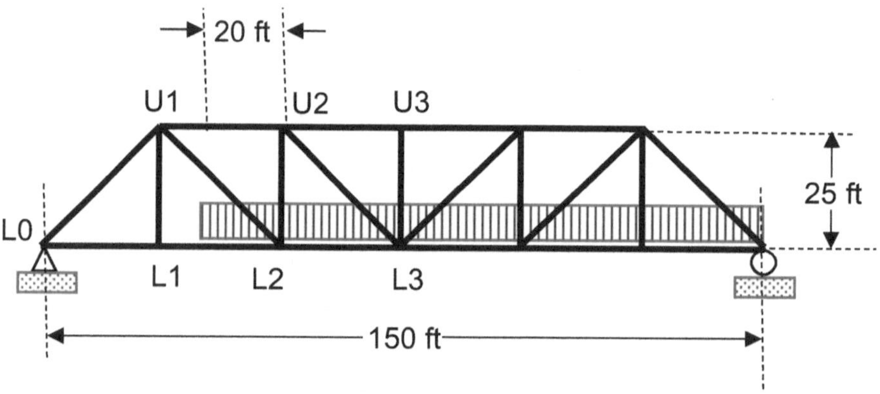

FIGURE E6.3 Pratt truss with partially distributed live load on statically determinate truss span.

The panel loads from the external live load distributed across 80% of the span are shown in Table E6.3a.

TABLE E6.3a
Panel Point Loads for Maximum Shear in Panel 1-2

Panel Point	Dead Load (kips)	Live Load (kips)	Dead + Live Load (kips)
0	18.75	0	18.75
1	37.5	32.0	69.5
2	37.5	98.0	135.5
3	37.5	100.0	137.5
4	37.5	100.0	137.5
5	37.5	100.0	137.5
6	18.75	50.0	68.75

To determine the displacement of a joint, j, due to only external live loads, Equation 6.13 is

$$\Delta_{Pi} = \Sigma F_{Qj}\left(\frac{F_{Pj}L_j}{A_jE}\right)$$

and statically determinate truss analyses are required to determine F_{Pj} and F_{Qj} for each member j of the truss with cross-sectional area shown in Table E6.3b. Q_i = 1 is applied vertically and horizontally at joint L3 to obtain F_{Qj} for each member. The results of these analyses for the vertical and horizontal displacement of joint L3 from external live load (distributed across 80% of the span at location of maximum shear force in panel 1-2) are shown in Table E6.3b.

TABLE E6.3b
Vertical and Horizontal Displacement of Joint L3

Live Load Vertical Displacement at L3—Partial Live Load (120 ft from right end of the truss span)

Member		F_{Pj} (kips)	Length (in)	A (in²)	F_{Qj} (kips)	$F_{Pj}L_jF_{Qj}/A_j$ (kip/in)
L0L1	Bottom chord	192	300	20	0.5	1440.0
L1L2	Bottom chord	192	300	20	0.5	1440.0
L2L3	Bottom chord	352	300	35	1	3017.1
L3L4	Bottom chord	376	300	35	1	3222.9
L4L5	Bottom chord	238.2	300	20	0.5	1786.5
L5L6	Bottom chord	238.2	300	20	0.5	1786.5
L0U1	End post (top chord)	−271.5	424	25	−0.71	3269.3
U1U2	Top chord	−352	300	25	−1	4224.0
U2U3	Top chord	−414	300	30	−1.5	6210.0
U3U4	Top chord	−414	300	30	−1.5	6210.0
U4U5	Top chord	−376	300	25	−1	4512.0
U5L6	End post (top chord)	−336.6	424	25	−0.71	4053.2
U1L1	Hanger	32	300	10	0	0.0
U2L2	Vertical	−62	300	5	−0.5	1860.0
U3L3	Vertical	0	300	5	0	0.0
U4L4	Vertical	−38	300	5	−0.5	1140.0
U5L5	Hanger	100	300	10	0	0.0
U1L2	Diagonal	226.3	424	10	0.71	6812.5
U2L3	Diagonal	87.68	424	5	0.71	5279.0
L3U4	Diagonal	53.74	424	5	0.71	3235.6
L4U5	Diagonal	195.2	424	10	0.71	5876.3
Σ (kip/in)						65374.9
Δ (in)						2.25

Live Load Horizontal Displacement at Joint L3—Partial Live Load (120 ft from right end of the truss span)

Member		F_{Pj} (kips)	Length (in)	A (in²)	F_{Qj} (kips)	$F_{Pj}L_jF_{Qj}/A_j$ (kip/in)
L0L1	Bottom chord	192	300	20	1	2880.0
L1L2	Bottom chord	192	300	20	1	2880.0
L2L3	Bottom chord	352	300	35	1	3017.1

(Continued)

TABLE E6.3b (*Continued*)
Vertical and Horizontal Displacement of Joint L3

Live Load Vertical Displacement at L3—Partial Live Load (120 ft from right end of the truss span)

Member		F_{Pj} (kips)	Length (in)	A (in²)	F_{Qj} (kips)	$F_{Pj}L_jF_{Qj}/A_j$ (kip/in)
L3L4	Bottom chord	376	300	35	0	0.0
L4L5	Bottom chord	238.2	300	20	0	0.0
L5L6	Bottom chord	238.2	300	20	0	0.0
L0U1	End post (top chord)	−271.5	424	25	0	0.0
U1U2	Top chord	−352	300	25	0	0.0
U2U3	Top chord	−414	300	30	0	0.0
U3U4	Top chord	−414	300	30	0	0.0
U4U5	Top chord	−376	300	25	0	0.0
U5L6	End post (top chord)	−336.6	424	25	0	0.0
U1L1	Hanger	32	300	10	0	0.0
U2L2	Vertical	−62	300	5	0	0.0
U3L3	Vertical	0	300	5	0	0.0
U4L4	Vertical	−38	300	5	0	0.0
U5L5	Hanger	100	300	10	0	0.0
U1L2	Diagonal	226.3	424	10	0	0.0
U2L3	Diagonal	87.68	424	5	0	0.0
L3U4	Diagonal	53.74	424	5	0	0.0
L4U5	Diagonal	195.2	424	10	0	0.0
Σ (kip/in)						8777.1
Δ (in)						0.30

Example 6.4

The 150 ft long six-panel horizontal parallel chord Pratt truss with h = 25 ft is subjected to a uniform live load of 4000 lb/ft distributed across 80% of the span (120 ft from the right end of the truss span). The truss has double diagonals in the center panel as shown in Figure E6.4. Using the method of virtual work, determine truss joint displacements due to external loads. Specifically, determine the vertical displacement of joint L3.

The panel loads from the external live load distributed across 80% of the span are in shown in Table E6.4a.

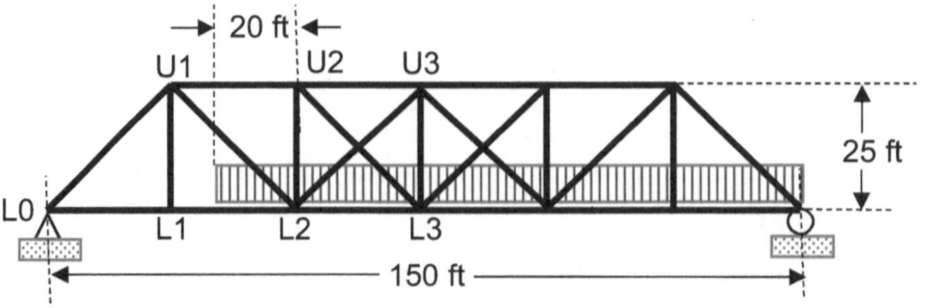

FIGURE E6.4 Pratt truss with partially distributed live load on statically indeterminate truss span.

TABLE E6.4a
Panel Point Loads for Maximum Shear in Panel 1-2

Panel Point	Dead Load (kips)	Live Load (kips)	Dead + Live Load (kips)
0	18.75	0	18.75
1	37.5	32.0	69.5
2	37.5	98.0	135.5
3	37.5	100.0	137.5
4	37.5	100.0	137.5
5	37.5	100.0	137.5
6	18.75	50.0	68.75

To determine the displacement of a joint, j, due to only external live loads, Equation 6.13 is

$$\Delta_{Pi} = \Sigma F_{Qj}\left(\frac{F_{Pj}L_j}{A_jE}\right)$$

and statically indeterminate truss analyses (see Chapter 5) are required to determine F_{Pj} and F_{Qj} for each j of the truss with cross-sectional area shown in Table E6.4b. $Q_i = 1$ is applied vertically at joint L3 to obtain F_{Qj} for each member. The results of these analyses for the vertical displacement of joint L3 from external live load (distributed across 80% of the span at location of maximum shear force in panel 1-2) are shown in Table E6.3b.

TABLE E6.4b
Vertical Displacement of Joint L3

Live Load Vertical Displacement at L3—Partial Live Load (120 ft from right end of the truss span)

Member		F_{Pj} (kips)	Length (in)	A (in²)	F_{Qj} (kips)	$F_{Pj}L_jF_{Qj}/A_j$ (kip/in)
LOL1	Bottom chord	192	300	20	0.5	1440.0
L1L2	Bottom chord	192	300	20	0.5	1440.0
L2L3	Bottom chord	382.1	300	35	1.225	4012.1
L3L4	Bottom chord	392.4	300	35	1.225	4120.2
L4L5	Bottom chord	238	300	20	0.5	1785.0
L5L6	Bottom chord	238	300	20	0.5	1785.0
LOU1	End post (top chord)	−271.5	424	25	−0.707	3255.5
U1U2	Top chord	−352	300	25	−1	4224.0
U2U3	Top chord	−383.9	300	30	−1.275	4894.7
U3U4	Top chord	−397.6	300	30	−1.275	5069.4
U4U5	Top chord	−376	300	25	−1	4512.0
U5L6	End post (top chord)	−336.6	424	25	−0.707	4036.1
U1L1	Hanger	32	300	10	0	0.0
U2L2	Vertical	−32	300	5	−0.275	528.0
U3L3	Vertical	46.5	300	5	0.451	1258.3
U4L4	Vertical	−21.6	300	5	−0.275	356.4
U5L5	Hanger	100	300	10	0	0.0
U1L2	Diagonal	226.3	424	10	0.707	6783.7
U2L3	Diagonal	45.1	424	5	0.388	1483.9

(Continued)

TABLE E6.4b (*Continued*)
Vertical Displacement of Joint L3

Live Load Vertical Displacement at L3—Partial Live Load (120 ft from right end of the truss span)

Member		F_{Pj} (kips)	Length (in)	A (in²)	F_{Qj} (kips)	$F_{Pj}L_jF_{Qj}/A_j$ (kip/in)
L2U3	Diagonal	−42.5	424	5	−0.319	1149.7
L3U4	Diagonal	30.5	424	5	0.388	1003.5
U3L4	Diagonal	−23.2	434	5	−0.319	642.4
L4U5	Diagonal	195.2	424	10	0.707	5851.5
Σ (kip/in)						59631.3
Δ (in)						2.06

The stiffness of the Pratt truss in Example 6.4 is increased by the double diagonal web members in the central panels, and the deflection at L3 is reduced to 92% of the 2.25 in determined for the truss in Example 6.3 with single diagonal web members in each panel.

Example 6.5

A statically determinate 150 ft long six-panel (d = 25 ft) horizontal parallel-chord Pratt truss with h = 25 ft is subjected to a Cooper's E80 live load. Using the method of virtual work, determine truss joint displacements due to external loads. Specifically, determine the maximum live load vertical and horizontal displacement of joint L3 for the Cooper's E80 live load position that creates maximum moment at joint L3. Example 4.11 reveals that Cooper's E80 live load with axle 11 at joint L3 creates the maximum live load moment of 14,088 kip-ft at L3. The panel point loads from Cooper's E80 live load with axle 11 at joint L3 are shown in Table E6.5a.

To determine the displacement of a joint due to only external loads, Equation 6.13 is $\Delta_{Pi} = \Sigma F_{Qj}\left(\frac{F_{Pj}L_j}{A_jE}\right)$ and statically determinate truss analyses are required to determine F_{Pj} and F_{Qj} for each member j of the truss with the cross-sectional areas shown in Table E6.5b. Q_i =1 is applied vertically and horizontally at joint L3 to obtain F_{Qj} for each member. The results of these analyses for the vertical and horizontal displacement of joint L3 from external live load (Cooper's E80 with axle 11 at L3 to create maximum live load moment at L3) are shown in Table E6.5b.

TABLE E6.5a
Joint Loads for Cooper's E80 Live Load with Axle 11 at L3

Panel Point	Live Load (kips)
L0	22.40
L1	146.16
L2	108.32
L3	140.16
L4	122.80
L5	98.16
L6	50.00

TABLE E6.5b
Vertical and Horizontal Displacement at Joint L3 for Cooper's E80 Live Load with Axle 11 at L3

Live Load Vertical Displacement at L3—Cooper's E80 Live Load

Member		F_{Pj} (kips)	Length (in)	A (in²)	F_{Qj} (kips)	$F_{Pj}L_jF_{Qj}/A_j$ (kip/in)
L0L1	Bottom chord	321.39	300	20	0.5	2410.4
L1L2	Bottom chord	321.39	300	20	0.5	2410.4
L2L3	Bottom chord	496.61	300	35	1	4256.7
L3L4	Bottom chord	490.27	300	35	1	4202.3
L4L5	Bottom chord	294.21	300	20	0.5	2206.6
L5L6	Bottom chord	294.21	300	20	0.5	2206.6
L0U1	End post (top chord)	−454.51	424	25	−0.71	5473.0
U1U2	Top chord	−496.61	300	25	−1	5959.3
U2U3	Top chord	−563.52	300	30	−1.5	8452.8
U3U4	Top chord	−563.52	300	30	−1.5	8452.8
U4U5	Top chord	−490.27	300	25	−1	5883.2
U5L6	End post (top chord)	−416.08	424	25	−0.71	5010.3
U1L1	Hanger	146.16	300	10	0	0.0
U2L2	Vertical	−66.91	300	15	−0.5	669.1
U3L3	Vertical	0	300	10	0	0.0
U4L4	Vertical	−73.25	300	15	−0.5	732.5
U5L5	Hanger	98.16	300	10	0	0.0
U1L2	Diagonal	247.81	424	15	0.71	4973.4
U2L3	Diagonal	94.62	424	10	0.71	2848.4
L3U4	Diagonal	103.6	424	10	0.71	3118.8
L4U5	Diagonal	277.26	424	15	0.71	5564.4
∑ (kip/in)						74831.0
Δ (in)						2.58

Live Load Horizontal Displacement at Joint L3—Cooper's E80 Live Load

Member		F_{Pj} (kips)	Length (in)	A (in²)	F_{Qj} (kips)	$F_{Pj}L_jF_{Qj}/A_j$ (kip/in)
L0L1	Bottom chord	321.39	300	20	1	4820.9
L1L2	Bottom chord	321.39	300	20	1	4820.9
L2L3	Bottom chord	496.61	300	35	1	4256.7
L3L4	Bottom chord	490.27	300	35	0	0.0
L4L5	Bottom chord	294.21	300	20	0	0.0
L5L6	Bottom chord	294.21	300	20	0	0.0
L0U1	End post (top chord)	−454.51	424	25	0	0.0
U1U2	Top chord	−496.61	300	25	0	0.0
U2U3	Top chord	−563.52	300	30	0	0.0
U3U4	Top chord	−563.52	300	30	0	0.0
U4U5	Top chord	−490.27	300	25	0	0.0
U5L6	End post (top chord)	−416.08	424	25	0	0.0
U1L1	Hanger	146.16	300	10	0	0.0
U2L2	Vertical	−66.91	300	15	0	0.0
U3L3	Vertical	0	300	10	0	0.0
U4L4	Vertical	−73.25	300	15	0	0.0

(Continued)

TABLE E6.5b (*Continued*)
Vertical and Horizontal Displacement at Joint L3 for Cooper's E80 Live Load with Axle 11 at L3

Live Load Horizontal Displacement at Joint L3—Cooper's E80 Live Load

Member		F_{Pj} (kips)	Length (in)	A (in²)	F_{Qj} (kips)	$F_{Pj}L_jF_{Qj}/A_j$ (kip/in)
U5L5	Hanger	98.16	300	10	0	0.0
U1L2	Diagonal	247.81	424	15	0	0.0
U2L3	Diagonal	94.62	424	10	0	0.0
L3U4	Diagonal	103.6	424	10	0	0.0
L4U5	Diagonal	277.26	424	15	0	0.0
Σ (kip/in)						13898.4
Δ (in)						0.48

Vertical impact = 16 + (600/(150 − 30)) = 21.0%

Rocking effect impact = 100/20 = 5.0% for trusses at 20 ft lateral spacing

Therefore, the live load plus impact vertical deflection = 1.26(2.58) = 3.25 in. AREMA Chapter 15 allowable deflection = 150(12)/640 = 2.81 in. Consequently, since the displacement is 16% greater than allowable, the truss members must be increased in size (gross cross-sectional area) to decrease vertical deflections by at least 0.44 in.

If some chord and diagonal member sizes (gross cross-sectional area) are increased, as shown in Table E6.5c, the live load plus impact deflection is reduced to 1.26(2.25) = 2.84, which is less than 1% above the AREMA Chapter 15 allowable deflection of 2.81 in. Adjusting member sizes in statically indeterminate trusses will also affect the forces in members F_{Pi} and F_{Qj}.

The method of virtual work is also a useful analytical tool for calculating truss displacements related to effects other than external loads. Examples 6.6 to 6.9 determine truss displacements due to temperature changes, length variations and support settlements.

TABLE E6.5c
Vertical Displacement at Joint L3 for Cooper's E80 Live Load with Axle 11 at L3

Live Load Vertical Displacement at L3—Cooper's E80 Live Load

Member		F_{Pj} (kips)	Length (in)	A (in²)	F_{Qj} (kips)	$F_{Pj}L_jF_{Qj}/A_j$ (kip/in)
L0L1	Bottom chord	321.39	300	30	0.5	1607.0
L1L2	Bottom chord	321.39	300	30	0.5	1607.0
L2L3	Bottom chord	496.61	300	35	1	4256.7
L3L4	Bottom chord	490.27	300	35	1	4202.3
L4L5	Bottom chord	294.21	300	30	0.5	1471.1
L5L6	Bottom chord	294.21	300	30	0.5	1471.1
L0U1	End post (top chord)	−454.51	424	25	−0.71	5473.0
U1U2	Top chord	−496.61	300	30	−1	4966.1
U2U3	Top chord	−563.52	300	30	−1.5	8452.8
U3U4	Top chord	−563.52	300	30	−1.5	8452.8
U4U5	Top chord	−490.27	300	30	−1	4902.7

TABLE E6.5c (*Continued*)
Vertical Displacement at Joint L3 for Cooper's E80 Live Load with Axle 11 at L3

Live Load Vertical Displacement at L3—Cooper's E80 Live Load

Member		F_{Pj} (kips)	Length (in)	A (in²)	F_{Qj} (kips)	$F_{Pj}L_jF_{Qj}/A_j$ (kip/in)
U5L6	End post (top chord)	−416.08	424	25	−0.71	5010.3
U1L1	Hanger	146.16	300	10	0	0.0
U2L2	Vertical	−66.91	300	15	−0.5	669.1
U3L3	Vertical	0	300	10	0	0.0
U4L4	Vertical	−73.25	300	15	−0.5	732.5
U5L5	Hanger	98.16	300	10	0	0.0
U1L2	Diagonal	247.81	424	20	0.71	3730.0
U2L3	Diagonal	94.62	424	15	0.71	1899.0
L3U4	Diagonal	103.6	424	15	0.71	2079.2
L4U5	Diagonal	277.26	424	20	0.71	4173.3
Σ (kip/in)						65155.8
Δ (in)						2.25

Example 6.6

A 150 ft long six-panel horizontal parallel chord Pratt truss with h = 25 ft is subjected to temperature changes of 60 °F at top chord, 45 °F at web and 30 °F at bottom chord members. Determine the vertical displacement of joint L3.

Considering displacements due to only temperature change, Equation 6.11 is

$$\Delta_{Pi} = \Sigma F_{Qj}\left(\alpha_t\left(\Delta T_j\right)L_j\right)$$

For the joint displacement required, a statically determinate truss analysis for $Q_i = 1$ applied vertically at joint L3 is required to obtain F_{Qj} for each member. The results of this analysis with $\alpha_T = 6.5 \times 10^{-6}$ are shown in Table E6.6.

TABLE E6.6
Vertical Displacement at Joint L3 due to Temperature Changes

Displacement of Joint L3 Due to Temperature Changes

Member		ΔT (°F)	L_j (in)	(L_jΔT) (in°F)	A_j (in²)	F_{Qj} (kips)	$L_j\Delta TF_{Qj}$ (in°F)
L0L1	Bottom chord	30	300	9000	20	0.5	4500
L1L2	Bottom chord	30	300	9000	20	0.5	4500
L2L3	Bottom chord	30	300	9000	35	1	9000
L3L4	Bottom chord	30	300	9000	35	1	9000
L4L5	Bottom chord	30	300	9000	20	0.5	4500
L5L6	Bottom chord	30	300	9000	20	0.5	4500
L0U1	End post (top chord)	45	424	19080	25	−0.71	−13547
U1U2	Top chord	60	300	18000	25	−1	−18000
U2U3	Top chord	60	300	18000	30	−1.5	−27000
U3U4	Top chord	60	300	18000	30	−1.5	−27000

(*Continued*)

TABLE E6.6 (*Continued*)
Vertical Displacement at Joint L3 due to Temperature Changes

Displacement of Joint L3 Due to Temperature Changes

Member		ΔT (°F)	L_j (in)	(L_jΔT) (in°F)	A_j (in²)	F_{Qj} (kips)	$L_j\Delta TF_{Qj}$ (in°F)
U4U5	Top chord	60	300	18000	25	−1	−18000
U5L6	End post (top chord)	45	424	19080	25	−0.71	−13547
U1L1	Hanger	45	300	13500	10	0	0
U2L2	Vertical	45	300	13500	5	−0.5	−6750
U3L3	Vertical	45	300	13500	5	0	0
U4L4	Vertical	45	300	13500	5	−0.5	−6750
U5L5	Hanger	45	300	13500	10	0	0
U1L2	Diagonal	45	424	19080	10	0.71	13547
U2L3	Diagonal	45	424	19080	5	0.71	13547
L3U4	Diagonal	45	424	19080	5	0.71	13547
L4U5	Diagonal	45	424	19080	10	0.71	13547
Σ (in°F)							−40406
Δ (in)							−0.26

Example 6.7

A 150 ft long six-panel horizontal parallel-chord Pratt truss with h = 25 ft is subjected to a shortening (length variation) of bottom chord panel lengths by 3/8 in. Determine the vertical displacement of joint L3.

Considering displacements due to only member length change, Equation 6.11 is

$$\Delta_{Pi} = \Sigma F_{Qj}\left(\Delta l_j\right)$$

For the joint displacement required, a statically determinate truss analysis for Q_i =1 applied vertically at joint L3 is required to obtain F_{Qj} for each member. The results of this analysis are shown in Table E6.7.

TABLE E6.7
Vertical Displacement of Joint L3 due to Bottom Chord Shortening

Vertical Displacement of L3 Due to Bottom Chord Panel Length Decrease of 3/8 in.

Member		d_c (in)	L_j (in)	A_j (in²)	F_{Qj} (kips)	(d_cF_{Qj}) (in)
L0L1	Bottom chord	−0.375	300	20	0.5	−0.19
L1L2	Bottom chord	−0.375	300	20	0.5	−0.19
L2L3	Bottom chord	−0.375	300	35	1	−0.38
L3L4	Bottom chord	−0.375	300	35	1	−0.38
L4L5	Bottom chord	−0.375	300	20	0.5	−0.19
L5L6	Bottom chord	−0.375	300	20	0.5	−0.19
L0U1	End post (top chord)	0	424	25	−0.71	0.00
U1U2	Top chord	0	300	25	−1	0.00
U2U3	Top chord	0	300	30	−1.5	0.00
U3U4	Top chord	0	300	30	−1.5	0.00

TABLE E6.7 (*Continued*)
Vertical Displacement of Joint L3 due to Bottom Chord Shortening

Vertical Displacement of L3 Due to Bottom Chord Panel Length Decrease of 3/8 in.						
Member		**d_c (in)**	**L_j (in)**	**A_j (in²)**	**F_{Qj} (kips)**	**(d_cF_{Qj}) (in)**
U4U5	Top chord	0	300	25	−1	0.00
U5L6	End post (top chord)	0	424	25	−0.71	0.00
U1L1	Hanger	0	300	10	0	0.00
U2L2	Vertical	0	300	5	−0.5	0.00
U3L3	Vertical	0	300	5	0	0.00
U4L4	Vertical	0	300	5	−0.5	0.00
U5L5	Hanger	0	300	10	0	0.00
U1L2	Diagonal	0	424	10	0.71	0.00
U2L3	Diagonal	0	424	5	0.71	0.00
L3U4	Diagonal	0	424	5	0.71	0.00
L4U5	Diagonal	0	424	10	0.71	0.00
Σ (in)						−1.50

Example 6.8

A 150 ft long six-panel horizontal parallel chord Pratt truss with h = 25 ft is subjected to a 1/2 in. increase (variation of d_c) of end post member lengths due to fabrication error. Determine the vertical displacement of joint L3.

Considering displacements due to only member length change, Equation 6.11 is

$$\Delta_{Pi} = \Sigma F_{Qj}\left(\Delta l_j\right)$$

For the joint displacement required, a statically determinate truss analysis for $Q_i = 1$ applied vertically at joint L3 is required to obtain F_{Qj} for each member. The results of this analysis are shown in Table E6.8.

TABLE E6.8
Vertical Displacement of Joint L3 due to End Post Member Lengthening

Vertical Displacement of L3 Due to End Post Lengthening of ½ in.						
Member		**d_c (in)**	**L_j (in)**	**A_j (in²)**	**F_{Qj} (kips)**	**(d_cF_{Qj}) (in)**
L0L1	Bottom chord	0	300	20	0.5	0.00
L1L2	Bottom chord	0	300	20	0.5	0.00
L2L3	Bottom chord	0	300	35	1	0.00
L3L4	Bottom chord	0	300	35	1	0.00
L4L5	Bottom chord	0	300	20	0.5	0.00
L5L6	Bottom chord	0	300	20	0.5	0.00
L0U1	End post (top chord)	0.5	424	25	−0.71	−0.36
U1U2	Top chord	0	300	25	−1	0.00
U2U3	Top chord	0	300	30	−1.5	0.00
U3U4	Top chord	0	300	30	−1.5	0.00

(*Continued*)

TABLE E6.8 (*Continued*)
Vertical Displacement of Joint L3 due to End Post Member Lengthening

Vertical Displacement of L3 Due to End Post Lengthening of ½ in.

Member		d_c (in)	L_j (in)	A_j (in²)	F_{Qj} (kips)	(d_cF_{Qj}) (in)
U4U5	Top chord	0	300	25	−1	0.00
U5L6	End post (top chord)	0.5	424	25	−0.71	−0.36
U1L1	Hanger	0	300	10	0	0.00
U2L2	Vertical	0	300	5	−0.5	0.00
U3L3	Vertical	0	300	5	0	0.00
U4L4	Vertical	0	300	5	−0.5	0.00
U5L5	Hanger	0	300	10	0	0.00
U1L2	Diagonal	0	424	10	0.71	0.00
U2L3	Diagonal	0	424	5	0.71	0.00
L3U4	Diagonal	0	424	5	0.71	0.00
L4U5	Diagonal	0	424	10	0.71	0.00
Σ (in)						−0.71

Example 6.9

A 150 ft long six-panel horizontal parallel chord Pratt truss with h = 25 ft is subjected to a 1.5 in vertical settlement at L0, a 0.75 in vertical settlement at L6 and a 0.5 in horizontal movement to the right at support L6. Determine the vertical displacement of joints L1, L3 and L5.

Considering displacements due to only settlements of supports, Equation 6.11 is

$$\Delta_{Pi} = -W_{QR}$$

W_{QR} is the work performed by the virtual reactions (due to Q_i) due to support movements Δ_r.

For the joint displacements required, a statically determinate truss analysis for Q_i =1 applied vertically at joints L1, L3 and L5 is required to obtain F_{Qj} for each member.

For a unit vertical load at L1, L0 = 0.833 and L6 = 0.167. Therefore, the work performed by the virtual reactions is $W_{QR} = -0.833(-1.5) - 0.167(-0.75) - 0(-0.5) = 1.375$ in vertical displacement of joint L1.

For a unit vertical load at L3, L0 = 0.50 and L6 = 0.50. Therefore, the work performed by the virtual reactions is $W_{QR} = -0.50(-1.5) - 0.50(-0.75) - 0(-0.5) = 1.125$ in vertical displacement of joint L3.

For a unit vertical load at L5, L0 = 0.167 and L6 = 0.833. Therefore, the work performed by the virtual reactions is $W_{QR} = -0.167(-1.5) - 0.833(-0.75) - 0(-0.5) = 0.875$ in vertical displacement of joint L5.

NOTES

1. Modern truss spans are unlikely to be designed with actual pin joints but are typically idealized as spans with pin connected members. If necessary, secondary forces related to this assumed behavior are calculated and superimposed on the primary axial forces obtained from the analysis assuming frictionless pin joints.
2. The change in member length is equal to the change in the distance between panel points at each end of the member.

3 See Chapter 7.
4 AREMA (2024) indicates that truss deflection must not exceed the span length divided by 640.
5 In AREMA (2024), the camber of trusses must equal to the deflection produced by the dead load plus a live load of 44 kN/m (3,000 lb/foot) of track.
6 Theorem of the Differential Coefficients of the Internal Work—Part 2 in Castigliano's book *The Theory of Equilibrium of Elastic Systems and Its Applications* (Castigliano, 1879).
7 Matrix methods involve either force (flexibility) or displacement (stiffness). Modern structural analysis is typically performed using the finite element method and computers. Finite element method software is typically coded based on systematic displacement (stiffness)method.
8 Statically determinate reactions are not independent as they can be found from the external loads P_i using static equilibrium. Therefore, Castigliano's Second Theorem is applicable to only trusses where displacements are caused by external loads.
9 Typically due to mechanical devices (turnbuckles) or fabrication errors.
10 Typically taken as $6.5 \times 10^{-6}\,^{\circ}F^{-1} = 1.1 \times 10^{-5}\,^{\circ}C^{-1}$.
11 Typically taken as 29×10^6 psi = 2.0×10^5 MPa.
12 Positive work is performed by Q_i reactions when they are in the same direction as the support movement. Negative work is performed by Q_i reactions when they are in the opposite direction as the support movement.
13 The number of structural analyses required depends on the positions of the moving loads. Only a single structural analysis is required for a uniform live load entirely on the span.

REFERENCES

American Railway Engineering and Maintenance-of-Way Association (AREMA), 2024, *Manual for Railway Engineering*, Chapter 15, Steel Structures, Lanham, MD.

Castigliano, C.A.P., 1879, *The Theory of Equilibrium of Elastic Systems and Its Application*, Dover Publications, New York, 1966.

BIBLIOGRAPHY

Armenakas, A.E., 1988, *Classical Structural Analysis—A Modern Approach*, McGraw-Hill, New York.

Hool, G.A., and Kinnie, W.S., 1923, *Stresses in Framed Structures*, McGraw-Hill, New York.

Utku, S., Norris, C.H., and Wilbur, J.B., 1991, *Elementary Structural Analysis*, 4th edition, McGraw-Hill, New York.

7 Secondary Flexural Stresses in Truss Members

7.1 INTRODUCTION

The primary axial forces in members of statically determinate and statically indeterminate trusses were determined in Chapters 4 and 5, respectively. The primary axial forces were determined under the assumption that externally applied loads and reactions are applied at the centroidal axes[1] of the truss members, coincide at a common point[2] at truss joints, are connected at their ends with frictionless pin joints and have cross-sections with straight centroidal axes that intersect at the truss joints.

However, in-plane[3] and out-of-plane[4] secondary flexural stresses in truss members can occur due to bending moments from

- eccentricities at connections[5] (when the centroidal axes of the member sections to not intersect at a common location) and from members that are not straight;
- transverse forces on members due to self-weight and appurtenances;[6]
- the distortion of vertical members;[7]
- rotationally fixed bearings;[8]
- elongation of chord members;[9]
- floorbeam end rotation from transverse frame action at vertical web members;[10] and/or,
- truss distortion and joint rigidity.[11]

Many typical short- and medium-span truss members are designed to resist only primary axial stresses[12] because secondary flexural stresses in the members are relatively small and implicitly considered within allowable axial stresses.[13] However, in some cases, explicit consideration of secondary bending moments in truss members is warranted, which must be combined with primary axial forces.[14]

Secondary bending moments of significance in truss spans are most often due to in-plane truss distortion with rigid joints, out-of-plane floorbeam distortion at vertical web members, in-plane eccentricities at joints, lack of member straightness and member beam behavior.[15] Large and complex truss spans may also require consideration of secondary bending moments in chord members associated with the distortion of floorbeam hanger and sub-vertical[16] members. Secondary flexural stresses can be considerable, and therefore, connection and member details must endeavor to limit secondary flexural stresses.[17]

Secondary bending moments provide some resistance to the external loads, thereby reducing the primary axial forces in members. Reducing the primary axial forces will subsequently precipitate a proportional reduction in secondary bending.[18] However, the primary axial force and related secondary bending moment reductions are not appreciable and typically negligible as shown in Example 7.1.

Example 7.1

Determine the reduction in primary axial forces and secondary bending moment in member L2L3 of a 150 ft long Pratt truss loaded by dead load of 1.5 kips/ft and uniformly distributed live load plus impact of 4.0 kips/ft (Figure E7.1).

DOI: 10.1201/9781003469971-8

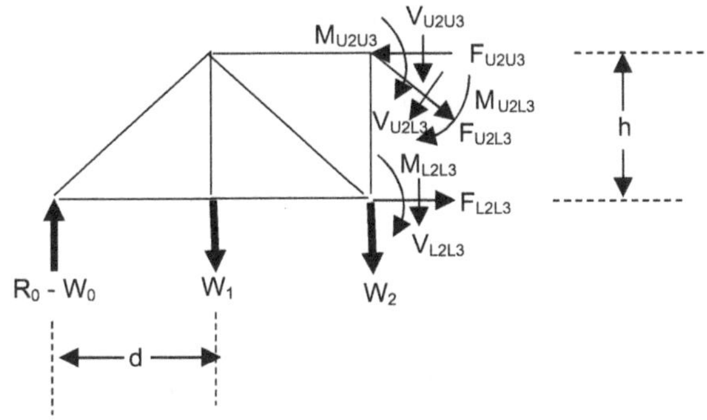

FIGURE E7.1 Pratt truss section.

$d = 25$ ft
$h = 25$ ft
$W_1 = W_2 = 37.5 + 100.0 = 137.5$ kips
$R_0 - W_0 = 137.5(5)/2 = 343.8$ kips

An analysis (see Example 7.2) yields the primary axial forces and secondary bending moments in truss members shown in Table E7.1.

TABLE E7.1
Member Forces

		Primary Axial Force (kips)			Secondary Bending Moment (kips-in)	
Member		DL	LL + I	DL + LL + I	Left Joint (lower number)	Right Joint (higher number or same number for vertical members)
L0L1	Bottom chord	93.8	250.0	343.8	595.8	680.9
L1L2	Bottom chord	93.8	250.0	343.8	−588.0	−236.0
L2L3	Bottom chord	150.0	400.0	550.0	446.7	2781.9
L0U1	End post	−132.6	−353.7	−486.3	−598.9	−468.8
U1U2	Top chord	−150.0	−400.0	−550.0	676.4	932.5
U2U3	Top chord	−168.8	−450.0	−618.8	−701.8	1432.6
U1L1	Hanger	37.5	100.0	137.5	−94.7	−97.6
U2L2	Vertical	−18.8	−50.0	−68.8	−135.1	−130.7
U3L3	Vertical	0.0	0.0	0.0	0.0	0.0
U1L2	Diagonal	79.6	212.1	291.7	−115.7	−84.4
U2L3	Diagonal	26.5	70.7	97.2	−87.8	12.5

From Figure E7.1,

$$\Sigma M_{U2} = 0 = (R_0 - W_0)(2d) - W_1(d) - F_{L2L3}(h) + M_{U2U3} + M_{U2L3} + M_{L2L3}$$

The moments due to secondary shear forces V_{U2U3}, V_{U2L3} and V_{L2L3} are negligible as secondary shear forces V_{U2U3}, V_{U2L3} and V_{L2L3} are close to joint U2. Therefore,

$$F_{L2L3} = \frac{(2R_0 - 2W_0 - W_1)d + M_{U2U3} + M_{U2L3} + M_{L2L3}}{h}$$
$$= \frac{(2(343.8) - 137.5)(25) - 701.8\text{-}87.8 + 446.7}{25}$$
$$= 536.3 \text{ kips}$$

The primary axial and secondary bending stresses are reduced by (1 − (536.3/550.0))100 = 2.5%. For most truss spans, the reduction of primary axial stress is less than 1%.

Truss members typically experience secondary flexural stresses proportional to the magnitude of the primary axial stresses. The secondary flexural stress in a member is also proportional to the distance to outer fiber in the plane of bending and inversely proportional to the length of the member. Therefore, when the members are symmetrical, the secondary flexural stresses will be proportional to the ratio of member width to length. Accordingly, AREMA (2024) recommends that secondary flexural stresses from truss distortion and joint rigidity can be neglected for slender truss members where the width of the member in the plane of the bending distortion is less than 10% of the member length.

This limitation on member stoutness provides that the secondary flexural stresses from truss distortion and joint rigidity for characteristically slender members of Pratt and Warren truss spans are typically less than 20% to 30% of the primary axial stresses. However, in some truss members,[19] the secondary flexural stresses are as high as 100% or more of the primary axial stress. Nonetheless, high secondary flexural stresses occur in only some members under some live load conditions and are localized at the extreme fibers of members.[20]

When required, secondary bending moments at the ends of members due to truss distortion and joint rigidity can be calculated by classical slope-deflection analysis[21] or by using modern numerical frame analysis software[22] assuming rigid or semi-rigid joints at the ends of members. With the exception of vertical web members directly loaded by floorbeam reactions, the location of the live load causing maximum secondary bending moments is not necessarily coincident with the live load location causing maximum primary axial force in a member. Consequently, the live load locations for maximum primary axial force in a member and for maximum secondary bending moment at the ends of the member can be developed using combined influence lines.[23] However, since secondary flexural stresses are typically considerably smaller than primary axial stresses, the live load location creating maximum primary axial force in the member is generally sufficient to determine the combined secondary flexural and primary axial stresses to be considered as the maximum combined stresses in a member.

Allowable stresses used for strength design include an allowance of 20% to 30% for the possibility of localized high secondary bending stresses at the extreme fibers of member cross-sections. However, localized secondary tensile flexural stresses and stress ranges combined with primary tensile axial stresses and stress ranges can precipitate strength or fatigue failure of a truss member where the maximum tensile stresses or stress ranges occur simultaneously with the maximum flexural stresses or stress ranges. Therefore, strength and fatigue are of particular concern for floorbeam hanger and subvertical members.[24]

Therefore, AREMA (2023) recommended that, for the strength of truss tension members other than floorbeam hanger members, only secondary bending stresses exceeding 27.6 MPa (4.0 ksi) need to be considered primary stresses.[25] AREMA (2024) recommends that secondary bending stresses[26] due to floorbeam end rotation in vertical truss members be considered primary stresses for strength design. For fatigue design of truss vertical members directly loaded by floorbeams, secondary bending stress ranges from all sources[27] are considered primary stress ranges. The most significant secondary tensile bending stresses and stress ranges in truss members are typically due to distortion in the plane of the truss with rigid joints. AREMA (2023) also recommended that for truss compression members, secondary bending stresses exceeding 20.7 MPa (3.0 ksi) be considered primary stresses.[28]

7.2 SECONDARY BENDING STRESSES IN MEMBERS DUE TO TRUSS DISTORTION AND JOINT RIGIDITY

Bolted gusset plates at truss joints provide restraint to member rotation[29] and cause members to bend due to distortion in the plane of the truss. Therefore, member length changes due to the primary axial forces and secondary bending moments are created in response to the constraint of angular deflections at the restrained joints. As shown in Figure 7.1, rigidity at joints n and p causes bending of member np, creating end bending moments M_n and M_p with corresponding shear forces V_n and V_p. Shear stresses created by secondary bending shear forces are relatively small and can be neglected. In addition, bending moments related to small eccentricities of the primary axial load when the member is bent can be neglected because the deformations are small. Deflection angles λ_{np} and λ_{pn} of member np at joints n and p are associated with M_n and M_p, respectively.

Figure 7.1 indicates

$$M_x = M_{pn} - V_{pn} x \quad (7.1a).$$

Taking moments about joint n provides

$$V_{pn} = \frac{M_{pn} + M_{np}}{l_{np}} \quad (7.1b),$$

and therefore,

$$M_x = M_{pn} - \frac{M_{pn} + M_{np}}{l_{np}} x \quad (7.1c).$$

Substituting Equation 7.1c into the equation for the elastic curve of a bending member provides

$$EI\frac{d^2y}{dx^2} - M = 0 = EI_{np}\frac{d^2y}{dx^2} - M_{pn} + \frac{M_{pn} + M_{np}}{l_{np}} x \quad (7.2a)$$

or

$$EI_{np}\frac{d^2y}{dx^2} = M_{pn} - \frac{M_{np} + M_{pn}}{l_{np}} x \quad (7.2b).$$

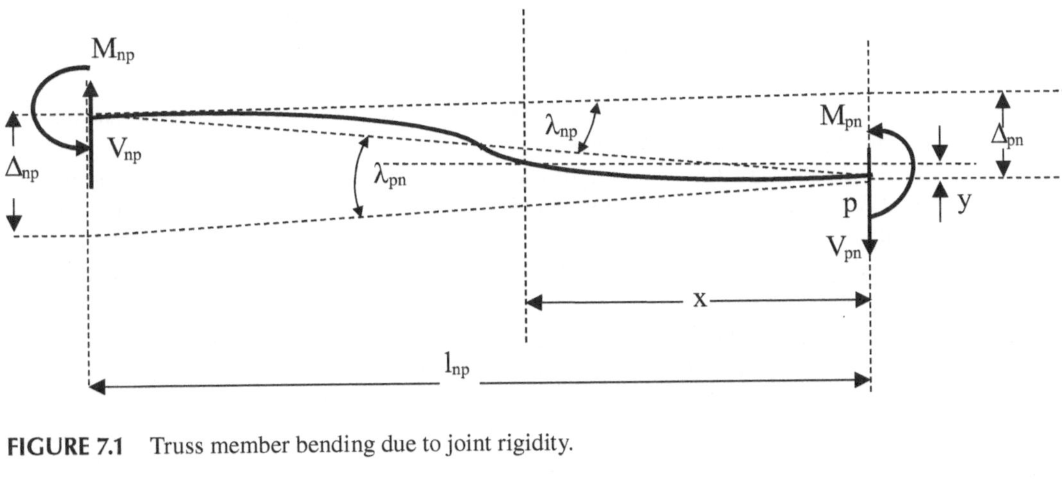

FIGURE 7.1 Truss member bending due to joint rigidity.

Integrating Equation 7.2b provides

$$EI_{np}\frac{dy}{dx} = M_{pn}x - \frac{M_{np}+M_{pn}}{2l_{np}}x^2 + C_1 \tag{7.3a}$$

Integrating Equation 7.3a provides

$$EI_{np}y = \frac{M_{pn}x^2}{2} - \frac{M_{np}+M_{pn}}{6l_{np}}x^3 + C_1x + C_2 \tag{7.3b}$$

At x = 0, $\frac{dy}{dx} = 0$ and from Equation 7.3a, $C_1 = 0$
At x = 0, y = 0 and from Equation 7.3b, $C_2 = 0$
At $x = l_{np}$, $y = \Delta_{np}$ and, therefore,

$$\Delta_{np} = \frac{l_{np}^2}{6EI_{np}}\left(2M_{pn} - M_{np}\right) \tag{7.4}$$

$$\lambda_{pn} = \frac{\Delta_{np}}{l_{np}} = \frac{l_{np}}{6EI_{np}}\left(2M_{pn} - M_{np}\right) \tag{7.5}$$

A similar analysis reveals

$$\lambda_{np} = \frac{\Delta_{pn}}{l_{np}} = \frac{l_{np}}{6EI_{np}}\left(2M_{np} - M_{pn}\right) \tag{7.6}$$

Solving Equations 7.5 and 7.6 for M_{np} and M_{pn} yields

$$M_{np} = \frac{2EI_{np}}{l_{np}}\left(2\lambda_{np} + \lambda_{pn}\right) \tag{7.7}$$

$$M_{pn} = \frac{2EI_{np}}{l_{np}}\left(2\lambda_{pn} + \lambda_{np}\right) \tag{7.8}$$

Equations 7.7 and 7.8 provide M_{np} and M_{pn} in member np created by the rigidity of the truss joints at n and p, respectively. Equations 7.7 and 7.8 also illustrate that reducing the moment of inertia of the member in the plane of the truss, $I_{np,}$[30] reduces the secondary bending stresses due to truss distortion and joint rigidity.

Figure 7.2a shows joint n in a truss span with design angles α, β and γ between members np_1 and np_2, np_2 and np_3 and np_3 and np_4, respectively.

Figure 7.2b shows joint n distorted due to primary axial forces in the members at the joint. The distortion at joint n creates changes in design angles dα, dβ and dγ between members np_1 and np_2, np_2 and np_3 and np_3 and np_4, respectively.

Figure 7.2c shows joint n distorted due to primary axial forces and secondary bending moments in members at the joint.

Figure 7.2c indicates the following relationships between the angular changes in the members at joint n due to primary axial forces and secondary bending moments:

$$\lambda_{np_2} + \alpha = \alpha + d\alpha + \lambda_{np_1} \tag{7.9a),}$$

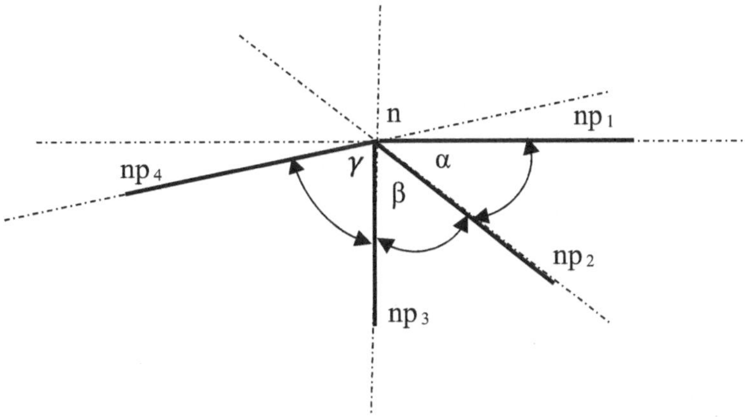

FIGURE 7.2a Truss joint.

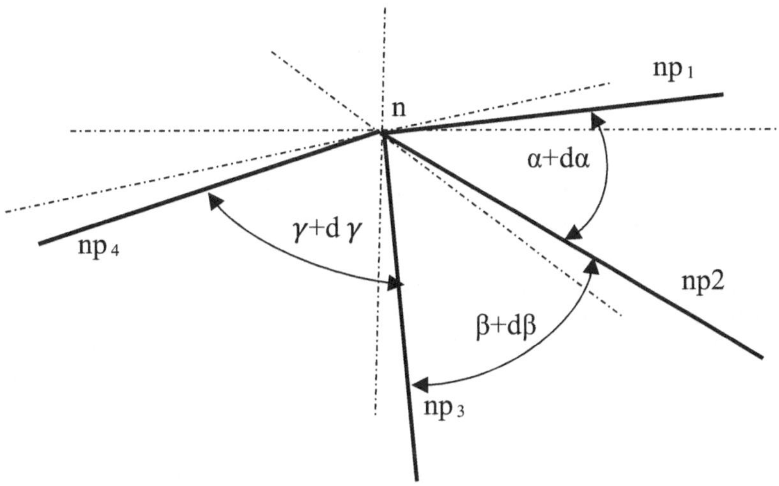

FIGURE 7.2b Truss joint distortion due to primary axial forces.

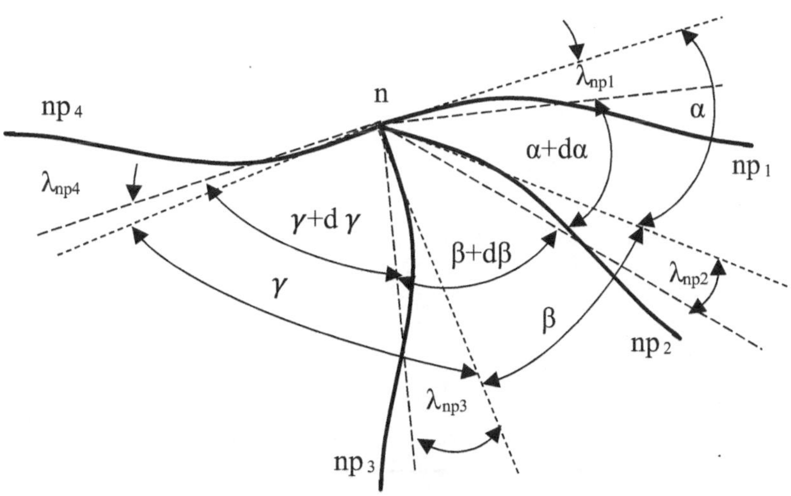

FIGURE 7.2c Truss joint distortion due to primary axial forces and secondary bending moments.

which provides

$$\lambda_{np_2} = \lambda_{np_1} + d\alpha \quad (7.9b).$$

Similarly,

$$\lambda_{np_3} = \lambda_{np_1} + d\alpha + d\beta \quad (7.10)$$

and

$$\lambda_{np_4} = \lambda_{np_1} + d\alpha + d\beta + d\gamma \quad (7.11).$$

Equations 7.9b, 7.10 and 7.11 can be expressed as

$$\lambda_{np_i} = \lambda_{np_1} + \sum d\theta_{np_i} = \lambda_{nref} + \sum d\theta_{npi} \quad (7.12)$$

where

λ_{np_i} = the secondary bending deflection angle of member np_i at joint n;

λ_{nref} = the secondary bending deflection angle of the first member encountered about joint n from outside the joint in a clockwise direction = the reference angle[31] for n ($\lambda_{nref} = \lambda_{np_1}$ in Figure 7.2c); and

$\sum d\theta_{npi}$ = the clockwise sum of the primary axial force angular changes from the joint n reference angle, λ_n, to member np_i.

End moments M_{ni} and M_{pi} for member np_i are determined from λ_{npi}, the secondary bending deflection angle of member np_i at joint n, and therefore, $\sum d\theta_{np_i}$ must be evaluated. Consequently, the angle changes at joint n, $d\theta_{np_i}$, due to the primary axial forces in truss members forming triangles with joints n = a, b and c, as shown in Figure 7.3, are determined by evaluating the geometrical changes due to changes in length of members ab, ac and bc.

Assuming the stresses σ_{ab}, σ_{ac} and σ_{bc} in members ab, ac and bc, respectively, are in axial tension, the members are straight and the joints are frictionless pins,[32] the change in length (elongation) of member np_i is

$$dl_{np_i} = \frac{\sigma_{np_i} l_{np_i}}{E} \quad (7.13).$$

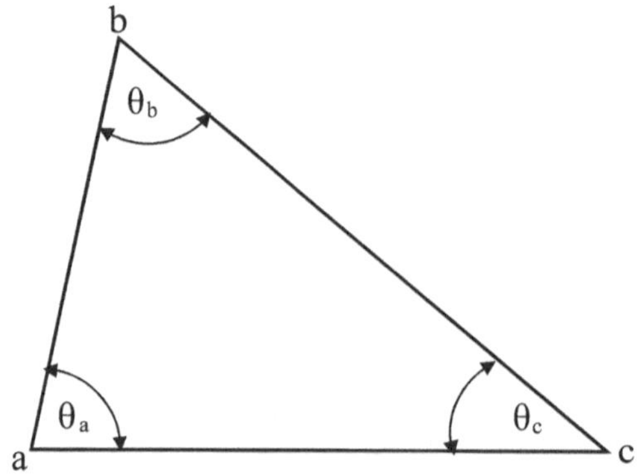

FIGURE 7.3 Truss member triangle.

The change in angular rotation $d\theta_a$ between members ab and ac is calculated with adjacent member ab elongated by $\frac{\sigma_{ab} l_{ab}}{E}$ and member ac considered not affected.[33] Therefore, the change in $d\theta_a$ between members ab and ac is calculated as

$$d\theta_a^{ab} = -\frac{\sigma_{ab}}{E}\cot\theta_b \tag{7.14a}$$

$d\theta_a$ is also calculated with adjacent member ac elongated by $\frac{\sigma_{ac} l_{ac}}{E}$ and member ab considered not affected, as follows:

$$d\theta_a^{ac} = -\frac{\sigma_{ac}}{E}\cot\theta_c \tag{7.14b}$$

or with the opposite member bc elongated by $\frac{\sigma_{bc} l_{bc}}{E}$ and member ab considered not affected, as follows:

$$d\theta_a^{bc} = \frac{\sigma_{bc}}{E}\left(\cot\theta_b + \cot\theta_c\right) \tag{7.14c}$$

Accordingly, the total change in $d\theta_a$ is

$$d\theta_a = d\theta_a^{ab} + d\theta_a^{ac} + d\theta_a^{bc} = -\frac{\sigma_{ab}}{E}\cot\theta_b - \frac{\sigma_{ac}}{E}\cot\theta_c + \frac{\sigma_{bc}}{E}\left(\cot\theta_b + \cot\theta_c\right) \tag{7.14d}$$

from which

$$d\theta_a = \frac{\sigma_{bc} - \sigma_{ba}}{E}\cot\theta_b + \frac{\sigma_{cb} - \sigma_{ca}}{E}\cot\theta_c \tag{7.15}$$

A similar evaluation of the change in $d\theta_b$ between members ab and bc indicates that

$$d\theta_b = \frac{\sigma_{ac} - \sigma_{ab}}{E}\cot\theta_a + \frac{\sigma_{ca} - \sigma_{cb}}{E}\cot\theta_c \tag{7.16}$$

and calculating the change in $d\theta_c$ between members ac and bc reveals

$$d\theta_c = \frac{\sigma_{ab} - \sigma_{ac}}{E}\cot\theta_a + \frac{\sigma_{ba} - \sigma_{bc}}{E}\cot\theta_b \tag{7.17}$$

Since the entire truss is in equilibrium, each joint must also be in equilibrium. Equilibrium of the free body diagram of joint n shown in Figure 7.4 requires that

$$\Sigma M_{np} = 0 = \sum_{i=1}^{nm} M_{npi} = M_{np1} + M_{np2} + M_{np3} + M_{np4} \tag{7.18a}$$

where nm = the number of members framing into joint n (nm = 4 in Figure 7.4).

Substituting Equation 7.7 and $K_{np_i} = \frac{I_{np_i}}{l_{np_i}}$ into Equation 7.18a and dividing by 2E provides,

$$K_{np_1}(2\lambda_{np_1} + \lambda_{p_1n}) + K_{np_2}(2\lambda_{np_2} + \lambda_{p_2n}) + K_{np_3}(2\lambda_{np_3} + \lambda_{p_3n}) + K_{np_4}(2\lambda_{np_4} + \lambda_{p_4n}) = 0 \tag{7.18b}$$

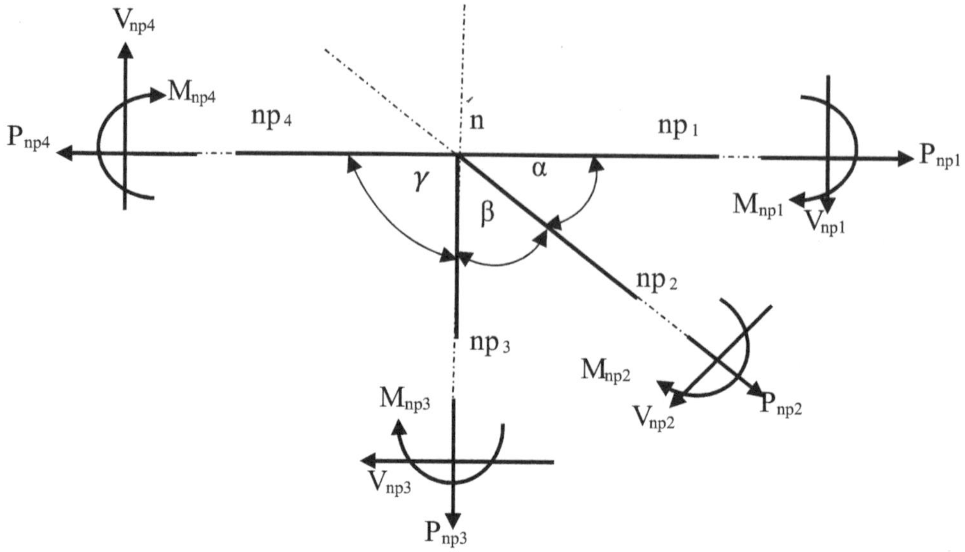

FIGURE 7.4 Free body diagram of joint n.

Substituting Equation 7.12 into Equation 7.18b yields

$$
\begin{aligned}
&2\lambda_{nref}(K_{np_1}+K_{np_2}+K_{np_3}+K_{np_4})+2(K_{np_1}\Sigma d\theta_{np_1}+K_{np_2}\Sigma d\theta_{np_2}+\\
&K_{np_3}\Sigma d\theta_{np_3}+K_{np_4}\Sigma d\theta_{np_4})+K_{p_1n}(\lambda_{p_1n}+\Sigma d\theta_{p_1n})+K_{p_2n}(\lambda_{p_2n}+\Sigma d\theta_{p_2n})+\\
&K_{p_3n}(\lambda_{p_3n}+\Sigma d\theta_{p_3n})+K_{p_4n}(\lambda_{p_4n}+\Sigma d\theta_{p_4n})=0
\end{aligned}
\tag{7.18c}
$$

Equation 7.18c is expressed as

$$
2\left[\left(\Sigma K_{np_i}\right)\lambda_{nref}+\Sigma\left(\left(K_{np_i}\right)\Sigma d\theta_{np_i}\right)\right]+\sum_{i=1}^{nm}\left(K_{p_in}\lambda_{p_in}+K_{p_in}\Sigma d\theta_{p_in}\right)=0 \tag{7.19}
$$

Equation 7.19 provides an effective method of calculating the secondary bending stresses in truss members np_i at joint n. Writing Equation 7.19 for all joints in the truss will provide a series of simultaneous equations that can be solved[34] for the secondary bending deflections λ_{nref}. Equation 7.12 is then used to calculate the deflection angles λ_{np_i} and λ_{p_in}, from which secondary bending stresses are calculated in each member at each joint. Example 7.2 illustrates the calculation of secondary bending stresses in horizontal chord Pratt truss members due to joint rigidity.

Example 7.2

Determine the maximum secondary bending stresses due to joint rigidity in the 150 ft long Pratt truss in Figure E7.2 for a uniformly distributed dead load of 1.5 kips/ft and a uniformly distributed live load plus impact of 4.0 kips/ft.

Member gross section properties, assuming a symmetrical section in the plane of bending, are as follows, and findings are presented in Table E7.2a.

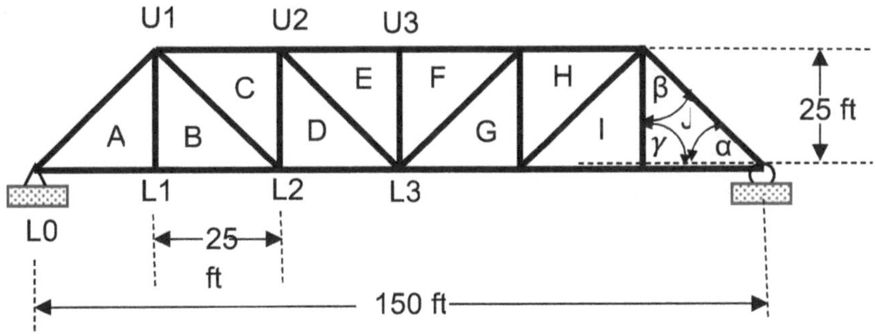

FIGURE E7.2 Pratt truss.

L = 6(25) = 150 ft
d = 25 ft
h = 25 ft (shallow for illustration purposes)

$$\cot\alpha = \frac{d}{h} = 1.00$$

$$\cot\beta = \frac{h}{d} = 1.00$$

$$\cot\gamma = 0$$

E = 29,000 ksi (steel)

TABLE E7.2a
Member Properties

Member np_i	Type	Length, l_{npi} (in)	Cross-Sectional Area, A_{npi} (in^2)	Moment of Inertia, I_{npi} (in^4)	Section Modulus, $S_{np} = I_{npi}/c_{npi}$ (in^3)	$K_{npi} = I_{npi}/l_{npi}$ (in^3)
L0L1	Bottom chord	300	20	2000	200	6.67
L1L2	Bottom chord	300	20	2000	300	6.67
L2L3	Bottom chord	300	35	3500	500	11.67
L3L4	Bottom chord	300	35	3500	500	11.67
L4L5	Bottom chord	300	20	2000	300	6.67
L5L6	Bottom chord	300	20	2000	200	6.67
L0U1	End post	424	25	2500	250	5.90
U1U2	Top chord	300	25	2500	250	8.33
U2U3	Top chord	300	30	3000	400	10.00
U3U4	Top chord	300	30	3000	400	10.00
U4U5	Top chord	300	25	2500	250	8.33
U5L6	End post	424	25	2500	250	5.90
U1L1	Hanger	300	10	100	20	0.33
U2L2	Vertical	300	5	100	20	0.33
U3L3	Vertical	300	5	100	20	0.33
U4L4	Vertical	300	5	100	20	0.33
U5L5	Hanger	300	10	100	20	0.33
U1L2	Diagonal	424	10	300	50	0.71
U2L3	Diagonal	424	5	200	50	0.47
L3U4	Diagonal	424	5	200	50	0.47
L4U5	Diagonal	424	10	300	50	0.71

I = maximum impact = 26% for all members (for illustration purposes includes floorbeam hanger members).

TABLE E7.2b
Primary Axial Stress in Members

Member			Primary Axial Force (kips)			
np_i	Type	Cross-Sectional Area, A_{npi} (in²)	Dead Load (DL)	Live Load + Impact (LL + I)	DL + LL + I	Primary Load Axial Stress (ksi)
L0L1	Bottom chord	20	93.8	250.0	343.8	17.2
L1L2	Bottom chord	20	93.8	250.0	343.8	17.2
L2L3	Bottom chord	35	150.0	400.0	550.0	15.7
L3L4	Bottom chord	35	150.0	400.0	550.0	15.7
L4L5	Bottom chord	20	93.8	250.0	343.8	17.2
L5L6	Bottom chord	20	93.8	250.0	343.8	17.2
L0U1	End post	25	−132.6	−353.7	−486.3	−19.5
U1U2	Top chord	25	−150.0	−400.0	−550.0	−22.0
U2U3	Top chord	30	−168.8	−450.0	−618.8	−20.6
U3U4	Top chord	30	−168.8	−450.0	−618.8	−20.6
U4U5	Top chord	25	−150.0	−400.0	−550.0	−22.0
U5L6	End post	25	−132.6	−353.7	−486.3	−19.5
U1L1	Hanger	10	37.5	100.0	137.5	13.8
U2L2	Vertical	5	−18.8	−50.0	−68.8	−13.8
U3L3	Vertical	5	0.0	0.0	0.0	0.0
U4L4	Vertical	5	−18.8	−50.0	−68.8	−13.8
U5L5	Hanger	10	37.5	100.0	137.5	13.8
U1L2	Diagonal	10	79.6	212.1	291.7	29.2
U2L3	Diagonal	5	26.5	70.7	97.2	19.4
L3U4	Diagonal	5	26.5	70.7	97.2	19.4
L4U5	Diagonal	10	79.6	212.1	291.7	29.2

Member primary maximum axial stresses are given in Table E7.2b.

Equations 7.15, 7.16 and 7.17 provide the angular changes at the joints of a triangle due to elongation of the members (assumed in axial tension). Due to the symmetry of loading and truss geometry, the angular changes in triangles F, G, H and I are the same as those in triangles E, D, C and B, respectively. Similarly, the angular changes in triangle J are the same as those in triangle A. For triangles A to J in Figure E7.2, the changes are given in Table E7.2c.

Equation 7.12 requires that the clockwise sum of the primary axial force angular changes from the joint n reference angle λ_n to member np_i $\Sigma d\theta_{np_i}$ be determined at each joint as shown in Table E7.2d.

Equation 7.19 requires that $\Sigma\left(\left(K_{np_i}\right)\Sigma d\theta_{np_i}\right)$ be determined, as shown in Table E7.2e.

Determine Equation 7.19 for each joint,

$$2\left[\left(\Sigma K_{np_i}\right)\lambda_n + \Sigma\left(\left(K_{np_i}\right)\Sigma d\theta_{np_i}\right)\right] + \sum_{i=1}^{nm}\left(K_{p_i n}\lambda_{p_i} + K_{p_i n}\Sigma d\theta_{p_i n}\right) = 0 = Q + R$$

where

$$Q = 2\left[\left(\Sigma K_{np_i}\right)\lambda_n + \Sigma\left(\left(K_{np_i}\right)\Sigma d\theta_{np_i}\right)\right]$$

$$R = \sum_{i=1}^{nm} \left(K_{p_in}\lambda_{p_i} + K_{p_in} \Sigma d\theta_{p_in} \right)$$

Joint n = L0 = L6:

$Q = 2(12.57)\lambda_{L0} + 2(7.67/1000) = 25.14\ \lambda_{L0} + (15.34/1000)$

Member np_i	Joint p_i	Member p_in	R
L0U1	U1	U1L0	$5.90\lambda_{U1} - (3.66/1000)$
L0L1	L1	L1L0	$6.67\lambda_{L1}$

$Q + R = 25.14\ \lambda_{L0} + 6.67\lambda_{L1} + 5.90\lambda_{U1} + (11.68/1000)$

Joint n = L1 = L5:

$Q = 2(13.67)\lambda_{L1} - 2(10.60/1000) = 27.34\ \lambda_{L1} - (21.20/1000)$

Member np_i	Joint p_i	Member p_in	R
L1L0	L0	L0L1	$6.67\lambda_{L0} + (7.67/1000)$
L1U1	U1	U1L1	$0.33\lambda_{U1} - (0.62/1000)$
L1L2	L2	L2L1	$6.67\lambda_{L2}$

$Q + R = 6.67\lambda_{L0} + 27.34\ \lambda_{L1} + 0.33\lambda_{U1} + 6.67\lambda_{L2} - (14.15/1000)$

Joint n = U1 = U5:

$Q = 2(15.27)\lambda_{U1} - 2(5.33/1000) = 30.54\ \lambda_{U1} - (10.66/1000)$

Member np_i	Joint p_i	Member p_in	R
U1U2	U2	U2U1	$8.33\lambda_{U2} + (20.41/1000)$
U1L2	L2	L2U1	$0.71\lambda_{L2} - (0.38/1000)$
U1L1	L1	L1U1	$0.33\lambda_{L1} - (0.80/1000)$
U1L0	L0	L0U1	$5.90\ \lambda_{L0}$

$Q + R = 5.90\ \lambda_{L0} + 0.33\lambda_{L1} + 30.54\ \lambda_{U1} + 0.71\lambda_{L2} + 8.33\lambda_{U2} + (8.57/1000)$

Joint n = L2 = L4:

$Q = 2(19.38)\lambda_{L2} - 2(13.16/1000) = 38.76\ \lambda_{L2} - (26.32/1000)$

Member np_i	Joint p_i	Member p_in	R
L2L1	L1	L1L2	$6.67\lambda_{L1} - (9.80/1000)$
L2U1	U1	U1L2	$0.71\lambda_{U1} - (1.05/1000)$
L2U2	U2	U2L2	$0.33\lambda_{U2} - (0.26/1000)$
L2L3	L3	L3L2	$11.67\ \lambda_{L3}$

TABLE E7.2c
Angular Changes due to Primary Axial Forces at Truss Joints

Triangle	Joint		$d\theta_{npi}$ (rad)		$E\,(d\theta_{npi})$ (ksi)	$d\theta_{npi}$ $\times 10^{-3}$ (rad)
A and J	L0	$d\alpha$	$\frac{\sigma_{U1L1}-\sigma_{L0U1}}{E}\cot(\beta)+\frac{\sigma_{U1L1}-\sigma_{L0L1}}{E}\cot(\gamma)$	$\frac{\sigma_{U1L1}-\sigma_{L0U1}}{E}$	13.8 − (−19.5) = 33.3	1.15
	U1	$d\beta$	$\frac{\sigma_{L0L1}-\sigma_{L0U1}}{E}\cot(\alpha)+\frac{\sigma_{L0L1}-\sigma_{U1L1}}{E}\cot(\gamma)$	$\frac{\sigma_{L0L1}-\sigma_{L0U1}}{E}$	17.2 − (−19.5) = 36.7	1.27
	L1	$d\gamma$	$\frac{\sigma_{L0U1}-\sigma_{U1L1}}{E}\cot(\beta)+\frac{\sigma_{L0U1}-\sigma_{L0L1}}{E}\cot(\alpha)$	$\frac{\sigma_{L0U1}-\sigma_{U1L1}}{E}+\frac{\sigma_{L0U1}-\sigma_{L0L1}}{E}$	−19.5 − (13.8) + (−19.5) − 17.2 = −33.3 − 36.7 = −70.0	−2.41
B and I	L2	$d\alpha$	$\frac{\sigma_{U1L1}-\sigma_{U1L2}}{E}\cot(\beta)+\frac{\sigma_{U1L1}-\sigma_{L1L2}}{E}\cot(\gamma)$	$\frac{\sigma_{U1L1}-\sigma_{U1L2}}{E}$	13.8 − (29.2) = −15.4	−0.53
	U1	$d\beta$	$\frac{\sigma_{L1L2}-\sigma_{U1L2}}{E}\cot(\alpha)+\frac{\sigma_{L1L2}-\sigma_{U1L1}}{E}\cot(\gamma)$	$\frac{\sigma_{L1L2}-\sigma_{U1L2}}{E}$	17.2 − (29.2) = −12.0	−0.41
	L1	$d\gamma$	$\frac{\sigma_{U1L2}-\sigma_{U1L1}}{E}\cot(\beta)+\frac{\sigma_{U1L2}-\sigma_{L1L2}}{E}\cot(\alpha)$	$\frac{\sigma_{U1L2}-\sigma_{U1L1}}{E}+\frac{\sigma_{U1L2}-\sigma_{L1L2}}{E}$	29.2 − (13.8) + 29.2 − (17.2) = 15.4 + 12.0 = 27.4	0.94
C and H	U1	$d\alpha$	$\frac{\sigma_{U2L2}-\sigma_{U1L2}}{E}\cot(\beta)+\frac{\sigma_{U2L2}-\sigma_{U1U2}}{E}\cot(\gamma)$	$\frac{\sigma_{U2L2}-\sigma_{U1L2}}{E}$	−13.8 − (29.2) = −43.0	−1.48
	L2	$d\beta$	$\frac{\sigma_{U1U2}-\sigma_{U1L2}}{E}\cot(\alpha)+\frac{\sigma_{U1U2}-\sigma_{U2L2}}{E}\cot(\gamma)$	$\frac{\sigma_{U1U2}-\sigma_{U1L2}}{E}$	−22.0 − (29.2) = −51.2	−1.77
	U2	$d\gamma$	$\frac{\sigma_{U1L2}-\sigma_{U2L2}}{E}\cot(\beta)+\frac{\sigma_{U1L2}-\sigma_{U1U2}}{E}\cot(\alpha)$	$\frac{\sigma_{U1L2}-\sigma_{U2L2}}{E}+\frac{\sigma_{U1L2}-\sigma_{U1U2}}{E}$	29.2 − (−13.8) + 29.2 − (−22.0) = 43.0 + 51.2 = 94.2	3.25
D and G	L3	$d\alpha$	$\frac{\sigma_{U2L2}-\sigma_{U2L3}}{E}\cot(\beta)+\frac{\sigma_{U2L2}-\sigma_{L2L3}}{E}\cot(\gamma)$	$\frac{\sigma_{U2L2}-\sigma_{U2L3}}{E}$	−13.8 − (19.4) = −33.2	−1.14
	U2	$d\beta$	$\frac{\sigma_{L2L3}-\sigma_{U2L3}}{E}\cot(\alpha)+\frac{\sigma_{L2L3}-\sigma_{U2L2}}{E}\cot(\gamma)$	$\frac{\sigma_{L2L3}-\sigma_{U2L3}}{E}$	15.7 − (19.4) = −3.7	−0.13
	L2	$d\gamma$	$\frac{\sigma_{U2L3}-\sigma_{U2L2}}{E}\cot(\beta)+\frac{\sigma_{U2L3}-\sigma_{L2L3}}{E}\cot(\alpha)$	$\frac{\sigma_{U2L3}-\sigma_{U2L2}}{E}$	19.4 − (−13.8) + 19.4 − (15.7) = 33.2 + 3.7 = 36.9	1.27

TABLE E7.2c (*Continued*)
Angular Changes due to Primary Axial Forces at Truss Joints

Triangle	Joint		$d\theta_{npi}$ (rad)		$E\,(d\theta_{npi})$ (ksi)	$d\theta_{npi}$ × 10^{-3} (rad)
E and F	U2	dα	$\frac{\sigma_{U3L3}-\sigma_{U2L3}}{E}\cot(\beta)+\frac{\sigma_{U3L3}-\sigma_{U2U3}}{E}\cot(\gamma)$	$\frac{\sigma_{U3L3}-\sigma_{U2L3}}{E}$	0 − (19.4) = −19.4	−0.67
	L3	dβ	$\frac{\sigma_{U2U3}-\sigma_{U2L3}}{E}\cot(\alpha)+\frac{\sigma_{L2L3}-\sigma_{U3L3}}{E}\cot(\gamma)$	$\frac{\sigma_{U2U3}-\sigma_{U2L3}}{E}$	−20.6 − (19.4) = −40.0	−1.38
	U3	dγ	$\frac{\sigma_{U2L3}-\sigma_{U3L3}}{E}\cot(\beta)+\frac{\sigma_{U2L3}-\sigma_{U2U3}}{E}\cot(\alpha)$	$\frac{\sigma_{U2L3}-\sigma_{U3L3}}{E}+\frac{\sigma_{U2L3}-\sigma_{U2U3}}{E}$	19.4 − (0) + 19.4 − (−20.6) = 19.4 + 40.0 = 59.4	2.05

TABLE E7.2d
Sum of the Angular Changes due to Primary Axial Forces at Truss Joints

Joint	Triangle	$d\theta_{npi} \times 10^{-3}$ (rad)	$\sum d\theta_{npi} \times 10^{-3}$ (rad)
L0 and L6	A	1.15	1.15
U1 and U5	C	−1.48	−1.48
	B	−0.41	−1.89
	A	1.27	−0.62
L1 and L5	A	−2.41	−2.41
	B	0.94	−1.47
L2 and L4	B	−0.53	−0.53
	C	−1.77	−2.30
	D	1.27	−1.03
U2 and U4	E	−0.67	−0.67
	D	−0.13	−0.80
	C	3.25	2.45
L3	D	−1.14	−1.14
	E	−1.38	−2.52
	F	−1.38	−3.90
	G	−1.14	−5.04
U3	F	2.05	2.05
	E	2.05	4.10

TABLE E7.2e
Relationships between the Angular Changes in Members at Joints due to Primary Axial Forces and Secondary Bending Moments

Joint, n	Member, np_i	K_{npi} (in³)	$\sum d\theta_{npi} \times 10^{-3}$ (rad)	$K\sum d\theta_{npi} \times 10^{-3}$ (in³)
L0 and L6	L0U1	5.90	0	0
	L0L1	6.67	1.15	7.67
	Σ	12.57	1.15	7.67
L1 and L5	L1L0	6.67	0	0
	L1U1	0.33	−2.41	−0.80
	L1L2	6.67	−1.47	−9.80
	Σ	13.67	−3.88	−10.60
U1 and U5	U1U2	8.33	0	0
	U1L2	0.71	−1.48	−1.05
	U1L1	0.33	−1.89	−0.62
	U1L0	5.90	−0.62	−3.66
	Σ	15.27	−3.96	−5.33
L2 and L4	L2L1	6.67	0	0
	L2U1	0.71	−0.53	−0.38
	L2U2	0.33	−2.30	−0.76
	L2L3	11.67	−1.03	−12.02
	Σ	19.38	−3.86	−13.16
U2 and U4	U2U3	10.00	0	0
	U2L3	0.47	−0.67	−0.31
	U2L2	0.33	−0.80	−0.26
	U2U1	8.33	2.45	20.41
	Σ	19.13	0.98	19.84
U3	U3U4	10.00	0	0
	U3L3	0.33	2.05	0.68
	U3U2	10.00	4.10	41.00
	Σ	20.33	6.15	41.68
L3	L3L2	11.67	0	0
	L3U2	0.47	−1.14	−0.54
	L3U3	0.33	−2.52	−0.83
	L3U4	0.47	−3.90	−1.83
	L3L4	11.67	−5.04	−58.82
	Σ	24.61	−12.60	−62.02

$$Q + R = 6.67\lambda_{L1} + 0.71\lambda_{U1} + 38.76\,\lambda_{L2} + 0.33\lambda_{U2} + 11.67\,\lambda_{L3} - (37.43/1000)$$

Joint n = U2 = U4:

$$Q = 2(19.13)\lambda_{U2} + 2(19.84/1000) = 38.26\,\lambda_{U2} + (39.68/1000)$$

Member np_i	Joint p_i	Member p_in	R
U2U3	U3	U3U2	$10.00\lambda_{U3} + (41.00/1000)$
U2L3	L3	L3U2	$0.47\lambda_{L3} - (0.54/1000)$
U2L2	L2	L2U2	$0.33\lambda_{L2} - (0.76/1000)$
U2U1	U1	U1U2	$8.33\,\lambda_{U1}$

$$Q + R = 8.33\,\lambda_{U1} + 0.33\lambda_{L2} + 38.26\,\lambda_{U2} + 10.00\lambda_{U3} + 0.47\lambda_{L3} + (79.38/1000)$$

The symmetrical loading and truss geometry results in no moment in member U3L3 and no angular deflection from secondary bending effects. This provides the expedients that

$$\lambda_{U3L3} = \lambda_{L3U3} = 0$$

From Equation 7.12,

$$\lambda_{U3L3} = \lambda_{U3} + \sum d\theta_{U3L3}$$

$$\lambda_{U3} = -\sum d\theta_{U3L3} = \frac{-2.05}{1000}$$

$$\lambda_{L3U3} = \lambda_{L3} + \sum d\theta_{L3U3}$$

$$\lambda_{L3} = -\sum d\theta_{L3U3} = \frac{2.52}{1000}$$

and the equilibrium equations are reduced to

$$25.14\,\lambda_{L0} + 6.67\lambda_{L1} + 5.90\lambda_{U1} = -(11.68/1000)$$

$$6.67\lambda_{L0} + 27.34\,\lambda_{L1} + 0.33\lambda_{U1} + 6.67\lambda_{L2} = (14.15/1000)$$

$$5.90\,\lambda_{L0} + 0.33\lambda_{L1} + 30.54\,\lambda_{U1} + 0.71\lambda_{L2} + 8.33\lambda_{U2} = -(8.57/1000)$$

$$6.67\lambda_{L1} + 0.71\lambda_{U1} + 38.76\,\lambda_{L2} + 0.33\lambda_{U2} = (8.02/1000)$$

$$8.33\,\lambda_{U1} + 0.33\lambda_{L2} + 38.26\,\lambda_{U2} = -(60.06/1000)$$

The solution of these equations is[35]

$\lambda_{L0} = -0.71/1000$ (secondary bending deflection of member L0U1)

$\lambda_{L1} = 0.66/1000$ (secondary bending deflection of member L1L0)

$\lambda_{U1} = 0.29/1000$ (secondary bending deflection of member U1U2)

$\lambda_{L2} = 0.10/1000$ (secondary bending deflection of member L2L1)

$\lambda_{U2} = -1.63/1000$ (secondary bending deflection of member U2U3)

$\lambda_{U3} = -2.05/1000$ (secondary bending deflection of member U3U4 and U3U2)

$\lambda_{L3} = 2.52/1000$ (secondary bending deflection of member L3L2 and L3L4)

Equation 7.12 gives the findings in Table E7.2f.

The secondary bending deflections (angular rotations) of greatest magnitude occur at members L1U1, U1L2, L2U2, L2L3, U2U3 and U2L3. Equation 7.7 gives the findings in Table E7.2g.

TABLE E7.2f
Angular Rotation due to Primary Axial Forces at Truss Joints

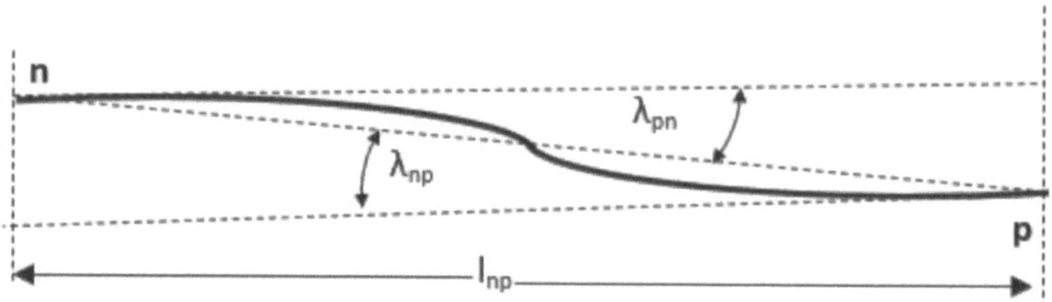

Joint n	Member np_i	$\lambda_n \times 10^{-3}$ (rad)	$\sum d\theta_{npi} \times 10^{-3}$ (rad)	$\lambda_{npi} \times 10^{-3}$ (rad)	$\lambda_{pin} \times 10^{-3}$ (rad)	$(2\lambda_{npi} + \lambda_{pin}) \times 10^{-3}$ (rad)
L0	L0U1	−0.71	0	−0.71	−0.33	−1.75
	L0L1		1.15	0.44	0.66	1.54
L1	L1L0	0.66	0	0.66	0.44	1.76
	L1U1		−2.41	−1.75	−1.60	−5.10
	L1L2		−1.47	−0.81	0.10	−1.52
U1	U1U2	0.29	0	0.29	0.82	1.40
	U1L2		−1.48	−1.19	−0.43	−2.81
	U1L1		−1.89	−1.60	−1.75	−4.95
	U1L0		−0.62	−0.33	−0.71	−1.37
L2	L2L1	0.10	0	0.10	−0.81	−0.61
	L2U1		−0.53	−0.43	−1.19	−2.05
	L2U2		−2.30	−2.20	−2.43	−6.83
	L2L3		−1.03	−0.93	2.52	0.66
U2	U2U3	−1.63	0	−1.63	2.05	−1.21
	U2L3		−0.67	−2.30	1.38	−3.22
	U2L2		−0.80	−2.43	−2.20	−7.06
	U2U1		2.45	0.82	0.29	1.93
U3	U3L3	−2.05	2.05	0	0	0
	U3U2		4.10	2.05	−1.63	2.47
L3	L3L2	2.52	0	2.52	−0.93	4.11
	L3U2		−1.14	1.38	−2.30	0.46
	L3U3		−2.52	0	0	0

TABLE E7.2g
Secondary Bending Stresses in Members

Member np_i	$K_{npi} = K_{pin}$ (in^3)	$(2\lambda_{npi} + \lambda_{pin}) \times 10^{-3}$	M_{npi} (kips-in)	$S_{npi} = S_{pin}$ (in^3)	σ^b_{npi} (ksi)	σ^a_{np} (ksi)	$(\sigma^a_{np}/\sigma^b_{np})100$ (%)
L0U1	5.90	−1.75	−598.9	250	−2.4	−19.5	12.3
L0L1	6.67	1.54	595.8	200	3.0	17.2	17.3
$\sum M_{L0}$			−3.10				
L1L0	6.67	1.76	680.9	200	3.4	17.2	19.8
L1U1	0.33	−5.10	−97.6	30	−3.3	13.8	23.6

TABLE E7.2g (*Continued*)
Secondary Bending Stresses in Members

Member np_i	$K_{npi} = K_{pin}$ (in^3)	$(2\lambda_{npi} + \lambda_{pin}) \times 10^{-3}$	M_{npi} (kips-in)	$S_{npi} = S_{pin}$ (in^3)	σ^b_{npi} (ksi)	σ^a_{np} (ksi)	$(\sigma^a_{np}/\sigma^b_{np})100$ (%)
L1L2	6.67	−1.52	−588.0	300	−2.0	17.2	11.4
ΣM_{L1}			−4.70				
U1U2	8.33	1.40	676.4	250	2.7	−22.0	12.3
U1L2	0.71	−2.81	−115.7	50	−2.3	29.2	7.9
U1L1	0.33	−4.95	−94.7	30	−3.2	13.8	22.9
U1L0	5.90	−1.37	−468.8	250	−1.9	−19.5	9.6
ΣM_{U1}			−2.80				
L2L1	6.67	−0.61	−236.0	300	−0.8	17.2	4.6
L2U1	0.71	−2.05	−84.4	50	−1.7	29.2	5.8
L2U2	0.33	−6.83	−130.7	30	−4.4	−13.8	31.6
L2L3	11.67	0.66	446.7	500	0.9	15.7	5.7
ΣM_{L2}			−4.40				
U2U3	10.00	−1.21	−701.8	400	−1.8	−20.6	8.5
U2L3	0.47	−3.22	−87.8	50	−1.8	19.4	9.0
U2L2	0.33	−7.06	−135.1	30	−4.5	−13.8	32.6
U2U1	8.33	1.93	932.5	250	3.7	−22.0	17.0
ΣM_{U2}			7.80				
U3L3	0.33	0	0.0	30	0.0	0.0	-
U3U2	10.00	2.47	1432.6	400	3.6	−20.6	17.4
L3L2	11.67	4.11	2781.9	500	5.6	15.7	35.4
L3U2	0.47	0.46	12.5	50	0.3	19.4	1.3
L3U3	0.33	0	0.0	30	0.0	0.0	-

Vertical members L1U1 (at joints L1 and U1), vertical members L2U2 (at joints L2 and U2) and bottom chord member L2L3 (at joint L3) have secondary flexural stresses that exceed 20%. $\sigma^a_{np}/\sigma^b_{np}$ can be reduced by reducing the stiffness[36] K_{np} by reducing the member's moment of inertia I_{np}.

7.3 SECONDARY BENDING STRESSES IN MEMBERS DUE TO ECCENTRICITIES AT CONNECTIONS

Where they exist, eccentricities at connections[37] create in-plane secondary bending stresses in members that are superimposed on the secondary bending stresses due to truss distortion and joint rigidity. Figure 7.5 is a free body diagram of joint n with eccentric forces P_{np4}, P_{np1} and P_{np3}.

The secondary bending moment at joint n, shown in Figure 7.5, due to joint eccentricity = M_{ne} = $P_{np4}(e_4) - P_{np1}(e_1) + P_{np3}(e_3)$. Equilibrium of the free body diagram of joint n shown in Figure 7.5 requires the following:

$$\Sigma M_n = 0 = \sum_{i=1}^{nm} M_{npi} + M_{ne} = M_{np1} + M_{np2} + M_{np3} + M_{np4} + M_{ne} \tag{7.20}$$

Substituting Equations 7.7 and 7.12 and $K_{np_i} = \frac{I_{np_i}}{l_{np_i}}$ into Equation 7.20 and dividing by 2E provides

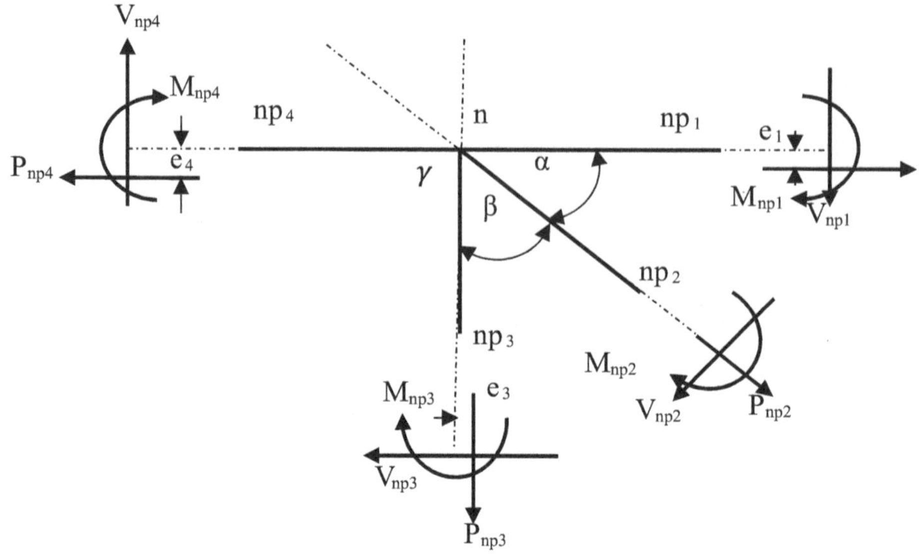

FIGURE 7.5 Free body diagram of joint with eccentric forces.

$$2\left[\left(\Sigma K_{np_i}\right)\lambda_n + \Sigma\left(\left(K_{np_i}\right)\Sigma d\theta_{np_i}\right)\right] + \sum_{i=1}^{nm}\left(K_{p_in}\lambda_{p_i} + K_{p_in}\Sigma d\theta_{p_in}\right) + \frac{M_{ne}}{2E} = 0 \tag{7.21}$$

Writing Equation 7.21[38] for all joints in the truss will provide a series of simultaneous equations which may be solved for the secondary bending deflections λ_n. Equation 7.12 is then used to calculate deflection angles λ_{np_i} and λ_{p_in} from which secondary bending stresses are calculated in each member at each joint. The secondary bending stresses can be considerable at the ends of members framing into a joint with eccentric forces. Therefore, connection eccentricities should be avoided by careful and proper detailing of the truss joints.

7.4 BENDING STRESSES IN VERTICAL WEB MEMBERS DUE TO FLOORBEAM DEFLECTION IN A TRANSVERSE FRAME

Out-of-plane bending stresses are induced in vertical web members from floorbeam end rotation in transverse frames. The transverse frames are considered either pinned/hinged or fixed at the top of the floorbeam hanger members depending on the stiffness of the lateral sway bracing at the top of the floorbeam hanger member, as shown in Figures 7.6 and 7.7.[39] If the lateral sway bracing at the top of the vertical web members is light and flexible, the top of the vertical web members are considered pinned/hinged as shown in Figure 7.6, with bending moment M_{fh} determined by slope deflection or other method of statically indeterminate frame analysis as follows:

$$M_{fh} = \frac{3Pa(L-a)I_p}{2hI_f + 3LI_p} \tag{7.22}$$

If the lateral sway bracing at the top of the vertical web members is stiff, the top of the vertical web members is considered fixed as shown in Figure 7.7, with bending moments M_{ff} and M_{tf} determined by slope deflection or other method of statically indeterminate frame analysis as follows:

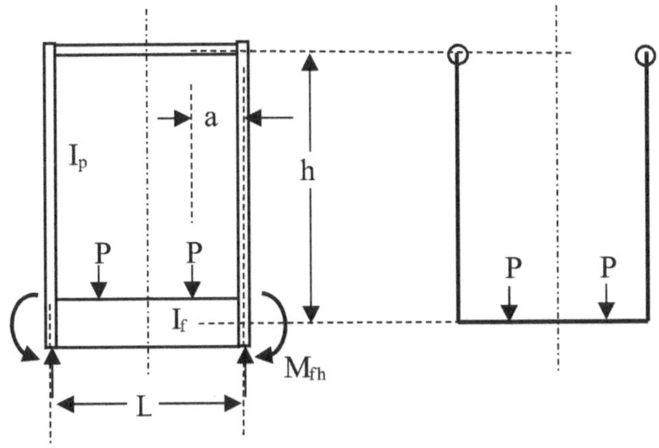

FIGURE 7.6 Transverse frame in a through-truss span with pinned/hinged ends at top of floorbeam hanger member.

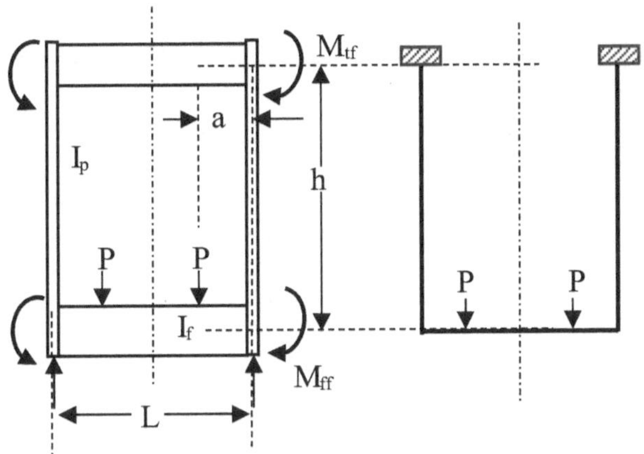

FIGURE 7.7 Transverse frame in a through-truss span with fixed ends at top of floorbeam hanger member.

$$M_{ff} = \frac{2Pa(L-a)I_p}{hI_f + 2LI_p} \tag{7.23a}$$

$$M_{tf} = \frac{Pa(L-a)I_p}{hI_f + 2LI_p} = \frac{M_{ff}}{2} \tag{7.23b}$$

Equations 7.22 and 7.23 indicate that reducing the moment of inertia of the vertical member in the plane transverse to the plane of the trusses I_p,[40] and increasing the moment of inertia of the floorbeam in the plane transverse to the plane of the trusses, I_f,[41] will reduce the bending stresses in vertical web members. Example 7.3 illustrates the use of Equations 7.22 and 7.23 to estimate the bending moments in floorbeam hanger members in a truss.

Example 7.3

Determine the secondary bending stresses due to transverse frame behavior in the floorbeam hanger member with truss dimensions h = 25 ft = 300 in, L = 20 ft = 240 in and a =

7.5 ft = 90 in and the following gross section properties: floorbeam hangers – A_p = 10 in^2; I_p = 520 in^4 and S_p = 65 in^3; floorbeams – I_f = 12,000 in^4; S_f = 500 in^3; P_{DL} = 1.5(25) = 37.5 kips; P_{LL+I} = 4.0(25) = 100 kips; P_{max} = 137.5 kips; P_{range} = 100 kips (In this case, instead of 65% of maximum impact, the designer has used 100% of maximum impact for the fatigue design of the hanger members. If 65% of maximum impact was considered, P_{range} = 92.8 kips); σ_{amax} = 137.5/10.0 = 13.75 ksi and σ_{arange} = 100/10.0 = 10.0 ksi.

Maximum bending moment and flexural stress in plane of transverse frame for pinned or hinged top of floorbeam hanger (see Figure E7.3a):

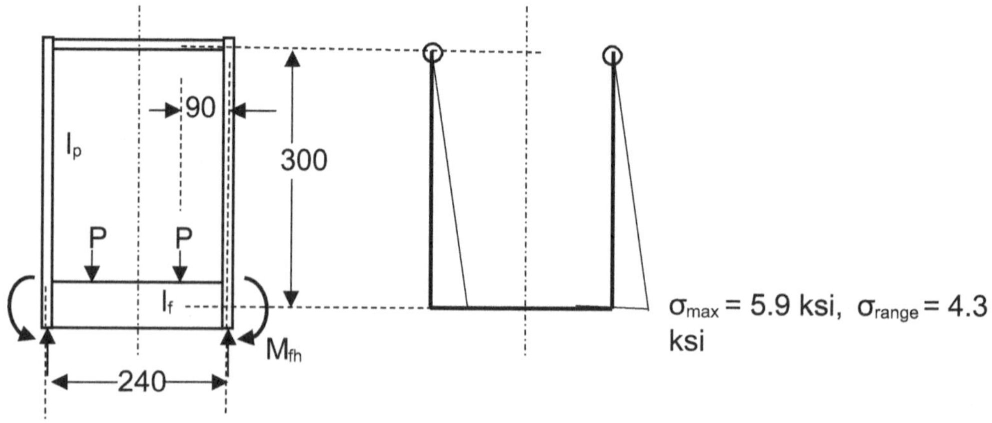

FIGURE E7.3a Bending stresses in floorbeam hanger members due to pinned/hinged transverse frame behavior.

$$M_{fh} = \frac{3Pa(L-a)I_p}{2hI_f + 3LI_p} = \frac{3(137.5)90(240-90)520}{2(300)12000 + 3(240)520} = 382.3\,\text{kip-in}$$

$$\sigma_{fh} = 382.3/65 = 5.9\text{ ksi}$$

Bending stress range in plane of transverse frame for pinned or hinged top of floorbeam hanger = (100/137.5)(5.9) = 4.3 ksi

Maximum bending moment and flexural stress in plane of transverse frame for fixed top of floorbeam hanger (see Figure E7.3b):

$$M_{ff} = \frac{2Pa(L-a)I_p}{hI_f + 2LI_p} = \frac{2(137.5)90(240-90)520}{300(12000) + 2(240)520} = 501.5\,\text{kip-in}$$

$$\sigma_{ff} = 501.5/65 = 7.7\text{ ksi}$$

$$M_{tf} = \frac{Pa(L-a)I_p}{hI_f + 2LI_p} = 250.7\,\text{kip-in}$$

$$\sigma_{tf} = 250.7/65 = 3.9\text{ ksi}$$

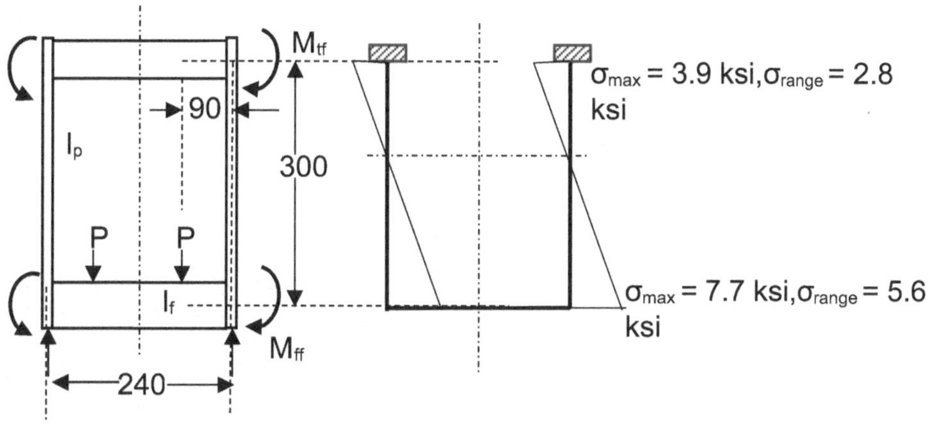

FIGURE E7.3b Bending stresses in floorbeam hanger members due to fixed transverse frame behavior.

Bending stress range at bottom of floorbeam hanger in plane of transverse frame for fixed top of floorbeam hanger = (100/137.5)(7.7) = 5.6 ksi

Bending stress range at top of floorbeam hanger in plane of transverse frame for fixed top of floorbeam hanger = 2.8 ksi

If all maximum bending stresses and bending stress ranges from Example E7.2 are considered:

Combined maximum stresses for pinned or hinged top of floorbeam hangers in transverse frame = 13.75 + 3.3 + 5.9 = 23.0 ksi
Combined stress ranges for pinned or hinged top of floorbeam hangers in transverse frame = 10.0 + 2.4 + 4.3 = 16.7 ksi
Combined maximum stresses at bottom of floorbeam hangers for fixed top of floorbeam hangers in transverse frame = 13.75 + 3.3 + 7.7 = 24.8 ksi
Combined stress ranges at bottom of floorbeam hangers for fixed top of floorbeam hangers in transverse frame = 10.0 + 2.4 + 5.6 = 18.0 ksi
Combined maximum stresses at top of floorbeam hangers for fixed top of floorbeam hangers in transverse frame = 13.75 + 3.2 + 3.9 = 20.9 ksi
Combined stress ranges at top of floorbeam hangers for fixed top of floorbeam hangers in transverse frame = 10.0 + 2.3 + 2.8 = 15.1 ksi
Maximum combined stress = 24.8 ksi at bottom of floorbeam hangers for fixed top of floorbeam hangers in transverse frame.
Maximum combined stress range = 18.0 ksi at bottom of floorbeam hangers for fixed top of floorbeam hangers in transverse frame.

If the AREMA (2023) recommendation is considered that only secondary maximum bending stresses in tension members from truss distortion and joint rigidity exceeding 4.0 ksi are to be considered primary maximum stresses:

Combined maximum stresses for pinned or hinged top of floorbeam hangers in transverse frame = 13.75 + 5.9 = 19.7 ksi

Combined maximum stresses at bottom of floorbeam hangers for fixed top of floorbeam hangers in transverse frame = 13.75 + 7.7 = 21.5 ksi
Combined maximum stresses at top of floorbeam hangers for fixed top of floorbeam hangers in transverse frame = 13.75 + 3.9 = 17.7 ksi
Maximum combined stress = 21.5 ksi at bottom of floorbeam hangers for fixed top of floorbeam hangers in transverse frame

---

Because of the superposition of the concurrent axial and flexural stress ranges in vertical web members that are directly loaded by floorbeam reactions, stress ranges can cause fatigue cracking at the top and bottom of floorbeam hanger members in through-truss spans and at the top of floorbeam hanger members in deck-truss spans. Figure 7.6, Figure 7.7 and Example 7.3 illustrate the bending stresses and stress ranges in floorbeam hanger members due to transverse frame behavior in through-trass spans. As shown in Figure 7.8, the bending stresses and stress ranges in vertical members of deck-truss spans above the floorbeam are very large when dimension h in Equation 7.22 is very small.[42]

7.5 SECONDARY BENDING STRESSES IN MEMBERS DUE TO TRANSVERSE FORCES BETWEEN TRUSS PANEL POINTS

Member fixed end moments M_{nf} can be determined for concentrated and uniformly distributed forces due to self-weight or other transverse loads. Considering the equilibrium of joint n:

$$\Sigma M_n = 0 = \sum_{i=1}^{nm} M_{npi} - M_{nf} \tag{7.24}$$

where

$M_{nf} = \sum_{i=1}^{nm} MF_{npi}$;

MF_{npi} = fixed end moment at joint n from transverse forces on member npi;
nm = the number of npi at joint n; and
M_{npi} = end moments at joint n due to joint rigidity at npi.

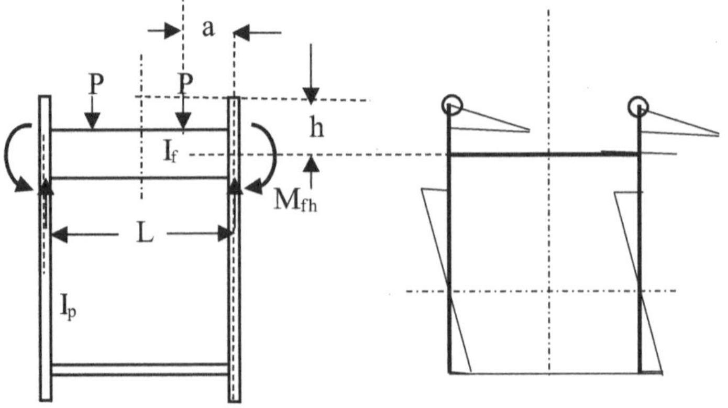

FIGURE 7.8 Transverse frame in a deck-truss span with pinned/hinged ends at top of floorbeam hanger member.

Substituting Equations 7.7 and 7.12 and $K_{np_i} = \frac{I_{np_i}}{l_{np_i}}$ into Equation 7.24 and dividing by 2E provides

$$2\left[\left(\Sigma K_{np_i}\right)\lambda_n + \Sigma\left(\left(K_{np_i}\right)\Sigma d\theta_{np_i}\right)\right] + \sum_{i=1}^{nm}\left(K_{p_in}\lambda_{p_i} + K_{p_in}\Sigma d\theta_{p_in}\right) - \frac{M_{nf}}{2E} = 0 \qquad (7.25)$$

Writing Equation 7.25[43] for all joints in the truss will provide a series of simultaneous equations that can be solved for the secondary bending deflections λ_n. Equation 7.12 is then used to calculate λ_{np_i} and λ_{p_in} from which secondary bending stresses are calculated in each member at each joint. The secondary bending stresses can be considerable at the ends of members in large trusses and with heavy appurtenances attached to the member between joints. The fixed end beam reactions also result in additional truss panel point forces.

7.6 SECONDARY BENDING STRESSES IN MEMBERS DUE TO ROTATIONALLY FIXED END SUPPORT OF THE TRUSS SPAN

Secondary bending stresses in members due to an external moment at a fixed support are calculated by replacing M_{ne} in Equation 7.21 for the joint at the fixed support with the external moment.[44]

7.7 SECONDARY BENDING STRESSES IN CHORD MEMBERS DUE TO ELONGATION OF FLOORBEAM HANGER MEMBERS

The in-plane secondary bending stresses in chord members due to the elongation of floorbeam hanger members is difficult to estimate without the aid of the appropriate software. Nevertheless, a simple model of the bottom chord as a fixed end beam provides some insight into the magnitude of the secondary bending stresses created by the elongation of floorbeam hanger members.

If the bottom chord L0L2[45] in a Pratt truss is assumed to behave as a fixed beam deflected at the center of the fixed beam (at L1) by the extension Δ_{U1L1} of floorbeam hanger member U1L1, the secondary bending stress due to the elongation of U1L1 is

$$\sigma_{L0L2} = \frac{24E\Delta_{U1L1}c}{L_{L0L2}^2} \qquad (7.26)$$

where c = distance from the neutral axis to the extreme fiber of bottom chord members (typically taken as half the depth of symmetrical cross-section bottom chord members) and L_{L0L2} = the length of the assumed fixed end beam (typically twice the panel length for symmetrical trusses).

Equation 7.26 reveals that decreasing Δ_{U1L1} by increasing the cross-sectional area is an effective means of reducing the secondary bending stress in the chord members L0L1 and L1L2 due to the elongation of member U1L1. This simplified model of bottom chord member behavior due to the extension of a floorbeam hanger member does not provide an accurate determination of the secondary bending stress but, as outlined in Example 7.4, illustrates the potential magnitude of the secondary bending stresses in chord members due to the elongation of floorbeam hanger members.

Example 7.4

Estimate the secondary bending stresses in 25 ft long and 16 in deep symmetrical bottom chord members L0L1 and L1L2 in a Pratt truss span due to the 0.25 in elongation of the floorbeam hanger member U1L1:

$$\sigma_{L0L2} = \frac{24E\Delta_{U1L1}c}{L_{L0L2}^2} = \frac{24(29000)(0.25)(8)}{(50(12))^2} = 3.9\,\text{ksi}$$

The secondary bending stresses in chord members due to the elongation of floorbeam hanger members in moderate length Pratt and Warren truss spans is typically quite small.[46] However, in subdivided Pratt (Baltimore) truss spans, the secondary bending stresses in bottom chord members due to floorbeam hanger elongation can be considerable and typically precipitate secondary bending stresses at top chord and vertical members where they are supported by no-load struts.[47]

7.8 SUMMARY OF SECONDARY BENDING IN TRUSS MEMBERS

7.8.1 Analysis of Secondary Bending Stresses in Truss Members

As outlined in Sections 7.2 to 7.7, classical slope-deflection methods are used to determine both in-plane and out-of-plane secondary bending stresses. However, secondary bending stresses are effectively analyzed by classical methods combined with modern structural analysis software. Secondary bending stresses and stress ranges due to in-plane effects (e.g., due to truss deformation considering rigid connections, eccentric connections, self-weight and/or floorbeam member extension) are readily calculated with modern 2D frame analysis software. Out-of-plane secondary bending stresses in vertical web members due to transverse frame behavior are also assessed by classical methods, 2D frame analysis software or, when applicable,[48] with modern 3D frame analysis software. AREMA (2024) acknowledges that calculating secondary stresses in truss members due to rigidity of joints, member eccentricity and transverse forces is cumbersome and typically does not appreciably affect the total (primary and secondary) stresses in members. It is also acknowledged that, where deemed necessary by the engineer, computer software enables calculating the total stresses in truss members.

7.8.2 Design for Secondary Bending Stresses in Truss Members

As indicated in Sections 7.2 to 7.7, secondary flexural stresses due to various causes are determined and considered for the design of truss members. However, secondary flexural stresses exceeding 20% to 30% of the primary axial stresses in truss spans can be precluded with certain design considerations,[49] such as using relatively narrow members to reduce in-plane secondary bending stresses. Secondary flexural stresses less than 25% of primary axial stresses are typical where the member width does not exceed 10% of member length. This geometric criterion is important in simple Pratt and Warren truss spans, but of particular significance in subdivided panel truss spans. Accordingly, AREMA (2024) recommends that secondary flexural stresses due to joint rigidity and truss deformation[50] do not require explicit consideration where member width does not exceed 10% of member length.

Connection eccentricities can be avoided where the centroidal axis of members is straight, coincident with the lines connecting joint centers and lies in the plane of the lines of action of loads and reactions. Markedly, end connections must be carefully detailed to avoid secondary bending in end post and bottom chord members. The detailing practices outlined in AREMA (2024) will preclude in-plane secondary flexural stresses due to eccentricities at joints.

Designers can also use relatively narrow vertical members and deep floorbeams. As illustrated in Section 7.4, out-of-plane secondary flexural stresses are reduced by using posts and floorbeam hangers of relatively narrow cross-section and deep floorbeams. Nevertheless, the secondary flexural stresses can easily exceed 25% if vertical member width is too great and floorbeam depth dimensions are too small. Consequently, AREMA (2024) recommends that out-of-plane secondary bending due to floorbeam end rotation be considered a primary effect for the strength and fatigue design of floorbeam hanger and subvertical members.[51]

Finally, designers can use relatively large cross-sections for hanger members in simple Pratt and Warren truss spans and in subvertical members in subdivided truss spans to reduce axial deformation of the members. Alternatively, since secondary flexural stresses due to member elongation is of greater concern in subdivided truss spans, the lengths of subvertical members can be modified to equalize displacement of chords under live load.

It is because of the inherent allowance for secondary stresses in the recommended allowable stresses and because localized yield stresses at the extreme fibers of a member will not precipitate failure that AREMA (2023) recommended considering only secondary tensile bending stresses that exceed 27.6 MPa (4.0 ksi) for members other than floorbeam hanger members and only secondary compressive bending stresses that exceed 20.7 MPa (3.0 ksi) as primary stresses. This recommendation was removed in AREMA (2024) because of the enhanced analysis available by modern software.

However, in addition to secondary flexural stresses in truss span members, truss span floor systems are also susceptible to secondary effects that cause bending. Out-of-plane bending of floorbeams can occur due to truss chord extension, which can be avoided with expansion joints for stringers and traction trusses (see Chapter 8).

NOTES

1 Gravity axis of the truss members.
2 The working point at a connection.
3 Parallel to the plane of the truss.
4 Perpendicular to the plane of the truss.
5 Connection eccentricities create in-plane secondary flexural stresses in truss members. Connection eccentricities can be avoided by proper detailing.
6 Transverse forces on truss members can create in-plane and out-of-plane bending. In-plane bending occurs due to the self-weight of members. Beam behavior for large transverse forces can be explicitly considered in design, such as for chord members in very long-span truss panels or where a superimposed uniformly distributed or concentrated weight is attached to a member. Beam behavior will result in additional fixed end bending moments and forces at panel points each side of the member.
7 The distortion of vertical members creates in-plane secondary flexural stresses in chord members. This is of particular concern in subdivided panel truss spans with secondary strut members supporting chord and vertical members.
8 The rotational restraint at the end joint will create external bending moments that precipitate secondary flexural stresses in the truss members. These stresses can be considerable for truss members at a rotationally fixed joint. Therefore, the end bearings of truss spans should be arranged to allow for free end joint rotation.
9 Elongation of chord members creates secondary lateral bending of floorbeams.
10 The maximum axial tensile stresses and stress ranges in vertical web members that are directly loaded by floorbeam reactions occur simultaneously with the maximum out-of-plane secondary tensile flexural stresses and stress ranges, suggesting that the out-of-plane secondary stresses and stress ranges are of greater concern for vertical web members directly loaded by floorbeam reactions than for vertical members that participate in truss action.
11 Bolted or welded gusset plates provide resistance to the free rotation of member ends. In-plane secondary bending results when the truss is distorted under load.
12 With members assumed straight and free to rotate (frictionless pins) at the ends.
13 Allowable axial stresses typically include an allowance of 20% to 30% for secondary flexural stresses.

14 Allowable stresses are typically reduced by about 30% to account for secondary stresses in truss members. Nevertheless, for long truss spans, it may be economical to determine the secondary flexural stresses and design the members with greater allowable stresses.
15 Beam behavior occurs due to member self-weight and/or externally applied loads between truss panel points.
16 The stresses in subvertical, substrut and subtie members of subdivided truss spans may create chord member deformations at the center that substantially exceed those at the ends of the chord members causing flexural stresses in the chord member. Vertical members in subdivided truss spans can also be susceptible to bending if supported by secondary struts at the center of the vertical member.
17 This is explicitly recommended by AREMA (2024) where trusses are to be designed and detailed such that secondary stresses from all sources are as small as is practical.
18 Because the secondary bending stresses are proportionally dependent on the primary axial stresses.
19 Because secondary flexural stresses are typically proportional to the ratio of member width to length, they can be great in trusses where the panels are subdivided, as in the Baltimore or Petit system (see Chapter 4).
20 Therefore, even if localized yield stress occurs at the extreme fibers of a member, the member will not fail. In addition, secondary bending stresses due to truss distortion and joint rigidity occur at the ends of members where gusset plates assist in resisting the flexure.
21 Moment distribution has also been used to approximate secondary bending moments in trusses with rigid joints.
22 Typically, based on the stiffness or displacement methods of structural analysis.
23 The superposition of member influence lines for maximum primary axial forces and the influence lines for maximum secondary bending moment in a member.
24 Floorbeam hanger and subvertical members are also subjected to many live load cycles in truss spans.
25 This recommendation is not present in AREMA (2024).
26 Typically, out-of-plane bending due to floorbeam end rotation from transverse frame behavior and in-plane bending due to truss distortion and joint rigidity.
27 Typically, in-plane secondary stresses due to joint rigidity and out-of-plane secondary stresses from floorbeam end rotation.
28 The recommended limit for secondary bending stress to be considered a primary stress was lower for compression members due to concern about overall member and local instability with combined secondary compressive bending and primary axial compression.
29 Because the geometric angles between members at a restrained joint cannot change.
30 Typically economically realized by reducing the width of the member in the plane transverse to the plane of the truss.
31 The number of unknown secondary bending deflection angles is 4j-6 since each member has a deflection at each end. However, with only j conditions, it is necessary to reduce the number of equations to j by referencing all the secondary bending deflection angles at each joint to a single reference angle.
32 In accordance with the methods of determining the primary axial forces in truss members.
33 Assuming small deformations and, therefore, additive changes in angles between straight members with frictionless joints.
34 Systems of linear simultaneous equations are typically solved by methods such as Gaussian or Gauss–Jordan elimination.
35 The linear simultaneous equations were solved by Gaussian elimination.
36 Long panels in trusses can also reduce the relative stiffness of members.
37 Eccentricities occur when the centroids of the member sections to not intersect at a single common location.
38 For connections without any eccentricities, $M_{ne} = 0$.
39 Alternatively, the bending moments can be calculated considering the lateral sway bracing moment of inertia in the plane of the transverse frame.
40 Typically economically realized by using narrow floorbeam hanger or subvertical member cross-sections.
41 Typically economically realized with deep floorbeam cross-sections.
42 When h = 0, Equation 7.22 yields the fixed end moment for a beam loaded as shown in Figure 7.8.
43 For members without considering transverse forces, $M_{nf} = 0$.
44 The solution is aided by the fact that the deflection angle at the joint with the fixed support is known.
45 Comprising bottom chord members L0L1 and L1L2.
46 Although typically greater for Warren truss than for Pratt truss chord members.
47 A similar effect occurs at the top chord of Warren trusses where no-load vertical members are connected.
48 3D models and analyses are typically only performed for large or complex truss spans.

49 Secondary flexural stresses less than 20% to 30% of the primary axial stresses are accommodated by the typical allowable stress allowance of 20% to 40% for secondary flexural stresses.

50 Additionally, secondary bending stresses due to truss distortion and joint rigidity occur at the ends of members where the presence of gusset plates assists in resisting the flexure.

51 For fatigue design, in-plane and out-of-plane secondary flexural stresses from all sources are considered primary flexural stresses.

REFERENCES

American Railway Engineering and Maintenance-of-Way Association (AREMA), 2023, *Manual for Railway Engineering*, Chapter 15, Steel Structures, Lanham, MD.

American Railway Engineering and Maintenance-of-Way Association (AREMA), 2024, *Manual for Railway Engineering*, Chapter 15, Steel Structures, Lanham, MD.

BIBLIOGRAPHY

Hool, G.A., and Kinne, B.S., 1923, *Stresses in Framed Structures*, McGraw-Hill, New York.

Johnson, J.B., Bryan, C.W., and Turneaure, F.E., 1916, *The Theory and Practice of Modern Framed Structures, Part III*, 9th edition, John Wiley & Sons, New York.

Johnson, J.B., Bryan, C.W., and Turneaure, F.E., 1917, *The Theory and Practice of Modern Framed Structures, Part II*, 9th edition, John Wiley & Sons, New York.

8 Lateral and Longitudinal Loads on Railway Truss Spans

8.1 INTRODUCTION

The internal forces in truss members due to external lateral and longitudinal forces must be considered when they exceed 25% of the member forces due to dead load (DL), live load (LL), impact (I) and centrifugal (CF) forces.[1] External lateral forces are from wind[2] (W_L and W_U), nosing[3] (N), and, on truss spans carrying curved track, continuous welded rail restraint (CWR_L) and centrifugal effects (CF). Lateral forces due to compression member bracing (B_R) must also be considered. External longitudinal forces due to live load (LF) are caused by locomotive traction (LF_T) and train braking (LF_B). External longitudinal forces also exist due to wind (LW_L and LW_U) and longitudinally anchored continuously welded rail (CWR).[4] In some cases, other forces (OF)[5] may also need to be anticipated. The magnitude of externally applied longitudinal and lateral forces due to LF_T, LF_B, W_U, W_L, LW_U, LW_L, N, CF, B_R, CWR_L and CWR are outlined in Chapter 3.

8.2 LATERAL FORCE ANALYSIS OF TRUSS SPANS

External lateral forces from wind on loaded truss spans (W_{LT} and W_{LB}),[6] wind on unloaded spans (W_{UT} and W_{UB}),[7] wind on the live load (W_{LL}),[8] N, CF and internal notional lateral forces due to compression member bracing (B_{RL} and B_{RU})[9] are transmitted to members of the main vertical trusses and to horizontal lateral and vertical lateral bracing systems. Horizontal lateral bracing systems are typically trusses in the planes of the loaded and unloaded chord members of the main trusses.[10] Vertical lateral bracing systems are typically braced frames between the main trusses (cross-braced, portal or sway frames).

The web members of horizontal lateral bracing trusses in the plane of the main truss loaded chord members are characteristically non-slender double horizontal diagonal members between panel points and strut members at panel points.[11] The horizontal diagonal members in the lateral bracing system trusses in the plane of the main truss unloaded truss chord members are typically relatively long and may be considered slender.[12] Slender members will be inactive in compression, and therefore, single horizontal diagonal members between panel points may be considered in tension regardless of lateral force direction. However, if the diagonal members of horizontal lateral bracing systems in the plane of the main truss chord members are non-slender,[13] both members between the panel points are considered active, and it is sufficiently accurate to assume that each diagonal at a section in the panel resists 50% of the panel shear force, with one diagonal member in tension and the other in compression.[14]

Vertical lateral bracing systems are typically in the plane of the main truss end and intermediate post members. Conventional cross-braced frames are used for the vertical lateral bracing systems in the plane of end members and intermediate vertical members of deck-truss spans. The end members and cross-bracing of deck-truss spans supported at the top (loaded) chord will be inclined, and those supported at the bottom (unloaded) chord will be vertical. Portal (end) and sway (intermediate) frames are used for the vertical lateral bracing systems in through-truss spans.[15] Portal bracing is typically in the plane of the inclined end post members of through-truss spans. The sway bracing is typically in the plane of the intermediate vertical post members of through-truss spans.

The bottom horizontal lateral bracing system reactions in through-truss spans are transferred to the supports in the plane of the main truss bottom chord members. The horizontal lateral bracing system

DOI: 10.1201/9781003469971-9

reactions in through-truss spans in the plane of the main truss top chord members are resisted at the top of the portal bracing and transferred to the supports by axial and flexural forces in the main truss end post members. The horizontal component of the axial forces in the inclined end post members creates tension or compression forces in the main truss bottom chord members of through-truss spans.

Sway bracing, as deep as the vertical clearance allows, is typically used in the plane of the main truss intermediate post members of through-truss spans with the top horizontal lateral bracing system struts as part of the sway bracing system. Since the lateral deflections of the top and bottom horizontal lateral bracing systems will be similar,[16] there is only a small magnitude force transferred between the top and bottom systems and, therefore, the sway bracing of through-truss spans.

The transfer of horizontal lateral bracing system forces in deck-truss spans depends on the location of the supports. The top lateral system forces in deck-truss spans supported in the plane of the main truss bottom chord are transferred through the cross-bracing to the supports. The bottom lateral system forces in deck-truss spans supported in the plane of the main truss bottom chord members are transferred to the supports. If the deck truss is supported in the plane of the main truss top chord members, the top horizontal lateral bracing system forces are transferred to the supports and the bottom horizontal lateral bracing system forces are transferred through the cross-bracing to the supports in the plane of the main truss top chord.

8.3 LONGITUDINAL FORCE ANALYSIS OF TRUSS SPANS

External longitudinal forces from LW_{LT} and LW_{LB},[17] LW_{UT} and LW_{UB},[18] LW_{LL}, live load traction (LF_T), live load braking (LF_B) and restraint of CWR are transmitted to members of the main trusses. Internal forces in the main truss loaded chord members due to the external longitudinal forces from LF_T and LF_B are typically significant. The internal forces in the main truss unloaded chord members and web members due to external longitudinal forces from LF_T and LF_B are typically trivial. External longitudinal forces from CWR restraint are also characteristically negligible.

8.4 LATERAL AND LONGITUDINAL FORCE ANALYSIS OF THROUGH-TRUSS SPANS

Through-truss spans that are loaded must consider external lateral forces from W_{LT}, W_{LB}, W_{LL}, N, CF, CWR_L and B_{RL} and external longitudinal forces from LW_{LT}, LW_{LB}, LW_{LL}, CWR, LF_T and LF_B. Through-truss spans that are unloaded must consider external lateral forces from W_{UT}, W_{UB}, B_{RU} and CWR_L and external longitudinal forces from CWR, LW_{UT} and LW_{UB}.

8.4.1 Lateral Force Analysis of Loaded Through-Truss Spans

As shown in Figure 8.1, W_L is the lateral wind pressure on the structure which is distributed to the panel points of the horizontal lateral bracing systems in the planes of the top (unloaded) and bottom (loaded) main truss chord members. AREMA (2024) recommends that W_L be established as 1.44 kPa (30 psf) on a loaded span.[19]

W_{LT} is the lateral wind force on the structure due to W_L at panel points in the plane of the main truss top chord members creating internal axial forces in the top horizontal lateral bracing system. The horizontal reactions at the end of the top horizontal lateral bracing system (at the top of the portal frame) have an overturning effect about axes in the plane of the bottom horizontal lateral bracing system members, creating uniform axial tension or compression in the bottom chord members of the main truss spans. Tension in the bottom chord members of the leeward main truss can be significant in design. AREMA (2024) recommends that W_{LT} be not less than 2.19 kN/m (150 lb/ft).

W_{LB} is the lateral wind force on the structure due to W_L at panel points in the plane of the main truss bottom chord members, creating only internal axial forces in the bottom horizontal lateral

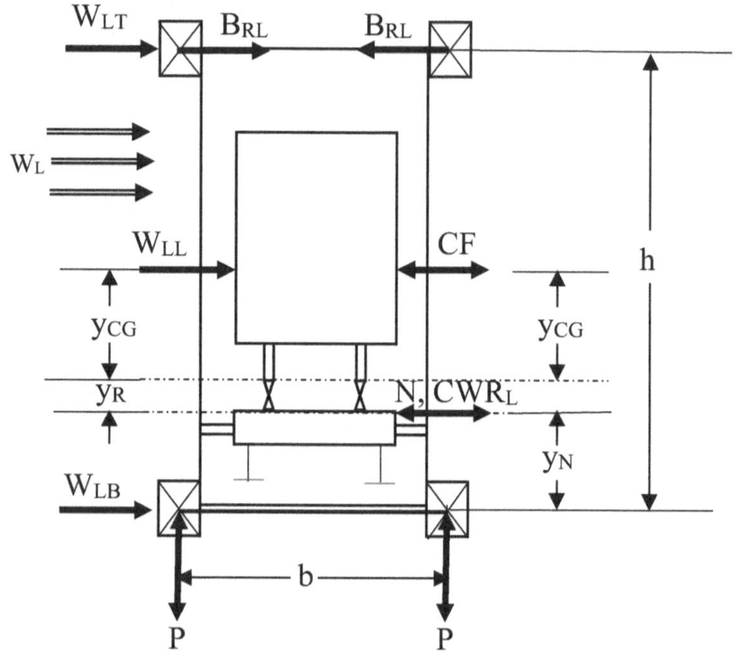

FIGURE 8.1 Lateral forces at the panel points of loaded through-truss spans.

bracing system because the plane of the bottom lateral horizontal bracing system is coincident with the plane of the main truss supports. AREMA (2024) recommends that W_{LB} be not less than 2.92 kN/n (200 lb/ft).

W_{LL} is the lateral wind force of 4.38 kN/m (300 lb/ft) on the live load due to W_L applied at $y_{CG} = 2.45$ m (8.0 ft) above the top of the rail. This force is resolved to create horizontal forces in the bottom horizontal lateral bracing system and vertical force couples, P_{WLL}, due to overturning through distance $y_{CG} + y_R + y_N$ at the panel points of the main truss bottom chord members. Therefore, vertical force $P_{WLL} = +/- W_{LL}$ (d)($y_{CG} + y_R + y_N$)/b, where d = the panel length, creates internal axial forces in the main truss members. Tension in the main truss bottom chord members of the leeward truss can be significant in design.

CF is the lateral centrifugal force due to track curvature applied at $y_{CG} = 2.45$ m (8.0 ft) above the top of the rail on the live load and resolved to create horizontal forces in the bottom horizontal lateral bracing system and vertical force couple P_{CF} due to overturning through distance $y_{CG} + y_R + y_N$ at the panel points of main truss bottom chord members. P_{CF} creates internal axial forces in the main truss members. Tension in the main truss bottom chord members of the leeward truss can be significant in design.

B_{RL} is the notional transverse shear force in a panel due to bracing of the main truss top chords, which is equal to 2.5% of the total axial force due to DL + LL + I in both main truss top chord members in the panel and creating internal axial forces in the top horizontal lateral bracing system.

CWR_L is the lateral force applied at the base of rail from CWR restraint on the curved track on the span.

Finally, N is the lateral force from equipment (due to horizontal sway of live load or axle group hunting at the interface between wheel and rail) applied as a single force of 25% of the maximum design axle load at the base of rail. This force is resolved to create internal axial forces in the horizontal lateral bracing system in the plane of the main truss bottom chord members and vertical force couple P_N due to overturning through distance y_N at the panel points of main truss bottom chord members. P_N creates internal axial forces in the main truss members. Tension in the bottom chord members of the leeward truss may be significant in design.

8.4.1.1 Lateral Force Analysis of Loaded Through-Truss Spans for W_{LT} (Lateral Wind Force on the Structure in the Plane of the Top Chord Members)

W_{LT} creates internal axial tension and compression forces in the horizontal lateral bracing system in the plane of the main truss top chord members.[20] W_{LT} also creates tension in the leeward main truss bottom chord members and compression in the windward main truss bottom chord members due to the transfer of the top horizontal lateral system end reactions, $P_T = \frac{W_{LT}(n-2)}{2}$, where there are n panels in the through-truss span. The magnitude of the tension and compression forces in the main truss bottom chord members depends on whether the base of the main truss end post members is considered pinned or fixed.[21]

If the portal frame posts (main truss end post members) are fixed at the base as shown in Figure 8.2,

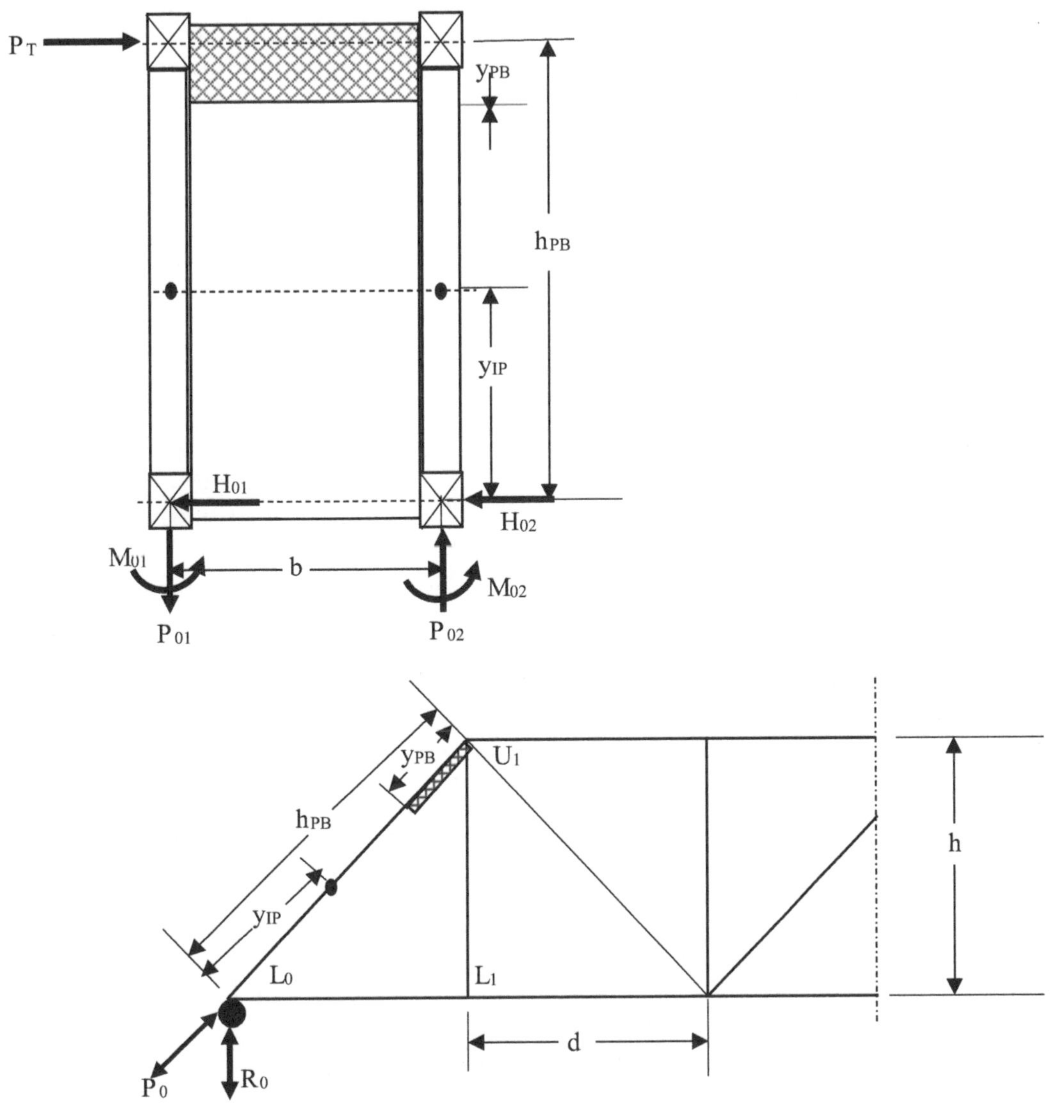

FIGURE 8.2 Overturning forces from wind on top chord of through-truss spans.

$$H_{01} = H_{02} = \frac{P_T}{2} \tag{8.1}$$

If the portal frame posts are fixed at the base and moments are taken about the assumed inflection points[22] in the plane of the end posts, the force in end post members below the inflection point is

$$P_{01} = P_{02} = \frac{P_T\left(h_{PB} - y_{IP}\right)}{b} \tag{8.2}$$

If the portal frame posts are fixed at the base and moments are taken about the base of the posts in the plane of the end posts,

$$M_{01} = M_{02} = \frac{P_T}{2} y_{IP} = \frac{P_T}{2}\left(\frac{h_{PB} - y_{PB}}{2}\right) = \frac{P_T}{4}\left(h_{PB} - y_{PB}\right) \tag{8.3}$$

If moments are taken about the longitudinal axis of the main truss bottom chord,

$$R_{01} = \frac{P_T h - 2M_{02}\left(\dfrac{h}{\sqrt{h^2 + d^2}}\right)}{b} = \frac{P_T}{2b}\left(2h - \left(h_{PB} - y_{PB}\right)\left(\frac{h}{\sqrt{h^2 + d^2}}\right)\right) \tag{8.4a}$$

$$R_{02} = \frac{P_T h - 2M_{01}\left(\dfrac{h}{\sqrt{h^2 + d^2}}\right)}{b} = \frac{P_T}{2b}\left(2h - \left(h_{PB} - y_{PB}\right)\left(\frac{h}{\sqrt{h^2 + d^2}}\right)\right) \tag{8.4b}$$

$$F_{LOL1} = \mp\frac{P_T}{2b}\left(2h - \left(h_{PB} - y_{PB}\right)\left(\frac{h}{\sqrt{h^2 + d^2}}\right)\right)\left(\frac{d}{h}\right) = \mp\frac{P_T d}{b}\left(1 - \frac{h_{PB} - y_{PB}}{2\sqrt{h^2 + d^2}}\right) \tag{8.5}$$

$$F_{LOU1} = \mp\frac{P_T}{b}\left(\sqrt{h^2 + d^2} - \frac{h_{PB} - y_{PB}}{2}\right) \tag{8.6}$$

If the portal frame posts are pinned at the base, $y_{IP} = 0$, the force in bottom chord and end post members is

$$P_{01} = P_{02} = \frac{P_T\left(h_{PB}\right)}{b} \tag{8.7}$$

$$M_{01} = M_{02} = 0 \tag{8.8}$$

$$R_{01} = R_{02} = \frac{P_T h}{b} \tag{8.9}$$

$$F_{LOL1} = \mp\frac{P_T d}{b} \tag{8.10}$$

$$F_{LOU1} = \mp\frac{P_T}{b}\left(\sqrt{h^2 + d^2}\right) \tag{8.11}$$

Example 8.1

A 150 ft long six-panel Pratt through-truss with a height of 35 ft between the centerline of the top and bottom chord members and floor system with a depth of 2.5 ft above the top of the bottom chord is subjected to an external lateral wind force on the structure of 30 psf. The trusses are 20 ft apart as shown in Figure E8.1. Determine the internal forces in the horizontal lateral truss bracing members in the plane of the top chord members due to W_{LT}. The diagonal members in the top horizontal lateral bracing system are considered slender. See Table E8.1a for the lateral wind forces.

TABLE E8.1a
Lateral Wind Forces at Joints of a Pratt Through-Truss Span

Truss Segment	Projected Area (ft^2)	Lateral Wind Force on End Joints (kips)	Lateral Wind Force on Interior Joints (kips)
Windward top chord	403	1.51	3.02
Windward bottom chord	478	1.20	2.39
Windward floor system	375	0.94	1.88
Leeward top chord	403	1.51	3.02
Leeward bottom chord	331	0.83	1.66

The lateral wind force at the plane of the top chord = 3.02(2)(1000)/25 = 241.6 lb/ft, which exceeds the AREMA (2024) recommended minimum of 150 lb/ft.

Maximum chord member forces (all joints loaded) are as follows:

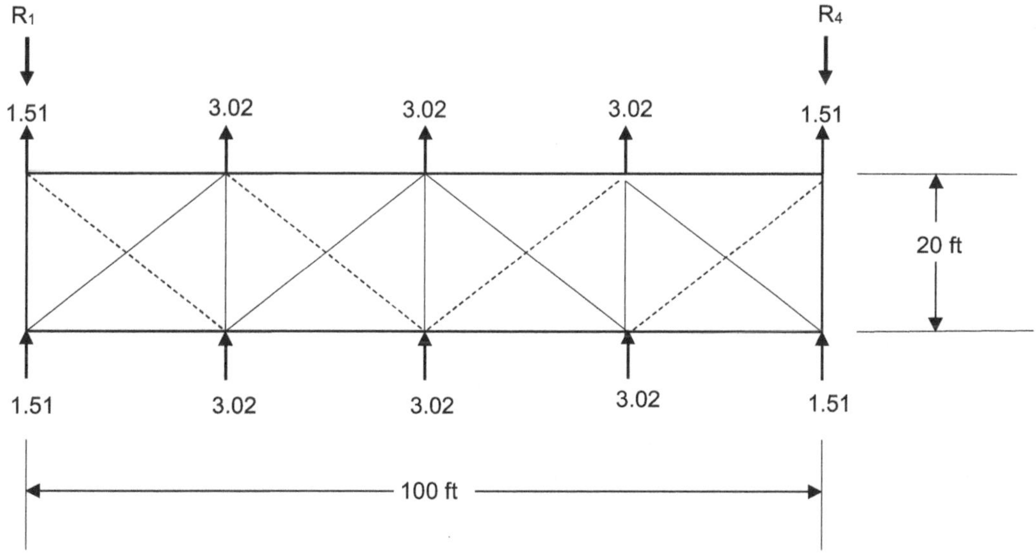

FIGURE E8.1 Plan of top horizontal lateral bracing system in through-truss spans.

$R_1 = R_4 = (1.51(4) + 3.02(6))/2 = 12.08$ kips

Force in windward member U1U2 = $F^W_{U1U2} = (2(1.51)(25) - 12.08(25))/20 = -11.33$ kips (C)

Force in windward member U2U3 = $F^W_{U2U3} = (3.02(2)(25) + 1.51(2)(50) - 12.08(50))/20 = -15.10$ kips (C)

Force in leeward member U1U2 = $F^L_{U1U2} = 0$
Force in leeward member U2U3 = $F^L_{U2U3} = (-1.51(2)(25)+12.08(25))/20 = 11.33$ kips (T)

Maximum web member forces considering wind as moving load with joints loaded for maximum panel shear by the approximate method (Table E8.1b):[23]

Shear in panel 1-2 = 3.02(2)(25)(3 + 2 + 1)/100 = 12.08 − 3.02 = 9.06 kips

$$\text{Force in diagonal member in panel 1-2} = 9.06\left(\frac{\sqrt{20^2+25^2}}{20}\right) = 14.50\,\text{kips}$$

Shear in panel 2-3 = 3.02(2)(25)(2 + 1)/100 = 4.53 kips (if wind not considered as moving load, shear in panel 2-3 = 12.08 − 2(1.51 + 3.02) = 3.02 kips)

$$\text{Force in diagonal member in panel 2-3} = 4.53\left(\frac{\sqrt{20^2+25^2}}{20}\right) = 7.25\,\text{kips}$$

TABLE E8.1b
Member Forces due to Lateral Wind in a Top Horizontal Lateral Bracing System

Member of Top Horizontal Lateral Bracing System	Internal Force due to W_{LT} (kips) Windward Truss	Leeward Truss
U1U2 & U4U5	−11.33	0
U2U3 & U3U4	−15.10	+11.33
Top bracing diagonal in panels 1-2 and 4-5	14.50	
Top bracing diagonal in panels 2-3 and 3-4	7.25	
Top bracing strut at panel points 1 & 5	Portal bracing top member (see Example 8.8 or 8.9)	
Top bracing strut at panel points 2 & 4	Sway bracing top member (see Example 8.10)	
Top bracing strut at panel point 3	Sway bracing top member (see Example 8.10)	

The end strut at panel points U1 and U5 is the upper member of the portal frame between the windward and leeward main trusses. Example 8.8 illustrates the analysis of this member as a component of a cross-braced portal frame (member AB in the cross-braced portal frame). Example 8.9 illustrates the analysis of this member as a component of a knee-braced portal frame (member ABCD in the knee-braced portal frame). The interior struts at panel points U2, U3 and U4 are the upper members of the sway frames between the windward and leeward main trusses. Example 8.10 illustrates the analysis of this member as a component of a sway frame (member ABC).

Example 8.2

A six-panel 150 ft long Pratt through-truss with h = 35 ft between the centerline of the top and bottom chord members and floor system with a depth of 2.5 ft above the top of the bottom chord is subjected to an external lateral wind force on the structure of 30 psf. The trusses are 20 ft apart. The portal bracing is 8 ft deep in the plane of the end post members. Determine the internal forces in the main truss end post and bottom chord members due to W_{LT} (see Table E8.2).

From Example 8.1 (analysis of top chord horizontal lateral truss bracing for W_{LT}),

P_T = 12.08 kips
h = 35 ft

$$h_{PB} = \sqrt{25^2 + 35^2} = 43.0\,\text{ft}$$

y_{PB} = 8 ft

If end post members are considered fixed at the base, Equations 8.5 and 8.6 provide

$$F_{LOL1} = \mp\frac{12.08(25)}{20}\left(1 - \frac{43.0 - 8.0}{2(43.0)}\right) = \mp 15.10(0.59) = \mp 8.95\,\text{kips}$$

(The force in the bottom chord members is uniform along the length of the bottom chord members.)

$$F_{LOU1} = \mp\frac{12.08}{20}\left(43.0 - \frac{43.0 - 8.0}{2}\right) = \mp 0.60(25.50) = \mp 15.40\,\text{kips}$$

If end post members are considered pinned at the base, Equations 8.10 and 8.11 provide

$$F_{LOL1} = \mp\frac{12.08(25)}{20} = \mp 15.10\,\text{kips}$$

(The force in the bottom chord members is uniform along the length of the bottom chord members.)

$$F_{LOU1} = \mp\frac{12.08}{20}(43.0) = \mp 25.97\,\text{kips}$$

TABLE E8.2
Member Forces due to Lateral Wind in Main Truss Members

	Uniform Axial Force due to W_{LT} (kips)	
Main Truss Members	**End Posts Fixed at Base**	**End Posts Pinned at Base**
Leeward bottom chord members	8.95	15.10
Windward bottom chord members	−8.95	−15.10
Leeward end post members L0U1 & U5L6	−15.40	−25.97
Windward end post members L0U1 & U5L6	15.40	25.97

The forces in the bottom chord and end post members are 69% greater if the base of the end posts are considered pinned.

8.4.1.2 Lateral Force Analysis of Loaded Through-Truss Spans for W_{LB} (Lateral Wind Forces on the Structure in the Plane of the Bottom Chord Members)

The truss is supported in the plane of the bottom chord members. Therefore, the lateral wind force on the structure at panel points in the plane of the bottom chord members creates only internal axial forces in the horizontal lateral bracing system in the plane of the bottom chord members.[24]

Example 8.3

Determine the internal forces in the chord members and web members of the bottom horizontal lateral bracing system in the truss of Example 8.1 due to W_{LB} applied at the bottom chord of the main trusses. The trusses are 20 ft apart as shown in Figure E8.3. The diagonal members are connected to the stringers and are, therefore, considered to be non-slender. Tables E8.3a and E8.3b provide the lateral wind forces.

TABLE E8.3a
Lateral Wind Forces at Joints of a Pratt Through-Truss Span

Truss Segment	Projected Area (ft^2)	Lateral Wind Force on End Joints (kips)	Lateral Wind Force on Interior Joints (kips)
Windward top chord	403	1.51	3.02
Windward bottom chord	478	1.20	2.39
Windward floor system	375	0.94	1.88
Leeward top chord	403	1.51	3.02
Leeward bottom chord	331	0.83	1.66

TABLE E8.3b
Lateral Wind Forces in Plane of Bottom Chord of a Pratt Through-Truss Span

	Lateral Wind Force on Structure in Plane of Bottom Chord Members, W_{LB}	
Truss Segment	Wind Force on Structure End Joints (kips)	Wind Force on Structure Interior Joints (kips)
Windward bottom chord	1.20 + 0.94 = 2.14	2.39 + 1.88 = 4.27
Leeward bottom chord	0.83	1.66

The lateral wind force at the plane of the bottom chord = (2.39 + 1.88 + 1.66)(1000)/25 = 237.2 lb/ft which exceeds the AREMA (2024) recommended minimum of 200 lb/ft.

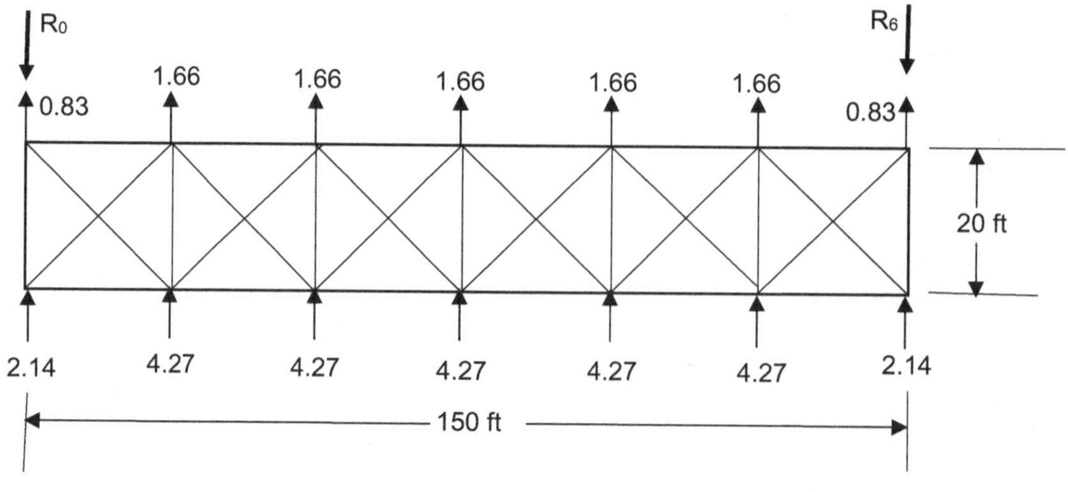

FIGURE E8.3 Plan of bottom horizontal lateral bracing system in through-truss spans.

The maximum chord member forces (all joints loaded) are

$R_0 = R_6 = (0.83(2) + 1.66(5) + 2.14(2) + 4.27(5))/2 = 17.80$ kips

Since the diagonal members are assumed to each resist 50% of the shear at a section, the forces in the diagonal members will be equal in magnitude but of opposite attitude, and forces in strut members will be nought. Therefore, the moments from forces in diagonal members will be null at the center of each chord member.

Force in windward chord member in panel 0-1 = (2.14(12.5) + 0.83(12.5) − 17.80(12.5))/20 = −9.27 kips

Force in leeward chord member in panel 0-1 = 9.27 kips

Force in windward chord member in panel 1-2 = (2.97(37.5) + 5.93(12.5) − 17.80(37.5))/20 = −24.10 kips

Force in leeward chord member in panel 1-2 = 24.10 kips

Force in windward chord member in panel 2-3 = (2.97(62.5) + 5.93(37.5) + 5.93(12.5) − 17.80(62.5))/20 = −31.52 kips

Force in leeward chord member in panel 2-3 = 31.52 kips

For maximum web member forces (joints loaded for maximum panel shear with approximate method),

Shear in panel 0-1 =+/−(5.93(25)(5 + 4 + 3 + 2 + 1)/150 = +/−14.83 kips

Force in diagonal member in panel 0-1 =

$$\mp 0.50(14.83)\left(\frac{\sqrt{20^2+25^2}}{20}\right) = \mp 0.50(14.83)(1.60) = \mp 11.86\,\text{kips}$$

Shear in panel 1-2 =+/−(5.93(25)(4 + 3 + 2 + 1)/150 = +/−9.88 kips

Force in diagonal member in panel 1-2 = +/−0.5(9.88)(1.60) = +/−7.91 kips

Shear in panel 2-3 = =+/−(5.93(25)(3 + 2 + 1)/150 = +/−5.93 kips

Force in diagonal member in panel 2-3 = +/−0.5(5.93)(1.60) = +/−4.74 kips

The struts at the panel points will have compressive forces (Table E8.3c); however, since the struts are the floorbeams of the main truss spans, they do not require further consideration.

TABLE E8.3c

Member Forces due to Wind in Plane of Bottom Chord of a Pratt Through-Truss Span

Member	Member Force due to W_{LB} (kips)
Windward truss L0L1 & L5L6	−9.27
Leeward truss L0L1 & L5L6	9.27
Windward truss L1L2 & L4L5	−24.10
Leeward truss L1L2 & L4L5	24.10
Windward truss L2L3 & L3L4	−31.52
Leeward truss L2L3 & L3L4	31.52
Bottom bracing diagonals in panels 0-1 and 5-6	+/−11.86
Bottom bracing diagonals in panels 1-2 and 4-5	+/−7.91
Bottom bracing diagonals in panels 2-3 and 3-4	+/−4.74

8.4.1.3 Lateral Force Analysis of Loaded Through-Truss Spans for W_{LL} (Lateral Wind Force on the Live Load)

W_{LL} applied at y_{CG} = 2.45 m (8.0 ft) above the top of the rail on the live load resolves to create internal axial forces in the horizontal lateral bracing system in the plane of the bottom chord members and internal axial forces in the main truss members due to overturning forces.

Example 8.4

A 150 ft long Pratt through-truss with h = 35 ft between the centerline of the top and bottom chord members and floor system with a depth of 2.5 ft above the top of the bottom chord is subjected to an external lateral wind force on the live load of 300 lb/ft at 8 ft above the top of the rail. The trusses are 20 ft apart. Determine the internal forces in the horizontal lateral bracing members in the plane of the main truss bottom chord members due to W_{LL}.

$R_0 = R_6$ = +/-0.30(150/2) = +/-22.50 kips

End panel point forces = 0.30(25)/2 = 3.75 kips

Interior panel point forces = 0.30(25) = 7.50 kips

From Example 8.3 with the ratio of 7.50/(4.27 + 1.66) = 1.26, the resulting member forces are shown in Table E8.4.

TABLE E8.4
Member Forces in Plane of Bottom Chord due to Wind on Live Load

Member	Member Force due to W_{LL} (kips)
Windward truss L0L1 & L5L6	−11.72
Leeward truss L0L1 & L5L6	11.72
Windward truss L1L2 & L4L5	−30.48
Leeward truss L1L2 & L4L5	30.48
Windward truss L2L3 & L3L4	−39.87
Leeward truss L3L3 & L3L4	39.87
Bottom bracing diagonals in panels 0-1 & 5-6	+/−15.00
Bottom bracing diagonals in panels 1-2 & 4-5	+/−10.00
Bottom bracing diagonals in panels 2-3 & 3-4	+/−6.00

Example 8.5

A 150 ft long Pratt through-truss with h = 35 ft between the centerline of the top and bottom chord members and floor system with a depth of 2.5 ft above the top of the bottom chord is subjected to an external lateral wind force on the live load of 300 lb/ft at 8 ft above the top of the rail. The trusses are 20 ft apart. Determine the internal forces in the main truss members due to W_{LL}:

End panel forces = 0.30(25)/2 = 3.75 kips

Interior panel forces = 0.30(25) = 7.50 kips

End panel point P_{WLL} = 3.75(8 + 4)/20 = +/−2.25 kips

Interior panel point P_{WLL} = 7.50(8 + 4)/20 = +/−4.50 kips

From Example 4.6 (maximum truss member forces for uniformly distributed live load of 5.5 kips/ft) with the ratio of 4.50/(5.5(25))) = 0.033 applied to the member forces calculated for the uniformly distributed live load, the member forces are given in Table E8.5.

TABLE E8.5
Member Forces due to Wind on Live Load

Member	Member Force due to W_{LL} (kips)	
	Leeward Truss	Windward Truss
L0L1 & L5L6	8.04	−8.04
L1L2 & L4L5	8.04	−8.04
L2L3 & L3L4	12.86	−12.86
L0U1 & U5L6	−13.82	13.82
U1U2 & U4U5	−12.86	12.86
U2U3 & U3U4	−14.46	14.46
U1L1 & U5L5	4.50	−4.50
U2L2 & U4L4	−4.05/1.80	4.05/−1.80
U3L3	0	0
U1L2 & L4U5	8.85/−0.55	−8.85/0.55
U2L3 & L3U4	4.98/−2.21	−4.98/2.21

8.4.1.4 Lateral Force Analysis of Loaded Through Truss Spans for CF (Centrifugal Force due to Curved Track)

CF applied at y_{CG} = 2.45 m (8.0 ft) above the top of the rail resolves to create internal axial forces in the horizontal lateral bracing system in the plane of the bottom chord members and internal axial forces in the main truss members due to overturning forces from centrifugal effects. For trusses with curved track, the analysis for lateral centrifugal force effects may be carried out in the same manner as for wind on the live load.

8.4.1.5 Lateral Force Analysis of Loaded Through Truss Spans for B_{RL} (Notional Force due to Compression Chord Member Bracing)

The notional transverse shear force, B_{RL}, in a horizontal lateral truss panel in the plane of the main truss compression (top) chord members, due to bracing of the compression chord members, creates internal axial forces in the horizontal lateral bracing system web members. AREMA (2024) recommends notional transverse shear force of 2.5% of the total axial compressive force in both members of a panel. This is a conservative approach intended to provide adequate strength and stiffness of horizontal bracing web members without need of a more onerous analytical approach.

Example 8.6

A 150 ft long Pratt through-truss with h = 35 ft between the centerline of the top and bottom chord members and floor system with a depth of 2.5 ft above the top of the bottom chord is subjected to B_{RL} from compression chord bracing under Cooper's E80 live load. The trusses are 20 ft apart. Determine the internal forces in the web members of the horizontal lateral bracing system in the plane of the top (compression) chords. From Example 4.11 (analysis of vertical truss members for Cooper's E80 live load; also see Table E8.6a):

Impact Percentage = 16+(600/(150-30)) +(100/20) = 26%

TABLE E8.6a
Compression Chord Bracing Forces

Through Pratt Truss Member	Dead Load Force (kips)	Maximum Live Load Force (kips)	Impact Load (kips)	Total Force (kips)	Compression Chord Member Bracing Force, B_{RL} (kips)
U1U2 & U4U5	−107.14	−354.96	−92.29	−554.39	−2(2.5/100)(554.39) = −27.72
U2U3 & U3U4	−120.54	−402.52	−104.66	−627.72	−2(2.5/100)(627.72) = −31.39

B_{RL} is the shear in the panels of the lateral bracing system in the plane of the top chord (see Table E8.6b).

TABLE E8.6b
Member Forces due to Compression Chord Bracing

Top Bracing System Web Members	Internal Force in Members due to B_{RL} (kips)
Diagonals in panels 1-2 & 4-5	27.72(32.02/20) = 44.37
Diagonals in panels 2-3 & 3-4	31.39(32.02/20) = 50.25
Top bracing strut at U1 & U5	−27.72/2 = −13.86
Top bracing strut at U2 & U4	(−27.72 − 31.39)/2 = −29.56
Top bracing strut at U3	−31.39

8.4.1.6 Lateral Force Analysis of Loaded Through-Truss Spans for N (Lateral Nosing Force from the Live Load)

The lateral force from equipment applied at the base-of-rail resolves to create internal axial forces in the horizontal lateral bracing system in the plane of the bottom chord members and internal axial forces in the main truss members due to overturning forces from equipment.

Example 8.7

A 150 ft long Pratt through-truss with h = 35 ft between the centerline of the top and bottom chord members and floor system with a depth of 2.5 ft above the top of the bottom chord is subjected to an external nosing force, N, from Cooper's E80 live load. The trusses are 20 ft apart. Determine the internal forces in the bottom horizontal lateral system and main truss members.

N = 80/4 = 20 kips
Interior panel P_N = +/−20(3.5)/20 = +/−3.50 kips
End panel P_N = +/−3.50/2 = +/−1.75 kips

From Example 8.4 (analysis for W_{LL}) with the ratio of 3.50/7.50 = 0.467, the member forces are given in Table E8.7a.

TABLE E8.7a
Lateral Bracing Member Forces due to Nosing Force

Lateral Bracing Member	Member Force due to N (kips)
Windward truss L0L1 & L5L6	−5.47
Leeward truss L0L1 & L5L6	5.47
Windward truss L1L2 & L4L5	−14.22
Leeward truss L1L2 & L4L5	14.22
Windward truss L2L3 & L3L4	−18.61
Leeward truss L3L3 & L3L4	18.61
Bottom bracing diagonals in panels 0-1 and 5-6	+/−7.00
Bottom bracing diagonals in panels 1-2 and 4-5	+/−4.67
Bottom bracing diagonals in panels 2-3 and 3-4	+/−2.80

From Example 8.5 (analysis for W_{LL}) with the ratio of 3.50/7.50 = 0.467, Table E8.7b gives the main member forces.

TABLE E8.7b
Main Truss Member Forces due to N

Main Truss Member	Member Force due to N (kips)	
	Leeward Truss	Windward Truss
L0L1 & L5L6	3.75	−3.75
L1L2 & L4L5	3.75	−3.75
L2L3 & L3L4	6.00	−6.00
L0U1 & U5L6	−6.45	6.45
U1U2 & U4U5	−6.00	6.00
U2U3 & U3U4	−6.75	6.75
U1L1 & U5L5	2.10	−2.10
U2L2 & U4L4	−1.05	1.05
U3L3	0	0
U1L2 & L4U5	3.87	−3.87
U2L3 & L3U4	1.29	−1.29

8.4.1.7 Lateral Force Analysis of Loaded Through Truss Spans for CWR_L (Lateral Force from Restraint of CWR on Curved Track)

Lateral forces from longitudinally restrained continuous welded rail are typically precluded by modern rail anchorage methods and, therefore, usually neglected in truss design.

8.4.1.8 Lateral Force Analysis of Vertical Lateral Bracing Systems in Through-Truss Spans

A supplemental structural system is required to transfer lateral forces from horizontal lateral bracing systems that are not located in the plane of the span supports to the support bearings. In through-spans, the lateral forces are transferred through a system of knee braces,[25] intermediate sway frames and end portal frames.

8.4.1.8.1 Portal Frame Bracing Analysis

Through-truss end portal frames must be designed to transfer the total wind force reaction, P_T, at the top horizontal lateral truss system to the supports via the end post members.[26] To accomplish this force

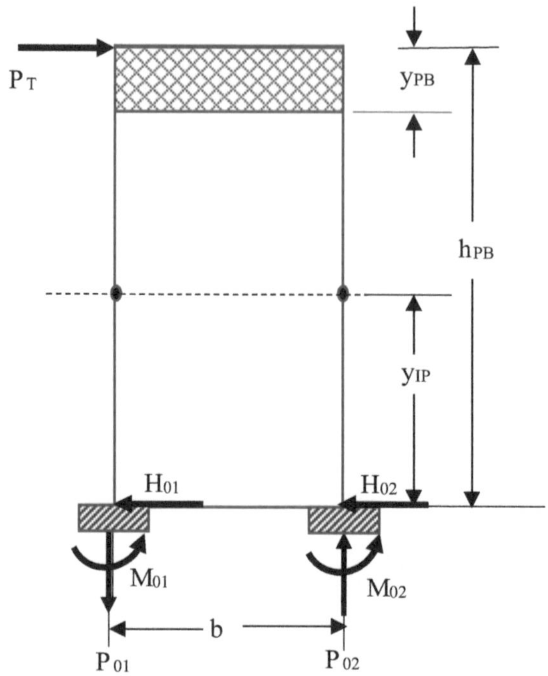

FIGURE 8.3 End portal forces from wind forces in the plane of the top chord members of through truss spans.

transfer, the end post members resist flexural and axial forces from portal frame action. To estimate the end portal frame effects with end post members considered fixed[27] at the base, it is assumed that inflection points are located midway between the bottom of the end post member and the bottom of the portal bracing frame at a distance, $y_{IP} = (h_{PB}\text{-}y_{PB})/2$ from the bottom of the end post member, as shown in Figure 8.3.

If the total wind force reaction, P_T, is resolved at the inflection points,

$$H_{01} = H_{02} = \frac{P_T}{2} \tag{8.12a}$$

$$P_{01} = P_{02} = \frac{P_T\left(h_{PB} - y_{IP}\right)}{b} = \frac{P_T\left(h_{PB} + y_{PB}\right)}{2b} \tag{8.12b}$$

$$M_{01} = M_{02} = \frac{P_T}{2} y_{IP} = \frac{P_T}{2}\left(\frac{h_{PB} - y_{PB}}{2}\right) = \frac{P_T}{4}\left(h_{PB} - y_{PB}\right) \tag{8.12c}$$

End portal bracing member internal forces can then be determined from free body equilibrium equations for one end post member of the portal frame as shown in Figures 8.4 and 8.5. Cross-bracing (Figure 8.4) and knee bracing (Figure 8.5) are the most common configurations of portal bracing used in modern through-truss construction.

8.4.1.8.1.1 Cross-braced Portal Frame Analysis A cross-braced portal frame can be analyzed by considering that the cross-bracing is slender such that the diagonal members are active only in tension.[28] The diagonal and horizontal members are considered pinned at the ends and, therefore, resist only axial forces. The posts are frame members that resist axial, bending and shear forces (see Figure 8.4).

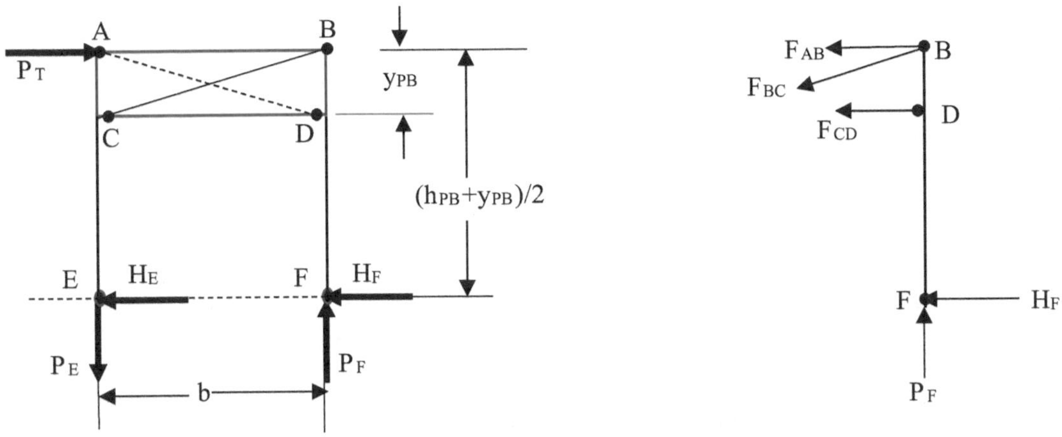

FIGURE 8.4 Cross-braced end portal forces from wind on top chord of a through-truss span.

With P_T resolved at the inflection points, as shown in Figure 8.4,

$$H_E = H_F = H_{01} = H_{02} = \frac{P_T}{2} \tag{8.13a}$$

$$P_E = P_F = P_{01} = P_{02} = P_T\left(\frac{h_{PB} + y_{PB}}{2b}\right) \tag{8.13b}$$

Vertical equilibrium of the free body diagram of post BDF indicates

$$F_{BC} = P_T\left(\frac{h_{PB} + y_{PB}}{2b}\right)\left(\frac{\sqrt{b^2 + y_{PB}{}^2}}{y_{PB}}\right) \tag{8.13c}$$

Moments about C of the free body diagram of post BDF indicate

$$F_{AB} = -P_T\left(\frac{h_{PB} + 3y_{PB}}{4y_{PB}}\right) \tag{8.13d}$$

Moments about B of the free body diagram of post BDF indicate

$$F_{CD} = -P_T\left(\frac{h_{PB} + y_{PB}}{4y_{PB}}\right) \tag{8.13e}$$

Post flexure is a maximum at C and D:

$$M_C = M_D = P_T\left(\frac{h_{PB} - y_{PB}}{4}\right) \tag{8.13f}$$

Flexure at base of posts G and H (Joint G is the windward post fixed base and Joint H is the leeward post fixed base):

$$M_G = M_H = -P_T\left(\frac{h_{PB} - y_{PB}}{4}\right) \tag{8.13g}$$

At pinned joints A, B, E and F:

$$M_A = M_B = M_E = M_F = 0 \quad (8.13h)$$

Axial force in posts CE and DF from equilibrium of isolated joints E and F:

$$F_{CE} = F_{EG} = P_E = P_T\left(\frac{h_{PB} + y_{PB}}{2b}\right) \quad (8.13i)$$

$$F_{DF} = F_{BD} = F_{FH} = -P_F = -P_T\left(\frac{h_{PB} + y_{PB}}{2b}\right) \quad (8.13j)$$

Axial force in post AC from equilibrium of isolated joint C:

$$F_{AC} = 0 \quad (8.13k)$$

Shear force in posts CE and DF above joints E and F:

$$V_{CE} = V_{DF} = V_{FH} = V_{EG} = H_E = H_F = \frac{P_T}{2} \quad (8.13l)$$

Shear force in the windward post AC below joint A:

$$V_{AC} = P_T + F_{AB} = \frac{P_T}{4}\left(1 - \frac{h_{PB}}{y_{PB}}\right) \quad (8.13m)$$

Shear force in the leeward post BD above joint D:

$$V_{BD} = \frac{P_T}{2} + F_{CD} = \frac{P_T}{4}\left(1 - \frac{h_{PB}}{y_{PB}}\right) \quad (8.13n)$$

Example 8.8

A 150 ft long Pratt through-truss with h = 35 ft between the centerline of the top and bottom chord members and floor system with a depth of 2.5 ft above the top of the bottom chord is subjected to an external lateral wind force on the structure of 30 psf. The trusses are 20 ft apart. The portal cross-bracing is 8 ft deep in the plane of the end post members. Determine the internal forces in the cross-braced portal frame members. (See Tables E8.8a and E8.8b.)

From Example 8.1 (analysis of top chord horizontal lateral truss bracing for W_{LT}),

$P_T = 12.08$ kips
$h = 35$ ft
$b = 20$ ft

$h_{PB} = \sqrt{25^2 + 35^2} = 43.0\,\text{ft}$

$y_{PB} = 8$ ft

From Equations 8.13 and Figure 8.4,

$$H_E = H_F = \frac{12.08}{2} = 6.04\,\text{kips}$$

$$P_E = P_F = 12.08\left(\frac{43.0 + 8.0}{2(20)}\right) = 15.40\,\text{kips}$$

$$F_{BC} = F_{AD} = 12.08\left(\frac{43.0 + 8}{2(20)}\right)\left(\frac{\sqrt{20^2 + 8^2}}{8}\right) = 41.47\,\text{kips}$$

$$F_{AB} = -12.08\left(\frac{43.0 + 3(8)}{4(8)}\right) = -25.29\ \text{kips}$$

$$F_{CD} = -12.08\left(\frac{43.0 + 8}{4(8)}\right) = -19.25\ \text{kips}$$

$$M_C = M_D = -M_G = -M_H = 12.08\left(\frac{43.0 - 8}{4}\right) = 105.70\ \text{kips} - \text{ft}$$

$$M_A = M_B = M_E = M_F = 0$$

$$F_{CE} = F_{GE} = P_E = 12.08\left(\frac{43.0 + 8}{2(20)}\right) = 15.40\ \text{kips}$$

$$F_{DF} = F_{BD} = F_{FH} = -P_F = -12.08\left(\frac{43.0 + 8}{2(20)}\right) = -15.40\ \text{kips}$$

$$F_{AC} = 0$$

$$V_{CE} = V_{DF} = V_{EG} = V_{FH} = H_E = H_F = \frac{12.08}{2} = 6.04\ \text{kips}$$

$$V_{AC} = V_{BD} = \frac{12.08}{4}\left(1 - \frac{43.0}{8}\right) = -13.21\,\text{kips}$$

TABLE E8.8a
Forces in Cross-Braced Portal Bracing Members due to Lateral Wind Forces in Plane of Top Chord

Member	Maximum Force in Member (kips)	
	Axial	Shear
End posts		
L0U1 & U5L6 leeward below cross-bracing	−15.40	6.04
L0U1 & U5L6 leeward above cross-bracing	−15.40	−13.21
L0U1 & U5L6.windward below cross-bracing	15.40	6.04
L0U1 & U5L6 windward above cross-bracing	0	−13.21
Cross-bracing		
Top lateral member at U1 & U5	−25.29	
Bottom lateral member	−19.25	
Diagonal members	41.47	

TABLE E8.8b
Moments in Cross-Braced Portal End Post Members due to Lateral Wind Forces in Plane of Top Chord

	Joint							
	A	B	C	D	E	F	G	H
End post L0U1 & U5L6 moment (kip-ft)	0	0	105.70	105.70	0	0	−105.70	−105.70

8.4.1.8.1.2 Knee-braced Portal Frame Analysis A knee-braced portal frame can be analyzed by considering that the knee-bracing is considered as pinned at the ends and, therefore, resist only axial forces. The posts and top strut are frame members that resist axial, bending and shear forces (see Figure 8.5).

$$H_H = H_G = H_{01} = H_{02} = \frac{P_T}{2} \tag{8.14a}$$

$$P_H = P_G = P_{01} = P_{02} = P_T\left(\frac{h_{PB} + y_{PB}}{2b}\right) \tag{8.14b}$$

Joints A, D, G and H are pinned:

$$M_A = M_D = M_G = M_H = 0 \tag{8.14c}$$

Taking moments about pinned joint A of the free body diagram of post AEG and about pinned joint D of the free body diagram of post DFH yields

$$F_{BE} = -F_{CF} = \frac{P_T\left(h_{PB} + y_{PB}\right)\sqrt{a^2 + y_{PB}^2}}{4ay_{PB}} \tag{8.14d}$$

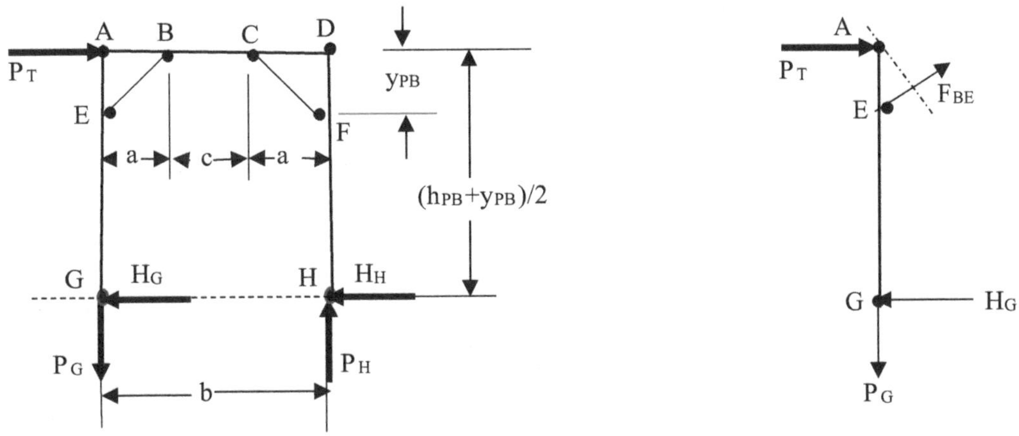

FIGURE 8.5 Knee-braced end portal forces from wind on top chord of a through-truss span.

Post flexure is a maximum at E, F, I and J:

$$M_E = M_F = -M_J = -M_I = P_T\left(\frac{h_{PB} - y_{PB}}{4}\right) \tag{8.14e}$$

Joints I and J are at the bases of windward and leeward posts, respectively.

Axial force in posts EG and FH from equilibrium of isolated joints G and H:

$$F_{EG} = F_{GI} = P_E = P_T\left(\frac{h_{PB} + y_{PB}}{2b}\right) \tag{8.14f}$$

$$F_{FH} = F_{HJ} = -P_F = -P_T\left(\frac{h_{PB} + y_{PB}}{2b}\right) \tag{8.14g}$$

Axial forces in posts DF and AE from vertical equilibrium of the free body diagram of posts DFH and AEG yield:

$$F_{DF} = -F_{AE} = P_T\left(\frac{h_{PB} + y_{PB}}{2b}\right)\left(\frac{b}{2a} - 1\right) \tag{8.14h}$$

Shear forces in the leeward post FH and windward post EG above joints H and G are

$$V_{FH} = V_{EG} = V_{GI} = V_{HJ} = H_G = H_H = \frac{P_T}{2} \tag{8.14i}$$

Shear force in the leeward post DF above joint F is

$$V_{DF} = \frac{P_T}{2} + F_{CF}\left(\frac{a}{\sqrt{a^2 + y_{PB}^2}}\right) = -\frac{P_T}{2}\left(\frac{h_{PB} + y_{PB}}{2y_{PB}} - 1\right) \tag{8.14j}$$

Shear force in the windward post AE above joint E is

$$-F_{BE}\left(\frac{a}{\sqrt{a^2+y_{PB}^2}}\right)=-\frac{P_T}{2}\left(\frac{h_{PB}+y_{PB}}{2y_{PB}}-1\right) \tag{8.14k}$$

Taking a section through isolated member ABCD between joints B and C and taking moments at joints B and C provides

$$M_B=-M_C=-F_{AE}a=P_T\left(\frac{h_{PB}+y_{PB}}{2b}\right)\left(\frac{b}{2}-a\right)=P_T\left(\frac{h_{PB}+y_{PB}}{4b}\right)c \tag{8.14l}$$

Horizontal equilibrium at joint D yields

$$F_{CD}=-V_{DF}=\frac{P_T}{2}\left(\frac{h_{PB}+y_{PB}}{2y_{PB}}-1\right) \tag{8.14m}$$

Horizontal equilibrium at joint A yields

$$F_{AB}=V_{AE}-P_T=\frac{-P_T}{2}\left(\frac{h_{PB}+y_{PB}}{2y_{PB}}-1\right)-P_T=-\frac{P_T}{2}\left(\frac{h_{PB}+y_{PB}}{2y_{PB}}+1\right) \tag{8.14n}$$

Horizontal equilibrium at joint C yields

$$F_{BC}=F_{CD}+F_{CF}\left(\frac{a}{\sqrt{a^2+y_{PB}^2}}\right)=\frac{P_T}{2}\left(\frac{h_{PB}+y_{PB}}{2y_{PB}}-1\right)+\frac{-P_T\left(h_{PB}+y_{PB}\right)}{4y_{PB}}=-\frac{P_T}{2} \tag{8.14o}$$

Vertical equilibrium of isolated member ABCD provides

$$V_{AB}=V_{CD}=-P_T\left(\frac{h_{PB}+y_{PB}}{2b}\right)\left(\frac{b}{2a}-1\right) \tag{8.14p}$$

$$V_{BC}=V_{AB}+F_{BE}\left(\frac{y_{PB}}{\sqrt{a^2+y_{PB}^2}}\right)=P_T\left(\frac{h_{PB}+y_{PB}}{2b}\right) \tag{8.14q}$$

Example 8.9

A 150 ft long Pratt through-truss with a height of 35 ft between the centerline of the top and bottom chord members and floor system with a depth of 2.5 ft above the top of the bottom chord is subjected to an external lateral wind force on the structure of 30 psf. The trusses are 20 ft apart. The portal knee-bracing is 8 ft deep in the plane of the end post members. Determine the internal forces in the knee-braced portal frame members. (See Tables E8.9a and E8.9b.)

From Example 8.1 (analysis of top chord horizontal lateral truss bracing for W_{LT}),

P_T = 12.08 kips
h = 35 ft
b = 20 ft

$a = 8$ ft
$c = 20 - 2(8) = 4$ ft
$h_{PR} = \sqrt{25^2 + 35^2} = 43.0$ ft
$y_{PB} = 8$ ft

Joints at A, D, G and H are pinned:

$$M_A = M_D = M_G = M_H = 0$$

From Equations 8.14 and Figure 8.5,

$$F_{BE} = -F_{CF} = \frac{12.08(51.0)\sqrt{8^2 + 8^2}}{4(8)(8)} = 27.23 \text{ kips}$$

$$M_E = M_F = -M_I = -M_J = 12.08\left(\frac{35.0}{4}\right) = 105.70 \text{ kips} - \text{ft}$$

Joints I and J are at the bases of windward and leeward posts, respectively:

$$F_{EG} = F_{GI} = P_E = 12.08\left(\frac{51.0}{2(20)}\right) = 15.40 \text{ kips}$$

$$F_{FH} = F_{HJ} = -P_F = -15.40 \text{ kips}$$

$$F_{DF} = -F_{AE} = 12.08\left(\frac{51.0}{2(20)}\right)\left(\frac{20}{2(8)} - 1\right) = 3.85 \text{ kips}$$

$$V_{FH} = V_{EG} = V_{GI} = V_{HJ} = \frac{12.08}{2} = 6.04 \text{ kips}$$

$$V_{AE} = V_{DF} = -\frac{12.08}{2}\left(\frac{51.0}{2(8)} - 1\right) = -13.21 \text{ kips}$$

$$M_B = -M_C = 12.08\left(\frac{51.0}{4(20)}\right)(20 - 2(8)) = 30.80 \text{ kips} - \text{ft}$$

$$F_{CD} = \frac{12.08}{2}\left(\frac{51.0}{2(8)} - 1\right) = 13.21 \text{ kips}$$

$$F_{AB} = -\frac{12.08}{2}\left(\frac{51.0}{2(8)} + 1\right) = -25.29 \text{ kips}$$

$$F_{BC} = -\frac{12.08}{2} = -6.04 \text{ kips}$$

$$V_{AB} = V_{CD} = -12.08\left(\frac{51.0}{2(20)}\right)\left(\frac{20}{2(8)} - 1\right) = -15.40(0.25) = -3.85 \text{ kips}$$

$$V_{BC} = 12.08\left(\frac{51.0}{2(20)}\right) = 15.40 \text{ kips}$$

TABLE E8.9a
Forces in Knee-Braced Portal Bracing Members due to Lateral Wind Forces in Plane of Top Chord

	Maximum Force in Member (kips)	
Member	**Axial**	**Shear**
End posts		
L0U1 & U5L6 leeward below knee-bracing	−15.40	6.04
L0U1 & U5L6 leeward above knee-bracing	3.85	−13.21
L0U1 & U5L6 windward below knee-bracing	+15.40	6.04
L0U1 & U5L6 windward above knee-bracing	−3.85	−13.21
Horizontal top struts		
Top lateral AB at U1 & U5	−25.29	−3.85
Top lateral BC at U1 & U5	−6.04	15.40
Top lateral CD at U1 & U5	13.21	−3.85
Portal bracing members		
Knee braces	+/− 27.23	

TABLE E8.9b
Moments in Knee-Braced Portal End Post Members due to Lateral Wind Forces in Plane of Top Chord

	Joint									
	A	**B**	**C**	**D**	**E**	**F**	**G**	**H**	**I**	**J**
Portal Frame Bending Moments (kip-ft)	0	30.80	−30.80	0	105.70	105.70	0	0	−105.70	−105.70

8.4.1.8.2 Sway Bracing Analysis

The magnitude of the force transferred between the top and bottom horizontal lateral bracing systems depends on the relative lateral deflections of the bracing systems. These deflections of the top and bottom horizontal lateral bracing systems are contingent upon whether the truss span is loaded or unloaded and on the relative stiffness of the lateral bracing systems. Therefore, since it is difficult to estimate the magnitude of the force transferred between the top and bottom horizontal lateral bracing systems and consequently resisted by the vertical sway bracing, structural engineers often design the sway bracing for a lateral force of about 50% of the lateral wind force at the panel points of the top horizontal lateral bracing system. This is intended to develop an adequately stiff vertical sway bracing system that permits the main truss truss top chord length for compression design as the panel length. An adequately stiff sway bracing system typically requires members that satisfy only maximum allowable slenderness criteria.

Example 8.10

A 150 ft long Pratt through-truss with h = 35 ft between the centerline of the top and bottom chord members and floor system with a depth of 2.5 ft above the top of the bottom chord is subjected to an external lateral wind force on the structure of 30 psf. The trusses are 20 ft apart. Determine the internal forces in the vertical sway bracing frame members of Figure E8.10.

From Example 8.1 (analysis of top chord horizontal lateral truss bracing for W_{LT}),

$P_T = 6.04$ kips (50% of the external force due to wind applied at the portal bracing frame)
$b = 20$ ft
$hs = 35$ ft
$ys = 7.5$ ft
$y_{IP} = (h_S - y_S)/2 = (35\text{-}7.5)/2 = 13.75$ ft

$$H_{01} = H_{02} = \frac{P_T}{2} = 3.02 \text{ kips}$$

$$P_{01} = P_{02} = \frac{P_T(h_S + y_S)}{2b} = \frac{6.04(35+7.5)}{2(20)} = 6.42 \text{ kips}$$ (41.6% of the external force due to wind applied at the portal bracing frame)

$$M_{01} = M_{02} = \frac{P_T}{2} y_{IP} = 3.02(13.75) = 41.53 \text{ kips-ft}$$ (39.2% of the external moment due to wind applied at the portal bracing frame)

$$H_F = H_G = \frac{P_T}{2} = 3.02 \text{ kips}$$

$$P_F = P_G = P_T\left(\frac{h_S + y_S}{2b}\right) = 6.04\left(\frac{35+7.5}{2(20)}\right) = 6.42 \text{ kips}$$

$$F_{BE} = -F_{BD} = -P_T\left(\frac{h_S + y_S}{2b}\right)\left(\frac{\sqrt{\frac{b^2}{4} + y_S^2}}{y_S}\right) = -6.42\left(\frac{12.5}{7.5}\right) = -10.70 \text{ kips}$$

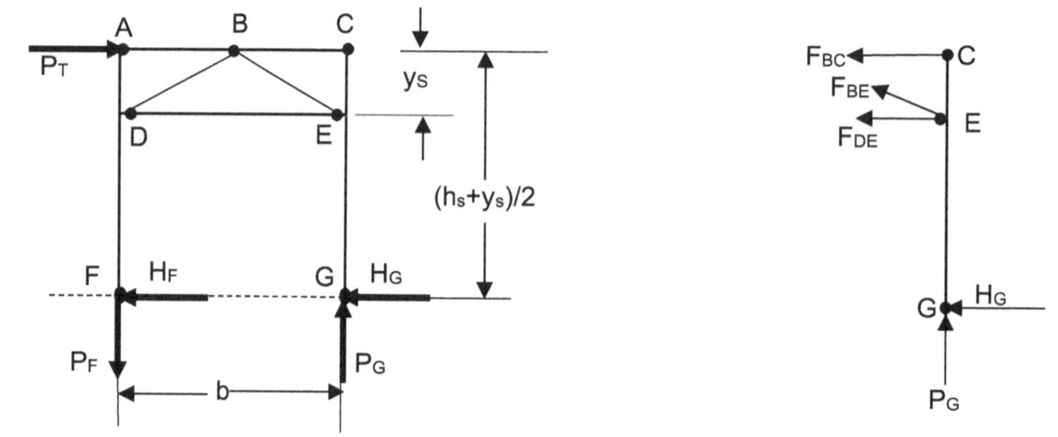

FIGURE E8.10 Sway bracing forces from wind in the plane of the top chord members of through-truss spans.

$$F_{DE} = \frac{P_G b}{2y_S} - H_G\left(\frac{h_S + y_S}{2y_S}\right) = P_T\left(\frac{h_S + y_S}{4y_S}\right) - \frac{P_T}{2}\left(\frac{h_S + y_S}{2y_S}\right) = 0$$

$$F_{BC} = \frac{P_T}{2}\left(\frac{h_S + y_S}{2y_S}\right) = 3.02\left(\frac{35 + 7.5}{2(7.5)}\right) = 8.56 \text{ kips}$$

$$F_{AB} = -\frac{P_T}{4}\left(3 + \frac{h_S}{y_S}\right) = -1.51\left(3 + \frac{35}{7.5}\right) = -11.58 \text{ kips}$$

The sway bracing forces are relatively small. Therefore, it is likely that members that satisfy maximum slenderness criteria will suffice to provide vertical lateral stability at truss intermediate panel points.

8.4.1.9 Pony Truss Knee Bracing Analysis

As shown in Figure 8.6, knee bracing connected to the top chord[29] and vertical post members is used to provide lateral stability to pony truss spans.[30] The external force at the top of knee braces is determined from W_{LT} and B_{RL}.[31] In accordance with AREMA (2024), W_{LT} is determined from the tributary area of the truss projection, and B_{RL} is 2.5% of the compression chord force.

The transverse spring constant K_{eq} provided by Figure 8.6 is (Galambos, 1988):[32]

$$K_{eq} = E/(h^2(h/(3I_v + 3I_D(h/l_D)^3) + (b/2I_{fb}))) \tag{8.15a}$$

If the floorbeams are of substantially greater flexural stiffness than the vertical members and knee braces[33] and, conservatively, the contribution of the diagonal members is neglected,

$$K_{eq} = \frac{3EI_V}{h^3} \tag{8.15b}$$

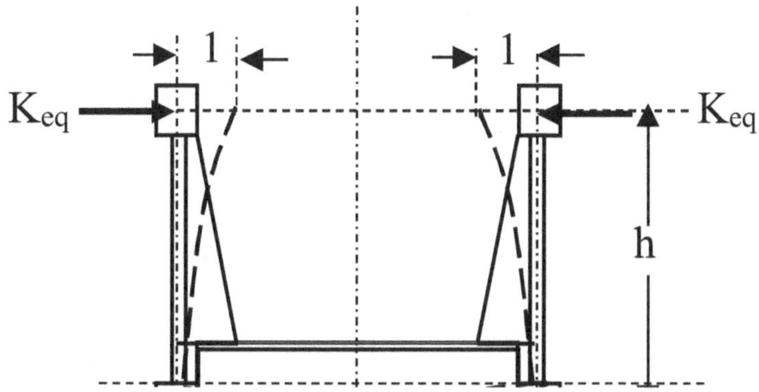

FIGURE 8.6 Transverse flexure of a pony truss cross-section.

where

I_V = the moment of inertia of the vertical member and knee brace;
I_{fb} = the moment of inertia of the floorbeam;
I_D = the moment of inertia of the diagonal member assumed fixed at the bottom chord member;
h = the distance between the centroid of the top and bottom chord members;
b = the distance between the centroid of the vertical web members; and
l_D = length of the diagonal web member.

The tranverse frame stiffness, K_{eq}, must be greater than the limiting transverse frame stiffness, K_c, that achieves buckling force P_c in chord members of length d and flexural stiffness EI_C. If K_c at each equally spaced panel point is considered equivalent and the half-wave shape of the buckled top chord is at least 1.8 times d, Engesser's solution (Galambos, 1988; Bleich, 1952) is[34]

$$K_c = \frac{\pi^2 P_C}{4K^2 d} = 1.46\frac{P_C}{d} = 1.46\frac{(FS)F}{d} \tag{8.16}$$

where

FS = 1.95 = the factor of safety applied to the axial design force, F, to obtain the critical P_C;
K = the effective length factor = 1.3 for the half-wave shape of the buckled top chord as at least 1.8 times d; and

---

Example 8.11

A pony truss span has the following characteristics:

Panel length = d = 25 ft = 300 in
Height of truss = l = 10 ft = 120 in
Cross-sectional area of top chord members = A_g = 61.5 in^2
Minimum radius of gyration of top chord members = r_{min} = 6.5 in
F_y = 50 ksi
Axial compressive force due to DL + LL + I in the top chord between vertical members = P_a = 1200 kips

$$K_c = 1.46\frac{1.95(1200)}{300} = 11.39 \text{ kips / in}$$

Try vertical member W12 × 35:

$A = 10.3\ in^2$
$I_x = 285\ in^4$
$r_x = 5.25\ in$
Assuming a floorbeam with much greater flexural rigidity than that the vertical member/knee brace,
$K_{eq} = 3(29{,}000)(285)/(120)^3 = 14.35$ kip/in ≥ 11.39 kips/in, OK.

8.4.1.10 Lateral Force Analysis for Through Trusses with Inclined Chords

If the lateral bracing system is in the plane of inclined main truss chord members,[35] it is typically sufficient to straighten (or flatten) the lateral bracing system in a horizontal plane and analyze the lateral forces at panel points from the inclined-chord member truss span.[36] The straightened lateral bracing system will be longer than the actual lateral bracing system in the inclined-chord member truss span because of the longer panels.[37]

8.4.1.11 Internal Horizontal Bracing Member Forces in Through Trusses due to External Lateral Forces

Example 8.12 summarizes the internal forces in the top and bottom horizontal lateral bracing systems in the through truss spans of Examples 8.1, 8.3, 8.4, 8.6, 8.7, 8.9 and 8.10 due to externally applied lateral forces. The internal forces in the top horizontal lateral bracing system are due to the external lateral forces from W_{LT} and B_{RL} as shown in Examples 8.1, 8.6, 8.9 and 8.10. The internal forces in the bottom horizontal lateral bracing system are due to the external lateral forces from W_{LB}, W_{LL} and N as shown in Examples 8.3, 8.4 and 8.7, respectively.

Example 8.12

A 150 ft long Pratt through-truss with h = 35 ft between the centerline of the top and bottom chord members and floor system with a depth of 2.5 ft above the top of the bottom chord carries tangent track and is subjected to external lateral forces from wind and live load. The trusses are 20 ft apart and have knee-braced end portals. Determine the internal forces in the top and bottom horizontal lateral bracing system web members from external lateral forces due to wind (Examples 8.1, 8.3, 8.4, 8.9 and 8.10) and live load (Examples 8.6 and 8.7).

From Examples 8.1, 8.6, 8.9 and 8.10, the calculated results are given in Table E8.12a.

TABLE E8.12a
Forces in Top Horizontal Bracing System due to Compression Member Bracing and Lateral Wind in the Plane of the Top Chord

Top Horizontal Lateral Bracing Web Member	Force in Top Horizontal Lateral Bracing Web Member of Leeward and Windward Trusses (kips)	
	Due to B_{RL}	Due to W_{LT}
Top bracing diagonals in panels 1-2 and 4-5	44.37	14.50
Top bracing diagonals in panels 2-3 and 3-4	50.25	7.25
Top bracing strut at U1 and U5	−13.86	−25.29/13.21
Top bracing strut at U2 and U4	−29.56	8.56/−11.58
Top bracing strut at U3	−31.39	8.56/−11.58

Forces in the top horizontal lateral bracing system diagonal web members are in tension due to the assumed tension only behavior in diagonal web members in the horizontal truss. Table E8.12b lists the calculated forces from Examples 8.3, 8.4 and 8.7.

TABLE E8.12b
Forces in Bottom Horizontal Bracing System due to Lateral Wind in Plane of the Bottom Chord, Wind on Live Load and Nosing Forces

	Force in Bottom Horizontal Lateral Bracing Web Members of Leeward and Windward Trusses (kips)			
Bottom Horizontal Lateral Bracing Web Member	W_{LB}	W_{LL}	N	Total
Bottom bracing diagonals in panels 0-1 and 5-6	+/−11.86	+/−15.00	+/−7.00	+/−33.86
Bottom bracing diagonals in panels 1-2 and 4-5	+/−7.91	+/−10.00	+/−4.67	+/−22.58
Bottom bracing diagonals in panels 2-3 and 3-4	+/−4.74	+/−6.00	+/−2.80	+/−13.54

Forces in the bottom horizontal lateral bracing system diagonal web members are in tension and compression due to the assumed sharing of panel shear in diagonal web members in the horizontal truss.

8.4.1.12 Internal Member Forces in the Main Through Truss Members due to External Lateral Forces

Example 8.13 summarizes the internal forces in the main truss members of Examples 8.1 through 8.5 and 8.7 due to externally applied lateral forces. The internal forces in the main truss members are due to the external lateral forces from W_{LT}, W_{LB}, W_{LL} and N.

Example 8.13

A 150 ft long Pratt through-truss with h = 35 ft between the centerline of the top and bottom chord members and floor system with a depth of 2.5 ft above the top of the bottom chord carries tangent track and is subjected to external lateral forces from wind and live load. The trusses are 20 ft apart and end post members are considered fixed at the base. Determine the internal forces in the main truss members due to external lateral forces from wind and equipment. Tables 8.13a and b give the calculated results from the examples.

TABLE E8.13a
Forces in Truss Members due to Wind and Nosing Forces

	Force in Leeward Main Truss Members (kips)							
Member	W_{LT} direct	W_{LT} overturning	W_{LB} direct	W_{LL} direct	W_{LL} overturning	N direct	N overturning	Total
Example	8.1	8.2, 8.9, 8.10	8.3	8.4	8.5	8.7	8.7	8.13
U1U2 & U4U5	0	0	0	0	−12.86	0	−6.00	−18.86
U2U3 & U3U4	11.33	0	0	0	−14.46	0	−6.75	−9.88
L0L1 & L5L6	0	8.95	9.27	11.72	8.04	5.47	3.75	47.20
L1L2 & L4L5	0	8.95	24.10	30.48	8.04	14.22	3.75	89.54
L2L3 & L3L4	0	8.95	31.52	39.87	12.86	18.61	6.00	117.81
L0U1 & U5L6	0	−15.40	0	0	−13.82	0	−6.45	−35.67
U1L1 & U5L5	0	−6.42	0	0	4.50	0	2.10	0.18

TABLE E8.13a (*Continued*)
Forces in Truss Members due to Wind and Nosing Forces

	Force in Leeward Main Truss Members (kips)							
Member	W_{LT} direct	W_{LT} overturning	W_{LB} direct	W_{LL} direct	W_{LL} overturning	N direct	N overturning	Total
U2L2 & U4L4	0	−6.42	0	0	−4.05/1.80	0	−1.05	−11.52
U3L3	0	−6.42	0	0	0	0	0	−6.42
U1L2 & L4L5	0	0	0	0	8.85/−0.55	0	3.87	12.72
U2L3 & L3U4	0	0	0	0	4.98/−2.21	0	1.29	6.27/−0.92

TABLE E8.13b
Forces in Truss Members due to Wind and Nosing Forces

	Force in Windward Main Truss Members (kips)							
Member	W_{LT} direct	W_{LT} overturning	W_{LB} direct	W_{LL} direct	W_{LL} overturning	N direct	N overturning	Total
Example	8.1	8.2	8.3	8.4	8.5	8.7	8.7	8.13
U1U2 & U4U5	−11.33	0	0	0	12.86	0	6.00	7.53
U2U3 & U3U4	−15.10	0	0	0	14.86	0	6.75	6.51
LOL1 & L5L6	0	−8.95	−9.27	−11.72	−8.04	−5.47	−3.75	−47.20
L1L2 & L4L5	0	−8.95	−24.10	−30.48	−8.04	−14.22	−3.75	−89.54
L2L3 & L3L4	0	−8.95	−31.52	−39.87	−12.86	−18.61	−6.00	−117.81
LOU1 & U5L6	0	15.40	0	0	13.82	0	6.45	35.67
U1L1 & U5L5	0	6.42	0	0	−4.50	0	−2.10	−0.18
U2L2 & U4L4	0	6.42	0	0	4.05/−1.80	0	1.05	11.52
U3L3	0	6.42	0	0	0	0	0	6.42
U1L2 & L4U5	0	0	0	0	−8.85/0.55	0	−3.87	−12.72
U2L3 & L3U4	0	0	0	0	−4.98/2.21	0	−1.29	-6.27/0.92

8.4.2 Longitudinal Force Analysis of Loaded Through Truss Spans

LF_T, LF_B, LW_{LT} and LW_{LB},[38] LW_{LL}, LW_U and restraint of CWR are transmitted to main truss members. Internal forces in the bottom (loaded) chord members due to external longitudinal forces from LF_T and LF_B are typically significant. The internal forces in the top (unloaded) chord members and web members due to external longitudinal forces from LF_T and LF_B are typically insignificant.

8.4.2.1 Longitudinal Forces from Live Load Traction and Braking

As shown in Figure 8.7, LF_B are inertia forces at the center of gravity of the live load (assumed at $y_{LF} = y_{LB} = 2.5$ m (8 ft) above the top of rail) and LF_T are transmitted at the rail car couplers of the live load (assumed at $y_{LF} = y_{LT} = 0.9$ m (3 ft) above the top of rail). L_{FT} at y_{LT} and L_{FB} at y_{LB} resolve to create a horizontal force at the top of rail and vertical force couples at the axles of each rail car.

Since the vertical force at the front axle of a car and the vertical force at the rear axle of a car are in opposite directions, the vertical forces within a panel tend to cancel each other and need not be explicitly considered. Therefore, the shear forces in each truss panel are essentially equal to the vertical reactions $R^V_0 = -R^V_6$.

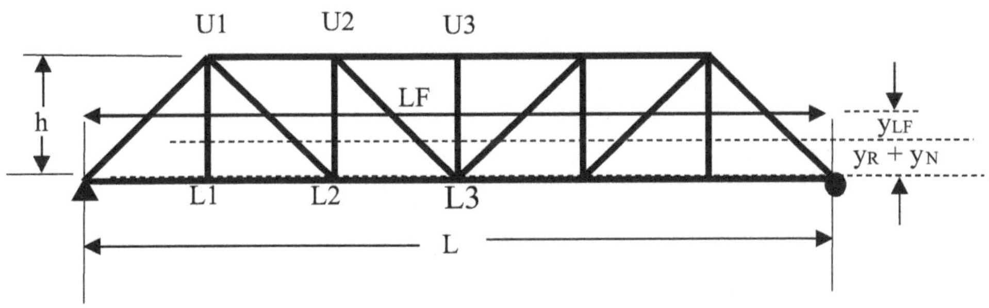

FIGURE 8.7 Longitudinal forces on through-truss spans.

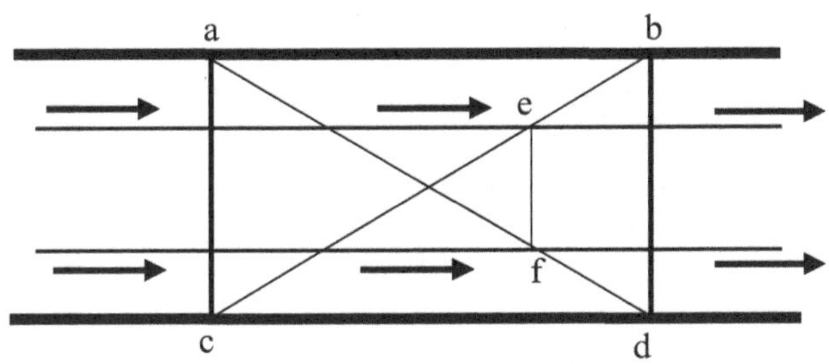

FIGURE 8.8 Traction bracing at the loaded chord of truss spans.

L_{FT} and L_{FB} are resisted by the fixed bearings of the main truss span by transmission through the floor system (floorbeams and stringers) at the bottom chord. Axial forces are transmitted through the stringers[39] causing lateral bending of floorbeams and, consequently, horizontal reactions at the panel points of the main truss bottom chord. Therefore, if the live load traverses the span from the fixed end, the chord stresses are tensile. Chord stresses are compressive when the live load traverses the span from expansion end. The forces in the chord members increase incrementally at each panel point from the expansion end to the fixed end.

Typically, the diagonal members of the bottom horizontal lateral bracing system are connected to the stringers at locations e and f as shown in in Figure 8.8 and therefore resist a portion of the LF. If diaphragms are installed between the stringers between locations e and f, traction bracing trusses are formed at befd that will assist in carrying LF to the main truss chord members at panel points b and d.[40] Truss floor systems in long spans can also contain contraction joints.[41]

Example 8.14

Determine the forces in the members of the truss in Figure 8.7 due to longitudinal forces from braking and traction. The through-truss span has the following dimensions:

$h = 35$ ft (centerline of top chord member to centerline of bottom chord member)
$L = 150$ ft
$y_{LFB} = 8$ ft
$y_{LFT} = 3$ ft
$y_R + y_N = 0.5 + 3.5 = 4.0$ ft (top of rail to centerline of bottom chord member)
$LF_B = 45 + 1.2(150) = 225$ kips at 8 ft above top of rail per track
$LF_T = 25\sqrt{150} = 306$ kips at 3 ft above top of rail per track
$R^V_{B0} = -R^V_{B6} = 225(8 + 4)/150 = 18.0$ kips per track

$R^V_{T0} = -R^V_{T6} = 306(3 + 4)/150 = 14.3$ kips per track
$R^H_{B0} = 225$ kips per track at fixed bearing
$R^H_{T0} = 306$ kips per track at fixed bearing
Longitudinal force at each intermediate panel point due to braking = 225/6 = 37.50 kips per track
Longitudinal force at each end panel point due to braking = 37.50/2 = 18.75 kips per track
Longitudinal force at each intermediate panel point due to traction = 306/6 = 51.00 kips per track
Longitudinal force at each end panel point due to traction = 51.00/2 = 25.50 kips per track

Internal member forces due to longitudinal force from live load (traction and braking) for web members:

Panel shears are equal since vertical axle forces within a panel tend to cancel.

$$\text{Diagonal web member force from braking} = \mp 18.0\left(\frac{\sqrt{25^2+35^2}}{35}\right) = \mp 22.1 \text{ kips per track}$$

$$\text{Diagonal web member force from traction} = \mp 14.3\left(\frac{\sqrt{25^2+35^2}}{35}\right) = \mp 17.6 \text{ kips per track}$$

Vertical web member force from braking = +/−18.0 kips
Vertical web member force from traction = +/−14.3 kips

Internal member forces due to longitudinal force from live load (traction and braking) for chord members:

Live load moving from right to left (from expansion bearing at L6 to fixed bearing at L0):
Member L0U1: Equilibrium at panel point L0 yields

$$F_B^{L0U1} = -18.0\left(\frac{\sqrt{25^2+35^2}}{35}\right) = -22.1 \text{ kips per track}$$

$$F_T^{L0U1} = -14.3\left(\frac{\sqrt{25^2+35^2}}{35}\right) = -17.6 \text{ kips per track}$$

Member U1U2: Take section through panel 1-2 and moments about panel point L2.
$F_B^{U1U2} = (18.0(50) - 18.75(8 + 4) - (37.50)(8 + 4))/35 = -6.43$ kips per track
$F_T^{U1U2} = (14.3(50) - 25.50(3 + 4) - (51.00)(3 + 4))/35 = -5.13$ kips per track
Member U2U3: Take section through panel 2–3 and moments about panel point L3.
$F_B^{U2U3} = (18.0(75) - 18.75(8 + 4) - 2(37.50)(8 + 4))/35 = -6.43$ kips per track
$F_T^{U2U3} = (14.3(75) - 25.50(3 + 4) - 2(51.00)(3 + 4))/35 = -5.13$ kips per track
Member U3U4: Take section through panel 3-4 and moments about panel point L3
$F_B^{U3U4} = -6.43$ kips per track
$F_T^{U3U4} = -5.13$ kips per track
Member U4U5: Take section through panel 4-5 and moments about panel point L4
$F_B^{U3U4} = (18.0(100) - 18.75(8 + 4) - 3(37.50)(8 + 4))/35 = -6.43$ kips per track

F_T^{U3U4} = (14.3(100) − 25.50(3 + 4) − 3(51.00)(3 + 4))/35 = −5.13 kips per track
Member L0L1: Equilibrium at L0 yields
F_B^{L0L1} = 18.0(25/35) − 225 = −212.1 kips per track
F_T^{L0L1} = 14.3(25/35) − 306 = −295.8 kips per track
Member L1L2: Take section through panel 1-2 and moments about panel point U1.
F_B^{L1L2} = (18.0(25) + 18.75(35 − 8-4) + 37.50(35 – 8 − 4) − 225(35))/35 = −175.2 kips per track
F_T^{L1L2} = (14.3(25) + 25.50(35 – 3 − 4) + 51.00(35 − 3 − 4) − 306(35))/35 = −245.5 kips per track
Member L2L3: Take section through panel 2-3 and moments about panel point U2.
F_B^{L2L3} = (18.0(50) + 18.75(35 – 8 − 4) + 2(37.50)(35 – 8 − 4) − 225(35))/35 = −137.7 kips per track
F_T^{L2L3} = (14.3(50) + 25.50(35 − 3 4) + 2(51.00)(35 – 3 − 4) − 306(35))/35 = −201.8 kips per track
Member L3L4: Take section through panel 3-4 and moments about panel point U4.
F_B^{L3L4} = (18.0(100) + 18.75(35 − 8-4) + 4(37.50)(35 − 8 − 4)– 225(35))/35 = −62.7 kips per track
F_T^{L3L4} = (14.3(100) + 25.50(35 − 3 − 4) + 4(51.00)(35 − 3 − 4) − 306(35))/35 = −114.3 kips per track
Member L4L5: Take section through panel 4-5 and moments about panel point U5.
F_B^{L4L5} = (18.0(125) + 18.75(35 − 8 − 4) + 5(37.50)(35 − 8 − 4)– 225(35))/35 = −25.2 kips per track
F_T^{L4L5} = (14.3(125) + 25.50(35 − 3 − 4) + 5(51.00)(35 − 3 − 4) − 306(35))/35 = −70.6 kips per track
Member L5L6: Equilibrium at L6 yields
F_B^{L5L6} = −18.0(25/35) = −12.9 kips per track
F_T^{L5L6} = −14.3(25/35) = −10.2 kips per track

When the live load moves from left to right (from fixed bearing at L0 to expansion bearing at L6) the forces in the top chord members and bottom chord members are in tension. It should be noted that the forces in the top chord members and web members are relatively small. The forces in the chord and web members are calculated per track and therefore must be divided by two for members in the trusses each side of the track. (See Table E8.14.)

TABLE E8.14
Longitudinal Forces in Truss Span Members

Main Truss Member	Braking Force LF_B (kips)	Traction Force LF_T (kips)	Longitudinal Force LF (kips)
Top chord members	+/- 3.22	+/−2.57	+/−3.22
End post members L0U1 & U5L6	+/−11.1	+/−8.8	+/−11.1
L0L1	+/−106.1	+/−147.9	+/−147.9
L1L2	+/−87.6	+/−122.8	+/−122.8
L2L3	+/−68.9	+/−100.9	+/−100.9
L3L4	+/−31.4	+/−57.2	+/−57.2
L4L5	+/−12.6	+/−35.3	+/−35.3
L5L6	+/−6.5	+/−5.1	+/−6.5
Diagonal members	+/−11.1	+/−8.8	+/−11.1
Vertical members	+/−18.0	+/−14.3	+/−18.0

The longitudinal forces due to braking and traction for the truss top chord and diagonal members are relatively small in comparison to the forces from dead, live and impact loads, and, therefore, usually neglected in design. Bottom chord members are often conservatively designed for the maximum horizontal reaction force = +/−306.0/2 = +/−153.0 kips (3.5% greater than the axial force in the end panel bottom chord members).

8.4.2.2 Longitudinal Forces from Restraint of CWR

Longitudinal forces from restraint of CWR also affect the members of a truss. However, as shown by analyses of the thermal forces from CWR on steel railway bridges (Fryba, 1996; Unsworth, 2018), horizontal fixed bearing forces are relatively small. Therefore, as vertical reactions at bearings are also small, the longitudinal forces from CWR in truss spans are typically unimportant for design.

8.4.2.3 Longitudinal Forces from Wind on the Structure and Live Load

AREMA (2024) recommends that 50% of the lateral wind force (W_L) be applied longitudinally to the structure at the top chord (LW_{LT}), at the bottom chord (LW_{LB}) and on the live load (LW_{LL}). As shown in Example 8.15, the resulting internal member forces are relatively small and may be regularly disregarded in design.

Example 8.15

A 150 ft long Pratt through truss with a height of 35 ft between the centerline of the top and bottom chord members and floor system with a depth of 2.5 ft above the top of the bottom chord is subjected to a lateral wind force on the structure of 30 psf and 300 lb/ft on the live load. The centerline of the trusses are 10 ft each side of the track centerline. Determine the internal forces in the main truss members due to the longitudinal component of the external wind forces on the structure and the live load. Table E8.15a provides the lateral wind forces on the structure.

TABLE E8.15a
Lateral Wind Forces at Joints of a Pratt Through-Truss Span

Truss Segment	Projected Area (ft²)	Lateral Wind Force on End Joints (kips)	Lateral Wind Force on Interior Joints (kips)
Windward top chord	403	1.51	3.02
Windward bottom chord	478	1.20	2.39
Windward floor system	375	0.94	1.88
Leeward top chord	403	1.51	3.02
Leeward bottom chord	331	0.83	1.66

Table E8.15b provides the longitudinal wind force on the truss span (50% of lateral wind force),

TABLE E8.15b
Longitudinal Wind Forces at Joints of a Pratt Through-Truss Span

Truss Segment	Longitudinal Wind Force on End Joints (kips)	Longitudinal Wind Force on Interior Joints (kips)
Windward top chord	+/−0.75	+/−1.51
Windward bottom chord	+/−1.07	+/−2.14
Leeward top chord	+/−0.75	+/−1.51
Leeward bottom chord	+/−0.42	+/−0.83

$R^V_0 = -R^V_6$ = (1.51)(4)(35)/150 = +/-1.41 kips per truss
R^H_0 = 4(1.51) + 6(2.14) = +/-18.88 kips per truss (fixed bearing)
Internal member forces due to longitudinal wind force on the structure,
Taking a section in panel 1-2 and moments about joint L2,
F_{U1U2} = (−1.41(50) + 0.75(35))/35 = −1.27 kips per truss
Taking a section in panel 2-3 and moments about joint L3,
F_{U2U3} = (−1.41(75) + 0.75(35) + 1.51(35))/35 = −0.77 kips per truss
Taking a section in panel 3-4 and moments about joint L3,
F_{U3U4} = (−1.41(75)+0.75(35)+2(1.51)(35))/35 = 0.77 kips per truss
Taking a section in panel 4–5 and moments about joint L4,
F_{U3U4} = (−1.41(100) + 0.75(35) + 3(1.51)(35))/35 = 1.27 kips per truss
Taking a section in panel 0-1 and moments about joint U1,
F_{L0L1} = (1.41(25) −18.88(35) + 1.07(35))/35 = −16.80 kips per truss
Taking a section in panel 1-2 and moments about joint U1,
F_{L1L2} = (2.14(35) + 1.07(35) + 1.41(25) − 18.88(35))/35 = −14.65 kips per truss
Taking a section in panel 2-3 and moments about joint U2,
F_{L2L3} = (2(2.14)(35) + 1.07(35) + 1.41(50) − 18.88(35))/35 = −11.51 kips per truss
Taking a section in panel 3-4 and moments about joint U4,
F_{L3L4} = (3(2.14)(35) + 1.07(35) + 1.41(100) − 18.88(35))/35 = −7.36 kips per truss
Taking a section in panel 4-5 and moments about joint U5,
F_{L3L4} = (4(2.14)(35) + 1.07(35) + 1.41(125) − 18.88(35))/35 = −4.21 kips per truss
Taking a section in panel 5-6 and moments about joint U5,
F_{L3L4} = (5(2.14)(35) + 1.07(35) + 1.41(125) − 18.88(35))/35 = −2.07 kips per truss
Taking a section through panels 0-1, 1-2 and 2-3 and vertical equilibrium provides
$-F_{L0U1} = F_{U1L2} = F_{U2L3}$ = 1.41(1.229) = 1.73 kips per truss
Lateral wind force on the live load of 300 lb/ft at 8 ft above the top of the rail:
End panel forces = 0.30(25)/2 = 3.75 kips
Interior panel forces = 0.30(25) = 7.50 kips
Longitudinal wind force on the live load of 300 lb/ft at 8 ft above the top of rail,
Force at bottom chord end panel points = +/−0.50(3.75)(12)/20 = +/−1.13 kips per truss
Force at bottom chord interior panel points = +/−0.50(7.50)(12)/20= +/−2.25 kips per truss
$R^V_0 = -R^V_6 = 0$
R^H_0 = 2.25(6) = 13.50 kips per truss (fixed bearing)
Internal member forces due to longitudinal wind force on the live load:
$F_{U1U2} = F_{U2U3} = 0$
Taking a section in panel 0-1 and moments about joint U1,
F_{L0L1} = −13.50 + 1.13 = = −12.37 kips per truss
Taking a section in panel 1-2 and moments about joint U1,
F_{L1L2} = −13.50 + 1.13 + 2.25 = −10.12 kips per truss
Taking a section in panel 2-3 and moments about joint U2,
F_{L2L3} = −13.50 + 1.13 − 2(2.25) = −7.87 kips per truss
F_{L3L4} = −13.50 + 1.13 + 3(2.25) = −5.62 kips per truss
F_{L4L5} = −13.50 +1.13 + 4(2.25) = −3.37 kips per truss
F_{L5L6} = −13.50 + 1.13 + 5(2.25) = −1.13 kips per truss

Table E8.15c gives a summary of internal member forces due to external longitudinal wind forces.

The internal force in bottom chord member L0L1 due to longitudinal wind forces is greatest at +/−29.17 kips. The internal force in bottom chord member L0L1 due to lateral wind forces is +/−47.20 kips (Example 8.9). The internal member forces in top chord and web members of the truss due to longitudinal wind forces are very small and typically not significant for design.

TABLE E8.15c
Main Truss Member Forces due to Longitudinal Wind Forces

Main Truss Member	Force due to Longitudinal Wind Load on the Structure, LW_L (kips)	Force due to Longitudinal Wind Load on the Live Load, LW_{LL} (kips)	Force due to Longitudinal Wind Load, $LW_L + LW_{LL}$ (kips)
L0U1 & U5L6	+/−1.73	0	+/−1.73
U1U2 & U4U5	+/−1.27	0	+/−1.27
U2U3 & U3U4	+/−0.77	0	+/−0.77
L0L1	+/−16.80	+/−12.37	+/−29.17
L1L2	+/14.65	+/−10.12	+/−24.77
L2L3	+/−11.51	+/−7.87	+/−19.38
L3L4	+/−7.36	+/−5.62	+/−12.98
L4L5	+/−4.21	+/−3.37	+/−7.58
L5L6	+/−2.07	+/−1.13	+/−3.20
U1L1 & U5L5	0	0	0
U2L2 & U4L4	+/−1.41	0	+/−1.41
U3L3	0	0	0
U1L2 & L4U5	+/−1.73	0	+/−1.73
U2L3 & L3U4	+/−1.73	0	+/−1.73

8.4.2.4 Internal Bracing Member Forces due to External Longitudinal Forces

Longitudinal forces do not create any forces in the horizontal lateral bracing systems in the plane of the top and bottom chord members.

8.4.2.5 Internal Member Forces in the Main Truss Members due to External Longitudinal Forces

Example 8.16 summarizes the internal forces in the main truss members due to externally applied longitudinal forces. The internal forces in the main truss members are due to the external longitudinal forces from LF, LW_{LT}, LW_{LB} and LW_{LL}.

Example 8.16

A 150 ft long Pratt through-truss with h = 35 ft between the centerline of the top and bottom chord members and floor system with a depth of 2.5 ft above the top of the bottom chord carries tangent track and is subjected to external longitudinal forces from wind and live load. The trusses are 20 ft apart, and end post members are considered fixed at the base. Determine the internal forces in the main truss members due to external longitudinal forces from wind and equipment. Table E8.16 gives the calculated results from Examples 8.14 and 8.15.

TABLE E8.16
Forces in Main Truss Members due to Longitudinal Forces from Wind and Equipment

Main Truss Member	Longitudinal Force in Main Truss Members (kips)		
	LF	LW_L	Total
U1U2 & U4U5	+/−3.22	+/−1.27	+/−4.49
U2U3 & U3U4	+/−3.22	+/−0.77	+/−3.99
L0L1	+/−147.9	+/−29.17	+/−177.07
L1L2	+/−122.8	+/−24.77	+/−147.57
L2L3	+/−100.9	+/−19.38	+/−120.28
L3L4	+/−57.2	+/−12.98	+/−70.18
L4L5	+/−35.3	+/−7.58	+/−42.88
L5L6	+/−6.5	+/−3.20	+/−9.70
L0U1 & U5L6	+/−11.1	+/−1.73	+/−12.83
U1L1 & U5L5	+/−18.0	0	+/−18.0
U2L2 & U4L4	+/−18.0	+/−1.41	+/−19.41
U3L3	+/−18.0	0	+/−18.0
U1L2 & L4U5	+/−11.1	+/−1.73	+/−12.83
U2L3 & L3U4	+/−11.1	+/−1.73	+/−12.83

8.4.3 Internal Member Forces due to External Lateral and Longitudinal Forces on Loaded Through Truss Spans

As outlined in Chapter 9, AREMA (2024) provides recommendations concerning allowable stress factors for load combinations that include internal member forces due to externally applied lateral and longitudinal forces. These load combinations primarily affect the bottom chord members of through-truss spans.

8.4.3.1 Internal Bracing Member Forces due to External Lateral Forces

The horizontal lateral bracing systems in the plane of the top and bottom chord members experience internal forces from external lateral forces only. Example 8.12 summarizes the internal forces in the top and bottom horizontal lateral bracing web members due to externally applied lateral forces.[42]

8.4.3.2 Internal Main Truss Member Forces due to External Lateral and Longitudinal Forces

The internal forces in the main truss members are due to the external lateral forces from W_{LT}, W_{LB}, W_{LL} and N. The internal forces in the main truss members are due to the external longitudinal forces from LF, LW_{LT}, LW_{LB} and LW_{LL}. Example 8.17 summarizes the internal forces in the main truss members due to externally applied lateral and longitudinal forces.

Example 8.17

A 150 ft long Pratt through-truss with h = 35 ft between the centerline of the top and bottom chord members and floor system with a depth of 2.5 ft above the top of the bottom chord carries tangent track and is subjected to external lateral and longitudinal forces from wind and live load. The trusses are 20 ft apart, and end post members are considered fixed at the base. Determine the internal forces in the main truss members due to external lateral and longitudinal forces from wind and live load. Tables E8.17a and b give the calculated results from Examples 8.13 and 8.16; Table E8.17c gives the design forces.

These tables illustrate that the leeward truss bottom chord members are most detrimentally affected by external lateral and longitudinal forces. The tables also illustrate that, particularly for light live loads, the bottom chord members of the windward truss require investigation for potential stress reversal.

TABLE E8.17a
Forces in Main Truss Members due to External Lateral and Longitudinal Forces

	Force in Leeward Truss Members (kips)		
Main Truss Member	Due to External Lateral Forces	Due to External Longitudinal Forces	Total
U1U2 & U4U5	−18.86	+/−4.49	−23.35
U2U3 & U3U4	−9.88	+/−3.99	−13.87
L0L1	47.20	+/−177.07	224.27/−129.87
L1L2	89.54	+/−147.57	237.11/−58.03
L2L3	117.81	+/−120.28	238.09/−2.47
L3L4	117.81	+/−70.18	187.99
L4L5	89.54	+/−42.88	132.42
L5L6	47.20	+/−9.70	56.90
L0U1 & U5L6	−35.67	+/−12.83	−48.50
U1L1 & U5L5	0.18	+/−18.0	18.18/−17.82
U2L2 & U4L4	−11.52	+/−19.41	7.89/−30.93
U3L3	−6.42	+/−18.0	11.58/−24.42
U1L2 & L4U5	12.72	+/- 12.83	25.55/−0.11
U2L3 & L3U4	6.27/−0.92	+/- 12.83	19.10/−13.75

TABLE E8.17b
Forces in Main Truss Members due to External Lateral and Longitudinal Forces

	Force in Windward Truss Members (kips)		
Main Truss Member	Due to External Lateral Forces	Due to External Longitudinal Forces	Total
U1U2 & U4U5	7.53	+/−4.49	12.02
U2U3 & U3U4	6.51	+/−3.99	10.50
L0L1	−47.20	+/−177.07	129.87/−224.27
L1L2	−89.54	+/−147.57	58.03/−237.11
L2L3	−117.81	+/−120.28	2.47/−238.09
L3L4	−117.81	+/−70.18	−187.99
L4L5	−89.54	+/−42.88	−132.42
L5L6	−47.20	+/−9.70	−56.90
L0U1 & U5L6	35.67	+/−12.83	48.50
U1L1 & U5L5	−0.18	+/−18.0	17.82/−18.18
U2L2 & U4L4	11.52	+/−19.41	30.93/−7.89
U3L3	6.42	+/−18.0	24.42/−11.58
U1L2 & L4U5	−12.72	+/- 12.83	0.11/−25.55
U2L3 & L3U4	−6.27/0.92	+/- 12.83	13.75/−19.10

TABLE E8.17c
Design Forces in Main Truss Members due to Lateral and Longitudinal Forces

	Force in Truss Members (kips)		
Main Truss Member	**Leeward Truss**	**Windward Truss**	**Design Force**
U1U2 & U4U5	−23.35	12.02	12.02/−23.35
U2U3 & U3U4	−13.87	10.50	10.50/−13.87
L0L1	224.27/−129.87	129.87/−224.27	∓224.27
L1L2	237.11/−58.03	58.03/−237.11	∓237.11
L2L3	238.09/−2.47	2.47/−238.09	∓238.09
L3L4	187.99	−187.99	∓187.99
L4L5	132.42	−132.42	∓132.42
L5L6	56.90	−56.90	∓56.90
L0U1 & U5L6	−48.50	48.50	∓48.50
U1L1 & U5L5	18.18/−17.82	17.82/−18.18	∓18.18
U2L2 & U4L4	7.89/−30.93	30.93/−7.89	∓30.93
U3L3	11.58/−24.42	24.42/−11.58	∓24.42
U1L2 & L4U5	25.55/−0.11	0.11/−25.55	∓25.55
U2L3 & L3U4	19.10/−13.75	13.75/−19.10	∓19.10

8.4.4 Lateral Force Analysis of Unloaded Through Truss Spans

Unloaded through-truss spans will be without lateral forces from W_{LT} and W_{LB}, W_{LL}, CF and N. However, W_{UT} and W_{UB} are developed from a static wind pressure of 2.39 kPa (50 psf), in accordance with AREMA (2024). Also, B_{RU} is severely reduced without the compression chord member forces from LL + I.

As shown in Figure 8.9, W_U is the lateral wind pressure on the structure distributed to the panel points of the horizontal lateral bracing systems in the planes of the main truss top and bottom chord members. AREMA (2024) recommends that W_U be established as 2.39 kPa (50 psf) on an unloaded span.[43]

W_{UT} is the lateral wind force on the structure due to W_U at panel points in the plane of the main truss top chord members, creating internal axial forces in the top horizontal lateral bracing system. The horizontal reactions at the end of the top horizontal lateral bracing system (at the top of the portal frame) have an overturning effect about axes in the plane of the bottom horizontal lateral bracing system members, causing uniform axial tension or compression in the bottom chord members of the main trusses.

W_{UB} is the lateral wind force on the structure due to W_U at panel points in the plane of the main truss bottom chord member, creating only internal axial forces in the bottom horizontal lateral bracing system because the plane of the bottom lateral horizontal bracing system is coincident with the plane of the main truss supports.

B_{RU} is the notional transverse shear force in a panel due to bracing of the top chords, which is equal to 2.5% of the dead load axial force in both top chord members in the panel and creates internal axial forces in the top horizontal lateral bracing system.

CWR_L is the lateral force due to continuous welded rail restraint on curved track.

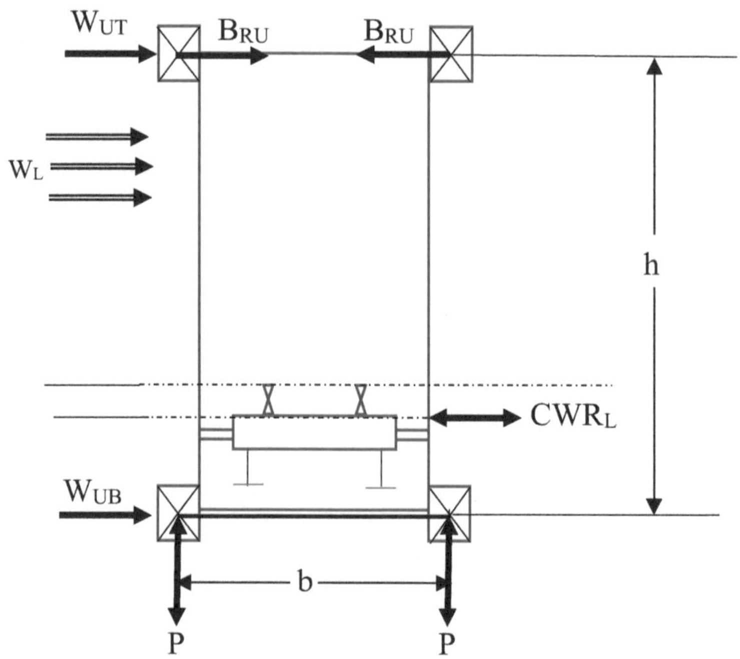

FIGURE 8.9 Lateral forces at the panel points of unloaded through-truss spans.

Example 8.18

A 150 ft long Pratt through-truss with a height of 35 ft between the centerline of the top and bottom chord members and floor system with a depth of 2.5 ft above the top of the bottom chord is subjected to an external lateral wind force on the structure of 50 psf. The trusses are 20 ft apart. The portal bracing is 8 ft deep in the plane of the end post members. Determine the internal forces in the top lateral bracing, bottom lateral bracing and main truss end post and chord members due to the external lateral wind forces, W_{UT}, W_{UB}, and compression chord member restraint force B_{RU}.

From Example 8.12 with a factor of 0.19 (DL/(DL + LL + I)) on B_{RL} and a factor of 1.67 on W_{LT}, Tables E8.18a and b give the calculated forces in the top and bottom horizontal lateral bracing systems. From Example 8.13 with a factor of 1.67 on W_{LT} and W_{LB}, Tables E8.18c and d give the calculated forces in the main truss members.

TABLE E8.18a
Forces in Top Horizontal Lateral Bracing Members due to Compression Member Bracing and Lateral Wind Force on an Unloaded Truss Span

Top Horizontal Lateral Bracing Web Member	Force in Top Horizontal Lateral Bracing Web Member of Leeward and Windward Trusses (kips)	
	Due to B_{RU}	Due to W_{UT}
Top bracing diagonals in panels 1-2 and 4-5	8.43	24.17
Top bracing diagonals in panels 2-3 and 3-4	9.55	12.08
Top bracing strut at U1 and U5	−2.63	−42.15/22.02
Top bracing strut at U2 and U4	−5.62	14.27/−19.30
Top bracing strut at U3	−5.96	−10.07

TABLE E8.18b
Forces in Bottom Horizontal Lateral Bracing Members due to Lateral Wind Force on Unloaded Truss Span

Bottom Horizontal Lateral Bracing Web Member	Force in Bottom Horizontal Lateral Bracing Web Members of Leeward and Windward Trusses (kips)
Bottom bracing diagonals in panels 0-1 and 5-6	+/−19.77
Bottom bracing diagonals in panels 1-2 and 4-5	+/−13.18
Bottom bracing diagonals in panels 2-3 and 3-4	+/−7.90

TABLE E8.18c
Forces in Main Truss Members due to Lateral Wind Forces on Unloaded Truss Span

	Force in Leeward Main Truss Members (kips)			
Member A	W_{UT} direct	W_{UT} overturning	W_{UB} direct	Total Unloaded Span
U1U2 & U4U5	0	0	0	0
U2U3 & U3U4	18.88	0	0	18.88
LOL1 & L5L6	0	14.92	15.45	30.37
L1L2 & L4L5	0	14.92	40.17	55.09
L2L3 & L3L4	0	14.92	52.53	67.45
LOU1 & U5L6	0	−25.67	0	−25.67
U1L1 & U5L5	0	−10.72	0	−10.72
U2L2 & U4L4	0	−10.72	0	−10.72
U3L3	0	−10.72	0	−10.72
U1L2 & L4U5	0	0	0	0
U2L3 & L3U4	0	0	0	0

TABLE E8.18d
Forces in Main Truss Members due to Lateral Wind Forces on Unloaded Truss Span

	Force in Windward Main Truss Members (kips)			
Member	W_{UT} direct	W_{UT} overturning	W_{UB} direct	Total Unloaded Span
U1U2 & U4U5	−18.88	0	0	−18.88
U2U3 & U3U4	−25.17	0	0	−25.17
LOL1 & L5L6	0	−14.92	−15.45	−30.37
L1L2 & L4L5	0	−14.92	−40.17	−55.09
L2L3 & L3L4	0	−14.92	−52.53	−67.45
LOU1 & U5L6	0	25.67	0	25.67
U1L1 & U5L5	0	10.72	0	10.72
U2L2 & U4L4	0	10.72	0	10.72
U3L3	0	10.72	0	10.72
U1L2 & L4U5	0	0	0	0
U2L3 & L3U4	0	0	0	0

8.4.5 Longitudinal Force Analysis of Unloaded Through-Truss Spans

Unloaded through-truss spans will be without LF_T and LF_B, LW_{LT} and LW_{LB} and LW_{LL}. Longitudinal forces from CWR restraint are considered negligible. However, LW_U, which affect only main truss members, are developed from a static wind pressure of 2.39 kPa (50 psf), in accordance with AREMA (2024).

Example 8.19

A 150 ft long Pratt through-truss with h = 35 ft between the centerline of the top and bottom chord members and floor system with a depth of 2.5 ft above the top of the bottom chord is subjected to an external lateral wind force on the structure of 50 psf. The trusses are 20 ft apart. The portal bracing is 8 ft deep in the plane of the end post members. Determine the internal forces in the main truss members due to the external LW_U.

Table E8.19 gives the results calculated from Example 8.16 with a factor of 50/30 on LW_L.

TABLE E8.19
Forces in Main Truss Members due to Longitudinal Wind Forces

Main Truss Member	Force in Main Truss Members, LW_U (kips)
U1U2 & U4U5	+/−2.12
U2U3 & U3U4	+/−1.28
L0L1	+/−48.62
L1L2	+/−41.28
L2L3	+/−32.30
L3L4	+/−21.63
L4L5	+/−12.63
L5L6	+/−5.33
L0U1 & U5L6	+/−2.88
U1L1 & U5L5	0
U2L2 & U4L4	+/−2.35
U3L3	0
U1L2 & L4U5	+/−2.88
U2L3 & L3U4	+/−2.88

8.4.6 Internal Forces due to External Lateral and Longitudinal Forces on Unloaded Through Truss Span

Example 8.20 summarizes the internal forces in the main truss members due to externally applied lateral and longitudinal forces. The internal forces in the main truss members are due to the external lateral forces from W_{UT} and W_{UB}. The internal forces in the main truss members are due to the external longitudinal forces from LW_{UT} and LW_{UB}.

Example 8.20

A 150 ft long Pratt through-truss with h = 35 ft between the centerline of the top and bottom chord members and floor system with a depth of 2.5 ft above the top of the bottom chord carries tangent track and is subjected to external lateral and longitudinal forces from wind on the unloaded span. The trusses are 20 ft apart and end post members are considered as fixed at the base. Determine the internal forces in the main truss members due to external lateral and longitudinal forces from wind.

From Examples 8.18 and 8.19, Tables E8.20a and b give the wind forces, and Table E8.20c gives the design forces.

TABLE E8.20a
Forces in Main Truss Members due to Lateral and Longitudinal Forces

Main Truss Member	Force in Leeward Truss Members (kips)		
	Due to External Lateral Forces	Due to External Longitudinal Forces	Total
U1U2 & U4U5	0	+/−2.12	+/−2.12
U2U3 & U3U4	18.88	+/−1.28	20.16
L0L1	30.37	+/−48.62	78.99/−18.25
L1L2	55.09	+/−41.28	96.37
L2L3	67.45	+/−32.30	99.75
L3L4	67.45	+/−21.63	89.08
L4L5	55.09	+/−12.63	67.72
L5L6	30.37	+/−5.33	35.70
L0U1 & U5L6	−25.67	+/−2.88	−28.55
U1L1 & U5L5	−10.72	0	−10.72
U2L2 & U4L4	−10.72	+/−2.35	−13.07
U3L3	−10.72	0	−10.72
U1L2 & L4U5	0	+/−2.88	+/2.88
U2L3 & L3U4	0	+/−2.88	+/−2.88

TABLE E8.20b
Forces in Main Truss Members due to Lateral and Longitudinal Forces

Main Truss Member	Force in Windward Truss Members (kips)		
	Due to External Lateral Forces	Due to External Longitudinal Forces	Total
U1U2 & U4U5	−18.88	+/−2.12	−21.00
U2U3 & U3U4	−25.17	+/−1.28	−26.45
L0L1	−30.37	+/−48.62	−78.99/18.25
L1L2	−55.09	+/−41.28	−96.37
L2L3	−67.45	+/−32.30	−99.75
L3L4	−67.45	+/−21.63	−89.08
L4L5	−55.09	+/−12.63	−67.72
L5L6	−30.37	+/−5.33	−35.70
L0U1 & U5L6	25.67	+/−2.88	28.55
U1L1 & U5L5	10.72	0	10.72
U2L2 & U4L4	10.72	+/−2.35	13.07
U3L3	10.72	0	10.72
U1L2 & L4U5	0	+/−2.88	+/−2.88
U2L3 & L3U4	0	+/−2.88	+/−2.88

TABLE E8.20c
Design Forces in Main Truss Members due to Lateral and Longitudinal Forces

Main Truss Member	Force in Truss Members (kips) Leeward Truss	Windward Truss	Design Force
U1U2 & U4U5	+/−2.12	−21.00	2.12/−21.00
U2U3 & U3U4	20.16	−26.45	20.16/−26.45
L0L1	78.99/−18.25	−78.99	∓78.99
L1L2	96.37	−96.37	∓96.36
L2L3	99.75	−99.75	∓99.75
L3L4	89.08	−89.08	∓89.08
L4L5	67.72	−67.72	∓67.67
L5L6	35.70	−35.70	∓35.70
L0U1 & U5L6	−28.55	28.55	∓28.55
U1L1 & U5L5	−10.72	10.72	+/−10.72
U2L2 & U4L4	−13.07	13.07	+/−13.07
U3L3	−10.72	10.72	+/−10.72
U1L2 & L4U5	+/−2.88	+/−2.88	+/−2.88
U2L3 & L3U4	+/−2.88	+/−2.88	+/−2.88

8.5 LATERAL AND LONGITUDINAL FORCE ANALYSIS OF DECK TRUSS SPANS

Deck-truss spans that are loaded must consider external lateral forces from W_{LT}, W_{LB}, W_{LL}, N, CF, CWR_L and B_{RL} and external longitudinal forces from LW_{LT}, LW_{LB}, LW_{LL}, CWR, LF_T and LF_B. Deck truss spans that are unloaded must consider external lateral forces from W_{UT}, W_{UB}, B_{RU} and CWR_L and external longitudinal forces from CWR, LW_{UT} and LW_{UB}.

8.5.1 Lateral Force Analysis of Loaded Deck-Truss Spans

Deck-truss spans have horizontal lateral bracing systems in the plane of the main truss top and bottom chord members and vertical lateral bracing systems in the plane of the main truss web members. Vertical lateral bracing system behavior depends on whether the main deck truss is supported in the plane of the top or bottom chord members. Deck trusses supported in the plane of the top chord will have the end vertical bracing system in the inclined plane of truss member U0L1. Pratt-type deck trusses supported in the plane of the bottom chord will have the end vertical bracing system in the plane of truss member U0L0. Warren-type deck trusses supported in the plane of the bottom chord will have the end vertical bracing system in the plane of truss member L0U1 with lighter bracing to resist external forces in the end panel at U0L0. These horizontal and vertical lateral bracing systems are subjected to the external lateral forces W_{LT}, W_{LB}, W_{LL}, CF, B_{RL} and N.

As shown in Figure 8.10, the horizontal lateral bracing system in the plane of the bottom (unloaded) chord members comprises the main truss bottom chord members and the bracing members (diagonal and strut members) between the bottom chord members.[44] The horizontal lateral bracing in the plane of the main truss bottom chord members is subject to W_{LB}, which creates compression in the windward bottom chord members and tension in the leeward bottom chord members. The main truss members are also subject to vertical forces from overturning effects of W_{LL}, CF and N. Assuming tangent track on the span, overturning provides $P_P = W_{LL}(y_{cg} + y_R + y_N)/b + N(y_N)/b$. The windward bottom chord members are in compression and the leeward bottom chord members are in tension due to overturning. Therefore, the internal forces in the main chord members are of the same attitude (tension or compression) when resisting the external lateral loads W_{LB}, W_{LL}, CF and N.

The horizontal lateral bracing system in the plane of the main truss top (loaded) chord members comprises the main truss top chord members and the bracing members (diagonal members and floorbeams) between the top chord members. The horizontal lateral bracing in the plane of the main truss top chord members is subject to the lateral forces W_{LT}, W_{LL}. CF, N and B_{RL}, which create compression in the windward top chord members and tension in the leeward top chord members. The overturning effects of W_{LL}. F_{CWR} and N create tension in the windward top chord members and compression in the leeward top chord members. As such, the internal forces in top chord members are usually relatively small due to the opposite attitude of the combined forces from external lateral forces and overturning.

For deck trusses supported in the plane of the top chord members, the vertical lateral bracing shown in Figure 8.10 is in the plane of the inclined end member U0L1 of the main truss. The diagonal members of the vertical lateral bracing are relatively long and can be considered tension-only members resisting an external reaction, P_{LB}, from W_{LB} such that the force in the diagonal member is $F_D = P_{LB}\left(\frac{\sqrt{h^2+b^2}}{b}\right)$ and the force in the bottom strut is P_{LB}. The internal forces from P_{LB} in the bracing system members in the plane of end member U0L1 are relatively small, and therefore,

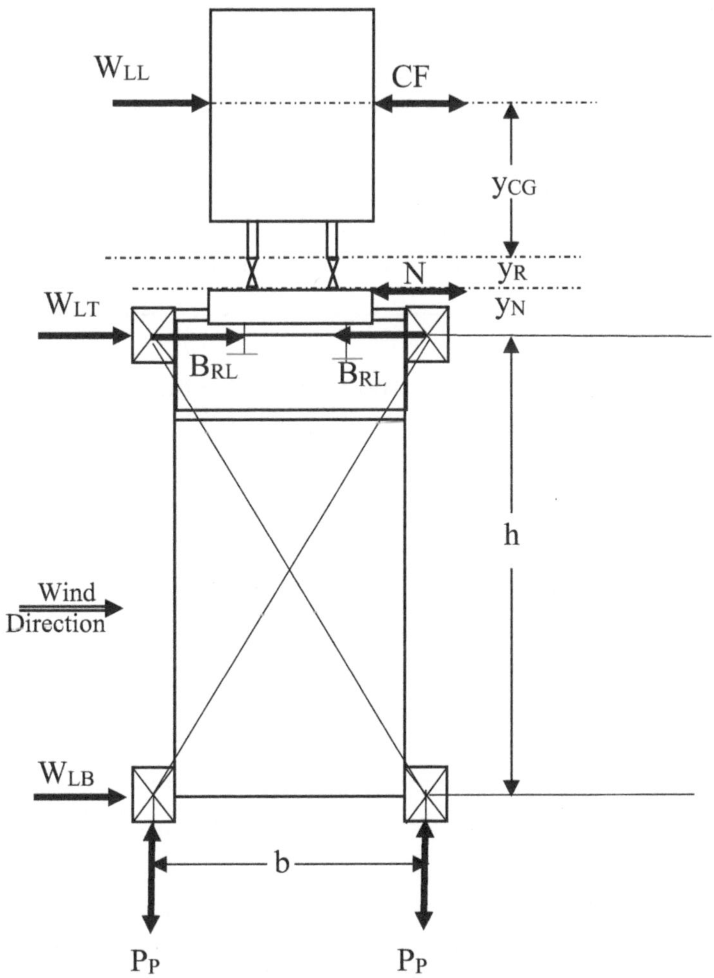

FIGURE 8.10 Lateral forces at panel points of loaded deck-truss spans.

the bracing is typically light like the sway bracing in the plane of the intermediate vertical web members of the deck truss. The internal forces in member U0L1 will create compression in the top chord members.

For deck trusses supported in the plane of the bottom chord members, the vertical lateral bracing shown in Figure 8.10 is in the plane of member U0L0 for Pratt-type trusses or L0U1 for Warren-type trusses. The diagonal members of the vertical lateral bracing are relatively long and can be considered tension-only members resisting external reaction P_{LT} from W_{LT} such that the force in the diagonal member is $F_D = P_{LT}\left(\frac{\sqrt{h^2+b^2}}{b}\right)$, and the force in the top strut is P_{LT}. It is noted that P_{LT} will be of greater magnitude than P_{LB} for deck-truss spans supported in the plane of the main truss bottom chord members. If the deck truss is supported in the plane of the main truss bottom chord and the diagonal member is U0L1, the reaction P_{LT} at U0 is carried to the supports by vertical lateral bracing in the plane of U0L0. If the deck truss is supported in the plane of the main truss bottom chord and the diagonal member is L0U1, the top lateral system reaction P_{LT} at U1 is transferred to the support by the inclined lateral bracing in the plane of member L0U1. The forces in the end panel at U0 are transferred by a lighter vertical lateral bracing system in the plane of end member U0L0.

8.5.2 Longitudinal Force Analysis of Loaded Deck-Truss Spans

If deck-truss spans are supported in the plane of the main truss top (loaded) chord members, the LF are resisted by the fixed bearings in the plane of the top chord members and create internal forces in the top chord members.[45] If deck-truss spans are supported in plane of the bottom chords, the LF create additional vertical reactions that create internal forces in all members of the truss. However, as shown in Example 8.21, the magnitude of these vertical reactions is relatively small in comparison with reactions from DL, LL and I. Therefore, the additional vertical reactions from LF have only a small effect on internal member forces and are typically neglected in design.

Example 8.21

A 225 ft long deck truss is 40 ft in height and supported in the plane of the bottom chord. Determine the vertical reactions created by the longitudinal braking force from Cooper's E80 live load.

LF_B = 45 +1.2(225) = 315 kips at 8 ft above top of rail per track
R_{LFB} = +/- 315(40 + 8)/225 = +/- 67.20 kips = LF_B/4.69
W_C = Weight of Coopers E80 LL on 225 ft span = 2(568 + (225-109)(4)) = 2064 kips per track
LF_B/W_C = (315/2064) = 0.153 (15.3% of the weight of the live load on the span)
R_{LFB} = LF_B/4.69 = 0.153W_c/4.69 = 0.0325W_C

Therefore, the additional reaction, R_{LFB}, is only 3.25% of the weight of the live load on the span.

8.5.3 Lateral Force Analysis of Unloaded Deck Truss Spans

The lateral wind forces on the unloaded structure (W_{UT} and W_{UB}), as shown in Figure 8.11, are developed from a static wind pressure of 2.39 kPa (50 psf), in accordance with AREMA (2024). The internal forces in the horizontal bracing and main truss members can be analyzed in the same way as for the lateral force analysis of loaded deck-truss spans with only the need to consider the external forces W_{UT}, W_{UB} and B_{RU}.

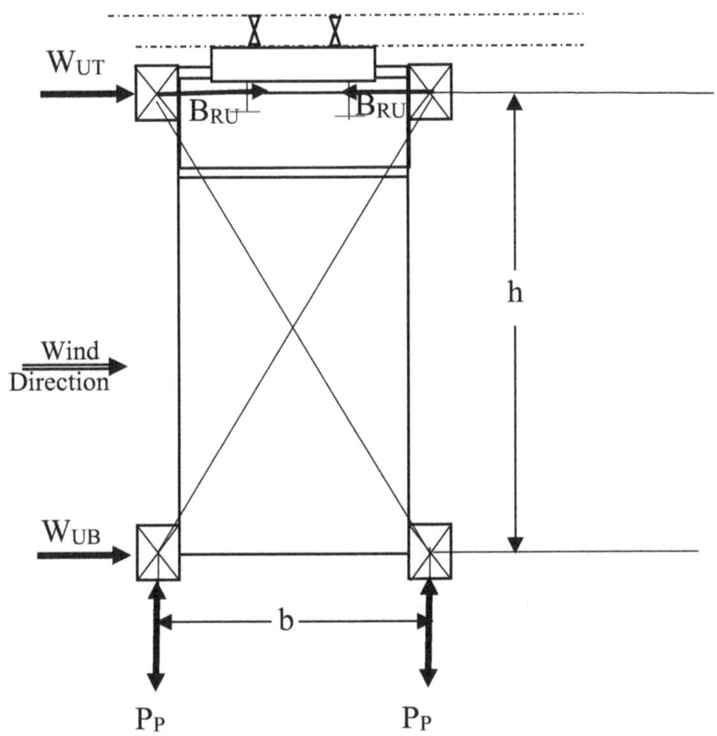

FIGURE 8.11 Lateral forces at panel points of unloaded deck-truss spans.

8.5.4 Longitudinal Force Analysis of Unloaded Deck-Truss Spans

As with the longitudinal force analysis of unloaded through-truss spans, the forces in the main truss members of unloaded deck-truss spans are relatively small and, typically, do not require consideration in design.

NOTES

1 Based on the Allowable Stresses for Combinations of Loads or Forces recommended in Chapter 15 of AREMA (2024).
2 AREMA (2024) recommends moving wind pressures of W_L = 1.44 kPa (30 psf) on the loaded bridge and W_U = 2.39 kPa on the unloaded bridge. The wind pressures are recommended to be moving loads to ensure that maximum member forces are considered. The wind pressure of 1.44 kPa (30 psf) is considered the maximum wind pressure in which trains would operate to remain stable (light trains can overturn under such pressures). The wind pressure of 2.39 kPa (50 psf) might exist in conjunction with rarely occurring hurricane-force wind speeds.
3 Nosing forces are lateral forces caused by locomotive and rail car truck (bolster) hunting at the wheel/rail interface.
4 Forces due to restrained CWR are typically neglected for rail anchoring systems that allow controlled longitudinal movement across the bridge (rail fasteners with Teflon® sliding surfaces are often used). For long spans, longitudinal rail movement across the bridge can also be controlled with rail expansion joints.
5 Such as member forces from temperature fluctuation or support settlements in statically indeterminate truss spans.
6 W_{LT} and W_{LB} are lateral wind forces at the top chord and bottom chord panel points, respectively, due to wind pressure, W_L, on the loaded span.
7 W_{UT} and W_{UB} are lateral wind forces at the top chord and bottom chord panel points, respectively, due to wind pressure, W_U, on the unloaded span.

8 W_{LL} is the wind force on the live load due to wind on the loaded bridge recommended by AREMA (2023) as 4.38 kN/m (300 lb/ft) applied at 2.4 m (8 ft) above the top of the rail. Some engineers will consider a higher location of application for rail lines that carry extensive double stack container trains, live loads with high vertical projections and/or large magnitude wind gusts.
9 B_{RU} is the notional bracing force due to dead load forces in compression flanges of unloaded spans. B_{RL} is the notional bracing force due to dead, live and impact forces in compression flanges of loaded spans.
10 In through-truss spans, the loaded chord members are the bottom chord members. In deck-truss spans, the loaded chord members are the top chord members. Typically, the horizontal bracing system in the plane of the loaded chord members carries greater lateral forces than the system in the plane of the unloaded chord members.
11 The diagonal web members in the horizontal lateral bracing truss in the plane of the loaded chord are considered non-slender if they are connected to the floor system stringers. The strut members (horizontal web members) are the floorbeams in the floor system and are typically capable of resisting lateral forces of considerable magnitude. Therefore, axial stresses in the floorbeams will be small and negligible in design. However, for narrow floorbeam bottom flange plates, it may be necessary to investigate the possibility of lateral buckling. Nevertheless, the bottom flange plates of most modern floorbeams are not narrow enough to be susceptible to lateral bucking.
12 The diagonal web members in the horizontal lateral bracing truss in the plane of the unloaded chord are considered slender as they are not connected to main truss members between the unloaded chord members. The strut members are the top horizontal members of the portal and sway frames.
13 This is typically the case for diagonal members of the horizontal lateral bracing trusses in the plane of the loaded chord. However, if diagonal members of the horizontal lateral bracing trusses in the plane of the unloaded chord are of robust cross-section, they may also be considered non-slender members.
14 This is because the approximate nature of the relatively small forces in lateral bracing systems do not warrant a more onerous structural analysis of the lateral bracing system as a statically indeterminate truss.
15 The portal and sway frame bracing in through-truss spans must provide adequate horizontal and vertical clearances. AREMA (2024) recommends a minimum horizontal clearance between posts of 5.5 m (18 ft) and minimum vertical clearance above top of rail of 7.0 m (23 ft).
16 Particularly when the span is loaded because the bottom lateral bracing system is designed for forces resulting from the loaded through truss span.
17 LW_{LT} and LW_{LB} are longitudinal wind forces at the top chord and bottom chord panel points, respectively, due to LW_L on the loaded bridge.
18 LW_{UT} and LW_{UB} are longitudinal wind forces at the top chord and bottom chord panel points, respectively, due to LW_U on the unloaded bridge.
19 AREMA (2024) recommends that W_U be established as 2.39 kPa (50 psf) on an unloaded span.
20 The horizontal lateral bracing system in the plane of the top chord members includes the main truss top chord members and the horizontal web members between the top chord members.
21 The base of the end posts is typically considered fixed due to the relatively stiff connection between the end post and bottom chord members at the support bearings.
22 The assumed inflection points are on the legs of the end portal frame midway between the base and the bottom of the portal bracing.
23 The approximate method conservatively locates panel point loads on each side of the panel to determine maximum positive and negative shear forces in the panel.
24 The lateral bracing system in the plane of the bottom chord members includes the bottom chord members and the horizontal web members between the bottom chord members.
25 Knee braces are used in pony truss spans that have no horizontal lateral bracing system in the plane of the top chord.
26 The portal frame is also subjected to an external distributed wind force on the windward post. However, this force is relatively very small and can be neglected in most cases. Nevertheless, in cases where this distributed load is considered it may be established, with sufficient accuracy, as a concentrated force in addition to P_T at the top of the post.
27 The connection at the base of the end posts is very stiff, approaching fixity.
28 In Figure 8.4, the inactive diagonal member in compression is shown with a dotted line.
29 Knee bracing connected to the top chord will preclude torsional deformation of the top chord members
30 Pony truss spans are through-truss spans without top lateral bracing; they are suitable for relatively short spans but rarely used in modern bridges because the relatively shallow depth of the trusses results in the truss members being of relatively large sections and, therefore, uneconomical. The top chord of pony truss spans is typically horizontal or parabolic in longitudinal profile.

31 W_{LT} and B_{RL} are the wind force at the top chord and notional compression bracing force for loaded spans, respectively. The wind force at the top chord and compression bracing force for unloaded spans are W_{UT} and B_{RU}, respectively. W_{LT} is determined from a wind pressure of 1.44 kPa (30 psf), and W_{UT} is determined from pressure of 2.39 kPa (50 psf).
32 Assuming the diagonal members are effectively fixed at the bottom chord members.
33 Typical of many pony truss spans.
34 This assumption is valid for the top chord members considered straight columns of uniform cross-section pinned and rigidly supported at both ends of the span with length greater than 1.3 times the panel length and an equal compressive axial force throughout all panels in the span. The assumptions of equal axial compressive force in each panel of the truss and rigid top chord end supports are inconsistent with actual truss behavior but provide K_{cr} with reasonable accuracy for preliminary assessment of knee brace requirements in pony truss spans.
35 Such as the top chord members of Parker or Pennsylvania truss spans.
36 The analysis is sufficiently conducted with panel point forces assuming equal horizontal panel lengths.
37 Panel length corresponds to the length of the inclined-chord member in the panel.
38 LW_{LT} and LW_{LB} are longitudinal wind forces at the top chord and bottom chord panel points, respectively, due to W_L on the loaded bridge.
39 The axial forces in stringers cause relatively small axial stresses and can be disregarded in floor system design.
40 For a floor system with two stringer lines between the main trusses, a diaphragm between stringers at locations e and f suffices to form a truss that will assist in carrying LF to the panel points at b and d. For a floor system with four stringer lines between the main trusses, a diaphragm between stringers at locations e and f and members between the stringers form a truss that will assist in carrying LF to the panel points at b and d. A traction bracing frame must also be installed at the other side of the panel to carry LF to panel points a and c when the live load traverses the track in the opposite direction to that shown in Figure 8.8.
41 Contraction joints are included in long-span floor systems to avoid lateral forces in floorbeams from bottom chord member extension. Longitudinal movement of stringers at contraction joints is typically enabled with the ends of stringers supported on a bracket bolted to the web plate of the floorbeam under the end of the stringer.
42 The horizontal struts in the bottom lateral bracing system are floorbeams with very small axial compressive forces and therefore are not considered in the analysis of the externally applied lateral forces.
43 AREMA (2024) recommends that W_L be established as 1.44 kPa (30 psf) on a loaded span.
44 Because the magnitude of W_{LB} is relatively small, in some cases the bottom lateral bracing diagonal web members are not required and only struts between the vertical members of the truss are installed for space frame stability.
45 The analysis is like that for the bottom (loaded) chord members of through-truss spans.

REFERENCES

American Railway Engineering and Maintenance-of-Way Association (AREMA), 2024, *Manual for Railway Engineering*, Chapter 15, Steel Structures, Lanham, MD.

Bleich, F., 1952, *Buckling Strength of Metal Structures*, McGraw-Hill, NY.

Fryba, L., 1996, *Dynamics of Railway Bridges*, Thomas Telford, NY.

Galambos, T. V., *Guide to Stability Design Criteria for Metal Structures*, John Wiley and Sons, NY.

Unsworth, J.F., 2018, *Design and Construction of Modern Steel Railway Bridges*, 2nd Edition, CRC Press, Boca Raton, FL.

BIBLIOGRAPHY

Grinter, L.E., 1942, *Theory of Modern Structures*, Vol. 1, Macmillan, New York.

Hool, G.S., and Kinne, W.S., 1923, *Stresses in Framed Structures*, McGraw-Hill, New York.

Part B

Structural Design of Railway Truss Bridges

9 Design of Railway Truss Members

9.1 INTRODUCTION

Truss chord and web members must be designed for stresses due to maximum axial tension; maximum axial compression; combined axial tension and flexure[1]; combined axial compression and flexure; secondary bending; and web member overload. Truss members must also be designed for effective stress ranges due to fluctuating tensile stresses from live load and live load stress reversals in web members.

9.2 STRENGTH DESIGN FOR MAXIMUM AXIAL TENSION

As outlined in Chapter 10, the maximum axial tension in truss members determined on the gross cross-sectional area must not exceed $0.55F_y$. and the maximum axial tension in truss members determined on the effective net cross-sectional area[2] must not exceed $0.47F_u$.

9.3 STRENGTH DESIGN FOR MAXIMUM AXIAL COMPRESSION

As outlined in Chapter 11, the maximum axial compression in centrally loaded[3] truss members determined on the gross cross-sectional area must not exceed F_{call}, which is determined considering the effective slenderness, Kl/r_{min}, of the compression member as shown in Table 9.1.

TABLE 9.1
Allowable Axial Compressive Stress in Truss Members

Kl/r_{min}	F_{call} (psi)	F_{call} (MPa)	Axial Compression
$\le 0.629\sqrt{\frac{E}{F_y}}$	$0.55F_Y$	$0.55F_y$	Yielding
$> 0.629\sqrt{\frac{E}{F_y}}$ but $< 5.034\sqrt{\frac{E}{F_y}}$	$0.60F_y - \left(17{,}500\frac{F_y}{E}\right)^{\frac{3}{2}}\left(\frac{Kl}{r_{min}}\right)$	$0.60F_y - \left(635\frac{F_y}{E}\right)^{\frac{3}{2}}\left(\frac{Kl}{r_{min}}\right)$	Inelastic buckling
$\ge 5.034\sqrt{\frac{E}{F_y}}$	$\frac{0.514\pi^2 E}{\left(\frac{Kl}{r_{min}}\right)^2}$	$\frac{0.514\pi^2 E}{\left(\frac{Kl}{r_{min}}\right)^2}$	Elastic (Euler) buckling

where

F_y = yield stress of the steel grade as specified in ASTM or other applicable specifications;
E = modulus of elasticity of steel = 200,000 MPa (29,000,000 psi);
l = length of the compression member;
Kl = effective length of the compression member;
K = 0.75 for compression members with bolted or welded end connections;[4]

DOI: 10.1201/9781003469971-11

r_{min} = the minimum radius of gyration of the compression member = the minimum value of $\sqrt{\frac{I_x}{A_g}}$ or $\sqrt{\frac{I_y}{A_g}}$;

I_x, I_y = the moment of inertia of the compression member in the longitudinal and lateral directions, respectively; and

A_g = the gross cross-sectional area of the compression member.

9.4 STRENGTH DESIGN FOR COMBINED AXIAL FORCES AND BENDING

As outlined in Chapters 10 and 11, the combined effects of the axial forces and bending must be considered in member design. Therefore, in addition to tensile and compressive axial stresses, truss members can be subjected to tensile and compressive primary and/or secondary flexural stresses.

9.4.1 Strength Design for Combined Axial Tension Forces and Bending

A tensile force reduces the bending effects on the member when tensile axial forces act simultaneously with bending. For the member shown in Figure 9.1,

$$M = \frac{wL^2}{8} - T\Delta \tag{9.1}$$

However, the deflection, Δ, is dependent on the bending moment, M, which is itself dependent on Δ, which is

$$\Delta = \frac{5wL^4}{384EI} - \frac{T\Delta L^2}{8EI} \tag{9.2}$$

and can be solved iteratively. The bending moment is

$$M = \frac{wL^2}{8} - \frac{T}{EI}\left(\frac{5wL^4}{384} - \frac{T\Delta L^2}{8}\right) \tag{9.3}$$

However, since the effect of tensile force T on Δ can be conservatively neglected in the analysis[5] of linear elastic members, the principal of superposition is applied to ensure that failure by tensile yielding or fracture does not occur.

The combined stresses due to axial tension on the gross cross-sectional area, σ_{ag}, and flexural tension on the net section modulus, σ_{bn}, must not exceed $0.55F_y$:

$$\sigma_{ag} + \sigma_{bn} \le 0.55F_y \tag{9.4a}$$

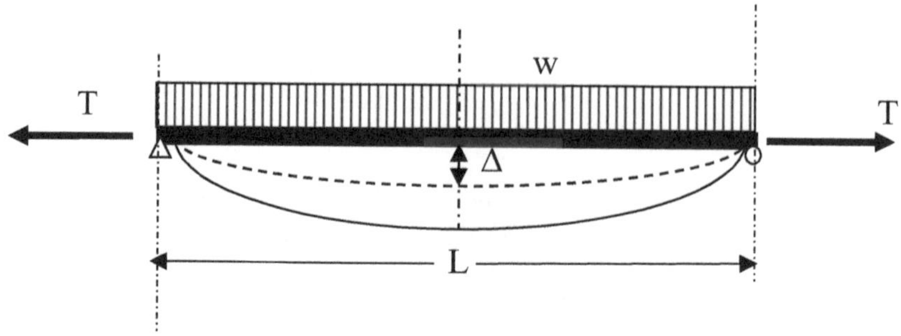

FIGURE 9.1 Axial tension combined with bending.

Correspondingly, since the axial stresses on the effective net cross-sectional area, σ_{aen}, must not exceed $0.47F_u$:

$$\frac{\sigma_{aen}}{0.47F_u} + \frac{\sigma_{bn}}{0.55F_y} \leq 1.0 \tag{9.4b}$$

If the combined stresses due to axial tension and flexure result in compressive stress, the flexural compressive stress must not exceed F_{bcall}, as shown in Table 9.2.

TABLE 9.2
Allowable Flexural Compressive Stress in Truss Members

Truss Member	F_{bcall}	Description
I shape bending perpendicular to the web	$0.55F_y$	
Welded or rolled members (other than box members) bending parallel to the web	$0.55F_y - \frac{0.55F_y^2}{6.3\pi^2 E}\left(\frac{l}{r_y}\right)^2$ or $\frac{0.13\pi Ebt_f}{ld\sqrt{1+\mu}}$	F_{bcall} is the larger of the values but not to exceed $0.55F_y$
Bolted built-up members (other than box members) bending parallel to the web	$0.55F_y - \frac{0.55F_y^2}{6.3\pi^2 E}\left(\frac{l}{r_y}\right)^2$	
Rolled channels	$\frac{0.13\pi Ebt_f}{ld\sqrt{1+\mu}}$	not to exceed $0.55F_y$
Box members	$0.55F_y - \frac{0.55F_y^2}{6.3\pi^2 E}\left(\frac{l}{r_e}\right)^2$	The effective slenderness ratio = $\frac{l}{r_e} = \sqrt{\frac{1.105\pi l S_x\sqrt{\Sigma\left(\frac{s}{t}\right)}}{\left(A\sqrt{\frac{I_y}{1+\mu}}\right)}}$

where

l = the distance between lateral supports of the compression flange of the member;

r_y = the radius of gyration of the compression flange and that portion of the web area on the compression side of the axis of bending, about an axis in the plane of the web;

A_f = the cross-sectional area of the smaller flange excluding any portion of the web;

d = overall depth of the member;

S_x = the section modulus of the box type member about its major axis;

A = total area enclosed within the center lines of the box member webs and flanges;

s/t = the ratio of the width of any flange or depth of any web component, s, to the thickness, t, of the component neglecting any portion of flanges which project beyond the box section; and

I_y = the moment of inertia of the box member about its minor axis.

9.4.2 Strength Design for Combined Axial Compression Forces and Bending

Axial compression combined with bending is of greater concern than axial tension combined with bending due to the potential for instability of characteristically slender compression members. The

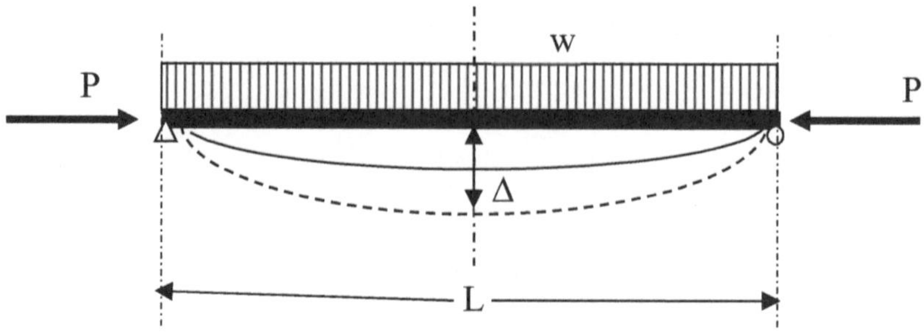

FIGURE 9.2 Axial compression combined with bending.

compressive force increases the bending effects on the member when compressive axial forces act simultaneously with bending. For the member shown in Figure 9.2,

$$M = \frac{wL^2}{8} + P\Delta \tag{9.5}$$

Deflection Δ depends on M, which itself depends on Δ, which is

$$\Delta = \frac{5wL^4}{384EI} + \frac{P\Delta L^2}{8EI} \tag{9.6}$$

and M is

$$M = \frac{wL^2}{8} + \frac{P}{EI}\left(\frac{5wL^4}{384} + \frac{P\Delta L^2}{8}\right) \tag{9.7}$$

Equation 9.6 indicates that the deflections build on themselves and instability can occur due to this $P - \Delta$ effect. Therefore, since an iterative solution to Equation 9.7 is required, which is not efficient for routine design work, semi-empirical interaction equations can be utilized to ensure that failure by combined compressive axial and flexural instability does not occur. The combined stresses due to axial compression and flexural compression must not exceed the interaction equations in Table 9.3.[6]

TABLE 9.3
Allowable Combined Axial and Flexural Compressive Stress in Truss Members

Ratio of Calculated to Allowable Axial Compressive Stress	Interaction Equation
$\frac{f_a}{F_{call}} \le 0.15$	$\frac{f_a}{F_{call}} + \frac{f_{bx}}{F_{bcallx}} + \frac{f_{by}}{F_{bcally}} \le 1.0$
$\frac{f_a}{F_{call}} > 0.15$	$\frac{f_a}{F_{call}} + \frac{f_{bx}}{F_{bcallx}\left(1 - \frac{f_a}{0.514\pi^2 E}\left(\frac{K_x l_x}{r_x}\right)^2\right)} + \frac{f_{by}}{F_{bcally}\left(1 - \frac{f_a}{0.514\pi^2 E}\left(\frac{K_y l_y}{r_y}\right)^2\right)} \le 1.0$
at locations braced in the planes of bending	$\frac{f_a}{0.55F_y} + \frac{f_{bx}}{F_{bcallx}} + \frac{f_{by}}{F_{bcally}} \le 1.0$

where

F_{call} = allowable axial compressive stress (Table 9.1);

F_{bcallx} and F_{bcally} = allowable compressive bending stress about the x and y axes, respectively, (Table 9.2);

f_a = calculated axial compressive stress;

f_{bx} and f_{by} = calculated compressive bending stress about the x and y axes, respectively; and

$\frac{K_x l_x}{r_x}$ and $\frac{K_y l_y}{r_y}$ = ratios of the effective length to the radius of gyration of the compression member about the x and y axes, respectively.

If the combined stresses due to axial compression and flexure result in tensile stress, the flexural tensile stress must not exceed $0.55F_y$ on the net section modulus.

9.5 SECONDARY FLEXURAL STRESSES IN TRUSS MEMBERS

As outlined in Chapter 7, secondary bending occurs in truss members due to eccentricities at connections, members that are not straight, transverse forces on the truss members, distortion of vertical truss members, floorbeam end rotation and/or, truss distortion and joint rigidity.

AREMA (2024) postulates that secondary flexural stresses in truss members can be precluded by careful design, detailing and fabrication to avoid eccentricities, out-of-straightness, transverse forces and distortion of vertical members. Furthermore, AREMA (2024) indicates that secondary stresses due to truss distortion and joint rigidity[7] need not be considered in truss members whose width parallel to the plane of distortion is less than 10% of the length of the member. Nevertheless, AREMA (2024) recommends that secondary flexural stresses due to floorbeam end rotation be considered primary flexural stresses for strength design of truss members directly loaded by floorbeam reactions. AREMA (2024) also recommends that secondary flexural stress ranges from all sources be considered for fatigue design of truss members directly loaded by floorbeam reactions. It should be noted that earlier than AREMA (2024), only secondary stresses exceeding 27.6 MPa (4,000 psi) in tension members and 20.7 MPa (3,000 psi) in compression members were recommended to be treated as primary flexural stresses for member strength design.[8]

9.6 STRENGTH OVERLOAD IN TRUSS WEB MEMBERS

To preclude the premature failure of web members,[9] AREMA (2024) recommends that web members[10] and their connections be designed such that an increase in the design live load that will amplify the total stress in the most highly stressed chord member by 33% will produce total stresses in the web members and their connections not greater than 1.33 times the allowable stresses. However, the most highly stressed chord member is unidentified until final member sizes are established.[11] Nonetheless, it is generally appropriate to consider the chord member with the greatest DL, LL and I forces to determine the web member overload design force, F_{ol}, at normal allowable stress as

$$F_{ol} = \frac{DL_w + K_c\left(LL + I + CF\right)_w}{1.33} \tag{9.8}$$

where

DL_W = dead load in web member under consideration;

$(LL+I+CF)_w$ = live load plus impact and centrifugal force in web member under consideration;

$$K_c = \frac{1.33\left(DL + LL + I + CF\right)_c - DL_c}{\left(LL + I + CF\right)_c} = \frac{133 - \%DL_c}{100 - \%DL_c};$$

$\%DL_c$ = the percentage of total load in the truss chord attributed to dead load = $100DL_c/(DL + LL + I + CF)_c$;

DL_c = dead load in most highly stresses chord member;
$(LL+I+CF)_c$ = live load plus impact and centrifugal force in most highly stressed chord member;
$(DL+LL+I+CF)_c$ = dead load, live load plus impact, and centrifugal force in most highly stressed chord member; and
$(LL+I+CF)_{in}$ = increased live load plus impact and centrifugal force in web member under consideration = $K_c\left(LL+I+CF\right)_w$.

Example 9.1

The design loads for the web members of the Pratt truss span of Example 4.11 are required considering the AREMA (2024) recommended web member overload stresses. The truss span carries tangent track.

For truss members except hangers, I = maximum impact = 16 + 600/(150-30) + (100/20) = 26.0%

For hanger members, I = 40 − 3(25)²/1600 + (100/20) = 43.8%

CF = 0 (tangent track on the truss span)

The dead load and live load forces in the chord member with greatest forces, U2U3, from Example 4.11 are given in Table E9.1a.

TABLE E9.1a
Forces in the Most Highly Stressed Chord Member

Pratt Through-Truss Member		DL_C (kips)	LL_C (kips)	$(LL + I + CF)_c$ (kips)	$(DL + LL + I + CF)_c$ (kips)	K_c
U2U3	Top chord	−120.54	−402.52	−507.18	−627.72	1.408

The dead load and live load forces in web members from Example 4.11 are given in Table E9.1b.

TABLE E9.1b
Overload Forces in Web Members

Pratt Through-Truss Member		DL_W (kips)	LL_W (kips)	$(LL + I + CF)_W$ (kips)	$(LL + I + CF)_{in}$ (kips)	F_{ol} (kips)
U1L1	Vertical	37.50	151.28	217.54	306.30	258.50
U2L2	Vertical	−18.75	−125.81/59.41	−158.52/74.86	−223.20/105.40	−181.92/65.15
U3L3	Vertical	0	0	0	0	0
U1L2	Diagonal	69.13	264.40/−20.69	333.14/−26.07	469.06/−36.71	404.65/24.38
U2L3	Diagonal	23.04	154.64/−73.01	194.85/−91.99	274.35/−129.52	223.60/−80.06

Table E9.1c, which compares the forces, shows that the increases in web member design forces due to the web overload criterion are small for this 150 ft long truss example. However, the web members of longer spans may entail greater increases in web member design forces due to the web overload criterion.

TABLE E9.1c
Comparison of Web Member Forces from DL + LL + I with Overload Forces

Member	DL_W + $(LL+I+CF)_W$ (kips)	F_{ol} (kips)	% difference
U1L1	255.04	258.50	+1.4
U2L2	−177.27/56.11	−181.92/65.15	+2.6/+16.1
U3L3	0	0	0
U1L2	402.27/43.06	404.65/24.38	+0.6
U2L3	217.89/−68.95	223.60/−80.06	+2.6/+16.1

9.7 FATIGUE DESIGN OF TRUSS MEMBERS

Tensile stress ranges initiate and propagate fatigue cracks. Therefore, fatigue design criteria are considered when the member is subject to a net applied tensile stress due to the combination of the member dead load (DL) and fatigue live load (LL + I_f).

Since DL does not contribute to stress range, only LL + I_f needs be considered when determining the effective stress range, S_R. S_R is the algebraic difference between the maximum and minimum calculated stress due to LL + I_f and, if the truss carries curved track, centrifugal force (CF). The fatigue design impact, I_f, is 65% of the maximum design impact, I.[12]

Where stress reversal occurs, S_R will include both the tensile and compressive stresses because members subject to cyclical loading at or near an initial flaw will be subject to an entire effective stress range cycle in tension due to the superposition of tensile residual stresses that may elevate the entire stress range cycle into the tensile stress region. Accordingly, fatigue design must consider effective stress range cycles in tension from LL + I_f and/or from LL tension and compression during stress reversal. S_R must not exceed the allowable fatigue (threshold) stress range, S_{Rfat}. S_{Rfat} for some typical tension member details are shown in Table 9.4.

TABLE 9.4
Allowable Fatigue Stress Ranges in Truss Members

Truss Member	Number of Constant Stress Cycles N	Detail Description	Fatigue Detail Category	S_{Rfat} (MPa)	S_{Rfat} (ksi)	S_R determined on
Truss chord members, end posts and connections	2,000,000[13]	Base metal of non-weathering steel away from welds	A	165	24	Gross section of member
		Base metal of weathering steel away from welds	B	124	18	Gross section of member
		Open fastener holes	D	69	10	Net section of member
		Hand holes and perforations	C	90	13	Net section of member
		Slip-critical connections	B	124	18	Gross section of connected elements
		Non-slip-critical connections	B	124	18	Net section of connected elements
Truss web members and connections in single track spans	>2,000,000[14]	Base metal of non-weathering steel away from welds	A	165	24	Gross section of member
		Base metal of weathering steel away from welds	B	110	16	Gross section of member
		Open fastener holes	D	48	7	Net section of member
		Hand holes and perforations	C	69	10	Net section of member
		Slip-critical connections	B	110	16	Gross section of connected elements
		Non-slip-critical connections	B	110	16	Net section of connected elements

The combined stress ranges due to axial tensile stress ranges, $\Delta\sigma_a$, and flexural tensile stress ranges due to primary and secondary bending, $\Delta\sigma_b$, must not exceed S_{Rfat},

$$\Delta\sigma_a + \Delta\sigma_b \leq S_{Rfat} \tag{9.9}$$

$\Delta\sigma_a$ and $\Delta\sigma_b$ are determined based on the cross-sectional area and section modulus of either the gross section or net section depending on the detail under consideration.

Example 9.2

Determine the effective force ranges from fluctuating live load and force reversal in members of the Pratt truss of Example 4.11:

I_f for members other than hangers = 0.65(26.0%) = 16.9%
I_f for hanger members (U1L1 & U5L5) = 0.65(43.8%) = 28.5%

The axial force ranges (LL + I_f) in tensile members and axial force reversal ranges from Example 4.11 are given in Table E9.2.

TABLE E9.2
Effective Force Ranges in Truss Members

Pratt Through-Truss Member with Tensile Forces		DL (kips)	LL (kips)	I_f (kips)	LL + I_f (kips)	DL + LL + I_f (kips)	Force Range (kips)
L0L1 & L5L6	Bottom chords	66.96	231.39	39.10	270.49	337.45	270.49
L1L2 & L4L5	Bottom chords	66.96	231.39	39.10	270.49	337.45	270.49
L2L3 & L3L4	Bottom chords	107.14	354.96	59.99	414.95	522.09	414.95
U1L1 & U5U5	Hangers	37.50	156.00	43.11	200.46	237.96	200.46
U2L2 & U4L4	Verticals	−18.75	−125.81/59.41	−21.26/10.04	−147.07/69.45	−165.82/50.70	216.52
U1L2 & L4U5	Diagonals	69.13	264.40/−20.69	44.68/−3.50	309.08/−24.19	378.21/44.94	333.27
U2L3 & L3U4	Diagonals	23.04	154.64/−73.01	26.13/−12.34	180.77/−85.35	203.81/−62.31	266.12

$\Delta\sigma_a$ is determined from the effective axial force ranges and the cross-sectional area of either the gross section or net section depending on the member detail under consideration.

9.8 STRESS REVERSAL IN THE DESIGN OF TRUSS MEMBERS

As shown in Example 9.2, the fatigue design stress range in the bottom chord members occurs due to the live load tensile force range, and in web members occurs due to changes in panel shear forces as the live load traverses the truss span.

However, as indicated in Chapter 8, the bottom chord members in windward truss end panels can experience strength design stress reversal due to load combinations including lateral forces from wind on the structure, wind on the live load, nosing and longitudinal live load forces. This is particularly conceivable for load combinations with light live loads, such as empty rail cars, or heavier live loads with considerable longitudinal forces from locomotive traction and train braking.[15]

9.9 LOAD COMBINATIONS FOR THE DESIGN OF TRUSS MEMBERS

Members subject to axial tensile stresses resulting from DL, LL, I and CF must be designed so that the maximum stresses do not exceed the allowable axial tension stresses in Section 9.2 or, if the axial tension member is combined with primary and/or secondary bending, Section 9.4.1. Members subject to axial compressive stresses resulting from DL, LL, I and CF must be designed so that the maximum stresses do not exceed the allowable axial compression stresses in Section 9.3 or, if the axial compression member is combined with primary and/or secondary bending, Section 9.4.2.

Members subject to stresses from DL, LL, I, CF, W_L, LF, N, CWR and other forces (OF)[16] must be designed with an allowable stress factor of 1.25 applied to the allowable axial tension stresses in Section 9.2, the allowable axial compression stresses in Section 9.3 or the allowable combined axial and flexural stresses of Section 9.4.[17] The allowable flexural stresses of Section 9.4 must be considered where secondary bending, as outlined in Section 9.5, occurs in members.

As depicted in Section 9.6, web members subject to axial stresses resulting from DL, LL, I and CF must also be designed so that the maximum stresses do not exceed the allowable axial stresses in Sections 9.2 through 9.4 with an allowable stress factor of 1.33. Members subject to only W_U must be designed for the allowable stresses in Sections 9.2 and 9.3 or, if applicable, Section 9.4. Members subject to S_R from LL plus I_f must be designed to not exceed S_{Rfat} as shown in Section 9.7. Table 9.5 indicates the allowable stress factors for the design of main truss members for some load combinations.

TABLE 9.5
Allowable Stress Factors for the Design of Truss Members

Load Combination		Truss Members	Allowable Stress Factor
LC1A	DL + LL + I + CF	Chord and web members	1.00
LC1B	LC1A + W_L + LF + N + CWR + OF	Chord, web and bracing members[18]	1.25
LC2	$(LL + I + CF)_{in}$	Web members	1.33
LC3	LL + I_f	Chord and web members with tensile force range from live load or force reversal[19]	1.00
LC4	DL + W_U	Chord and bracing members	1.00
LC5	B_{RL} + N + W_L + CWR_L + CF	Bracing members between compression chord members for loaded truss spans[20]	1.25

The member forces in Table 9.5 are due to,

DL = dead load (vertically applied external gravity load)
LL = live load (concentrated and/or uniformly distributed vertically applied external load)
I = maximum impact or dynamic increment for strength (vertically applied external load)
I_f = mean impact or dynamic increment for fatigue (vertically applied external load)
CF = the centrifugal effect of live load on curved track (laterally applied external load)
W_L = wind on the loaded truss span (laterally and longitudinally applied external forces from wind on the structure (W_{LT}, W_{LB}, LW_{LT}, LW_{LB}) and wind on the live load (W_{LL}, LW_{LL})). LW_{LT}, LW_{LB} and LW_{LL} apply to main truss chord and web members only.
W_{LT} = lateral wind load at top chord of loaded truss span
W_{LB} = lateral wind load at bottom chord of loaded truss span
LW_{LT} = longitudinal wind load at top chord of loaded truss span
LW_{LB} = longitudinal wind load at bottom chord of loaded truss span
W_{LL} = lateral wind load on the live load
LW_{LL} = longitudinal wind load on the live load
W_U = wind on the unloaded truss span (laterally and longitudinally applied external forces from wind on the structure (W_{UT}, W_{UB}, LW_{UT}, LW_{UB})). LW_{UT} and LW_{UB} apply to main truss chord and web members only.
W_{UT} = lateral wind load at top chord of unloaded truss span
W_{UB} = lateral wind load at bottom chord of unloaded truss span
LW_{UT} = longitudinal wind load at top chord of unloaded truss span

LW_{UB} = longitudinal wind load at bottom chord of unloaded truss span

LF = the effect of live load traction (LF_T) or braking (LF_B) (longitudinally applied external loads)

N = nosing or truck hunting on loaded truss span (laterally applied external load)

CWR = the forces due to restraint of continuously welded rail on tangent track (longitudinally applied external force)

CWR_L = the forces due to restraint of continuously welded rail on curved track (laterally applied external force)

OF = other forces such as those related to statical indeterminacy

$(LL + I + CF)_{in}$ = the live load + maximum impact load + centrifugal effect forces in web members for the web member strength overload condition outlined in Section 9.6

B_R = notional force for compression chord bracing members (laterally applied notional force B_{RL} for loaded truss spans and B_{RU} for dead load only on truss spans)

In addition, members of the truss stressed during construction have allowable stress factors of 1.25 for erection loads (EL) and 1.33 for erection loads combined with wind on the unloaded structure ($EL + W_U$). Members of the truss stressed during maintenance jacking have allowable stress factors of 1.50 for DL.

Example 9.3

Determine the load combinations for the design of the main Pratt truss members of Example 4.11. The load combinations that apply to the main truss chord and web members are LC1A, LC1B, LC2, LC3 and LC4.

Example 4.11 provides DL + LL +I for load combination LC1A (strength load combination for external vertical forces on loaded trusses).

Example 4.11 and Example 8.17 provide LC1A + W_L + LF + N for load combination LC1B (strength load combination for external vertical, lateral and longitudinal forces on loaded trusses).

Example 9.1 provides $DL_W + K_c(LL + I)_W$ for load combination LC2 (strength load combination for web member overload).

Example 9.2 provides LL + I_f for load combination LC3 (fatigue load combination for external vertical live load forces).

Example 8.18 and Example 8.19 provide W_U for load combination LC4 (strength load combination for external lateral and longitudinal forces on unloaded trusses).

From Example 4.11,

For truss members except hangers, $I = 16 + 600/(150 - 30) + (100/20) = 26.0\%$
For hanger members, $I = 40 - 3(25)^2/1600 + (100/20) = 43.8\%$
Tangent track on bridge, CF = 0

Examples 4.11 and 8.17 give the forces in Table E9.3a.

From Example 9.1, the calculated member forces are given in Table E9.3b.

From Example 9.2, Table E9.3c gives the calculated member force ranges.

From Example 8.18 and Example 8.19, Table E9.3d gives the calculated member forces.

From Examples 4.11, 8.18 and 8.19, Table E9.3e gives the calculated member forces.

Table E9.3f summarizes the design forces.

Table E9.3f indicates the axial tension and compression design forces for strength and the design force range for fatigue. The tables in Example 9.3 are typically shown on truss design drawings as components of a tabulation of stresses (stress sheet).

TABLE E9.3a
Member Forces for Strength Load Combinations LC1A and LC1B

Pratt Through-Truss Member	LC1A (DL + LL + I) (kips)	(W_L + LF + N) (kips)	1.25(LC1B) = LC1A+W_L+LF+N (kips)	LC1B (kips)	Design Force LC1A or LC1B (kips)
L0L1	358.51	∓224.27	582.78	466.22	466.22
L1L2	358.51	∓237.11	595.62	476.50	476.50
L2L3	554.39	∓238.09	792.48	633.98	633.98
L3L4	554.39	∓187.99	742.38	593.90	593.90
L4L5	358.51	∓132.42	490.93	392.74	392.74
L5L6	358.51	∓56.90	415.41	332.33	358.51
L0U1 & U5L6	−616.78	∓48.50	−665.28	−532.22	−616.78
U1U2 & U2U3	−554.39	12.02/−23.35	−577.74	−462.19	−554.39
U2U3 & U3U4	−627.72	10.50/−13.87	−641.59	−513.27	−627.72
U1L1 & U5L5	255.04	∓18.18	273.22	218.58	255.04
U2L2 & U4L4	−177.26/56.11	∓30.93	−208.19/87.04	−166.55/69.63	−177.26/69.63
U3L3	0	∓24.42	∓24.42	∓19.54	∓24.42
U1L2 & L4U5	402.27/43.06	∓25.55	427.82	342.26	402.37
U2L3 & L3U4	217.89/−68.95	∓19.10	236.99/−88.05	189.59/−70.44	217.89/−70.44

TABLE E9.3b
Member Forces for Strength Load Combinations LC1A, LC1B and LC2

Pratt Through-Truss Member	LC2 (kips)	Strength Design Force LC1A, LC1B or LC2 on Loaded Span (kips)
L0L1	n/a	466.22
L1L2	n/a	476.50
L2L3	n/a	633.98
L3L4	n/a	593.90
L4L5	n/a	392.74
L5L6	n/a	358.51
L0U1 & U5L6	n/a	−616.78
U1U2 & U2U3	n/a	−554.39
U2U3 & U3U4	n/a	−627.72
U1L1 & U5L5	258.50	258.50
U2L2 & U4L4	−181.92/65.15	−181.92/69.63
U3L3	0	+/−24.42
U1L2 & L4U5	404.65	404.65
U2L3 & L3U4	223.60/−80.06	223.60/−80.06

TABLE E9.3c
Member Force Ranges for Fatigue Load Combination LC3

Pratt Through-Truss Member	Fatigue Design Force Range LC3 (kips)	Force Range Type
L0L1	270.49	Tensile force range
L1L2	270.49	Tensile force range

(Continued)

TABLE E9.3c (*Continued*)
Member Force Ranges for Fatigue Load Combination LC3

Pratt Through-Truss Member	Fatigue Design Force Range LC3 (kips)	Force Range Type
L2L3	414.95	Tensile force range
L3L4	414.95	Tensile force range
L4L5	270.49	Tensile force range
L5L6	270.49	Tensile force range
U1L1 & U5L5	200.46	Tensile force range
U2L2 & U4L4	216.52	Force reversal range
U1L2 & L4U5	333.27	Force reversal range
U2L3 & L3U4	266.12	Force reversal range

TABLE E9.3d
Member Forces from Wind on Unloaded Truss for Strength Load Combination LC4

Pratt Through-Truss Member	Lateral Wind Load on Leeward Truss (kips)	Lateral Wind Load on Windward Truss (kips)	Longitudinal Wind Load (kips)	Wind Force on Unloaded Truss Members (kips)
L0L1	30.37	−30.37	+/−48.62	+/−78.99
L1L2	55.09	−55.09	+/−41.28	+/−96.37
L2L3	67.45	−67.45	+/−32.30	+/−99.75
L3L4	67.45	−67.45	+/−21.63	+/−89.08
L4L5	55.09	−55.09	+/−12.63	+/−67.72
L5L6	30.37	−30.37	+/−5.33	+/−35.70
L0U1 & U5L6	−25.67	25.67	+/−2.88	+/−28.55
U1U2 & U2U3	0	−18.88	+/−2.12	+/−21.00
U2U3 & U3U4	18.88	−25.17	+/−1.28	20.16/−26.45
U1L1 & U5L5	−10.72	10.72	0	+/−10.72
U2L2 & U4L4	−10.72	10.72	+/−2.35	+/−13.07
U3L3	−10.72	10.72	0	+/−10.72
U1L2 & L4U5	0	0	+/−2.88	+/−2.88
U2L3 & L3U4	0	0	+/−2.88	+/−2.88

TABLE E9.3e
Member Forces from Dead Load and Wind on Unloaded Truss for Strength Load Combination LC4

Pratt Through-Truss Member	Dead Load Force, DL (kips)	Wind Force on Unloaded Truss Members (kips)	Strength Design Force on Unloaded Truss Members (kips)
L0L1	66.96	+/−78.99	145.95/−12.03
L1L2	66.96	+/−96.37	163.33/−29.41
L2L3	107.14	+/−99.75	206.89
L3L4	107.14	+/−89.08	196.22
L4L5	66.96	+/−67.72	134.68/−0.76
L5L6	66.96	+/−35.70	102.66

TABLE E9.3e (*Continued*)
Member Forces from Dead Load and Wind on Unloaded Truss for Strength Load Combination LC4

Pratt Through-Truss Member	Dead Load Force, DL (kips)	Wind Force on Unloaded Truss Members (kips)	Strength Design Force on Unloaded Truss Members (kips)
L0U1 & U5L6	−115.17	+/−28.55	−143.72
U1U2 & U2U3	−107.14	+/−21.00	−128.14
U2U3 & U3U4	−120.54	20.16/−26.45	−146.99
U1L1 & U5L5	37.50	+/−10.72	48.22
U2L2 & U4L4	−18.75	+/−13.07	−31.82
U3L3	0	+/−10.72	+/−10.72
U1L2 & L4U5	69.13	+/−2.88	72.01
U2L3 & L3U4	23.04	+/−2.88	25.92

TABLE E9.3f
Main Truss Member Strength and Fatigue Design Forces

Truss Member	Strength Design Tension Force (kips)	Strength Design Compression Force (kips)	Strength Member Design	Fatigue Design Force (kips)
L0L1 & L5L6	466.22	−12.03	Tension member design with negligible compression (2.6%) due to wind on unloaded span	270.49
L1L2 & L4L5	476.50	−29.41	Tension member design with negligible compression (6.2%) due to wind on unloaded span	270.49
L2L3 & L3L4	633.98	0	Tension member design	414.95
L0U1 & U5L6	0	−616.78	Compression member design	0
U1U2 & U2U3	0	−554.39	Compression member design	0
U2U3 & U3U4	0	−627.72	Compression member design	0
U1L1 & U5L5	258.50	0	Tension member design	194.39
U2L2 & U4L4	69.63	−181.92	Compression and tension member design	216.52
U3L3	24.42	−24.42	Negligible forces due to wind on loaded span and nominal cross-section only	0
U1L2 & L4U5	404.65	0	Tension member design	333.27
U2L3 & L3U4	223.60	−80.06	Tension and compression member design	266.12

Example 9.4

Determine the load combinations for the design of the bracing members of the Pratt truss in Chapter 8. The load combinations that apply to the bracing members are LC1B, LC4 and LC5.

Example 8.12 provides W_L+ N for LC1B and B_{RL} + W_{LT} for LC5 (strength load combination for external lateral forces on loaded trusses).

Example 8.18 provides W_U for LC4 and B_{RU} + W_{UT} for LC5 (strength load combination for external lateral forces on loaded trusses).

From Examples 8.12 and 8.18, Tables E9.4a and b give the top and bottom bracing lateral forces.

TABLE E9.4a
Top Lateral Bracing Member Strength Design Forces

Top Horizontal Lateral Bracing Web Member	Force in Top Horizontal Lateral Bracing Web Member of Leeward and Windward Trusses (kips)		
	1.25(LC1B) = 1.25 (B_{RL} + W_{LT})	1.25(LC5) = 1.25 (B_{RU} + W_{UT})	Design force
Top bracing diagonals in panels 1-2 and 4-5	58.87	32.60	47.10
Top bracing diagonals in panels 2-3 and 3-4	57.50	21.63	46.00
Top bracing strut at U1 and U5	−39.15	−44.78/19.39	−35.82/15.51
Top bracing strut at U2 and U4	−41.14	−24.92/8.65	−32.91/6.92
Top bracing strut at U3	−42.97	−16.03	−34.38

TABLE E9.4b
Bottom Lateral Bracing Member Strength Design Forces

Bottom Horizontal Lateral Bracing Web Member	Force in Bottom Horizontal Lateral Bracing Web Members of Leeward and Windward Trusses (kips)		
	1.25(LC1B) = 1.25(W_{LB} + N)	LC4 (W_{UB})	Design Force
Bottom bracing diagonals in panels 0-1 and 5-6	+/−33.86	+/−19.77	+/−27.09
Bottom bracing diagonals in panels 1-2 and 4-5	+/−22.58	+/−13.18	+/−18.06
Bottom bracing diagonals in panels 2-3 and 3-4	+/−13.54	+/−7.90	+/−10.83

NOTES

1 The flexural stresses are tensile at the extreme fibers on one side of the member and compressive at the extreme fibers on the other side of the member.
2 The effective net section is determined by considering the net section from the removal of holes and perforations and the distribution of tensile stresses at connections (shear lag).
3 If the truss members are not centrally loaded, secondary stresses due to the eccentricity of the force at the end connection might need to be considered depending on their magnitude.
4 In accordance with AREMA (2024).
5 In usual structures the effect is small (Bresler, Lin and Scalzi, 1968).
6 The development of these interaction equations is shown in Chapter 11.
7 Truss distortion and joint rigidity are a common source of secondary flexure.
8 This recommendation is consistent with allowable stress design practice, which considers that allowable stresses include about 20% to 30% for secondary flexural stresses.
9 This recommendation by AREMA is because of a historical concern with respect to web members failing prior to chord members.
10 Some engineers consider the end posts of through-truss spans diagonal web members for calculating web member overload forces.
11 Typically based on the allowable stresses for strength and allowable fatigue stress ranges.
12 Maximum design impact must be applied for the fatigue design of some members and machinery in movable bridges, bridges with a discontinuity in the rail and members supporting special trackwork.
13 For truss chord and end post members, and their connections, 2,000,000 constant stress range cycles are appropriate for typical truss spans with length exceeding 30 m (100 ft).
14 For truss web members and their connections in truss spans loaded on one track, more than 2,000,000 constant stress range cycles are appropriate. For truss web members and their connections in truss spans loaded on two tracks, 2,000,000 constant stress range cycles are appropriate.
15 Stress reversal in bottom chord members may not be an issue for modern truss member sections, such as box-section members.
16 Other forces such as those from temperature and settlement of supports.
17 However, the section of the member must not be less than that required to meet the requirements of the member subject to DL + LL + I + CF.

18 For bracing members, LC1A loads are not applicable. Typically, LC1B loads that apply to lateral bracing members are W_L and N. LF and CWR loads are transmitted to main truss chord and web members and do not affect lateral bracing members.
19 The allowable fatigue stress range, S_{Rfat}, is the threshold stress range.
20 For loaded through-truss spans, LC5 load N will not apply. For unloaded truss spans, LC5 load is $B_{RU} + W_U + CWR_L$.

BIBLIOGRAPHY

American Railway Engineering and Maintenance-of-way Association (AREMA), 2023, *Manual for Railway Engineering (MRE) Volume 2*, Chapter 15, Steel Structures, Lanham, MD.

American Railway Engineering and Maintenance-of-way Association (AREMA), 2024, *Manual for Railway Engineering (MRE) Volume 2*, Chapter 15, Steel Structures, Lanham, MD.

Bresler, B., Lin, T.Y., and Scalzi, J.B., 1968, *Design of Steel Structures*, 2nd edition, John Wiley & Sons, New York.

10 Design of Truss Tension Members

10.1 INTRODUCTION

Axial tension members in railway truss bridge spans are frequently nonredundant steel tension members (NSTM).[1] Therefore, the strength (yielding and ultimate) and fatigue limits require attention in design. AREMA (2024) recommends load combinations (see Chapter 9) for the strength and fatigue design of truss members subjected to primary axial tension and secondary[2] flexural stresses and stress ranges.

Accordingly, Table 10.1 indicates the recommended load combinations and primary and secondary stresses for the strength design of truss tension members. Table 10.2 designates the recommended load combinations and primary and secondary stress ranges for the fatigue design of main truss tension members.[3] Brittle fracture is typically mitigated by appropriate material selection, design detailing and fabrication quality control. Serviceability criteria relating to axial tension members[4] must also be considered in truss tension member design.

TABLE 10.1
Requirements for Tension Member Strength Design

Truss Tension Element	Strength Design			
	Load combinations	Primary axial stresses	Primary flexural stresses due to floorbeam distortion	Secondary flexural stresses[5]
Chord members	LC1A or LC1B	On net cross-sectional area of member or effective net cross-sectional area of connection[6]	N/A	On net section modulus if member width parallel to plane of distortion greater than 10% of member length
Diagonal web members	LC1A or LC1B or LC2	On net cross-sectional area of member or effective net cross-sectional area of connection	N/A	On net section modulus if member width parallel to plane of distortion greater than 10% of member length
Vertical web members (post, hanger, sub-vertical members)	LC1A or LC1B or LC2	On net cross-sectional area of member or effective net cross-sectional area of connection	On net section modulus[7]	On net section modulus if member width parallel to plane of distortion greater than 10% of member length

DOI: 10.1201/9781003469971-12

TABLE 10.2
Requirements for Tension Member Fatigue Design

Tension Element	Fatigue Design			
	Load Combination	Primary Axial Stress Ranges	Primary Flexural Stress Ranges due to Floorbeam Distortion	Secondary Flexural Stress Ranges
Truss members other than vertical members directly loaded directly by floorbeam reactions[8]	LC3	On net cross-sectional area of member or effective gross or net cross-sectional area of the member elements that are directly connected	N/A	On net section modulus if member width parallel to plane of distortion greater than 10% of member length[9]
Vertical web members directly loaded by floorbeam reaction (hanger, subvertical members)	LC3	On net cross-sectional area of member or effective gross or net cross-sectional area of the member elements that are directly connected	On net section modulus[10]	On net section modulus[11]

10.2 SECONDARY BENDING IN THE STRENGTH AND FATIGUE DESIGN OF TRUSS TENSION MEMBERS

AREMA (2024)[12] and Table 10.1 indicate that secondary flexural stresses must be considered for strength design load combinations LC1A, LC1B or LC2 of members with width parallel to the plane of distortion and greater than 10% of member length. However, for strength design, vertical web members must also consider secondary flexural stresses from floorbeam end rotation as primary stresses.[13] AREMA (2024) and Table 10.2 indicate that secondary flexural stress ranges from all sources, including floorbeam end rotation, which is considered a primary stress range, must be determined for fatigue design LC3 of vertical members directly loaded by floorbeam reactions.

The shear stresses associated with the secondary bending of ordinary truss members are typically negligible. In-plane secondary flexural stresses will slightly, but not appreciably, reduce primary axial stresses, and therefore, the total combined primary axial and secondary flexural stresses are normally greater than those due to axial tension[14] alone. As indicated in Chapter 7 and Chapter 9, secondary flexural stresses in truss tension members may be due to eccentricities at connections,[15] transverse forces on members,[16] vertical distortion of members,[17] rotational restraint at end joints,[18] floorbeam end rotation from transverse frame action at vertical members[19] and/or truss distortion and joint rigidity.[20]

The secondary flexural stresses due to eccentricities at connections are generally precluded by careful detailing of working points. However, if eccentricities at joints exist, they are accounted for as shown in Chapter 7. Transverse bending of truss members due to self-weight and forces between panel points is typically negligible for usual truss designs but if present are also determined as shown in Chapter 7.

Secondary bending from vertical member distortion is avoided by providing vertical members with robust cross-sectional areas or allowing for such extension in design. Nevertheless, secondary bending due to vertical member distortion (such as supporting struts in subdivided truss spans) can be estimated as shown in Chapter 7. Secondary flexural stresses from rotational restraint at bearings and truss joints are also typically small for members with width parallel to the plane of distortion less than 10% of the member length,[21] but if necessary, can also be determined as shown in Chapter 7.

The out-of-plane bending due to floorbeam end rotation is often considerable and must be determined as a primary flexural stress for all vertical members[22] when considering strength design. For fatigue design of members directly loaded by floorbeam reactions, the out-of-plane bending range due to floorbeam end rotation must also be determined as a primary stress range in addition to any other relevant secondary flexural stress ranges. The determination of flexural stresses and stress ranges due to floorbeam end rotation is also outlined in Chapter 7.

For most simple Pratt and Warren truss spans with relatively slender members, the secondary flexural stresses due to joint rigidity, self-weight and member distortion seldom exceed 25% of the primary axial stress.[23] Therefore, in many cases, secondary flexural stresses other than those due to out-of-plane floorbeam end rotation need not be explicitly considered in design because localized yield stresses at the extreme fibers of a member will not precipitate failure. Nevertheless, some members in long or complex truss spans require explicit consideration of secondary flexural stresses from joint rigidity, transverse forces and member distortion.

10.3 NON-REDUNDANT STEEL TENSION MEMBERS

NSTMs in trusses require particular attention to material toughness, member details, fabrication and welding practices, shop inspection and testing. For improved material toughness, more stringent Charpy V-notch impact test results are required for NSTMs, and to avoid tensile stress concentrations that may precipitate cracking, Fatigue Detail Category D, E and E' details[24] must be avoided on NSTMs. AREMA (2024) also recommends that the submerged arc welding process be used for butt splices, flange plate to web plate welds and box member corner welds in NSTM truss members.

10.4 STRENGTH DESIGN OF TRUSS TENSION MEMBERS

As outlined in Table 10.1, strength design entails considering combined axial tension forces and bending moments for members whose width exceeds 10% of the member length and for vertical members in truss spans.

The maximum axial tensile force, T_{max}, due to dead load (DL), live load (LL), maximum impact (I) and centrifugal force (CF), which is load combination[25] LC1A, is determined with a load combination factor of 1.00 applied to allowable stresses that can be increased to 1.25 when also considering lateral, longitudinal and other forces,[26] which is load combination LC1B. Load combination LC2 with a factor of 1.33 is applied to web member overload forces $(LL + I + CF)_{in}$.

10.4.1 Strength Design of Truss Tension Members for Axial Forces

The axial strength of a tension member is contingent upon yielding of the gross cross-sectional area, A_g, occurring prior to failure (defined at ultimate strength) on the effective net cross-sectional area,[27] A_{en}, which is expressed mathematically as

$$F_y A_g \leq \varphi F_U A_{en} \tag{10.1}$$

where

F_y = the tensile yield stress of the steel;
F_u = the ultimate tensile stress of the steel; and
$\phi = 0.85$ = the connection strength capacity reduction factor (Salmon and Johnson, 1980).

The connection strength capacity reduction factor reflects that yielding will occur on the gross cross-sectional area of a tension connection before the capacity of the net cross-section is attained if the ratio of net cross-sectional area to the gross cross-sectional area is greater than 0.85. AREMA (2024) uses a safety factor of 9/5, which when substituted into Equation 10.1 provides

$$0.55 F_y A_g \leq 0.47 F_u A_{en} \tag{10.2}$$

Therefore, the axial strength of tension member T is

$$T = 0.55F_yA_g \tag{10.3a}$$

or

$$T = 0.47F_uA_{en} \tag{10.3b}$$

The design of the tension member should be established based on the lesser of Equations 10.3a or 10.3b where $T \geq T_{max}$ (tension strength is greater than the maximum tensile force in the member).

10.4.1.1 Net Cross-Sectional Area of Tension Members

The gross cross-sectional area of an axial tension member, A_{gm}, is reduced by the greatest cross-sectional area of perforations or open holes to obtain the net cross-sectional area, A_{nm}, of the axial tension member, which is also the effective net cross-sectional area of the axial tension member,[28] A_{enm}.

However, bolted connections at the ends of an axial tension member require consideration of potential failure lines to determine the net cross-sectional area of the connection, A_{nc}, and further consideration of the distribution of tensile stress at the connection to determine the effective net cross-sectional area at the connection, A_{enc}. The gross cross-sectional area, A_{gc}, across a bolted tension member connection is reduced to the net cross-sectional area, A_{nc}, by the cross-sectional area of the fastener holes along the line of potential tensile failure. The potential tensile failure line, w_{nc}, of least length at connections is (Cochrane, 1922),

$$w_{nc} = w_{gc} - \sum\nolimits_{i=1}^{N_b} d_b + \sum\nolimits_{j=1}^{N_{b-1}} \frac{s_j^2}{4g_j} \leq 0.85w_{gc} \tag{10.4}$$

where

w_{gc} = the gross length across the connection;
N_b = the number of bolt holes in the failure line;
d_b = the effective diameter of the bolt holes = bolt diameter + 3 mm (+ 1/8 in);
s = the hole stagger or pitch (the hole spacing in the direction parallel to the load); and
g = the hole gage (the hole spacing in the direction perpendicular to the load).

Therefore, A_{nc} at a connection is

$$A_{nc} = w_{nc}(t_{mc}) = A_{gc} - \left(\sum\nolimits_{i=1}^{N_b} d_b - \sum\nolimits_{j=1}^{N_{b-1}} \frac{s_j^2}{4g_j}\right)(t_{mc}) \tag{10.5a}$$

where t_{mc} = the thickness of the member at the connection.

Where the bolt holes are not staggered, s = 0, and

$$A_{nc} = A_{gc} - \left(\sum\nolimits_{i=1}^{N_b} d_b\right)(t_{mc}) \tag{10.5b}$$

10.4.1.2 Effective Net Cross-Sectional Area of Tension Members

A_{enm} is equal to A_{nm} since there are no shear lag effects away from connections. Similarly, at the end connections of axial tension members where all elements of a member are connected[29] to transfer the tensile force, $A_{enc} = A_{nc}$.

However, shear lag occurs at connections when the tension force in the member is not transmitted by all member elements in the connection.[30] Therefore, A_{enc} is determined to reflect that the tensile force is not uniformly distributed across the entire A_{nc} at the connection, as shown in Figure 10.1.

Shear lag effects are related to the length of the connection[31] and the efficacy of the tension member with respect to the transfer of forces on the shear plane between the member and the connection plate[32] (Munse and Chesson, 1963).

The connection efficiency or shear lag reduction factor, U_c, is described by the ratio of the eccentricity, e_x, which is the distance between the center of gravity of connected member elements and the shear plane as shown in Figure 10.1 and Figure 10.2, and the connection length, L_c, as[33]

$$U_c = \left(1 - \frac{e_x}{L_c}\right) \tag{10.6}$$

Consequently, where not all elements of a member at the connection are fastened with bolts, A_{enc} is

$$A_{enc} = U_c A_{nc} \tag{10.7}$$

Angle members connected by only one leg[34] are particularly susceptible to shear lag effects, and AREMA (2024) recommends U_c between 0.60 and 0.80 depending on the number of bolts per fastener line in the connection.[35]

For members that are continuous through a joint (such as truss chord members continuous across several panels) or connections where load transfer between the chord segments is efficient, AREMA (2024) indicates that shear lag may not be of concern and U_c can be taken as 1.00. However, in such circumstances, consideration of $A_{enc} = 0.90A_{nc}$ could be appropriate for design (Bowles, 1980). Nevertheless, at joints with continuous chord members, U_c is considered for the part of the tension force transferred from diagonal members to chord members at the gusset plate connection.

10.4.2 Strength Design of Truss Tension Members and Connections for Combined Axial Tensile Forces and Tensile Bending

As indicated in Section 9.4.1, the combined stresses due to primary axial tension on the gross cross-sectional area, σ_{ag}, and secondary flexural tension on the net section modulus, σ_{bn}, must not exceed the allowable yield stress, $0.55F_y$. The secondary flexural stresses on the net section modulus may occur parallel (in-plane), σ_{bnx}, and perpendicular (out-of-plane), σ_{bny}, to the plane of the truss span, such that

$$\sigma_{ag} + \sigma_{bnx} + \sigma_{bny} \le 0.55F_y \tag{10.8a}$$

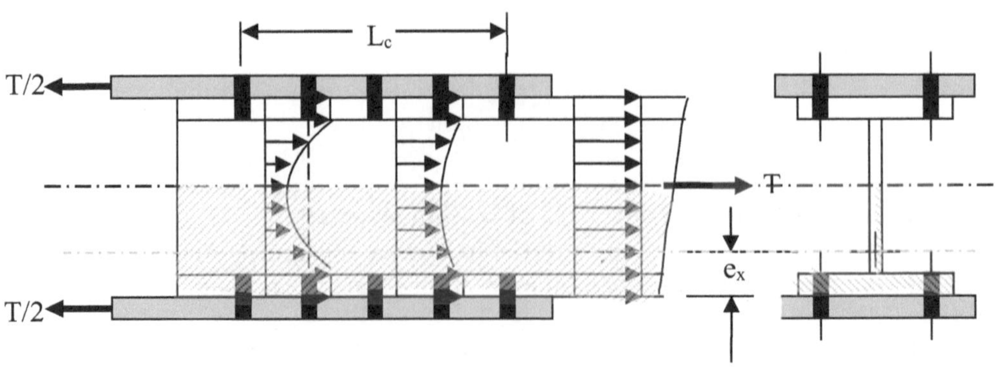

FIGURE 10.1 Shear lag at a tension connection.

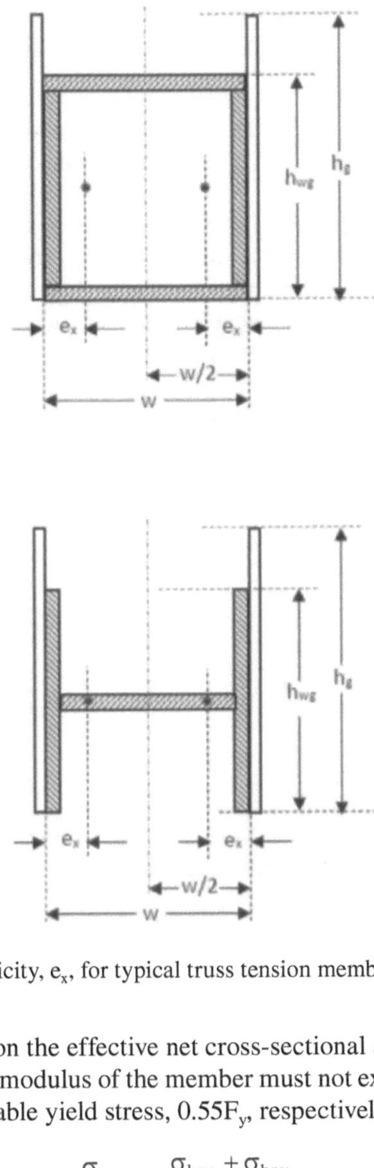

FIGURE 10.2 Shear lag eccentricity, e_x, for typical truss tension member connections.

Also, the primary axial stress on the effective net cross-sectional area, σ_{aen}, and secondary flexural stresses, σ_{bn}, on the net section modulus of the member must not exceed the allowable fracture (ultimate) stress, $0.47F_u$, and allowable yield stress, $0.55F_y$, respectively, such that

$$\frac{\sigma_{aen}}{0.47F_u} + \frac{\sigma_{bnx} + \sigma_{bny}}{0.55F_y} \leq 1 \tag{10.8b}$$

where

σ_{ag} = either the primary axial tension on the gross cross-sectional area of the member, σ_{agm}, or the primary axial tension on the gross cross-sectional area at the connection, σ_{agc}. Typically, $\sigma_{agm} = \sigma_{agc}$.

σ_{bnx} = the in-plane secondary flexural tension on the net section modulus of the member, σ_{bnmx}, or the in-plane secondary flexural tension on the net section modulus at the connection, σ_{bncx}.

σ_{bny} = the out-of-plane secondary flexural tension on the net section modulus of the member, σ_{bnmy}, or the out-of-plane secondary flexural tension on the net section modulus at the connection, σ_{bncy}. Secondary out-of-plane bending from floorbeam end rotation must be considered primary out-of-plane bending for tension member design.

σ_{aen} = either the primary axial tension on the effective net cross-sectional area of the member, σ_{aenm}, or the primary axial tension on the effective net cross-sectional area at the connection, σ_{aenc}.

10.4.3 Strength Design of Truss Tension Members and Connections for Combined Axial Tensile Forces and Compressive Bending

As indicated in Section 9.4.2, members subjected to the minimum primary axial tensile stress and concurrent secondary flexural stress that result in a net flexural compressive stress must not exceed the allowable flexural compressive stress, F_{bcall}, which depends on the member type and propensity for lateral-torsional instability.[36]

H-sections are often used for truss members with the weak axis in the plane of the truss. Consequently, the H-section truss member is subject to strong axis bending from out-of-plane secondary flexure (e.g., bending in vertical member I-sections due to floorbeam end rotation and transverse forces such as wind) and weak-axis bending from in-plane secondary effects (e.g., bending due to connection rigidity and bending of chord members from the extension of vertical tension members). For rolled and welded built-up H-section truss members in strong axis bending, the allowable compressive bending stress, F_{bcall}, is the larger[37] of the two elastic critical lateral torsional buckling stresses, F'_{cr} of Equation 10.9a, or F''_{cr} of Equation 10.9b:[38]

$$F'_{cr}=\left(\frac{0.131\pi E}{ld\left(\sqrt{1+\upsilon}\right)/A_f}\right) \tag{10.9a}$$

$$F''_{cr}=0.55F_y-\frac{0.55F_y^2}{6.3\pi^2 E}\left(\frac{l}{r_y}\right)^2 \tag{10.9b}$$

where

F'_{cr} and F''_{cr} are less than F_y;

l = the length of the H-section member between lateral supports for the truss member in strong axis flexural compression;

d = width of the H-section member (depth if used as I-section);

ν = Poisson's ratio for steel = 0.3;

A_f = the cross-sectional area of the flange plate with the smallest cross-sectional area for the truss member in strong axis flexure; and

r_y = minimum radius of gyration of the compression flange and portion of web in compression about an axis in the plane of the web for the truss member in flexure.

AREMA (2024) also restricts H-section member strong axis slenderness (to preclude Euler buckling) to

$$L/r_y \leq 5.55\sqrt{E/F_y} \tag{10.10}$$

A rolled or built-up H-section truss member bent about the weak axis is not susceptible to lateral–torsional instability because the out-of-plane moment of inertia greatly exceeds the in-plane moment of inertia and, consequently, $F_{bcall} = 0.55F_y$.

Box sections are also frequently used for truss axial force members[39] and provide beneficial lateral–torsional stiffness for both weak and strong axis bending. Accordingly, since the strong and weak axis slenderness of truss box-section members is typically small, $F_{bcall} \approx 0.55F_y$ for both weak- and strong-axis bending. However, if a box-section member bent about the strong axis is appreciably slender,[40] lateral–torsional buckling should be considered with Equation 10.9b and an equivalent slenderness ratio, $(L/r)_e$, yielding

$$F_{bcall}=0.55F_y\left(1-\frac{0.064F_y l S_x\sqrt{\Sigma\left(s/t\right)}}{EA\sqrt{I_y}}\right) \tag{10.11}$$

where S_x = the section modulus of the box-section member about the strong axis and A = the area enclosed within the centerlines of the box-section member web (side) and cover (flange) plates. For cover plates, s/t = the ratio of the cover plate width to thickness neglecting portions that project beyond the web plates, and for web plates, s/t = the ratio of the web plate height to thickness. I_y = the moment of inertia of the box-section member about the weak axis.

10.5 FATIGUE DESIGN OF TRUSS TENSION MEMBERS

As outlined in Table 10.2, fatigue design must consider primary axial force ranges and, in the case of tension members directly loaded by floorbeam reactions, combined primary axial force and secondary[41] bending moment ranges.

The effective axial tensile force range, ΔT, due to LL is the difference between the maximum and minimum axial tensile forces due to LL, fatigue design impact (I_f) and CF. Since DL does not contribute to stress range, for truss members in axial tension under DL, the effective axial tensile force range is LL + I_f + CF. For truss members in axial compression under DL and in axial tension with the application of LL + I_f + CF, the effective axial tensile stress range is determined by the superposition of forces due to the axial compressive DL on A_{gm} or A_{gc} and axial tension from LL + I_f + CF on A_{nm} of the member or the effective net cross-sectional area for fatigue, A_{efat} at connections.

Truss members can also experience live load stress reversal,[42] which can include both an axial tensile and axial compressive component, from the application of LL + I_f + CF. However, since tensile residual stresses in the member[43] can locally raise an entire live load axial stress range cycle into the tensile stress region, ΔT due to LL force reversal in a member is determined as the algebraic difference between the maximum tensile and compressive forces in the member due to LL + I_f + CF.

The allowable fatigue stress range, S_{Rfat} for tension members depends on the Fatigue Detail Category for load-induced fatigue[44] provided by AREMA (2024). Allowable S_{Rfat} need not be considered when the combined axial DL and axial LL + I_f result in a net axial compressive stress in the member. However, members that experience ΔT due to repeated fluctuations in axial tension or force reversal must consider S_{Rfat}.

10.5.1 Fatigue Design of Truss Tension Members for Axial Force Ranges

Axial stress ranges from fluctuating LL and force reversal in members must not exceed S_{Rfat}. Therefore, the fatigue resistance, ΔT_{fat}, of an axial tension member is

$$\Delta T_{fat} = S_{Rfat} A_{efat} \tag{10.12}$$

where A_{efat} = the effective gross or net cross-sectional area of the member elements that are directly connected. This reduction accounts for shear lag effects for fatigue design, which occurs at stress levels below fracture.[45] Where all elements of the member are connected, A_{efat} = A_{nc}; however, where only some of the elements are connected,[46] A_{efat} is either the gross cross-sectional area of only the member elements connected or the net cross-sectional area of only the member elements connected depending on the detail under consideration. For example, slip-critical[47] (or friction-type) bolted connections have an A_{efat} equal to the gross cross-sectional area of only the member elements that are directly connected.

S_{Rfat} depends on the number of design cycles and details of the tension member. For chord and end post members, and their connection design, the number of design cycles is 2,000,000 for truss spans with influence lines exceeding 30.5 m (100 ft).[48] For web member and connection design,[49] the number of design cycles is 2,000,000 for multiple-track truss spans and greater than 2,000,000 for single-track spans. Details such as transitions, apertures or other discontinuities in axial tension members subjected to fluctuating stresses must be carefully considered to ensure an allowable fatigue stress range based on the appropriate detail category.

Tension members without open holes, perforations or other openings may be designed as Fatigue Detail Category A with S_{Rfat} = 165 MPa (24 ksi). However, the allowable fatigue stress range depends on the number of fatigue stress cycles at connections. AREMA (2024) recommends S_{Rfat} based on Fatigue Detail Category B, of 125 MPa (18 ksi) on the gross cross-sectional area of the base metal of connected elements of slip-critical connections subjected to 2,000,000 cycles or fewer or 110 MPa (16 ksi) on the gross cross-sectional area of the base metal of connected elements of slip-critical connections subjected to greater than 2,000,000 stress range cycles. Bolts will generally not experience fatigue failure prior to the base metal, and therefore, AREMA (2024) includes no recommendations concerning allowable shear stress ranges for bolts.

Tension members with large perforations, such as cover plate perforations and hand holes[50] in box-section members, behave as Fatigue Category C details (Bonachera and Connor, 2017). However, where small open holes, perforations, slots, or other openings exist in a tension member, S_{Rfat} may be severely reduced to Fatigue Detail Category D with S_{Rfat} of 48 MPa (7 ksi).

10.5.2 Fatigue Design of Truss Tension Members for Combined Axial Force and Bending Ranges

Tensile stress ranges in members subjected to fluctuating axial forces and bending moments must not exceed S_{Rfat}.

As indicated in Table 10.2, fluctuating stresses in truss tension members that are floorbeam hanger or subvertical members[51] must include all out-of-plane and in-plane secondary flexural stress ranges in the member for fatigue design. Out-of-plane bending ranges from floorbeam end rotation is a secondary stress range that must be considered as a primary flexural stress range for hanger or sub-vertical member fatigue design.

The axial stress range, $\Delta\sigma_{an}$, and flexural stress ranges,[52] $\Delta\sigma_{bn}$, on the net cross-sectional area of the member must not exceed the allowable fatigue stress range

$$\Delta\sigma_{an} + \Delta\sigma_{bnx} + \Delta\sigma_{bny} + \Delta\sigma_{bnfy} \leq S_{Rfat} \tag{10.13}$$

where

$\Delta\sigma_{an}$ = axial stress range on effective net cross-sectional area for fatigue = $\Delta T/A_{efat}$;

$\Delta\sigma_{bnx}$ = in-plane secondary bending stress range on net section modulus of the member;

$\Delta\sigma_{bny}$ = out-of-plane secondary bending stress range on net section modulus of the member; and

$\Delta\sigma_{bnfy}$ = out-of-plane primary bending stress range due to floorbeam end rotation on net section modulus of the member.

10.6 DESIGN OF TRUSS TENSION MEMBERS

Tension members in steel railway truss spans are typically fabricated with structural shapes and plates. Due to large-magnitude live loads, secondary bending, connection geometry and stress reversals, it is often necessary to fabricate built-up tension members using several structural shapes. Accordingly, the most common built-up tension member sections use three welded plates to fabricate H-section members[53] and four welded plates (top cover plate, bottom cover plate and two side or web plates) or two channel shapes and two cover plates to create box-section members. H-section members are typically the most economical members for truss tension members. However, box-section members are often used for tension members carrying large-magnitude loads and forces. The top and bottom cover plates in box-section members may be solid[54] or perforated.

10.6.1 Box-Section Tension Members with Perforated Cover Plates

Closed tension members, such as box-section members, use perforated cover plates, as shown in Figures 10.3 and 10.4, to create an open member for egress of water and to reduce member weight. Perforations in cover plates must be ovaloid or elliptical with a length, c, not greater than twice the perforation width, b, and have a clear distance between perforations, l_p–c, not less than the distance between the nearer lines of connections, b_p.[55] Ovaloid perforations are the most common in modern truss box-section cover plates.

Where perforated cover plates in tension members are not spliced, open perforation can be used at the end of cover plates provided the open length does not exceed half the distance between the nearer lines of connections. Where perforated cover plates in tension members are spliced, the clear distance between the end perforation and the end of the cover plate shall not be less than the distance between the nearer lines of connections.

Shear deformation in tension members, which is primarily due to self-weight and wind loads, is relatively small, and AREMA (2024) recognizes this by providing nominal recommendations for the thickness of perforated cover plates as a minimum of 2% of the length between the nearer connection lines (typically the distance between longitudinal welds to web plates in box members). The transverse shear force, V, associated with the shear deformation is distributed equally to all parallel planes. However, since V is generally small in tension members,[56] perforated cover plate thickness is primarily dependent on the requirements for axial tension, bending and recommended minimum material thickness.[57] Nevertheless, the thickness of perforated cover plates must be considered based on V under LC1A or LC1B. The maximum transverse shear stress, τ_V, at the center of the cover plate is

$$\tau_V = \frac{3V}{2b_p t_{pc}} \tag{10.14}$$

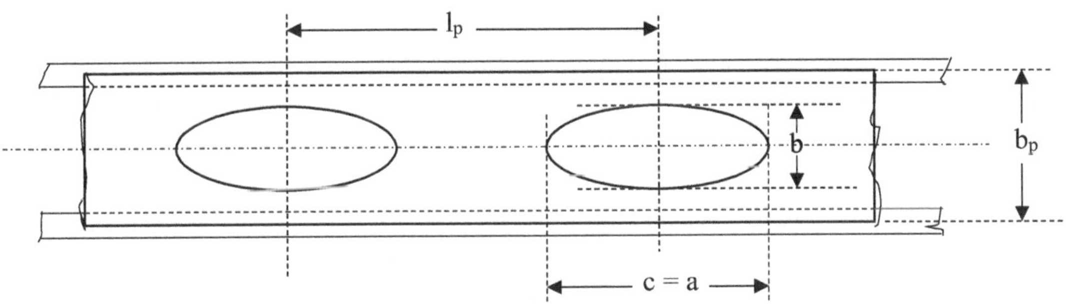

FIGURE 10.3 Elliptical perforations in cover plates.

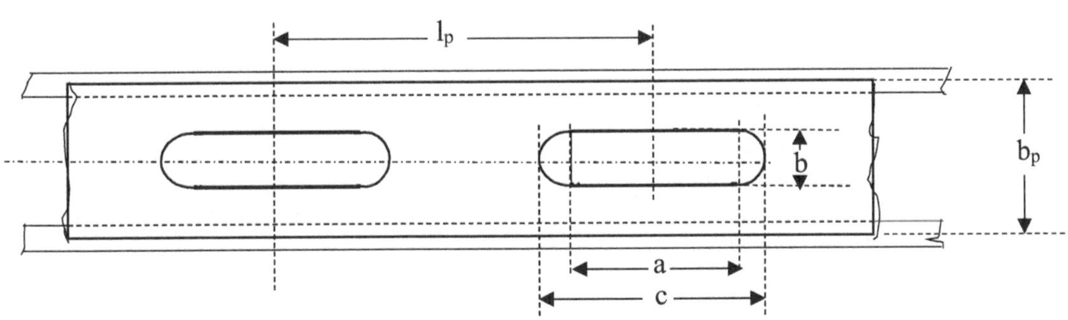

FIGURE 10.4 Ovaloid perforations in cover plates

where b_p = the width of the perforated cover plate and t_{pc} = the thickness of the perforated cover plate.

Therefore, the longitudinal shear force, V' over the distance between the centers of perforations or apertures, l_p, is

$$V' = \tau_V \left(l_p t_{pc} \right) = \frac{3Vl_p}{2b_p} \tag{10.15}$$

and the shear stress over the net area of the plate between the centers of perforations, τ_{pc}, is

$$\tau_{pc} = \frac{V'}{\left(l_p - c \right) t_{pc}} = \frac{3Vl_p}{2b_p t_{pc} \left(l_p - c \right)} \tag{10.16}$$

where c = the length of the elliptical or ovaloid perforation, which must not be more than twice the perforation width, b

Rearranging Equation 10.16 and considering $\tau_{pc} \le \tau_{all} = 0.35F_y$ yields

$$t_{pc} \ge \frac{3Vl_p}{2\tau_{all} b_p \left(l_p - c \right)} \ge \frac{3Vl_p}{2\left(0.35F_y \right) b_p \left(l_p - c \right)} \ge \frac{4.29Vl_p}{F_y b_p \left(l_p - c \right)} \tag{10.17}$$

For tensile axial and bending stresses, the net section through the perforation of the plate is included in member net cross-sectional area A_n and net section modulus S_n. Perforated cover plate tension members behave as Fatigue Category C details.[58]

The effect of perforations in box-section member cover plates must also be considered for serviceability design. Elastic truss joint displacements are the result of elastic changes in member length (axial extension or contraction), Δ_a caused by internal forces in the truss members. The axial extension or contraction, Δ_a for box-section members with solid cover plates is determined on the gross cross-sectional area:

$$\Delta_a = \frac{Pl}{A_g E} \tag{10.18}$$

where P = axial force in the member, l = length of the member, A_g = gross cross-sectional area of the member and E = modulus of elasticity of the member (200,000 MPa (29,000 ksi) for steel).

However, the axial extension or contraction of box-section members with perforated cover plates are determined based on an effective cross-sectional area, A_{ed}, based on the axial stiffness of the box-section member with perforated plates:

$$\Delta_a = \frac{Pl}{A_{ed} E} \tag{10.19}$$

where $A_{ed} = K_r A_g$ and K_r = the axial stiffness factor for perforated box-section members. K_r for a box-section member with perforated cover plates can be determined by the reciprocal theorem of elasticity (Greenspan, 1943). The axial stiffness factor for box-section members with load parallel to the major axis of cover plates with ovaloid perforations is

$$K_{ro} = \frac{1}{1 + \dfrac{A_g V_r}{A_n V_g} + \dfrac{3V_c}{C(n) V_g}} \tag{10.20a}$$

where

A_g = the gross cross-sectional area of the box-section member

A_n = the net cross-sectional area of the box-section member = $A_g - A_p$

A_p = the cross-sectional area of perforations in the box-section member

V_r = the volume of rectangular portions of ovaloid perforations in the box-section member cross-section = $A_{pr}t_{pc}$

A_{pr} = 2ba = 2b (c-b) for ovaloid perforations in each cover plate of box-section members (ba for only one perforated cover plate)

V_c = the volume of circular portions of ovaloid perforations in the box-section member cross section = $A_{pc}t_{pc}$

$A_{pc} = 2\pi b^2/4 = \pi b^2/2$ for ovaloid perforations in each cover plate of box-section members ($\pi b^2/4$ for only one perforated cover plate)

$V_g = A_g l_p$ = the gross volume of segment of the box-section member with perforated cover plates

$C(n) = 1 - 1/(2n^2) - 1/(2n^4)$

$n = A_g/(A_g - A_n)$

The axial stiffness factor for box-section members with load parallel to the major axis of cover plates with elliptical perforations is

$$K_{re} = \frac{1}{1+\dfrac{\left(1+2\dfrac{b}{a}\right)V_e}{C(n)V_g}} \qquad (10.20b)$$

where

a = c = the length of major axis of elliptical perforation;

b = the length of minor axis of elliptical perforation;

V_e = the volume of elliptical perforations in the box-section member cross-section = $A_{pe}t_{pc}$; and

$A_{pe} = \pi ba/2$ = for elliptical perforations in each cover plate of box-section members ($\pi ba/4$ for only one perforated cover plate).

Based on the axial stiffness of box-section members with ovaloid and elliptical perforations in the cover plates, the effective cross-sectional area for calculating axial extension or contraction, A_{ed}, and truss joint displacements is recommended by AREMA (2024) as

$$A_{ed} = A_g - (V_p)/l_p \qquad (10.21)$$

where

V_p = volume of the perforations in the box-section member

$= (A_{pr} + A_{pc})t_{pc}$ for ovaloid perforations in each cover plate of the box-section member

$= A_{pe}t_{pc}$ for elliptical perforations in each cover plate of the box section member.

The shape or plate components of tension members must be adequately fastened together to ensure integral behavior of the member. This is accomplished by strength and fatigue design of the fillet welds in open members to transfer the load from the connected side plates to the web plate over the length of the member's connection to the gusset plate. The strength and fatigue design of the four welds[59] in box-section members considers the transfer of load from the connected side plates to the cover plates over the length of the member's connection to the gusset plate. In many instances, minimum size welds in accordance with the applicable welding code[60] suffice to transfer the maximum and tensile range connection loads.

10.6.2 Tension Members with Lacing Bars and Batten Plates

Due to uneconomical fabrication, modern truss spans do not typically use tension members with lacing bars and/or batten plates (Unsworth, 2018).

Example 10.1

Design chord member L2L3 from Example 9.3 using a welded built-up H-section. The secondary bending stress on the gross section from joint rigidity is determined to be 20% of the primary axial stress at joint L3 and 10% of the primary axial stress at joint at joint L2. ASTM A709 Grade 50W steel is used with F_y = 50 ksi and F_U = 70 ksi.

From Example 9.3,

T_{DL} = 107.14 kips
T_{LL} = 354.96 kips
I_m = 26.0%
I_f = 0.65(26.0) = 16.9 %
T_{LC1A} = 107.41 + 1.26(354.96) = 107.41 + 447.25 = 554.39 kips
T_{LC1B} = (554.39 + 238.09)/1.25 = 792.48/1.25 = 633.98 kips
T_{max} = T_{LC1B} = 633.98 kips
T_{LC3} = 1.169(354.96) = S_R = 414.95 kips
L_{L2L3} = length of member L2L3 = 25 ft = 300 in

Strength Design:

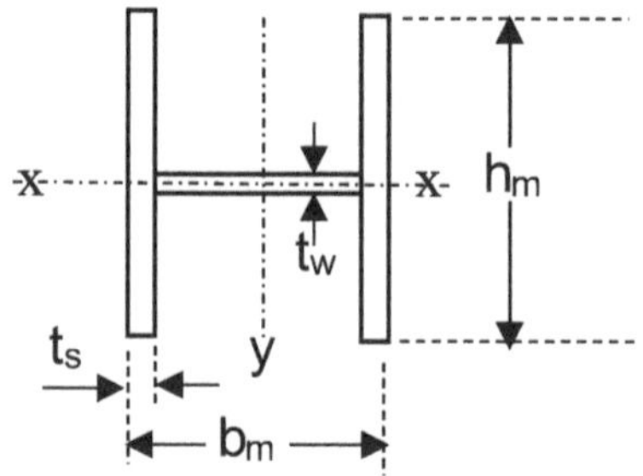

FIGURE E10.1a Gross cross-sectional area of member L2L3.

a) Gross section properties of member (Figure E10.1a): Try built-up section of 2 – 12" × 1.25" vertical side plates and 1 – 9.5" × 1/2" horizontal web plate.
   h_m = 12 in
   b_m = 12 in
   t_s = 1.25 in
   t_w = 0.50 in
   b_m/L_{L2L3} = 12/300 = 0.04 ≤ 0.10, therefore no need to consider secondary flexural stresses in accordance with AREMA (2024) or secondary flexural stresses due to joint rigidity and truss distortion in accordance with AREMA (2023)
   Gross cross-sectional area of member = A_{gm} = 2(12)(1.25) + (9.5)(0.5) = 30 + 4.75 = 34.75 in^2

Gross vertical (in-plane) moment of inertia of member = I_{gmx} = $(2(1.25)(12)^3 + 9.50(0.50)^3)/12 = 360.00 + 0.10 = 360.10$ in^4

Gross vertical section modulus of member = S_{gmx} = $2I_{gmx}/h_m$ = 2(360.10)/12 = 60.02 in^3

Gross horizontal (out-of-plane) moment of inertia of member = I_{gmy} = $0.5(9.5)^3/12 + 2(12)(1.25)^3/12 + 2(1.25)(12)(5.375)^2 = 35.72 + 3.91 + 866.72 = 906.35$ in^4

Gross horizontal section modulus of member = S_{gmy} = $2I_{gmy}/b_m$ = 2(906.35)/12 = 151.06 in^3

b) Effective net section properties of member at connection (Figure E10.1b):

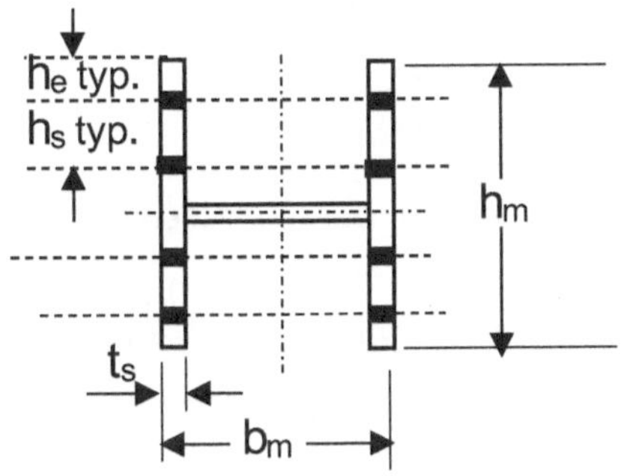

FIGURE E10.1b Net cross-sectional area of member L2L3 at connection.

No of bolts = 633.98/12.9 = 24.6

Consider four bolt holes in each side plate at connection to gusset plates,

h_e = 1.5 in

h_s = 3 in

Width of gross cross-sectional area of connecting elements = w_{gc} = h_m = 12 in

Width of net cross-sectional area of connecting elements = w_{nc} = 12 − 4(1) = 8 in

w_{nc}/w_{gc} = 8/12 = 0.67 ≤ 0.85, OK

Gross cross-sectional area of connecting elements = A_{gcc} = 2(1.25)(12) = 30.0 in^2

Net cross-sectional area of connecting elements = A_{ncc} = 2(1.25)(12) − 8(1.25)(1.0) = 20.0 in^2

Net cross-sectional area at connection = A_{nc} = 2(8)(1.25) + 9.5(0.5) = 20.0 + 4.75 = 24.75 in^2

A_{nc}/A_{gc} = 24.75/34.75 = 0.71 ≤ 0.85, OK

Allowable bolt shear force for 7/8 in. diameter bolts with Class A or Class C faying surface = 12.9(0.60) = 7.74 kips (see Chapter 12)

Allowable bolt shear stress for 7/8 in. diameter bolts with Class B faying surface = 21.5(0.60) = 12.90 kips (see Chapter 12)

No. bolts with Class A or Class C faying surface = 633.98/7.74 = 82 bolts

No. bolt rows for Class A or Class C faying surface = 82/8 = 11 rows

No. bolts with Class B faying surface = 633.98/12.90 = 50 bolts

No. bolt rows for Class B faying surface = 50/8 = 7 rows

Consider a non-staggered bolted connection with 3 in bolt spacing parallel to force,

Length of connection for Class A or Class C faying surface = L_c = 11(3) = 33 in
Length of connection for Class B faying surface = L_c = 7(3) = 21 in

c) Eccentricity for shear lag (Figure E10.1c):

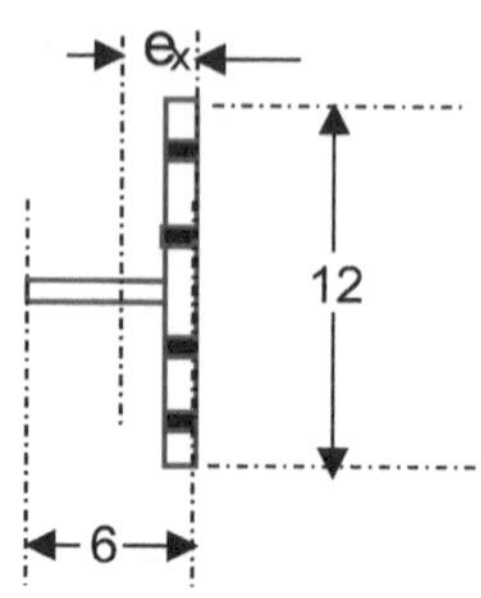

FIGURE E10.1c Cross-section for determination of eccentricity for shear lag.

e_x = (4.75(0.50)(4.75/2+1.25) + 12(1.25)(1.25/2))/(12(1.25) + 4.75(0.5)) = 1.035 in.
For Class A or Class C faying surface, $U_c = 1 - e_x/L_c$ = 1 − 1.035/33 = 0.97, use maximum U_c of 0.90.
For Class B faying surface $U_c = 1 - e_x/L_c$ = 1 − 1.035/21 = 0.95, use maximum U_c of 0.90.
Effective net cross-sectional area of member at connection = $A_{enc} = U_c\,(A_{nc})$ = 0.90(24.75) = 22.28 in^2
Net vertical moment of inertia at connection = I_{ncx} = 360.10 − (8(1.25)(1)^3)/12 − 4(1.25)(1)(4.5)^2 − 4(1.25)(1)(1.5)^2 = 360.10 − 113.33 = 246.77 in^4
Net vertical section modulus at connection = 246.77/6 = 41.13 in^3
Net horizontal moment of inertia at connection = I_{ncy} = 906.35 − 8(1)(1.25)^3/12 − 8(1)(1.25)(5.375)^2 = 906.35 −1.30 − 288.91 = 616.14 in^4
Net horizontal section modulus at connection = 616.14/6 = 102.69 in^3

d) Primary axial stresses:
Axial stress on gross cross-sectional area of member = $\sigma_{agm} = T_{max}/A_{gm}$ = 633.98/34.75 = 18.24 ksi
Axial stress on effective net cross-sectional area of connection = $\sigma_{anc} = T_{max}/A_{enc}$ = 633.98/22.28 = 28.46 ksi

In accordance with AREMA (2024), the secondary flexural stresses need to be considered only if $b_m/L_{L2L3} > 0.10$. In accordance with AREMA (2023), the secondary flexural stresses due to joint rigidity need to be considered only if $b_m/L_{L2L3} > 0.10$, and only secondary flexural stresses exceeding 4.0 ksi need to be considered primary stresses. Therefore, since $b_m/L_{L2L3} < 0.10$, secondary flexural stresses need not be considered. However, for illustrative purposes, this example will assume that $b_m/L_{L2L3} > 0.10$, and only excess secondary flexural stresses in accordance with AREMA (2023) are considered.

e) In-plane secondary flexural stresses due to joint rigidity and truss distortion:
Secondary flexural stress due to joint rigidity and truss distortion on gross section of member at joint L3 = σ_{bgm} = 0.20(18.24) = 3.65 ksi
Secondary bending moment on member gross section due to joint rigidity and truss deformation = 3.65(60.02)/12 = 18.26 kip-ft
Secondary bending stress on net section at connection due to joint rigidity and truss distortion = σ_{bnc} = 18.26(12)/41.13 = 5.33 ksi

f) In-plane secondary flexural stresses due to member eccentricities: There are no eccentricities due to connection detailing.
g) In-plane secondary flexural stresses due to member weight:
Weight of member L2L3 = 34.75(490)/(144(1000)) = 0.118 kip/ft
Vertical fixed end moment due to self-weight at L3 = $0.118(25)^2/12$ = 6.15 kip-ft
Secondary vertical flexural stress on gross section of member due to self-weight at joint L3 = 6.15(12)/60.02 = 1.23 ksi
Secondary vertical flexural stress on net section at connection due to self-weight at joint L3 = 6.15(12)/41.13 = 1.79 ksi
h) Out-of-plane secondary flexural stresses due to wind on member:
Wind on member L2L3 = 30(1.0)/1000 = 0.030 kip/ft (include for load case LC1B)
Horizontal fixed end moment due to wind at L3 = $0.03(25)^2/12$ = 1.56 kip-ft
Secondary horizontal flexural stress on gross section of member due to wind at joint L3 = 1.56(12)/151.06 = 0.12 ksi
Secondary horizontal flexural stress on net section at connection due to wind at joint L3 = 1.56(12)/102.69 = 0.18 ksi
i) Secondary flexural stresses:
On gross section of member = σ_{bgm} = 3.65 + 1.23 + 0.12 = 5.00 ksi
On net section of member at connection = σ_{bnc} = 5.33 + 1.79 + 0.18 = 7.30 ksi
j) Excess secondary flexural stresses in accordance with AREMA (2023):
On gross section of member = σe_{bgm} = 5.00 − 4.0 = 1.00 ksi
On net section at connection = σe_{bnc} = 7.30 − 4.0 = 3.30 ksi
k) Combined primary axial and excess secondary bending stresses for strength design:
$\sigma_{agm} + \sigma_{ebgc}$ = 18.24 + 1.00 = 19.24 ksi ≤ 0.55(50) ≤ 27.5 ksi, OK
$\sigma_{anc}/(0.47F_U) + \sigma_{ebnc}/(0.55F_y)$ = (28.46/32.90)+ (3.30/27.50) = 0.87 + 0.12 = 0.99 ≤ 1, OK

Fatigue Design:

l) Category B with S_{Rfat} = 18 ksi on gross section of member:
Combined axial and flexural stress ranges on member gross section:
(414.95/633.98)(18.24 + 1.00) = 0.65(19.24) = 12.59 ksi ≤ 18 ksi, OK
m) Category B with S_{Rfat} = 18 ksi on effective section of slip-critical connection:
Effective cross-sectional area for fatigue = gross section of connected elements = A_{efat} = 2(1.25)(12) = 30.0 in^2
$I_{efatx} = 2(1.25)(12)^3/12$ = 360.00 in^4
S_{efatx} = 360.00/6 = 60.00 in^3
$I_{efaty} = 2(12)(1.25)^3/12 + 2(12)(1.25)(5.375)^2$ = 870.63 in^4
S_{efaty} = 870.63/6 = 145.10 in^3
Combined axial and flexural stress ranges on effective section at connection:
(414.95/633.98)(633.98/30.0 + 3.30(41.13/60.0)) = 13.8 + 2.3 = 16.1 ksi ≤ 18 ksi, OK
n) Connection of H-section member plates:
Fillet welds connecting the vertical side and horizontal web plates of the member must transfer the connection force over the length of the connection to the gusset plate.
Maximum fillet weld = 0.50 − (5/64) = 0.42 in., use 3/8 in fillet welds
Allowable stress for electrodes with 70 ksi tensile strength = 19.0 ksi
Allowable stress on base metal= 0.35(50) = 17.5 ksi
Allowable fatigue stress on weld = 16 ksi (Category B)
Total weld length required for strength = 633.98/(0.707(17.5)(0.375)) = 136.6 in

Total weld length required for fatigue = 414.95/(0.707(16.0)(0.375)) = 97.8 in
Each weld length required = 136.6/4 = 34.16 in
Use a minimum connection length, L_c = 36 in

o) Connection of member side plates to gusset plates:
U_c = 1-1.035/36 = 0.97, maximum of 0.90 used
For Class A or Class C faying surface, non-staggered bolt spacing = 36/10 = 3.60 in, use 3.75 in bolt spacing with L_C = 3.75(10) = 37.50 in
For Class B faying surface, non-staggered bolt spacing = 36/6 = 6.00 in, L_C = 36.00 in

Connection dimensions (Figure E10.1d):

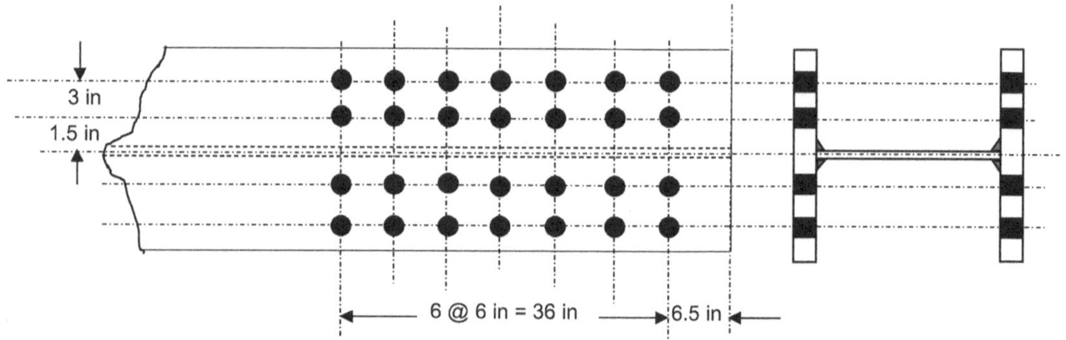

FIGURE E10.1d H-section member L2L3 end connection.

Shear stress on bolts = 633.98/56 = 11.32 kips per bolt
Allowable shear stress on bolts = 21.50(0.60) = 12.90 kips, Ok
Bearing stress on bolts = 633.98/(56(7/8)(1.25) = 10.35 ksi per bolt
Allowable bearing stress on bolt = (6 − 0.875/2))(70)/((2)(0.875)) = 222.5 ksi or 1.2(70) = 84.0 ksi, OK
Bolt spacing ≥ 2(7/8)(10.35)/70 + 7/16 ≥ 0.26 + 0.44 ≥ 0.70 in, OK
Edge distance ≥ 1.5(7/8) ≥ 1.31 in, OK
End distance ≥ 1.5 + 4(1.25) ≥ 6.5 in, OK
Maximum bolt spacing for sealing ≤ $4t_{min}$ + 4 ≤ 7 in, OK
t_{min} ≥ (6 − 4)/4 ≥ 0.50 in (gusset plate t ≥ 0.50 in), OK

Check block shear (see Chapter 12):
$T_{max} \le 0.30F_UA_{nv} + 0.50F_UA_{nt}$
or
$T_{max} \le 0.30F_yA_{gv} + 0.50F_UA_{nt}$
A_{gv} = 2(6.5+36)(1.25) = 106.25 in^2
$A_{nv} = A_{gv}$ − 2(7)(1)(1.25) = 106.25 − 17.50 = 88.75 in^2
A_{nt} = 2(4.5)(1.25) − 4(1)(1.25) = 11.25 − 5.0 = 6.25 in^2
F_y = 50 ksi
F_u = 70 ksi
633.98 ≤ 0.30(70)(88.75) + 0.50(70)(6.25) = 1863.8 + 218.8 = 2082.5 kips, OK
633.98 ≤ 0.30(50)(106.25) + 0.50(70)(6.25) = 1593.8 + 218.8 = 1812.5 kips, OK

Because the side plates are 1.25 in thick, block shear at the member connection is not an issue. However, block shear of the gusset plate connection must also be investigated as shown in Chapter 12.

p) Member and connection summary:
Use built-up section of two 12" × 1.25" vertical side plates and one 9.5" × 1/2" horizontal web plate with 3/8 in. fillet welds and Class B connection with 56-7/8 in. diameter slip-critical bolts.

Example 10.2

Design chord member L2L3 of Example 9.3 using a welded built-up box-section with ovaloid perforated cover plates. The secondary bending stress on the effective section (perforated cover plates) from joint rigidity is determined to be 20% of the primary axial stress at joint L3 and 10% of the primary axial stress at joint at joint L2. ASTM A709 Grade 50W steel is used with F_y = 50 ksi and F_U = 70 ksi.

From Example 9.3,

T_{DL} = 107.14 kips
T_{LL} = 354.96 kips
I_m = 26.0%
I_f = 0.65(26.0) = 16.9 %
T_{LC1A} = 107.41 + 1.26(354.96) = 107.41 + 447.25 = 554.39 kips
T_{LC1B} = (554.39 + 238.09)/1.25 = 792.48/1.25 = 633.98 kips
T_{max} = T_{LC1B} = 633.98 kips
T_{LC3} = 1.169(354.96) = S_R = 414.95 kips
L_{L2L3} = length of member L2L3 = 25 ft = 300 in

Since the allowable fatigue stress range is only 10 ksi on the net section for a perforated cover plate box-section, a considerably large cross-sectional area is required. The estimated net cross-sectional area required is 414.95/10.0 = 41.50 in^2

Strength Design:

a) Gross section properties of member: AREMA (2024) indicates that the net section of the cover plate through the perforation for tension members shall be included as a part of the area of the member (Figure E10.2a). Try built-up section of two 11" × 1.75" vertical side plates and two 10.25" × 1/2" horizontal cover plates.
h_m = 12 in.
b_m = 12 in.
t_s = 1.75 in.
t_c = 0.50 in
h_p = height of side plate = h_m − 2(0.5) = 12 − 1.0 = 11 in

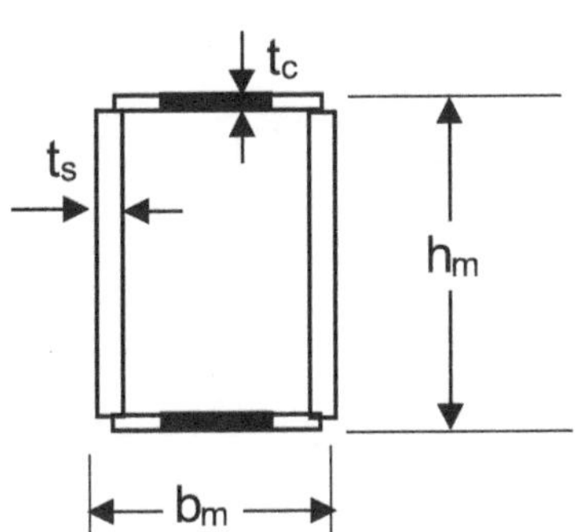

FIGURE E10.2a Cross-sectional area of member L2L3 at perforation.

b_p = width of cover plates = $b_m - t_s$ = 12 − 2.00 = 10.25 in
b_m/L_{L2L3} = 12/300 = 0.04, therefore no need to consider secondary flexural stresses in accordance with AREMA (2024) or secondary flexural stresses due to joint rigidity and truss distortion in accordance with AREMA (2023)
b = width of perforation = 6 in
c = length of perforation = 12 in ≤ 2b = 2(6) = 12 in, OK
l_p = center-to-center of perforations = 30 in
l_p–c = 30 − 12 = 18 in ≥ b_p = 10.25 in, OK

Cross-sectional area of member = A_m = 2(11)(1.75) + 2(10.25)(0.5) − 2(6)(0.5) = 38.75 + 10.25 − 6.00 = 42.75 in^2
Self-weight of L2L3 = ((18/30)42.75 + (12/30)(48.75)(490)/(1000(144)) = 0.154 kip/ft
Wind pressure on L2L3 = 0.030 kip/ft

$$V = \sqrt{\left(\frac{0.154(25)}{2}\right)^2 + \left(\frac{0.03(25)}{2}\right)^2} = \sqrt{3.71 + 0.14} = 1.96 \text{ kips}$$

From Equation 10.17,

$$t_{pc} \geq \frac{4.29(1.96)(30)}{50(11.25)(30-12)} = 0.03 \text{ in, OK (insignificant, as expected for tension member)}$$

Vertical (in-plane) moment of inertia of member = I_{mx} = 2(1.75)(11)^3/12 + 2(10.25)(0.50)^3/12 + 2(10.25)(0.50)(5.75)^2 − 2(6.00)(0.50)^3/12 − 2(6.00)(0.50)(5.75)^2 = 388.21 + 0.21 + 338.89 − 0.13 − 198.38 = 727.31 − 198.51 = 528.80 in^4
Vertical section modulus of member = S_{mx} = 528.80/6 = 88.13 in^3

Horizontal (out-of-plane) moment of inertia of member = I_{my} = 2(0.50)(10.25)^3/12 + 2(11)(1.75)^3/12 + 2(11)(1.75)(5.125)^2 − 2(0.50)(6.0)^3/12 = 89.74 + 9.83 + 993.89 − 18.00 = 1093.5 − 18.00 = 1075.5 in^4
Horizontal section modulus of member = S_{my} = 1075.5/6 = 179.24 in^3

b) Effective net section properties of member at connection: Consider four bolt holes in each side plate at connection (Figure E10.2b).
Width of gross cross-sectional area at connection = w_{gc} = h_m = 11 in.
Width of net cross-sectional area at connection = w_{nc} = 11 − 4(1) = 7 in.
w_{nc}/w_{gc} = 7/11 = 0.64 ≤ 0.85, OK
Gross cross-sectional area of connecting elements = A_{gcc} = 2(1.75)(11) = 38.50 in^2
Net cross-sectional area of connecting elements = A_{ncc} = 38.50 − 8(1.75)(1.00) = 24.50 in^2
A_{ncc}/A_{gcc} = 24.50/38.50 = 0.64 ≤ 0.85, OK
Net cross-sectional area at connection = A_{nc} = 42.75 − 8(1.75)(1.0) = 42.75 − 14.0 = 28.75 in^2

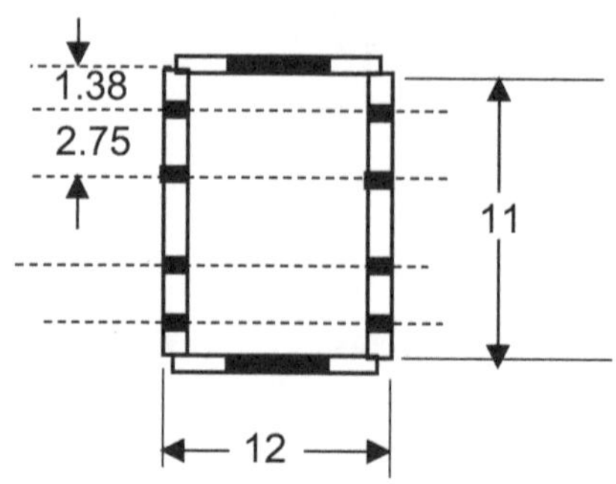

FIGURE E10.2b Net cross-sectional area of member L2L3 at connection.

$A_{nc}/A_m = 28.75/42.75 = 0.67 \le 0.85$, OK

Allowable bolt shear stress for 7/8 in. diameter bolts with Class B faying surface = 21.5(0.60) = 12.90 kips (see Chapter 12)

No.bolts with Class B faying surface = 633.98/12.90 = 50 bolts

No.bolt rows for Class B faying surface = 50/8 = 7 rows

Consider a non-staggered bolted connection with 3 in bolt spacing parallel to force; length of connection for Class B faying surface = L_c = 7(3) = 21 in

c) Eccentricity for shear lag (Figure E10.2c):

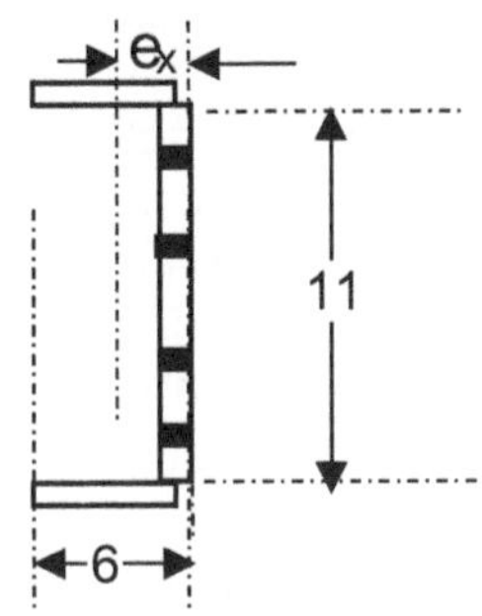

FIGURE E10.2c Cross-section for determining eccentricity for shear lag.

e_x = 2(5.125)(0.50)(3.438) + 11(1.75)(0.875))/(11(1.75) + 2(5.125)(0.5)) = 34.46/24.38 = 1.414 in

For Class B faying surface $U_c = 1 - e_x/L_c = 1 - 1.414/21 = 0.93$, use maximum U_c of 0.90

Effective net cross-sectional area at connection = $A_{enc} = U_c\,(A_{nc}) = 0.90(28.75) = 25.88$ in²

Net vertical (in-plane) moment of inertia at connection = I_{ncx} = 528.80 − 8(1.75)(1.00)³/12 − 4(1.75)(1.0)(4.125)² − 4(1.75)(1.0)(1.375)² = 528.80 − 1.17 − 119.11 − 13.23 = 528.80 − 133.51 = 395.29 in⁴

Net vertical section modulus at connection = 395.29/6 = 65.88 in³

Net horizontal (out-of-plane) moment of inertia at connection = I_{ncy} = 1075.5 − 8(1.0)(1.75)³/12 − 8(1.0)(1.75)(5.125)² = 1075.5 − 3.57 − 367.72 = 704.21 in⁴

Net horizontal section modulus at connection = 704.21/6 = 117.37 in³

d) Primary axial stresses:

On cross-sectional area of member = $\sigma_{am} = T_{max}/A_m$ = 633.98/42.75 = 14.83 ksi

On effective net cross-sectional area of connection = $\sigma_{anc} = T_{max}/A_{enc}$ = 633.98/25.88 = 24.50 ksi

e) In-plane secondary flexural stresses due to joint rigidity and truss distortion:

Gross cross-sectional area for displacement calculations = $A_{ed} = A_{gm} - (V_p)/I_p$

$$V_p = \left(2(6)\left(12 - 2(3)\right) + \frac{\pi 6^2}{2}\right)(0.50) = (72.00 + 56.55)0.50 = 64.27 \text{ in}^3$$

$A_{ed} = A_m - (V_p)/I_p$ = 42.75 − 64.27/12 = 42.75 − 5.36 = 37.39 in²

Axial stress on cross-sectional area of member for deflection calculations = $\sigma_{amd} = T_{max}/A_{ed}$ = 633.98/37.39 = 16.96 ksi

Secondary vertical flexural stress on section due to joint rigidity and truss distortion at joint L3 based on displacement calculations = $\sigma_{bmL3} = \sigma_{bm}$ = 0.20(16.96) = 3.39 ksi

Secondary bending moment on member section due to joint rigidity and truss deformation = 3.39(88.13)/12 = 24.91 kip-ft

Secondary bending stress on net section at connection due to joint rigidity and truss deformation = σ_{bnc} = 24.91(12)/65.88 = 4.54 ksi

f) In-plane secondary flexural stresses due to member eccentricities: Connections are detailed to avoid secondary flexural stresses due to eccentricities.

g) Secondary flexural stresses due to member weight:
Self-weight of L2L3 = 0.154 kip/ft
Vertical fixed end moment due to self-weight = $0.154(25)^2/12$ = 8.02 kip-ft
Secondary vertical flexural stress on member due to self-weight at joint L3 = 8.02(12)/88.13 = 1.09 ksi
Secondary vertical flexural stress on net section at connection due to self-weight at joint L3 = 8.02(12)/65.88 = 1.46 ksi

h) Secondary flexural stresses due to wind on member:
Wind on L2L3 = 30(1)/1000 = 0.030 kip/ft (include for load case LC1B)
Horizontal fixed end moment due to wind at L3 = $0.030(25)^2/12$ = 1.56 kip-ft
Secondary horizontal flexural stress on member due to wind at joint L3 = 1.56(12)/179.24 = 0.10 ksi
Secondary horizontal flexural stress on net section of connection due to wind at joint L3 = 1.56(12)/117.37 = 0.16 ksi

i) Secondary flexural stresses:
On member = σ_{bm} = 3.39 + 1.09 + 0.10 = 4.58 ksi
On net section at connection = 4.54 + 1.46 + 0.16 = 6.16 ksi

In accordance with AREMA (2024), the secondary flexural stresses need to be considered only if $b_m/L_{L2L3} > 0.10$. In accordance with AREMA (2023), the secondary flexural stresses due to joint rigidity need to be considered only if $b_m/L_{L2L3} > 0.10$, and only secondary flexural stresses exceeding 4.0 ksi need to be considered primary stresses. Therefore, since $b_m/L_{L2L3} < 0.10$, secondary flexural stresses need not be considered. However, for illustrative purposes, this example will assume $b_m/L_{L2L3} > 0.10$ and only consider excess secondary flexural stresses in accordance with AREMA (2023).

j) Excess secondary flexural stresses:
On member = σ_{ebm} = 4.58 − 4.0 = 0.58 ksi
On net section of connection = 6.16 − 4.0 = 2.16 ksi

k) Combined primary axial and excess secondary flexural stresses for strength design:
$\sigma_{am} + \sigma_{ebm}$ = 14.83 + 0.58 = 15.41 ksi ≤ 0.55(50) ≤ 27.50 ksi, OK
$\sigma_{anc}/(0.47Fu) + \sigma_{ebm}/(0.55Fy)$ = (24.50/32.90) + (2.16/27.50) = 0.74 + 0.08 = 0.82, ≤ 1.0, Ok

Fatigue Design:

l) Category C with S_{Rfat} = 10 ksi on member:
C combined axial and flexural stress ranges on member:
Primary stress range on cross-sectional area of member = $\Delta T/A_m$ = 414.95/42.75 = 9.71 ksi
9.71 + (414.95/633.98)(0.58) = 10.09 ksi, OK 1% overstress

m) Category B with S_{Rfat} = 18 ksi on effective section of slip-critical connection:
Effective cross-sectional area for fatigue = A_{efat} = 2(1.75)(11) = 38.50 in^2
S_{Rm} = 414.95/38.50 + (414.95/633.98)(2.16) = 12.18 ksi ≤ 18 ksi, OK

n) Connection of box-section member plates and member to gusset plates (see Example10.1):

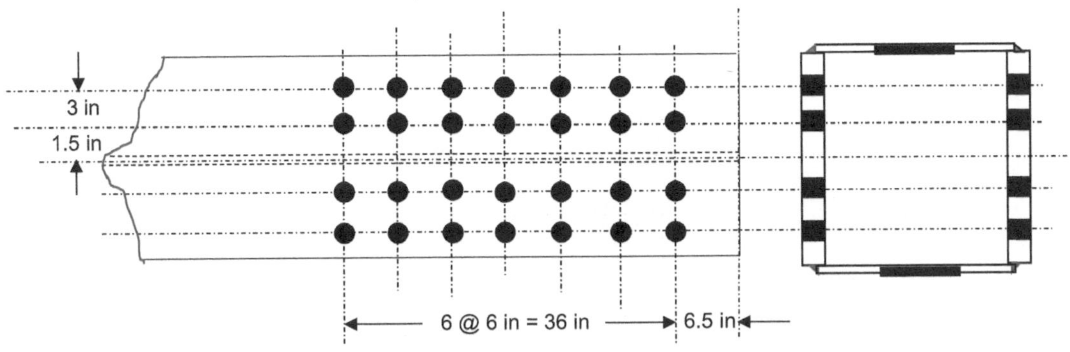

FIGURE E10.2d Box-section member L2L3 end connection.

Block shear will not be an issue for 1.75 in thick side plates (see Example 10.1).

o) Member and connection summary (Figure E10.2d):
Use built-up section of two 11" × 1.75" vertical side plates and two 10.25" × 1/2" horizontal cover plates with 6" × 12" ovaloid perforations using 3/8 in. fillet welds and Class B connection with 56-7/8 in. diameter slip-critical bolts.

Example 10.3

Design chord member L2L3 of Example 9.3 using a welded built-up box-section with solid cover plates. The secondary bending stress on the effective section from joint rigidity is determined to be 20% of the primary axial stress at joint L3 and 10% of the primary axial stress at joint at joint L2. ASTM A709 Grade 50W steel is used with F_y = 50 ksi and F_U = 70 ksi.

From Example 9.3,

T_{DL} = 107.14 kips
T_{LL} = 354.96 kips
I_m = 26.0%
I_f = 0.65(26.0) = 16.9 %
T_{LC1A} = 107.41 + 1.26(354.96) = 107.41 + 447.25 = 554.39 kips
T_{LC1B} = (554.39 + 238.09)/1.25 = 792.48/1.25 = 633.98 kips
T_{max} = T_{LC1B} = 633.98 kips
T_{LC3} = 1.169(354.96) = S_R = 414.95 kips
L_{L2L3} = length of member L2L3 = 25 ft = 300 in

Strength Design:

a) Gross section properties of member (Figure E10.3a):
Member L2L3 is a section built-up with side plates and, except for hand holes at the bottom cover plate at the end of the member, solid cover plates. The hand holes are single 10" × 16" elliptical holes through the bottom cover plate and horizontal bracing gusset plate at the end of the member. The location and size of the hand holes are such that they do not affect the gross section properties. Try built-up section of two 11" × 1.00" vertical side plates and two 11" × 1/2" solid horizontal cover plates.

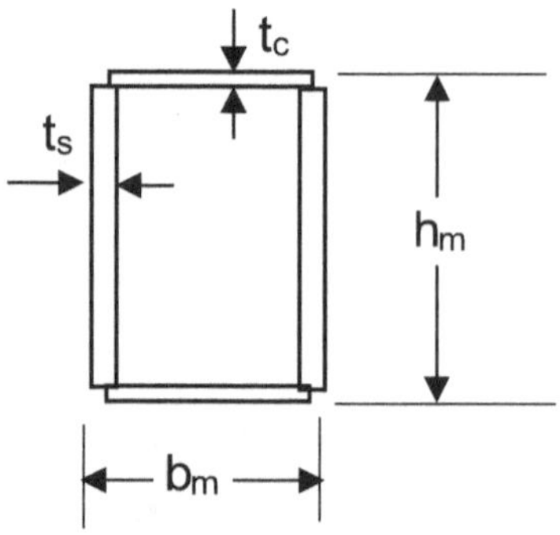

FIGURE E10.3a Cross-sectional area of member L2L3.

h_m = 12 in.
b_m = 12 in.
t_s = 1.00 in.
t_c = 0.50 in
h_p = height of side plate = h_m − 2(0.5) = 12 − 1.0 = 11 in
b_p = width of cover plates = b_m − t_s = 12 − 1.0 = 11 in
b_m/L_{L2L3} = 12/300 = 0.04, therefore no need to consider secondary flexural stresses in accordance with AREMA (2024) or secondary flexural stresses due to joint rigidity and truss distortion in accordance with AREMA (2023)
Cross-sectional area of member = A_m = 2(11)(1.00) + 2(11)(0.5) = 22.00 + 11.00 = 33.00 in^2
Vertical (in-plane) moment of inertia of member = I_{mx} = 2(1.00)(11)^3/12 + 2(11)(0.50)^3/12 + 2(11)(0.50)(5.75)^2 = 221.83 + 0.23 + 363.69 = 585.75 in^4
Vertical section modulus of member = S_{mx} = 585.75/6 = 97.63 in^3
Horizontal (out-of-plane) moment of inertia of member = I_{my} = 2(0.50)(11)^3/12 + 2(11)(1.00)^3/12 + 2(11)(1.00)(5.50)^2 = 110.92 + 1.83 + 665.50 = 778.25 in^4
Horizontal section modulus of member = S_{my} = 778.25/6 = 129.71 in^3

b) Effective net section properties of member at connection; consider four bolt holes in each side plate (Figure E10.3b).

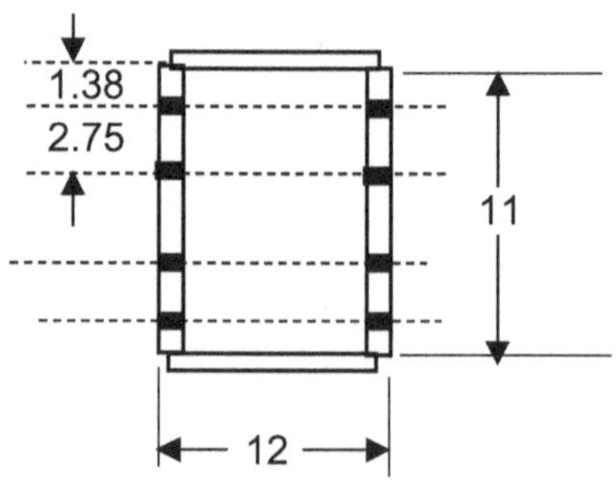

FIGURE E10.3b Net cross-sectional area of member L2L3 at connection.

Width of gross cross-sectional area at connection = w_{gc} = h_m = 11 in
Width of net cross-sectional area at connection = w_{nc} = 11 − 4(1) = 7 in
w_{nc}/w_{gc} = 7/11 = 0.64 ≤ 0.85, OK

Gross cross-sectional area of connecting elements = A_{gcc} = 2(1.00)(11) = 22.00 in^2
Net cross-sectional area of connecting elements = A_{ncc} = 22.00 − 8(1.00)(1.00) = 14.00 in^2
A_{ncc}/A_{gcc} = 14.0/22.0 = 0.64 ≤ 0.85, OK
Net cross-sectional area at connection = A_{nc} = 33.00 − 8(1.00)(1.0) = 25.00 in^2
A_{nc}/Am = 25.0/33.0 = 0.76 ≤ 0.85, OK
Allowable bolt shear stress for 7/8 in. diameter bolts with Class B faying surface = 21.5(0.60) = 12.90 kips (see Chapter 12)
No. bolts with Class B faying surface = 633.98/12.90 = 50 bolts
No. bolt rows for Class B faying surface = 50/8 = 7 rows
Consider a non-staggered bolted connection with bolt spacing parallel to force:
Length of connection for Class B faying surface = L_c = 7(3) = 21 in

c) Eccentricity for shear lag (Figure E10.3c):

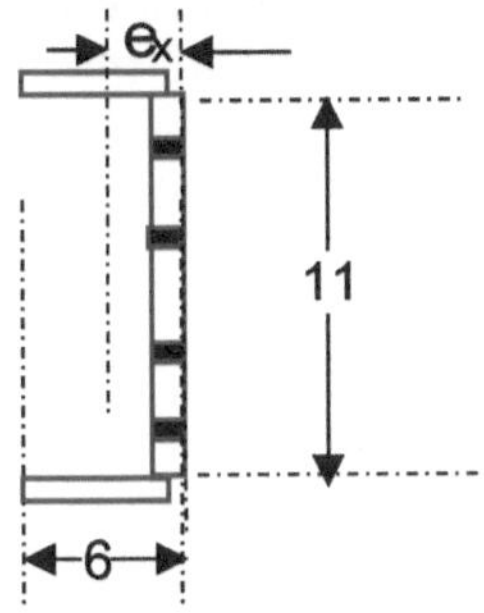

FIGURE E10.3c Cross-section for determination of eccentricity for shear lag.

e_x = 2(5.5)(0.50)(3.25) + 11(1.00)(0.50))/(11(1.00) + 2(5.5)(0.5)) = 23.38/16.50 = 1.417 in
For Class B faying surface $U_c = 1 - e_x/L_c$ = 1 − 1.417/21 = 0.93, use maximum U_c of 0.90
Effective net cross-sectional area at connection = $A_{enc} = U_c\,(A_{nc})$ = 0.90(25.00) = 22.50 in^2
Net vertical (in-plane) moment of inertia at connection = I_{ncx} = 585.75− 8(1.00)(1.00)^3/12 − 4(1.00)(1.0)(4.125)^2 − 4(1.00)(1.0)(1.375)^2 = 585.75 − 0.67 − 68.06 − 7.56 = 585.75 − 76.29 = 509.46 in^4
Net vertical section modulus at connection = 509.46/6 = 84.91 in^3
Net horizontal (out-of-plane) moment of inertia at connection = I_{ncy} = 778.25 − 8(1.0)(1.00)^3/12 − 8(1.0)(1.00)(5.5)^2 = 778.25 − 0.67 − 242.00 = 535.58 in^4
Net horizontal section modulus at connection = 535.58/6 = 89.26 in^3

d) Primary axial stresses
On cross-sectional area of member = $\sigma_{am} = T_{max}/A_m$ = 633.98/33.00 = 19.21 ksi
On effective net cross-sectional area of connection = $\sigma_{anc} = T_{max}/A_{enc}$ = 633.98/22.50 = 28.18 ksi

e) In-plane secondary flexural stresses due to joint rigidity and truss distortion:
Secondary vertical flexural stress on section due to joint rigidity and truss distortion at joint L3 = $\sigma_{bmL3} = \sigma_{bm}$ = 0.20(19.21) = 3.84 ksi
Secondary bending moment on member section due to joint rigidity and truss deformation = 3.84(97.63)/12 = 31.26 kip-ft

Secondary bending stress on net section at connection due to joint rigidity and truss deformation = σ_{bnc} = 31.26(12)/84.91 = 4.42 ksi

f) In-plane secondary flexural stresses due to member eccentricities: Connections are detailed to avoid secondary flexural stresses due to eccentricities.

g) Secondary flexural stresses due to member weight:
Self-weight of L2L3 = 33.00(490)/(1000(144)) = 0.112 kip/ft
Vertical fixed end moment due to self-weight = $0.112(25)^2/12$ = 5.85 kip-ft
Secondary vertical flexural stress on member due to self-weight at joint L3 = 5.85(12)/97.63 = 0.72 ksi
Secondary vertical flexural stress on net section at connection due to self-weight at joint L3 = 5.85(12)/84.91 = 0.83 ksi

h) Secondary flexural stresses due to wind on member:
Wind on L2L3 = 30(1)/1000 = 0.030 kip/ft (include for load case LC1B)
Horizontal fixed end moment due to wind at L3 = $0.030(25)^2/12$ = 1.56 kip-ft
Secondary horizontal flexural stress on member due to wind at joint L3 = 1.56(12)/129.71 = 0.14 ksi
Secondary horizontal flexural stress on net section of connection due to wind at joint L3 = 1.56(12)/89.26 = 0.21 ksi

i) Secondary flexural stresses:
On member = σ_{bm} = 3.84 + 0.72 + 0.14 = 4.70 ksi
On net section at connection = 4.42 + 0.83 + 0.21 = 5.46 ksi

In accordance with AREMA (2024), the secondary flexural stresses need to be considered only if $b_m/L_{L2L3} > 0.10$. In accordance with AREMA (2023), the secondary flexural stresses due to joint rigidity need to be considered only if $b_m/L_{L2L3} > 0.10$, and only secondary flexural stresses exceeding 4.0 ksi need to be considered primary stresses. Therefore, since $b_m/L_{L2L3} < 0.10$, secondary flexural stresses need not be considered. However, for illustrative purposes, this example will assume $b_m/L_{L2L3} > 0.10$ and only consider excess secondary flexural stresses in accordance with AREMA (2023).

j) Excess secondary flexural stresses:
On member = σ_{ebm} = 4.70 − 4.0 = 0.70 ksi
On net section of connection = 5.46 − 4.0 = 1.46 ksi

k) Combined primary axial and excess secondary flexural stresses for strength design:
$\sigma_{am} + \sigma_{ebm}$ = 19.21 + 0.70 = 19.91 ksi ≤ 0.55(50) ≤ 27.50 ksi, OK
$\sigma_{anc}/(0.47Fu) + \sigma_{ebm}/(0.55Fy)$ = (28.18/32.90) + (1.46/27.50) = 0.86 + 0.05 = 0.91 ≤ 1.0, OK

Fatigue Design:

l) Category B with S_{Rfat} = 16 ksi on member:
Combined axial and flexural stress ranges on member:
Primary stress range on cross-sectional area of member = $\Delta T/A_m$ = 414.95/33.00 = 12.57 ksi
12.57 + (414.95/633.98)(0.70) = 13.03 ksi, OK

m) Category B with S_{Rfat} = 18 ksi on effective section of slip-critical connection:
Effective cross-sectional area for fatigue = A_{efat} = 2(1.00)(11) = 22.00 in^2
S_{Rm} = 414.95/22.00 + (414.95/633.98)(1.46) = 19.81 ksi, NG, increase side plate width to (414.95/22.00)/(18.0 − (414.95/633.98)(1.46)) = 1.11 in, use 1.125 in thick side plates

n) Connection of box-section member plates and member to gusset plates (see Example 10.1):

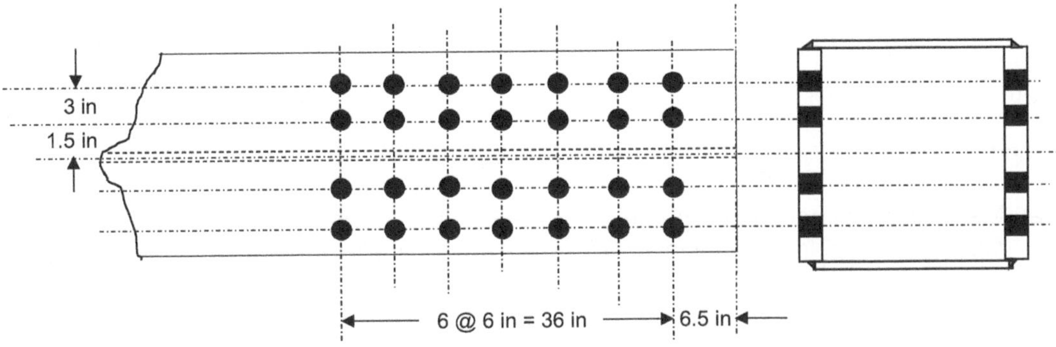

FIGURE E10.3d Box-section member L2L3 end connection.

Block shear will not be an issue with 1.125 inch thick side plates.

n) Member and connection summary (Figure E10.3d):

Use built-up section of two 11" × 1.125" vertical side plates and two 10.875" × 1/2" horizontal web plate with 3/8 in. fillet welds and Class B connection with 56-7/8 in. diameter slip-critical bolts.

Table 10.3 and Examples 10.1, 10.2 and 10.3 illustrate that box-section members, particularly those with perforated cover plates, are not typically economical for tension chord members.

TABLE 10.3
Tension Member L2L3 Designs in Examples 10.1, 10.2 and 10.3

Example	Side Plates (in)	Web or Cover Plates (in)	Gross Cross-sectional area (in^2)	Weight (lbs)
10.1	2 12 × 1.25	1 9.5 × 0.50	34.75	2950
10.2	2 11 × 1.75	2 10.25 × 0.50 with 6 × 12 ovaloid perforations	42.75	3850
10.3	2 11 × 1.125	2 11 × 0.50	35.75	3025

Example 10.4 outlines the design of hanger tension member U1L1 where AREMA (2024) indicates that out-of-plane flexure due to floorbeam rotation is considered a primary stress for the strength and fatigue designs, and all secondary flexural stresses related to live load are considered in fatigue design.

Example 10.4

Design hanger member U1L1 of Example 9.3 using a built-up H-section. The secondary bending stress from joint rigidity is determined to be 20% of the gross axial stress in member U1L1 at joint U1 and 25% of the gross axial stress in member U1L1 at joint L1. The floorbeam at joint L1 has a moment of inertia of 12,000 in^4. ASTM A709 Grade 50W steel is used with F_y = 50 ksi and F_U = 70 ksi.

From Example 9.3,

T_{DL} = 37.50 kips
T_{LL} = 156.00 (alternate live load)

$I_m = 43.8\%$
$I_f = 0.65(43.8) = 28.5\%$
$T_{max} = T_{LC1A} = 37.50 + 1.438(156.00) = 37.50 + 224.3 = 261.82$ kips (255.04 kips excluding alternate live load)
$T_{LC3} = S_R = 1.285(156.00) = 200.46$ kips (194.39 kips excluding alternate live load)
$L_{U1L1} = 35$ ft = 420 in

Strength Design:

a) Member gross section properties (Figure E10.4a): Try built-up section of 2 12" × 1.00" vertical side plates and 1 10.0" × 1/2" horizontal web plate.

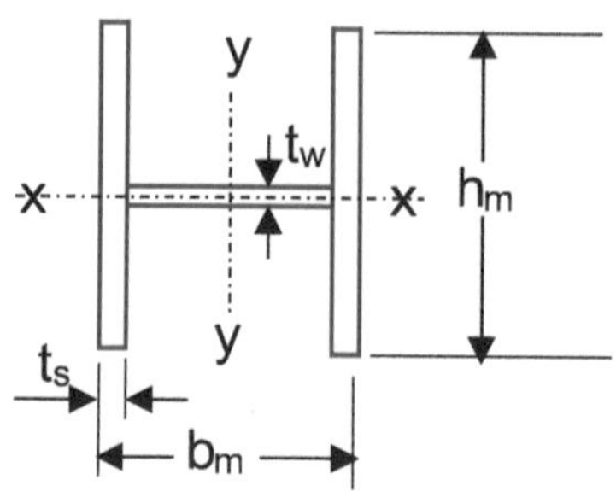

FIGURE E10.4a Gross cross-sectional area of member U1L1.

$h_m = 12$ in.
$b_m = 12$ in.
$t_s = 1.00$ in.
$t_w = 0.50$ in
$b_m/L_{U1L1} = 12/420 = 0.03$, OK
Gross cross-sectional area of member = $A_{gm} = 2(12)(1.00) + (10.0)(0.5) = 24.0 + 5.00 = 29.00$ in^2
In-plane gross moment of inertia of member = $I_{gmx} = 2(1.00)(12)^3/12 + 10.0(0.50)^3/12 = 288.10$ in^4
In-plane gross section modulus of member = $S_{gmx} = 2I_{gmx}/h_m = (288.10)/6 = 48.02$ in^3
Out-of-plane moment of inertia of member = $I_{gmy} = 2(12)(1.00)^3/12 + 2(12)(1.00)(6 - 0.50)^2 + 0.50(10.0)^3)/12 = 2.00 + 726.0 + 41.67 = 769.67$ in^4
Out-of-plane section modulus of member = $S_{gmy} = 2I_{gmy}/b_m = 2(769.67)/12 = 128.28$ in^3

b) Member net section properties: The member net section properties are equal to the member gross section properties as no perforations or holes exist in the member other than at connections at joints U1 and L1.
$A_{nm} = A_{gm} = 29.00$ in^2
$S_{nmx} = S_{gmx} = 48.02$ in^3
$S_{nmy} = S_{gmy} = 128.28$ in^3

c) Connection gross section properties: The connection gross section properties are equal to the member gross section properties as no there is change in cross section of member U1L1.
$A_{gc} = A_{gm} = 29.00$ in^2
$S_{gcx} = S_{gmx} = 48.02$ in^3
$S_{gcy} = S_{gmy} = 128.28$ in^3
Gross width of connecting elements = $w_{gc} = 12$ in

d) Connection effective net section properties: Consider two bolt holes in each side plate to gusset plate at connection (Figure E10.4b).

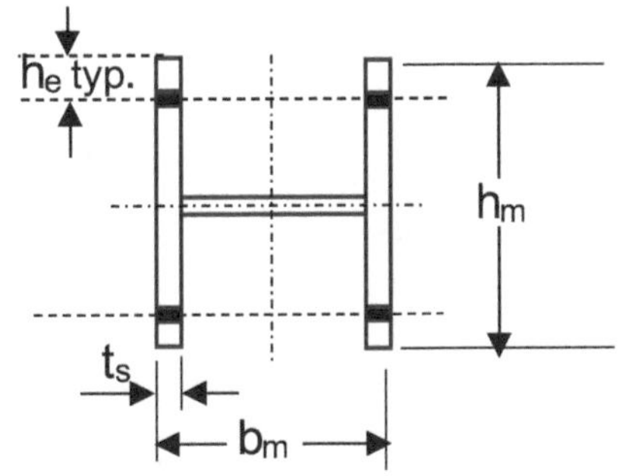

FIGURE E10.4b Net cross-sectional area at member U1L1 end connection.

h_e = 1.5 in.
Net width of connecting elements = w_{nc} = 12 − 2(1) = 10 in.
w_{nc}/w_{gc} = 10/12 = 0.83 ≤ 0.85, OK
Net cross-sectional area of connecting elements = A_{ncc} = 2(1.00)(12) − 4(1.00)(1.00) = 20.0 in^2
A_{ncc}/A_{gc} = 20.0/29.0 = 0.69 ≤ 0.85, OK
Net cross-sectional area at connection = A_{nc} = 2(10)(1.00) + 10.0(0.5) = 20.0 + 5.00 = 25.00 in^2
A_{nc}/A_{gc} = 25.0/29.0 = 0.86 ≈ 0.85, OK
Allowable bolt shear stress for 7/8 in. diameter bolts with Class B faying surface = 21.5(0.60) = 12.90 kips
No. bolts with Class B faying surface = (1.25)(261.82)/12.90 = 327.28/12.90 = 25.4 bolts, use 26 bolts (see Chapter 12)
No. bolt rows for Class B faying surface = 26/4 = 6.5 rows, use 7 rows (28 bolts)
Consider a non-staggered bolted connection with 3.5 in bolt spacing parallel to force.
Length of connection for Class B faying surface = L_c = 6(3.5) = 21 in

e) Eccentricity for shear lag (Figure E10.4c):

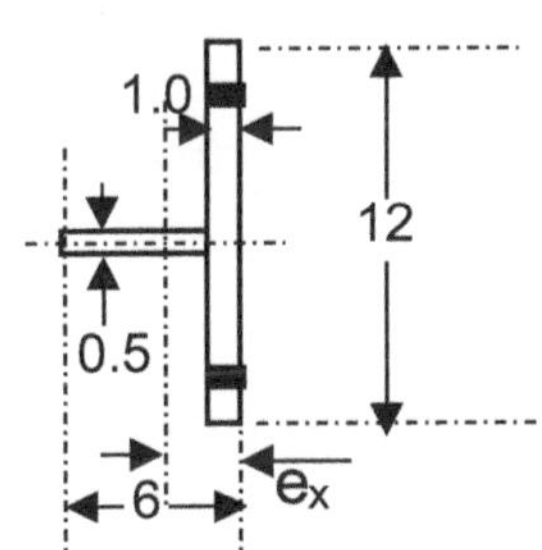

FIGURE E10.4c Cross-section for determining eccentricity for shear lag.

e_x = (5.00(0.50)(5.00/2 + 1.00) + 12(1.00)(1.00/2))/(12(1.00) + 5.00(0.5)) = 1.017 in.
For Class B faying surface $U_c = 1 - e_x/L_c$ = 1 − 1.017/21 = 0.95, use maximum U_c of 0.90
Effective net cross-sectional area at connection = $A_{enc} = U_c\ (A_{nc})$ = 0.90(25.0) = 22.50 in^2
Net out-of-plane moment of inertia at connection = I_{ncy} = 769.67 − 4(1)(1)^3/12 − 4(1.00)(1)(6-0.50)^2 = 769.67- 0.33 − 121.0 = 648.34 in^4

Net out-of-plane section modulus at connection = S_{ncy} = 648.34/6 = 108.06 in^3
Net in-plane moment of inertia at connection = I_{ncx} = 288.10 − 4(1)(1)³/12 − 4(1)(1.00)(4.50)² = 288.10 − 0.33 − 81.0 = 206.77 in^4
Net in-plane section modulus at connection = S_{ncx} = 206.77/6 = 34.46 in^3

f) Primary axial stresses:
$\sigma_{agm} = T_{max}/A_{gm}$ = 261.82/29.00 = 9.03 ksi
$\sigma_{anc} = T_{max}/A_{enc}$ = 261.82/22.50 = 11.64 ksi

g) In-plane secondary flexural stresses due to joint rigidity and truss distortion:
Secondary flexural stress due to joint rigidity and truss distortion of member gross section at joint U1 = σ_{bgmU1} = 0.2(9.03) = 1.81 ksi
Secondary bending moment due to joint rigidity and truss distortion of member gross section at joint U1 = 1.81(48.02)/12 = 7.24 kip-ft
Secondary flexural stress due to joint rigidity and truss distortion of member gross section at joint L1 = σ_{bgmL1} = 0.25(9.03) = 2.26 ksi
Secondary bending moment due to joint rigidity and truss distortion of member gross section at joint L1 = 2.26(48.02)/12 = 9.04 kip-ft
Secondary bending stress on net section of connection due to joint rigidity and truss distortion at U1 = 7.24(12)/34.46 = 2.52 ksi
Secondary bending stress on net section of connection due to joint rigidity and truss distortion at L1 = 9.04(12)/34.46 = 3.15 ksi

h) Secondary flexural stresses due to member eccentricities: There are no eccentricities at connections due to detailing.

i) Out-of-plane secondary flexural stresses due to wind on member:
Wind on U1L1 = 0.030 kip/ft
Out-of-plane fixed end moment due to wind at U1 and L1 = 0.030(35)²/12 = 3.06 kip-ft
Secondary out-of-plane flexural stress on member = 3.06(12)/128.28 = 0.29 ksi
Secondary out-of-plane flexural stress at connection net section at U1 and L1 = 3.06(12)/108.06 = 0.34 ksi

j) Out-of-plane primary flexural stresses due to floorbeam rotation: In accordance with Equations 7.29a and 7.29b, the strength design bending moment due to floorbeam rotation, assuming fixity of joints U1 and L1,

$$M_{L1} = \frac{2Pa(L-a)I_p}{hI_f + 2LI_p} = \frac{2(261.82)90(240-90)769.67}{420(12000)+2(240)769.67} = 1005.8 \text{ kip} - \text{in}$$

$$M_{U1} = \frac{M_{L1}}{2} = 502.9 \text{ kip} - \text{in}$$

Out-of-plane primary bending stress in member at L1 = $\sigma_{pbgmL1} = M_{L1}/S_{nmy}$ = 1005.8/128.28 = 7.84 ksi
Out-of-plane primary bending stress in member at U1 = $\sigma_{pbgmU1} = M_{U1}/S_{nmy}$ = 3.92 ksi
Out-of-plane primary bending stress at connection at L1 = $\sigma_{pbgmL1} = M_{L1}/S_{ncy}$ = 1005.8/108.06 = 9.31 ksi
Out-of-plane primary bending stress at connection at U1 = $\sigma_{pbgmU1} = M_{U1}/S_{ncy}$ = 4.65 ksi

k) In accordance with AREMA (2023) and AREMA (2024), secondary flexural stresses other than out-of-plane primary flexural stresses due to floorbeam rotation need not be considered, and combined stresses for strength design are
9.03 + 9.31 = 18.34 ksi ≤ 0.55(50) ≤ 27.5 ksi, OK
(11.64/32.90) + (9.31/27.5) = 0.35 + 0.34 = 0.69 ≤ 1.0, OK

Fatigue Design:

l) Member effective section properties: Category B with S_{Rfat} = 16 ksi on member gross section properties.
m) Slip-critical connection effective section properties:
Category B with S_{Rfat} = 16 ksi on gross section of connected elements
$A_{efat} = 2(12)(1.00) = 24.0\ in^2$
$I_{efatx} = 2(1.0)(12)^3/12 = 288.00\ in^4$
$S_{efatx} = 288.00/6 = 48.00\ in^3$
$I_{efaty} = 2(12)(1.0)^3/12 + 2(12)(1.0)(5.50)^2 = 728.00\ in^4$
$S_{efaty} = 728.00/6 = 121.00\ in^3$
n) Combined axial and flexural stress ranges on the member gross section (stress ranges at L1 govern):
(200.46/261.82)(9.03 + 7.84 + 2.26) = 0.77(19.13) = 14.65 ksi ≤ 16 ksi, OK
o) Combined axial and flexural stress ranges on the slip-critical connection effective section for fatigue (stress ranges at L1 govern):
(200.46/261.82)((261.82/24.0) + (1005.8/121.0) + (9.04(12)/48.00)) = 0.77(10.91 + 8.31 + 2.26) = 0.77(21.48) = 16.45 ksi, OK at less than 3% over
p) Connection of H-section member plates:
Fillet welds connecting the side and web plates of member U1L1 must transfer the connection force over the length of the connection to the gusset plate.
Maximum fillet weld = 0.50 − (5/64) = 0.42 in., use 3/8 in fillet welds
Allowable stress for electrodes with 70 ksi tensile strength = 19.0 ksi
Allowable stress on base metal= 0.35(50) = 17.5 ksi
Total weld length required for strength = 261.82/(0.707(17.5)(0.375)) = 56.4 in
Total weld length required for fatigue = 200.46/(0.707(16.0)(0.375)) = 47.3 in
Each weld length required = 56.4/4 = 14.1 in
Use a minimum connection length, L_c = 14 in
p) Connection of member side plates to gusset plates: For Class B faying surface, non-staggered bolt spacing = 21/6 = 3.50 in.

Connection dimensions (Figure E10.4d):

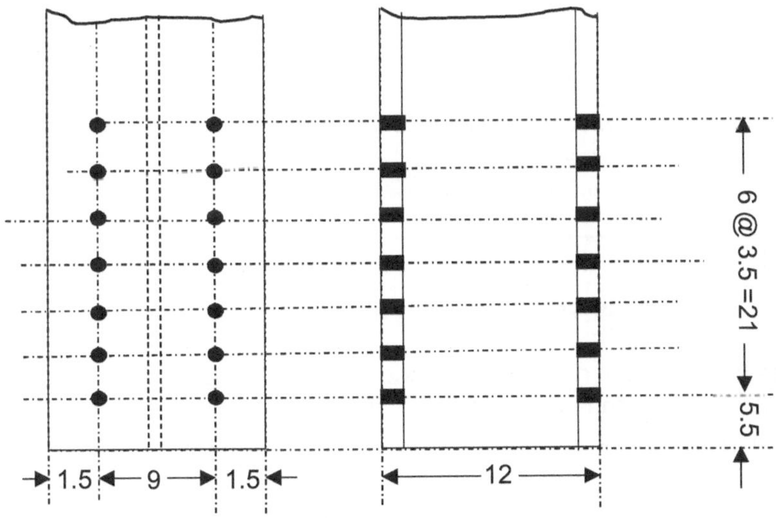

FIGURE E10.4d End connection of member U1L1.

Shear stress on bolts = 261.82/28 = 9.35 kips per bolt
Bearing stress on bolts = 261.82/(28)(7/8)(1.00) = 10.69 ksi per bolt
Allowable bearing stress on bolt = (3.5 − 0.875/2)(70)/(2(0.875) = 122.50 ksi or 1.2(70) = 84 ksi, OK
Bolt spacing ≥ 2(7/8)(14.96)/70 + 7/16 ≥ 0.37 + 0.44 ≥ 0.81 in, OK
Edge distance ≥ 1.5(7/8) ≥ 1.31 in, OK
End distance ≥ 1.5 + 4(1.00) ≥ 5.5 in, OK
Check block shear (see Chapter 12):
$T_{max} \le 0.30F_UA_{nv} + 0.50F_UA_{nt}$

or

$T_{max} \le 0.30F_yA_{gv} + 0.50F_UA_{nt}$
$A_{gv} = 2(5.5+21)(1.00) = 53.00\ in^2$
$A_{nv} = A_{gv} - 2(7)(1)(1.00) = 53.00 - 14.00 = 39.00\ in^2$
$A_{nt} = 2(4.5)(1.00) - 2(1)(1.00) = 9.00 - 2.0 = 7.00\ in^2$
$F_y = 50$ ksi
$F_u = 70$ ksi
261.82 ≤ 0.30(70)(39.00) + 0.50(70)(7.00) = 819.0 + 245.0 = 1064.0 kips, OK
261.82 ≤ 0.30(50)(53.00) + 0.50(70)(7.00) = 795.0 + 245.0 = 1040.0 kips, OK
q) Member and connection summary:
Use built-up section of two 12" × 1.00" side plates and one 10.0" × 1/2" web plate with 3/8 in. fillet welds and Class B connection with 28-7/8 in. diameter slip-critical bolts.

10.7 SERVICEABILITY OF AXIAL TENSION MEMBERS

The flexibility of axial tension members must be limited to preclude excessive deflections or vibrations. The maximum slenderness between points of support, L_u/r_{min}, is recommended by AREMA (2024) as

$$\frac{L_u}{r_{min}} \le 200 \tag{10.22}$$

where

$$r_{min} = \sqrt{I_{min}/A_g};$$

I_{min} = the minimum moment of inertia of the axial tension member in bending about an axis over a length, L_u, between supports; and
A_g = the gross cross-sectional area of the truss tension member.

Example 10.5

Investigate the slenderness for serviceability of the hanger member U1L1 of Example 10.3:

A_g = Gross cross-sectional area of member = A_{gm} = 29.00 in²
I_{min} = Gross in-plane moment of inertia of member = I_{xgm} = 288.10 in⁴

$$r_{min} = \sqrt{288.10/29.00} = 3.15\ \text{in}$$

L_u = 35 ft = 420 in
L_u/r_{min} = 420/3.15 = 133.3 ≤ 200, OK

NOTES

1 Also referred to as fracture-critical members. These are members whose failure would be expected to result in collapse of the truss or inability of the truss to perform its intended design function.

2 The secondary flexural stresses and stress ranges are those not determined via typical classical truss analyses assuming axial forces applied at the centroidal axis of members that terminate at frictionless pin connections. Classical methods of determining secondary bending effects are onerous. However, modern analytical methods based on direct stiffness are incorporated into software that can readily provide secondary bending moments in truss members such those created by rigidity of joints, member eccentricities and transverse frame behavior.

3 Tables 10.1 and 10.2 refer to load combinations LC1A, LC1B, LC2 and LC3 outlined in Chapter 9.

4 Slenderness limitations to mitigate vibrations of tension members.

5 These secondary flexural stresses can be due to member eccentricity, transverse forces on members, distortion of vertical members, rotational restraint at end joints, elongation of chord members and/or truss distortion and joint rigidity. Truss distortion and joint rigidity is typically the most prevalent source of secondary flexural stresses.

6 The net cross-sectional area of a member is the gross cross-sectional area of the member with the greatest cross-sectional area of open holes or perforations in the member removed. The effective net cross-sectional area at a connection is the gross cross-sectional area at the connection with the greatest cross-sectional area of bolt holes on potential failure lines removed to obtain the net cross-sectional area at the connection and, thereby, further reduced to account for shear lag to obtain the effective net cross-sectional area at the connection.

7 These are out-of-plane secondary flexural stresses due to floorbeam end rotation; however, they are considered primary flexural stresses for strength design of vertical members in truss spans.

8 Vertical web members that are loaded only by floorbeam reactions such as hanger and sub-vertical web members.

9 AREMA (2024) indicates that secondary stresses (strength design) must be considered on net section modulus if member width parallel to plane of distortion greater than 10% of member length. AREMA (2024) does not explicitly indicate that secondary stress ranges (fatigue design) must be considered on net section modulus if member width parallel to plane of distortion greater than 10% of member length, but it is implied by the reference to secondary stresses.

10 These are secondary flexural stress ranges due to floorbeam end rotation, but they are considered primary flexural stress ranges for fatigue design of vertical members directly loaded by floorbeam reactions.

11 AREMA (2024) recommends that secondary flexural stress ranges from all sources be considered for fatigue design of vertical members directly loaded by floorbeam reactions.

12 From about 1919 to 1934, AREMA Chapter 15 recommended that primary axial and, if necessary, secondary flexural stresses be combined with an allowable stress factor of 1.33. From 1934 to 2023, AREMA Chapter 15 recommended that, if necessary, only secondary tensile stresses exceeding 27.6 MPa (4,000 psi) be considered primary tensile flexural stresses. In 2021, AREMA Chapter 15 recommended that secondary flexural stress ranges be considered primary flexural stress ranges for the fatigue design of members directly loaded by floorbeam reactions. In 2022, AREMA Chapter 15 indicated explicitly that secondary flexural stresses also be considered primary flexural stresses for the strength design of members directly loaded by floorbeam reactions. In 2024, AREMA Chapter 15 indicated that secondary flexural stresses be considered primary flexural stresses for the strength design of all vertical members. From about 1919 to 2023, AREMA Chapter 15 recommended that secondary flexural stresses due to connection rigidity and truss distortion were necessary to determine if members with width parallel to the plane of distortion exceeded 10% of member length. In 2024, AREMA Chapter 15 recommended that secondary flexural stresses were necessary to determine if members with width parallel to the plane of distortion exceeded 10% of member length.

13 The flexural stresses due to out-of-plane bending from floorbeam end rotation are exacerbated in vertical web members directly loaded by floorbeam reactions (hanger and subvertical members) as they are subjected to simultaneous maximum axial and flexural tensile stresses.

14 Truss members are subjected to primary axial, secondary axial and secondary bending stresses. Secondary axial forces are small and can be neglected.

15 Eccentricity of member centroidal axes and out-of-straightness of members. These secondary bending stresses are in the plane of the truss.

16 Beam action of truss members due to self-weight or other forces between joints where the forces are transferred in typical truss analysis. These secondary bending stresses are in-plane (the plane of the truss) for vertically applied loads and out-of-plane for laterally applied forces.

17 Restraint from members such as collision struts or bending of chord members from the extension of vertical members may create in-plane secondary bending. Tension in stringer and lateral bracing system connections due to chord extension may also precipitate secondary stress in floor system and bracing system members. Due to out-of-plane bending stresses, floorbeam hangers and subvertical members will typically require larger cross-sectional areas, which is beneficial in reducing the tensile axial deformation causing secondary bending stresses in chord members due to the extension of vertical members.
18 External bending moments causing in-plane secondary flexural stresses in truss members are created due bearings that do not accommodate rotation.
19 Floorbeam end rotation creates considerable out-of-plane bending moments in vertical web members and, particularly, in truss hanger and sub-vertical members loaded directly by floorbeam reactions.
20 Bolted gusset plates provide resistance to member end rotation thereby creating secondary flexural stresses in the plane of the truss. Since the magnitude of bending created in truss members is typically proportional to the intensity of the primary axial forces, because of primary axial stress deformation, the secondary flexural stresses are proportional to the primary axial stresses.
21 Also, secondary bending stresses at the ends of members are typically small due to the presence of gusset plates, which effectively increase the moment of inertia of the member near connections.
22 The bending effect is greater for vertical members directly loaded by floorbeam reactions as the primary axial stresses and secondary flexural stresses occur simultaneously. Vertical members indirectly loaded by floorbeam reactions are also engaged in truss action.
23 Allowable stresses usually include an allowance of about 30% of primary axial stress for secondary flexural stress.
24 For example transverse welds at attachments to a tension member.
25 Load combinations and associated allowable stresses are recommended in AREMA Chapter 15. These load combinations and associated allowable stresses are typically presented as 'stress sheets' on truss design drawings.
26 Lateral forces from wind on the loaded bridge and nosing (truck hunting). Longitudinal forces from wind on the loaded bridge, braking, traction and anchored continuous welded rail. Other forces such as those from temperature and support settlements of statically indeterminate truss spans.
27 The net cross-sectional area is the reduced cross-sectional area due to relatively small openings in members such as holes for bolts. The reduced cross-sectional area due to large openings in members, such as perforations in cover plates of box-section members, is the net cross-sectional area at a section taken through the perforation.
28 Away from the member connections.
29 Typically, only two elements of a member are connected at panel point gusset plates.
30 This is typical of truss member connections where fastening to gusset plates is made at the side plates of built-up members or flanges of rolled members.
31 The length of the connection is the distance between the first and last fastener holes in the connection element.
32 Gusset plates in truss spans.
33 Where all elements of a member are connected to transfer the tensile force in the member, the connection is 100% efficient and $U_c = 1.00$.
34 Typical of bracing member connections.
35 The larger value of 0.80 is used for angles when there are four or more bolts per fastener line in the connection (i.e., a relatively long connection).
36 Strong axis bending can result in weak-axis lateral-torsional instability affecting the allowable compressive bending stress.
37 Equation 10.9a is the critical elastic buckling stress where warping torsion is negligible, such as in torsionally strong H-sections and Equation 10.9b is the critical elastic buckling stress where pure (St. Venant) torsion is negligible, such as in torsionally weak H-sections.
38 Due to fastener slip between the components of bolted built-up members (rarely used in modern truss member design), only Equation 10.9b needs consideration for strong axis bending.
39 Particularly to reduce the slenderness of truss members subject to primary axial compression forces.
40 Rectangular sections with much greater in-plane depth than out-of-plane width, which are uncommon for truss box-section member dimensions.
41 For fatigue, AREMA (2024) essentially designates the secondary bending moment ranges as primary bending moment ranges.
42 Particularly the web members of trusses near the center of a truss span.
43 Typically due to hot rolling, cutting and welding of plates and shapes.

44 Distortion-induced fatigue may occur due to secondary stresses not typically determined from analysis and design procedures, particularly at non-rigid details. Distortion-induced fatigue is precluded by careful design of details and connections that are stiff enough to preclude secondary strains.
45 Shear lag for strength design is evaluated at stress levels near fracture.
46 The flanges of truss H-members (I-members rotated 90°) or web (side) plates of truss box members connected to gusset plates are typical examples of indirectly loaded member connections with shear lag effects considered by determining the live load stress range with the effective gross or net cross-sectional area, A_{efat}, of only the connected elements.
47 Pre-tensioned connections with ASTM F3125. Grade A325M (A325) or Grade A490M (A490) bolts. The number of bolts in a connection is determined based on the allowable stresses for slip-critical connections provided by AREMA (2024). For example, a Class B connection with Grade A325M (A325) bolts provides an allowable stress of 150 MPa (21.5 ksi). Slip-critical connections are recommended for main member connections and all connections subject to stress reversal and/or cyclical loading.
48 Most railway truss span chord members have influence lines that exceed 30.5 m (100 ft).
49 Including hanger and other members resisting only floorbeam reactions.
50 However, hand holes of reasonable size (say, about 250 mm (10 in) × 400 mm (16 in.)) in the cover plates of box sections within gusset plate connections are typically not a concern for maximum tensile axial and bending stresses or for cyclical axial and bending stress ranges due to proximity of the hand holes to the gusset plate connection and some reinforcement provided by horizontal gusset plates bolted to the bottom cover plate of the box-section member for bottom lateral bracing.
51 Subvertical members are directly loaded by floorbeam reactions in subdivided truss spans.
52 Flexural stresses can occur parallel (in-plane) and perpendicular (out-of-plane) to the plane of the truss span.
53 Rolled W and HP sections are also used for members with relatively small tension forces.
54 Solid cover plates have hand holes for access to bolted joint connections. Hand holes are typically fabricated in the bottom cover plates at the end of members to allow for access to bolted connections at joints.
55 The distance between the nearer lines of connections is typically the distance between the longitudinal welds connecting cover plates to web plates of box members, which is essentially the cover plate width, b_p.
56 V at the centerline of the cover plate is generally larger for compression members.
57 In accordance with AREMA (2024), the minimum steel thickness is 9.5 mm (3/8 in), except for gusset plates where the minimum thickness is 12.7 mm (1/2 in).
58 Engineers use solid cover plates for truss tension members to establish Fatigue Detail Category A or B members. Hand holes in the bottom cover plate within gusset plate connections are typically not a net section concern for strength and/or fatigue design of tension members.
59 Cover plate to web plate welds in box section members can require CJP or large fillet welds to develop the weld within the gusset plate length. Large fillet welds can be avoided in large gusset plates with a long connection for developing the weld for strength and fatigue loadings.
60 For example, American Welding Society ANSI/AASHTO/AWS D1.5 and Canadian Standards Association W59 in North America.

REFERENCES

American Railway Engineering and Maintenance-of-Way Association (AREMA), 2023, *Manual for Railway Engineering*, Chapter 15, Steel Structures, Lanham, MD.

American Railway Engineering and Maintenance-of-Way Association (AREMA), 2024, *Manual for Railway Engineering*, Chapter 15, Steel Structures, Lanham, MD.

Bonachera Martin, F.J., and Connor, R.J., 2017, *Load-induced Fatigue Category of Hand Holes and Manholes*, Purdue University, West Lafayette, IN.

Bowles, J.E., 1980, *Structural Steel Design*, McGraw-Hill, New York.

Cochrane, V.H., 1922, *Rules for Rivet Hole Deductions in Tension Members*, November, Engineering News Record, New York.

Greenspan, M., 1943, *Axial Rigidity of Perforated Structural Members*, Journal of Research of the National Bureau of Standards, Washinton, DC.

Munse, W.H., and Chesson, E., 1963, "Riveted and Bolted Joints: Net Section Design", *ASCE Journal of the Structural Division*, 83(ST1).

Salmon, C.G., and Johnson, J.E., 1980, *Steel Structures Design and Behavior*, Harper & Row, New York.

Unsworth, J.F., 2018, *Design and Construction of Modern Steel Railway Bridges*, 2nd Edition, CRC Press, Boca Raton, FL.

11 Design of Truss Compression Members

11.1 INTRODUCTION

Non-slender compression members fail at a load that generates stress levels in the member that exceed the limiting material strength of yielding. However, slender compression members fail by instability because at a critical buckling force, less work is required by the load to bend than to shorten[1] the member.

Axial compression members in steel railway truss spans are typically slender and therefore subject to instability[2] at critical buckling stress, which is less than the yield stress of the material.[3] Therefore, slender axial compression members must be designed considering the critical buckling force, which depends on member dimensions,[4] support conditions[5] and material properties[6] rather than solely material strength. Accordingly, Table 11.1 outlines the recommended load combinations, primary stresses and secondary stresses for the strength and stability design of truss compression members. Serviceability criteria relating to axial compression members must also be considered in truss compression member design.

11.2 SECONDARY BENDING IN THE STRENGTH AND STABILITY DESIGN OF TRUSS COMPRESSION MEMBERS

As indicated in Section 9.4.2, secondary bending can be of considerable concern for axial compression members. AREMA (2024),[8] and Table 11.1 indicates that secondary flexural stresses must be considered for strength and stability design (load combinations LC1A, LC1B or LC2) of members with width parallel to the plane of distortion greater than 10% of member length.

The shear stresses associated with the secondary bending of typical truss members are typically negligible. In-plane secondary flexural stresses will slightly, but not appreciably, reduce primary axial stresses, and therefore, the total combined primary axial and secondary flexural stresses are normally greater than those due to axial compression[9] alone. As indicated in Chapter 7 and Chapter 9, secondary flexural stresses in truss compression members are caused by eccentricities

TABLE 11.1
Requirements for Compression Member Strength and Stability Design

	Strength and Stability Design		
Truss Compression Element	**Load combinations**	**Primary axial stresses**	**Secondary flexural stresses[7]**
Chord members	LC1A or LC1B	On gross cross-sectional area of member	On gross section modulus if member width parallel to plane of distortion greater than 10% of member length
Web members	LC1A or LC1B or LC2	On gross cross-sectional area of member	On gross section modulus if member width parallel to plane of distortion greater than 10% of member length

DOI: 10.1201/9781003469971-13

at connections, transverse forces on members, the distortion of tension members that frame into compression members, rotational restraint at end joints, and/or truss distortion and joint rigidity. The secondary flexural stresses due to eccentricities at connections are generally precluded by careful detailing of working points. However, if eccentricities at joints exist, they are accounted for as shown in Chapter 7. Transverse bending of truss members due to self-weight and forces between panel points is typically negligible for usual truss designs but if present can be calculated as shown in Chapter 7. Secondary bending from vertical member distortion is avoided by providing vertical members with robust cross-sectional areas or allowing for such extension in design. Nevertheless, if present, secondary bending due to vertical member distortion (such as supporting struts in subdivided truss spans) can be estimated as shown in Chapter 7. Secondary flexural stresses from rotational restraint at bearings and truss joints are also typically small for members with width parallel to the plane of distortion less than 10% of the member length[10] but can also be determined as shown in Chapter 7.

For most simple Pratt and Warren truss spans with relatively slender members, the secondary flexural stresses due to joint rigidity, self-weight and member distortion seldom exceed 25% of the primary axial stress.[11] Therefore, in many cases, secondary flexural stresses need not be explicitly considered in design. Still, some compression members, in long or complex truss spans, may require explicit consideration of secondary flexural stresses from joint rigidity, transverse forces and member distortion.

11.3 STRENGTH AND STABILITY DESIGN OF TRUSS COMPRESSION MEMBERS

Strength and stability design must consider combined primary axial compression forces and secondary bending moments in the case of members whose width exceeds 10% of the member length.

The axial compressive force, P, due to dead load (DL), live load (LL), maximum impact (I) and centrifugal force (CF), which is load combination[12] LC1A, is determined with a load combination factor of 1.00 applied to allowable stresses that can increase to 1.25 when also considering lateral, longitudinal and other forces,[13] which is load combination LC1B. LC2 with a factor of 1.33 is applied to web member overload forces $(LL + I + CF)_{in}$.

11.3.1 Strength and Stability Design of Truss Compression Members for Axial Forces

As shown in Figure 11.1, linear elastic analysis is appropriate when the force in slender compression members reaches the critical buckling load prior to the compressive stress in the member attaining the proportional limit. However, inelastic analysis is required for members that are non-slender because the axial compressive stress in the member will exceed the proportional limit prior to attaining the critical buckling force.

11.3.1.1 Elastic Critical (Euler) Buckling Stress

Slender members will buckle at loads generating compressive stresses below the proportional limit stress. The magnitude of this elastic critical buckling stress depends on the stiffness, length and end conditions of the compression member, as well as imperfections in loading and geometry. Nevertheless, assuming that the external force on the member is applied along the centroidal axis of the member, the member has no geometric imperfections (perfectly straight), plane sections remain plane after deformation, flexural deflection is small and is the only deformation considered (shear deflection is neglected), and Hooke's law is applied, then the differential equation of the compression member deflection curve is

$$\frac{d^2y(x)}{dx^2}+k^2y(x)=U \tag{11.1}$$

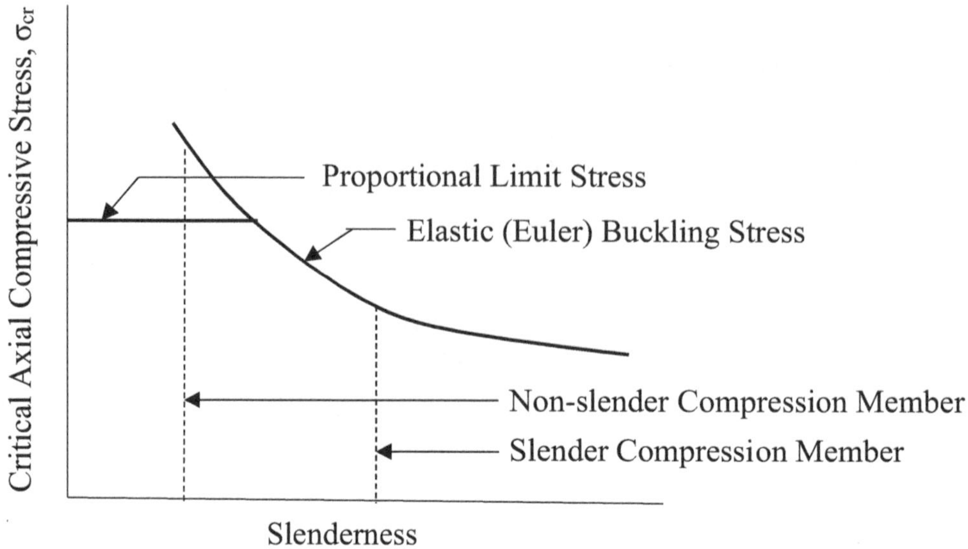

FIGURE 11.1 Elastic compression member behavior.

where

$y(x)$ = the lateral deflection of the compression member;
U depends on the effects of the compression member end conditions on P;
$k^2 = \frac{P}{EI}$; and
P = the force applied at the end and along the centroidal axis of the compression member.

The solution of Equation 11.1 for the elastic critical buckling force, P_{cr}, is readily accomplished by considering the appropriate boundary conditions (Wang, Wang and Reddy, 2005). P_{cr} for various member end conditions is shown in Table 11.2.

Figures 11.2 and 11.3 illustrate the various compression member end conditions shown in Table 11.2 where L = length of member between end supports, V = shear force in member, M = bending moment at end of member and Δ = lateral deflection at free or guided end of member with other end fixed.

As illustrated by Table 11.2, the elastic critical (Euler) buckling force can be expressed as

$$P_{cr} = \frac{\pi^2 EI}{(KL)^2} \tag{11.2}$$

Substituting $I = A_g r^2$ into Equation 11.2 provides the critical elastic buckling stress, σ_{cr}, as

$$\sigma_{cr} = \frac{P_{cr}}{A_g} = \frac{\pi^2 E}{(KL/r)^2} \tag{11.3}$$

Rearranging Equation 11.3 yields

$$K = \text{effective length factor} = \frac{\pi}{L}\sqrt{\frac{EI}{P_{cr}}} \tag{11.4}$$

TABLE 11.2
Critical Buckling Force for Various Compression Member End Conditions

End Condition	U	P_{cr}
Both ends pinned (Figure 11.2a)	0	π^2EI/L^2
Both ends fixed (Figure 11.2b)	$-\left(V/EI\right)x+M/EI$	$4\pi^2EI/L^2$
One end fixed and other end free (Figure 11.2c)	$k^2\Delta$	$\pi^2EI/4L^2$
One end hinged and other end fixed (Figure 11.2d)	$\left(M/EIL\right)x$	$2.046\pi^2EI/L^2$
One end guided and other end fixed (Figure 11.3a)	$P\Delta/2EI$	π^2EI/L^2
One end hinged and other end guided (Figure 11.3b)	0	$\pi^2EI/4L^2$

Values of K from Equation 11.4 for the various end conditions and corresponding critical buckling force, P_{cr}, in Table 11.2 are shown in Table 11.3.

However, due to the ideal conditions of the mathematical model, which may not be representative of actual end conditions, the effective length factors in Table 11.3 are generally reduced for the allowable stress design of compression members. For railway trusses with moving live loads, the forces in other members framing into the end of a member under consideration will be less than maximum when the member under consideration is subject to maximum force. Therefore, the ideally pinned end condition is not established because this force arrangement imposes rotational restraints and secondary bending stresses at the end of the member under consideration. Consequently, AREMA (2024) recommends K = 0.75 for bolted member end connections.

A safety factor must be applied to Equation 11.3 to account for small load eccentricities and geometric imperfections as shown in Figure 11.3c and Figure 11.3d. Accordingly, AREMA (2024) uses a factor of safety, FS, of 1.95 to arrive at the allowable axial compressive strength, F_{call}, of

$$F_{call}=\frac{\sigma_{cr}}{1.95}=\frac{0.514\pi^2E}{(KL/r)^2}=\frac{5.073E}{(KL/r)^2} \tag{11.5}$$

As shown in Figure 11.1, Equation 11.5 provides the allowable compressive stress for members with slenderness, KL/r, greater than the slenderness where the elastic buckling stress is equal to the proportional limit or yield stress. However, members with intermediate slenderness exhibit inelastic behavior due to geometric imperfections,[14] material nonlinearities and residual stresses[15] introduced during fabrication and erection of steel truss spans.

11.3.1.2 Inelastic Critical Buckling Stress

As illustrated by Figure 11.1, compression members of intermediate slenderness[16] will buckle at loads above the proportional limit, F_p, when some cross-section fibers have already yielded before

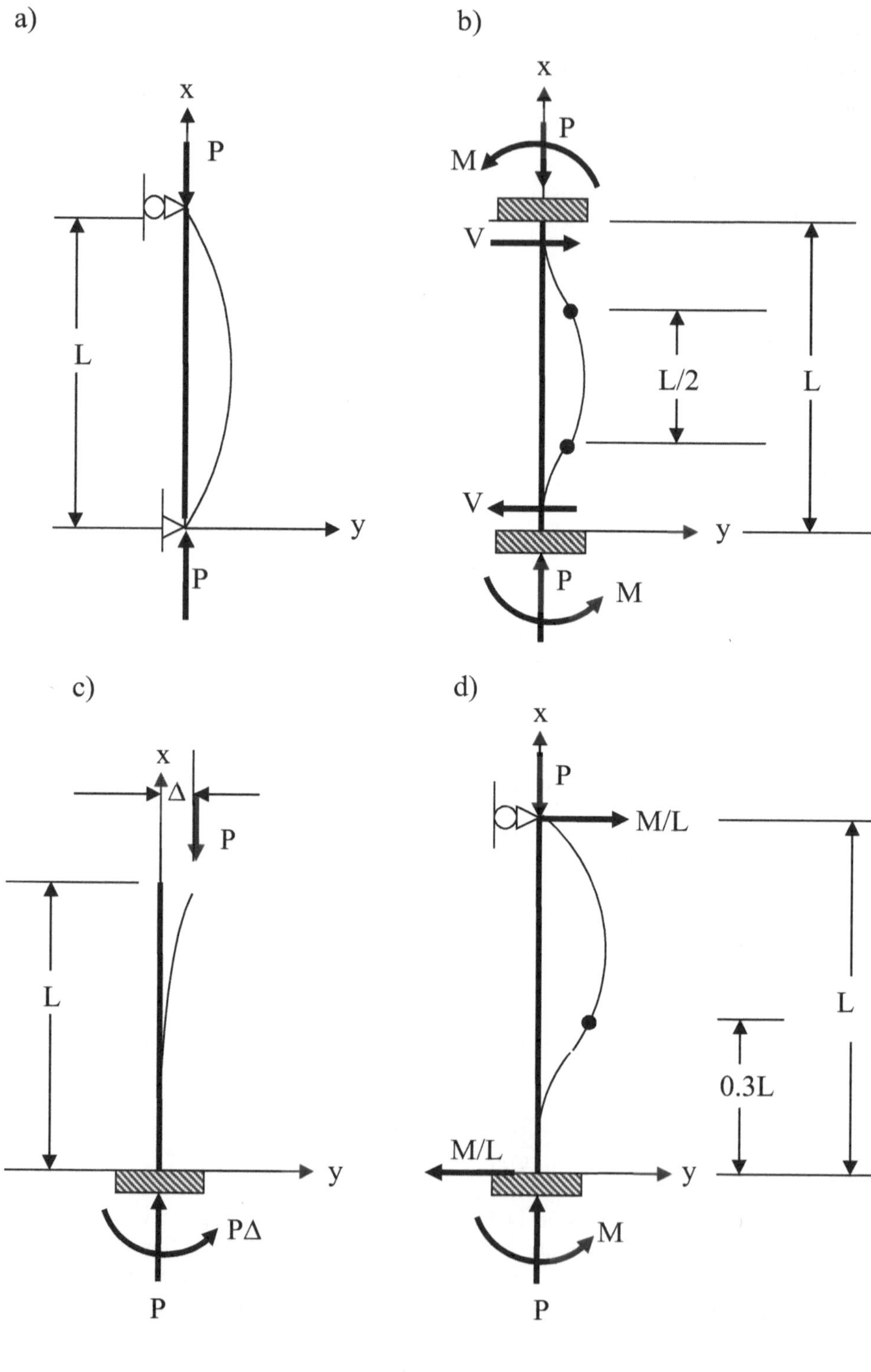

FIGURE 11.2 Compression member end conditions. (a) pinned–pinned, (b) fixed–fixed, (c) fixed–free and d) fixed–hinged.

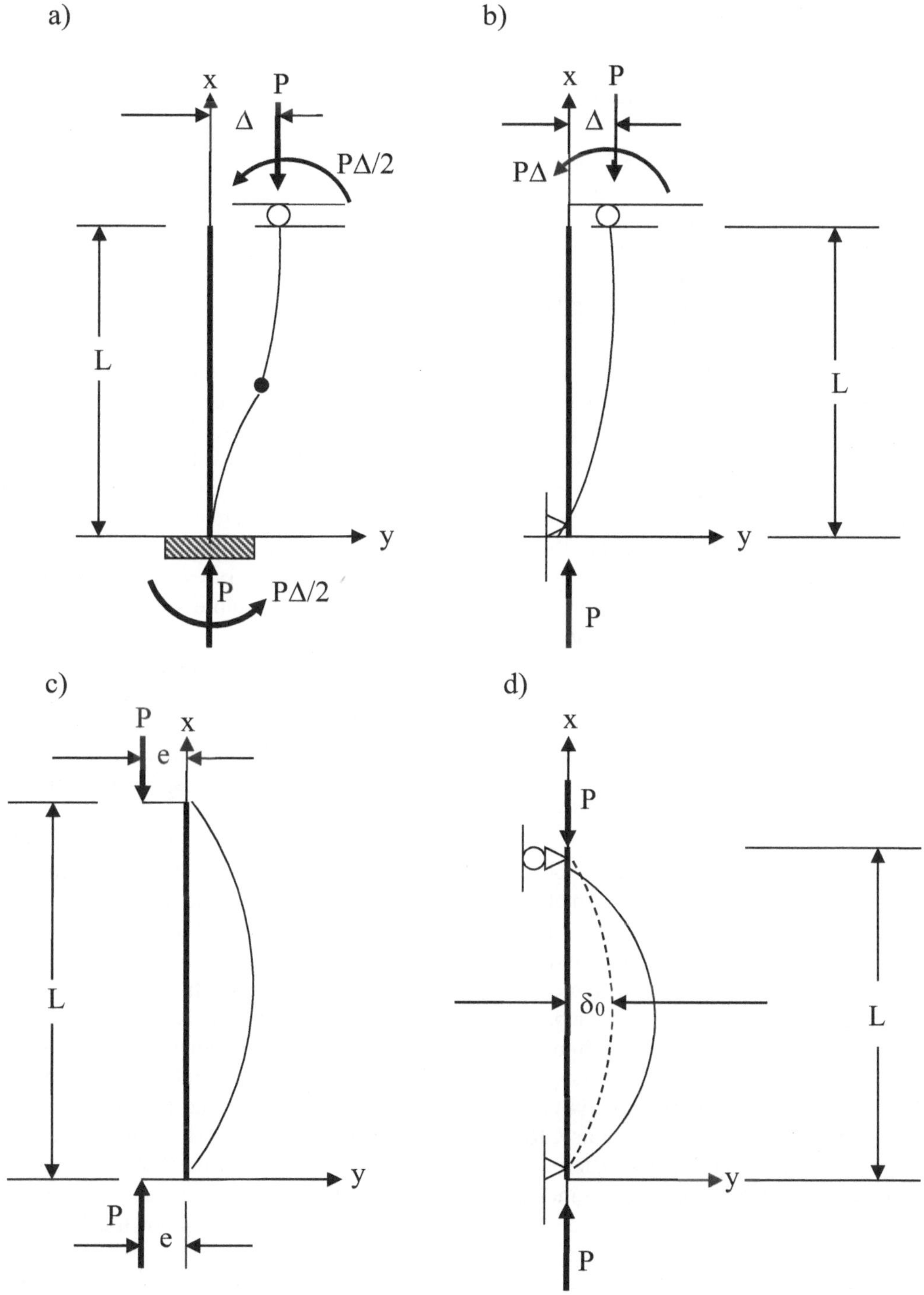

FIGURE 11.3 Compression member end and initial conditions. (a) fixed–guided, (b) hinged–guided, (c) eccentric load and d) geometric imperfection.

TABLE 11.3
Effective Lengths for Various Compression Member End Conditions

Effective Length Factors for Various Compression Member End Conditions

End condition	K
Both ends pinned (Figure 11.2a)	1.00
Both ends fixed (Figure 11.2b)	0.50
One end fixed and other end free (Figure 11.2c)	2.00
One end hinged and other end fixed (Figure 11.2d)	0.70
One end guided and other end fixed (Figure 11.3a)	1.00
One end hinged and other end guided (Figure 11.3b)	2.00

the initiation of instability. Therefore, the effective modulus of elasticity is less than the initial value. Consequently, the nonlinear behavior of intermediate slenderness compression members due to material nonlinearities may be modelled by replacing the elastic modulus, E, with a lesser effective modulus, E_{eff}, in Equation 11.5. Engesser proposed the effective modulus as both the tangent modulus and the reduced modulus (Timoshenko and Gere, 1961), and Shanley also proposed an inelastic compression member theory (Tall, 1974). The theory indicates that actual inelastic compression member behavior lies between that of the tangent and reduced modulus curves.

However, because test results are closer to the tangent modulus curve values (Chen and Lui, 1987), the tangent modulus has often been used to develop inelastic compression curves and equations. Nevertheless, the tangent modulus theory considers material nonlinearities but not geometric imperfections. AREMA (2024) implicitly considers geometric imperfections with an increased safety factor of 1.95.[17] However, test observations[18] indicated discrepancies in the strength of compression members determined in accordance with tangent modulus theory. It was determined that residual stresses in the test specimens were responsible for the inconsistencies in compression member strength. Therefore, in addition to material and geometric nonlinearities, residual stresses in steel members require consideration for compression member design.

The rolling of structural steel plates and shapes and fabrication bending, cutting and/or welding procedures characteristically create residual stresses that affect the inelastic critical buckling stress in a compression member. The pattern of compressive and tensile residual stresses is very dependent on member cross section and dimensions. The presence of varying residual stresses will affect the material compressive stress–strain curve and establish a different effective modulus of elasticity in each direction across a compression member cross section. Consequently, the effective modulus of elasticity will differ depending on the buckling direction (weak or strong axis) and must be considered independently to determine compression member strength when allowing for residual stresses.

Subsequent to the Lehigh University studies that illustrated residual stresses are the principal reason for the discrepancies between compression member test results and the tangent modulus theory, the Column Research Council (CRC) conducted further tests and performed analytical studies of weak and strong axis inelastic buckling behavior with linear and parabolic residual stress distributions across the compression member cross-section. These studies revealed that within the inelastic range, the compression member curves were parabolic[19] and of the following form:

$$\sigma_{cr} = F_y - B(KL/r)^2 \tag{11.6}$$

The residual stresses used in the CRC studies were $0.3F_y$. However, $0.5F_y$ is used to conservatively represent the residual stresses and provide a smooth transition to the elastic buckling curve. The constant B with $F_p = 0.5F_y$ and $F_r = F_y - F_p$ (Bleich, 1952) is

$$B = \frac{F_y^2}{4\pi^2 E} \tag{11.7},$$

and the inelastic critical buckling stress is

$$\sigma_{cr} = F_y\left(1-\frac{F_y}{4\pi^2 E}\left(\frac{KL}{r}\right)^2\right) \quad (11.8).$$

From Equation 11.8, the slenderness ratio at $\sigma_{cr} = 0.5F_y$ is

$$C_c = \frac{KL}{r} = \sqrt{\frac{2\pi^2 E}{F_y}} = 4.443\sqrt{\frac{E}{F_y}} \quad (11.9).$$

Therefore, from Equation 11.3, for $KL/r > C_c$,

$$\sigma_{cr} = \frac{\pi^2 E}{(KL/r)^2} \quad (11.10a),$$

and from Equation 11.8, for $KL/r \le C_c$,

$$\sigma_{cr} = F_y\left(1-\frac{F_y}{4\pi^2 E}\left(\frac{KL}{r}\right)^2\right) = F_y\left(1-\frac{\left(\frac{KL}{r}\right)^2}{2C_c^2}\right) \quad (11.10b).$$

Equations 11.10a and 11.10b are plotted in Figure 11.4.

AISC allowable stress design uses Equation 11.10a with a constant safety factor of 1.92 and Equations 11.10b with a variable safety factor:

$$FS = \frac{5}{3}+\frac{3}{8}\frac{(KL/r)}{C_c}-\frac{1}{8}\left(\frac{(KL/r)}{C_c}\right)^3 \text{ for } (KL/r) \le C_c \quad (11.11)$$

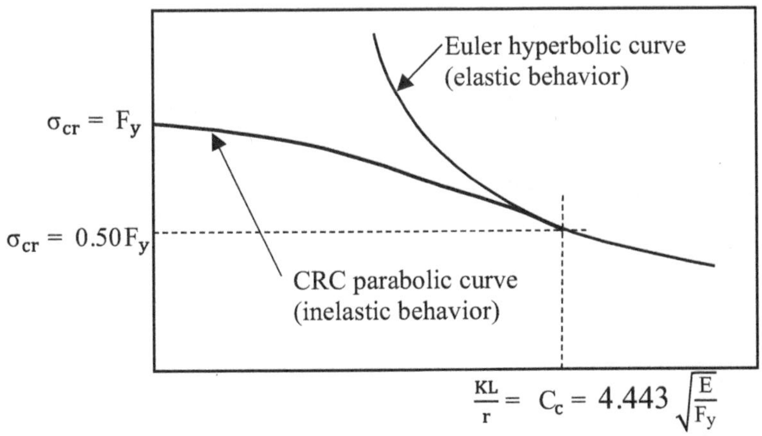

FIGURE 11.4 CRC critical elastic and parabolic inelastic compression member behavior.

Therefore, from Equations 11.3 with a safety factor of 1.92, for KL/r > C_c,

$$F_{call} = \frac{\sigma_{cr}}{1.92} = \frac{\pi^2 E}{1.92(KL/r)^2} = 5.140\frac{E}{(KL/r)^2} \tag{11.12a}$$

and, from Equation 11.8, for KL/r ≤ C_c,

$$F_{call} = \frac{\sigma_{cr}}{\frac{5}{3}+\frac{3}{8}\frac{(KL/r)}{C_c}-\frac{1}{8}\left(\frac{(KL/r)}{C_c}\right)^3} = \frac{F_y\left(1-\frac{F_y}{4\pi^2 E}\left(\frac{KL}{r}\right)^2\right)}{\frac{5}{3}+\frac{3}{8}\frac{(KL/r)}{C_c}-\frac{1}{8}\left(\frac{(KL/r)}{C_c}\right)^3}$$

$$= \frac{F_y\left(1-\frac{\left(\frac{KL}{r}\right)^2}{2C_c^2}\right)}{\frac{5}{3}+\frac{3}{8}\frac{(KL/r)}{C_c}-\frac{1}{8}\left(\frac{(KL/r)}{C_c}\right)^3} \tag{11.12b}$$

However, the uncertainties involved in calculating K will result in overestimates for σ_{cr}, particularly when KL/r is between 40 and 100, and to mitigate this, AREMA (2024) adopts the conservative straight line approximation of

$$\sigma_{cr} = F_y - B(KL/r) \tag{11.13}$$

The constant B (the slope of the line) in Equation 11.13 can be established by calculating compression member curves using variable and constant safety factors. In this manner, using an appropriate safety factor, AREMA (2024) recommends an allowable axial compression stress of

$$F_{call} = \frac{\sigma_{cr}}{FS} = 0.60F_y - \left(635\frac{F_y}{E}\right)^{\frac{3}{2}}\left(\frac{KL}{r}\right), \text{ in SI units} \tag{11.14a}$$

or

$$F_{call} = \frac{\sigma_{cr}}{FS} = 0.60F_y - \left(17500\frac{F_y}{E}\right)^{\frac{3}{2}}\left(\frac{KL}{r}\right), \text{ in US customary or imperial units} \tag{11.14b}$$

The straight-line transition curve for the allowable compression stress is made to intersect the Euler elastic curve at $0.20F_y$ to represent the effects of eccentricities, initial imperfections and residual stresses. Inserting $F_{call} = 0.20F_y$ into Equations 11.14a and 11.14b yields

$$\frac{KL}{r} = 5.034\sqrt{\frac{E}{F_y}} \tag{11.15}$$

Equation 11.15 is the limiting slenderness ratio or critical buckling coefficient, C_c, to preclude elastic buckling of compression members with the AREMA (2024) straight-line approximation for allowable compression stress.

When $F_{call} = 0.55F_y$, KL/r from Equations 11.14a and 11.14b is

$$\frac{KL}{r} \leq 0.629\sqrt{\frac{E}{F_y}} \tag{11.16}$$

Compression members that are very short and without the potential for instability have slenderness below that given by Equation 11.16. The Euler buckling curve (Equation 11.12a with FS = 1.95 instead of 1.92) and Equation 11.14 are plotted in Figure 11.5.

A comparison of the AISC allowable elastic (Equation 11.12a) and inelastic (Equation 11.12b) compressive stress and the AREMA allowable elastic (Equation 11.12a with FS = 1.95) and inelastic (Equation 11.14) compressive stress is shown in Figure 11.6.

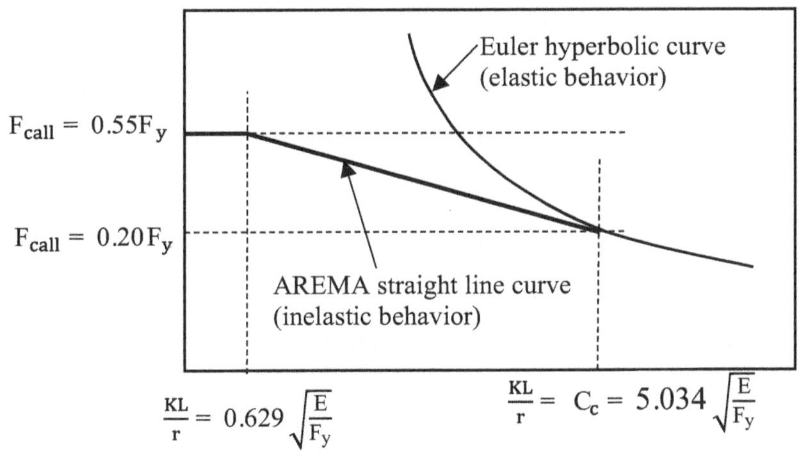

FIGURE 11.5 AREMA allowable elastic and straight-line inelastic compression member behavior.

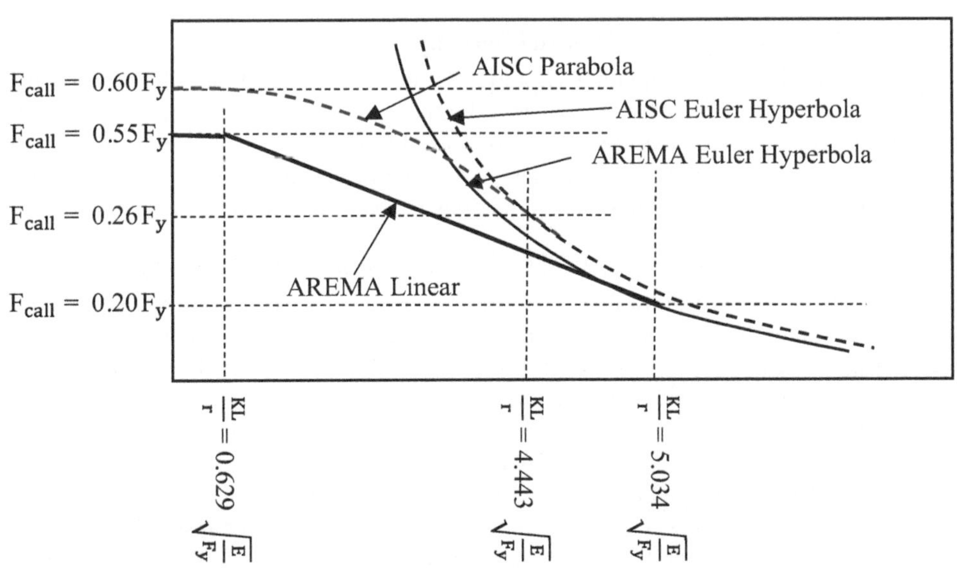

FIGURE 11.6 AISC (CRC) and AREMA allowable elastic and inelastic compression member behavior.

Therefore, the AREMA allowable axial compression force, $P_{call} = F_{call}A_g$, is

$$P_{call} = (0.55F_y)A_g, \text{ when } \frac{KL}{r} \leq 0.629\sqrt{\frac{E}{F_y}} \tag{11.17}$$

$$P_{call} = \left(0.60F_y - \left(635\frac{F_y}{E}\right)^{\frac{3}{2}}\left(\frac{KL}{r}\right)\right)A_g, \text{ when } \frac{KL}{r} \geq 0.629\sqrt{\frac{E}{F_y}} \text{ and },$$

$$\frac{KL}{r} \leq 5.034\sqrt{\frac{E}{F_y}}, \text{ in SI units} \tag{11.18a}$$

$$P_{call} = \left(0.60F_y - \left(17500\frac{F_y}{E}\right)^{\frac{3}{2}}\left(\frac{KL}{r}\right)\right)A_g, \text{ when } \frac{KL}{r} \geq 0.629\sqrt{\frac{E}{F_y}} \text{ and}$$

$$\frac{KL}{r} \leq 5.034\sqrt{\frac{E}{F_y}}, \text{ in US Customary or Imperial units} \tag{11.18b}$$

$$P_{call} = \left(\frac{0.514\pi^2 E}{(KL/r)^2}\right)A_g, \text{ when } \frac{KL}{r} > 5.034\sqrt{\frac{E}{F_y}} \tag{11.19}$$

where A_g = the gross cross-sectional area of the member.

11.3.2 Strength and Stability Design of Truss Compression Members for Axial Forces and Bending

Secondary tensile bending stresses seldom exceed the axial compressive stress in railway truss members. However, combined axial compression and secondary compressive bending requires consideration in the design of truss compression members.

11.3.2.1 Axial Compression and Uniaxial Bending

As indicated in Section 9.4.2, Equation 9.7 requires an iterative solution, which is not efficient for routine design work. Alternatively, for some boundary conditions and loads, the differential equation for axial compression and flexure can be solved (Unsworth, 2018).[20] However, for regular design work, limitations on combined stresses or semi-empirical interaction equations have been developed. AREMA (2024) uses interaction equations for both the yielding and stability criteria.

11.3.2.1.1 Interaction Equations for Axial Compression and Uniaxial Bending

Interaction equations used in allowable stress design are determined from interaction equations developed for ultimate loads. When members remain elastic, the axial compressive force, P, has no effect on the moment-curvature-axial compression relationship. However, in the inelastic range, where partial yielding has occurred, the moment–curvature–axial compression relationship is dependent on the axial compression. The nonlinear relationship between bending moment, M, and axial compressive force, P depends on the curvature, Φ, of the member. The ultimate axial compressive load is attained when yielding under combined bending and axial compression reduces the member stiffness due to partial yielding of the cross-section and creates instability. Additionally, in the range of inelastic behavior, the superposition of effects due to M and P is not applicable. The

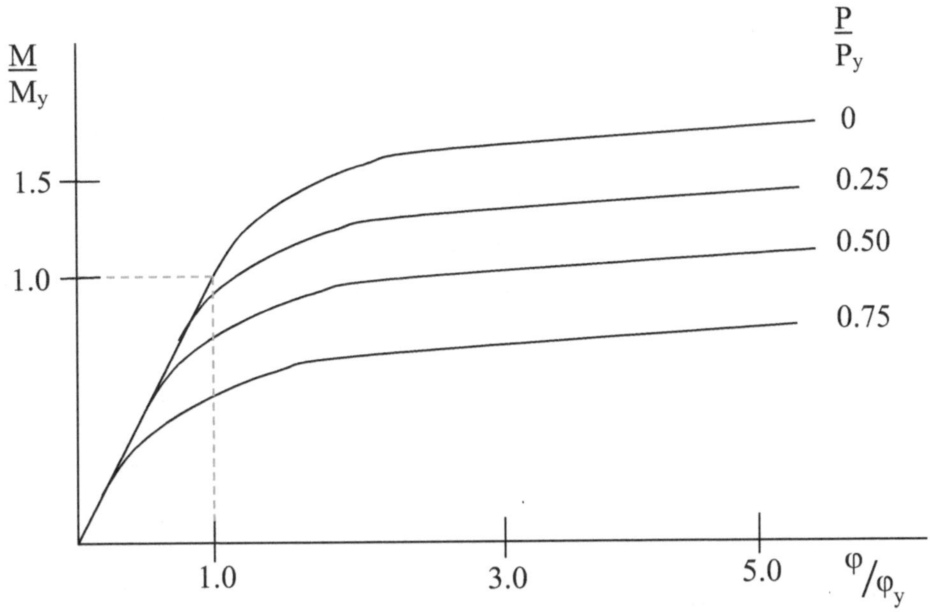

FIGURE 11.7 Typical moment–curvature–axial compression relationship curves.

moment–curvature–axial compression relationships for members are determined analytically for various degrees of member yielding.[21] A typical moment-curvature-axial compression relationship is shown in Figure 11.7.

Once the moment–curvature–axial compression relationship is established, each value of P/P_y with KL/r (where r is the radius of gyration in the plane of bending) is combined with various values of M/M_y until instability occurs (at M_u). Since M/M_y creates deflection, Δ, which creates an additional bending moment $(P/P_y)\Delta$, iterative analysis such as numerical integration by Newmark's method or another step-by-step numerical integration technique is often used. Once the appropriate P/P_y and M/M_u are determined (when deflections calculated in two successive iterations are in sufficient agreement) for various values of KL/r, interaction diagrams such as that shown in Figure 11.8 are produced. These interaction curves are approximated by interaction equations for various values of L/r.

The curve for L/r = 0 is approximated as

$$\frac{P}{P_y} + \frac{M}{1.18M_u} = 1.0 \tag{11.20a}$$

or

$$\frac{\sigma_a}{0.55F_y} + \frac{\sigma_b}{1.18F_u} = 1.0 \tag{11.20b}$$

For allowable stress design, Equation 11.20b is conservatively expressed as

$$\frac{\sigma_a}{0.55F_y} + \frac{\sigma_b}{F_{bcall}} = 1.0 \tag{11.21}$$

where L/r = 0, $\sigma_a = P/A_g$ = the normal stress due to applied axial compression, $\sigma_b = M/S_g$ = the normal stress due to applied bending moment, S_g = gross section modulus of member, F_u = the ultimate bending stress, and F_{bcall} = the allowable compressive stress for bending alone. This interaction

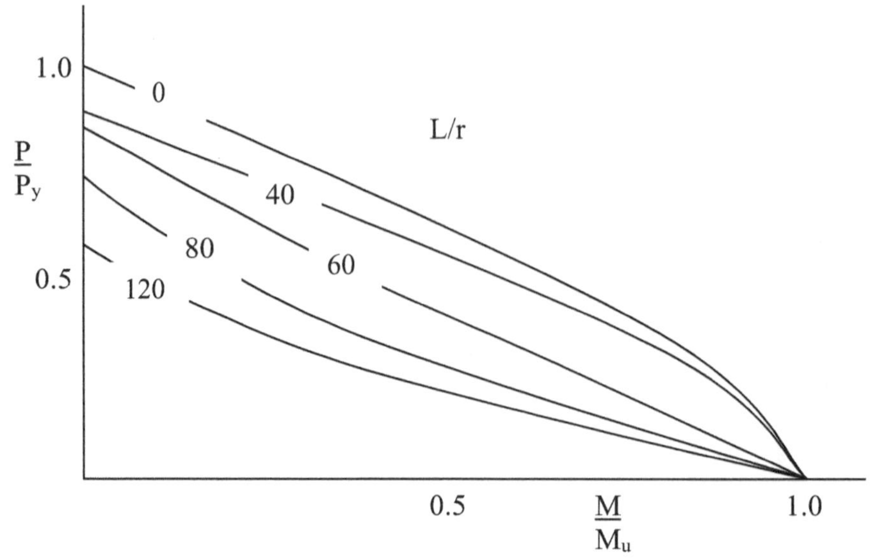

FIGURE 11.8 Typical interaction curves.

equation is applicable to members with low slenderness, such as locations that are braced in the plane of bending, where yielding will be the failure criterion. However, for members with larger slenderness, the stability criterion must also be investigated.

The yield criterion for members of larger slenderness is established as Equation 11.21 but considering the allowable axial compression stress, F_{call}, instead of $0.55F_y$ due to the potential for allowable axial compressive stresses to be controlled by instability when $\frac{KL}{r} \geq 0.629\sqrt{\frac{E}{F_y}}$. Equation 11.21 is then

$$\frac{\sigma_a}{F_{call}} + \frac{\sigma_b}{F_{bcall}} = 1.0 \tag{11.22}$$

The interaction curves are approximated for various values of L/r by interaction equations as

$$\frac{P}{P_{cr}} + \frac{M}{M_u(1 - P/P_e)} = 1.0 \tag{11.23a}$$

or

$$\frac{\sigma_a}{F_{cr}} + \frac{\sigma_b}{F_u(1 - \sigma_a/\sigma_e)} = 1.0 \tag{11.23b}$$

where

P_{cr} = the critical axial buckling load;

P_e = the Euler buckling load = $\frac{\pi^2 EI}{L^2}$;

F_{cr} = the critical axial buckling stress; and

σ_e = the Euler buckling stress = $\frac{\pi^2 E}{\left(KL/r\right)^2}$.

Equation 11.23b, using FS = 1.95 for allowable stress design for axial buckling, is conservatively expressed as

$$\frac{\sigma_a}{F_{call}} + \frac{\sigma_b}{F_{bcall}(1 - \sigma_a/(\sigma_e/1.95))} = 1.0 \tag{11.24a}$$

or

$$\frac{\sigma_a}{F_{call}} + \frac{\sigma_b}{F_{bcall}\left(1 - \frac{\sigma_a}{0.514\pi^2 E}\left(\frac{KL}{r}\right)^2\right)} = 1.0 \tag{11.24b}$$

The yield criterion of Equation 11.22 and the stability criterion of Equation 11.24b should be investigated for all members subject to simultaneous bending and axial compression. Equations 11.22 and 11.24b, where $\sigma_{bx} = 0$ or $\sigma_{by} = 0$ (uniaxial bending), are shown in Section 9.4.2, which outlines the interaction equations for biaxial bending.

11.3.2.1.2 Interaction Equations for Axial Compression and Biaxial Bending

The strength of members subjected to axial compression and biaxial bending is complex. Theoretical procedures have been developed for short members and longer members to produce interaction curves (Culver, 1966; Chen and Astuta, 1977) and confirmed as reasonable by experiment and in computer studies for typical members by Birnstiel (1968), Pillai (1980) and others (Galambos, 1988).

Since a design methodology for axial compression and biaxial bending must include the case of axial compression and uniaxial bending, it would appear reasonable to extend the interaction Equations 11.22 and 11.24b to the case of biaxial bending with axial compression. Therefore, the interaction formula relating to the stability criterion is,

$$\frac{\sigma_a}{F_{call}} + \frac{\sigma_{bx}}{F_{bcallx}\left(1 - \frac{\sigma_a}{0.514\pi^2 E}\left(\frac{K_x L_x}{r_x}\right)^2\right)} + \frac{\sigma_{by}}{F_{bcally}\left(1 - \frac{\sigma_a}{0.514\pi^2 E}\left(\frac{K_y L_y}{r_y}\right)^2\right)} = 1.0 \tag{11.25}$$

Equation 11.25 for biaxial bending is analogous to Equation 11.24b for uniaxial bending.

For members with low slenderness (where yielding controls), at locations of supports or when braced in the plane of bending, Equation 11.25 becomes

$$\frac{\sigma_a}{F_{call}} + \frac{\sigma_{bx}}{F_{bcallx}} + \frac{\sigma_{by}}{F_{bcally}} = 1.0 \tag{11.26}$$

Equation 11.26 for biaxial bending is analogous to Equation 11.22 for uniaxial bending

where

$F_{call} = 0.55F_y$ when $L/r = 0$ (at locations of bracing and at supports);
σ_{bx} = the normal bending stress about the x axis;
σ_{by} = the normal bending stress about the y axis;
F_{bcallx} = the allowable compressive bending stress about the x axis;
F_{bcally} = the allowable compressive bending stress about the y axis; and
$\frac{K_x L_x}{r_x}, \frac{K_y L_y}{r_y}$ = effective slenderness ratio of the member about axes x and y, respectively.

Equations 11.25 and 11.26 are shown in Section 9.4.2 for biaxial bending where $\sigma_{bx} \neq 0$ or $\sigma_{by} \neq 0$.

11.3.2.2 AREMA Recommendations for Combined Axial Compression and Biaxial Bending

AREMA (2024) recommends that members subjected to axial compression and biaxial bending be designed in accordance with Equations 11.21 extended for biaxial bending, 11.25 and 11.26. However, AREMA (2024) recognizes that for members with relatively small axial compressive forces, the secondary effects are negligible. Therefore, when $\frac{\sigma_a}{F_{call}} \leq 0.15$, Equation 11.25 is expressed as

$$\frac{\sigma_a}{F_{call}} + \frac{\sigma_{bx}}{F_{bcallx}} + \frac{\sigma_{by}}{F_{bcally}} \leq 1.0 \tag{11.27}$$

Both yielding and stability effects must be considered when $\frac{\sigma_a}{F_{call}} > 0.15$. Consequently, AREMA (2024) recommends that when $\frac{\sigma_a}{F_{call}} > 0.15$, Equations 11.22 (the yield criterion), extended for biaxial bending, and Equations 11.25 (the stability criterion) be used for design as

$$\frac{\sigma_a}{0.55F_y} + \frac{\sigma_{bx}}{F_{bcallx}} + \frac{\sigma_{by}}{F_{bcally}} \leq 1.0 \tag{11.28}$$

and

$$\frac{\sigma_a}{F_{call}} + \frac{\sigma_{bx}}{F_{bcallx}\left(1 - \frac{\sigma_a}{0.514\pi^2 E}\left(\frac{K_x L_x}{r_x}\right)^2\right)} + \frac{\sigma_{by}}{F_{bcally}\left(1 - \frac{\sigma_a}{0.514\pi^2 E}\left(\frac{K_y L_y}{r_y}\right)^2\right)} \leq 1.0 \tag{11.29}$$

where

σ_a = the normal axial compressive stress;
σ_{bx} = the normal flexural stress about the x axis;
σ_{by} = the normal flexural stress about the y axis;
F_{call} = the allowable axial compressive stress for axial compression only;
F_{bcallx} = the allowable flexural compressive stress for bending only about x axis; and
F_{bcally} = the allowable flexural compressive stress for bending only about y axis.

Equation 11.28 relates to the yield criterion, which is appropriate to consider at support locations, for members with very low slenderness ratios and at locations braced in the planes of bending. The design of typical members subjected to axial compression and bending is usually governed by the stability criterion of Equation 11.29.

11.3.3 Design of Truss Compression Members

Compression members in steel railway truss spans are typically fabricated with structural shapes and plates. However, due to large-magnitude live loads, secondary bending and connection geometry, it can be necessary to fabricate built-up compression members with several structural shapes. Some built-up compression member sections use three welded plates to fabricate H-sections (I-sections rotated 90°), and many use four welded plates to generate commonly used box-section compression members.[22] In cases where a closed box-section member is undesirable, such as where the ingress of water is difficult to preclude, open compression members are used.[23] In such cases, open built-up box-section members are typically fabricated with perforated cover plates.[24]

11.3.3.1 Design of Built-up Compression Members

Compression members can be built up with lacing bars, batten plates and perforated cover plates, as shown in Figure 11.9. Nevertheless, modern truss compression members are typically built-up box-sections with perforated cover plates as shown in Figure 11.9f.

Only bending deformations, as illustrated by Equation 11.1, were considered in developing Equation 11.2 for P_{cr}. The effect of shear forces was neglected. For solid-section compression members, this is appropriate. However, for built-up compression members, the shear forces can

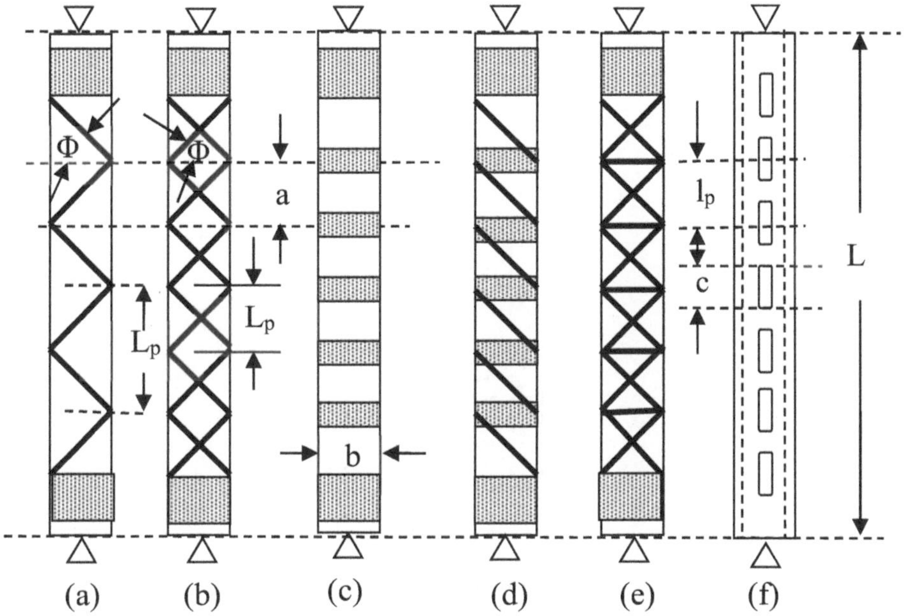

FIGURE 11.9 Typical configurations of built-up compression members.

create deformations of the open section that reduce the overall stiffness and thereby reduce the elastic buckling strength. Therefore, the curvature of the compression member due to shear must be included in Equation 11.1 to determine the critical buckling load, $\overline{P}_{cr}$ for built-up compression members. As shown in Figure 11.10, the shear force is

$$V(x) = P\sin\phi(x) = P\frac{dy(x)}{dx} \tag{11.30}$$

with curvature γ_v as follows:

$$\gamma_v = \frac{\beta}{A_g G_{eff}}\frac{dV(x)}{dx} = \frac{\beta P}{A_g G_{eff}}\frac{d^2y(x)}{dx^2} \tag{11.31}$$

where

- β = a numerical factor to correct for non-uniform stress distribution across the cross section of the compression member;
- A_g = the gross cross-sectional area of the compression member;
- G_{eff} = the effective shear modulus = $E_{eff}/(2(1+v))$; and
- v = Poisson's ratio = 0.3 for steel.

Substiting Equation 11.31 into Equation 11.1 yields the differential equation

$$\frac{d^2y(x)}{dx^2} + \frac{k^2y(x)}{\left(1+\frac{\beta P}{A_g G_{eff}}\right)} = \frac{U}{\left(1+\frac{\beta P}{A_g G_{eff}}\right)} \tag{11.32}$$

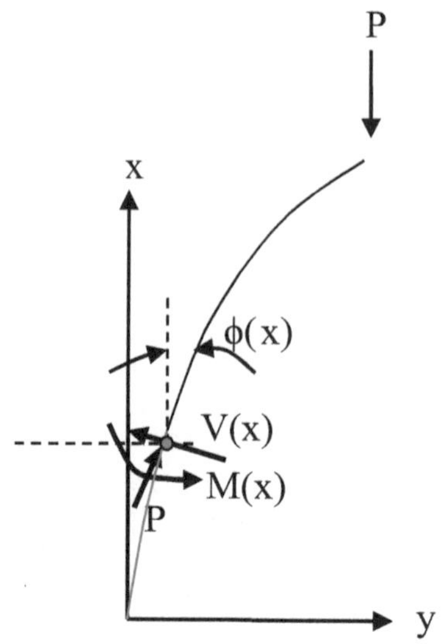

FIGURE 11.10 Bending moment and shear force at compression member cross-section.

with solution analogous to Equation 11.2 of

$$\bar{P}_{cr} = \frac{\pi^2 E_{eff} I}{(\alpha KL)^2} = \frac{P_{cr}}{\alpha^2} \tag{11.33}$$

where P_{cr} = the critical buckling load for the compression member with gross cross-sectional area, A_g, moment of inertia, I, and length, L, and

$$\alpha = \sqrt{\left(1 + \frac{\beta P_{cr}}{A_g G_{eff}}\right)} = \sqrt{1 + \Omega P_{cr}} \tag{11.34}$$

$$\Omega = \frac{\beta}{A_g G_{eff}} = \frac{2\beta(1+\nu)}{A_g E_{eff}} \tag{11.35}$$

Ω is determined through investigation of the deformations of the lacing bars, batten plates and/or perforated cover plates caused by lateral displacements from the shear force, V (x), (Bleich, 1952; Timoshenko and Gere, 1961).

11.3.3.1.1 Critical Elastic Buckling of Built-Up Compression Members with Lacing Bars Only

Shear is resisted by pin-connected truss behavior of lacing bars (for double lacing, consider tension resistance only):

$$\Omega = \frac{1}{A_{lb} E_{eff} \sin\varphi \cos^2\varphi} \tag{11.36a}$$

$$\alpha = \sqrt{1 + \Omega P_{cr}} = \sqrt{1 + \left(\frac{\pi^2 I}{(KL)^2}\right) \frac{1}{A_{plb} \sin\varphi \cos^2\varphi}} = \sqrt{1 + \left(\frac{\pi^2 I}{(KL)^2}\right) \frac{\left(a^2 + b^2\right)^3}{A_{plb} a b^2}} \tag{11.36b}$$

where

A_{plb} = the cross-sectional area of the diagonal lacing bars in each panel of the member
For single lacing $A_{plb} = A_{lb} = t_{lb}w_{lb}$
For double lacing $A_{plb} = 2A_{lb} = 2\ t_{lb}w_{lb}$
t_{lb} = thickness of lacing bar
w_{lb} = width of lacing bar
a = the panel length, L_P, for members with double diagonal lacing (Figures 11.9b and 11.9e) or 50% of the panel length, $0.5L_P$, for members with single diagonal lacing (Figure 11.9a)
b = the distance between the centroid of the main compression elements of the member
I = moment of inertia of the compression member in the plane of flexural buckling
Φ = the angle of the lacing bar from the line perpendicular to the member axis (should be about 30° for single lacing and 45° for double lacing)

11.3.3.1.2 Critical Elastic Buckling of Built-up Compression Members with Batten Plates Only

Shear is resisted by flexure of batten plates and main member elements (Figure 11.9c):

$$\Omega = \frac{ab}{12E_{eff}I_b} + \frac{a^2}{24E_{eff}I_f} \tag{11.37a}$$

$$\alpha = \sqrt{1+\Omega P_{cr}} = \sqrt{1+\left(\frac{\pi^2 I}{(KL)^2}\right)\left(\frac{ab}{12I_b}+\frac{a^2}{24I_f}\right)} \tag{11.37b}$$

where I_b = the moment of inertia of the batten plates in a panel in the plane of flexural buckling and I_f = the moment of inertia of the main elements of the member each side of the longitudinal axis through the center of the member about the centroidal axis of half the member.

If the shear rigidity of the batten plates is small, reduction of the built-up compression member critical buckling force will result. Inclusion of the batten plate shearing strain into Equation 11.37a yields

$$\Omega = \frac{ab}{12E_{eff}I_b} + \frac{a^2}{24E_{eff}I_f} + \frac{\beta a}{2A_bG_{eff}b} \tag{11.37c}$$

$$\alpha = \sqrt{1+\left(\frac{\pi^2 I}{(KL)^2}\right)\left(\frac{ab}{12I_b}+\frac{1.2aE}{2A_bbG}+\frac{a^2}{24I_f}\right)} = \sqrt{1+\left(\frac{\pi^2 I}{(KL)^2}\right)\left(\frac{ab}{12I_b}+\frac{3.12a}{2A_bb}+\frac{a^2}{24I_f}\right)} \tag{11.37d}$$

where A_b = cross sectional area of a batten plate.

11.3.3.1.3 Critical Elastic Buckling of Built-up Compression Members with Lacing Bars and Batten Plates[25]

Shear is resisted by pin-connected truss behavior of lacing bars (for double lacing consider tension resistance only) (Figure 11.9d):

$$\Omega = \frac{1}{A_{plb}E_{eff}\sin\varphi\cos^2\varphi} + \frac{b}{A_bE_{eff}a} \tag{11.38a}$$

$$\alpha = \sqrt{1+\left(\frac{\pi^2 I}{(KL)^2}\right)\left(\left(\frac{1}{A_{plb}\sin\varphi\cos^2\varphi}\right)+\left(\frac{1}{A_b\tan\varphi}\right)\right)} = \sqrt{1+\left(\frac{\pi^2 I}{(KL)^2}\right)\left(\frac{\left(a^2+b^2\right)^3}{A_{plb}ab^2}+\frac{b}{A_b a}\right)} \tag{11.38b}$$

11.3.3.1.4 *Critical Elastic Buckling of Built-up Compression Members with Perforated Cover Plates*

Shear is resisted by flexure of the main member elements because the perforated cover plates act effectively as rigid batten plates between the perforations (Figure 11.9f):

$$\Omega = \frac{9c^3}{64 l_p E_{eff} I_f} \tag{11.39a}$$

$$\alpha = \sqrt{1+\Omega P_{cr}} = \sqrt{1+\frac{\pi^2 I}{(KL)^2}\left(\frac{9c^3}{64 l_p I_f}\right)} \tag{11.39b}$$

where l_p = the distance between the center of perforations and c = the length of the perforation. The gross section of the plate through the perforation must be considered part of the area of the member.

Most modern steel railway superstructure compression members are assumed to be pin connected at each end (K = 0.75). To avoid the fabrication cost of built-up laced bar and batten plate compression members and the bending inefficiencies of using batten plates, perforated cover plates are often used for built-up compression members in modern railway steel superstructures.

11.3.3.1.5 *Box-section Compression Members with Perforated Cover Plates*

Closed compression members, such as box-section members, use perforated cover plates, as shown in Figures 11.9f, 11.11 and 11.12, to create an open member for egress of water and to reduce member weight.[26] Perforations in cover plates must be ovaloid or elliptical[27] with a length, c, not greater than twice the perforation width, b, and have a clear distance between perforations, l_p–c, not less than the distance between the nearer lines of connections, b_p.[28] In addition, to ensure stability of the main member elements at perforations, AREMA (2024) recommends that the effective slenderness ratio, C_{pc}, which is the ratio of the length of the perforation, c to the radius of gyration, r_{pc}, of the half member about its own axis, as shown in Figure 11.13, not exceed 20 nor 33% of the member slenderness ratio, L/r, about its axis perpendicular to the plane of the perforation.

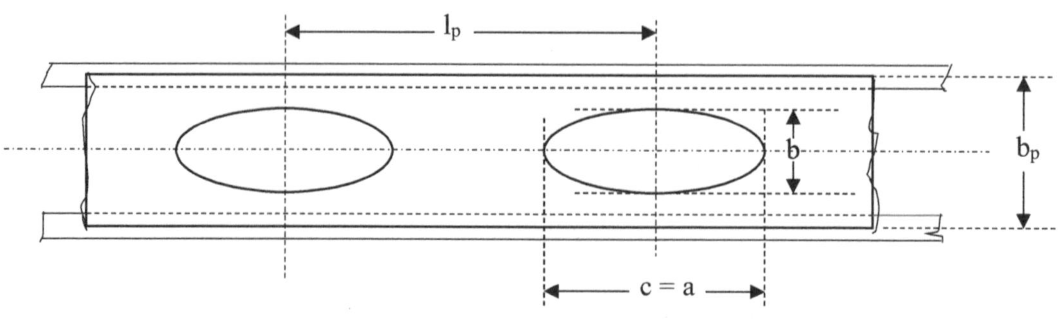

FIGURE 11.11 Elliptical perforations in cover plates.

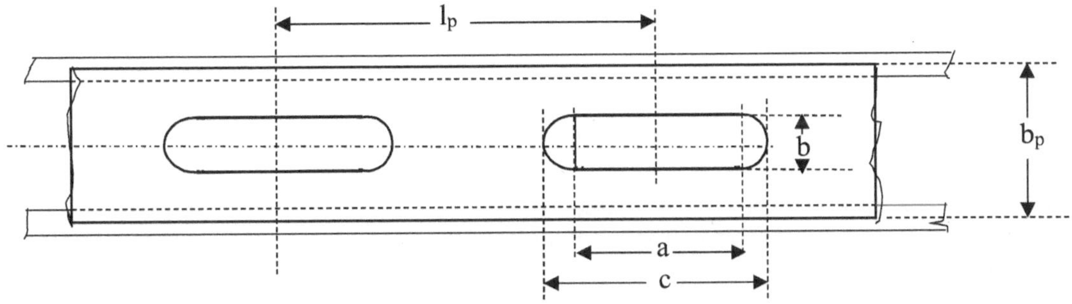

FIGURE 11.12 Ovaloid perforations in cover plates.

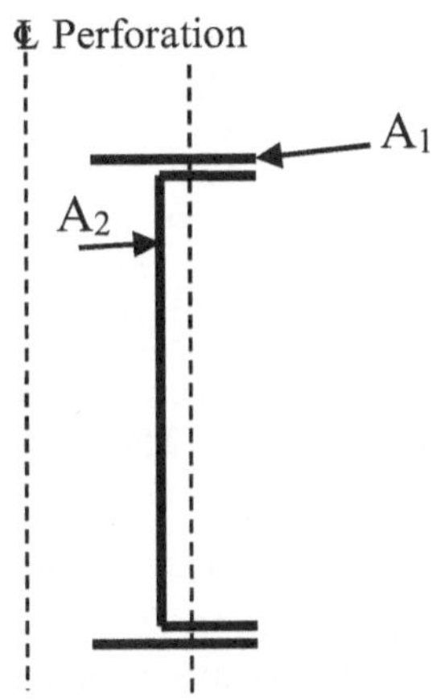

FIGURE 11.13 Half of perforated plate compression member.

Therefore, the effective slenderness ratio is

$$C_{pc} = \frac{c}{r_{pc}} \leq 20 \tag{11.40a}$$

and

$$C_{pc} = \frac{c}{r_{pc}} \leq \frac{L}{3r} \tag{11.40b}$$

where

c = the length of the perforation

$r_{pc} = \sqrt{I_{pc}/A_{pc}}$

I_{pc} = the moment of inertia of the main elements of the member each side of the longitudinal axis through the center of the member about the centroidal axis of the member

A_{pc} = the area of half of the member ($A_{pc} = 2A_1 + A_2$ in Figure 11.13)

L = length of the compression member

r = radius of gyration of the member about its axis perpendicular to the plane of the perforation

The thickness of perforated cover plates, t_{pc}, should be governed by the largest of the following expressions relating to local plate stability criteria:

$$t_{pc} \geq {}^{b_p}\!/_{50} \tag{11.41}$$

$$t_{pc} \geq 1.17\left(b_p - b\right)\sqrt{{}^{F_y}\!/_{E}} \tag{11.42}$$

Perforated cover plate thickness is also based on the transverse shear,[29] V due to LC1A or LC1B at the centerline of the cover plate. The minimum perforated cover plate thickness (see Equation 10.17) is

$$t_{pc} \geq \frac{4.29Vl_p}{F_y b_p\left(l_p - c\right)} \tag{11.43}$$

V in compression members is primarily due to self-weight, wind loads and 2.5% of the compressive axial force in the member. The shear force due to 2.5% of the compressive axial force in the member must not be less than

$$V = \frac{PF_y}{150F_{call}} \tag{11.44}$$

where P = the axial compressive force on the member, F_y = the yield stress, and F_{call} = the allowable axial compressive stress for axial compression only. The shear forces related to self-weight and wind are generally small and may be neglected in many cases.

Figure 11.14 illustrates that the AREMA (2024) recommendation for minimum shear force (Equation 11.44) ensures that shear forces of relatively greater proportion to the axial compressive force are used for the design of very slender compression members that approach the Euler elastic critical buckling load.[30]

Where perforated cover plates in compression members are not spliced, an open perforation is used at the end of cover plates provided that the open length does not exceed half the distance

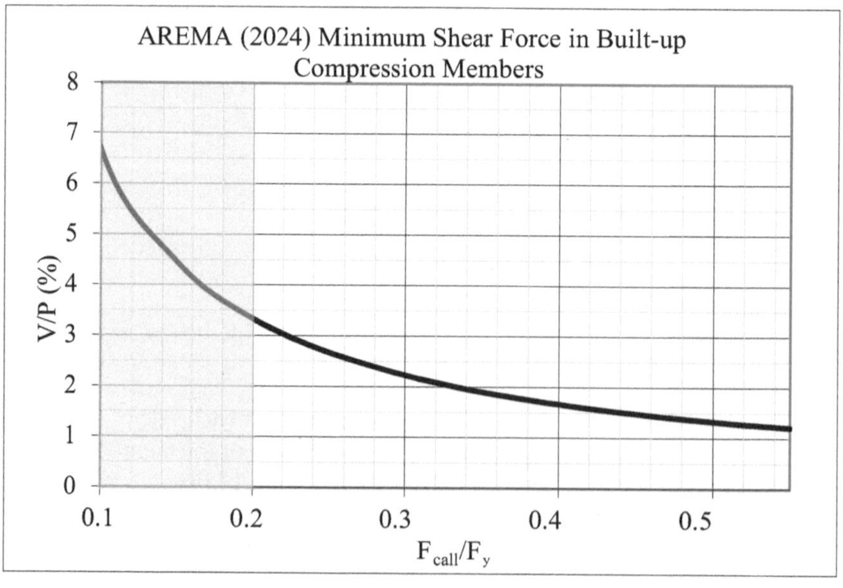

FIGURE 11.14 AREMA (2024) recommendation for minimum shear force.

between the nearer lines of connections. Where perforated cover plates in compression members are spliced, the clear distance between the end perforation and the end of the cover plate should not be less than the distance between the nearer lines of connections, except that 50% of this distance can be used for compression members that are faced for bearing.

Perforated cover plate thickness must also meet the requirements for axial compression, bending and recommended minimum material thickness. For compressive axial and bending stresses, the gross section through the perforation of the plate (with only the perforated area removed) is included in the member gross cross-sectional area, A_g and gross section modulus, S_g. The effect of perforations in box-section member cover plates must also be considered for serviceability design. Based on the axial stiffness of box-section members with ovaloid and elliptical perforations in the cover plates (see Equation 10.20a and 10.2b) the effective cross-sectional area for the calculation of axial extension or contraction, A_{ed}, and truss joint displacements is recommended by AREMA (2024) as

$$A_{ed} = A_g - (V_p)/l_p \qquad (11.45)$$

where V_p = volume of the perforations in the box-section member
= $(A_{pr} + A_{pc})t_{pc}$ for ovaloid perforations in each cover plate of the box-section member
= $A_{pe}t_{pc}$ for elliptical perforations in each cover plate of the box section member

The shape or plate components of built-up compression members must be adequately fastened together to ensure integral behavior of the member. This is accomplished by strength design of the four flange-to-web fillet welds in H-section members to transfer the load from the connected flange plates to the web plate over the length of the member's connection to the gusset plate. Similarly, the strength design of the four welds[31] in box-section members considers the transfer of load from the connected web plates to the cover plates over the length of the member's connection to the gusset plate. In many instances, the minimum size welds in accordance with the applicable welding code[32] suffice to transfer the maximum compressive connection loads.

P_{call} (Equations 11.17, 11.18 and 11.19) is contingent upon the main element web plates and cover plates being of adequate strength and stability as individual components to preclude local buckling. To ensure that the webs of main elements of built-up compression members do not buckle prior to the member, a minimum side plate thickness, t_w, is recommended by AREMA (2024)[33] as

$$t_w \geq \frac{0.90 b_p \sqrt{F_y / E}}{\sqrt{F_{call} / \sigma_a}} \qquad (11.46)$$

where

$$F_{call} / \sigma_a \leq 4$$

F_{call} = allowable compressive stress, P_{call}/A_g
$\sigma_a = P/A_g$, the calculated compressive stress in the member
b_p = unsupported distance between the nearest lines of fasteners or welds of the cover plate

Additionally, to ensure that cover plates for main elements of built up compression members do not buckle locally prior the member, AREMA (2024) recommends a minimum cover plate thickness, t_{pc}, of

$$t_{pc} \geq \frac{0.72 b_p \sqrt{F_y / E}}{\sqrt{F_{call} / \sigma_a}} \qquad (11.47)$$

11.4 LOCAL BUCKLING OF OUTSTANDING ELEMENTS OF COMPRESSION MEMBERS

Compression members are often box sections without outstanding elements. However, for cases where compression members have outstanding elements such as side plates of H-sections the local elastic stability of the outstanding elements requires consideration. Based on elastic stability of a plate with an infinitely long unsupported edge, the width of outstanding elements in compression with thickness t_{co} must not exceed $0.43t_{co}\sqrt{\frac{E}{F_y}}$ for main truss members and $0.50t_{co}\sqrt{\frac{E}{F_y}}$ for bracing and secondary members. Example 11.1 outlines the design of a perforated cover plate built-up compression member.

Example 11.1

Design chord member U2U3 of Example 9.3. The secondary bending stress from joint rigidity is determined to be 20% of the primary axial stress at joint U3 and 10% of the primary axial stress at joint at joint U2. ASTM A709 Grade 50W steel is used with F_y = 50 ksi and F_U = 70 ksi. Use a built-up perforated plate box-section for U2U3.

From Example 9.3,

P_{max} = −627.72 kips

L_{U2U3} = length of member U2U3 = 25 ft = 300 in

Try built-up section of two 11" × 1.25" vertical side plates and two 10.75" × 1/2" horizontal cover plates (Figure E11.1a).

h_m = 12 in.

b_m = 12 in.

t_s = 1.25 in.

b_p = 12 − 1.25 = 10.75 in.

t_c = 0.50 in

From Equation 11.41,

$t_c \geq 10.75/50 \geq 0.22$ in, OK

b_m/L_{U2U3} = 12/300 = 0.04

AREMA (2024) indicates that the gross section of the plate through the perforation shall be considered as a part of the gross cross-sectional area of the member.

Consider 6" wide × 12" long ovaloid perforations spaced at 30" in member cover plates (Figure E11.1b).

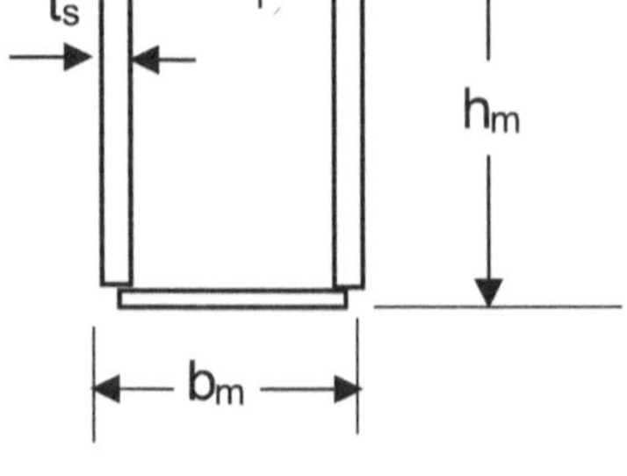

FIGURE E11.1a Section of member at solid cover plate section.

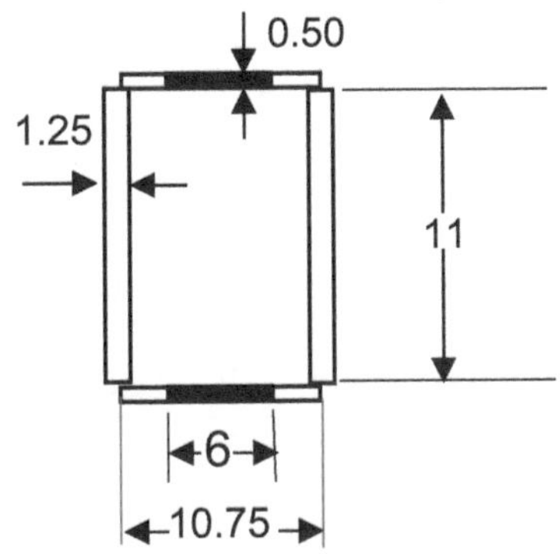

FIGURE E11.1b Section of member at cover plate perforation.

b = 6 in
From Equation 11.42,

$$t_c \geq 1.17(10.75 - 6)\sqrt{50/29000} = 0.23 \text{ in, OK}$$

c = 12 in ≤ 2b = 2(6) = 12 in, OK
$l_p = 30$
$l_p - c = 30 - 12 = 18$ in ≥ $b_p = 10.75$ in, OK

Gross cross-sectional area of member = $A_g = 2(11)(1.25) + 2(10.75)(0.5) - 2(6)(0.50) = 27.50 + 10.75 - 6.00 = 38.25 - 6.00 = 32.25$ in^2

Gross moment of inertia of member resisting vertical flexure = $I_{gx} = 2(1.25)(11)^3/12 + 2(10.75)(0.5)^3/12 + 2(0.5)(10.75)(5.75)^2 - 2(0.5)(6)(5.75)^2 - 2(6)(0.5)^3/12 = 277.29 + 0.22 + 355.42 - 198.38 - 0.13 = 434.42$ in^4

Gross section modulus of member resisting vertical flexure = $S_{gx} = (434.55)/6 = 72.43$ in^3

$$\text{Radius of gyration of member resisting vertical flexure} = r_{gx} = \sqrt{\frac{I_{gx}}{Ag}} = \sqrt{\frac{434.42}{32.25}} = 3.67 \text{ in}$$

Gross moment of inertia of member resisting horizontal (lateral) flexure = $I_{gy} = 2(0.5)(10.75)^3/12 + 2(11)(1.25)^3/12 + 2(11)(1.25)(5.375)^2 - 2(0.5)(6)^3/12 = 103.52 + 3.58 + 794.49 - 18.00 = 883.59$ in^4

Gross section modulus of member resisting horizontal (lateral) flexure = $S_{gy} = (883.59)/6 = 147.27$ in^3

$$\text{Radius of gyration of member resisting horizontal flexure} = r_{gy} = \sqrt{\frac{I_{gy}}{Ag}} = \sqrt{\frac{883.59}{32.25}} = 5.23 \text{ in}$$

Axial stress on gross cross-sectional area of member = $\sigma_{ag} = P_{max}/A_g = -627.72/32.25 = -19.46$ ksi

Secondary vertical flexural stress due to joint rigidity and truss distortion at joint L3 = $\sigma_{bg} = 0.20(-19.46) = -3.89$ ksi

Self-weight = −((12/30(38.25) + (18/30(32.25))490/(144(1000)) = 0.118 kip/ft

Fixed end bending moment due to self-weight = $-0.118(25)^2/12 = 6.14$ kip-ft

Secondary vertical flexural stress due to self-weight = −(6.14)(12)/72.43 = −1.02 ksi

Wind force on member = 30(1.0)/1000 = 0.030 kip/ft

Fixed end bending moment due to wind = $0.030(25)^2/12 = 1.56$ kip-ft

Secondary flexural stress due to wind = −(1.56)(12)/72.43 = −0.26 ksi
Secondary flexural stress = −3.89 − 1.02 − 0.26 = −5.17 ksi
In accordance with AREMA (2024), the secondary flexural stress needs to be considered only if $b_m/L_{L2L3} > 0.10$; however, for this example the secondary flexural stress will be considered.
In accordance with AREMA (2023), the excess secondary bending stress = σ_{ebg} = −5.17 − (−3.0) = −2.17 ksi.
Excess secondary vertical flexural stress = −2.17(3.89+1.02)/5.17 = −2.06 ksi
Excess secondary horizontal flexural stress = −2.17(0.26)/5.17 = −0.11 ksi

Allowable axial compressive stress:

$KL/r_{gx} = 0.75(300)/3.67 = 61.31$
$KL/r_{gy} = 0.75(300)/5.23 = 43.02$
$KL/r_{min} = 61.31$

Considering shear deformation of the perforated cover plate box-section, $\bar{y}$ = (2(2.375)(0.5)(2.375/2 + 0.625) + 11(1.25)(0.625))/(2(2.375)(0.5) + 11(1.25) = 0.800 in. Additionally, I_f = the moment of inertia of the main elements of the member each side of the longitudinal axis through the center of the member about the centroidal axis of half the member = $11(1.25)^3/12 + 11(1.25)(0.800 - 0.625)^2 + 2(0.5)(2.375)^3/12 + 2(0.5)(2.375)(2.375/2 + 0.625 - 0.800)^2 = 5.762$ in^4 (Figure E11.1c).

From Equation 11.39b,

$$\alpha = \sqrt{1 + \frac{300\pi^2}{(225)^2}\left(\frac{9(10)^3}{64(30)(5.762)}\right)} = 1.024 \text{ (2.4\% increase in effective length of member)}$$

$K_v = \alpha K = 1.024(61.31) = 62.752$

$$0.629\sqrt{\frac{29000}{50}} = 15.1 < \frac{KL}{r_{min}} = 62.75 < 5.034\sqrt{\frac{29000}{50}} = 121.2$$

As shown by Equations 11.17–11.19 (also shown in Table 9.1 of Section 9.3), the maximum axial compression in centrally loaded[34] truss compression members determined on the

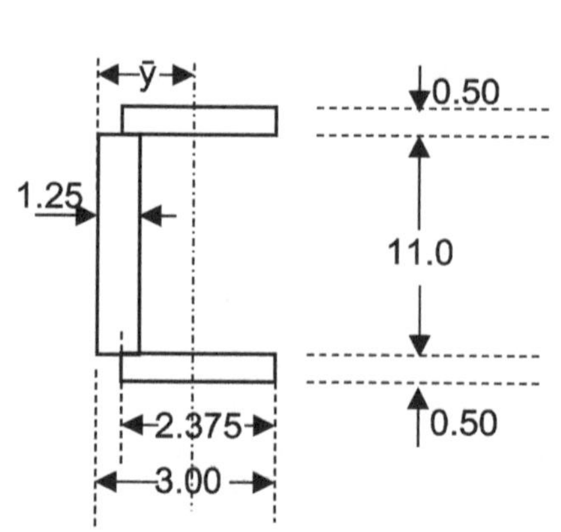

FIGURE E11.1c Section to determine the moment of inertia about centroid of half of member.

gross cross-sectional area must not exceed F_{call}, which is determined considering the effective slenderness, KL/r_{min}, of the compression member. Therefore,

$$F_{call} = 0.60F_y - \left(17{,}500\frac{F_y}{E}\right)^{\frac{3}{2}}\left(\frac{K_V L}{r_{min}}\right) = 30.0 - \frac{165.7}{1000}(62.75) = 30.0 - 10.40 = 19.60 \text{ ksi}$$

(19.84 ksi not considering shear deformation)

From Equations 11.46 and 11.47 for side plate and cover plate buckling,

$$\sqrt{F_{call}/\sigma_\alpha} = \sqrt{19.60/19.46} = 1.00 \le 2, \text{ OK}$$

$$t_w \ge \frac{0.90(10.75)\sqrt{50/29000}}{1.00} \ge 0.40 \text{ in, OK}$$

$$t_{pc} \ge \frac{0.72(10.75)\sqrt{50/29000}}{1.00} \ge 0.32 \text{ in, OK}$$

The transverse shear force in U2U3 due to self-weight, wind loads and 2.5% of the compressive axial force is

V = 0.118(25/2) + 30(1)(25/2)/1000 + 0.025(627.72) = 1.30 + 0.38 + 15.69 = 17.37 kips

From Equations 11.44,

$$V = 17.37 \text{ kips} \ge \frac{627.72(50)}{150(19.60)} = 10.68 \text{ kips, OK}$$

From Equations 11.43,

$$t_c \ge \frac{4.29(17.37)(30)}{50(10.75)(30-12)} \ge 0.23 \text{ in, OK}$$

As shown in Table 9.2, the maximum flexural compression determined on the gross section modulus of a box-section must not exceed F_{bcall}, which is determined considering the slenderness of the compression member:

$$F_{bcall} = 0.55F_y - \frac{0.55F_y^2}{6.3\pi^2 E}\left(\frac{L}{r_e}\right)^2$$

$$\frac{L}{r_e} = \sqrt{\frac{1.105\pi L S_x\sqrt{\Sigma\left(\frac{s}{t}\right)}}{\left(A\sqrt{\frac{I_y}{1+\mu}}\right)}}$$

where

$L = 300$ in

$S_x = 72.43$ in^3

$A = 10.75(11.50) = 123.63$ in^2

$\sum s/t = 2(10.75)/0.5 + 2(11)/1.25 = 43.00 + 17.60 = 60.60$

$I_y = 883.59$ in^4

$$\frac{L}{r_e} = \sqrt{\frac{1.105\pi(300)72.43\sqrt{60.60}}{\left(123.63\sqrt{\dfrac{883.59}{1+0.3}}\right)}} = 13.50$$

$$F_{bcall} = 0.55(50{,}000) - \frac{0.55(50{,}000)^2}{6.3\pi^2\left(29x10^6\right)}(13.50)^2 = 27500 - 138.9 \text{ psi} = 27.36 \text{ ksi}$$

Compressive Axial and Flexural Interaction

As shown by Equations 11.25 and 11.26 (also shown in Table 9.3), the axial and flexural stress interaction equation must be satisfied:

$f_a/F_a = 19.46/19.60 = 0.99$

$f_{bx} = 2.06$ ksi

$f_{by} = 0.11$ ksi

$$0.99 + \frac{2.06}{27.36\left(1 - \dfrac{19.60}{0.514\pi^2 E}(62.75)^2\right)} + \frac{0.11}{27.36\left(1 - \dfrac{19.60}{0.514\pi^2 E}\left(1.024(43.02)\right)^2\right)}$$

$= 0.99 + 0.16 + 0.01 = 1.15$

15% overstress considering secondary flexural stresses. However, because $b_m/L_{U2U3} = 12/300 = 0.04 \le 0.10$ and AREMA (2023) does not require consideration of secondary bending due to connection rigidity and truss distortion,

Secondary vertical flexural stress = −2.17(1.02)/5.17 = −0.43 ksi

Secondary horizontal flexural stress = −2.17(0.26)/5.17 = −0.11 ksi

$$0.99 + \frac{0.43}{27.36\left(1 - \dfrac{19.60}{0.514\pi^2 E}(62.75)^2\right)} + \frac{0.11}{27.36\left(1 - \dfrac{19.60}{0.514\pi^2 E}\left(1.024(43.02)\right)^2\right)}$$

$= 0.99 + 0.03 + 0.01 = 1.03,$

OK with 3% overstress due to secondary flexure from self-weight and wind considered.

However, because $b_m/L_{U2U3} = 12/300 = 0.04 \le 0.10$, AREMA (2024) does not require consideration of secondary bending stresses.

Considering stability at the perforations,

I_{pc} = the moment of inertia of the main elements of the member each side of the longitudinal axis through the center of the member about the centroidal axis of half the member = $I_f = 5.762$ in^4

$A_{pc} = A_f = 1.25(11) + 2(2.375)(0.50) = 16.13$ in^2

$$r_{pc} = \sqrt{\frac{5.762}{16.13}} = 0.60 \text{ in}$$

Effective slenderness ratio = $C_c = c/r_{pc} = 12/0.60 = 20.1$
$C_c \leq L/(3r_{gy}) = 300/(3(5.23)) = 19.12$
$C_c \leq 20$
C_c exceeds the recommended allowable effective slenderness ratio by 5%. This is likely acceptable.

11.5 SERVICEABILITY OF TRUSS COMPRESSION MEMBERS

The slenderness ratio, L/r must be limited to values that will preclude excessive vibration or deflection, which is of particular concern for compression member stability (e.g., to avoid excessive secondary flexural compressive stress due to member curvature). Therefore, AREMA (2024) recommends that

$$\frac{L}{r_{min}} \leq 100 \text{ for main compression members,} \tag{11.49a}$$

$$\frac{L}{r_{min}} \leq 120 \text{ for wind and sway bracing compression members} \tag{11.49b}$$

NOTES

1 The member is shortened when stresses in the member from applied loads exceed the compressive yield stress, which is essentially equal to the tensile yield stress.
2 If the compression member is nonredundant, failure of the member can result in collapse of the truss span.
3 Strength design is defined by yield stress, and stability design is defined by the critical buckling stress.
4 Particularly compression member dimensions that result in initially bent or eccentrically loaded members.
5 Pinned, fixed, free or elastically restrained supports at the ends of the compression member.
6 Notably steel residual stresses due to manufacturing and fabrication.
7 These secondary flexural stresses can be due to member eccentricity, transverse forces on members, distortion of vertical members, rotational restraint at end joints, elongation of chord members, and/or truss distortion and joint rigidity. Truss distortion and joint rigidity are typically the most prevalent sources of secondary flexural stresses.
8 From about 1919 to 1934, AREMA Chapter 15 recommended that primary axial and, if necessary, secondary flexural stresses be combined with an allowable stress factor of 1.33. From 1934 to 2023, AREMA Chapter 15 recommended that if necessary, only secondary compression stresses exceeding 20.7 MPa (3,000 psi) be considered primary compressive flexural stresses. Since about 1919, AREMA Chapter 15 has recommended that secondary flexural stresses were necessary to determine if members with width parallel to the plane of distortion exceeded 10% of member length.
9 Truss members are subjected to primary axial, secondary axial and secondary bending stresses. Secondary axial forces are small and can be neglected.
10 Additionally, secondary bending stresses at the ends of members are typically small due to the presence of gusset plates, which effectively increase the moment of inertia of the member near connections.
11 Allowable stresses usually include an allowance of about 30% of primary axial stress for secondary flexural stress.
12 Load combinations and associated allowable stresses are recommended in AREMA Chapter 15. These are typically presented as ‘stress sheets’ on truss design drawings.
13 Lateral forces from wind on the loaded bridge and nosing (truck hunting). Longitudinal forces from wind on the loaded bridge, braking, traction and anchored continuous welded rail. Other forces such as those from temperature and support settlements of statically indeterminate truss spans.
14 Geometric nonlinearities due to initial curvature and/or force eccentricity.
15 Since residual stresses are most affected by size, the use of high-strength steel can reduce their effects. Annealing to reduce residual stresses (heat treatment) increases the strength of a compression member.
16 The length and dimensions of typical steel railway truss compression members precipitate intermediate slenderness.
17 The safety factor for tension members is 1.82 as indicated in Chapter 10.
18 These tests on short columns were performed at Lehigh University in the 1950s.
19 Commonly referred to as Johnson parabolas and first reported in the CRC *Guide to Design Criteria for Metal Compression Members* in 1960.

20 Solved for simply supported uniformly distributed, concentrated and end moments.
21 Indicated by depth of yielded material from the fibers with the largest compressive strain.
22 Box-section compression members are often used due to the decreased slenderness of such sections in comparison with H-sections.
23 This is generally the case for railway compression members exposed to rain, ice and snow.
24 Built-up open compression members using lacing bars and batten plates were commonly used in riveted construction. However, modern welded built-up compression members such as box-section members typically use perforated cover plates to avoid the fabrication cost of laced bar built-up compression members and the bending inefficiencies of using batten plates.
25 This equation was developed and then later used by Engesser in connection with investigations into the collapse of the Quebec Bridge.
26 Where required, the perforations also allow for the internal painting of box-section members.
27 Typically, ovaloid perforations are preferred.
28 The distance between the nearer lines of connections is typically the distance between the longitudinal welds connecting cover plates to web plates of box members, which is essentially the cover plate width, b_p.
29 The transverse shear is equally distributed to the transverse planes of the compression member.
30 As shown in Figure 11.5, Euler elastic critical buckling occurs when $F_{call}/F_y \leq 0.20$.
31 Cover plate to web plate welds in box-section members can require CJP or large fillet welds to develop the weld within the gusset plate length. Large fillet welds can be avoided in large gusset plates with long connections for developing the weld for strength loadings.
32 For example, AWS D1.5 in the United States and CSA W59 in Canada
33 Based on uniform compression of elastic plates theory (Timoshenko and Gere, 1961)
34 If the truss members are not centrally loaded, secondary stresses due to the eccentricity of the force at the end connection may need to be considered, depending on their magnitude. However, proper detailing at plate joints will preclude secondary stresses due to eccentricity.

REFERENCES

American Railway Engineering and Maintenance-of-Way Association (AREMA), 2023, *Manual for Railway Engineering*, Chapter 15, Steel Structures, Lanham, MD.

American Railway Engineering and Maintenance-of-Way Association (AREMA), 2024, *Manual for Railway Engineering*, Chapter 15, Steel Structures, Lanham, MD.

Birnstiel, C., 1968, "Experiments on H-Columns under Biaxial Bending", *Journal of Structural Engineering, ASCE*, 94(ST10).

Bleich, F., 1952, *Buckling Strength of Metal Structures*, McGraw Hill, New York.

Chen, W.F., and Astuta, T., 1977, *Theory of Beam Columns*, Vols. 1 and 2, McGraw-Hill, New York.

Chen, W.F., and Lui, E.M., 1987, *Structural Stability*, Elsevier, New York.

Culver, C.G., 1966, "Exact Solution of the Biaxial Bending Equations", *Journal of Structural Engineering, ASCE*, 92(ST2).

Galambos, T.V., 1988, *Guide to Stability Design Criteria for Metal Structures*, John Wiley & Sons, New York.

Pillai, U.S., 1980, *Comparison of Test Results with Design Equations for Biaxially Loaded Steel Beam Columns*, Civil Engineering Research Report No. 80–2, Royal Military College of Canada, Kingston, Canada.

Tall, L., ed., 1974, *Structural Steel Design*, 2nd edition, John Wiley & Sons, New York.

Timoshenko, S.P., and Gere, J.M., 1961, *Theory of Elastic Stability*, McGraw-Hill, New York.

Unsworth, J.F., 2018, *Design and Construction of Modern Steel Railway Bridges*, 2nd edition, CRC Press, Boca Raton, FL.

Wang, C.M., Wang, C.Y., and Reddy, J.N., 2005, *Exact Solutions for Buckling of Structural Members*, CRC Press, Boca Raton, FL.

12 Design of Truss Member Connections and Splices

12.1 INTRODUCTION

Russ connection design must consider the forces at the end of members, in gusset plates at the confluence of members and at splices in members.[1]

The design forces due to the load combinations of Chapter 9 are outlined in Chapter 10 and Chapter 11 for tension member and compression member design, respectively. However, the design forces due to the load combinations of Chapter 9 differ for member end connections and splices to ensure, where necessary, that the end connections and splices are not weak elements of the member. In such cases, the connections and splices are designed for member capacity or augmented computed forces. Table 12.1 outlines the member end connection and splice strength design criteria recommendations by AREMA (2024).

Gusset plates at joints at the ends of members that are not components of splices are designed for the computed forces in the members.[2] The computed axial tensile force range is used for the fatigue design of member end connections and splices, other than end connections of vertical web members directly loaded by floorbeam reactions.[3] The fatigue design of vertical web members directly loaded by floorbeam reactions must consider the computed combined axial force range and bending moment range.

All connections must be slip critical per AREMA (2024) and must not have fewer than three bolts per plane of connection or equivalent strength in welding. The weld is preferably a fillet weld that is parallel and symmetrical to the direction of force. The minimum distance between the center of bolts in a connection must not exceed $3d_b$. In accordance with AREMA (2024), the distance from a free edge to the center of a bolt must not exceed the values in Table 12.2.

TABLE 12.1
Requirements for Member End Connection and Splice Strength Design

Truss Member Connection	Strength Design: Load Combinations	Strength Design: Connection Design Force
Chord member end connections	LC1A or LC1B	Maximum computed axial force
Diagonal web member end connections	LC1A or LC1B or LC2	Maximum computed axial force
Vertical web member end connections not directly loaded by floorbeam reaction (post members)	LC1A or LC1B or LC2	Axial strength of member
Vertical web member end connections directly loaded by floorbeam reaction (hanger, subvertical members)	LC1A or LC1B or LC2	1.25 times maximum computed axial force
Bracing and secondary member end connections	LC1B or LC4	Lesser of axial strength of member or 1.5 times maximum computed axial force
Bracing member end connections in members to reduce the unsupported length of another member	LC5	2.5% of the axial force in the member being braced
Tension splices and compression splices not faced for bearing	LC1A or LC1B or LC2	Axial strength of spliced member
Compression splices faced for bearing and spliced on four sides adjacent to panel point[4]	LC1A or LC1B or LC2	50% of computed axial force

DOI: 10.1201/9781003469971-14

TABLE 12.2
AREMA (2024) Bolted Connection Edge Distance Requirements

Free Edge Type	Minimum Distance from Free Edge to the Center of a Bolt
Sheared	$1.75(d_b)$
Rolled, planed or thermally cut	$1.5(d_b)$
Rolled to meet clearances	$1.25(d_b)$

The distance from an outside free edge to the first line of bolts should not exceed 4(10 + t) or 150 mm (SI units) or 1.5 + 4t or 6 in (US customary and imperial units). The distance between the center of the nearest bolt and the end of the connected member in line with the force in the member must not exceed $\frac{2d_b\sigma_p}{F_U}$, where[5]

d_b = the bolt diameter
σ_p = the calculated bearing stress from the design force
F_U = the specified ultimate tensile strength of the connection

Yielding will occur on the gross cross-sectional area of a tension connection before the capacity of the net cross-section is attained if the ratio of net cross-sectional area to the gross cross-sectional area is greater than 0.85 since the length of the gusset plate or splice is small in comparison to member length, and therefore, inelastic deformation of the gross cross-section is restricted (Kulak, Fisher and Struik, 1987). Accordingly, AREMA (2024) recommends that the net cross-sectional area of the connecting element not exceed 85% of the gross cross-sectional area of the connecting element.

12.2 MEMBER END CONNECTION AND SPLICE DESIGN

The number of slip-critical bolts in chord and diagonal member end connections is determined from the maximum computed axial tension or compression force in the chord or diagonal member and the allowable effective shear strength of the bolts. The number of slip-critical bolts in vertical member end connections not directly loaded by floorbeam reactions (subject to combined floor system and truss action forces) and splices is determined based on member axial yield capacity and the allowable effective shear strength of the bolts. The number of slip-critical bolts for design of vertical member end connections directly loaded by floorbeam reactions considers combined axial and bending forces by increasing the maximum computed axial force by 25% and considering the allowable effective shear strength of the bolts. The bearing stress on the member or splice at the bolted connection must also be investigated. In addition, tension member end connections and splices require that block shear be investigated. Fatigue design of slip-critical end connections in chord and web members is based on the gross area of the connected elements with an allowable fatigue stress range of 124 MPa (18 ksi) and 110 MPa (16 ksi), respectively.[6]

12.3 ALLOWABLE STRESSES AT MEMBER END CONNECTIONS AND SPLICES

Allowable bolt shear, bolt bearing and, for tension member end connections and splices, block shear rupture stresses are required for design of slip-critical member end connections and splices.

12.3.1 Allowable Shear Force of Slip-Critical Bolted Connections

High-strength steel bolts are specified by ASTM Standard F3125, which includes Grades A325M (A325) and A490M (A490) heavy hex head bolts, and Grades F1852 and F2280 twist-off bolts (RCSC, 2014). The minimum tensile strength, F_{bU}, for Grade A325M (A325), A490M (A490), F1852 and F2280 bolts is shown in Table 12.3.

TABLE 12.3
Minimum Tensile Strength of ASTM F3125 Bolts

Bolt Grade	F_{bU} (MPa)	F_{bU} (ksi)
A325M/A325	830	120
A490M/A490	1040	150
F1852	-	120
F2280	-	150

TABLE 12.4
Recommended Minimum Bolt Pre-Tension Force

	Minimum Required Bolt Pre-Tension, T_{bP} kN (kips)	
Bolt Diameter, d_b, mm (in)	Grade A325 Bolts	Grade A490 Bolts
12 (1/2)	50 (12)	62 (15)
16 (5/8)	88 (19)	111 (24)
20 (3/4)	138 (28)	173 (35)
22 (7/8)	167 (39)	210 (49)
24 (1)	199 (51)	249 (64)
36 (1½)	448 (103)	561 (148)

Slip-critical connections have effective shear strength based on the magnitude of the prestress force and the shear slip coefficient of the steel connection elements. The shear slip force, P_{bv}, is

$$P_{bv} = mnf'_{bv}\left(A_b\right) = k_s m\alpha \sum_{i=1}^{n} T_{bPi} \tag{12.1}$$

where

f'_{bv} = the effective allowable bolt shear stress
k_s = the shear slip coefficient of the steel connection
A_b = cross sectional area of the bolts
m = the number of slip planes (faying surfaces)
n = the number of bolts in the connection
T_{bPi} = the specified pre-tension in bolt i. AREMA (2024) recommends the minimum bolt pre-tension shown in Table 12.4 for slip-critical connections.
$\alpha = T_{bi}/T_{bPi}$
T_{bi} = actual pre-tension in bolt i

From Equation 12.1, f'_{bv} is

$$f'_{bv} = \frac{k_s\alpha \sum_{i=1}^{n} T_{bPi}}{nA_b} \tag{12.2}$$

Using a factor of safety (FS) of 1.5 and, conservatively, $\alpha = 1$ (in practical connections, α is typically between 1.13 and 1.35 depending on the method of bolt installation) with $A_{st} = 0.767(A_b)$, Equation 12.2 for Grade A325M (A325) bolts when the specified pre-tension in each bolt, T_{bPi}, is equal is

$$f'_{bv} = \frac{k_s\alpha T_{bPi}}{1.5\left(A_b\right)} = \frac{k_s\left(0.70F_{bU}A_{st}\right)}{1.5\left(A_b\right)} = \frac{k_s\left(0.70\left(830\right)0.767\right)}{1.5} = 297k_s \text{ MPa} \tag{12.3a}$$

$$f'_{bv} = \frac{k_s \alpha T_{bPi}}{1.5(A_b)} = \frac{k_s(0.70F_{bU}A_{st})}{1.5(A_b)} = \frac{k_s(0.70(120)0.767)}{1.5} = 43.0k_s \text{ ksi} \tag{12.3b}$$

AREMA (2024) outlines four slip-critical connection faying surface conditions for design and the allowable shear stress for Grade A325M (A325) bolts in standard holes[7] in slip-critical connections (Table 12.5).

Most modern steel railway truss fabrication processes include bolted joint preparation of surfaces with slip coefficients of about 0.50 resulting in an effective allowable shear force of 57.0 kN (12.9 kips) for a 22 mm (7/8 in) diameter bolt.[8] Shear on the bolt shank does not need to be evaluated as the slip values in Table 12.5 provide FS against bolt material shear greater than 2, even with the bolt threads in the shear plane. If shear failure of a slip-resistant connection occurs, the connection will behave as a bearing type connection.

12.3.2 Allowable Bearing Stress of Bolted Connections

Bearing failures are manifested as either yielding due to bearing of the connection element (member, splice or gusset plate) material against the bolt shank and/or tear out (block shear) of the connection element material near an edge.

The ultimate bearing strength, F_B, of the connection element material bearing on the bolt shank is related to the connection element material ultimate tensile strength, F_u, by the following linear relationship (Kulak, Fisher and Struik, 1987):

$$F_B = \left(\frac{l_e}{d_b}\right) F_U \tag{12.4}$$

where l_e = the distance from the centerline of the bolt to the nearest edge in the direction of the force and d_b = the diameter of the bolt. Using FS = 2.5 against bearing on the plate material and the AREMA (2024) recommendation that $l_e \geq 3d_b$ results in an allowable bolt-bearing stress, f_B, of

$$f_B = \frac{F_B}{FS} = 1.2F_u \tag{12.5}$$

TABLE 12.5
Allowable Effective Shear Stress for Grade A325M (A325) Bolts in Standard Holes in Slip-Critical Connections

Class	Surface Description	Mean Slip Coefficient, k_{sm}	Allowable Effective shear stress, f'_{bv}, MPa (ksi)
A	Uncoated clean mill scale and blast-cleaned surface before Class A coating	0.30	90 (12.9)
B	Uncoated blast-cleaned surface to SSPC-SP6 or better, blast-cleaned surfaces with Class B coatings and unsealed thermal-sprayed (i.e., metallized) surfaces	0.50	150 (21.5)
C	Hot-dip galvanized surfaces	0.30	90 (12.9)
D	Blast-cleaned surfaces with Class D coatings	0.45	135 (19.3)

The yield strength in pure shear, F_v, is[9]

$$F_v = \frac{F_y}{\sqrt{3}} \tag{12.6}$$

and the yield strength, P_y, of the shear block failure shown in Figure 12.1 is

$$P_y = 2t_p\left(l_e - \frac{d_b}{2}\right)\frac{F_y}{\sqrt{3}} = 1.15t_p\left(l_e - \frac{d_b}{2}\right)F_y \tag{12.7}$$

where t_p = the plate thickness.

The bearing strength of the bolt, F_{bB}, is

$$F_{bB} = f_B d_b t_p \tag{12.8}$$

which must not exceed the yield strength of the shear block given by Equation 12.7. Therefore,

$$P_y = 1.15t_p\left(l_e - \frac{d_b}{2}\right)F_y \geq F_{bB} \geq f_B d_b t_p \tag{12.9}.$$

Rearranging Equation 12.9 yields

$$\frac{l_e}{d_b} \geq \frac{0.87f_B}{F_y} + 0.5 \tag{12.10a},$$

which is conservatively simplified (for $l_e/d_b \geq 1.4$[10]) as

$$\frac{l_e}{d_b} \geq \frac{f_B}{F_y} \tag{12.10b}$$

Rearranging Equation 12.10b and using FS = 2.0[11] provides the allowable bearing stress as

$$f_B \leq \frac{l_e F_y}{d_b} \leq \frac{l_e F_u}{d_b(FS)} = \frac{l_e F_u}{2d_b} \tag{12.11a}$$

Equations 12.5 and 12.11a are the allowable bearing stresses on bolts recommended by AREMA (2024).

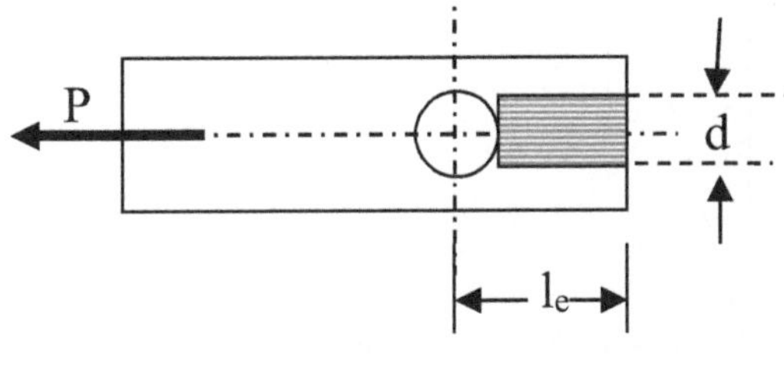

FIGURE 12.1 Bolt tear out (block shear) failure.

The minimum bolt spacing, s_b, is developed from Equation 12.11a as

$$s_b = l_e + \frac{d_b}{2} = \frac{2d_b\sigma_{br}}{F_U} + \frac{d_b}{2} \tag{12.11b}$$

Accordingly, the minimum distance between the center of the nearest bolt and end of the connection in the direction of the bolt pressure, e_b, is

$$e_b = \frac{2d_b\sigma_{br}}{F_U} \tag{12.11c}$$

where σ_{br} = the computed bearing stress due to the connection design force.

12.3.3 Allowable Tension Force in Bolted Connections

Figure 12.2, a typical curve from bolt tension testing, illustrates that the behavior of a bolt under tensile load is elastic for small elongations.

The strength of a bolt loaded in direct tension is not affected by pre-tension stresses from torquing during installation (Kulak, 2002). This is because the pre-tension load is readily dissipated as a direct tensile load is applied to a connection. Therefore, bolt strength is determined based on ultimate tensile strength. The allowable tensile stress, f_{bt}, in a Grade A325M (A325) bolt using FS = 2.0 and $A_{st} = 0.75A_b$ is

$$f_{bt} = \frac{0.75F_{bU}}{FS} = 310 \text{ MPa (45 ksi)} \tag{12.12}$$

However, as shown in Figure 12.3, the bolts in tension connections are subjected to additional tensile forces, T_Q, created by the prying action resulting from flexibility of the connection leg. From Figure 12.3, the bending moment at the bolt line, M_b, is

$$T(b) = M_f + \left(\frac{A_{nb}}{A_{gf}}\right)M_b = M_f + \left(\frac{A_{nb}}{A_{gf}}\right)\alpha M_f = M_f(1+\alpha\eta) \tag{12.13}$$

and with $M_f = \dfrac{w_b t_p^2}{4} F_y$

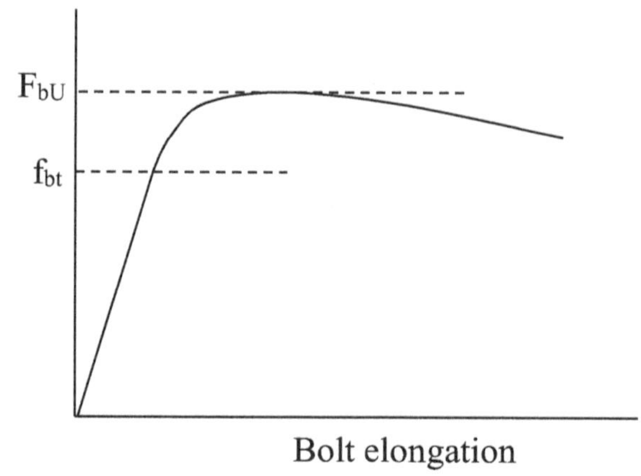

FIGURE 12.2 Bolt direct tension stress–strain curve.

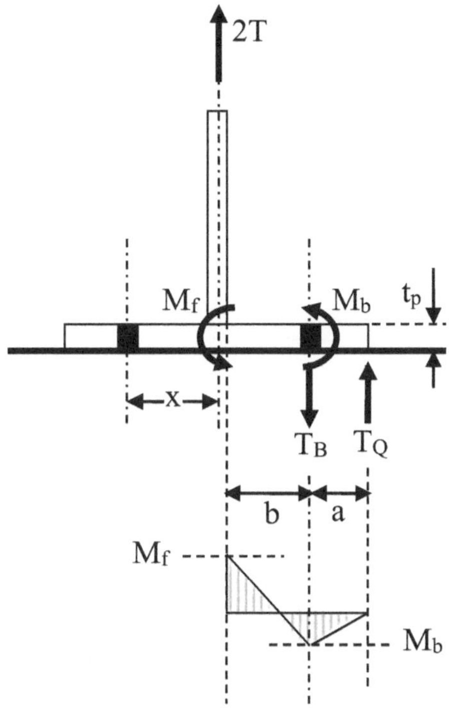

FIGURE 12.3 Prying action on bolted joint in tension.

$$T_Q(a) = M_b = \alpha\eta M_f = \frac{\alpha\eta w_b t_p^2 F_y}{4} \tag{12.14}$$

where

- T = the applied tensile force per bolt
- T_Q = the prying tensile force per bolt
- $T_B = T + T_Q$ = the total tensile force per bolt
- $h = A_{nb}/A_{gf}$
- A_{nb} = the net area of the flange at the bolt line
- A_{gf} = the gross area of the flange at the intersection with the web plate
- $a = M_b/M_f$ and depends on T_Q/T
- w_{bi} = tributary area for prying of each bolt i

The bolt tension, $T_B = T + T_Q$, with substitution of Equations 12.13 and 12.14 is

$$T_B = T + T_Q = \frac{M_f(1+\eta\alpha)}{b} + \frac{\alpha\eta w_b t_p^2 F_y}{4a} = \frac{M_f(1+\eta\alpha)}{b} + \frac{\alpha\eta M_f}{a} = T\left(1 + \frac{\alpha\eta b}{(1+\alpha\eta)a}\right) \tag{12.15}$$

Therefore,

$$T_Q = T\left(\frac{\alpha\eta b}{(1+\alpha\eta)a}\right) \tag{12.16}$$

Further manipulation of these equations provides thickness, t_p, as

$$t_p \le \sqrt{\frac{4T_B ab}{w_b F_y \left(a + \alpha\eta\left(a+b\right)\right)}} \tag{12.17}$$

However, Equations 12.15–12.17 are difficult to use in routine design. Consequently, analytical and experimental studies have provided a semi-empirical equation (Douty and McGuire, 1965):

$$T_Q = T\left(\frac{0.50 - \left(\frac{w_b t_p^4}{30ab^2 A_b}\right)}{\frac{a}{b}\left(\left(\frac{a}{3b}\right)+1\right)+\left(\frac{w_b t_p^4}{6ab^2 A_b}\right)}\right) \tag{12.18}$$

Equation 12.18 can be further simplified (Kulak, Fisher and Struik, 1987) as

$$\text{in SI units (mm): } T_Q = T\left(\frac{3b}{8a} - \frac{t_p^3}{328\text{x}10^3}\right) \tag{12.19a}$$

$$\text{and in US customary and imperial units (in): } T_Q = T\left(\frac{3b}{8a} - \frac{t_p^3}{20}\right) \tag{12.19b}$$

Further analytical and empirical studies (Nair, Birkemoe and Munse, 1974) have provided other empirical equations for the prying force, but Equation 12.19 is simple and conservative for use in routine design of bolted connections subjected to tension.

AREMA (2024) recommends the allowable tensile stress on fasteners, including the effects of prying, as 300 MPa (44 ksi) and 370 MPa (54 ksi) for Grade A325M (A325) and A490M (A490) bolts, respectively. Connections with bolts subjected to direct tension should be avoided in the main members of steel truss spans. However, bolts in tension and prying can occur combined with shear in connections.

12.3.4 Allowable Combined Shear and Tension at Slip-Critical Bolted Connections

Connections in steel railway truss spans may be subjected to combined shear and tension forces (e.g., floorbeam hangers). An ultimate strength interaction equation developed from tests (Chesson, Faustino and Munse, 1965) is shown in Figure 12.4.

The solid line in Figure 12.4 can be expressed as

$$\left(\frac{F_{bt}}{F_{bU}}\right)^2 + \left(\frac{F_{bv}}{F_{vU}}\right)^2 = 1.0 \tag{12.20}$$

where

F_{bt} = the ultimate tensile stress under combined shear and tension
F_{bU} = the ultimate tensile stress under tension only
F_{bv} = the ultimate shear stress under combined shear and tension
F_{vU} = the ultimate shear stress under shear only = 0.62 F_{bU} (Kulak, Fisher and Struik, 1987)

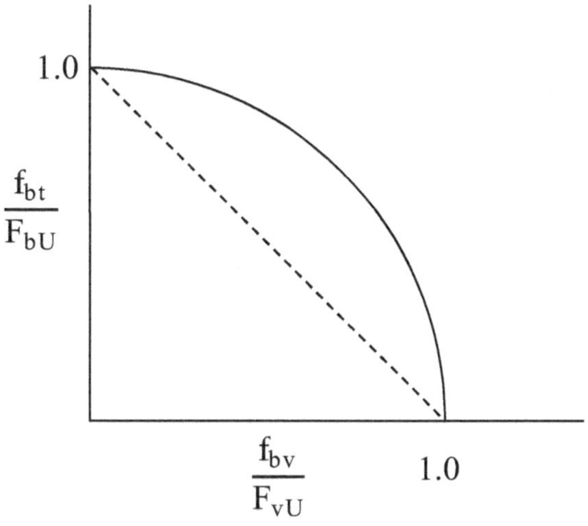

FIGURE 12.4 Bolt shear and tension interaction diagram.

Therefore, Equation 12.20 can be expressed as

$$\left(\frac{F_{bt}}{F_{bU}}\right)^2 + 2.60\left(\frac{F_{bv}}{F_{bU}}\right)^2 = 1.0 \tag{12.21}$$

Equation 12.20 in terms of allowable stresses is

$$\left(\frac{\sigma_{bt}}{f_{bt}}\right)^2 + \left(\frac{\tau_{bv}}{f_{bv}}\right)^2 = 1.0 \tag{12.22}$$

where

σ_{bt} = the tensile stress in the bolt (including prying action)
f_{bt} = the nominal tensile stress from pre-tension = T_{bP}/A_b
t_{bv} = the shear stress in bolt
f_{bv} = the allowable effective bolt shear stress = f'_{bv} (Table 12.5)
T_{bP} = the bolt pre-tension force (Table 12.4)
A_b = the nominal area of the bolt ~ 1.33 A_{st}
A_{st} = the effective bolt area through the threaded portion of the bolt

Therefore, for slip-resistant connections,

$$\left(\frac{\sigma_{bt}}{\left(T_{bP}/A_b\right)}\right)^2 + \left(\frac{\tau_{bv}}{f'_{bv}}\right)^2 = 1.0 \tag{12.23}$$

The elliptical Equation 12.23 can be simplified by a straight-line approximation (dashed line for ultimate stress values in Figure 12.4) as

$$\left(\frac{\sigma_{bt}}{\left(T_{bP}/A_b\right)}\right) + \left(\frac{\tau_{bv}}{f'_{bv}}\right) = 1.0 \tag{12.24}$$

and can be rearranged as

$$\tau_{bv} \leq f'_{bv}\left(1-\frac{\sigma_{bt}}{\left(T_{bP}/A_b\right)}\right) \tag{12.25}$$

If t_{bv} is taken as f_{bv} when combined with tension, Equation 12.25 becomes

$$f_{bv} \leq f'_{bv}\left(1-\frac{\sigma_{bt}}{\left(T_{bP}/A_b\right)}\right) \tag{12.26}$$

which is the allowable shear stress for combined shear and tension recommended by AREMA (2024).

12.3.5 Allowable Fatigue Stress Range for Bolted Slip-Critical Connections

Connections subjected to repeated fluctuations of stress must be designed considering allowable fatigue stress ranges. AREMA (2024) recommends the allowable tensile stress range, based on Fatigue Detail Category B, of 125 MPa (18 ksi) for the base metal of slip-critical connections subjected to 2 million cycles or less, or 110 MPa (16 ksi) for the base metal of slip-critical connections subjected to greater than 2 million stress range cycles. Bolts will generally not experience fatigue failure prior to the base metal and, therefore, AREMA (2024) includes no recommendations concerning allowable shear stress ranges for bolts.

12.3.6 Block Shear Strength at Bolted End Connections and Splices of Tension Members

Block shear must be evaluated at the end connections and splices of tension members. Block shear failure at the end connection or splice of a tension member occurs at ultimate shear stress on planes along the bolt lines (lines a-a in Figure 12.5) and ultimate tensile stress on planes between bolt lines

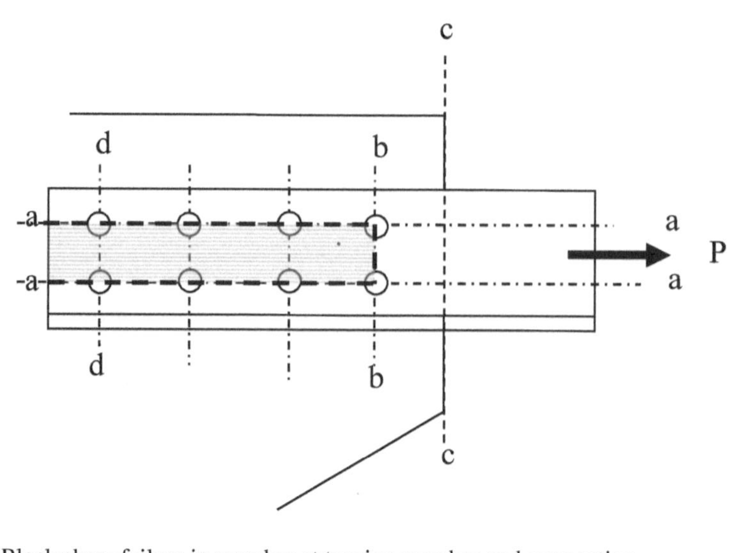

FIGURE 12.5 Block shear failure in member at tension member end connection.

(line b-b in Figure 12.5). The failure ultimately results in the tear-out of a section (shaded area of the member in Figure 12.5).

Allowable block shear strength is the sum of the allowable tensile fracture strength of area, A_{nt}, perpendicular to the tension force and either the allowable shear fracture strength of area, A_{nv}, or the allowable shear yield strength of area, A_{gv}, parallel to the tension force. Therefore, with FS = 2.0 for the allowable tensile fracture, shear fracture and shear yielding strengths,

$$\text{The allowable tensile fracture strength} = 0.50U_{bs}F_uA_{nt} \quad (12.27)$$

$$\text{The allowable shear fracture strength} = 0.30F_uA_{nv} \quad (12.28)$$

$$\text{The allowable shear yielding strength}^{12} = 0.30F_yA_{gv} \quad (12.29)$$

where

F_u = the ultimate tensile stress of the tension member or splice steel;
F_y = the tensile yield stress of the tension member or splice steel;
A_{gv} = the gross area of the tension member end connection or splice subject to shear stress;
A_{nv} = the net area of the tension member end connection or splice subject to shear stress;
A_{gt} = the gross area of the tension member end connection or splice subject to tension stress;
A_{nt} = the net area of the tension member end connection or splice subject to tension stress; and
U_{bs} = a reduction coefficient based on the tensile stress distribution. U_{bs} = 1.0 when the tension stress is uniform and U_{bs} = 0.5 when the tension stress is non-uniform.

$$\text{Accordingly, the allowable block shear strength based on tension fracture combined with shear fracture is} = 0.50U_{bs}FuA_{nt} + 0.30F_uA_{nv} \quad (12.30)$$

$$\text{and the allowable block shear strength based on tension fracture combined with shear yielding is} = 0.50U_{bs}F_uA_{nt} + 0.30F_yA_{gv}. \quad (12.31)$$

In accordance with AREMA (2024), the allowable block shear strength of tension member end connections or splices is the lesser of Equations 12.30 and 12.31.

12.4 MEMBER END CONNECTIONS AT GUSSET PLATES

The ends of truss chord and web members meet at joints that are typically made with steel gusset plates and high strength steel bolts in slip-critical connections. The member end connections must consider the allowable bearing stresses on the gusset plate.[13]

The stress distribution in gusset plates from tension and compression forces transferred from truss members is complex. Gusset plate connections have traditionally been designed using beam theory to determine axial, bending and shear stresses at various critical sections of the gusset plate. However, a slender beam model does not provide an accurate analysis of the gusset plate stress state[14] and considering axial, shear and block shear stresses in gusset plates from the transferred member forces is appropriate for the design of ordinary gusset plates (Ocel, 2013).

The gusset plates at tension member end connections must be designed for allowable yield, fracture, block shear and fatigue. Gusset plate yielding and fracture can also result from shear stresses transferred to the gusset plate by tension and compression in diagonal web members framing into the gusset plate. Block shear failures in gusset plates, due to forces transferred from tension members, can transpire because of tensile fracture stresses combined with shear yield or shear fracture stresses. The gusset plates at compression member end connections must be designed for allowable compression stresses from gusset plate buckling and partial shear yielding. Gusset plates at the ends of compression members also experience instability due to partial shear yielding. The transfer of web member tensile and compressive axial stress in gusset plates typically occurs over an effective section width (Whitmore, 1952). The Whitmore section width does not properly characterize the axial stress transmission in gusset plates at chord members in tension or compression (Ocel, 2013).

The chord members at gusset plates at truss panel points are continuous or discontinuous (spliced) at the panel point. Gusset plates at truss panel points connected to continuous chord members transfer the difference in the chord member forces across the panel point. Truss panel points with spliced chord members must transfer the minimum chord member yield capacity at the panel point through the gusset plate. The difference in the continuous chord member forces will be of lesser magnitude than discontinuous (spliced) chord member forces at gusset plate connections. Consequently, it is often preferable to use continuous chord members through truss panel points and splice the chord members for member yield capacity near the panel point in the least stressed member.

The dimensions of gusset plates typically depend on the dimensions and angles of the web members framing into the gusset plate. Nevertheless, gusset plates should be designed to be as compact as possible. This not only reduces gusset plate material consumption but also reduces slenderness ratios and free edge distances for greater buckling strength of the gusset plate. A detailed finite element analysis to determine the stress types, distributions, and magnitudes in complex geometry[15] gusset plate connections is usually required.[16]

12.4.1 Gusset Plate Axial Tensile Strength at Member End Connections

Gusset plate tension design is controlled by allowable yielding, fracture and block shear stresses.

12.4.1.1 Gusset Plate Axial Tensile Yield, Fracture and Fatigue Design at Member End Connections

The distribution of axial tension stress in a gusset plate that is transferred from a tension member connected to the gusset plate occurs over an effective or Whitmore[17] width. As shown in Figure 12.6, the Whitmore width, w_e, is the length of a line perpendicular to the member axis at the last row of

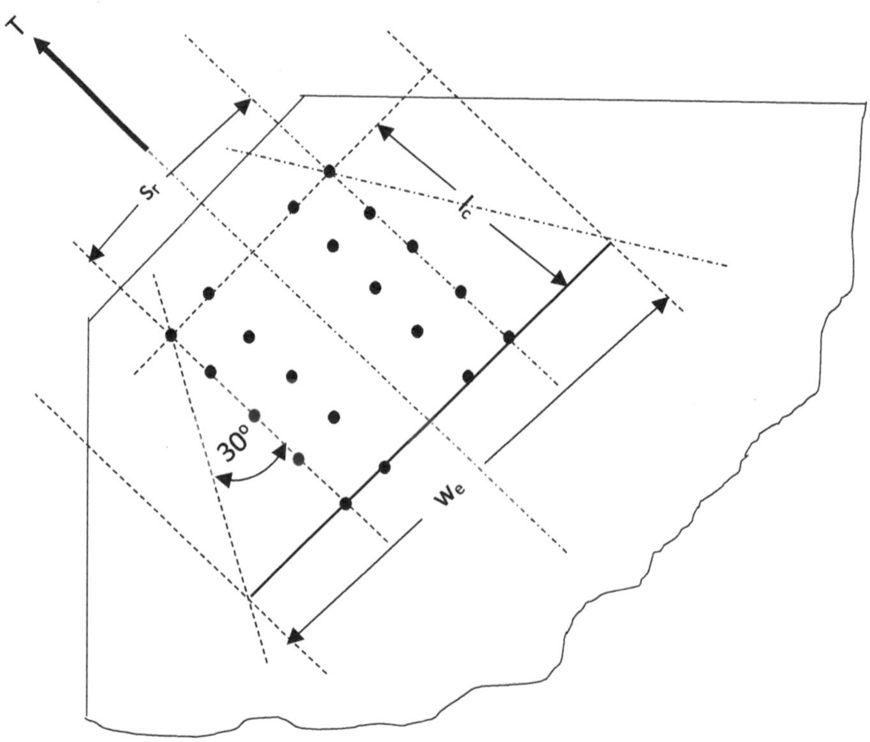

FIGURE 12.6 Whitmore effective width in gusset plates at tension members.

fasteners in the member connection; w_e bounded by lines diverging away from the member at 30° to the axis of the member from the exterior fasteners in the first row of fasteners in the connection. The near edge of a gusset plate may bound w_e.[18]

As illustrated in Figure 12.6, w_e is

$$w_e = 2l_c \tan(30^\circ) + s_r = 1.151_c + s_r \tag{12.32}$$

The allowable stress for yielding on the gross w_e of the gusset plate and the allowable stress for fracture on the net w_e of the gusset plate must be investigated. The allowable stress based on yielding of the Whitmore effective gross section, $w_e t_p$, immediately below the last line of bolts is[19]

$$\sigma_{aty} = \frac{P}{w_e t_p} \le 0.55F_y \tag{12.33}$$

where w_e = the Whitmore effective gross width, σ_{aty} = the axial tension stress on the Whitmore effective gross area, $w_e t_p$; P = the axial tension force in the member; t_p = the thickness of the gusset plate and F_y = the tensile yield stress of gusset plate steel.

The allowable of tensile stress based on fracture of the Whitmore effective net section, $w_{ne} t_p$, through the last line of bolts is

$$\sigma_{atU} = \frac{P}{w_{ne} t_p} \le 0.47F_U \tag{12.34}$$

where w_{ne} = the Whitmore effective net width, σ_{atU} = the axial tension stress on the Whitmore effective net area, $w_{ne} t_p$ and F_U = the ultimate tensile stress of gusset plate steel. The allowable stress range based on fatigue of $w_e t_p$ through the last line of bolts is[20]

$$S_R = \frac{\Delta P}{w_e t_p} \le S_{Rfat} \tag{12.35}$$

where ΔP = the axial tension force range in the member and S_{Rfat} = the allowable fatigue stress range based on the number of stress range cycles and detail.

12.4.1.2 Gusset Plate Tensile Block Shear Strength at Member End Connections

Block shear must be evaluated at tension connections utilizing gusset plates. Block shear failure in the gusset plate at tension member end connections occurs due to the yield or ultimate shear stress through the cross-sectional areas on planes along the bolt lines (areas A_{gv} and A_{nv} in Figure 12.7) and the ultimate tensile stress through the cross-sectional area on planes between bolt lines (area A_{nt} in Figure 12.7). The failure ultimately results in the tear-out of a section (shaded area of the member in Figure 12.7).

Allowable block shear strength is the sum of the allowable tensile fracture strength of area, A_{nt}, perpendicular to the tension force and either the allowable shear fracture strength of area, A_{nv}, or the allowable shear yield strength of area, A_{gv}, parallel to the tension force.

In accordance with AREMA (2024), the allowable block shear strength of gusset plates at tension member end connections or splices is the lesser of Equation 12.30 or Equation 12.31 where

F_u = the ultimate tensile stress of the gusset plate steel;
F_y = the tensile yield stress of the gusset plate steel;
A_{gv} = the gross area of the gusset plate subject to shear stress;
A_{nv} = the net area of the gusset plate subject to shear stress;
A_{gt} = the gross area of the gusset plate subject to tension stress; and
A_{nt} = the net area of the gusset plate subject to tension stress.

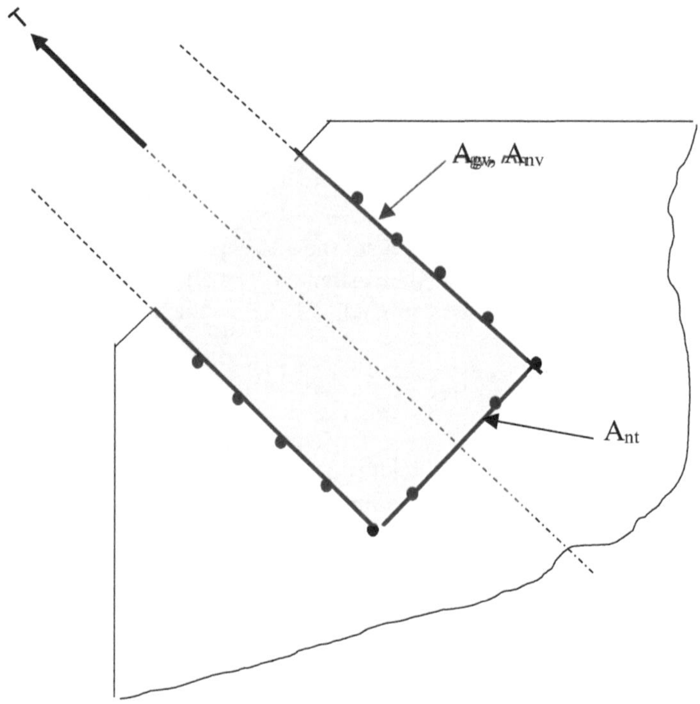

FIGURE 12.7 Block shear failure in gusset plates at tension members.

12.4.2 Gusset Plate Axial Compression Strength at Member End Connections

Gusset plate compression strength is controlled by allowable axial compression stresses and instability from partial shear yielding.

12.4.2.1 Gusset Plate Axial Compressive Strength and Stability at Member End Connections

As shown in Figure 12.8, the distribution of axial compression stress in a gusset plate that is transferred from a compression member to the gusset plate occurs over w_e. The allowable axial compression stress on the effective gross section, $w_e t_p$, immediately below the last line of bolts is

$$\sigma_{ac} = \frac{P}{w_e t_p} \le F_{call} \tag{12.36}$$

where

σ_{ac} = the axial compression stress on $w_e t_p$ of the gusset plate

P = the axial compression force in the member

F_{call} = the allowable axial compression stress considering instability with an effective slenderness ratio, Kl_w/r_w

Kl_w = the effective buckling length of the gusset plate

l_w = the distance from the center of the last line of bolts[21] in a compression member to the nearest gusset plate restraint. l_w may extend to the first row of bolts (restraint line) at an interface member such as a chord member.

$r_w = t_p / \sqrt{12}$

t_p = the thickness of the gusset plate

K = the effective length factor, typically between 0.50 and 0.65 for properly braced gusset plates (Thornton and Kane, 1999).[22] Typical gusset plate connections for railway truss main members are well braced against lateral sway instability and K is often 0.50.

The allowable axial compression stress, F_{call}, is

$$\text{when } \frac{Kl_w}{r_w} \le 0.629\sqrt{\frac{E}{F_y}} \text{ then } F_{call} = 0.55F_y \tag{12.37}$$

$$\text{when } 0.629\sqrt{\frac{E}{F_y}} < \frac{Kl_w}{r_w} < 5.034\sqrt{\frac{E}{F_y}} \text{ then } F_{call} = 0.60F_y - \left(635\frac{F_y}{E}\right)^{\frac{3}{2}}\left(\frac{Kl_w}{r_w}\right) \text{MPa} \tag{12.38a}$$

$$\text{when } 0.629\sqrt{\frac{E}{F_y}} < \frac{Kl_w}{r_w} < 5.034\sqrt{\frac{E}{F_y}} \text{ then } F_{call} = 0.60F_y - \left(175\frac{F_y}{E}\right)^{\frac{3}{2}}\left(\frac{Kl_w}{r_w}\right) \text{ksi} \tag{12.38b}$$

$$\text{when } \frac{Kl_w}{r_w} \ge 5.034\sqrt{\frac{E}{F_y}} \text{ then } F_{call} = \frac{0.514\pi^2 E}{\left(\frac{Kl_w}{r_w}\right)^2} = \frac{5.073E}{\left(\frac{Kl_w}{r_w}\right)^2} \tag{12.39}$$

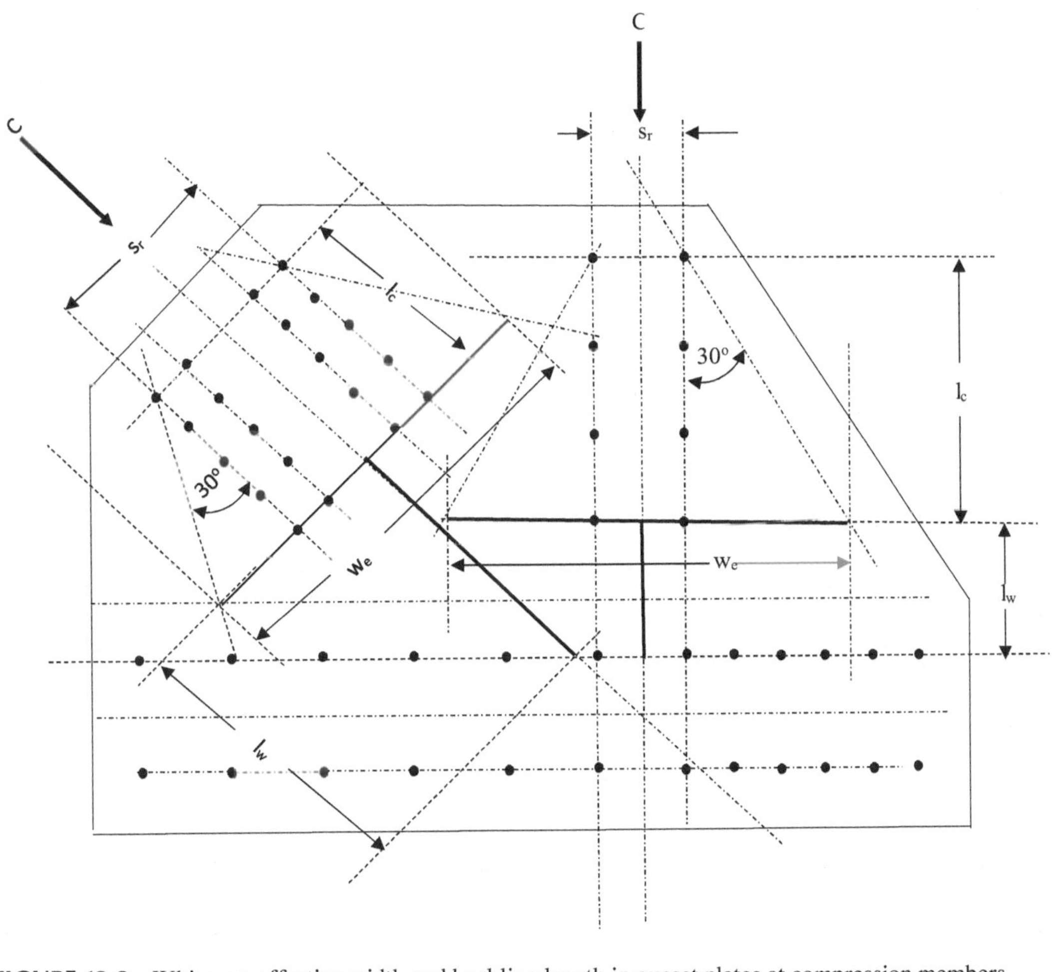

FIGURE 12.8 Whitmore effective width and buckling length in gusset plates at compression members.

12.4.2.2 Gusset Plate Partial Shear Strength at Compression Member End Connections

Gusset plate instability can also result from the formation of partial shear stress planes (Ocel, 2013). Possible partial shear planes on the gross section of the gusset plate along a compression member are of interest because as the partial shear plane yields, the associated decrease in the gusset plate elastic modulus reduces the out-of-plane rotational restraint of the gusset plate acting in compression as an idealized column. Therefore, the allowable stress for yielding on partial shear planes is also investigated to fully understand gusset plate axial compression strength. Figure 12.9 illustrates two possible shear planes adjacent to a compression member consisting of a vertical partial shear plane with strength $0.30F_yA_{gvv}/\sin\alpha$ and a horizontal partial shear plane with strength $0.30F_yA_{gvh}/\cos\alpha$ where A_{gvv} = vertical plane gross area subject to partial shear stress and A_{gvh} = horizontal plane gross area subject to partial shear stress.

In general, partial plane shear yielding is not required to be checked for chord or vertical members because there is not an admissible shear plane that would cause out-of-plane instability of the gusset plate.

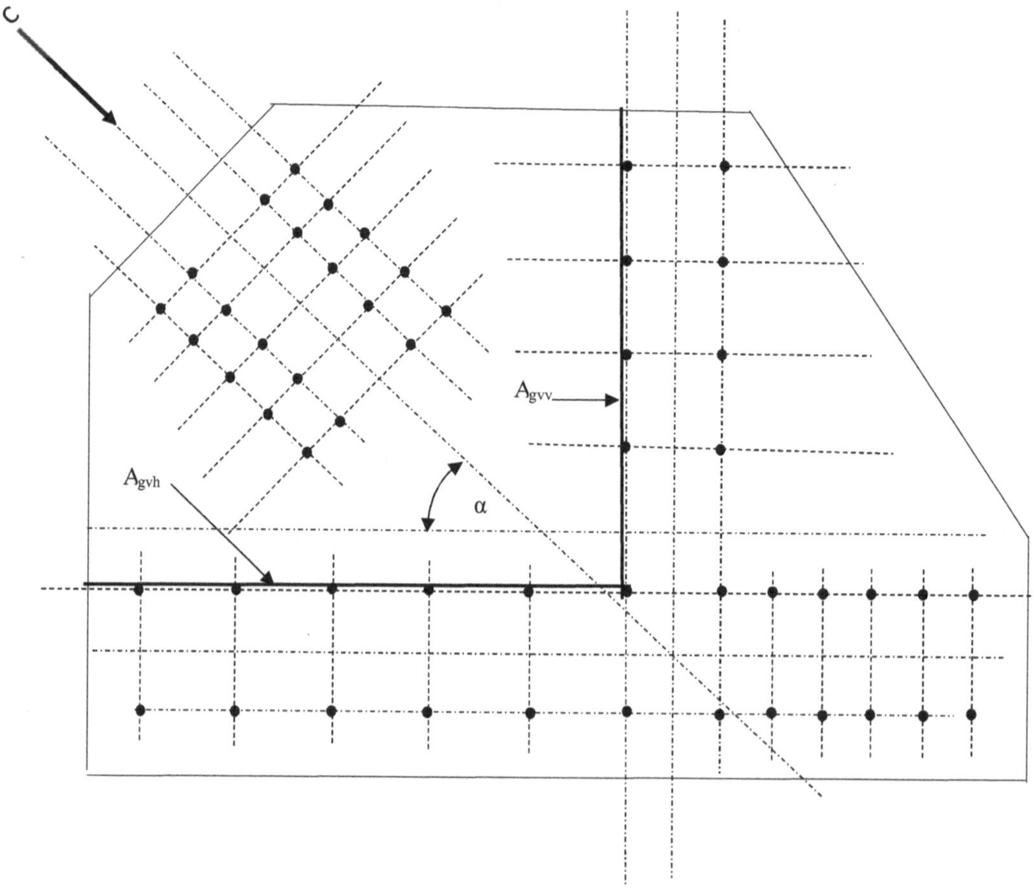

FIGURE 12.9 Partial shear planes in gusset plates at compression members.

12.4.3 Gusset Plate Shear Strength

Gusset plates must also be designed considering the allowable shear yield stresses on the gross and net sections of possible full horizontal and vertical shear planes through the gusset plate. The full shear planes consist of horizontal shear planes with strength $0.30F_yA_{gvh}$ or $0.30F_uA_{nvh}$ and vertical shear planes with strength $0.30F_yA_{gvv}$ or $0.30F_uA_{nvv}$ where

A_{gvv} = full vertical plane gross area subject to shear stress
A_{nvv} = full vertical plane net area subject to shear stress
A_{gvh} = full horizontal plane gross area subject to shear stress
A_{nvh} = full horizontal plane net area subject to shear stress

However, full shear planes will not develop at the intersection with members, thereby reducing the number of possible full shear planes in the gusset plate. For example, as shown in Figure 12.10, horizontal full shear planes develop through the gross area, A_{gvh}, and net area, A_{nvh}, of the gusset plate, but a vertical full shear plane is not admissible through the bottom chord member.

12.4.4 Free Edge of Gusset Plates

Based on elastic stability, some design engineers limit the slenderness, b/t_p, where b is the length of a free edge, to less than $2.06\sqrt{\frac{E}{F_y}}$ and if it is exceeded, use edge-stiffening angles proportioned such

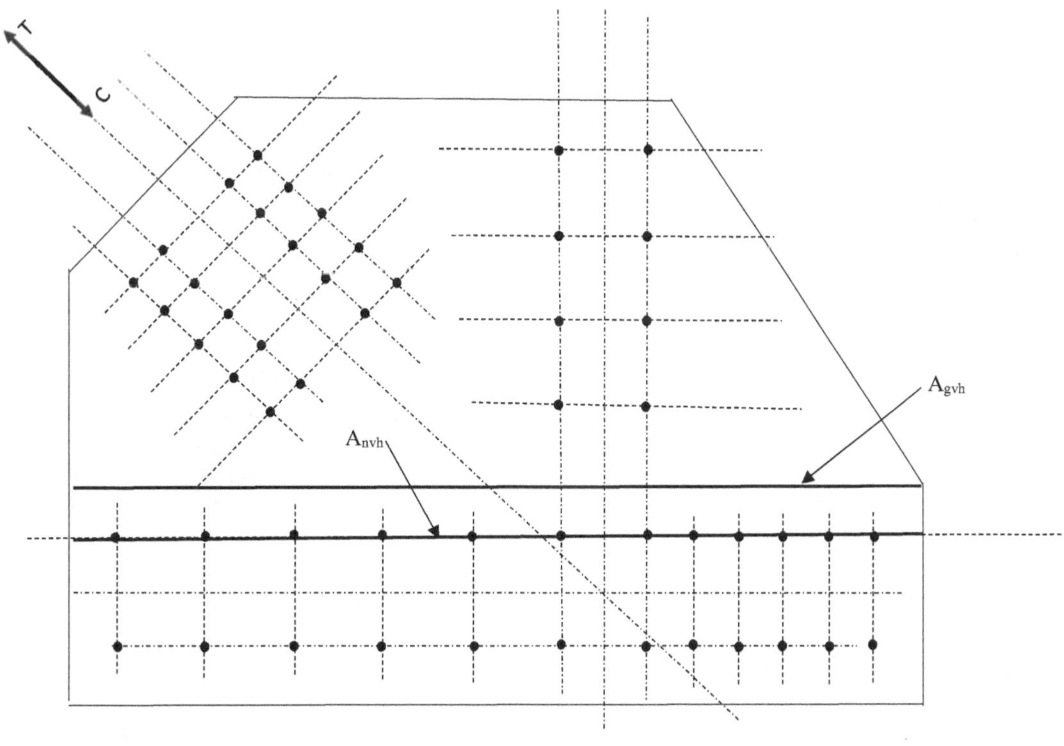

FIGURE 12.10 Full horizontal shear planes in gusset plate.

that b/r_w, where r_w is the radius of gyration of the stiffening angle, is less than 120. However, this plate buckling limitation is valid for only very thin plates, and since AREMA (2024) recommends a minimum gusset plate thickness of 0.5 in, the free edge slenderness criterion is not required by AREMA (2024).[23]

12.4.5 Gusset Plates at Continuous Chord Members

Gusset plates at truss panel points connected to continuous chord members transfer the difference in the chord member forces across the panel point. Therefore, for continuous chord members, the bolts in the chord member must be arranged symmetrically about the centerline of the chord member.

The largest difference in continuous chord member forces is equal to the greatest horizontal resultant forces from the diagonal members framing into the gusset plate, provided member forces all meet at a common working point. As shown in Example 12.1, the continuous chord member forces transferred to gusset plates in truss spans with diagonal web members oriented in a single direction, such as Pratt and Howe truss spans, are determined from the load position for maximum force in the diagonal member. As shown in Example 12.2, the continuous chord member forces transferred to gusset plates in truss spans with diagonal web members oriented in each direction, such as Warren truss spans, are determined from the load position for maximum force in each diagonal member with the concurrent force in the other diagonal member framing into the gusset plate. For truss chord members that are continuous through a joint, AREMA (2024) indicates shear lag may not be of concern and U_c may be effectively taken as 1.00. However, in some circumstances, engineering judgment may indicate that consideration of $U_c = 0.90$ may be appropriate for chord member design (Bowles, 1980).[24]

Example 12.1

Design the gusset plate for joint L2 of the Pratt truss with the dead load and maximum live load forces shown in Table E12.1a. Members L1L2, L2L3, U1L2 and U2L2 frame into the gusset plates at L2. Member U1L2 frames into the gusset plate at a 45° angle. The gusset plate is not a component of a bottom chord splice as the H-section bottom chord is continuous through panel point L2.

TABLE E12.1a
Dead Load and Maximum Live Load Forces in Members at Joint L2

Pratt Through-Truss Member		Dead Load Force, DL (kips)	Maximum Live Load Force, LL_{max} (kips)	Maximum Live Load Plus Impact Force, LL_{max} + I (kips)	Dead Plus Maximum Dynamic Live Load, DL + LL_{max} + I (kips)	Tension or Reversal Force Range, ΔF (kips)
L1L2	Bottom chord	93.8	323.9	391.9	485.7	368.1
L2L3	Bottom chord	150.0	496.9	601.3	751.3	564.7
U2L2	Vertical positive shear	−18.8	−125.8	−152.2	−171.0	0
	Vertical negative shear	−18.8	59.4	71.9	53.1	90.7
U1L2	Diagonal positive shear	79.6	303.4	367.1	446.7	344.8
	Diagonal negative shear	79.6	−23.7	−28.9	50.7	108.5

For slip-critical connections, try a gusset plate with

t_p = 0.50 in
F_y = 50 ksi
F_u = 70 ksi
7/8 in diameter ASTM F3125 Grade A325 bolts are used in standard holes with connection surfaces prepared for a mean slip coefficient, k_{sm}, of 0.50.
Bolt effective shear strength in SC connection = 21.5(0.60) = 12.9 kips.
Single shear connection between member and each gusset plate.

The forces in Table E12.1a are maximum forces and not concurrent forces at L2. Since the bottom chord is continuous through joint L2, the gusset place experiences only the difference of the forces in L1L2 and L2L3. A free body diagram of joint L2 indicates that the greatest difference of the forces in L1L2 and L2L3 occurs in conjunction with the maximum force in diagonal member U1L2.
The length of the influence line for the chord members at joint L2 is greater than 100 ft and, therefore, the allowable fatigue stress is 18 ksi on the gross section of the connected elements. The length of the influence line for the web members at joint L2 is less than 100 ft and, therefore, the allowable fatigue stress on the gross section of the connected elements is 16 ksi.

Since U2L2 experiences forces from both floor system and truss action, the U2L2 end connection is designed for the strength of the member (see Table E12.1b for minimum bolts):

Use H-section with 12 × 0.5 flange plates and 11 × 0.5 web plate
L_{U2L2} = 300 in
P_{U2L2} = −171.0 kips
A = 17.5 in^2
$I_x = 2(0.5)(12)^3/12 + 11(0.5)^3/12 = 144.1$ in^4

$$r_x = \sqrt{\frac{144.1}{17.5}} = 2.9 \text{ in}$$

$I_y = 2(12)(0.5)^3/12 + 2(12)(0.5)(5.75)^2 + 0.5(11)^3/12 = 452.5$ in^4

$$r_y = \sqrt{\frac{452.5}{17.5}} = 5.1 \text{ in}$$

$KL/r_{min} = 0.75(300)/2.9 = 78.4$
$F_{call} = 30 - (165.7(78.4))/1000 = 17.0$ ksi

Capacity of member U2L2 = 17.0(17.5) = 297.5 kips, OK ≥ 171.0 kips.
Allowable bolt-bearing stress = 1.2(70) = 84 ksi or = 1.09(70)/(2(0.875)) = 43.75 ksi (with minimum edge spacing of 1.25(0.875) = 1.09 in)
Bolt-bearing stress at bottom chord member on gusset plate = 157.9/(12(1)(0.5)) = 26.3 ksi OK
Bolt-bearing stress at diagonal member on gusset plate = 223.4/(20(1)(.5))= 22.3 ksi OK
Bolt-bearing stress at vertical member on gusset plate = 148.8/(12(1)(0.5)) = 24.8 ksi OK

a) Gusset plate tension strength controlled by allowable yielding and fracture stresses: At connection to member U1L2, T = 446.7kips and 5 rows of 4 bolts in each connection (Figure E12.1a).

TABLE E12.1b
Minimum Bolts for Member Connections to Gusset Plate at joint L2

Pratt Through-Truss Member		Maximum Design Force (kips)	Minimum Bolts for DL + LL_{max} + I with f'_{bv} = 21.5 ksi	Bolts in Each Gusset Plate Connection
L1L2 L2L3	Bottom chord	(446.7cos 45°)/ (2(12.9)) = 157.9/12.9	12.2	Use minimum of 6 bolts each side of working point
U2L2	Vertical	297.5/2 = 148.8	11.5	Use 6 rows of 2 bolts in each connection
U1L2	Diagonal	446.7/2 = 223.4	17.3	Use 5 rows of 4 bolts in each connection

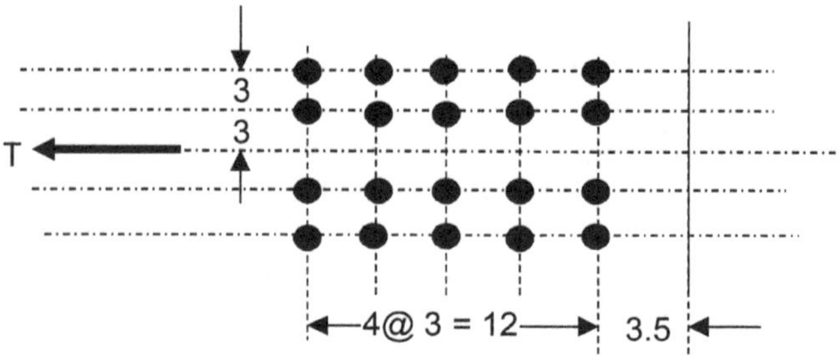

FIGURE E12.1a Connection to gusset plate at U1L2.

l_c = 4(3.0) = 12.0 in
s_r = 4(3) = 12 in
$w_e = 2l_c\tan(30^o) + s_r = 1.15l_c + s_r$ = 1.15(12)+34 = 25.86 in (see Figure 12.6 and Figure E12.1c)
w_{ne} = 25.86 − (4)(1) = 21.86 in
w_{ne}/w_e = 21.86/25.86 = 0.85

$$\sigma_{aty} = \frac{\left(\frac{446.7}{2}\right)}{(25.86)(0.5)} = 17.3 \text{ ksi} \le 0.55(50) \le 27.5 \text{ ksi, OK}$$

Assuming U_c = 0.90 (see Chapter 10),

$$\sigma_{atU} = \frac{\left(\frac{446.7}{2}\right)}{0.90(21.86)(0.5)} = 22.7 \text{ ksi} \le 0.47(70) \le 32.9 \text{ ksi, OK}$$

At connection to member L1L2 and L2L3:
T = 446.7cos 45° = 315.9 kips
Minimum of 6 bolts each side of working point
w_e = 12.0 in
w_{ne} = 12.0 − 2(1.0) = 10.0 in (see Figure E12.1c)
w_{ne}/w_e = 10.0/12.0 = 0.83

$$\sigma_{aty} = \frac{\left(\frac{315.9}{2}\right)}{(12)(0.5)} = 26.3 \text{ ksi} \le 0.55(50) \le 27.5 \text{ ksi, OK}$$

Assuming U_c = 0.90 (bottom chord member forces transferred from diagonal members)

$$\sigma_{atU} = \frac{\left(\frac{315.9}{2}\right)}{0.90(10)(0.5)} = 35.1\,\text{ksi} \le 0.47(70) \le 32.9\text{ ksi, say OK at 7\% overstress}$$

or increase gusset plate thickness by minimum of 1/32 in

At connection to member U2L2

T = 53.1 kips

l_c = 5(3.0) = 15.0 in

s_r = 2(3) = 6 in

$w_e = 2l_c \tan(30^o) + s_r = 1.15l_c + s_r$ = 1.1515 + 6 = 23.25 in (see Figure 12.6 and Figure E12.1c)

w_{ne} = 23.25 − (2)(1) = 21.25 in

w_{ne}/w_e = 21.25/23.25 = 0.91

Maximum w_{ne} = 0.85(23.25) = 19.76 in

$$\sigma_{aty} = \frac{\left(\frac{53.1}{2}\right)}{(23.25)(0.5)} = 2.3\ \text{ksi} \le 0.55(50) \le 27.5\ \text{ksi, OK}$$

Assuming U_c = 0.90 (see Chapter 10)

$$\sigma_{atU} = \frac{\left(\frac{53.1}{2}\right)}{0.90(19.76)(0.5)} = 3.0\ \text{ksi} \le 0.47(70) \le 32.9\ \text{ksi, OK}$$

b) Gusset plate tension strength controlled by allowable block shear

At connection to member U1L2:

T = 446.7 kips

5 rows of 4 bolts in each connection

Shear plane length = 3.5 + 4(3) = 15.5 in

Tension plane length = 4(3) = 12 in

A_{gv} = 2(15.5)(0.5) = 15.5 in² (see Figure 12.7 and Figure E12.1c)

A_{nv} = 2(15.5 − 5(1))(0.5) = 10.5 in²

A_{gt} = (3)(4)(0.5) = 6.0 in² (see Figure 12.7 and Figure E12.1c)

A_{nt} = 6.0 − 4(1)(0.5) = 4.0 in²

U_{bs} = 1.0

Allowable block shear strength based on tension fracture strength combined with shear fracture strength = $0.50U_{bs}F_uA_{nt} + 0.30F_uA_{nv}$ = 0.50(1.0)(65)(4.0) + 0.30(65)(10.5) = 130.0 + 204.8 = 334.8 kips

Allowable block shear strength based on tension fracture strength combined with shear yield strength = $0.50U_{bs}F_uA_{nt} + 0.30F_yA_{gv}$ = 0.50(1.0)(65)(4.0) + 0.30(50)(15.5) = 130.0 + 232.5 = 362.5 kips

Allowable block shear strength = 334.8 kips ≥ 223.4 kips, OK

c) Gusset plate compression strength controlled by allowable compression stresses

At connection to member U2L2 (Figure E12.1b):

C = 171.0 kips

C_c = 297.5 kips (capacity of member)

No. bolts = 297.5/(2(12.9)) = 11.5 bolts each side (6 rows of 2 bolts each side)

l_w = 7.75 in (see Figure 12.8 and Figure E12.1b)

$$r_w = \frac{t_p}{\sqrt{12}} = \frac{0.5}{\sqrt{12}} = 0.144\,\text{in}$$

K = 0.5 (lateral sway restrained)

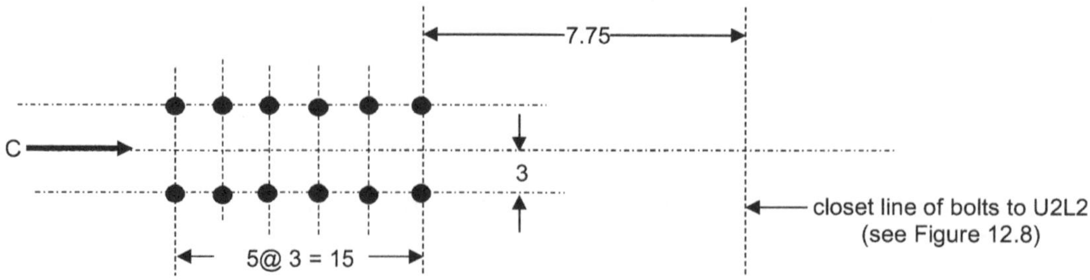

FIGURE E12.1b Connection to gusset plate at U2L2.

$$0.629\sqrt{\frac{E}{F_y}} = 0.629\sqrt{\frac{29{,}000}{50}} = 15.15 \le Kl_w/r_w = 0.5(7.75)/0.144 = 26.91 \le$$

$$5.034\sqrt{\frac{E}{F_y}} = 5.034\sqrt{\frac{29{,}000}{50}} = 121.2$$

$$F_{call} = 0.60(50) - \left(175\left(\frac{50}{29000}\right)\right)^{\frac{3}{2}}(26.91) = 30 - 0.166(26.91) = 25.5 \text{ ksi}$$

l_c = 5(3.0) = 15.0 in
s_r = 2(3) = 6.0 in
$w_e = 2l_c \tan(30^o) + s_r = 1.15l_c + s_r$ = 1.15(15) + 3(2) = 23.25 in (see Figure 12.6 and Figure E12.1c)

$$\sigma_{ac} = \frac{C}{w_e t_p} = \frac{\left(\frac{171.0}{2}\right)}{23.25(0.50)} = 7.4\text{ksi} \le F_{call} = 25.5\text{ksi, OK}$$

d) Gusset plate compression strength controlled by allowable partial shear: At connection to member U2L2, There is not an admissible partial shear plane that would cause out-of-plane instability of the gusset plate.
e) Gusset plate shear strength controlled by allowable full shear: A vertical full shear plane is not admissible through the bottom chord member.
Horizontal shear force = (446.7/2)cos 45° = 157.9 kips
Length of gusset plate = 52.5 in (see Figure 12.10 and Figure E12.1c)
A_{gvh} = 52.5(0.5) = 26.25 in^2
A_{nvh} = (52.5-10(1))0.5 = 21.25 in^2 (see Figure 12.10 and Figure E12.1c)
Full horizontal shear plane strength on gross area = $0.30F_yA_{gvh}$ = 0.30(50)(26.25) = 393.8 kips ≥ 157.9 OK.
Full horizontal shear plane strength on net area = $0.30F_uA_{nvh}$ = 0.30(65)(21.25) = 414.4 kips ≥ 157.9 OK
f) Gusset plate fatigue
At connection to member U1L2:
ΔT = 344.8 kips
S_R = (344.8/2)/((25.86)(0.5)) = 13.3 ksi ≤ 16 ksi, OK
At connection to U2L2:
ΔT = 90.7 kips
S_R = (90.7/2)/(23.25(0.50)) = 3.9 ksi ≤ 16 ksi, OK
At connection to bottom chord:
ΔT = (315.9)(344.8/446.7) = 243.8 kips

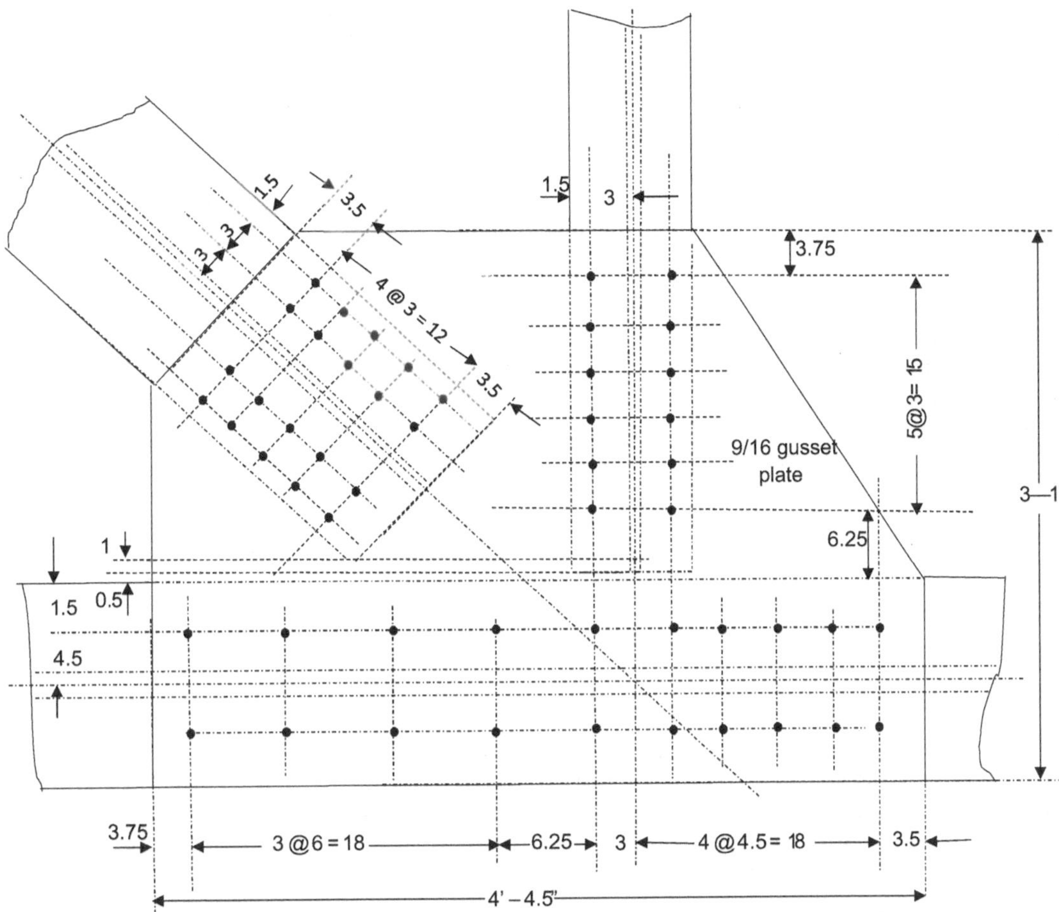

FIGURE E12.1c Gusset plate at L2 of Pratt truss.

S_R = (243.8/2)/(12.0(0.5)) = 20.3 ksi, > 18 ksi, NG, increase gusset plate thickness to 9/16

Use 9/16 in thick gusset plate with a minimum of 6 bolts each side of working point in chord members.

Spacing along line of force not less than $2d_b\sigma_{bc}/F_u + d_b/2$ = 2(0.875)(22.3)/70 + 0.875/2 = 0.56 + 0.44 = 1.00 in

Spacing of bolts not less than $3d_b$ = 2.63 in

Use minimum spacing of 3.0 in

Use maximum spacing of 6.0 in for sealing connection.

Edge distance not less than $1.5d_b$ = 1.31 in

Edge distance not greater than $1.5 + 4t_g$ = 1.5 + 4(0.5) = 1.5 + 2 = 3.5 in or 6 in

Loaded edge distance not less than $2d_b\sigma_{bc}/F_u$ = 0.56 in

Use minimum edge distance of 1.5 in

Use maximum edge distance of 3.5 in

Example 12.2

Design the gusset plate for joint L2 of the Warren truss with the dead load and maximum live load forces shown in Table E12.2a. Members L1L2, L2L3, U1L2, U2L2 and L2U3 frame

TABLE E12.2a
Dead Load and Maximum Live Load Forces in Members at Joint L2

Warren Through-Truss Member		Dead Load Force, DL (kips)	Maximum Live Load Force, LL_{max} (kips)	Maximum Live Load Plus Impact Force, LL_{max} + I (kips)	Dead Plus Maximum Dynamic Live Load, DL + LL_{max} + I (kips)	Tension or Reversal Force Range, ΔF (kips)
L1L2	Bottom chord	93.8	323.9	391.9	485.7	368.1
L2L3	Bottom chord	168.8	563.2	681.5	850.3	640.1
U2L2	Vertical positive shear	0	0	0	0	0
	Vertical negative shear	0	0	0	0	0
U1L2	Diagonal positive shear	79.6	303.4	367.1	446.7	344.8
	Diagonal negative shear	79.6	−23.7	−28.9	50.7	108.5
L2U3	Diagonal positive shear	−26.8	−178.9	−216.5	−243.3	-
	Diagonal negative shear	−26.8	83.8	101.4	74.6	128.2

into the gusset plates at L2. Members U1L2 and L2U3 frame into the gusset plate at a 45° angle. The gusset plate is not a component of a bottom chord splice as the H-section bottom chord is continuous through panel point L2.

For member slip-critical connections, try gusset plate with,

$t_p = 0.50$ in
$F_y = 50$ ksi
$F_u = 70$ ksi

7/8 in diameter ASTM F3125 Grade A325 bolts are used in standard holes with connection surfaces prepared for a mean slip coefficient, k_{sm}, of 0.50.

Bolt effective shear strength in SC connection = 21.5(0.60) = 12.9 kips.

Single shear connection between member and each gusset plate.

The forces in Table E12.2a are maximum forces and not concurrent forces at L2. Since the bottom chord is continuous through joint L2, the gusset place experiences only the difference of the forces in L1L2 and L2L3, which is determined from the maximum force in U1L2 with the concurrent force in U2L3 or the maximum force in U2L3 with the concurrent in U1L2. The length of the influence line for the chord members at joint L2 is greater than 100 ft, and therefore, the allowable fatigue stress on the gusset plate is 18 ksi. The length of the influence line for the web members at joint L2 is less than 100 ft, and therefore, the allowable fatigue stress on the gusset plate is 16 ksi.

Since member U2L2 experiences forces from floor system and truss action, the connection is designed for the strength of the member. For the connection for strength of U2L2, use H-section with 12 × 0.5 flange plates and 11 × 0.5 web plate:

$L_{U2L2} = 300$ in
$P_{U2L2} = 0$
$A = 17.5$ in^2

$I_x = 2(0.5)(12)^3/12 + 11(0.5)^3/12 = 144.1\ in^4$

$r_x = \sqrt{\frac{144.1}{17.5}} = 2.9\ in$

$I_y = 2(12)(0.5)^3/12 + 2(12)(0.5)(5.75)^2 + 0.5(11)^3/12 = 452.5\ in^4$

$r_y = \sqrt{\frac{452.5}{17.5}} = 5.1 in$

$KL/r_{min} = 0.75(300)/2.9 = 78.4$

$F_{call} = 30 - (165.7(78.4))/1000 = 17.0$ ksi

Capacity of member U2L2 = 17.0(17.5) = 297.5 kips

An analysis of the live load location for the maximum forces in U1L2 and L2U3 and the concurrent forces in members L2U3 and U1L2, respectively, indicates the forces in Table E12.2b.

TABLE E12.2b
Maximum and concurrent forces in members U1L2 and L2U3.

Warren Through-Truss Member		Dead Load Force, DL (kips)	Maximum Live Load Force, LL_{max} (kips)	Maximum Live Load Plus Impact Force, LL_{max} + I (kips)	Dead Plus Maximum Dynamic Live Load, DL + LL_{max} + I (kips)	Tension or Reversal Force Range, ΔF (kips)
U1L2	Maximum diagonal positive shear	79.6	303.4	367.1	446.7	344.8
	Concurrent force in U2L3	−26.8	−99.37	−120.24	−147.0	-
L2U3	Diagonal positive shear	−26.8	−178.9	−216.5	−243.3	-
	Concurrent force in U1L2	79.6	203.95	246.78	326.4	231.8

Maximum difference in forces in members L1L2 and L2L3 based on maximum force in member U1L2 = (446.7 + 147.0)cos 45 = 419.8 kips

Maximum difference in forces in members L1L2 and L2L3 based on maximum force in member L2U3 = (243.3 + 326.4)cos 45 = 402.8 kips

Therefore, maximum difference in continuous chord member forces at L2 = 419.8 kips (Table E12.2c).

TABLE E12.2c
Minimum Bolts for Member Connections to L2 Gusset Plate

Warren Through Truss Member		Minimum Bolts for DL + LL_{max} + I with f'_{bv} = 21.5 ksi	Minimum Bolts in Each Gusset Plate Connection
L1L2 L2L3	Bottom chord	(446.7 + 147.0)cos45°/(2(12.9)) = 209.9/12.9 = 16.3	Use minimum of 8 bolts each side of working point
U2L2	Vertical	11.5	6 rows of 2 bolts each side
U1L2	Diagonal	17.3	5 rows of 4 bolts each side of connection
L2U3	Diagonal	9.4	5 rows of 2 bolts each side of connection

Allowable bolt-bearing stress = 1.2(70) = 84 ksi or = 1.09(70)/(2(0.875)) = 43.75 ksi (with edge spacing of 1.25(0.875) = 1.09 in)

Bolt-bearing stress at bottom chord member on gusset plate = (419.8/2)/(20(1)(0.5)) = 21.0 ksi OK

Bolt-bearing stress at diagonal member U1L2 on gusset plate = 223.4/(20(1)(0.5))= 22.3 ksi OK

Bolt-bearing stress at vertical member = 0

Bolt-bearing stress at diagonal member L2U3 on gusset plate = 121.7/(10(1)(0.5))= 24.3 ksi OK

a) Gusset plate tension strength controlled by allowable yielding and fracture stresses
   At connection to member U1L2 (Figure E12.2a),
   T = 446.7 kips
   5 rows of 4 bolts in each connection

   l_c = 4(3.0) = 12.0 in
   s_r = 4(3) = 12 in
   $w_e = 2l_c\tan(30°) + s_r = 1.15l_c + s_r$ = 1.1512 + 34 = 25.86 in (see Figure 12.6 and Figure E12.2d)
   w_{ne} = 25.86 − (4)(1) = 21.86 in
   w_{ne}/w_e = 21.86/25.86 = 0.85

$$\sigma_{aty} = \frac{(446.7/2)}{(25.86)(0.5)} = 17.3\,\text{ksi} \le 0.55(50) \le 27.5\,\text{ksi, OK}$$

   Assume U_c = 0.90 (see Chapter 10).

$$\sigma_{atU} = \frac{\left(\frac{446.7}{2}\right)}{0.90(21.86)(0.5)} = 22.7\,\text{ksi} \le 0.47(70) \le 32.9\,\text{ksi, OK}$$

   At connection to member L2U3 (Figure E12.2b),
   T = 74.6 kips (very small tension force)
   5 rows of 2 bolts each side of connection
   l_c = 4(3) = 12 in
   s_r = 2(3) = 6 in
   $w_e = 2l_c\tan(30°) + s_r = 1.15l_c + s_r$ = 1.1512 + 32 = 19.80 in (see Figure 12.6 and Figure E12.2d)
   w_{ne} = 19.80 − (2)(1) = 17.80 in
   w_{ne}/w_e = 17.80/19.80 = 0.90

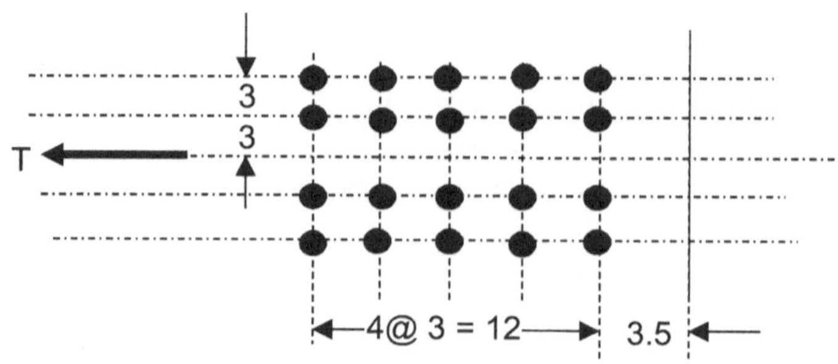

FIGURE E12.2a Connection to gusset plate at U1L2.

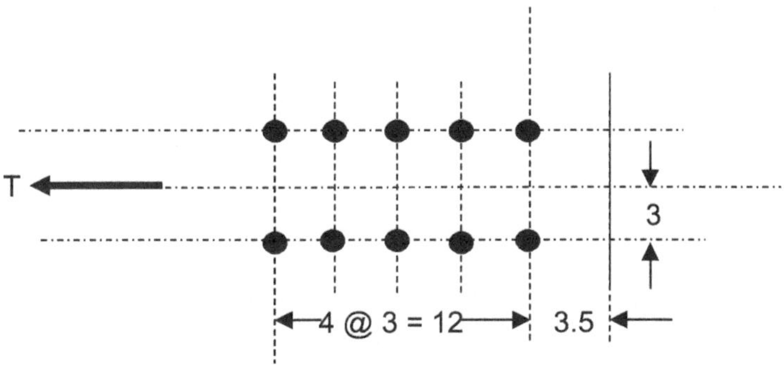

FIGURE E12.2b Connection to gusset plate at L2U3 in tension.

maximum $w_{ne} = 0.85(19.80) = 16.83$ in

$$\sigma_{aty} = \frac{\left(\frac{74.6}{2}\right)}{(19.80)(0.5)} = 3.8\,\text{ksi} \le 0.55(50) \le 27.5\,\text{ksi, OK}$$

Assume $U_c = 0.90$ (see Chapter 10):

$$\sigma_{atU} = \frac{\left(\frac{74.6}{2}\right)}{0.90(16.83)(0.5)} = 4.9\,\text{ksi} \le 0.47(70) \le 32.9\,\text{ksi, OK}$$

At connection to member L1L2 and L2L3,

$T = 419.8$ kips

Minimum of 8 bolts each side of working point:

$w_e = 12.0$ in

$A_g = 12(0.5) = 6.0\ \text{in}^2$

$w_{ne} = 12.0 - 2(1.0) = 10.0$ in

$w_{ne}/w_e = 10.0/12.0 = 0.83$

$A_n = (10.0)(0.5) = 5.0\ \text{in}^2$

$$\sigma_{aty} = \frac{\left(\frac{419.8}{2}\right)}{(12)(0.5)} = 35.0\,\text{ksi} > 0.55(50) > 27.5\,\text{ksi}$$, NG, increase gusset plate thickness to minimum of 5/8 in

Assume $U_c = 0.90$ (bottom chord member forces transferred from diagonal members)

$$\sigma_{atU} = \frac{\left(\frac{419.8}{2}\right)}{(0.90)\,5.0} = 46.6\,\text{ksi} > 0.47(70) > 32.9\,\text{ksi}$$, NG, increase gusset plate thickness to minimum of 3/4 in.

b) Gusset plate tension strength controlled by allowable block shear

At connection to member U1L2 with ½ in thick gusset plate,

$T = 446.7$ kips

5 rows of 4 bolts in each connection

$A_{gv} = 2(3.5+12.0)(0.5) = 15.5\ \text{in}^2$ (see Figure 12.7 and Figure E12.2d)

$A_{nv} = 2(15.5 - 5(1))(0.5) = 10.5\ \text{in}^2$

$A_{gt} = 12(0.5) = 6.0\ \text{in}^2$ (see Figure 12.7 and Figure E12.2d)

$A_{nt} = 6.0 - 4(1)(0.5) = 4.0\ \text{in}^2$

$U_{bs} = 1.0$

Allowable block shear strength based on tension fracture strength combined with shear fracture strength = $0.50U_{bs}F_uA_{nt} + 0.30F_uA_{nv}$ = 0.50(1.0)(70)(4.0) + 0.30(70)(10.5) = 140.0 + 220.5 = 360.5 kips

Allowable block shear strength based on tension fracture strength combined with shear yield strength = $0.50U_{bs}F_uA_{nt} + 0.30F_yA_{gv}$ = 0.50(1.0)(70)(10.5) + 0.30(50)(15.5) = 140.0 + 232.5 = 372.5 kips

Allowable block shear strength = 360.5 kips ≥ 223.4 kips, OK.

At connection to member L2U3,

T = 74.6 kips (very small tension force and block shear will not govern)

c) Gusset plate compression strength controlled by allowable compression stresses

At connection to member L2U3 with ½ in thick gusset plate (Figure E12.2c):

C = 243.4 kips

5 rows of 2 bolts each side of connection

l_w = 3.5 + 4.5 + 1.5/sin 45° = 10.12 in (see Figure 12.8 and Figure E12.2d)

$$r_w = \frac{t_p}{\sqrt{12}} = \frac{0.5}{\sqrt{12}} = 0.144\,\text{inches}$$

K = 0.5 (lateral sway restrained)

Kl_w/r_w = 0.5(10.12)/0.144 = 35.14

$$F_{call} = 0.60(50) - \left(175\left(\frac{50}{29000}\right)\right)^{\frac{3}{2}}(52.33) = 30 - 0.166(35.14) = 24.2\,\text{ksi}$$

$$\sigma_{ac} = \frac{C}{w_e t_p} = \frac{\left(\frac{243.3}{2}\right)}{19.80(0.50)} = 12.3\,\text{ksi} \le F_{call} = 23.4\,\text{ksi, OK}$$

At connection to U2L2 with ½ in thick gusset plate:

C_c = 297.5 kips

C = 0

No. bolts = 297.5/(2(12.9)) = 11.5 bolts each side (6 rows of 2 bolts each side)

l_w = 7.75 in (see Figure 12.8 and Figure E12.2d)

$$r_w = \frac{t_p}{\sqrt{12}} = \frac{0.5}{\sqrt{12}} = 0.144\,\text{inches}$$

K = 0.5 (lateral sway restrained)

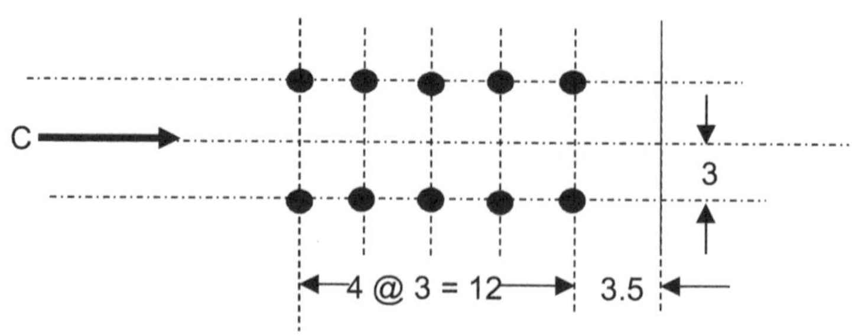

FIGURE E12.2c Connection to gusset plate at L2U3 in compression.

$$0.629\sqrt{\frac{E}{F_y}} = 0.629\sqrt{\frac{29{,}000}{50}} = 15.15 \quad \le \ Kl_w/r_w \ = \ 0.5(7.75)/0.144 \ = \ 26.91 \ \le$$

$$5.034\sqrt{\frac{E}{F_y}} = 5.034\sqrt{\frac{29{,}000}{50}} = 121.2$$

$$F_{call} = 0.60(50) - \left(175\left(\frac{50}{29000}\right)\right)^{\frac{3}{2}}(26.91) = 30 - 0.166(26.91) = 25.5\,\text{ksi}$$

$l_c = 5(3.0) = 15.0$ in
$s_r = 2(3) = 6.0$ in
$w_e = 2l_c\tan(30^\circ) + s_r = 1.15l_c + s_r = 1.15(15) + 3(2) = 23.25\,\text{inches}$ (see Figure 12.6 and Figure E12.2d)

$$\sigma_{ac} = \frac{C}{w_e t_p} = 0$$

d) Gusset plate compression strength controlled by allowable partial shear
   At connection to member L2U3 with ½ in thick gusset plate:
   - C = 243.3 kips
   - A_{gvv} = (1.5 + 5 + 15 + 5)(0.5) = 26.5(0.5) = 13.25 in² (see Figure 12.9 and Figure E12.2d)
   - A_{gvh} = 28.0(0.5) = 14.00 in² (see Figure 12.9 and Figure E12.2d)
   - Vertical partial shear plane with strength = $0.30F_yA_{gvv}/\sin\alpha$ = 0.30(50)(13.25)/sin 45° = 281.1 kips
   - Horizontal partial shear plane with strength = $0.30F_yA_{gvh}/\cos\alpha$ = 0.30(50)(14.00)/cos 45° = 297.0 kips
   - Allowable partial shear strength = 281.1 kips ≥ 121.7 kips, OK

e) Gusset plate shear strength controlled by allowable full shear A vertical full shear plane is not admissible through the bottom chord member.
   - With ½ in thick gusset plate,
   - A_{gvh} = 61.50(0.5) = 30.75 in² (see Figure 12.10 and Figure E12.2d)
   - A_{nvh} = (61.50-10(1))0.5 = 25.75 in²
   - Full horizontal shear plane strength on gross area = $0.30F_yA_{gvh}$ = 0.30(50)(30.75) = 461.3 kips
   - Full horizontal shear plane strength on net area = $0.30F_uA_{nvh}$ = 0.30(65)(25.75) = 502.1 kips
   - Horizontal shear force = 419.8/(2) = 209.9 kips ≤ 461.3 kips, OK

f) Gusset plate fatigue
   At connection to U1L2 with ½ in thick gusset plate:
   - S_R = (344.8/2)/(25.86(0.5)) = 13.3 ksi ≤ 16 ksi, OK

   At connection to L2U3 with ½ in thick gusset plate:
   - S_R = (128.2/2)/(19.80(0.5)) = 6.5 ksi ≤ 16 ksi, OK

   At connection of continuous bottom chord (L1L2 and L2L3):
   - S_R = (419.8/2)(344.8/446.7)/(12(0.5)) = 27.0 ksi, > 18 ksi, NG, increase gusset plate thickness to 3/4 in
   - Use ¾ in thick gusset plate with minimum of 8 bolts each side of working point in chord members.
   - Spacing along line of force not less than $2d_b\sigma_{bc}/F_u + d_b/2$ = 2(0.875)(24.3)/70 + 0.875/2 = 0.61 + 0.44 = 1.05 in

Spacing of bolts not less than $3d_b = 2.63$ in
Use minimum spacing of 3.0 in
Use maximum spacing of 6.0 in for sealing connection.
Edge distance not less than $1.5d_b = 1.31$ in
Edge distance not greater than $1.5 + 4t_g = 1.5 + 4(0.5) = 1.5 + 2 = 3.5$ in or 6 in
Loaded edge distance not less than $2d_b\sigma_{bc}/F_u = 0.61$ in
Use minimum edge distance of 1.5 in
Use maximum edge distance of 3.5 in

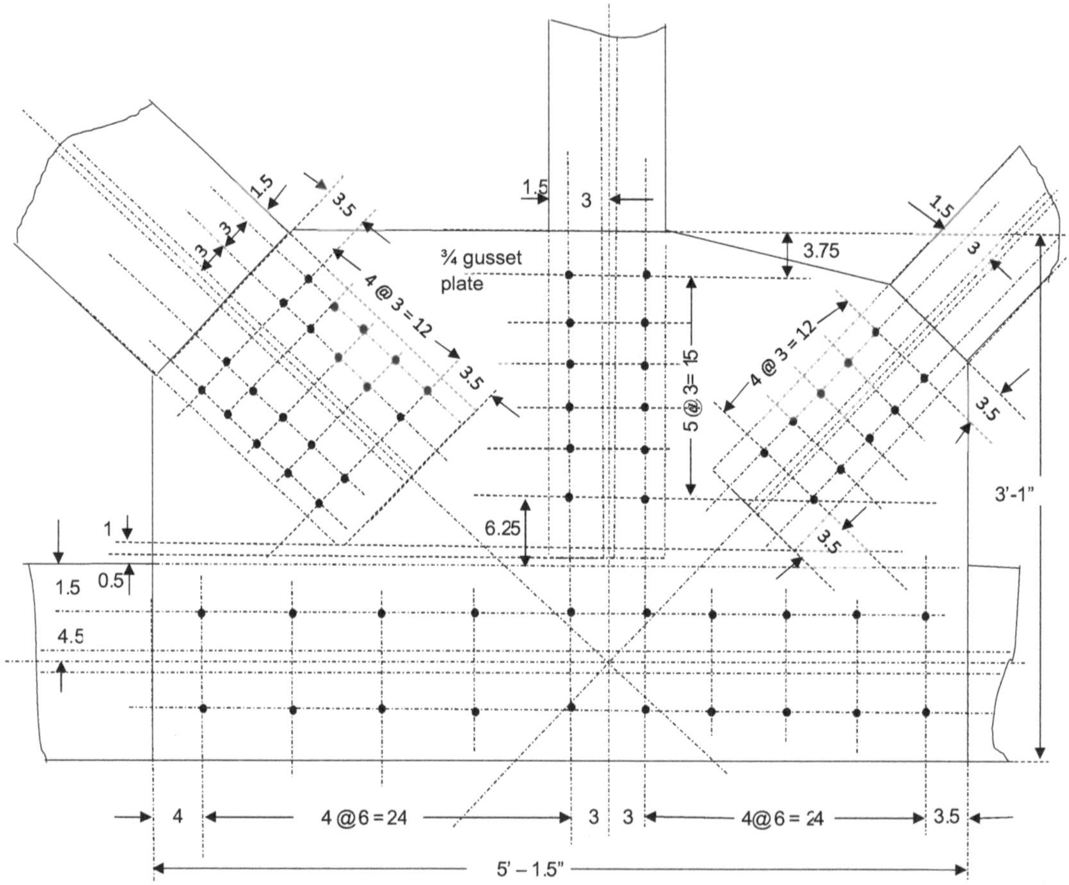

FIGURE E12.2d Gusset plate at L2 of Warren truss.

12.4.5.1 Members Spliced within a Panel

When chord members are continuous through truss panel points, as in Examples 12.1 and 12.2, the chord member may require splicing within a truss panel. Splices in chord members must be designed for a minimum strength equal to the full capacity of the member being spliced (AREMA, 2024). Bolts in member splices must be designed as slip-critical connections considering allowable effective shear and, in case of joint failure by slippage, bearing stresses.

A common axial member bolted splice involves the use of lap joints on the member elements. Splices located in the center of truss members must also be sufficiently rigid to resist bending from self-weight and other lateral forces. Therefore, splices in members within panels should be as close as practical to the gusset plate and in the member with the lower axial forces. Generally, all elements of members are spliced on each side of the element to avoid eccentricities[25] and shear lag effects.

12.4.5.1.1 Tension Member Splices

Rolled or built-up tension members are spliced by bolted plates and therefore designed as net area tension members. The splice components[26] of tension members must be designed for the yield capacity of the tension member considering the allowable fracture and block shear stresses in the splice components.

12.4.5.1.2 Compression Member Splices

The splice components of compression members must be designed for the capacity of the compression member. Compression member splices faced for bearing can be designed for 50% of the capacity of the member being spliced (AREMA, 2024).

12.4.6 Gusset Plates at Chord Members Spliced at Panel Points

When chord members are spliced at a panel point, the gusset plates must be designed as components of the splice to transfer the capacity of the chord members.[27] Chord members in modern railway truss spans are typically H-sections, as shown in Example 12.3 and Example 12.4, or box sections, as shown in Example 12.5. Typical arrangements of H-section, I-section and box-section bottom chord members[28] spliced at panel points are shown in Figure 12.11.

H-section[29] member splices consist of splice plates on the horizontal component (web plate) of the member and, in conjunction with the gusset plates, splice plates on the two vertical components (side or flange plates) of the member. I-section[30] member splices consist of splice plates on the horizontal components (flange plates) of the member and, in conjunction with the gusset plates, splice plates on vertical component (web plates) of the member. Box-section[31] member splices consist of splice plates on the two horizontal components (cover plates) of the member and, in conjunction with the gusset plates, splice plates on the two vertical components (side plates) of the member.

The splice design should endeavor to balance the forces in the gusset plates and other splice material at the splice location. Accordingly, the force in each horizontal splice of a chord member[32] is determined in proportion to the gross cross-sectional area of the horizontal component of the member to the gross cross-sectional area of the member. Similarly, the force in each vertical splice of a chord member[33] is determined in proportion to the gross cross-sectional area of the vertical component of the member to the gross cross-sectional area of thc member. However, as shown in Figures 12.11a and 12.11c, the vertical splices include the gusset plate as a component of the splice and the forces distributed to each component of the vertical splice must be determined by balancing the splice such that the resultant moment from forces in the splice components each side of the vertical plate being spliced is zero. The forces in diagonal members connected to gusset plates that are additive[34] to the splice forces must also be considered when balancing the splice.[35] If there are multiple splice plates each side of the splice, including the gusset plate on the outside of the vertical splice, the forces in each splice component may be determined in proportion to the thickness of each splice component to the thickness of the multiple splice plates each side of the splice.

The forces in the various splice components and gusset plates will determine the number of fasteners (based on connection shear and bearing), along with the yield, fracture, fatigue, and block shear rupture stresses in the splice components. The gusset plate is also subject to flexural stresses due to the eccentric force in the chord members, which must be considered when determining the yield, fracture and fatigue stresses in the gusset plate as a splice component.

The eccentric application of the chord force to the gusset plate may be considered by evaluation of the axial stress in the gusset plate overlap[36] due to the chord member force and the normal stress in the gusset plate due to the eccentric chord member force as shown in Figures 12.12 and 12.13, respectively. Accordingly, an effectiveness factor, E_f, that indicates the effectiveness of the gusset plate as a splice component was developed (Kulicki and Reiner, 2011)[37] to account for the eccentric chord member force in the gusset plate. E_f is the ratio of the uniform axial stress on the gusset plate overlap,

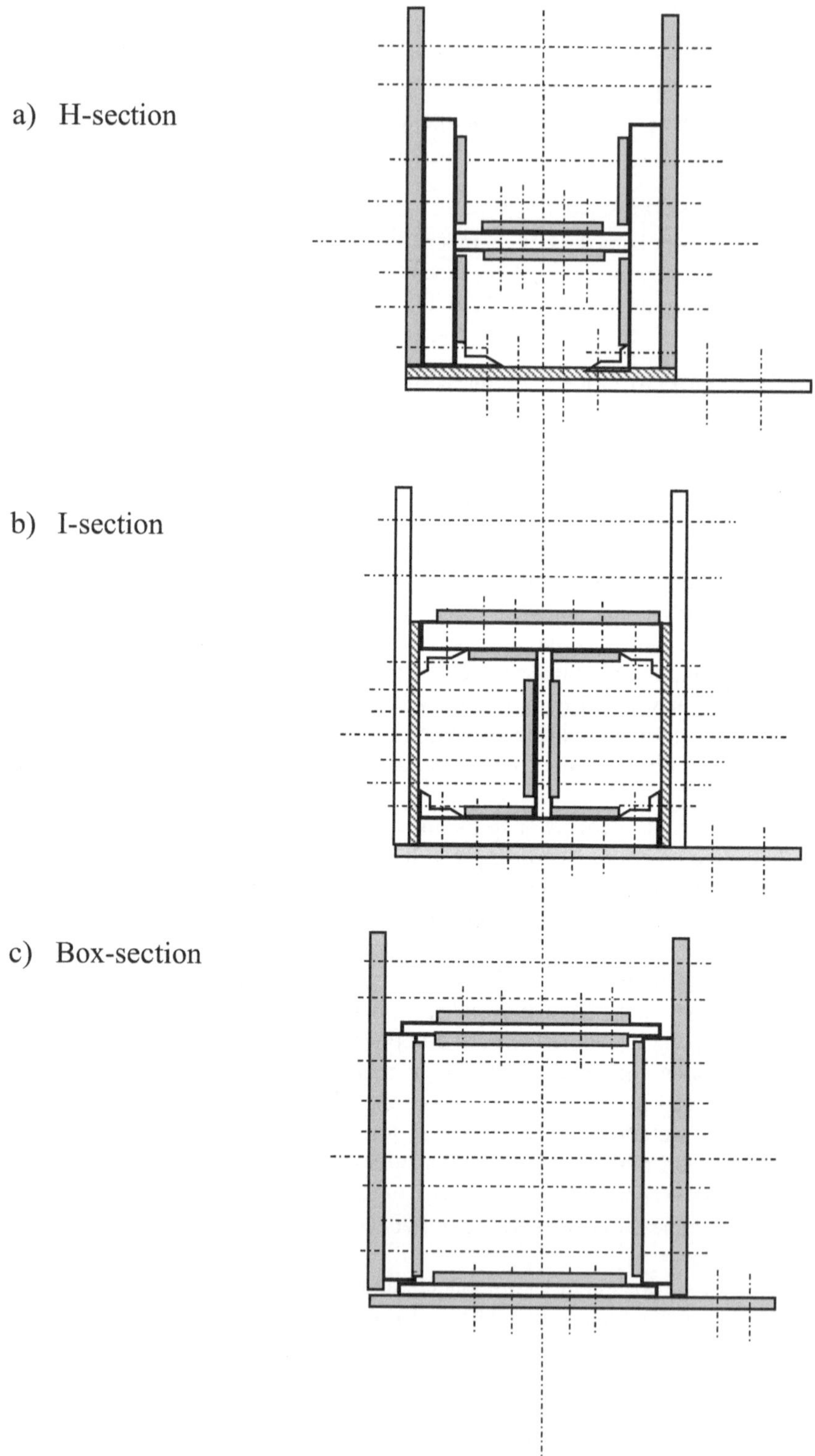

FIGURE 12.11 Typical arrangements of sections at bottom chord member splices.

h_{wg}, due to the chord member force to the maximum normal stress on the gusset plate full section area, $h_{gg}t_p$, due to the eccentric chord load. Since the maximum normal stress on the full gusset plate height, h_{gg}, due to the eccentric chord load is the sum of the axial normal stress and the flexural normal stress due to the eccentricity of the chord member relative to the centroid of the gusset plate, E_f is

$$E_f = \frac{\frac{F}{A_{wg}}}{\frac{F}{A_{gg}} + \frac{Fe_c}{S_g}} = \frac{\frac{1}{A_{wg}}}{\frac{1}{A_{gg}} + \frac{e_c}{S_g}} \tag{12.39}$$

where

F = the tension or compression force in the splice
A_{wg} = the gross cross sectional area of the overlapping portion of the gusset plate = $h_{wg}t_p$
h_{wg} = the overlapping portion of the gusset plate = the depth of the chord member
A_{gg} = the gross cross sectional area of the gusset plate = $h_{gg}t_p$
h_{gg} = the height of gusset plate at splice location
e_c = the eccentricity of the force in the chord member = $(h_{gg} - h_{wg})/2$
S_g = section modulus of the entire gusset plate at the splice = $h_{gg}^2(t_p)/6$

For truss chord members where load transfer between the chord segments is efficient, AREMA (2024) indicates that shear lag is not of concern and U_c can be effectively taken as 1.00 for tension members. However, in some circumstances, $U_c = 0.90$ can be appropriate for design (Bowles, 1980).

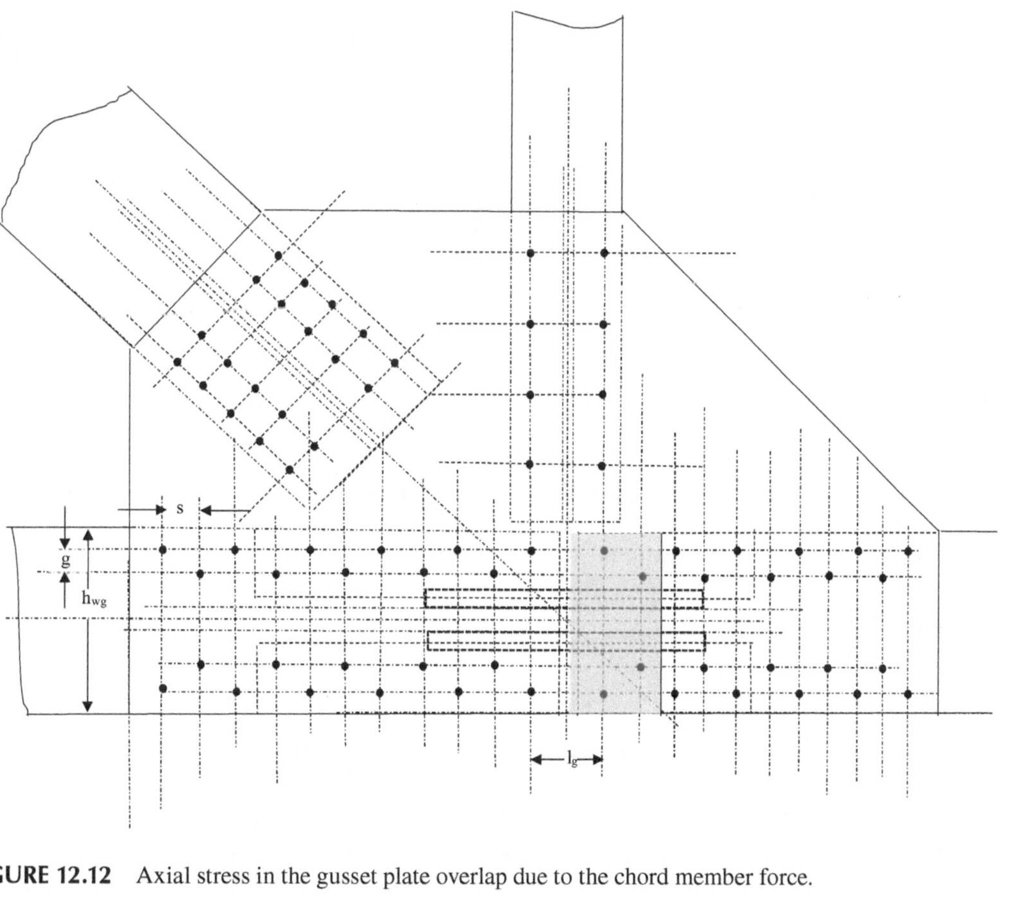

FIGURE 12.12 Axial stress in the gusset plate overlap due to the chord member force.

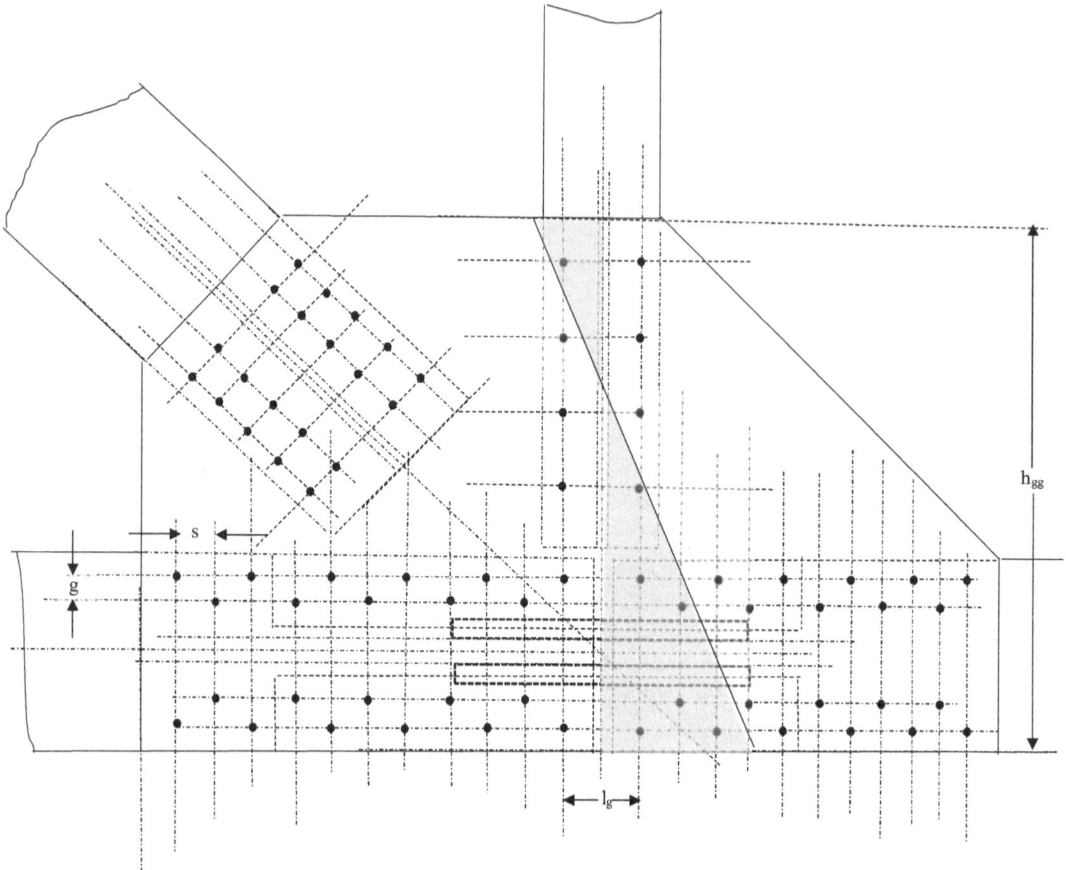

FIGURE 12.13 Axial and flexural stress in the gusset plate due to the eccentric chord member force.

Example 12.3

Design the gusset plate for joint L2 of the Warren truss with the dead load and maximum live load forces shown in Table E12.3a. Members L1L2, L2L3, U1L2, U2L2 and L2U3 frame into the gusset plates at L2. Members U1L2 and L2U3 frame into the gusset plate at a 45° angle. The H-section bottom chord is not continuous through panel point L2 and the splice between members L1L2 and L2L3 consists of inside splice plates, the gusset plates and outside splice plates.

For member slip-critical connections, try gusset plate and splice plate with,

$t_p = 0.50$ in
$F_y = 50$ ksi
$F_u = 70$ ksi

7/8 in diameter ASTM F3125 Grade A325 bolts are used in standard holes with connection surfaces prepared for a mean slip coefficient, k_{sm}, of 0.50.

Bolt effective shear strength in SC connection = 21.5(0.60) = 12.9 kips.

Single shear connection between member and each gusset plate

TABLE E12.3
Dead Load and Maximum Live Load Forces in Members at Joint L2

Warren Through-Truss Member		Dead Load Force, DL (kips)	Maximum Live Load Force, LL_{max} (kips)	Maximum Live Load Plus Impact Force, LL_{max} + I (kips)	Dead Plus Maximum Dynamic Live Load, DL + LL_{max} + I (kips)	Tension or Reversal Force Range, ΔF (kips)
L1L2	Bottom chord	93.8	323.9	391.9	485.7	368.1
L2L3	Bottom chord	168.8	563.2	681.5	850.3	640.1
U2L2	Vertical positive shear	0	0	0	0	0
	Vertical negative shear	0	0	0	0	0
U1L2	Diagonal positive shear	79.6	303.4	367.1	446.7	344.8
	Diagonal negative shear	79.6	−23.7	−28.9	50.7	108.5
L2U3	Diagonal positive shear	−26.8	−178.9	−216.5	−243.3	-
	Diagonal negative shear	−26.8	83.8	101.4	74.6	128.2

Since the bottom chord is not continuous through joint L2, the gusset place is part of a splice that must be designed for the capacity of members L1L2 and L2L3 and the additive horizontal force components of the diagonal members U1L2 and L2U3. The forces in Table E12.3 are maximum and not concurrent forces. The influence line for the chord members at joint L2 is longer than 100 ft, and therefore, the allowable fatigue stress on the gusset plate is 18 ksi. The influence line for the web members at joint L2 is shorter than 100 ft, and therefore, the allowable fatigue stress on the gusset plate is 16 ksi.
Design gusset plate for force in connected members
Gusset plate at connection to U1L2 (Figure E12.3a):

T = 446.7 kips
ΔT = 344.8 kips

Connection fasteners:

No. bolts for strength = (446.7/2)/12.9 = 17.3, use 20 bolts each side of splice (4 rows of 5 bolts)
Allowable bolt bearing stress on gusset plate = 1.2(70) = 84 ksi or = 1.09(70)/2(.875) = 43.8 ksi (with edge spacing of $1.25d_b$ in).
Bolt bearing stress at diagonal member U1L2 on gusset plate = 223.4/(20(1)(0.5))= 22.3 ksi OK.

Yield, fracture and fatigue:

l_c = 4(3.0) = 12.0 in
s_r = 4(3) = 12 in

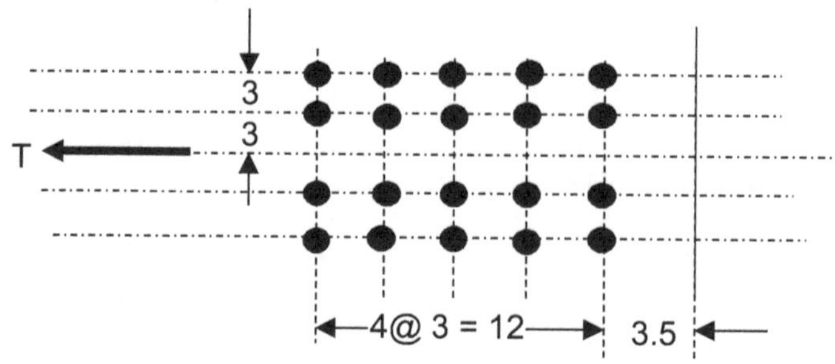

FIGURE E12.3a Connection to gusset plate at U1L2.

$w_e = 2l_c \tan(30°) + s_r = 1.15l_c + s_r = 1.1512 + 34 = 25.86$ in (see Figure 12.6 and Figure E12.3h)
$w_{ne} = 25.86 - (4)(1) = 21.86$ in
$w_{ne}/w_e = 21.86/25.86 = 0.85$

$$\sigma_{aty} = \frac{\left(\frac{446.7}{2}\right)}{(25.86)(0.5)} = 17.3\,\text{ksi} \le 0.55(50) \le 27.5\,\text{ksi, OK}$$

Assume $U_c = 0.90$

$$\sigma_{atU} = \frac{\left(\frac{447.6}{2}\right)}{0.90(21.86)(0.5)} = 22.7\,\text{ksi} \le 0.47(70) \le 32.9\,\text{ksi, OK}$$

$S_R = (344.8/2)/(25.86(0.5)) = 13.3$ ksi ≤ 16 ksi, OK

Block shear:

$A_{gv} = 2(3.5+12.0)(0.5) = 15.5$ in^2
$A_{nv} = 2(15.5 - 5(1))(0.5) = 10.5$ in^2
$A_{gt} = 12(0.5) = 6.0$ in^2
$A_{nt} = 6.0 - 4(1)(0.5) = 4.0$ in^2
$U_{bs} = 1.0$
Allowable block shear strength based on tension fracture strength combined with shear fracture strength = $0.50U_{bs}F_uA_{nt} + 0.30F_uA_{nv} = 0.50(1.0)(70)(4.0) + 0.30(70)(10.5) = 140.0 + 220.5 = 360.5$ kips
Allowable block shear strength based on tension fracture strength combined with shear yield strength = $0.50U_{bs}F_uA_{nt} + 0.30F_yA_{gv} = 0.50(1.0)(70)(10.5) + 0.30(50)(15.5) = 140.0 + 232.5 = 372.5$ kips
Allowable block shear strength = 360.5 kips ≥ 446.7/2 ≥ 223.4 kips, OK.

Gusset plate at connection to U2L2:

U2L2 experiences forces from floorbeam flexure and truss behavior. Therefore, design connection bolts for strength of the member.
Use H-section with 12 x 0.5 flange plates and 11 X 0.5 web plate
$L_{U2L2} = 300$ in
$P_{U2L2} = 0$
$A = 17.5$ in^2

I_x = 2(0.5)(12)^3/12 + 11(0.5)^3/12 = 144.1 in^4

$$r_x = \sqrt{\frac{144.1}{17.5}} = 2.9\,\text{in}$$

I_y = 2(12)(0.5)^3/12 + 2(12)(0.5)(5.75)^2 + 0.5(11)^3/12 = 452.5 in^4

$$r_y = \sqrt{\frac{452.5}{17.5}} = 5.1\,\text{in}$$

KL/r_{min} = 0.75(300)/2.9 = 78.4

F_{call} = 30– (165.7(78.4))/1000 = 17.0 ksi

Capacity of member U2L2 = 17.0(17.5) = 297.5 kips (design force)

Connection fasteners

No. bolts = 297.5/(2(12.9)) = 11.5 bolts each side (6 rows of 2 bolts each side of connection)

Compression strength:

l_w = 8.0 in (see Figure 12.8 and Figure E12.3h)

$$r_w = \frac{t_p}{\sqrt{12}} = \frac{0.5}{\sqrt{12}} = 0.144\,\text{inches}$$

K = 0.5 (lateral sway restrained)

$$0.629\sqrt{\frac{E}{F_y}} = 0.629\sqrt{\frac{29{,}000}{50}} = 15.15 \le Kl_w/r_w = 0.5(8.0)/0.144 = 27.78 \le$$

$$5.034\sqrt{\frac{E}{F_y}} = 5.034\sqrt{\frac{29{,}000}{50}} = 121.2$$

$$F_{call} = 0.60(50) - \left(175\left(\frac{50}{29000}\right)\right)^{\frac{3}{2}}(27.78) = 30 - 0.166(27.78) = 25.4\,\text{ksi}$$

l_c = 5(3.0) = 15.0 in

s_r = 2(3) = 6 in

w_e = 2l_c tan(30°) + s_r = 1.15l_c + s_r = 1.1515 + 32 = 23.25 in (see Figure 12.8 and Figure E12.3h)

$\sigma_{ac} = 0$

Gusset plate at connection to L2U3 (Figure E12.3b):

C = −243.3 kips

T = 74.6 kips

ΔT = 128.2 kips (force reversal)

Connection fasteners

No. bolts strength = (243.3/2)/12.9 = 9.4 (5 rows of 2 bolts)

Bolt bearing stress at diagonal member L2U3 on gusset plate = 121.7/(10(1)(.5))= 24.3 ksi OK.

Yield, fracture and fatigue:

l_c = 4(3) = 12 in

s_r = 2(3) = 6 in

w_e = 2l_c tan(30°) + s_r = 1.15l_c + s_r = 1.1512 + 32 = 19.80 in (see Figure 12.6 and Figure E12.3h)

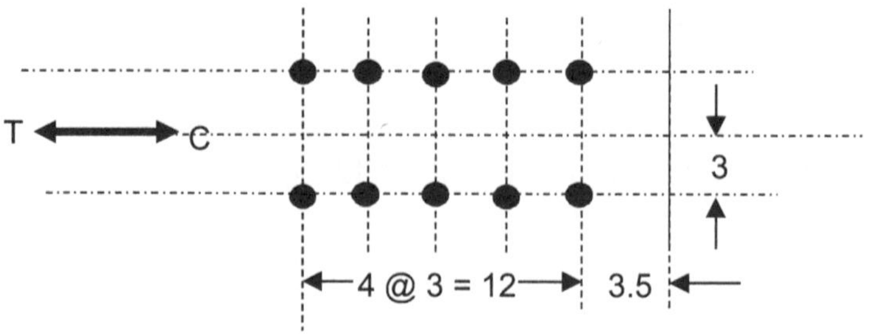

FIGURE E12.3b Connection to gusset plate at L2U3.

$w_{ne} = 19.80 - (2)(1) = 17.80$ in
$w_{ne}/w_e = 17.80/19.80 = 0.90$
maximum $w_{ne} = 0.85(19.80) = 16.83$ in

$$\sigma_{aty} = \frac{\left(\frac{74.6}{2}\right)}{(19.80)(0.5)} = 3.8\,\text{ksi} \le 0.55(50) \le 27.5\,\text{ksi, OK}$$

$$\sigma_{atU} = \frac{\left(\frac{74.6}{2}\right)}{0.90(16.83)(0.5)} = 4.9\,\text{ksi} \le 0.47(70) \le 32.9\,\text{ksi, OK}$$

$S_R = (128.2/2)/(19.80(0.5) = 6.5$ ksi ≤ 16 ksi, OK

Block shear:

$A_{gv} = 2(3.5 + 12.0)(0.5) = 15.5$ in^2 (see Figure 12.7 and Figure E12.3h)
$A_{nv} = 2(15.5 - 5(1))(0.5) = 10.5$ in^2
$A_{gt} = 6.0(0.5) = 3.0$ in^2
$A_{nt} = 3.0 - 2(1)(0.5) = 2.0$ in^2
$U_{bs} = 1.0$
Allowable block shear strength based on tension fracture strength combined with shear fracture strength = $0.50U_{bs}F_uA_{nt} + 0.30F_uA_{nv} = 0.50(1.0)(70)(2.0) + 0.30(70)(10.5) = 70.0 + 220.5 = 290.5$ kips
Allowable block shear strength based on tension fracture strength combined with shear yield strength = $0.50U_{bs}F_uA_{nt} + 0.30F_yA_{gv} = 0.50(1.0)(70)(2.0) + 0.30(50)(15.5) = 70.0 + 232.5 = 302.5$ kips
Allowable block shear strength = 290.5 kips ≥ 37.3 kips, OK

Compression strength
$l_w = 3.5 + 4.5 + 1.375/\sin 45° = 9.94$ in (see Figure 12.8 and Figure E12.3h)

$$r_w = \frac{t_p}{\sqrt{12}} = \frac{0.5}{\sqrt{12}} = 0.144\,\text{inches}$$

K = 0.5 (lateral sway restrained)
$Kl_w/r_w = 0.5(9.94)/0.144 = 34.53$

$$F_{call} = 0.60(50) - \left(175\left(\frac{50}{29000}\right)\right)^{\frac{3}{2}}(34.53) = 30 - 0.166(34.53) = 24.3\,\text{ksi}$$

$l_c = 4(3.0) = 12.0$ in

$s_r = 2(3) = 6$ in

$w_e = 2l_c \tan(30^\circ) + s_r = 1.15l_c + s_r = 1.15(12) + 3(2) = 19.80 \text{inches}$ (see Figure 12.8 and Figure E12.3h)

$$\sigma_{ac} = \frac{P}{w_e t_p} = \frac{\left(\frac{243.3}{2}\right)}{19.80(0.50)} = 12.3 \text{ksi} \leq F_{call} = 24.3 \text{ksi, OK}$$

Compression partial shear planes

$A_{gvv} = 26.5(0.5) = 13.25$ in^2 (see Figure 12.9 and Figure E12.2)

$A_{gvh} = 28.0\,(0.5) = 14.00$ in^2 (see Figure 12.9 and Figure E12.2)

Vertical partial shear plane with strength = $0.30F_yA_{gvv}/\sin\alpha = 0.30(50)(13.25)/\sin 45^\circ = 281.1$ kips

Horizontal partial shear plane with strength = $0.30F_yA_{gvh}/\cos\alpha = 0.30(50)(14.0)/\cos 45^\circ = 297.0$ kips

Allowable partial shear strength = 281.1 kips ≥ 121.7 kips, OK.

Splice analysis:

Design splice for least yield capacity of chord members L1L2 and L2L3

Yield capacity of member L1L2

T = 485.7 kips

ΔT = 368.1 kips

Vertical plates: 2- 0.625 in × 12 in

Horizontal plate: 0.375 x 10.75"

$A_{gm} = 2(0.625)(12) + (0.375)(10.75) = 19.03$ in^2

$\sigma_{ay} = 485.7/19.03 = 25.5$ ksi ≤ 0.55(50) ≤ 27.5 ksi, OK

Yield capacity of L1L2 = 27.5(19.03) = 523.4 kips

Yield capacity/T = 523.4/485.7 = 1.08

Yield capacity/ΔT = 523.4/368.1 = 1.42

Yield capacity of L2L3

T = 850.3 kips

ΔT = 640.1 kips

Vertical plates: 2- 1.25" x 12"

Horizontal plate: 0.375 x 9.5"

$A_{gm} = 2(1.25)(12) + (0.375)(9.5) = 33.56$ in^2

$\sigma_{ay} = 850.3/33.56 = 25.3$ ksi ≤ 0.55(50) ≤ 27.5 ksi, OK

Yield Capacity of L2L3 = 27.5(33.56) = 922.9 kips

Yield capacity/T = 922.9/850.3 = 1.09

Yield capacity/ΔT = 922.9/640.1 = 1.44

Horizontal splice: The top and bottom horizontal splice plates are bolted to the horizontal parts of members L1L2 and L2L3 (Figure E12.3c).

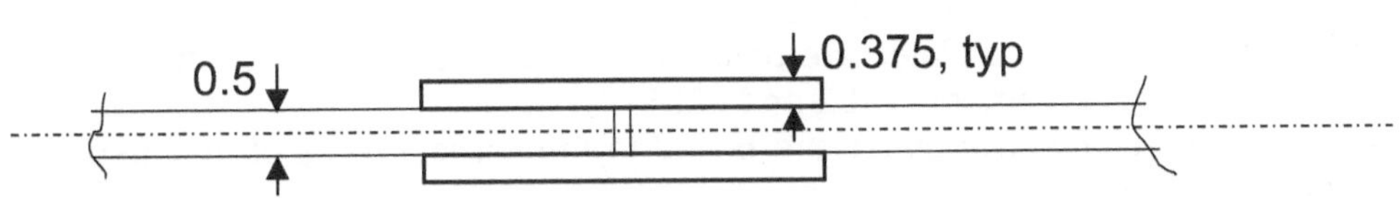

FIGURE E12.3c Elevation of horizontal splice at L2.

	Chord Member Gross Cross-Sectional Area (in²)	
Member Part	**L1L2**	**L2L3**
Vertical plates	2(12)(0.625) = 15.00	2(12)(1.25) = 30.00
Horizontal plate	0.50(10.75) = 5.38	0.50(9.5) = 4.75
Total	20.38	34.75

Since splices must be designed based on capacity of the members,

Force in each Horizontal Splice Plate (kips)	
L1L2	**L3L3**
(523.4/2)(5.38/20.38) = 69.04	(922.9/2)(4.75/34.75) = 63.08

Least yield capacity of each horizontal splice plate = 63.08 kips (L2L3 side of splice)
No. bolts strength = 63.08/12.9 = 5.0 (use 2 rows of 3 bolts each side of splice)
Bearing stress on splice = 63.08/(6(0.375)(1.0)) = 28.0 ksi
Length of each horizontal splice plate = 2(3.5 + 2(3.0) + 3.5) = 26 in
Width of each horizontal splice plate = 7.0 in
Thickness of each horizontal splice plate = 0.375 in
$A_{gs} = 7.0(0.375) = 2.63$ in²
$A_{ns} = 2.63 - 2(1.0)(0.375) = 1.88$ in²
$A_{ns}/A_{gs} = 1.88/2.63 = 0.71$
$A_{efat} = 7.0(0.375) = 2.63$ in²
$\sigma_{ay} = 63.08/2.63 = 24.0$ ksi $\leq 0.55(50) \leq 27.5$ ksi, OK
$\sigma_{aU} = 63.08/1.88 = 33.6$ ksi $> 0.47(70) > 32.9$ ksi, say OK 2% overstress
$S_{Rfat} = (640.1/922.9)63.08/2.63 = 16.6$ ksi ≤ 18, OK

Block shear:

$A_{gv} = 2(3.5+6.0)(0.375) = 7.13$ in²
$A_{nv} = 2(9.5 - 3(1))(0.375) = 4.88$ in²
$A_{gt} = 5.5(0.375) = 2.06$ in²
$A_{nt} = (5.5 - 2(1)(0.375) = 1.31$ in²
$U_{bs} = 1.0$
Allowable block shear strength based on tension fracture strength combined with shear fracture strength = $0.50U_{bs}F_uA_{nt} + 0.30F_uA_{nv}$ = 0.50(1.0)(70)(1.31) + 0.30(70)(4.88) = 45.9 + 102.4 = 148.2 kips
Allowable block shear strength based on tension fracture strength combined with shear yield strength = $0.50U_{bs}F_uA_{nt} + 0.30F_yA_{gv}$ = 0.50(1.0)(70)(1.31) + 0.30(50)(7.13) = 45.9 + 107.0 = 152.9 kips
Allowable block shear strength = 148.2 kips ≥ 63.08 kips, OK.

Vertical splices:

The inside vertical splice plates are bolted to the vertical parts on each side of the horizontal part of members L1L2 and L2L3.
The outside vertical splice plates are bolted to the vertical parts of members L1L2 and L2L3.
The gusset plate is a component of the outside vertical splice shown in Figure E12.3d.

Splice component	Chord member L1L2	Chord member L2L3
Chord depth (in)	12.0	12.0
Chord vertical plate thickness (in)	0.625	1.25
Gusset plate thickness (in)	0.50	0.50
Inside splice plate thickness	0.50	0.50
Outside splice plate thickness	0.50	0.50

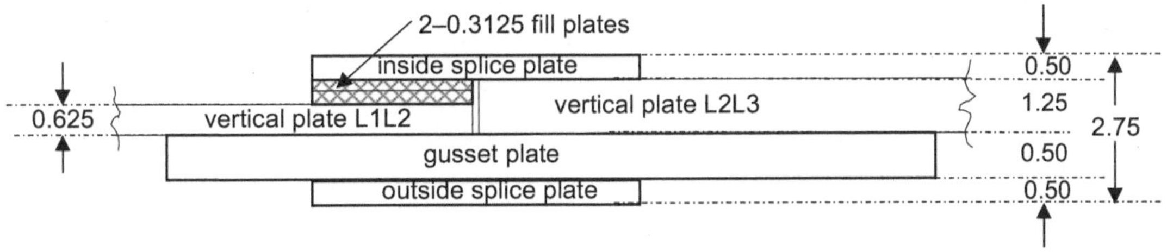

FIGURE E12.3d Plan of vertical splice at L2.

	Force in each Vertical Splice (kips) L1L2	Force in each Vertical Splice (kips) L2L3
Chord	(523.4/2)(15.00/20.375) = 192.7	(922.9/2)(30.00/34.75) = 398.4
Diagonal	(446.7/2) cos 45° = 157.9	(74.6/2) cos 45° = 26.4
Total	350.6	424.8

Least yield capacity of each vertical splice = 350.6 kips (L1L2 side of splice)
Determine forces in vertical splice components (inside splice, gusset and outside splice plates) by balancing the splice on the L1L2 side as shown in Figure E12.3e,

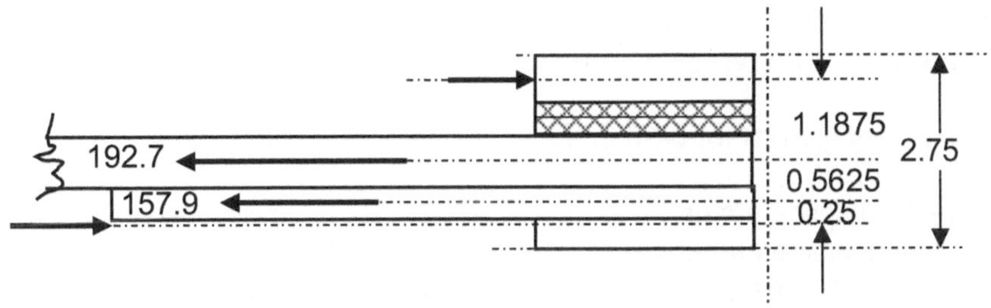

FIGURE E12.3e Plan of vertical splice forces from L1L2.

Splice component	Force (kips)	Distance to Center Of Gusset Plate and Outside Splice Plate (in)	Moment at Center Of Gusset Plate and Outside Splice Plate (kips-in)
Chord vertical plate	192.7	0.8125	156.6
Gusset plate diagonal force	157.9	0.25	39.48
Total	350.6	-	196.0

Splice Component	Force (kips)	Distance to Center Of Gusset Plate and Outside Splice Plate (in)	Moment at Center of Gusset Plate and Outside Splice Plate
Inside splice plate	−196.0/2.00 = −98.00	2.00	−196.0
Gusset plate and outside splice plate	-(350.6−98.0) = −252.6	0	0
Total	−350.6	-	−190.6

Splice Component	L1L2	L2L3
Gusset plate	−252.6(0.50/1.0) + 157.9 = 31.60	−252.6(0.50/1.0) = −126.3
Outside splice plate	−252.6(0.50/1.0) = −126.3	−126.3

Inside splice plate force = 98.00 kips
Outside splice plate force = 126.3 kips
L1L2 side of gusset plate force = 31.60 kips
L2L3 side of gusset plate force = 126.3 kips
L1L2 chord force = 98.00 + 126.3 − 31.60 = 192.7 kips
L1L2 + L2L3 chord force = 192.7 + 126.3 = 350.6 − 31.6 = 319.0 kips
L1L2 side of outside splice plate–gusset plate connection force = 126.3 − 31.60 = 94.70 kips
L2L3 side of outside splice plate–gusset plate connection force = 126.3 + 126.3 = 252.6 kips

Inside vertical splice plates:

No. bolts = 98.0/12.9 = 7.6 (8 bolts each side of vertical splice) (4 bolts in each inside splice plate on each side of horizontal splice)
Use two (12 − 1.25)/2 = 5.375 in inside splice plates either side of horizontal splice
$A_{gis} = 2(5.375)(0.50) = 5.38\ in^2$
$A_{nis} = 5.38 - 4(1.0)(0.50) = 3.38\ in^2$ (see Figure E12.3h)
$A_{nis}/A_{gis} = 3.38/5.38 = 0.63$
$A_{efat} = 5.38\ in^2$
$\sigma_{sy} = 98.0/5.38 = 18.2\ ksi \le 27.5\ ksi$, OK
$\sigma_{sU} = 98.0/3.38 = 29.0\ ksi \le 32.9\ ksi$, OK
$S_{Rfat} = (368.1/523.4)98.0/5.38 = 12.8\ ksi \le 18\ ksi$, OK

Block shear will not govern for 0.50 in thick inside vertical splice plate.

Outside vertical splice plates:
Outside vertical splice plates are 12 in in vertical direction
No. bolts = 126.3/12.9 = 9.8 (use 10 bolts each side of vertical splice)
$A_{gos} = 12(0.50) = 6.00\ in^2$
$A_{nos} = 6.00 - 4(1.0)(0.50) = 4.00\ in^2$ (see Figure E12.3h)
$A_{nos}/A_{gos} = 0.67$
$A_{efat} = 6.00\ in^2$
$\sigma_{sy} = 126.3/6.00 = 21.0\ ksi \le 27.5\ ksi$, OK
$\sigma_{sU} = 126.3/4.00 = 31.6\ ksi \le 0.47(70) \le 32.9\ ksi$, OK
$S_{Rfat} = (368.1/523.4)126.3/6.00 = 14.8\ ksi \le 18\ ksi$, OK

Gusset plate at connection to L1L2
Connection fasteners
No. bolts strength = (94.7)/12.9 = 7.3 (use 10 bolts, see Figure E12.3h)

Bolt bearing stress at chord member L1L2 on gusset plate = 94.7/(10(1)(0.5)) = 18.9 ksi OK (Figure E12.3f)

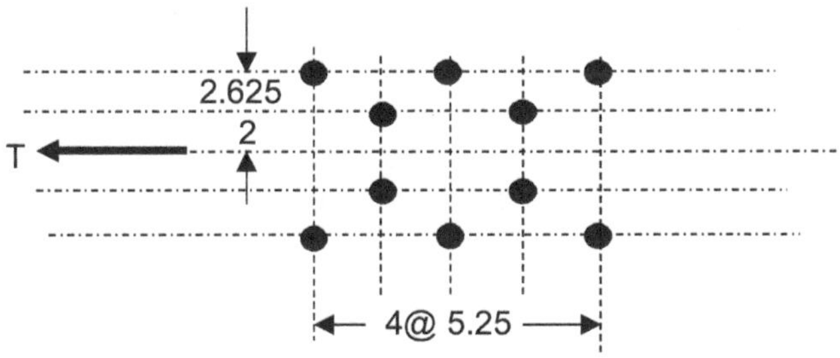

FIGURE E12.3f Connection to gusset plate at L1L2.

Yield, fracture and fatigue:

Calculate E_f to account for eccentric bottom chord force (see Figures 12.13 and 12.14 and Figure E12.3).

h_{gg} = 37.0 in
h_{wg} = 12.0 in
A_{gg} = 37.0(0.5) = 18.5 in^2
A_{wg} = 12.0(0.5) = 6.0 in^2
$S_g = h_{gg}^2 t_p/6 = (37.0)^2(0.5)/6 = 114.1$ in^3
$e_c = h_g/2 - h_{wg}/2 = 37.0/2 - 12.0/2 = 18.5 - 6.0 = 12.5$ in

$$E_f = \frac{\frac{1}{A_{wg}}}{\frac{1}{A_{gg}} + \frac{e_c}{S_g}} = \frac{\frac{1}{6.0}}{\frac{1}{18.5} + \frac{12.5}{114.1}} = \frac{0.167}{0.054 + 0.110} = 0.982$$

A_{ggp} = 12(0.5) = 6.0 in^2
A_{ngp} = 6.0 − 2(1.0)(0.5) = 5.0 in^2 (see Figure E12.3h)
A_{ngp}/A_{ggp} = 0.83
σ_{aty} = (31.6)/((6.0)(0.982)) = 5.3 ksi ≤ 27.5 ksi, OK
σ_{atU} = (31.6)/(5.0)(0.982)) = 6.4 ksi ≤ 32.95 ksi, OK
S_{Rfat} = (368.1/523.4)(31.6)/((12)(0.98)(0.50)) = 3.8 ksi ≤ 18 ksi, OK
Block shear will not govern for the relatively small connection force of 94.7 kips.

Gusset plate at connection to L2L3 (Figure E12.3g)
Connection fasteners
No. bolts strength = (252.6)/12.9 = 19.6 bolts (use 20 bolts, see Figure E12.3h)

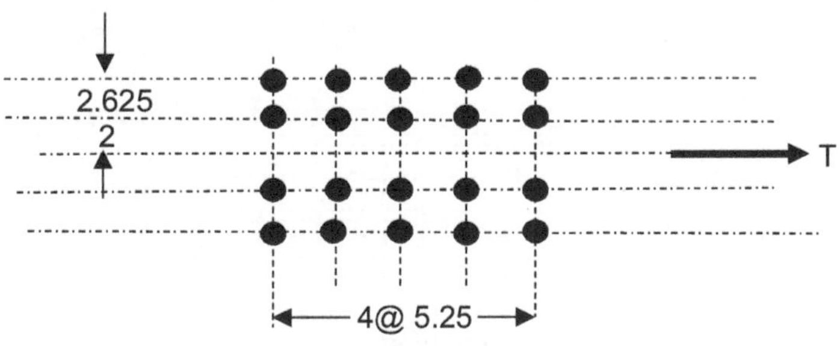

FIGURE E12.3g Connection to gusset plate at L2L3.

Yield, fracture and fatigue

$A_{ggp} = 6.0\ in^2$
$A_{ngp} = 4.0\ in^2$
$A_{ngp}/A_{ggp} = 0.67$
$\sigma_{gy} = (126.3)/(12(0.98)(0.50)) = 21.5\ ksi \le 27.5\ ksi$, OK
$\sigma_{gU} = (126.3)/(8.0(0.98)(0.50)) = 32.2\ ksi \le 32.9\ ksi$, OK
$S_{Rfat} = (640.1/922.9)(126.3)/(12(0.98)(0.50)) = 14.9\ ksi \le 18\ ksi$, OK
Block shear (see Figure E12.3h):
$A_{gv} = 2(5.75+21.0)(0.50) = 26.75\ in^2$
$A_{nv} = 2(26.75 - 5(1))(0.50) = 21.75\ in^2$
$A_{gt} = 2(2.625 + 2.0)(0.50) = 4.63\ in^2$
$A_{nt} = (2(2.625 + 2.0) - 4(1.0))0.50 = 2.63\ in^2$
$U_{bs} = 1.0$
Allowable block shear strength based on tension fracture strength combined with shear fracture strength = $0.50U_{bs}F_uA_{nt} + 0.30F_uA_{nv} = 0.50(1.0)(70)(2.63) + 0.30(70)(21.75) = 91.88 + 456.8 = 548.6$ kips
Allowable block shear strength based on tension fracture strength combined with shear yield strength = $0.50U_{bs}F_uA_{nt} + 0.30F_yA_{gv} = 0.50(1.0)(70)(2.63) + 0.30(50)(26.75) = 91.88 + 401.3 = 493.1$ kips
Allowable block shear strength = 493.1 kips ≥ 252.6 kips, OK.

Use 0.50 in thick gusset plate.

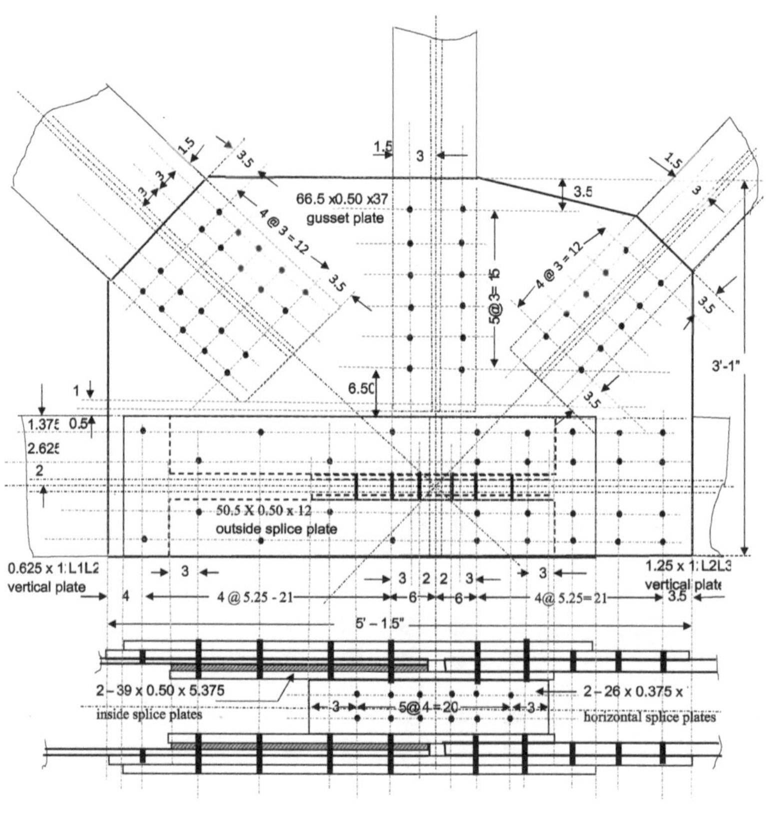

FIGURE E12.3h Gusset plate at L2 of Warren truss.

In many cases, outside splice plates, such as those in Example 12.3, are not used in lieu of thicker gusset plates as shown in Examples 12.4 and 12.5.

Example 12.4

Design the gusset plate at joint L2 of the Warren truss with the dead load and maximum live load forces shown in Table E12.3a. Members L1L2, L2L3, U1L2, U2L2 and L2U3 frame into the gusset plates at L2. Members U1L2 and L2U3 frame into the gusset plate at a 45° angle. The H-section bottom chord is not continuous through panel point L2 and the splice between members L1L2 and L2L3 consists of inside splice plates and the gusset plates. The gusset plate design for the connection of web members U1L2, U2L2 and L2U3 is shown in Example 12.3. The horizontal splice at panel point L2 between members L1L2 and L2L3 is also shown in Example 12.3. Design the vertical splices at panel point L2 consisting of inside splice plates and the gusset plates.

Vertical splice analysis
Design splice for least yield capacity of chord members L1L2 and L2L3
Yield capacity of L1L2
$T = 485.7$ kips
$\Delta T = 368.1$ kips
Vertical plates: 2- 0.625 in × 12 in
Horizontal plate: 0.375 × 10.75 in
$A_{gm} = 2(0.625)(12) + (0.375)(10.75) = 19.03\ in^2$
$\sigma_{ay} = 485.7/19.03 = 25.5\ ksi \le 0.55(50) \le 27.5\ ksi$, OK
Yield capacity of L1L2 = 27.5(19.03) = 523.4 kips
Yield capacity/T = 523.4/485.7 = 1.08
Yield capacity/ΔT = 523.4/368.1 = 1.42

Yield capacity of L2L3
$T = 850.3$ kips
Vertical plates: 2- 1.25 in × 12 in
Horizontal plate: 0.375 × 9.5 in
$A_{gm} = 2(1.25)(12) + (0.375)(9.5) = 33.56\ in^2$
$\sigma_{ay} = 850.3/33.56 = 25.3\ ksi \le 0.55(50) \le 27.5\ ksi$, OK
Yield Capacity of L2L3 = 27.5(33.56) = 922.9 kips
Yield capacity/T = 922.9/850.3 = 1.09
Yield capacity/ΔT = 922.9/640.1 = 1.44

The inside vertical splice plates are bolted to the vertical parts on each side of the horizontal part of members L1L2 and L2L3. The gusset plate is the outside vertical splice shown in Figure E12.4a.

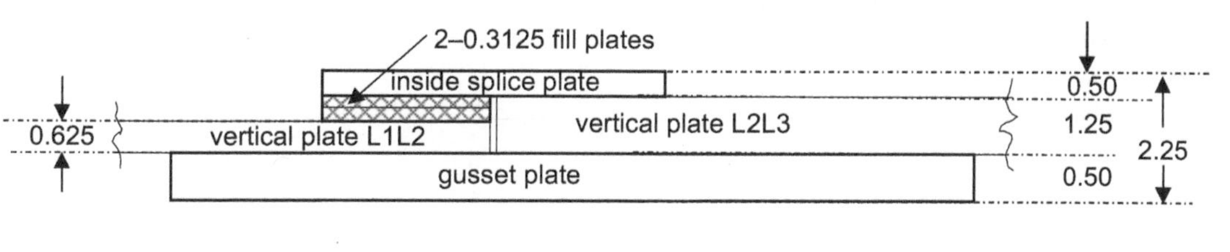

FIGURE E12.4a Plan of vertical splice at L2.

Splice component	Chord member L1L2	Chord member L2L3
Chord depth (in)	12.0	12.0
Chord vertical plate thickness (in)	0.625	1.25
Gusset plate thickness (in)	0.50	0.50
Inside splice plate thickness	0.50	0.50

	Force in Each Vertical Splice (kips) L1L2	Force in Each Vertical Splice (kips) L2L3
Chord	(523.4/2)(15.00/20.375) = 192.7	(922.9/2)(30.00/34.75) = 398.4
Diagonal	(446.7/2) cos 45° = 157.9	(74.6/2) cos 45° = 26.4
Total	350.6	424.8

For least yield capacity of each vertical splice = 350.6 kips (L1L2 side of splice), determine forces in vertical splice components (inside splice and gusset plate) by balancing the splice on the L1L2 side as shown in Figure E12.4b.

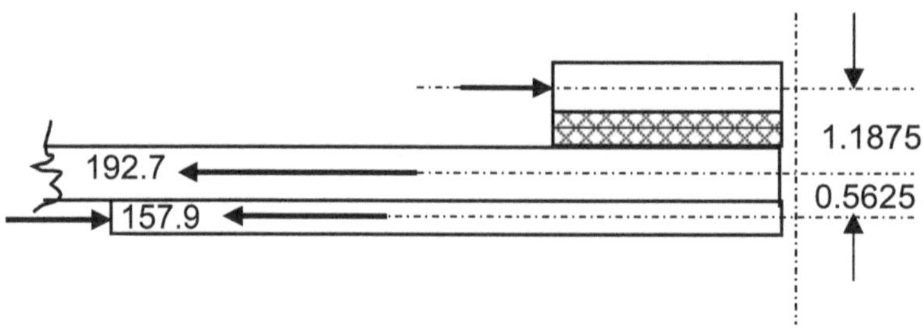

FIGURE E12.4b Plan of vertical splice forces from L1L2 with 0.50 in thick gusset plate.

Splice component	Force (kips)	Distance to center of gusset plate (in)	Moment at center of gusset plate and outside splice plate (kips-in)
Chord vertical plate	192.7	0.5625	108.4
Gusset plate diagonal force	157.9	0	0
Total	350.6	-	108.4

Splice component	Force (kips)	Distance to center of gusset plate (in)	Moment at center of gusset plate (kip-in)
Inside splice plate	−108.4/1.75 = −61.94	1.75	−108.4
Gusset plate	-(350.6−61.94) = −288.7	0	0
Total	−350.6	-	−108.4

Splice component	L1L2	L2L3
Gusset plate	−288.7 + 157.9 = −130.8	−288.7

Inside splice plate force = 61.94 kips
L1L2 side of gusset plate force = 130.8 kips
L2L3 side of gusset plate force = 288.7 kips

Inside vertical splice plates:

No. bolts = 61.94/12.9 = 4.8 (6 bolts each side of vertical splice) (3 bolts in each inside splice plate on each side of horizontal splice)
Use two (12-1.25)/2 = 5.375 in inside splice plates either side of horizontal splice
A_{gis} = 2(5.375)(0.50) = 5.38 in^2
A_{nis} = 5.38 − 4(1.0)(0.50) = 3.38 in^2
A_{nis}/A_{gis} = 3.38/5.38 = 0.63
A_{efat} = 5.38 in^2
σ_{sy} = 61.94/5.38 = 11.5 ksi ≤ 27.5 ksi, OK
σ_{sU} = 61.94/3.38 = 18.3 ksi ≤ 32.9 ksi, OK
S_{Rfat} = (368.1/523.4)61.94/5.38 = 8.1 ksi ≤ 18 ksi, OK
Block shear will not govern for 0.50 in thick inside vertical splice plate

Gusset plate at connection to L1L2
Connection fasteners
No. bolts strength = (130.8)/12.9 = 10.1 (use min. 10 bolts)
Bolt bearing stress at chord member L1L2 on gusset plate = 130.8/(10(1)(0.5)) = 26.2 ksi OK.
Yield, fracture and fatigue:
Calculate E_f to account for eccentric bottom chord force (see Figure E12.3h).
h_{gg} = 37.0 in
h_{wg} = 12.0 in
A_{gg} = 37.0(0.5) = 18.5 in^2
A_{wg} = 12.0(0.5) = 6.0 in^2
$S_g = h_g^2 t_p/6$ = (37.0)^2(0.5)/6 = 114.1 in^3
$e_c = h_g/2 - h_{wg}/2$ = 37.0/2 − 12.0/2 = 18.5 − 6.0 = 12.5 in

$$E_f = \frac{\dfrac{1}{A_{wg}}}{\dfrac{1}{A_{gg}} + \dfrac{e_c}{S_g}} = \frac{\dfrac{1}{6.0}}{\dfrac{1}{18.5} + \dfrac{12.5}{114.1}} = \frac{0.167}{0.054 + 0.110} = 0.982$$

A_{ggp} = 12(0.5) = 6.0 in^2
A_{ngp} = 6.0 − 2(1.0)(0.5) = 5.0 in^2 (see Figure E12.3h)
A_{ngp}/A_{ggp} = 5.0/6.0 = 0.83
σ_{aty} = (130.8)/((6.0)(0.982)) = 22.2 ksi ≤ 27.5 ksi, OK
σ_{atU} = (130.8)/(5.0)(0.982)) = 26.6 ksi ≤ 32.95 ksi, OK
S_R = (368.1/523.4)(130.8)/((12)(0.98)(0.50)) = 15.6 ksi ≤ 18 ksi, OK
Block shear will not govern for the relatively small force of 130.8 kips.

Gusset plate at connection to L2L3
Connection fasteners
No. bolts strength = (288.7)/12.9 = 22.4 bolts (use min 24 bolts)
Yield, fracture and fatigue:
σ_{gy} = (288.7)/(12(0.98)(0.50)) = 49.1 ksi > 27.5 ksi, NG
σ_{gU} = (288.7)/(8.0(0.98)(0.50)) = 73.5 ksi > 32.9 ksi, NG (4 bolts across section)
S_{Rfat} = (640.1/922.9)(288.7)/(12(0.98)(0.50)) = 34.0 ksi > 18 ksi, NG
Try gusset plate thickness = 1.125 in
σ_{gy} = (288.7)/(12(0.98)(1.125)) = 21.8 ksi ≤ 27.5 ksi, OK
σ_{gU} = (288.7)/(8.0(0.98)(1.125)) = 32.7 ksi ≤ 32.9 ksi, OK (4 bolts across section)
S_{Rfat} = (640.1/922.9)(288.7)/(12(0.98)(1.125)) = 15.1 ksi ≤ 18 ksi, OK
Block shear will not govern design for 1.125 in thick gusset plate

Determine forces in vertical splice components (inside splice and gusset plate) by rebalancing the splice on the L1L2 side with a 1.125 in thick gusset plate as shown in Figure E12.4c.

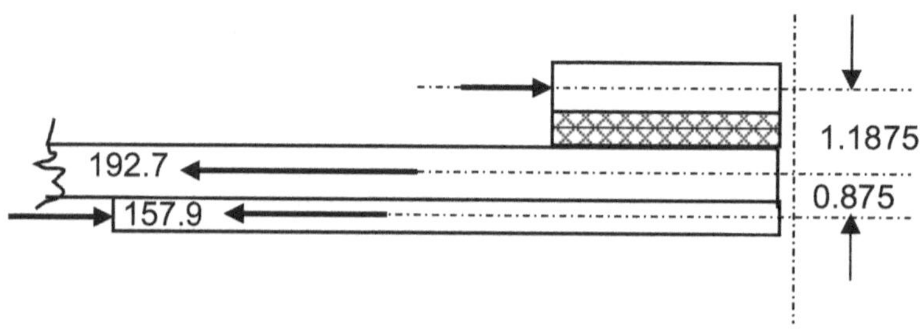

FIGURE E12.4c Plan of vertical splice forces from L1L2 with 1.125 in thick gusset plate.

Splice component	Force (kips)	Distance to center of gusset plate (in)	Moment at center of gusset plate and outside splice plate (kip-in)
Chord vertical plate	192.7	0.875	168.6
Gusset plate diagonal force	157.9	0	0
Total	350.6	-	168.6

Splice component	Force (kips)	Distance to center of gusset plate (in)	Moment at center of gusset plate (kip-in)
Inside splice plate	−168.6/2.0625 = −81.75	2.0625	−168.6
Gusset plate	-(350.6−81.75) = −268.9	0	0
Total	−350.6	-	−168.6

Splice component	L1L2	L2L3
Gusset plate	−268.9 + 157.9 = −111.0	−268.9

Inside splice plate force = 81.75 kips
L1L2 side of gusset plate force = 111.0 kips
L2L3 side of gusset plate force = 268.9 kips

Inside vertical splice plates:

No. bolts = 81.75/12.9 = 6.3 (8 bolts each side of vertical splice) (4 bolts in each inside splice plate on each side of horizontal splice)
σ_{sy} = 81.75/5.38 = 15.2 ksi ≤ 27.5 ksi, OK
σ_{sU} = 81.75/3.38 = 24.2 ksi ≤ 32.9 ksi, OK
S_{Rfat} = (368.1/523.4)81.75/5.38 = 10.7 ksi ≤ 18 ksi, OK
Block shear will not govern for 0.50 in thick inside vertical splice plate

Gusset plate at connection to L1L2
Connection fasteners
No. bolts strength = (111.0)/12.9 = 8.6 (use min. 10 bolts)
Bolt bearing stress at chord member L1L2 on gusset plate = 111.0/(10(1)(1.125)) = 9.9 ksi OK.
Yield, fracture and fatigue:

$$E_f = \frac{\dfrac{1}{A_{wg}}}{\dfrac{1}{A_{gg}} + \dfrac{e_c}{S_g}} = \frac{\dfrac{1}{6.0}}{\dfrac{1}{18.5} + \dfrac{12.5}{114.1}} = \frac{0.167}{0.054 + 0.110} = 0.982$$

A_{ggp} = 12(1.125) = 13.50 in^2
A_{ngp} = 13.50 − 2(1.0)(1.125) = 11.25 in^2 (2 bolts across section)(see Figure E12.3h)
A_{ngp}/A_{ggp} = 11.25/13.50 = 0.83
σ_{aty} = (111.0)/((13.50)(0.982)) = 8.4 ksi ≤ 27.5 ksi, OK
σ_{atU} = (111.0)/(11.25)(0.982)) = 10.0 ksi ≤ 32.95 ksi, OK
S_R = (368.1/523.4)(111.0)/((13.50)(0.98)) = 5.9 ksi ≤ 18 ksi, OK
Block shear will not govern for the relatively small force of 111.0 kips.

Gusset plate at connection to L2L3
Connection fasteners
No. bolts strength = (268.9)/12.9 = 20.8 bolts (use min 22 bolts)
Yield, fracture and fatigue:
A_{ggp} = 12(1.125) = 13.50 in^2
A_{ngp} = 13.50 − 4(1.0)(1.125) = 9.00 in^2 (4 bolts across section)(see Figure E12.3h)
A_{ngp}/A_{ggp} = 9.00/13.50 = 0.67
σ_{gy} = (268.9)/(13.50(0.98)) = 20.3 ksi ≤ 27.5 ksi, OK
σ_{gU} = (268.9)/(9.00(0.98)) = 30.4 ksi ≤ 32.9 ksi, OK
S_{Rfat} = (640.1/922.9)(268.9)/(13.50(0.98)) = 14.1 ksi ≤ 18 ksi, OK
Use gusset plate thickness of 1.125 in

Example 12.5

Design the gusset plate at joint L2 of the Warren truss with the dead load and maximum live load forces shown in Table E12.3a. Members L1L2, L2L3, U1L2, U2L2 and L2U3 frame into the gusset plates at L2. Members U1L2 and L2U3 frame into the gusset plate at a 45° angle. The box-section bottom chord is not continuous through panel point L2 and the splice between members L1L2 and L2L3 consists of inside splice plates and the gusset plates. The gusset plate design for the connection of web members U1L2, U2L2 and L2U3 is shown in Example 12.3. The horizontal splices at panel point L2 between members L1L2 and L2L3 are determined similarly to that shown in Example 12.3. The top horizontal element (top cover plate) is spliced with two splice plates (one each side of the top horizontal element). The bottom horizontal element (bottom cover plate) is spliced with one splice plate on the topside of the bottom horizontal element and the horizontal gusset for the lateral bracing on the underside of the bottom horizontal element. Design the vertical splices at panel point L2 consisting of inside splice plates and the gusset plates.

Splice analysis
Design splice for least yield capacity of chord members L1L2 and L2L3
Yield capacity of L1L2
T = 485.7 kips
ΔT = 368.1 kips
Vertical plates: 2- 0.625" x 12"
Horizontal plates: 2- 0.50 x 10.75"
A_{gm} = 2(0.625)(12) + 2(0.50)(10.75) = 25.75 in^2
σ_{ay} = 485.7/25.75 = 18.9 ksi ≤ 0.55(50) ≤ 27.5 ksi, OK
Yield capacity of L1L2 = 27.5(18.9) = 518.7 kips
Yield capacity/T = 518.7/485.7 = 1.07
Yield capacity/ΔT = 518.7/368.1 = 1.41

Yield capacity of L2L3
T = 850.3 kips
ΔT =640.1 kips
Vertical plates: 2- 1.00" x 12"
Horizontal plate: 2—0.50 x 9.5"
A_{gm} = 2(1.00)(12) + 2(0.50)(9.5) = 33.50 in^2
σ_{ay} = 850.3/33.50 = 25.4 ksi ≤ 0.55(50) ≤ 27.5 ksi, OK
Yield Capacity of L2L3 = 27.5(33.50) = 921.3 kips
Yield capacity/T = 921.3/850.3 = 1.08
Yield capacity/ΔT = 921.3/640.1 = 1.44

Vertical splices:

The inside vertical splice plates are bolted to the vertical parts of members L1L2 and L2L3. The gusset plate is the outside vertical splice shown in Figure E12.5a.

	Chord Member	
Splice Component	**L1L2**	**L2L3**
Chord depth (in)	12.0	12.0
Chord vertical plate thickness (in)	0.625	1.00
Gusset plate thickness (in)	1.00	1.00
Inside splice plate thickness	0.50	0.50

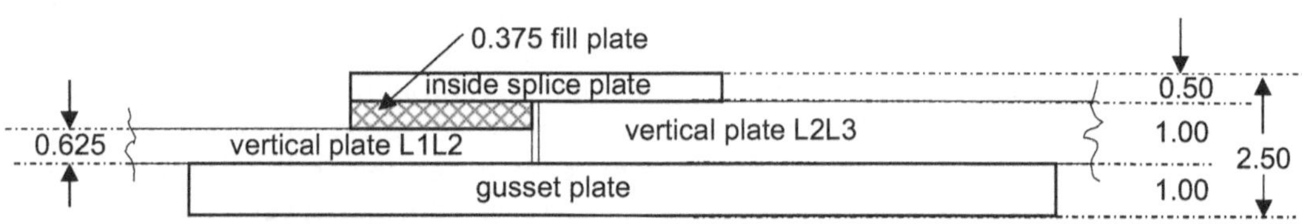

FIGURE E12.5a Plan of vertical splice at L2.

	Force in each Vertical Splice (kips)	
	L1L2	L2L3
Chord	(518.7/2)(15.00/25.75) = 151.1	(921.3/2)(24.00/33.50) = 330.0
Diagonal	(446.7/2) cos 45° = 157.9	(74.6/2) cos 45° = 26.4
Total	309.0	356.4

Least yield capacity of each vertical splice = 309.0 kips (L1L2 side of splice)

Determine forces in vertical splice components (inside splice and gusset plate) by balancing the splice on the L1L2 side as shown in Figure E12.5b.

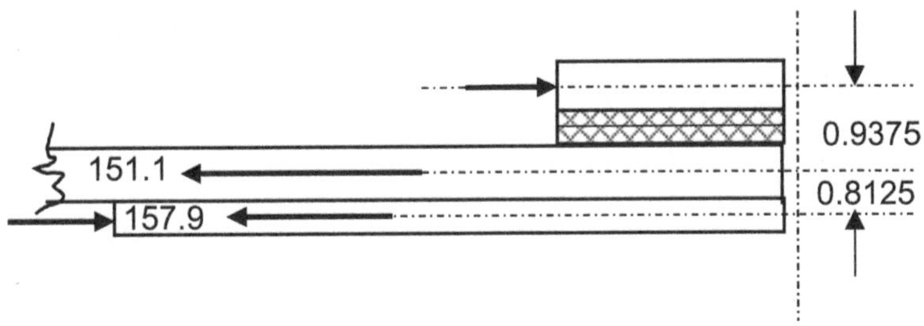

FIGURE E12.5b Plan of vertical splice forces from L1L2.

Splice component	Force (kips)	Distance to center of gusset plate (in)	Moment at center of gusset plate and outside splice plate (kips-in)
Chord vertical plate	151.1	0.8125	122.8
Gusset plate diagonal force	157.9	0	0
Total	309.0	-	122.8

Splice component	Force (kips)	Distance to center of gusset plate (in)	Moment at center of gusset plate (kip-in)
Inside splice plate	−122.8/1.75 = −70.15	1.50	−122.8
Gusset plate	-(309.0−70.15) = −238.9	0	0
Total	−309.0	-	−122.8

Splice component	L1L2	L2L3
Gusset plate	−238.9 + 157.9 = −80.95	−238.9

Inside splice plate force = 70.15 kips
L1L2 side of gusset plate force = 80.95 kips
L2L3 side of gusset plate force = 238.9 kips

Inside vertical splice plates:

No. bolts = 70.15/12.9 = 5.4 (6 bolts each side of vertical splice)
Use 10” deep inside splice plate

$A_{gis} = (10.0)(0.5) = 5.00\ in^2$
$A_{nis} = 5.00 - 3(1.0)(0.50) = 3.50\ in^2$ (2 rows of 3 bolts)
$A_{nis}/A_{gis} = 3.50/5.00 = 0.70$
$A_{efat} = 5.00\ in^2$
$\sigma_{sy} = 70.15/5.0 = 14.0$ ksi ≤ 27.5 ksi, OK
$\sigma_{sU} = 70.15/3.5 = 20.0$ ksi ≤ 32.9 ksi, OK
$S_{Rfat} = (368.1/518.7)70.15/5.0 = 10.0$ ksi ≤ 18 ksi, OK
Block shear will not govern for 0.50 in thick inside vertical splice plate

Gusset plate at connection to L1L2
Connection fasteners
No. bolts strength = (80.95)/12.9 = 6.3 (use min. 8 bolts)
Bolt bearing stress at chord member L1L2 on gusset plate = 80.95/(8(1)(1.0)) = 10.1 ksi OK.

Yield and fracture:

Calculate E_f to account for eccentric bottom chord force (see Figure E12.3)
$h_{gg} = 37.0$ in
$h_{wg} = 12.0$ in
$A_{gg} = 37.0(0.5) = 18.5\ in^2$
$A_{wg} = 12.0(0.5) = 6.0\ in^2$
$S_g = h_g^2 t_p/6 = (37.0)^2(0.5)/6 = 114.1\ in^3$
$e_c = h_g/2 - h_{wg}/2 = 37.0/2 - 12.0/2 = 18.5 - 6.0 = 12.5$ in

$$E_f = \frac{\frac{1}{A_{wg}}}{\frac{1}{A_{gg}} + \frac{e_c}{S_g}} = \frac{\frac{1}{6.0}}{\frac{1}{18.5} + \frac{12.5}{114.1}} = \frac{0.167}{0.054 + 0.110} = 0.982$$

$A_{ggp} = 12(1.0) = 12.0\ in^2$
$A_{ngp} = 12.0 - 2(1.0)(1.0) = 10.0\ in^2$ (see Figure E12.3h)
$A_{ngp}/A_{ggp} = 10.0/12.0 = 0.83$
$\sigma_{aty} = (80.95)/((12.0)(0.982)) = 6.9$ ksi ≤ 27.5 ksi, OK
$\sigma_{atU} = (80.95)/(10.0)(0.982)) = 8.2$ ksi ≤ 32.95 ksi, OK
$S_{Rfat} = (368.1/518.7)(80.95)/((12)(0.98)(1.0)) = 4.9$ ksi ≤ 18 ksi, OK
Block shear will not govern for the relatively small force of 80.95 kips.

Gusset plate at connection to L2L3
Connection fasteners
No. bolts strength = (238.9)/12.9 = 18.5 bolts (use min 20 bolts)
Yield, fracture and fatigue:
$\sigma_{gy} = (238.9)/(12(0.98)(1.0)) = 20.3$ ksi ≤ 27.5 ksi, OK
$\sigma_{gU} = (238.9)/(8.0(0.98)(1.0)) = 30.4$ ksi ≤ 32.9 ksi, OK (4 bolts across section)
$S_{Rfat} = (640.1/921.3)(238.9)/(12(0.98)(1.0)) = 14.1$ ksi ≤ 18 ksi, OK
Block shear will not govern design for 1.0 in thick gusset plate
Use a gusset plate with 1 in thickness

Examples 12.1 to 12.5 outline the design of gusset plates for continuous and discontinuous chord members using the Whitmore section at web member connections for the determination of yield,

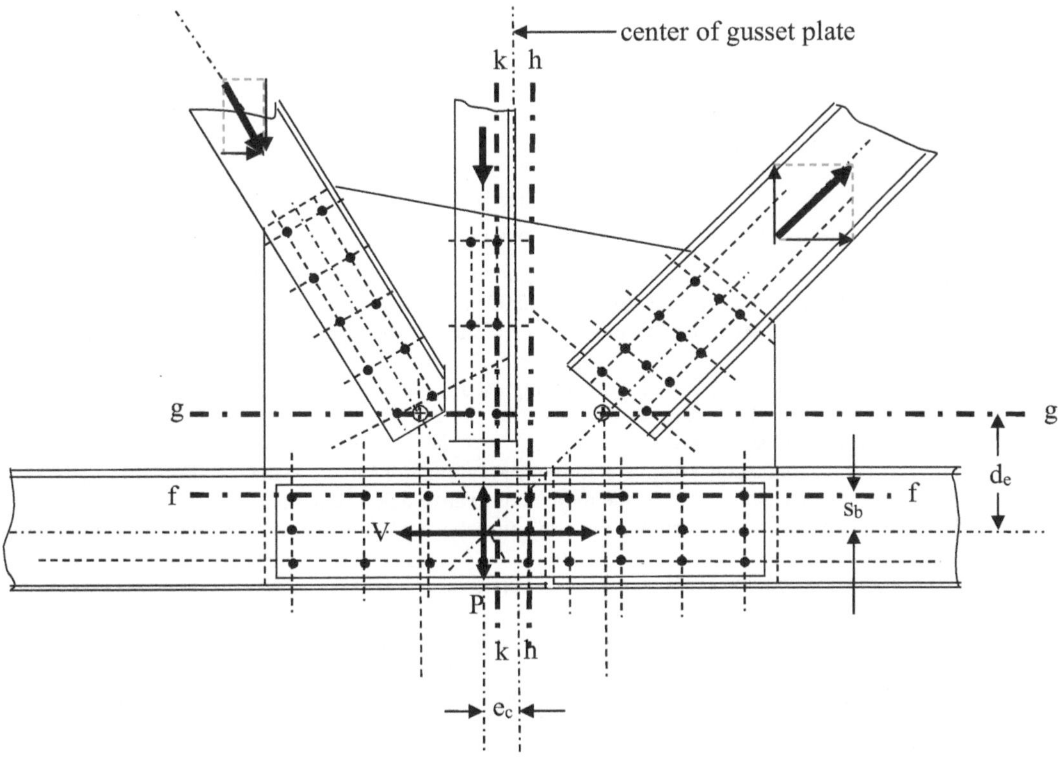

FIGURE 12.14 Critical sections for gusset plate analysis by slender beam theory.

fracture, fatigue and compressive forces. Block shear is also determined for tension connections. However, for heavily loaded railway truss members, it may be appropriate to also investigate gusset plate stresses from shear forces, bending moments and axial forces with beam theory[38] at critical sections (e.g., lines f-f, g-g, h-h and k-k in Figure 12.14).

The critical sections, such as lines g-g and h-h in Figure 12.14, may be reviewed for combined stresses from the following:

- shear yielding on the gross section of the gusset plate from the resultant horizontal, V, (on section g-g) and vertical, P, (on section h-h) forces in the members
- axial tension or compression on the gross section of the gusset plate from the resultant horizontal, V, (on section h-h) and vertical, P, (on section g-g) forces in the members
- bending moments, for example, $M = \pm V\ (d_e) \pm P\ (e_c)$ at section g-g.

Critical sections such as lines f-f and k-k in Figure 12.14 may be reviewed for combined stresses from shear fracture on the net section of the gusset plate from the resultant horizontal, V, (on section f-f) and vertical, P, (on section k-k) forces in the members; axial tension on the net section or compression on the gross section of the gusset plate from the resultant horizontal, V, (on section k-k) and vertical, P, (on section f-f) forces in the members or bending moments at sections f-f and k-k. Depending on connection geometry, other critical sections and interfaces may require review for combined stresses due to slender beam theory forces. Nonetheless, for typical truss span gusset plates it is typically sufficient to disregard slender beam theory for analysis and design.

NOTES

1 Chord members are spliced within panels or at panel points. Web members are typically not spliced unless they are members of very large truss spans.
2 Gusset plates that are components of chord member splices at panel points must be designed for the capacity of the spliced chord members.
3 Vertical members directly loaded by floorbeam reactions are hanger and subvertical members.
4 The compression member splice must be as close to the panel point as practicable.
5 Equation $\frac{2d_b\sigma_p}{F_U}$ is based on allowable bearing stresses at bolted connections as developed in Section 12.3.2.
6 Based on their respective influence line lengths, chord members are designed for more than 2,000,000 stress range cycles and web members for fewer than 2,000,000 stress range cycles.
7 AREMA (2024) also provides effective allowable bolt shear stresses for slotted and oversize holes for F3125 Grade 325 and 490 bolts.
8 22 mm (7/8 in) diameter bolts are the most commonly used bolts in railway bridges.
9 Based on the von Mises yield criterion.
10 This is the case in practical structures.
11 A lower FS is used due to the conservative nature of the assumptions made to develop Equation 12.10.
12 The coefficient for shear yield is taken as the more conservative 0.30 instead of 0.35.
13 The bolt bearing stresses on the member material must be considered during member end connection design.
14 For example, tests show that shear stresses in gusset plates are closer to V/A than 1.5 V/A as predicted by beam theory.
15 Complex geometry gusset plate connections may occur in very large trusses, in heavily loaded trusses, at chord member splices and where heavily chamfered members frame into the gusset plates.
16 A supplementary benefit of finite element analysis is the determination of gusset plate deformations, which are, otherwise, very difficult to estimate.
17 Whitmore (1952)
18 Some engineers will also truncate the Whitmore section at adjacent member fasteners lines contacted by the Whitmore lines diverging from the exterior fasteners in the first row of fasteners in the connection. It is likely that this is too conservative.
19 Equation 12.33 is slightly conservative as the Whitmore effective section is taken through the center of the last line of bolts.
20 The Whitmore gross width is used for slip-critical bolted connections.
21 Previous recommended practice considered l_w as the average of the the distances from the center and ends of the Whitmore width, w_e, to the nearest gusset plate restraint.
22 If gusset plates are not braced against lateral movement, K may be greater than 1.0 and is often taken as 1.2.
23 This criterion was removed from AREMA recommended practice in 2020.
24 Nevertheless, the shear lag factor, U_c, must be determined to calculate the Whitmore effective net width for the web members framing into the gusset plate.
25 Eccentricities at splices are particularly detrimental in compression members.
26 Typically, splice plates.
27 In contrast to panel points with continuous chord members that are spliced within panels adjacent to the panel point in the lower stressed member.
28 The arrangement of H-section, I-section and box-section top chord members at panel points is similar with the sections rotated 180 degrees.
29 H-sections can be fabricated from three welded plates or by using rolled H or W sections with strong axis in horizontal direction.
30 I-sections can be fabricated from three welded plates or by using rolled H or W sections with weak axis in horizontal direction.
31 Box sections can be fabricated from four welded plates.
32 There will be one horizontal splice in H-section members and two horizontal splices in box-section members.
33 There will be two vertical splices in H-section members and box-section members.
34 For example, tension diagonals connected to gusset plates at the bottom chord of truss spans or compression members that are connected to gusset plates at the top chord of truss spans.
35 The vertical splice must transmit the chord capacity and the horizontal component of the additive diagonal force.

36 The gusset plate overlap is typically the portion of the gusset plate fastened to the chord member.

37 The method typically provides a conservative estimate of the stresses due to chord member force eccentricity.

38 An alternative to slender beam theory, the uniform force method, which is strongly dependent on connection geometry, has been used for building design (Thornton and Kane, 1999).

REFERENCES

American Railway Engineering and Maintenance-of-Way Association (AREMA), 2024, *Manual for Railway Engineering*, Chapter 15, Steel Structures, Lanham, MD.

Bowles, J.E., 1980, *Structural Steel Design*, McGraw-Hill, New York.

Chesson, E., Faustino, N.L., and Munse, W.H., 1965, "High-Strength Bolts Subjected to Tension and Shear", *Journal of the Structural Division, ASCE*, 91(ST5).

Douty, R.T., and McGuire, W., 1965, "High Strength Bolted Moment Connections", *Journal of the Structural Division, ASCE*, 91(ST2).

Kulak, G., 2002, *High Strength Bolts*, AISC Steel Design Guide 17, Chicago, IL.

Kulak, G., Fisher, J.W., and Struik, J.H.A., 1987, *Guide to Design Criteria for Bolted and Riveted Joints*, 2nd edition, John Wiley & Sons, New York.

Kulicki, J.M., and Reiner, B.M., 2011, *Chapter 13—Truss Bridges, Structural Steel Designer's Handbook*, McGraw-Hill, New York.

Nair, R.S., Birkemoe, P.C., and Munse, W.H., 1974, "High Strength Bolts Subjected to Tension and Prying", *Journal of the Structural Division, ASCE*, 100(ST2).

Ocel, J., 2013, *Guidelines for the Load and Resistance Factor Design and Rating of Riveted and Bolted Gusset-Plate Connections for Steel Bridges*, The National Academies Press, National Academies of Sciences, Engineering, and Medicine, Washington, DC.

Research Council on Structural Connections (RCSC), 2014, *Specification for Structural Joints Using High-Strength Bolts*, AISC, Chicago, IL.

Thornton, W.A., and Kane, T., 1999, "Design of Connections for Axial, Moment and Shear Forces", Chapter 2 in *Handbook of Structural Steel Connection Design and Details* (Tamboli, A.R., ed.), McGraw-Hill, New York.

Whitmore, R.E., 1952, *Experimental Investigation of Stresses in Gusset Plates*, University of Tennessee Experiment Station Bulletin 16.

Index

Note: Page numbers in *italics* indicate figures and page numbers in **bold** indicate tables on the corresponding page. Numbers following "n" refer to notes.

K

L

M

T

U

V

W